COMPLETELY REVISED AND UPDATED EDITION!

WELDER'S

H A N D B O O K

A COMPLETE GUIDE TO MIG, TIG, ARC & OXYACETYLENE WELDING

RICHARD FINCH

HPBooks

Most HPBooks are available at special quantity discounts for bulk purchases for sales promotions, premiums, fund-raising or educational use. Special books, or book excerpts, can also be created to fit specific needs.

For details, write: Special Markets, The Berkley Publishing Group, 375 Hudson Street, New York, New York 10014.

HPBooks
are published by
The Berkley Publishing Group
a division of Penguin Putnam Inc.
375 Hudson Street
New York, New York 10014

This book is a fully revised and updated edition of *Welder's Handbook*
by Richard Finch and Tom Monroe, originally published in 1985,
copyright © 1985 by Price Stern Sloan, Inc.

First Printing of Revised Edition: February 1997

©1997 Richard Finch
30   29   28   27   26   25

The Penguin Putnam Inc. World Wide Web site address is
http://www.penguinputnam.com.

Library of Congress Cataloging-in-Publication Data

Finch, Richard. 1935-
    Welder's handbook / Richard Finch. — Rev. ed.
        p.   cm.
    Includes index.
    ISBN 1-55788-264-9
    1. Welding—Handbooks, manuals, etc. I. Title.
TS227.F5   1997
671.5'2—dc20   96-44794
                                                         CIP

Book Design & Production by Bird Studios
Interior photos by the author unless otherwise noted
Cover photos courtesy Lincoln Electric

Acknowledgments

This second edition of *Welder's Handbook* was written at a time when welding technology was literally exploding compared to the state of the art that existed when the first edition was published. And many of the companies listed in the first edition merged, went out of business, and changed names at a surprising rate. Therefore, the names and companies that contributed to the publication of this second edition will possibly be unfamiliar to the reader in many cases. But go to their business and get acquainted with them. They will be most helpful to you in your welding activities. Here are the names of the people and the companies who helped with this book:

Mr. Rick Wuchner of So-Cal Air Gas, Ventura, CA; Mr. Jeff Noland of HTP America, Inc., Arlington Heights, IL; Mr. Bill Berman of Daytona Mig, Inc., Daytona Beach, FL; Mr. Ed Morgan of Lincoln Electric Co., Santa Fe Springs, CA; Mr. Ray Snowden of Allan Hancock College, Santa Maria, CA; Mr. Seth Hammond and Mrs. Tannis Hammond of Specialty Welding, Goleta, CA; Mr. Tom Giffen and Mr. Ron Chase of Western Welding, Goleta, CA; Mr. Dave Williams of Williams Lo-Buck Tools, Norco, CA; Mr. Michael Reitman of United States Welding Corporation, Carson City, NV; Ms. Darlene Tardiff of Harbor Freight Tools, Inc., Camarillo, CA; Mr. Phil Pilsen of Pro-Tools, Tampa, FL; Mr. Henry Hauptfuhrer of The Eastwood Co., Malvern, PA; Ms. Lisa West of Smith Equipment, Watertown, SD; Mr. Dale Wilch of Dale Wilch Sales, Kansas City, MO; Mr. Hal Olcutt of Victor-Thermadyne, Denton, TX; Mittler Bros Tools, Huntsman Welding Helmets, Salt Lake City, UT; Mr. Dick Casperson of Miller Electric Co., Appleton, WI; and two friends who helped, Mr. Dale Johnson and Mr. Jerry Jones, both of Goleta, CA. ■

Contents

INTRODUCTION . . . . **V**

1 METAL BASICS & HEAT CONTROL **1**

2 WELDING, BRAZING & CUTTING **8**

3 WELDING EQUIPMENT **16**

4 WELDING SAFETY **24**

5 FITTING & JIGGING **31**

6 CLEANING BEFORE WELDING **44**

7 GAS WELDING & HEAT FORMING **48**

8 TORCH CUTTING **69**

9 GAS BRAZING & SOLDERING **76**

10 ARC WELDING **82**

11 MIG WELDING **94**

12 TIG WELDING **105**

13 PLASMA CUTTING & WELDING **123**

14 SPECIAL WELDING PROCESSES **128**

15 WELDING RODS, WIRES & FLUXES **136**

16 WELDING PROJECTS **145**

17 WELDING CERTIFICATION & TRAINING **155**

GLOSSARY **161**

INDEX **165**

Introduction

THE WORLD DEPENDS ON WELDING

Welding is an art, and it is also a means to achieve an end result. For instance, welding is a process used to build automobiles and airplanes, and it is also a process used to build stronger buildings, highway bridges, hydroelectric power plants, and probably the chair you are sitting in while you read this page.

If you possess a television set and a refrigerator, spot welding was used to make those appliances. If you are reading this page in a bookstore or a library, the bookshelves probably utilize welded brackets and bases to hold the shelves in place. If every weld in the world should give way at once, our world would fall apart. Automobiles would fall into piles of metal in the parking lots, airplanes would fall from the sky in pieces, ships would come apart and sink, and even the bed you sleep in would fall to the floor.

But welding is a satisfying talent that almost anyone can develop. I really enjoy building things out of metal. I probably spend more time admiring my welded projects than I spend in welding them. The welding table that I designed and built for a project in this book (see page 146) is one of the most useful welding projects that I have ever completed. Regardless of your reason for wanting to learn to weld, or to be a better weldor, you will surely want to build

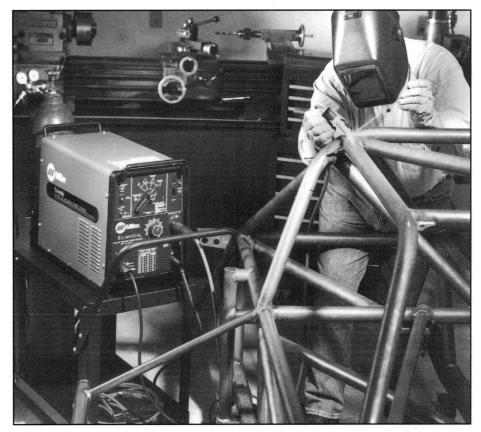

this table before you try any other projects because this welding table will support many of your future projects.

Learning to weld is similar to learning to play a piano. You can be playing "chopsticks" with just a few minutes' instruction, but it takes a lot of practice to play Mozart on the piano. Likewise, you can learn to run a weld bead in just a few minutes, but you will have to spend time practicing before you build a real race car, airplane, or an ocean-going aluminum yacht. The projects that I have laid out for you in Chapter 16 of this book will provide a low-pressure way to practice your welding before you

tackle the really important projects.

So, read this book, study the pictures and captions, follow my suggestions, and learn to weld at your own pace.

WELDER and WELDOR

Webster's New Collegiate Dictionary defines "Welder" as the machine that is used to weld with, and "Weldor" as the person who uses the machine. However, the American Welding society prefers "Welder" for the person as well as the machine. I rather like to call myself a "Weldor"! ∎

Metal Basics & Heat Control

T here are many factors involved in producing good welds in metal: the equipment used, the filler metal used, the preparation of the parts to be welded, and especially, the correct application of heat to the weld. In manual (not automatic) welding, the weldor is the artist who knows how much heat to apply to produce good welds. Before you can weld, you must learn how to control heat, and have some fundamental knowledge of the basic types of metals and their properties.

Solid, Liquid, Gas

My high school physics teacher, Larry Grundy, taught me that almost all matter on this earth is in three basic forms: (1) Solid (frozen), (2) Liquid (molten), and (3) Gas (vapor). At the time, I don't think that I believed him, but as I got older and tried things my own way, I began to see that he was right. I learned by experience that temperature, and especially heat control, plays a very important part in how the earth exists. And it really plays a very important part in welding.

Example: Metal can exist in these three forms, just like ice. Steel is solid at or below its 2700° F melting point. Ice is solid at or below its 32° F melting point. Heat the ice to 32° F and it becomes liquid water, then heat

This totally awesome Cobra replica, built by Butler Racing, Inc., incorporates several types of welding processes in building the frame (see photo next page). Photo by Bill Kerian, Butler Racing, Inc.

it to 212° F and it begins to boil and vaporize.

Heat steel to 2,786° F and it becomes liquid or molten, then continue to heat it to over 5,500° F and it begins to vaporize. In welding, when the liquid/molten weld puddle begins to solidify, we say it is *freezing*. Certain kinds of arc welding rods are called "fast freeze rods."

While you are learning to weld, keep the picture example of ice cubes on the kitchen stove in mind. If you get the metal too hot, it will vaporize, just like the ice cubes do. Study the charts in this chapter to get a better idea of the melting and boiling points

(temperatures) of the many kinds of metals that are weldable.

Color Changes of Metals

As you practice your welding, brazing and soldering skills, you will recognize the color changes in metals as the metal is heated and cooled. Study the charts in this chapter to better understand the temperatures that result in specific colors when you are heating steel.

Aluminum does not exhibit the same kind of color changes that steel does, but it goes from shiny to dull as it is heated, then to shiny again as it

Lift off the body of the Cobra and you see the real beauty behind the machine. MIG, TIG, and oxyacetylene welding were used to produce this true work of metal-sculpture art. Photo by Bill Kerian, Butler Racing, Inc.

Keep these working temperatures in mind. Each process—welding, brazing or soldering—is different. If you overheat the weld bead, you could "vaporize" your project! Master temperature control, and you will become a much better weldor.

begins to melt.

Stainless steel does not go through the number of color changes that mild steel and carbon steel go through in increasing temperatures, but stainless steel does turn red just before it melts.

Similarly, brass and copper do not show the same color changes that steel does. Brass just gets lighter in color, then shiny as it melts; copper just gets red, then dark as it reaches melting temperature.

CHART TEMPERATURES

First, study the chart that gives "Temperatures of Soldering, Brazing and Welding Processes." You will see a temperature range for each process. Soldering is the lowest temperature range, 250° F to 800° F. Brazing is the next highest temperature range at 800° F to 1600° F, and Fusion welding at the melting point of the metal to be welded.

Next, study the chart that shows "Weights, Melting Points & Boiling Points of Metals." From this chart, you will begin to understand why you

must control the heat in order to produce the results you want, which is good, strong metal joining techniques.

Control the Heat

In every welding, brazing and soldering process that I describe and instruct in this book, I tell you to control the heat. Too cold will produce a weak bond or no bond at all, and too much heat will boil or vaporize the bond and even ruin your metal.

TYPES OF METAL

Before you can begin welding, you must know what kind of metal you are going to weld. Usually, but not always, you should weld the same kind of metals together. There will be cases where you will want to weld stainless steel to low alloy carbon steel, or solder brass to aluminum. In the applicable chapters, I will show you how to do it. But first, study the following paragraphs and photos to find out how to identify metals.

Ferrous & Non-Ferrous Metals

Ferrous indicates that a metal has iron content, and that it is attracted to a magnet. Non-ferrous means that the metal doesn't have any iron content and that it is not magnetic. Cast iron, mild steel and chrome molybdenum steel are ferrous metals. Aluminum,

WEIGHTS, MELTING POINTS & BOILING POINTS OF METALS

Metal	Wt. Per Cu Ft in Lb	Melting Point in Degrees F (C)	Boiling Point in Degrees F (C)
Aluminum	166	1217 (658)	4442 (2450)
Bronze	548	1566-1832 (850-1000)	
Brass	527	1652-1724 (900-940)	
Carbon	219	6512 (3600)	
Chromium	431	3034 (1615)	
Copper	555	1981 (1083)	4703 (2595)
Gold	1205	1946 (1063)	5380 (2971)
Iron	490	2786 (1530)	5430 (2999)
Lead	708	621 (327)	3137 (1725)
Magnesium	109	1100 (593)	
Manganese	463	2300 (1260)	
Mild Steel	490	2462-2786 (1350-1530)	5450 (3049)
Nickel	555	2645 (1452)	4950 (2732)
Silver	655	1761 (960)	4010 (2210)
Tin	455	449 (231)	4120 (2271)
Titanium	218	3263 (1795)	
Tungsten	1186	5432 (3000)	10706 (5930)
Zinc	443	786 (419)	1663 (906)
4130 Steel	495	2550	5500 (3051)

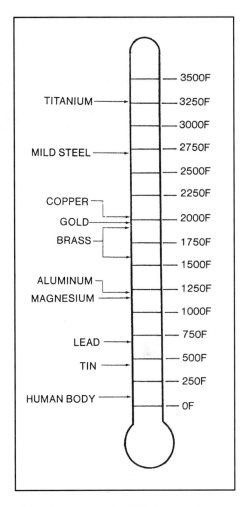

This "Thermometer" will help you visualize the melting points of various metals. It will also give you an idea of how much heat you will need to weld various metals.

Melting points of metals vary widely. Use this chart to determine how hot you will need to get the metal in order to fusion weld it. If available, I have also included boiling temperatures for each metal, where they are available, a temperature you should surely avoid.

brass, copper, gold, silver, lead and magnesium are non-ferrous metals and they are not magnetic and are not attracted to a magnet. In welding terms, ferrous and non-ferrous metals are considered to be "dissimilar metals" but they can be joined by various welding, brazing or soldering processes.

For instance, I'll show you a special Cronatron brand of solder that will join any and all metals if a tensile strength of the bond at 7,000 psi would be acceptable. In this book, I will also show you how to braze copper to cast iron and even how to do repair welds without heat! More on that later. Now, let's become more familiar with the various types of metals.

Cast Iron—This metal is usually rough-textured because of how it is manufactured. A cast-iron part is formed by pouring molten iron into a sand mold, thus giving it the form and texture of the interior of the mold. Typical cast-iron parts include automotive engine blocks, exhaust manifolds, manual-transmission cases, older lawn mower and garden-tractor engines, and early-style farm equipment.

Where it is cut with a lathe, saw, grinder or whatever, cast iron usually has a gray, grainy appearance. When ground with a high-speed grinding wheel, red sparks are generated.

Cast iron is ferrous, or magnetic. It can be arc-welded with stick-type electrodes or brazed or fusion-welded with an oxyacetylene torch.

Forged Steel—This is a rough metal, but smoother than cast iron. It's used for most engine connecting rods, some crankshafts, axle shafts and

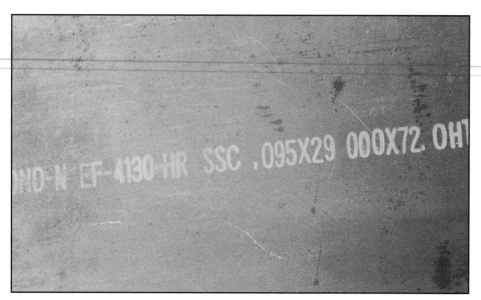

Chrome moly steel is factory-marked for ease of identification. Here you see condition N (normalized), 4130-HR (4130 hot rolled), .095 x 29 (.095" thick x 29" wide) x 72 (72" long), OHT 144610 (heat number 144610 certification).

some chassis components. Forging steel is done by hammering a red hot steel billet into the desired shape in a forging press. Machined, cut or ground, forged steel is light gray or silver inside. Grinding forged steel creates yellow or white sparks.

Forged steel is a ferrous metal. It can be welded with gas, arc, TIG, MIG or plasma-arc methods. But, because steel forgings are intended for high-load/high-fatigue applications, such parts should be welded using the best methods available—TIG, plasma or DC arc—if welded at all. Usually, damaged forged-steel parts should be replaced rather than welded.

Stainless Steel—Stainless is very smooth and hard. It's usually found in sheets, but can be cast. As the name implies, it will not stain or corrode easily. Stainless steel is used to make kitchen cutlery, pots and pans, exhaust systems for airplanes or autos and, in the early '80s, the DeLorean sports car outer skin. When ground with a high-speed abrasive stone, it does not give off sparks. Instead, stainless steel

turns black where ground.

Stainless steel is an exception to the ferrous rule. It is steel, but it is usually non-magnetic. However, a few of the hundred or so different stainless steel alloys are magnetic. Stainless steel welds best with TIG, but arc and gas welding can be used as well. It can also be gas-brazed.

Mild Steel—This is the most common of all metals, used in automobile bodies and chassis parts, lawn-mower handles, bicycle fenders, house furniture, filing cabinets–the list is endless. Cut or ground, it looks bright gray to silver. Grinding with a stone wheel generates a shower of yellow sparks.

Mild steel is ferrous. It can be welded by every method described in this book: gas, arc, TIG, MIG, plasma-arc and spot welding. It is very easy to work with.

Chrome Moly Steel—Chrome moly is also called low carbon steel because it contains some carbon for greater strength than mild steel but less than 0.04%. It is ferrous and can

be heat treated after welding to 190,000 psi tensile strength. It welds best with TIG or oxyacetylene and should not be brazed because it has grains which can open to molten brass and often cracks as a result of brazing.

Cast Brass—Brass is often rough-textured because it is usually cast in sand molds. Cast brass is used for water-valve housings and, with bronze alloy, for boat propellers. If you cut or machine cast brass or bronze—a brass alloy—it appears smooth and yellow or gold in color, depending on the specific alloy. Brass should not be ground. Soft metals such as brass load up, or clog, abrasive grinding wheels, quickly rendering them useless. Cast brass is non-ferrous and non-magnetic. It can be welded with flux-coated brass rod and an oxyacetylene torch.

Cast Aluminum—This metal is usually rough, but because it's cast at a much lower temperature than cast iron, it can be cast in molds with a smoother finish. Cast aluminum is used for late-style lawnmower engines, motorcycle crankcases, intake manifolds for automobile engines and, more recently, automobile engine blocks and cylinder heads. As with brass, don't grind aluminum with a stone wheel. It also clogs the abrasive. Cast aluminum is non-ferrous. It can be welded with TIG, plasma and gas, and brazed with aluminum filler. Cast aluminum can also be arc-welded.

Sheet Aluminum—This metal is smooth and shiny. It can even be polished to a mirror finish. Sheet aluminum comes in thicknesses as thin as kitchen foil or as thick as 2" or 3" plates. Sheet aluminum is used for screen-door frames, airplane wings, race car chassis, lawn furniture, siding

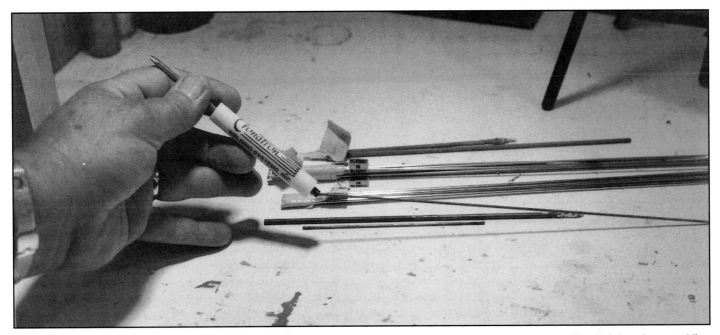

Small magnetic screwdriver is a handy way to test for ferrous or non-ferrous metals. In this picture, the magnet will not pick up brass welding rod because brass is non-ferrous, thus non-magnetic.

on buildings, and many other common applications. Sheet aluminum is non-ferrous. Certain alloys can be welded with gas or TIG. You can also arc-weld some sheet aluminum.

Titanium—This metal looks similar to stainless steel, but it's much shinier when welded or filed. Although relatively light, it is very strong. Titanium-alloy forgings, tubes and sheets are used in aircraft and race car construction. It is very expensive. Titanium doesn't give off sparks when ground with an abrasive wheel. A ferrous metal, titanium can be TIG-welded, but shielding is critical.

Identification Markings

Most metal alloys have identification markings. For example, chrome-moly steel is usually marked "4130 Cond. N," for 4130 steel, normalized condition. The condition of metal indicates how its temper, or

hardness, was achieved. For instance, it may be in an as-fabricated condition, work-hardened or heat-treated. The same applies to tubing and sheet chrome-moly steel.

Sheet aluminum is usually marked to indicate the basic alloy and its conditions. Typical examples of aluminum alloys are 2024-T4, 3003-H14 and 6061-0. Refer to the chart of aluminum alloys, page 120 for a more detailed explanation.

Sheet stainless steel is marked "301, 308, 316 or 347," for example, depending on the specific alloy. The markings are similar to those for sheet aluminum except that STAINLESS may also be printed on the sheet.

For an indepth discussion of popular steel and aluminum alloys, read HPBooks' *Metal Fabricator's Handbook*. Not only are important properties of metals discussed, but the specifics of each of the popular alloys are covered.

Permanent Metal Markers—In my own metal shop, I keep a paint-

type metal marker handy to mark cut-off pieces of metal so I can identify them later. I mark the cut-off piece with the same identification that was used on the full sheet, such as 6061-0, .-90" or 4130-N-.025" and so forth.

Ferrous or Non-Ferrous?—I also keep a small magnetic screwdriver in my shirt pocket for quick identification of metals. Why? Because I cannot always visually identify the metal I'm working with. As already discussed, a magnet attracts ferrous metals—those

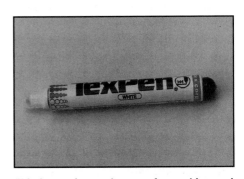

This is a paint marker pen for marking and writing on metal. Buy one and use it to mark scraps of metal for future ease of identification.

TEMPERATURE OF WELDING FUELS

FUEL	AIR° F	w/OXYGEN° F
Acetylene (C2H2)	4800° F	6300° F
Hydrogen (H2)	4000° F	5400° F
Propane (CaHa)	3800° F	5300° F
Butane	3900° F	5400° F
Mapp Gas	2680° F	5300° F
Natural Gas (CH4+H2)	3800° F	5025° F

COLOR OF STEEL AT VARIOUS TEMPERATURES

In Fahrenheit (Centigrade)

Faint Red	900 (482)
Blood Red	1050 (566)
Dark Cherry Red	1075 (579)
Medium Cherry Red	1250 (677)
Cherry Red	1375 (746)
Bright Red & Scaling	1550 (843)
Salmon and Scaling	1650 (899)
Orange	1725 (941)
Lemon	1825 (996)
Light Yellow	1975 (1079)
White	2200 (1204)
Dazzling White	2350 (1288)

Colors as viewed in medium light, not bright sunlight.

COLORS FOR TEMPERING CARBON STEEL

Color	Metal Temp
Pale Yellow	428 (220)
Straw	446 (230)
Gold Yellow	469 (243)
Brown	491 (255)
Brown & Purple	509 (265)
Purple	531 (277)
Dark Blue	550 (287)
Bright Blue	567 (297)
Pale Blue	610 (321)

Carbon steel assumes colors when heated.

Such dissimilar metals must be joined by brazing or soldering.

Mistakes Will Happen

Of course, there's more to identifying metals than determining if they're ferrous or non-ferrous.

Even though cast iron and mild steel are both magnetic and ferrous, the two cannot be joined using conventional welding methods. Special care must be taken to choose the proper welding process and welding rod.

Several years ago I was in a big rush to weld several aluminum stove hoods for a local restaurant. My supply of welding rod included short pieces that had been put back in the holder by a previous weldor. By mistake, a piece of 316 stainless rod had been put back in the aluminum rod holder.

I did not notice the heavier weight of the stainless rod compared to the aluminum rod, and I proceeded to weld at least 12 inches of .090" aluminum stove hood seam with the stainless rod!

The dissimilar metal weld did not fuse together. Stainless steel will not mix with aluminum. I had to cut the stainless rod out of the seam and make a patch to cover my mistake.

Stainless steel rod is not magnetic and aluminum rod is not magnetic, but the aluminum rod is only one-third the weight of stainless rod. I made a mistake!

WELDING TEMPERATURE

With the frozen state of water in mind, consider the same condition for a piece of steel plate. When the steel is "frozen," below about 2700° F, its melting point, it is solid. When heated above that temperature, steel changes to a liquid. And if you heat it to a

containing iron. Magnets are not attracted to non-ferrous metals—metals that contain little or no iron, such as aluminum.

So, I touch my magnetic pocket screwdriver to the metal I want to weld. If it sticks, the metal has iron

content. If the magnet doesn't stick, the metal has little or no iron content. But this is only a preliminary test. Some non-magnetic metals, such as stainless steel, cannot be fusion-welded to other non-magnetic metals, such as aluminum or magnesium.

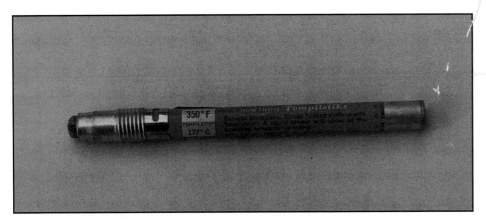

This is a heat-measuring Tempil stick. You can see 350° F printed near the tip. Mark the chalk-like stick on metal and the mark will melt at 350° F plus or minus 2°.

temperature considerably above the melting point, steel can boil and vaporize into the atmosphere. You obviously shouldn't vaporize a weld bead, so keep the weld-puddle temperature below the boiling point.

Overheat at First

Most beginning weldors don't heat the base metal enough. They fail to get it to the melting point and keep it there. I usually tell the beginner to go ahead and burn up, or overheat a few practice pieces. Doing this gives the beginner a feel for temperature control. Melting temperatures vary considerably from metal to metal. If you can master temperature control, you have a head start on welding successfully.

Temperature Control

If you can learn to correctly control the temperature of the weld puddle—the molten pool that forms the weld bead—you can do a good job of welding. To control the weld puddle, you must learn how to judge and control the temperature of the metal you're welding.

Getting back to the water example, water will pour and flow at room temperature. However, if you lower its temperature to below its melting point, or 32° F (0° C), it becomes a solid—ice. You can then handle it just like a block of steel; saw it, drill it or sit it on the freezer shelf without the need to contain it.

If you put this block of solid water in a pan and heat it to 212° F (100° C), it begins to boil and vaporize as it changes from a solid, to a liquid, then to a gas. Steam boils off and escapes into the atmosphere. Water, therefore, has a freezing point, a melting point

and a boiling point.

Once you've mastered temperature control, the next thing to do is to become familiar with which metals can be welded and by which methods. The accompanying charts give the melting points of metals that can be welded, brazed, and soldered.

Make photocopies—you have the author's and publisher's permission—of the five charts in this chapter. Post them in your welding shop for quick reference. You'll be amazed how helpful they'll be. ■

Welding, Brazing & Cutting 2

Historical records tell us that iron, gold, silver and bronze were commonly used as far back as 6,000 years ago. Biblical records indicate that metal swords existed even in the time of Adam and Eve. However, the significant joining of metal became common practice with the invention of the forge, which superheated fire by blowing air into a bed of red-hot coals.

Oxygen mixed with acetylene gas and burned in a blowpipe was first used for metal fusion welding at the turn of the 20th century (1900). The invention of oxyacetylene welding did a lot to promote the aircraft industry in World War I.

Arc welding was in its infancy in the early 1900's. My earliest memories as a child were of watching an arc weldor at work and getting my eyes burned from the arc. That was in 1938. I was 3 years old then. But the first arc welding rods were merely bare steel rods.

Oxyacetylene really became a well-developed art in building the WWII steel tube fuselage airplanes. The two gases, oxygen and acetylene, did not change but the regulators, hoses, and welding torches improved. My first torch was a 1950 vintage Victor Aircraft torch, and it did a very good job of building go-karts for me from 1958–1962.

TIG (Heli-Arc) welding was

Oxyacetylene fuel is used in a torch handle to gas weld on aluminum bracket. Flux cored filler rod makes this process very easy to do.

developed during WWII to improve the ability to weld aluminum parts for airplanes. Heli-Arc welding began to improve in the 1970's and 1980's when it utilized inverter technology. Inverter technology allows much smaller and lighter transformers than with straight AC or DC welding equipment.

In 1970, a typical AC-DC Heli-Arc welding machine weighed 400 to 500 pounds. Today a machine with even greater capabilities will weigh less than 100 pounds because of the use of inverters, rectifiers and electronic circuitry.

OVERVIEW OF WELDING PROCESSES

In this chapter, I will provide an overview of the 10 most common methods of welding, brazing, and soldering metals together. This book is written to be a handbook for the beginning weldor and as a source

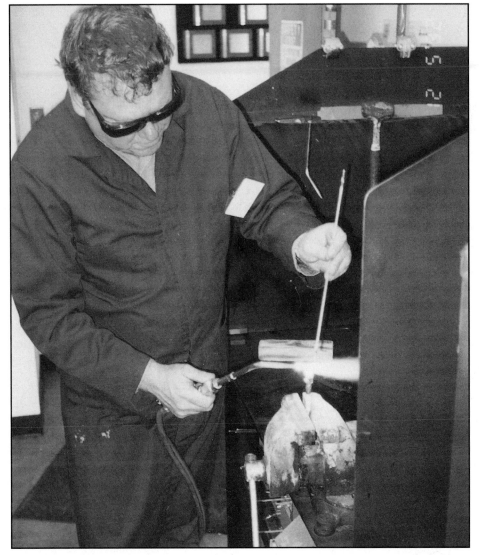

Oxyacetylene fuel is used to braze two pieces of mild steel together. The brazing rod is not called filler rod because it does not mix with the steel parts, it adheres to them to give a 90,000-pound-per-square-inch bond.

book for the more experienced weldor. Schools and factory welding departments will find that memorizing pages of welding hints and tips is not necessary if you have *Welder's Handbook* handy. The processes described in this chapter are: (1) Gas (oxyacetylene) welding and heating, (2) torch (oxyacetylene) cutting, (3) gas brazing and soldering, (4) arc (stick) welding, (5) MIG (wire feed) welding, (6) TIG (Heli-Arc) welding, (7) plasma cutting, (8) spot welding,

(9) plasma welding, and (10) cold welding with epoxies.

Gas Welding & Heating

This process is more accurately called oxy(oxygen)acetylene welding and heating. The heat produced by an oxyacetylene welding or heating torch is 6,300° F (3,482° C), which is very adequate for most welding projects. The heat control techniques used in gas welding are very similar to the heat-control techniques required in

most other kinds of welding, therefore, gas welding is usually taught first in most formal welding schools.

The size of the torch is directly related to the capacity or thickness of the metal it will weld or heat-form. A small torch set up for welding aircraft parts would usually weld steel no thicker than 3/16", .1875" or 4.7mm. A large torch set up for use in oil field work might be capable of fusion welding steel up to 1" (25mm) thick.

If I could only afford one welding machine for my welding shop, I would pick a gas welding setup. Next would likely be a wire-feed (MIG) unit that could be adapted to do TIG (Heli-Arc) welding. Current, modern welding equipment is readily added to by use of modular units that have specific purposes.

Oxyacetylene Brazing & Soldering

Any torch that will weld can also be used for brazing and for soldering. An oil-field-sized torch that will fusion-weld 1/2" thick steel can be used to braze and solder by reducing the gas and oxygen pressures at the regulators, by using a small tip, and by adjusting the flame to a very soft, low heat output flame. However, if accuracy is important in your project, such as jewelry making, a torch setup that is suited to the required heat output will be the best.

Flux—When torch brazing and soldering, it is necessary to use flux to clean the parts while heating. This may be done by using flux-filled or flux-coated rods and solders. Flux is a cleaning agent that comes in many forms, each suited to the base metal and the joining process used.

When preparation is done, brazing

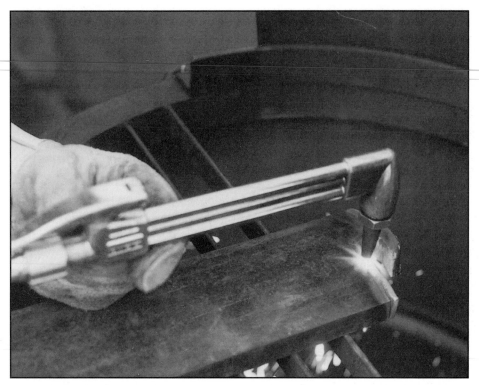

Oxygen and acetylene are mixed and burned by this cutting torch to cut through a thick steel bar in just a few seconds.

and soldering are the easiest of all metal heat-joining processes. First, the metal is prepared, cleaned and fitted for joining. Next, the appropriate amount of heat is applied to the parts with the torch flame, and then the filler metal and flux are applied. When the seam or joint is completed, the heat is removed by taking the torch flame away from the part. When the part cools, water is used to wash off the flux. That is it. Most people can learn to braze and to solder in just a few minutes. But preparation is the key to success in brazing and soldering.

Oxy-Flame Cutting

Only certain metals can be oxidized (rusted) and therefore only those metals can be cut with a cutting torch. Stainless steel and aluminum do not rust, and therefore the oxygen from a cutting torch will not effectively cut those metals even though the oxy-fuel cutting torch has enough heat to melt them.

To effectively demonstrate the oxygen-oxidizing process used in torch cutting, I describe a trick of cutting steel plate with oxygen only on page 74. To better understand the cutting torch process, read that special sidebar. In addition to plasma and laser beam cutting processes described in this book, there are several other cutting processes, including *air-arc gouging, magnesium arc gouging, abrasive wheel cutting* and, of course, *saw blade cutting.*

Oxyacetylene cutting requires more practice to perfect than other welding processes, but it is a very handy skill to have when you are building a metal project. Auto dismantlers also make good use of oxygen-propane cutting torches when salvaging reusable parts for cars and trucks.

Arc Welding

For many years, the accepted method for putting big things together was by arc welding, also called stick welding in the trade. The term "stick welding" was used because the coated welding rod was a "stick" of wire about 12 inches long, coated with flux to provide for strong, defect-free welds. Weldors also called it stick welding to differentiate from "wire" feed MIG welding.

Many of the world's tallest skyscraper buildings were put together with hot rivets and strengthened by arc welding. Most of the nuclear powerplants in the world were welded with low hydrogen welding rod and arc welding machines.

Arc welding is somewhat difficult to learn, but the small shop will find that many things can be welded with the average arc welding machine. A small 220 amp/220 volt buzz box arc welder can build 2- and 4-wheel trailers very adequately.

SMAW—Shield Metal Arc Welding (SMAW) is the oldest and simplest form of electric welding. Many small shops have a "buzz box" arc welder put away in the corner somewhere. From 1940 through 1990, many hot rods, race cars, utility trailers, farm repairs and auto repairs were successfully done with transformer-powered arc welders.

The stick electrode is usually a piece of steel, stainless steel, or aluminum rod that is coated with a clay-like mixture of fluorides, carbonates, oxides, metal alloys and binders to provide a gas shield for the weld puddle. When the rod is used in

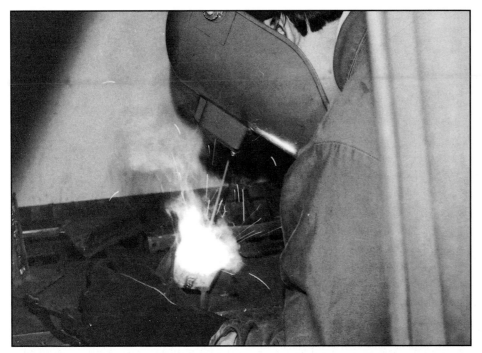

Arc welding is being practiced by this welding student at Allan Hancock College. A stick of flux covered 1/8" diameter steel rod/wire is called a "consumable" electrode, because it provides the spark heat for the weld, then becomes the filler metal as it melts into the part being welded.

welding, the coating cleans and provides a protective cover for the hot weld head to protect it until it cools below the critical point where the atmosphere could degrade the weld bead.

Arc Cutting

The arc welding machine can be used to sever steel, cast iron, aluminum, stainless steel and even the most exotic metals. Two primary methods can be used. The industry standard for many years has been the *air-arc* that uses air pressure combined with the 6,000° F to 10,000° F arc temperature to blow away the molten metal in a somewhat wide kerf. A 3/16" air-arc rod will usually cut a 1/2" wide kerf, and the kerf is usually more jagged than when the metal is cut with an oxyacetylene torch.

Because of the way the air-arc works, most weldors call it the *air-arc gouging* process. The primary use of this process is to cut out defective welds on thick sections, usually over 1/2" thick. The arc is started as in arc welding and as soon as a puddle of molten metal is established, the arc rod is laid nearly parallel with the metal surface, and high pressure air is triggered to blow through the hollow arc rod, blowing away metal as soon as the 10,000° F arc makes the puddle. It is a quick-and-dirty method of metal removal.

And it requires a special electrode holder that has provisions for forcing 90 to 150 pounds of air pressure through the arc rod. A fireproof blanket must be used to catch the sparks blown out of the kerf. Another product of air-arc gouging is noise, caused by the high pressure air. Ear protection is a requirement when using the air-arc process.

Magnesium Arc Rod—A somewhat newer process for cutting and gouging metal is the magnesium-based arc rod produced by specialty manufacturers. The best feature of this cutting rod is that it requires no special equipment such as air pressure and special rod holders. It can be used with any standard stick electrode holder.

This rod can be kept in a tool box and used for cutting off rusty bolts and nuts, severing metal where accessibility is a problem, and where oxyacetylene cutting would be impossible, such as when cutting aluminum plate and castings, stainless steel, and even cast iron. The method of use is similar to air-arc gouging. The arc is struck and then the rod is laid nearly parallel to the surface and pushed forward to extrude the metal from the kerf. Several passes must be made to cut through 1/2" metal plates. This cutting/gouging rod is economical in terms of time and money. In a particular case where NASA required an exotic metal to be cut into blocks for machining, the cut for each block took over one hour and expended a $100 bandsaw blade. The same cut could be done with Cronatron 1100 rod in 5 minutes at a cost of $5.00 for the cutting rod.

MIG Welding

Wire feed welding, officially known as GMAW (gas metal arc welding), has really become popular since 1985. It is possible to drive to your local hardware store, buy a $300 MIG welder, take it home and plug it into a 110V outlet in your shop and immediately begin welding.

But, as with most things in life, one machine just will not do everything. The least expensive MIG welder must

Here, I am MIG/wire feed welding a bracket to a large framework. MIG is also an arc welding process, but much smaller electrode wire is used, usually .030" diameter, and a large roll of the electrode wire is fed into the weld puddle.

use flux-cored filler wire, and it is limited to 3/16" and thinner metals, and it produces gobs of smoke and spatter. But when you are building a barbecue grill or making shop equipment, or ornamental iron, you can make the least expensive machines work for you.

On the opposite side of the picture, robotic MIG welders that assemble new car bodies and components cost tens of thousands of dollars and make welds that are so good that the parts appear to be one piece, with little or no evidence of welding.

Although MIG welding is the easiest kind of welding to get started with, it is also one of the more difficult processes to master. Even in the cutting-edge race car fabrication business, many people find that accurate weld seams are hard to master because of at least three reasons:

• You can't see the puddle easily because the nozzle of the MIG gun

blocks your view. A partial solution to this is to look at the weld puddle from the side rather than from behind the MIG gun.

• The smoke and sparks that are a result of the flux-cored MIG process

make seeing the puddle very hard to do. A partial solution to this problem would be to use inert gas and solid wire to reduce the smoke.

• The weld begins when you squeeze the MIG gun trigger and continues until you release the trigger. You are committed to weld, ready or not. A partial solution to this problem is to pulse weld, either manually or by squeezing the trigger on and off every second or so, and the other solution to this problem is practice, practice, practice.

TIG Welding

The Cadillac or Rolls Royce of all welding processes. You can operate a TIG (Heli-Arc) GTAW (gas tungsten arc welding) torch while dressed in a white Sunday suit, tie and long sleeves. And the weld you do will be neater, probably stronger, and easier to do than with all other welding processes that require fusion of two or more pieces of metal.

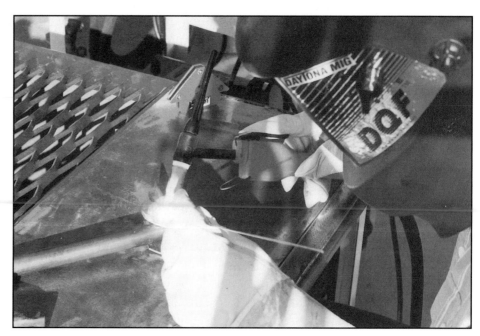

In this photo, I am TIG welding a handle for my welding table. TIG is also called Heli-Arc welding, and it is a very clean, but expensive, process.

hold the puddle while the heat spreads into the part, hold the puddle while you reposition yourself or the filler rod, and they allow for a cool-down of the puddle before shutting off. Automatic TIG welders can be programmed to make perfect welds on things such as the fuel system plumbing on the space shuttle and in nuclear power plants.

An ideal feature of many of the top-end TIG machines is that they can be used to stick weld and to MIG weld, with the appropriate additions of auxiliary equipment.

Plasma Cutting

This is another welding/cutting process that became very popular in the late 1980s, to the point that almost every sheet metal fabricating shop and metal fabricating shop can afford one. Plasma cutters are extremely easy to operate. Anyone can be given one minute of instruction and then can make a cut in any metal. Setting up the machine and replacing the consumable nozzles and electrodes takes just a little more instruction, but still, any shop helper can do it.

Cutting stainless steel sheet and tubing is one of the more difficult things to do in a metal fabrication shop, but a plasma cutter can just as easily cut stainless steel as any other metal, including mild steel. All metals can be cut with a plasma cutter, including aluminum, brass, copper, titanium, but in sheet thickness only. When cutting over 1/2", other cutting methods should be considered.

Spot Welding, Laser Cutting, EBW

EBW—Electron beam welding (EBW) is not a process for small fabrication shops, and is mentioned

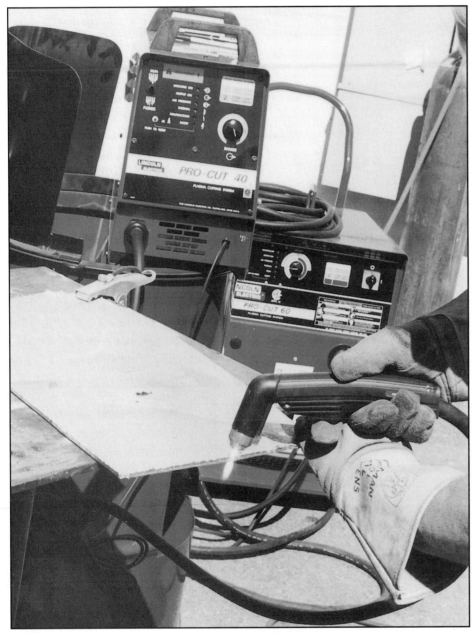

A Lincoln Welding Equipment weldor is using a plasma cutter and dry nitrogen to easily cut a thick aluminum plate. The plasma machine is the one to the left.

The least expensive TIG welders make an inert gas protected weld puddle, but the heat is not adjustable while welding. You are committed to whatever heat setting you decided on before you started, or you must stop and readjust the heat. Test welds before welding the valuable part will partly solve this problem.

The best but most expensive TIG welding machine can be programmed to make nearly perfect welds with only moderate input from the operator/weldor. Pulsed TIG can give you short or long applications of heat with cooling times between application of heat so that your dipping the filler metal into the molten puddle will coincide with the pulses of heat from the machine.

Foot-pedal-operated TIG machines allow the weldor to start a puddle,

This large, expensive spot welding machine is being tested for spot welding aluminum. Once they are properly adjusted, spot welding machines are very easy to operate.

Natural gas, super-heated by injecting large amounts of air into the flame, is used in this ornamental ironworks oven to heat steel bars for forming into fireplace tools.

here only to familiarize you with this process. As the name implies, EBW uses a thin electron beam to fusion-weld two or more parts together when other welding methods would not be possible. Large manufacturers use this process to join automotive transmission gears to a common shaft, missile component manufacturers use the process to join intricate parts that must be made in pieces and then joined as if the final product were one strong, complicated part. Most EBW is done in vacuum chambers with inert gas to prevent atmospheric contamination. Specialty fabricators in most large cities can provide EBW services in small lots of one or more parts projects. Laser beam welding (LBW) is similar to EBW, and can be performed by specialty shops.

Laser Beam Cutting—Laser beam cutting is a more popular cutting method. Many fabrication shops contract with specialty shops to laser cut metal parts that are ready to use as cut. Lasers can easily cut through most metals, up to 1" thick. Thinner materials, .025" through .375" are ready-made for super cuts by laser beam. Laser beam cutting leaves no slag on the backside of the cut as other methods do. One LBC specialty shop hands out key rings with their logo cut in .08" stainless steel, and the key ring is as smooth on the back side as it is on the top side. Look in the yellow pages of your phone book for laser cutting.

Spot Welding—Spot welding is within the capabilities of any metal fabrication shop. Mild steel and low carbon steel can be spot welded with the least expensive equipment. Aluminum spot welding is very common in aerospace and in mass-produced commercial equipment. A body shop spot welder can be operated from most MIG, TIG and stick welding power supplies.

Aluminum spot welding machines are far more complex. The average spot welder for aluminum is twice as large as a refrigerator and new costs are around $100,000. Aluminum requires much more precision to weld than does steel. ∎

METALS & WELDING PROCESSES

The following chart will help you determine which filler materials are available for which metals, and the welding process that can be used to weld each respective metal.

Material	*Thickness	PROCESS							
		Arc Weld.	MIG Weld.	TIG Weld.	Spot Weld.	Gas Weld.	Electron Beam	Oxyacet. Brazing	Lead Solder
CARBON STEEL	S	X	X	X	X	X	X	X	X
	I	X	X	X	X	X	X	X	X
	M	X	X	-	-	X	X	X	-
	T	X	X	-	-	X	X	-	-
LOW ALLOY STEEL	S	X	X	X	X	X	X	X	X
	I	X	X	X	X	X	X	X	X
	M	X	X	-	-	-	X	-	-
	T	X	X	-	-	-	X	-	-
STAINLESS STEEL	S	X	X	X	X	X	X	X	X
	I	X	X	X	X	-	X	X	X
	M	X	X	-	-	-	X	-	-
	T	X	X	-	-	-	X	-	-
CAST IRON	I	X	X	X	-	X	-	X	X
	M	X	X	-	-	X	-	X	-
	T	X	X	-	-	X	-	X	-
ALUMINUM	S	X	X	X	X	X	X	X	X
	I	X	X	X	X	X	X	X	X
	M	X	X	X	-	-	X	X	-
	T	X	X	-	-	-	X	-	-
TITANIUM	S	-	X	X	-	X	X	X	X
	I	-	X	X	-	-	X	-	-
	M	-	X	X	-	-	X	-	-
	T	-	X	-	-	-	X	-	-
COPPER & BRASS	S	-	X	X	-	X	X	X	X
	I	-	X	-	-	-	X	X	X
	M	-	X	-	-	-	X	X	-
	T	-	X	-	-	-	X	-	-
MAGNESIUM	S	-	X	X	X	-	X	X	-
	I	-	X	X	X	-	X	X	-
	M	-	X	-	-	-	X	-	-
	T	-	X	-	-	-	X	-	-

* S=sheet up to 3mm, 1/8".   I= 3–6mm, 1/8-1/4"   M= medium, 6–19mm, 1/4–3/4".   T=thick, 19mm, 3/4" and up.

Welding Equipment

3

For welding shops ready to upgrade their welding equipment, the best process is to shop around for the best equipment that suits their purpose. In almost every instance, the smartest thing to do is to try the equipment before you buy it unless it is a welder with known capabilities. If your friend has a welder made by a specific company and he is happy with it, then your choice of welder is narrowed down significantly.

But with steady improvements in technology, it is advisable to shop around and ask to be introduced to the latest and best the industry has to offer. Electronic welding helmets were not even invented five years ago, and now they are on every weldor's shopping list. Pulsed MIG and TIG welders are new developments, and they offer better control of the weld, both for the novice weldor and the professional weldor. Inverter technology has really opened up the welding machine industry to provide portable welders that will do more than the big, heavy machines made in the 1960s and 1970s.

WHERE TO BUY

Local welding supply dealers usually sell at retail except for sales promotions. Large department stores usually have tool departments, and most of these tool departments have

The first welding setup for any shop should include a portable gas welding (oxyacetylene) torch, carrier and bottles such as this Harris torch, that sells for about $300.00.

welding equipment for sale. Mail-order catalogs are other good sources for welding equipment. Most of the automotive and aircraft monthly magazines feature advertisements from reputable welding supply firms, and their ads provide good information about prices and features. Home improvement centers are also good sources of welders and welding supplies.

For occasional use, check with tool rental firms, and don't overlook used welding equipment listed in newspaper want ads. However, buying used welding equipment is

about as risky as buying a used car or used lawn mower. In every case, trying the equipment before you buy is a very good practice.

GAS WELDING EQUIPMENT

For sure, a gas welding torch setup is strongly advised in any beginner's shop. Being able to heat, bend, form and torch-cut steel parts is essential. Even large, well-established fabrication shops can always benefit from a lightweight, portable gas welding and cutting setup. The most

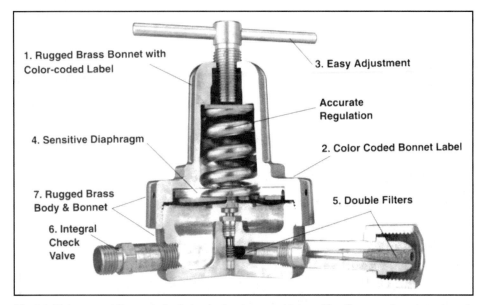

1. Rugged Brass Bonnet with Color-coded Label

3. Easy Adjustment

Accurate Regulation

4. Sensitive Diaphragm

2. Color Coded Bonnet Label

7. Rugged Brass Body & Bonnet

5. Double Filters

6. Integral Check Valve

Most initial gas welding sets include single stage regulators like the one show here in cut-away. Tightening the top screw will compress the large spring that acts on the diaphragm to provide pressure to the torch. Victor Equip. Co.

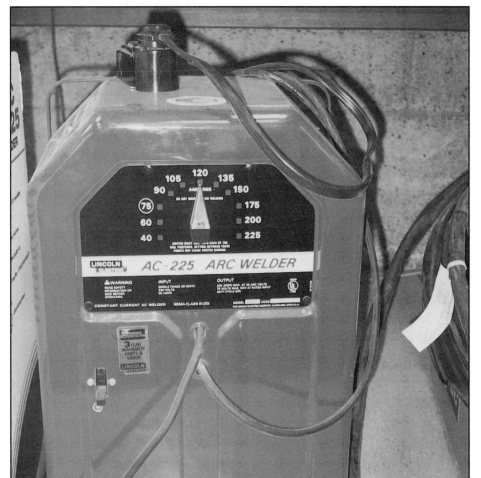

This Lincoln Electric AC-225 "Buzz Box" arc welder will last a lifetime and it sells for about $300.00. It will stick weld and cut metals with magnesium cutting electrodes. It will build trailers and similar equipment, and even dirt track race cars.

important point to decide on when purchasing gas welding equipment is whether you will be working with very thin material—0.49" and thinner —or with thicker material, 0.50" to .250" thick and thicker. For thin material, choose an aircraft-type torch and possibly add a jeweler's torch to your tool box. If your projects will be race cars, trailer or farm equipment, a standard-sized torch is recommended.

ARC WELDING EQUIPMENT

Several different brands of stick electrode arc welding equipment are available. Usually, these 220 volt, 225 amp to 250 amp buzz box welders are priced about the same as a small portable oxyacetylene torch setup.

The pros and cons of buzz box welders are similar to the same for MIG-wire feed welders:

•Buzz box welders are only useable with coated stick electrodes. Stick electrodes have a shorter useful life than bare welding rod and welding wire because moisture affects the flux coating and reduces the arc rod's welding ability. Old, wet stick welding rod is almost useless and must be disposed of.

•Buzz box welders are not production welding machines because they have a duty cycle of 10% to 40%, meaning that they must rest 6 to 9 minutes in every 10 minutes of welding.

Positive aspects of buzz-box welders are:

•You can purchase coated stick welding rod in small (1 to 5 lb.)

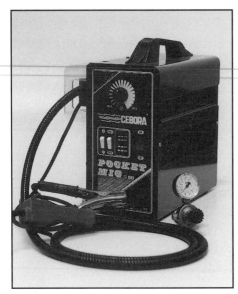

This Pocket MIG wire feed welder by Daytona MIG was the first small 110-volt powered welder to be sold in America. It weighs 42 pounds, will plug in and operate from 110-volt house current, and will weld steel up to 3/16" thick. Daytona MIG.

packages, and that way you have fresh rod for each project. And you can purchase small amounts of several kinds of welding rod.

• Even though 20% duty cycle means that you can weld 2 minutes, and that you must let the welder cool for 8 minutes before resuming, that is not a serious problem for most small projects. Changing used-up rods, tack welding and moving around to the other side will average out the 20% factor.

Larger (up to 500 amp) arc welders are available, but most shop welding jobs are done with 1/8" rod at 130 amps or less, usually at 100 amps.

Welding schools, fabrication shops, production welding shops and construction shops are users of the higher amp arc welding machine.

Another feature of large arc welding machines is that they are usually able to be switched from AC to DC

Another portable MIG welder from Daytona MIG is this 220-volt, single-phase machine that can be plugged into your clothes dryer outlet. It can weld with gas or flux-cored wire. Daytona MIG.

positive and DC negative polarity for special welding jobs.

MIG WELDING EQUIPMENT

By far the most popular small shop welder is the 110-Volt MIG/wire feed

machine. With a 5-pound roll of flux-cored wire installed, these machines are all ready to plug into house current and weld immediately. There are two drawbacks to the 110-Volt MIG welders:

•Flux-cored wire feed welding

Miller Electric makes these four sizes of portable MIG/wire feed welding machines. The weights range from 53 pounds to 125 pounds. Much larger models are also available from Miller Electric. Miller Electric.

produces lots of smoke and spatter, and the spatter would be objectionable in appearance in most welds.

•The bottom-of-the-line MIG welders cannot be converted to gas and they cannot be converted to weld aluminum. You can only weld steel with them.

Some of the positive aspects of portable MIG welders are:

• They are almost always ready to weld. You simply adjust the voltage knob, adjust the wire speed knob, turn on the power switch, and you are ready to tack weld or to weld.

• It is possible to weld for several hours without ever changing the electrode (wire roll).

• Higher capacity (and higher cost) MIG welders are so trouble-free that they are used in robots to weld automobiles in auto manufacturing.

• Lower- to medium-priced MIG welders are available that can be switched to gas, and the polarity can also be switched for gas and aluminum welding.

TIG WELDING EQUIPMENT

TIG welders are available in more configurations than most other welders. All TIG welders must be supported with argon gas or other shielding gas, but as one manufacturer states in its welding booklet, TIG welding can be done with a regulated argon flow and 24 volts from two car batteries wired in series! Safe to say, there would be no current control in a setup like that, so you need to decide which welding setup suits your needs.

Scratch-start TIG torches do a pretty good job if you are welding thin to

This Combination TIG-MIG Power Supply, arc (stick) welder will do just about any job in a commercial welding shop. The foot pedal amp control is on top of the machine. Photo: So-Cal Airgas.

The portable, gasoline engine powered welder has a TIG module and foot pedal amp control for portable field welding of steel and stainless steel. It will also power a MIG/wire feed module and will do stick welding (Lincoln Electric Co.).

I cut my TIG welding teeth on a machine just like this one. It was made in late 1969 or 1970, but it still does high-quality welding. Look for a machine like this at bankruptcy sales, school surplus sales.

medium thickness steel tubing and steel sheet. There are several brands of DC-only TIG welders that use scratch start and pulse current to weld steel. You cannot weld aluminum with a DC-only welder. Before buying one of these welders, try it out, or at least ask for a satisfaction-or-your-money-back guarantee. They do excellent work on 4130 steel and mild steel. They cost about 2 1/2 times more than buzz box welders.

Foot-pedal-operated, variable remote current control AC/DC, high frequency TIG welding machines are a dream to operate. The bottom-of-the-line, full-feature TIG welder sells for about 4 to 5 times the cost of a bottom-of-the-line buzz box AC welder or DC-only MIG welder, but they will do many kinds of fusion welding on most materials, and the welds are aircraft-quality welds.

The mid-price TIG machines cost 4

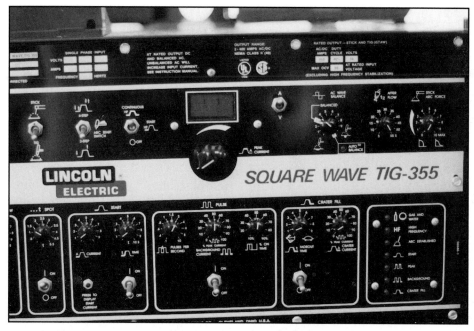

TIG welding is much easier to do with this almost top-of-the-line square wave machine. Indy race car builders love to use this machine because it does aircraft-quality welds every time. Lincoln Electric Co.

This small shop TIG welding/plasma cutting set, complimented by a welding/cutting table and a mechanic's tool box is a very good start for a small fabricating operation. The TIG welder operates on 220-volt, 30 amp house current, and the plasma cutter operates on 110-volt house current and 70 psi compressed air.

to 5 times as much as the entry-level TIG welder, but those machines can be used in trade schools, aircraft factories, and other production welding facilities.

The top-of-the-line TIG welders feature pulsed, square wave current, pre-flow adjustments, post-weld timer, and repeatable weld programs. Many of them have digital readouts, and will cost about as much as a small economy car. But they make an average weldor look like a welding talent!

PLASMA CUTTING EQUIPMENT

Good, low-priced plasma cutters can be found in any mail-order ads, in car magazines, in aircraft magazine ads, and in large department stores, as well as in local welding supply outlets.

The best thing about plasma cutters is that they will cut stainless steel as

easily as they cut mild steel. If you have ever tried to hacksaw or bandsaw cut a sheet of .080" thick stainless steel sheet, you know that it will ruin your blade in just a couple of inches of cut. With the plasma arc cutter, you can cut a 4' x 8' sheet of stainless steel in half in about 2 minutes, and do it again and again before you have to replace the consumable electrode in the torch handle.

Plasma cutters come in many sizes and prices, just like other welding equipment. The bottom-of-the-chart plasma cutting machine costs $500, and a top-of-the-line, BIG plasma cutter costs $5,000 or more. The differences are in what the machine will do. They all will cut all metals—steel, aluminum, copper, stainless steel, brass, and titanium—but some plasma cutters do the job better and cleaner than others do.

Even the higher-priced plasma cutters have duty cycles. If you cut for

10 minutes without stopping, their "fire" will die out and not come back on until the unit cools for 10 minutes. If possible, try the plasma cutter before you buy it, or get a satisfaction-or-your-refund guarantee.

SPOT WELDING EQUIPMENT

These sheet metal assembly welders come in two classifications. The transformer-type will spot weld steel, galvanized steel and, sometimes, stainless steel. The second type will spot weld aluminum, magnesium and titanium by special programmed weld sequences. These electronics-controlled, air- or hydraulic-operated capacitive discharge welders are very large and very expensive. Never,

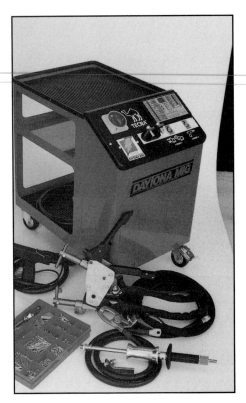

Daytona MIG sells this Techna spot/stud welder. It has 5,000 amps output, will do both pinch spot welds and one-side spot welds. It will also do seam welds and will spot weld studs for body shop metal pulling.

under any circumstances, should you buy an aluminum spot welder unless the sales contract provides for setup and hundreds of samples performed and certified to prove that the aluminum spot welder will perform as you want it to. See page 131 for a picture.

One of the best methods for deciding on which welding equipment to buy is to visit several shops, welding shops, airplane or race car fabrication shops or ornamental iron shops, and ask the people there for their recommendations. Then make up your own mind and buy or rent the equipment that best suits you. In Chapter 7 through Chapter 17, this book will tell and teach you how to use the equipment.

WELDING AREA

Chair—Get a lightweight, comfortable chair. I prefer a chair like that used by a draftsman—one that can be raised or lowered to suit the work height. The chair should have rollers so it can be moved without having to pick it up. Don't spend a lot of money. Check a used-furniture store or a flea market first.

Table—Build a metal-top welding table. The table top should be steel—not aluminum or wood—at least 1/4" thick and 2' square. Make the top larger if you plan many big, heavy welding jobs. A good all-purpose welding table uses a 3/8" thick, 3' x 4' steel plate for the top. The frame to hold the welding table can be anything sturdy enough.

Keep the table top fairly rust-free, especially if you are using it for arc welding or TIG welding. Rust is a poor electrical conductor and will even contaminate gas welds. I power-sand the top of my weld table occasionally to expose shiny steel. Never paint the top of a weld table because it blocks current flow from the ground electrode to the workpiece.

Drill a 1/2" or larger hole in the back side of the table top or frame and put a bolt and nut there for a more secure ground connector attachment. See page 146 for a welding table design you can build yourself.

Keep an assortment of C-clamps, weights, metal clothespin clamps and holding fingers nearby to hold the pieces in position while you weld. Never ask anyone to hold a small part while you weld it. They won't like it when the sparks fly and the metal heats up!

CONVENIENCE EQUIPMENT

Sandblaster—A great addition to any welding shop is a sandblaster. You can buy a small portable sandblaster that holds 50 pounds of sand for about the price of 3 tanks of gasoline. I used one to completely sandblast the frame of a 1956 Ford 1/2-ton truck I was restoring. The job took about 6 hours, but the result was a like-new frame that was ready to accept a coat of polyurethane paint.

A sandblaster is also handy for cleaning small parts and numerous projects. But it does take several minutes to get the sand ready. And after sandblasting, you'll need a bath to wash all the sand off. It gets everywhere—even where the sun doesn't shine!

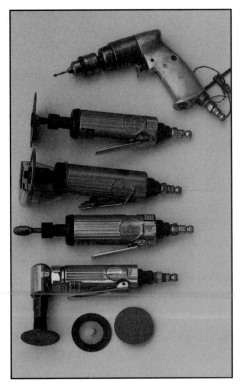

This handy assortment of air-operated tools, a 1/4" drill, 2 cut-off wheels, a die grinder and an angle sander, really make fitting and welding a lot easier. The average cost was $20 per tool.

For once-a-year welding projects, consider a portable gasoline-powered arc welder from your local rental yard. The one shown here will weld right out of the back of your pickup truck.

The most economical way to heat large quantities of steel parts for ornamental iron projects is to use air-charged natural gas ovens like the one shown here. The heat output is over 3,000° F, but in large amounts of BTU's. Photo: Kummer Ornamental Iron.

If you buy a 4–8 cfm siphon-type sandblaster, you'll need at least a 1- or 2-hp air compressor. A 1/2-HP compressor doesn't have sufficient capacity. Pressure-type sandblasters work two or three times faster, but will cost up to ten times more. The siphon-type sandblaster will do a good job of cleaning metal for the occasional, small welding job. For more frequent weldors, and more complex welded assemblies, consider investing in a pressure-type sandblaster.

Metal Marker—A handy item to have around the welding shop is a metal marker. It's a tube of paint with a ball-point end that marks almost anything. It will mark smooth metal, rough metal, oily metal, wood, glass or even plastic. You'll find these markers at most welding supply shops.

I use a metal marker for identifying scraps of metal that are suitable for future use. A tube costs about as much as a hamburger, but lasts about two years in normal use.

Never use a scribe for marking metal. Scribe lines in metal act the same as perforations in a piece of paper–ready-made for tearing, breaking or cracking. If you do scribe-mark a piece of metal, the scribed area should be discarded because of the potential for cracks. Such a case would be marking a cut line.

Temperature Indicator—As discussed in Chapter 1, when welding or heating metal, it's crucial to get the metal to just the right temperature—no more. Excess heat will often ruin the metal. This can happen even when loosening rusted or otherwise stuck parts, too.

The way to assure yourself of exact temperatures is with a temperature-indicating crayon or paint. Temperature indicator is applied to the area to be heated or welded. When the indicated temperature is reached, the crayon or paint melts like wax on a hot surface.

Temperature-indicating crayons and paints come in more than 100 different temperatures, from 100°F (38C) to 2500°F (1371C). Claimed accuracy is plus or minus 3°F. These and other types of temperature indicators by Omega Engineering and Tempil can be purchased at most welding-supply stores. ■

Welding Safety

4

W elding is more hazardous than most other shop processes. The dangers of fire and explosion, burned hands or metal in the eyes are always present. Additionally, high-pressure oxygen, acetylene, argon, CO2 or helium tanks present a potential work hazard. Hot metal is always a danger if it comes in contact with anything flammable or meltable, or your skin.

To start with, I'll relate some horror stories–welding explosions caused by carelessness.

First, an explosion that caused a fatality occurred several years ago when a local weldor was assigned the job of cutting several 55-gallon oil drums in half. These were to be made into cattle-feeding troughs. He had cut several empty drums in two, splitting them lengthwise with his cutting torch. On the last drum he cut open, he was straddling it as if riding a horse. Just as his cutting torch pierced the metal, the drum exploded, blowing the weldor through the shop's corrugated roof. It was estimated that he was blown 50-feet straight up.

I don't know what the oil drums contained, but the one that exploded obviously contained a fatal combination of explosive vapors.

Small Weld Job—Very Big Explosion

Then there was the story one of my

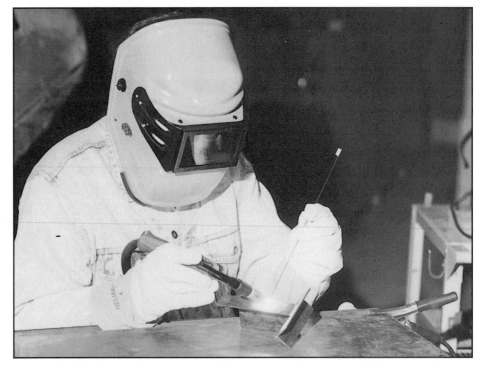

Welding is dangerous business! Make sure you take all safety precautions. This includes proper protective clothing, and the appropriate welding head gear and eye protection required by the type of welding you are doing. Photo: Gayle Finch

auto-shop students related to me about an explosion in my neighborhood.

This accident occurred in the driveway of a private residence near my house. The destruction was so great that the police and fire departments blocked off three streets to preserve order while they surveyed the damage.

A local oil-trucking company had hired two neighborhood weldors to weld taillight brackets on the rear bumper of some new trucks. The trucks were really fancy, with

fiberglass cabs, chrome-plated dual exhaust stacks, and a large stainless-steel tank on the back for hauling crude oil. The tank was a two-layer assembly with fiberglass insulation between the inner tank and the outer cover. The welding had been completed on one truck. The second truck was backed into the driveway so the weldor could get the welding cables to the rear bumper. The truck was so big that it occupied the entire driveway.

The auto-shop student had been

Flammable Containers & Welding

The lesson to be learned from these accidents is that containers of flammable products should *never be welded* on or cut with a torch. It is safest to refuse to weld on or near any such tank or container, even if it has been completely drained and sitting empty for a long time. The vapors and flammable materials can permeate the metal, and even though it smells clean, you may not be able to smell it. Even vapors from non-flammable liquids can be explosive under certain conditions. And for sure, vapors from flammable liquids are explosive.

As a weldor, people will ask you to weld many different types of containers. You may even want to weld something similar yourself, but the safe thing is not to do it.

If you absolutely must have a tank welded, such as a classic gas tank for a restoration that simply can't be replaced, take it to a professional welding shop that specializes in welding gas, oil and similar flammable tanks. They'll know what preparation and welding procedures are needed to prevent accidents such as those just described. Typically, a professional welding shop will "boil out" the tank in radiator cleaning acid to remove any oil or gas residue, then purge the tank with an inert gas such as argon while welding it. Clearly, these are techniques above and beyond the skills of even talented do-it-yourself weldors.

watching, but went home for lunch. He lived in a house one block from where the welding was being done. He heard a tremendous explosion and ran outside just in time to see a huge stainless-steel tank fall on top of the house behind his. At first he thought a missile from nearby Vandenberg AFB had exploded and was falling on the neighborhood. Then he realized the debris had come from where the trucks were being welded, so he ran down the street to find the new truck had been totally demolished! One of the weldors had been blown across the street, bruised but alive. The other fellow was found inside the kitchen of the house, also bruised but alive.

The truck didn't fare as well. The oil tank, of course, had been blown off the frame, straight up 100 feet or more and had parked itself on the house roof. The cab was shattered. The two chrome-plated exhaust pipes had been flattened to the ground. The frame of the truck was warped and twisted. Pieces of the truck landed on cars and the rooftops of several nearby houses. The new truck was totaled! Damage to the house was 50%. The two men welding on the truck miraculously escaped with only minor injuries.

The explosion was caused by a welding spark igniting vapors escaping from the crude-oil tank.

ARC WELDING SAFETY

Radiation Burn

The primary safety hazard in arc welding is from ultraviolet-light (radiation) burns to the eyes and, to a lesser degree, the skin. This hazard applies to TIG, plasma-arc and wire-feed welding as well.

The burns received from electric-arc welding are similar to sunburn, except usually deeper into the skin or eyes. One reason for the deeper and potentially more severe burns may be the ultraviolet light source–it's much closer to the body than the sun is. Therefore, arc-welding radiation is more intense. And there's less atmospheric dust to filter the rays of the arc welder.

Clothing

The solution for preventing burns is simply to shield the skin and eyes from ultraviolet light. You should wear a long-sleeve, close-weave shirt and trousers of material least likely to ignite from sparks. Regular work clothes are acceptable. Wool or heavy denim, such as blue jeans, works just fine for shielding radiation. Many professional weldors use leather aprons, jackets or pants for burn protection. And don't forget welding gloves. Your hands are the closest to the heat and light source. Consequently, they are most vulnerable to burns from radiation, sparks and hot metal.

For maximum burn protection from your long-sleeve shirt, tuck the sleeves inside the top of the gloves and button the collar at the top. Many beginning weldors neglect to button up their shirt collar and end up with a nasty ultraviolet burn on their neck. Wearing a leather flap on the helmet to shield the neck and throat is the best way to avoid neck and throat burns.

Gloves—Protect your hands and wrists from burns with leather gloves. Although leather doesn't burn easily, it will char and shrink if it contacts hot metal or a flame. Choose the most

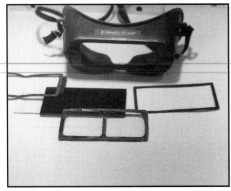

Special corrective lenses (lower left) can be used in place of corrective eyeglasses. Such lenses are available at welding supply stores. This lens will fit either gas welding goggles or an arc welding helmet.

flexible gloves possible. For light TIG welding, I prefer deerskin gloves. For heavy arc welding on structural steel, heavy cowhide or horsehide gloves give you the protection needed.

Shoes—If you weld bridges professionally, you should wear leather, high-top, safety boots. Regardless of whether you weld for a hobby or professionally, never wear low-cut shoes, particularly the slip-on type such as loafers. Sparks and molten metal can slip down inside your shoes. Regardless of your pain threshold, you'll drop everything so you can get your shoe off and the hot metal out while you are dancing the jig. And while you are dancing the jig, the dropped welding torch could start a fire or explosion! Never wear nylon jogging shoes. If exposed to high heat, they could melt to your foot! For hobby welding, an old pair of leather lace-up shoes will be OK.

Arc-Welding Lenses

I prefer the rectangular 2 x 4 1/4-in. lens. It's available in various shades and tints. The standard tint is green #10. A darker tint for welding in bright sunlight is green #11 or #12. A #9 lens is OK for low-hydrogen welding or TIG-welding steel. TIG or

RECOMMENDED WELDING LENSES

Application	Base Metal Thickness (inches)	Suggested Shade No.
Arc welding with 1/16, 3/32, 1/8, 5/32" electrodes	1/8 to 1/4"	10
3/16, 7/32, 1/4" electrodes	1/4 to 1"	12
5/13/8-in. electrodes	Over 1	14
TIG welding with non-ferrous 1/6, 3/32, 1/8, 5/32-in. electrodes*	Up to 1/4	11
TIG welding with ferrous 1/16, 3/32, 1/8, 5/32-in. electrodes*	Up to 1/4	12
Soldering	All	2
Brazing	Up to 1/4	3 or 4
Light cutting	Up to 1/4	3 or 4
Medium cutting	1-6	4 or 5
Heavy cutting	Over 6	5 or 6
Gas welding (light)	Up to 1/8	4 or 5
Gas welding (medium)	1/8-1/2	5 or 6
Gas welding (heavy)	Over 1/2	6 or 8

* For MIG welding, decrease shade no. by one.
Make sure your eyes are protected with the right lens when welding.

Clockwise from top left: (1) HTP chin-up arc welding helmet is inexpensive and makes welding easy; just move your chin to open and close the lens. (2) Fiber helmet retrofitted with electronic lens. (3) Lightweight, large lens Huntsman helmet with #10 lens. (4) Daytona MIG/D.Q.F. helmet with adjustable electronic lens. Pick one that suits your budget and that works for you.

wire-feed welding aluminum require a darker-tint lens, such as #10, #11, or even #12. You should pick the darkest lens that still allows you to see the weld puddle.

Don't use gold- and silver-plated lenses. Although they are pretty, one tiny scratch in the gold or silver plating could admit enough ultraviolet light to burn an eye.

A tinted lens should always be protected from weld spatter, scratches and breakage. Cover it with a clear, disposable plastic or glass lens. Change the lens protector as often as necessary to ensure distortion-free sight. Likewise, clean the lens as often as you would a pair of eyeglasses.

CHOICE OF HELMETS

In the late 1990s, electronic helmets are the rage. Every weldor who doesn't own one, wants one. They really do work, but there are drawbacks. If you strike a new arc 50 or so times in a day, your eyes will itch at the end of the day if you are using an electronic helmet. That is because the fraction of a second that it takes the lens to darken from #2 to #10 or #12 allows ultraviolet light to strike your eyes. The effect is cumulative. It adds up by the end of the day. The time lag for darkening is noticeable. Another negative is the cost of electronic helmets, which run

about 10 times or more the price of a standard arc-welding helmet.

Another special arc welding helmet is a "chin-up" model that allows the weldor to lift the lens only, just enough to see the weld area immediately before striking the arc. To close the lens, the weldor simply moves his chin about 1/2". For me, this is easier than the other methods of protecting my face, neck and eyes from welding flash burns.

The good thing about electronic helmets is that you can see where the arc will strike immediately before you weld. As with many things in welding, "try before you buy."

There are about 100 different safety-approved welding helmets, or hoods. I have two different helmets for myself.

Get the lightest helmet that covers your face completely. Whether you are welding bridges or ships for a living, or welding a lawn mower or child's bicycle, the lightest helmet will distract you the least. You need all the concentration you can muster to do a good welding job, so you certainly don't want a heavy helmet pulling down on your neck and head. Regardless of weight, though, you need maximum protection.

Safety Glasses

If you can't afford safety glasses, you can't afford to weld. Don't think safety glasses are too much bother—they don't compare to the bother of impaired eyesight. Your eyes will not tolerate hot metal in them. I wear my safety glasses when grinding, chipping, filing or anytime there's a chance of metal flying around.

Safety for Bystanders

When arc-welding, you have the

ARC WELDING EYE BURNS

Common courtesy dictates that the weldor is responsible for shielding his welding from bystanders, helpers, and dogs and cats. If a pet watches you arc weld for 60 seconds—one minute—the pet can have severe eye radiation burns for 12 hours or more. Be thoughtful. Don't let pets or people watch you arc weld, TIG, or MIG weld, or plasma cut. Shield your arc welding.

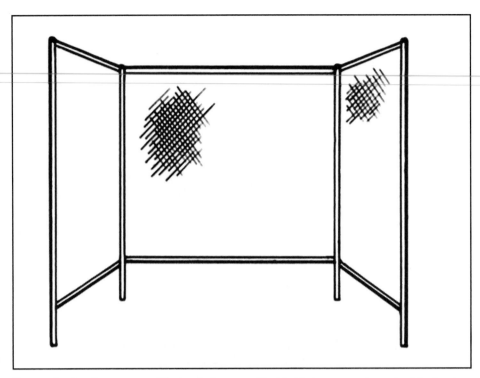

Be sure to do all your arc welding and plasma cutting behind a screen to protect bystanders and pets from eye damage from ultraviolet radiation. You can buy screens at your local welding supply, or you can make a temporary screen from three pieces of 1/8" to 1/4" plywood, hinged together at the edges.

obligation to protect bystanders from ultraviolet burns to their eyes. First of all, you should provide a suitable screen around the weld area so direct flash can't be seen by anybody not wearing correct eye protection. For a temporary screen, place a sheet of plywood or corrugated tin so it shields those nearby.

If you can see arc-weld flash from any distance, it can burn your eyes. The closer you are to the flash, the more severe the burn will be. The same rule applies to pets and other animals. Within reason, keep them from looking at the welding flash by using shielding. If anyone is in the welding area, always say "watch your eyes" before you strike the arc. This gives the person a chance to turn the other way or close their eyes. If you weld regularly in a certain area, permanent or portable screens should be used. These can be built or purchased from a weld-supply outlet.

GAS-WELDING SAFETY

Oxyacetylene welding presents its own set of safety considerations.

Everything I said about shoes, gloves and clothing for arc welding also applies to gas welding.

Goggles

Goggles for gas welding or cutting are different than arc-welding helmets because they don't have to shield the weldor from hostile light. Oxyacetylene welding doesn't produce ultraviolet light, so it can't cause flash burns. The only burn you can get from oxyacetylene is from the heat of the flame, and from sparks from the oxidized metal being welded or cut.

The best goggles for oxyacetylene welding are fiberglass and have a 2 x 4 1/4" lens. You can wear most conventional vision-correcting eyeglasses under them. In addition, they are easier to put on and adjust than the separate-eyepiece goggles.

Bottled Welding Gases

Welding gases come in high-strength, high-pressure cylinders. When full, they contain the following pressures at room temperature:

Oxygen—2200 psi
Argon—2200 psi
Hydrogen—2200 psi
Acetylene—375 psi

If a pressurized cylinder of oxygen or argon welding gas falls and breaks off a valve, the resulting release of high-pressure gas can turn it into a rocket. I know of one high-pressure cylinder with a broken-off valve that went completely through four automobiles and the brick wall of a welding shop. As with fire, bottled welding gas is useful when contained and controlled. But also like fire, pressurized gas can be destructive. Always chain welding gas cylinders in an upright position to prevent them

BOOM!

Although acetylene is bottled at comparatively low pressure, the acetylene itself is highly explosive. Like leaking gas from a natural gas pipeline, nothing happens if acetylene leaks into a confined space unless there is a spark or flame to ignite it. Then, it can explode just like a stick of dynamite. Never carry flammable gas such as acetylene, hydrogen or MPS gas in the closed trunk of a car. If vapors escape, a major explosion could result. Instead, haul acetylene or MPS gas cylinders in an open truck or trailer.

Don't let these safety tips worry you unnecessarily. Every time you drive a car, you are carrying around enough gasoline to do about the same damage as the contents of any welding gas cylinder. Therefore, if you observe proper safety precautions, you should have no trouble.

SOLDERING SAFETY

You may have heard stories about the dangers of lead poisoning from soldering. Here's how lead-poisoning dangers can be avoided:

•Smoking, eating or drinking while handling lead solder can eventually cause lead poisoning. Be sure to wash your hands and face with soap and water after handling lead solder.
•Do not breathe fumes from overheated lead solder. If you keep the temperature between 250° F (121° C) and 900° F (482° C), there is almost no danger of lead poisoning. But lead heated to over 1000° F (538° C) will give off fumes that are hazardous to your lungs and body.
•The harmful effects of absorbing lead into the human body are largely cumulative. We all remember the Mad Hatter from *Alice in Wonderland*. Years ago, hatters used lead to make hats. Also, printers used hot lead to make printing plates. In time, lead accumulated in their bodies, resulting in various nervous disorders, anemia, sometimes paralysis and death. If you experience colic, muscular cramps or constipation when using lead, stop immediately and contact a physician.

from falling. An acetylene cylinder must be kept upright to keep acetone, a stabilizing gas, at the bottom of the cylinder. If the heavier acetone gets into the cylinder regulator, it will damage its rubber parts.

GENERAL SAFETY TIPS

Work Area

The most common welding accident is burned hands and arms. Keep first-aid equipment for burns in the work area. And, as I already discussed, eye injuries can occur if you get careless. So keep handy a phone number for emergency medical treatment for unexpected injuries, particularly eye injuries. A medical doctor is the only one who can properly treat any eye injuries. Don't try to treat an eye injury yourself. Get to a doctor immediately.

The ideal welding shop should have bare concrete floors, and bare metal walls and ceilings to reduce the possibility of fire. Although you probably don't have such a building, the important thing is to keep flammables, and rags, wood scraps, piles of paper and other combustibles out of the welding area. The same goes for wood floors; never weld or cut over one.

If you must weld in an enclosed garage, make every effort to eliminate anything that could trap a spark. Sparks can smolder for hours and then burst into flames. So, regardless of where you're welding, be sure to have a fire extinguisher nearby. Also, keep a 5-gallon bucket of water handy to cool off hot metal and quickly douse small fires.

Never use a cutting torch inside your workshop. Take whatever you're going to cut outside, away from flammables. Also be aware that welding sparks can ignite gasoline fumes in a confined area.

Grinding Sparks

When grinding with a portable grinder or bench grinder, the resulting high-velocity sparks are tiny pieces of metal. These tiny projectiles are in an oxidized state and at the same temperature as metal being cut with a gas torch. Therefore, be as careful of where grinding sparks fall as you would if you were using a cutting torch.

Eyeglasses

An often-overlooked cause of less-than-perfect welding is not being able to see the weld puddle clearly. Frequently, people who don't have good vision attempt to weld without

Inexpensive drug store-purchased reading glasses will do in many cases to prevent you from having to wear your expensive prescription glasses in the welding shop.

This weldor is almost perfectly set up for MIG welding. She is using a suction vent to take away smoke and fumes, she has new clean gloves, a full-face helmet and a pair of Nomex fireproof coveralls. Buttoning up the collar would keep sparks from going down her blouse! Lincoln Electrical Co.

eyeglasses. If you want to do a good welding job, but need glasses, have your eyes checked and corrected first. Or, wear your glasses.

If you don't want to subject your expensive bifocals to welding damage, there are other solutions. The least-expensive solution is to buy a pair of reading glasses at a drugstore or supermarket. These come in standard magnification increments of +1.25, +1.50 and so on. Welding supply shops also sell eyesight-correcting lenses that fit the 2 x 4 1/4" opening in helmets or goggles. These are available in +1.25, +1.50 and higher increments, too.

Comfort

I suppose most people think a weldor at work is not comfortable. That is somewhat true, but the more comfortable you are while welding, the better your work will be. I prefer to sit when welding. Why stand and have leg fatigue interfere with concentration?

Just as you would sit down and get situated to write a letter, get yourself comfortable before starting to weld. Think about it: You wouldn't squat down and write a letter on the floor. Likewise, you wouldn't stand up with your arms stretched out to write a letter with the paper against a wall. And you certainly wouldn't lay on your back, writing a letter using the side of the desk! If you tried writing in these awkward positions, your handwriting would not look good. That's why you can't weld as well lying on your back, or squatting on the floor.

There are exceptions. For instance, you can't flip a car on its roof just to weld on a new muffler. However, you can get yourself as comfortable as possible if you have to weld in such a position. ■

Fitting & Jigging

5

Believe it or not, this is the most important chapter in this book! You can be the world's best weldor, but if the parts fit poorly, you simply cannot weld them properly. If you have a 1/4" wide gap in a piece of .030" sheet metal, you simply cannot do a nice, strong weld across that gap. If you have a rough, jagged angle iron (steel) frame to weld for a two-wheel trailer, the welds will be weak and could crack when least desirable, like when you are towing a heavy load on the highway.

Fitting is the process of cutting and shaping metal pieces so they fit together without large gaps. Big welding operations usually employ a weldor-and-fitter team–two people who fit and complete the welds. If you were to time and apportion the operation, fitting takes about 80% of the time; welding takes the balance. These percentages illustrate how important fitting is to making good welds.

You must fit pieces properly before welding to make sure they stay in place during welding and subsequent cooling. The idea is to avoid large gaps that must be filled with weld bead.

Jigging is assembling a project in a fixture to ensure that the welded assembly conforms to design specifications. Jigs are used in large-scale production to assure consistent

Simple plywood-and-nail jig for 4130 steel tube airplane frame. Fit tubing so 1/16-in. welding rod cannot be inserted in weld-seam gap. Remember, if the parts don't fit exactly right, you won't be able to weld them properly, no matter how good you are as a weldor.

quality. You must plan ahead of doing the actual welding in order to get good welds.

FITTING

Thin-Wall Tubing

When welding thin-wall steel tubing, such as for a race-car roll cage or an airplane fuselage, I fit the tubing joints so there are no gaps—the joints are almost watertight before welding! At the very least, your welding project's joints should fit close enough

so a welding rod cannot be inserted between them. See the photo above.

One of the toughest fitting jobs is building a set of tubular engine-exhaust headers. For welding, fit the weld joints as closely as possible. It's not easy. Even though I've had a lot of practice, I sometimes overtrim a curved section of exhaust. If I overtrim a tube, I have to splice in a section or start over with a new piece. Practice is still the best way to get good fits.

Always cut the pieces a little longer

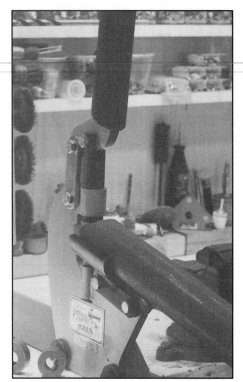

Roll bar and race car frames need to have the tubing ends notched as is being done here with a Williams Lo-buck tubing and pipe notcher.

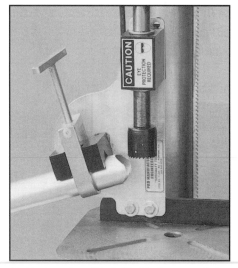

ProTools makes this tubing hole saw notcher that fits in a drill press. Tubing fitting is now much easier and even fun with tools like these. ProTools photo.

than necessary so you'll have plenty of metal to file and fit into the proper shape. As one of my welding supervisors used to say," ... you cut it off twice and the piece is still too

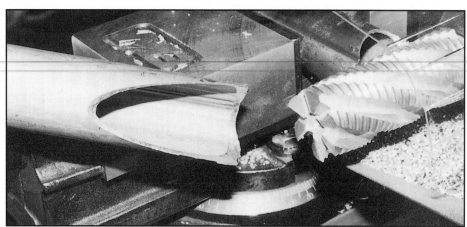

Mittler Bros. Tool Co. makes this tubing notcher that can notch up to 70 degree cut.

Up to 2-inch diameter tubing can be fishmouthed in a matter of seconds. This notcher bolts to the front of your work bench and is operated with a 1/2-inch drill motor. Dale Wilch Sales.

short!" This means that if you start with a metal piece that's too small, you'll never be able to fit it properly by trimming. So, be sure to rough-trim a part first to provide extra metal for final fitting. The resulting fit will be much more accurate.

Fishmouth Joints

Recent improvements to long-established fishmouth fitting methods have greatly simplified this tubing fitting process. In this chapter, I have illustrated several different tools that make fishmouthing and fitting tubing a lot easier than it was even as recently as 1990. If you have lots of

tubing to weld, by all means try one of the tools I have illustrated here.

The fishmouth tool that uses bi-metal hole saws will work very well on all kinds of tubing from 1/2" diameter up to 2 1/2" diameter. Larger sizes could be cut by making a larger tool and using larger diameter hole saws. I have cut 4" diameter tubing by making special setups in a vertical mill. There is one limitation to the hole saw method, and that limitation is the angle of cut. Most hole saw cutters will do 90° cuts down to 45° cuts, but no steeper angles.

The milling cutter shown in this chapter will mill cut angles down to

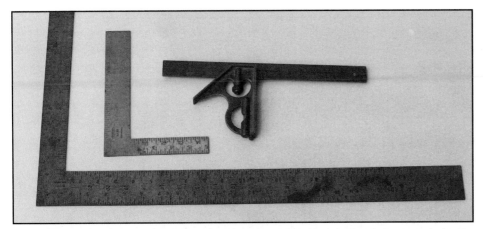 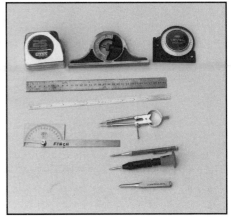

Tools that are necessary for weld parts measuring and fitting: (Left Photo) Carpenter's framing square, cabinet square and sliding 45° and 90° scale. (Right Photo) 25 ft. steel tape, machinist's angle finder, carpenter's angle finder, 12" steel rule (inches and mm), 12" steel rule (in 100ths), angle protractor, circle scribe, 2 spring punches, steel punch.

28°, but the cost of the tool is a lot more than the cost of a hole saw cutter.

If you don't have a lot of tubing fits to make, you can still do the job by hand. I use a half-round mill file to hand-fit most tubing. A small air die-grinder and rotary file will also work in many cases, but this method is very slow compared to the power cutter method.

Angle & Plate Steel—Even when fitting angle and plate steel, you need a tight fit. Any wide gaps will vary weld quality, making the finished joint weaker.

Heavy-Gauge Material—When fitting thick-wall tubing, heavy pipe or steel plate, you must bevel the edges to provide proper penetration, page 88. Some certified welding specifications call for the gap to be less than one-third the diameter of the welding rod being used, regardless of base-metal thicknesses. That's a tight fit!

Tools

Tools for fitting range from the ubiquitous hacksaw to a heavy-duty power shear. The hacksaw is obviously a hand tool; the shear a power tool. You'll need some tools from each category to do a quick and accurate job of fitting. You'll also need some tools to complement your cutting tools.

Measuring Tools—These are needed to establish where a cut will be made. One of the handiest measuring tools is a retractable steel tape. A 10' tape will usually do; however, big projects can require a 50' tape. For smaller jobs, you can use a 12" or 18" rigid steel rule. A 6" flexible rule in your shirt pocket can come in handy.

Also, a carpenter's framing square is necessary for laying out those square cuts. One of these durable tools is great for marking parts to be cut or fitted, or setting them up to be welded.

In addition to a framing square, you should also get a machinist's combination square. Use this tool for fine layout work and for work involving angles other than 90°.

Marking Tools—These are needed to provide a line for making cuts according to your measurements. When making a cut with a torch, you'll need a mark that won't be obliterated by the flame. These

marking tools include a scribe and a center punch or prick punch. A center punch will work, but a prick punch is made specifically for this purpose—making a line of closely spaced prick marks.

Another method for marking metal to be cut with a cutting torch is with a soapstone marker. It marks like chalk, but will not burn at high cutting temperatures. By all means, have several pieces of soapstone marker in your toolbox.

For marking arcs or circles, you'll need a divider—a special scribe that's similar to a compass, but with two sharpened steel points. One place you might want to use a scribe mark and bluing is for flame-cutting flanges for exhaust manifolds. Here, the cut must be precise, and the scribe line is better than soapstone for accuracy.

Remember that you should only use a scribe for marking a cut line. This is particularly true when marking sheet metal because of the tendency of a tear or break to start at a scribe line—the scribe line creates a stress riser.

Hand Tools—As mentioned previously, the hacksaw is the first to consider. Just make sure you have plenty of replacement blades. A

For schools and fabrication shops, Scotchman makes this multi-purpose metal shear, punch and notcher, great for fast, accurate fittings of plate, tubing and angles. It even makes spear-point fence posts.

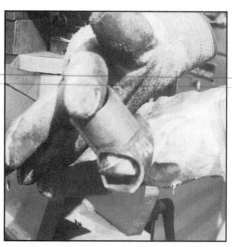

The Scotchman metal shear can notch heavy black pipe for ready-to-weld fit-ups.

The proper way to make 90° fit-ups in angle steel. This fit-up was made in just seconds with the Scotchman metal shear.

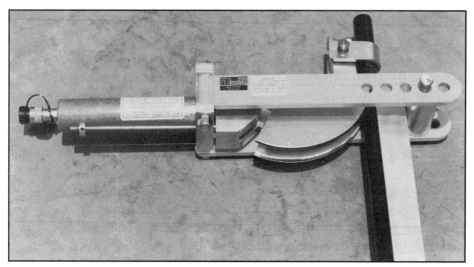

This hydraulic bender handles tubing sizes 1/2" diameter, up to 2" diameter. ProTools photo.

Tubing will flatten if you try to bend it without the proper tool. This hand-operated mechanical bender will make correct bends in tubing up to 2" and .139" wall thickness. ProTools photo.

hacksaw is great for cutting tubing and small structural shapes such as angle iron and channels. Tinner's snips are great for making straight cuts on sheet metal; make curved cuts with aviation snips. Aviation snips can also be used for trimming thin-wall tubing.

For final fitting and smoothing, you'll need an assortment of files. Start with three basic styles: flat, half-round and round. Get double-cut, coarse-tooth files for removing the most material with each pass. As for the size of the files, bigger projects require bigger files, and vice versa.

Power Tools—Professional weldors prefer power tools because time is money, and these tools remove material quickly. Even though the initial tool cost is higher, time is worth more. If you weld much angle steel, a 4" rotary, hand-held grinder is invaluable. Your welds can look totally professional if you dress them with a grinder. A grinder is also great

For tubing bends that require angle accuracy, this 20-ton upright hydraulic bender will make bends that are repeatable by observing the bubble level at the right on the tubing. ProTools photo.

For 90° weld joints, this angle clamp provides repeatable and accurate fit-ups.

Magnetic angle clamps really come in handy for making fit-ups of steel parts. They don't work on aluminum, copper, brass, stainless and other non-ferrous metals.

for fitting and dressing after using the cutting torch.

There are many power tools from which to choose. One of the most common is the disc grinder. It's great for smoothing cuts, especially those made by a torch. A disc sander can be fitted with a cup-type stone, which removes metal quickly. For smaller fitting and smoothing jobs, a 2" grinder works particularly well. It's easier to control, thus, a more accurate fitting job can be done.

Die grinders—pneumatic or electric—are useful for making accurate fits. Abrasive stones or carbide cutters can be used, depending on the material being fitted. For instance, a carbide-steel cutter is useful when fitting soft metals, such as aluminum, because of the tendency of an abrasive to load up—clog with metal particles. When coated with paraffin wax, a cutter has less tendency to load up.

For making fast cuts, a saber saw can be used in place of the hacksaw. The powered, oscillating saw rapidly eats through thin-gauge metal, and must be used with care. It takes more effort to follow the cut line than to make the actual cut.

For making cuts on heavy-gauge steel, it's frequently best to use a cutting torch, page 69. Because flame cutting leaves a relatively rough edge—how rough depends on the skill of the operator—it's always necessary to do the final fitting and smoothing with a grinder.

Safety—When using cutting and trimming tools, protect yourself against flying metal chips. This is especially important when working with power tools, such as high-speed grinders and cutters. Protect your eyes. Clear goggles are OK, but a full face shield is much better.

Don't forget your arms and legs. Wear a shirt and pants with full-length sleeves and legs. Wear gloves when using a disc grinder or torch.

WELDING JIGS

A welding jig is a fixture designed for holding parts in position during welding. The term scares many would-be weldors because they envision a very complicated device designed by a welding engineer. This is true only if it was designed to weld hundreds of parts on a production-line basis. Jig-welded usually means that all weldments, or welded assemblies, come out looking exactly the same,

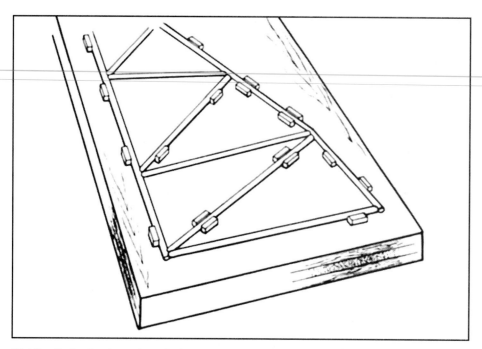

Welding jig for fuselage of steel-tube airplane: I use 5/8"-thick particle board and nail 1 X 1-1/2 X 2-in. pine blocks to it to hold tubing in place. Position blocks about 3 in. from each weld joint to avoid fire hazard.

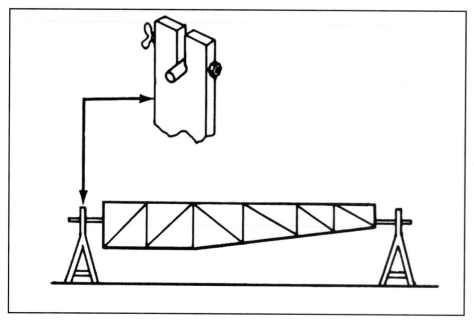

Simple rotating fixture brings the weld seam to the weldor. Using one saves a lot of time and makes all of the welds easy to see and to weld.

with consistent quality. For example, you should use a welding jig if you're building 1000 airliner seats.

For a one-time project, a jig can be as simple as Vise-Grip pliers or just a piece of angle iron used to prop up a part while it's welded in place. The

more sophisticated welding jig can be like that shown above for welding race-car frames and roll cages. Fortunately, you probably won't have to concern yourself with such a device. The most frequently used jig in my welding shop is a simple three-

legged finger. The metal finger takes the place of your finger so you won't get burned while holding a part in place during welding.

Remember, a jig is whatever it takes to hold parts in place until tack welding and finish-welding can be done. Tack welds are a series of very short welds spaced at even intervals. The tack welds hold two pieces of metal together so they can be finish-welded.

Wooden Jigs—A wooden jig is just that—plywood or particle board with small wood blocks nailed to it to hold metal pieces in place. When welding one or two assemblies, such a simple wooden jig is sufficient. See the illustrations, page 36. I have welded airplane parts, race-car parts, even factory production parts such as turbocharger wastegates, turbocharger control shafts and seat headrests on wooden jigs.

Permanent Steel Jigs—Factories use heavy steel welding jigs to ensure consistent sizes and fit of parts. You wouldn't want to buy a new exhaust system for your car and discover that the factory had welded the muffler inlet pipe on the wrong side. Factory welding jigs to ensure that welded parts will be interchangeable. Even small race-car factories have welding jigs to assure interchangeability and to improve production rates.

Welding jigs have their drawbacks, though. For example, when welding 4130 steel in a heavy welding jig, the parts sometimes must be *stress-relieved* to remove internal loads. The jig doesn't allow the parts to twist and conform to stresses from warpage. It holds the weldments in position, regardless of how they want to move. Therefore, internal stresses develop in the welded assembly. These stresses

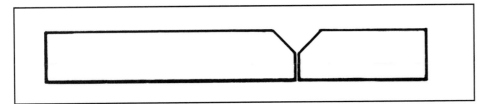

It is not possible to weld thick materials with one pass. You must bevel the edges as shown here, then weld two, three or more passes to obtain 100% weld fusion and strength.

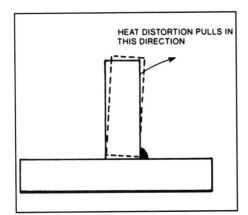

Metal tends to shrink when welded. As weld bead cools, the vertical piece in this drawing is pulled toward the weld. Allow for distortion by tack-welding both sides, then welding on alternate sides as illustrated in stitch-welding, page 58.

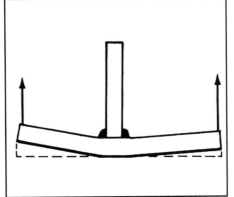

Horizontal piece warps as weld bead cools. Such warpage is normal, and can be corrected by heating and straightening after the weld is completed. In this illustration, dull red heat should be applied to the bottom of the horizontal plate to straighten it.

must be relieved, or the part may fail when it's put into use, or service.

Stress-Relieving Welded Assemblies

When a complicated, rigidly braced structure such as an airplane engine mount, power plant high-pressure steam pipe or a race car suspension member is welded, stresses remaining in the metal can cause premature fatigue cracking—caused by many loading and unloading cycles—unless they are relieved. Stress-relieving is accomplished by heating part or all of the structure to about two-thirds of the melting point and then holding the structure at that temperature for two hours or longer to assure that all the residual stresses are relieved. Then the structure is allowed to cool to room

temperature in still air.

It ought to be obvious that stress-relieving is not possible except in an oven. However, many long-time weldors falsely believe they are stress-relieving structures or parts of structures when they heat it with an oxyacetylene torch and then immediately let the structure cool. They are ANNEALING, softening the part, taking away tensile strength, but they are not stress-relieving the part. In most cases, parts heated to blood red with a torch and then allowed to cool immediately are weaker and more prone to cracking than if they were just welded and left to cool. If you still don't believe me, then check out what I have to say about the actual process in Chapter 7, page 68, then decide if you have the

tools, equipment and skill to stress-relieve in your own shop.

Or, if you have real need to stress-relieve a part or an assembly that you have welded, by all means, contact a MIL-SPEC, certified heat treater in your locality for advice.

Thankfully, most 4130 steel welded structures are pretty strong and won't crack in spite of how badly they are treated by improper post-heating.

Avoiding Warpage—Study the sketches in this chapter and you'll learn to cope with warpage in welding. I said cope with because you cannot stop warpage, just limit it.

When building a large tubular structure such as an airplane fuselage, I usually start welding at the front and work toward the back, alternating from side to side. See the above sketch. Even better would be to have two people welding symmetrically opposite sides simultaneously, but that is not done easily. So do the next best thing when welding a large structure: Weld one joint, then the opposite one to cancel the effects of warpage from welding the first joint.

Check alignment after each pair of welds. Repeat this welding-and-checking process until the structure is completely welded. It would be a shame to get a frame or fuselage 80% complete and discover that it's 1/2" out of square. A frame, fuselage or large assembly that far out of square is scrap metal.

Tack Weld First—Almost every structure should be tack-welded prior to finish welding. As mentioned, tack welds are a series of small welds between two adjacent pieces. Spaced about 1 1/2" apart, they serve to align the two pieces, hold them together and help prevent warpage. When the final bead is made, the tack welds are

37

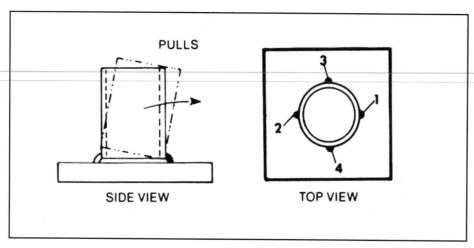

Example of controlling warpage from tack welding: Space tube up from flat plate. Make tack-weld 1. Then square tube to plate; make tack-weld 2. Make tack-weld 3. then square tube to plate; make tack-weld 4. Finish weld can now be made.

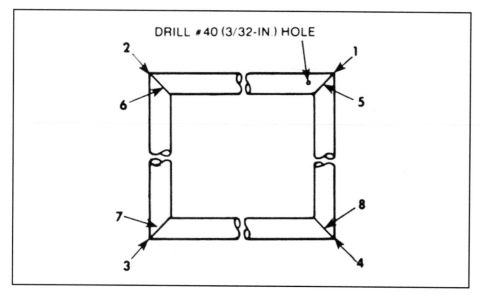

Sequence used to tack-weld tubing in rectangular pattern. Setup is similar to aircraft-fuselage or race car frame bulkhead. When welding closed tubing, vent by drilling hole as shown to prevent weld from blowing out as it's finished.

remelted and become a part of it.

Only where the designer calls for complete welding of a joint before welding another section should you bypass the tack-welding rule.

Gas Welded Structures—When torch-welding an airplane fuselage or race car frame, I stress-relieve the just-completed weld by slowly pulling the torch away from the work. I do this over a period of about 60 seconds. I never just finish a weld and jerk the torch away. That would surely cause

cracking.

Another way to minimize stress cracks in welded assemblies is controlling the air temperature in your workshop. Never weld in a cold or drafty workshop. A weld is more sensitive to cold and drastic changes than the human body! You'll have the best results welding in a room temperature of 90° F (32C), but 70–80° F (21–27C) works OK, too. All you'll have to do is become accustomed to working at above-

average room temperature.

Don't try to weld in extremely cold weather. The chances of 4130 steel cracking after welding at 40°F (4C) are 20 times greater than at 80° F (27C). Mild steel is less prone to cracking from cold-air shock after welding.

TIG Welded Structures—When I am in a hurry to TIG weld something, it is often hard for me to remember to let the post-flow timer do its thing, which is to protect the weld from air until it has cooled to a certain temperature. But that is exactly what the Argon post-flow timer is for, to protect the weld just completed. Let it work for you. Don't weld and jerk the torch away.

DESIGNING AND BUILDING WELDING JIGS AND FIXTURES

Much is said about welding out of position. You've probably seen bumper stickers noting the various positions weldors are capable of performing in. Although position in welding terms means the position of the weld surface—flat, vertical or overhead—all experts agree that the weldor should be in as comfortable a position as possible. So, avoid welding while laying on your back or standing on your head if there's an easier, more comfortable position.

The way to get in the best welding position—for both you and the electrode—is to make a fixture for rotating the assembly. If you're welding an airplane fuselage, make a fixture such as that shown nearby. Similar fixtures can be fabricated for welding other assemblies. With this fixture, you have better access to all welds. If you're welding a large trailer,

I use inexpensive electrical conduit—E.M.T.—to mock up engine mounts for auto engine-to-airplane installations. The three following photos show this engine mount development and installation.

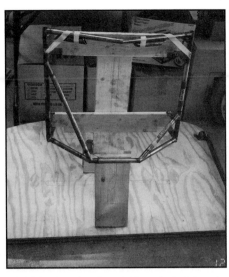

Once the E.M.T. tubing mock-up is completed, the real 4130 steel tubing parts are cut and fitted to the plywood firewall mock-up.

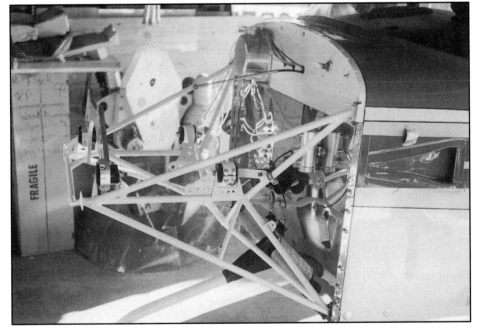

After 20 hours of cutting, fitting, welding and painting, the Buick V-6 engine mount is bolted to the airplane firewall, ready for the engine.

hoist it up on its side and turn it over to get to the other side of the weld joints. The "fixture" doesn't have to be exotic. The secret is, make the work accessible to you.

Obviously, it is not possible to make a weld position fixture to flip an ocean liner up on its side for welding, but every effort should be made to position your parts for FLAT position welding of all seams.

Almost anything you use to hold parts in the proper fit-ups for welding would be considered to be a welding fixture. Making a fixture for one-time weld projects is the bottom of the

scale, and building massive weld jigs to produce hundreds of identical welded assemblies is at the opposite end of the scale. Most production weld jigs weigh 10 to 20 times more than the parts they produce. In the next several paragraphs I will tell you how to design and build five different kinds of welding jigs.

Plywood and Nails Jig

Look at the photo on page 39. You can see that I drew a simple outline on the plywood of the tubing shape I wanted. Next, I drove small finishing nails into the plywood to help position the two pieces of 4130 steel tubing. Because the plywood will catch fire easily, I wet the plywood with water in the tack weld area, and then I only tack-welded the tubing together while it is on the plywood. The plywood often chars if I am gas welding, and it will slightly char if I am TIG or MIG welding. When you are using plywood or particle board for welding jigs, just make sure you thoroughly wet the charred area to prevent the board from catching fire and burning.

Jess Meyers and his associates have built a temporary engine support framework to hold this 4.3-liter Chevrolet V-6 engine in place while they develop the engine mount for the RV-6A experimental airplane. Belted Air Power photo.

I use a lot of corrugated cardboard to make patterns for 1/4" thick steel parts, as shown here. The photo at right shows the finished part.

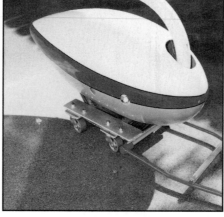

The 1/4" steel weld assembly in the previous photo is this wheel cradle that makes hangar parking of my airplane much easier to do. The low bar is mild steel, welded.

Particle Board and Blocks

Many airplane and many race car frames have been welded together on a wooden welding table like the one shown in the drawing on page 36.

Particle board is good for building large welding jigs because it is naturally flatter, smoother and less expensive than similarly sized sheets of plywood. The framework under the particle board should be 1" x 4" lumber, as straight and knot-free as possible. Remember that a crooked jig will produce a crooked part.

The particle board welding jig table should be made slightly larger than the framework you are building. You will need at least one inch extra on all sides so there will be room to nail the 1" x 1" x 2" positioning blocks. The race car frame shown in the picture on page 78 was built on a single 4' x 8' x 3/4" sheet of particle board. Building the welding jig took about 4 hours and fitting and tack welding the tubular race car frame took another 12 hours. This table method sure beats trying to build a race car or airplane frame on a concrete shop floor.

Plywood and Bolts Jig

Building a dimensionally accurate tubular framework such as an airplane engine mount requires a welding jig that will maintain the mounting dimensions and bolt pattern of the airplane fuselage. In the pictures shown on pages 39, you can see how I designed and built a plywood-and-bolts welding jig for my airplane project.

First, I drew an exact shape of the airplane firewall on a piece of 3/8" plywood and cut out the plywood firewall shape with a saber saw. Next, I made four exact replicas of the four engine mount brackets that are part of the airplane firewall. Then I bolted the four brackets to the plywood and backed them up with 2 lengths of 1" x 1" angle steel for stiffness. The 3/8" plywood would twist if not braced with the angles.

Then, I fitted and tack-welded the actual engine mount brackets and, with the engine in place, I fitted galvanized EMT tubing to the firewall jig and the engine. Once I was satisfied with the engine mount design, I removed the EMT framework and fitted 4130 tubing to make the actual mount. Total time invested in building the firewall jig, fitting the engine to the firewall, and tack-brazing the EMT mock-up

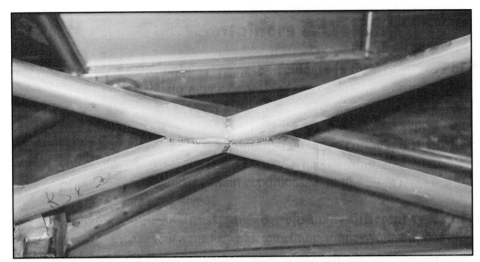

The angle of this tubing fit-up is 55°, too steep for most of the hole saw type notchers, but a notcher cut can be made, then enlarged by hand filing and hand grinding.

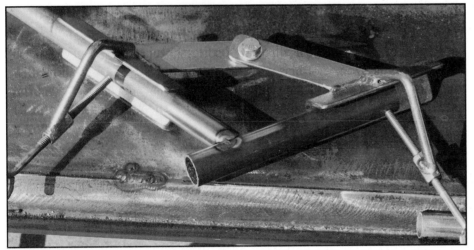

Williams Lo-Buck Tools makes this handy little angle holder for welding. It can be used on the welding table and on the car or airplane fuselage fit-up. Neat!

mount was about 25 hours. Building the actual 4130 engine mount took about 12 hours.

Once the mount was completely welded on the plywood-and-bolt jig, it was removed from the jig and fit-checked on the airplane firewall. Three of the four mounting bolt holes slid into place and fit perfectly the first time. The lower right side mounting point was off by about 1/8", probably because my plywood jig was not 100% accurate. Good fitting procedure, combined with TIG welding, produced a strong accurate

engine mount.

Surface Plate

A favorite way of making welding jigs for many welding shops is to buy a large, heavy sheet of mild steel. Most shops use 4' x 8' x 1" thick steel plates that weigh 1,325 pounds. A 4' x 8' x 1/2" thick sheet of mild steel weighs 662 pounds.

Most fabrication shops support the steel plate on a framework made of 1" beams and 2" heavy wall tubing, with adjustments on each of 6 or 8 legs so that the table can be adjusted to make

it perfectly level. Having a level weld table makes it easy to level and fit parts on top of the table.

Some weld shops allow the weldor to tack-weld his assembly to the top of the table. When the assembly is ready to be removed from the table, the tack welds are cut off with a thin abrasive wheel, and then the steel tabletop is dressed smooth with a #80 grit flap wheel.

Other welding shops do not allow tack welding on the table top, so parts must be clamped to the top with C-clamps and magnetic holding devices. You may want to have a dozen or more of these welding aids in your shop for making welding fixtures.

Thick steel plate can also be purchased in larger slabs, up to 8' x 16'. A one-foot-square piece of 1" thick steel plate weighs 42 pounds. It is also possible to buy special cast iron welding tables that are made with up to 50% of the table top open for jigging and clamping.

Tubular and Angle Jigs

If your plan is to make a large number of welded assemblies, you will certainly want to build one or more permanent weld jigs. A typical application of this weld jig might be for building replica car frames like the AC Cobra frame shown on page 2.

The easiest way to build a welding jig of this complexity is to first build a prototype race car frame and attach all the components and brackets that would be included in the finished car. Next, take the car apart and set the frame up at a good, workable height. From this prototype frame, you can develop a welding jig that will allow you to place each part in it for welding.

Butler Race Cars of Goleta, CA, uses this strong steel jig to position parts for welding AC Cobra car replica frames.

Butler Race Cars' fit-up of rectangular tubing is done by saw-cutting the tubing, de-burring the cut edges. The fit-up is then tack-welded as shown here.

Often, the jig to build a 200-pound car frame in will weigh 2,000 pounds by itself. Each jig design will be unique to the parts you need to build, but most important of all:

MAKE VERY SURE THAT THE WELDED ASSEMBLY CAN BE REMOVED FROM THE JIG WHEN THE WELDING IS ALL DONE

We have all heard about building boats in basements, then they won't fit through the exit doorway after they are finished! You will likely need to make the jig a bolt-together fixture so that you can take it apart and put it back together each time you weld up the assembly in it.

For more accurate welding jigs, install locating pins, 1/4" diameter, into each removable connection. You want any and all bolt-together parts on your weld jig to fit exactly the same every time you reassemble them.

CONVERT YOUR WOOD-CUTTING BANDSAW TO CUT STEEL & ALUMINUM

Almost any bandsaw can be converted to cut metal for fitting and building jigs. There are two secrets to cutting steel on a bandsaw. No. 1, buy a high-quality BI-METAL bandsaw blade with lots of teeth per inch for cutting steel. No. 2, slow the blade down to a very slow cutting speed, usually 5% of the wood cutting speed.

Correct metal cutting speed for bandsaws is measured in "feet per minute" of blade travel. If your bandsaw has 12" diameter drive wheels, it travels 3.14 x 12" each revolution (12" x 3.14 = 37.68" travel per revolution). 4130 chrome moly steel should be cut at 270-feet-per-minute blade travel. Multiply 270 feet per minute x 12" and you get 3,240 inches per minute. Divide 37.68" bandsaw drive wheel circumference into 3,240 inches, and you get 86 rpm. That is what rpm you need to have your drive wheel turning.

Your bandsaw motor probably turns 1,725 rpm. Therefore, you need to

Almost any belt-driven bandsaw can be converted to cut metal. The secret is a quality saw blade and a speed reduction drive that slows the blade down to the right speed for the particular metal. See chart on next page.

gear your motor to the drive wheel down with a 20-to-1 reduction of some sort. I have three different speed ratios on my bandsaw. One is 2-to-1 for sawing wood at 862 rpm. x 37.68" per revolution = 32,480 inches per minute 12 = 2,706 fpm. To effectively saw aluminum plate up to 6" thick, gear your bandsaw to 3.4-to-1 reduction ratio. This gives you a blade speed of 1,593 fpm. To effectively

CHART FOR BANDSAW SPEEDS, METAL CUTTING
1725 RPM MOTOR, 12" DIAMETER DRIVE WHEEL

Metal	Blade Speed in fpm	Motor to Drive Wheel Reduction Ratio	Blade Teeth Per Inch	Blade Width
Stainless Steel Titanium	Not Recom.	N/A	N/A	N/A
T1-6A-14V	45 fpm.	14 rpm, 125-to-1	18 teeth	1/2"
Bronze 4130	80 fpm	25 rpm, 70-to-1	12 teeth	12"
Chrome Moly Carbon	270 fpm	86 rpm, 20-to-1	14 teeth	1/2"
Steel 1020	330 fpm	105 rpm, 16.4-to-1	14 teeth	1/2"
Aluminum	1,600 fpm	509 rpm, 3.4-to-1	9 teeth	14" to 1/2"
Wood	2,700 fpm	2-to-1	6 teeth	1/4" to1/2"

FOR ALL METALS, LUBRICATE THE BLADE WITH DRESSING WAX STICK
GIVEN: 1725 motor rpm 12" Dia. Drive Wheel 37.68" Circumference ∏ 12 = 3.14 feet, 1 Revolution, blade travel is 3.14 feet

HOW TO "GEAR" YOUR BANDSAW TO CUT STEEL

The misnomer "gear" should be clarified to explain that belts, pulleys, chains and sprockets are an easy and inexpensive way to slow down the bandsaw blade. You could "gear" down the bandsaw drive wheel by putting a 14-tooth sprocket on the motor and a 280-tooth sprocket on the bandsaw shaft to saw 4130 steel, but you would also need a 1,725 tooth sprocket on the bandsaw shaft to saw titanium, and that many teeth on a sprocket would be very bulky and very expensive. So, you "gear" your bandsaw down in steps.

Employing 3 jack shafts with reduction ratios of approximately 3-to-1 will provide a final drive speed of 65 rpm, more than slow enough to saw chrome moly steel, which requires 86 rpm or less. The first jack shaft should be a V-belt and pulleys to prevent chain and sprocket noise at the speeds that electric motors turn (1,725). A 2" pulley on the motor and a 6" pulley on the first jack shaft will give a 3-to-1 reduction ratio and a speed of 575 rpm.

A 12-tooth #35 pitch chain sprocket on the first jack shaft and a 48-tooth #35 pitch chain sprocket on the second jack shaft, connected to a 24-tooth sprocket on the bandsaw drive wheel shaft, will give 72 output rpm, a final ratio of 24-to-1.

saw mild steel and chrome moly steel up to 2" thick, you need to gear your bandsaws to a 34-to-1 reduction ratio. This ratio gives you a blade speed of 159 fpm. That is somewhat slow for production steel sawing, but your blades will last longer. A 1/2" wide blade works best for steel, a 1/4" wide blade will do for aluminum sawing. ■

TEMPIL TEMPERATURE INDICATORS

F	C	F	C	F	C
100	38	306	152	977	525
103	39	313	156	1000	538
106	41	319	159	1022	550
109	43	325	163	1050	566
113	45	331	166	1100	593
119	48	338	170	1150	621
125	52	344	173	1200	649
131	55	350	177	1250	677
138	59	363	184	1300	704
144	62	375	191	1350	732
150	66	388	198	1400	760
156	69	400	204	1425	774
163	73	413	212	1450	788
169	76	425	218	1480	804
175	79	438	226	1500	816
182	83	450	232	1550	843
188	87	463	239	1600	871
194	90	475	246	1650	899
200	93	488	253	1700	927
206	97	500	260	1750	954
213	101	525	274	1800	982
219	104	550	288	1850	1010
225	107	575	302	1900	1038
231	111	600	316	1950	1066
238	114	625	329	2000	1093
244	118	650	343	2050	1121
250	121	675	357	2100	1149
256	124	700	371	2150	1177
263	128	725	385	2200	1204
269	132	750	399	2250	1232
275	135	800	427	2300	1260
282	139	850	454	2350	1288
288	142	900	482	2400	1316
294	146	932	500	2450	1343
300	149	950	510	2500	1371

Use Temperature Indicator chart for converting Fahrenheit to Centigrade or vice versa. Chart indicates availability of Tempilstick and Tempilaq temperature indicators. Chart courtesy Tempil Division, Big Three Industries, Inc.

Cleaning Before Welding

6

Beginning weldors think that the heat from welding will burn away any dirt, oil, and paint, but that is not so. In fact, dirt, rust, oil, paint, mill scale, oxides and other contaminants will do just that: contaminate the weld, even on farm and ranch equipment.

As a kid on the farm, I believed that welds were expected to break, because they often did if the repairs were done in a hurry, with little or no pre-weld cleaning and preparation. Later in life, as a welding inspector in Atomic Energy Nuclear Power Plants, I learned that good welds were directly dependent on good fitting practices and on proper cleaning of the metal AND the filler rod before the welds were done.

CLEANING METHODS

A good rule of thumb is that if it is worth your time to weld something, it is surely worth your time to prepare the weld area before welding to ensure that the weld will be as strong as possible.

Sandpaper Cleaning

For structures that are mostly tubes welded together, sandpaper makes the best cleaning agent. In my shop, I keep a large roll of emery cloth #80-

My helper, Jerry Jones, uses a strip of open-weave abrasive to clean aluminum tubing before I weld it.

grit to prepare both steel and aluminum tubing for welding. Tear off a length of cloth and use it like shining shoes, usually a clean area about 1" back from the weld area is necessary for preparing the welding. Yes, it takes time to deoxidize aluminum tubing prior to welding, but the weld will be much easier to do with clean metal, and the integrity of the weld will be much better with all the dirt and contaminants removed prior to welding.

If I am gas- or TIG-welding a complete airplane fuselage, I will expect to spend about four hours cleaning all the tubing ends and connection areas prior to welding. Just before welding a particular joint, I spray the tubes in the weld area with acetone and wipe off the acetone with a clean, white, lint-free cloth. I also wipe the welding rod with acetone just before I use it to weld with.

Yes, if I get in a hurry and don't take time to clean the 4130 steel tubing before welding, I can still make a strong weld, but it takes quite a bit longer to make the weld because I really have to watch out for porosity, slag and cracks in the weld. Lack of fusion also becomes a problem if I

Siphon-type sandblaster holds about 40 pounds of sand. It can also use other abrasives such as walnut shells. Homemade screen is for sifting sandblaster sand for reuse. After third sandblasting, silica 30 sand becomes to fine and will no longer clean parts adequately.

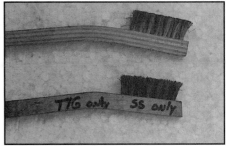

I keep a set of small stainless steel bristle wire brushes for cleaning parts before I weld them and while I am welding them. Notice that one of the brushes is marked "S.S. Only."

don't clean the parts prior to welding.

For parts that are not easy to sandpaper clean, try the small angle sander shown on page 22. For larger parts, such as angle steel for trailer building, use a 4" angle grinder.

Sandblasting

The dirtiest and least desirable way to clean parts before and after welding is to open-air sandblast them. But there are times when you have no other choice. For instance, you may want to modify an old boat trailer, or repair an old auto frame, or you may want to use up some steel or aluminum material that has been out in the weather for many months. The only way to clean large parts is to sandblast them.

For the times when this becomes necessary, I lay down a large sheet of plastic in an area where it will be easy to contain the sand when I am sandblasting. For instance, when I restored a 1956 Ford pickup, I sandblasted the frame in the 8-foot wide area between my garage and the 6-foot high side fence. Covering the area with a big sheet of plastic made it much easier to clean up the sand after the rust removal job was complete.

Air Compressor—You will need a minimum 1-horsepower air compressor for sandblasting. A 2- or 3-horsepower compressor will work even better. And after the sandblasting is done, you can blow off the frame/assembly with the air nozzle. The air compressor will also be necessary for operating one of the glass bead cabinets shown on page 46. If you don't own an air compressor, you can usually rent one for a day or two and save making such a large investment.

Power Sanding

It is very tempting to just sand the thin-wall tubing on a power sanding disc or on a sanding belt, but don't do it. Power sanders are designed to remove a lot of material quickly, and you will find that your .049" wall tubing is only .010" thick on the ends where you power sanded it. Unfortunately, hand sanding is the only safe way to prepare thin tubing for welding, brazing, or soldering.

Wire Brushes

I do keep an assortment of small stainless steel bristle wire brushes for cleaning the welds between passes. But I never mix the metals that I use them on. For instance, one brush is marked "Stainless Steel Only," one brush is marked "Aluminum Only," and one brush is marked "Steel Only." That is because small amounts of one metal could stick to the wire brush bristles and be transferred to a dissimilar kind of metal and cause defects in the weld. For instance, I certainly do not want aluminum, copper, brass, lead, or magnesium in my 4130 steel weld structure. The very small amounts of dissimilar metal would cause cracks in the chrome moly weld.

Don't try to remove oxidation from aluminum on a motor-driven wire wheel. The wire wheel will erode the aluminum and it will also transfer steel into the aluminum and contaminate your weld.

The same thing is true when trying

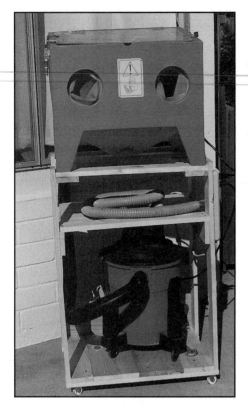

A minimum cabinet for glass bead cleaning also requires the use of a canister-type shop vacuum to suck out the dust and to help the suction-type glass bead gun pick up glass beads for blasting.

When I was growing up on a farm, I thought all welds were eventually supposed to break because little thought was given to pre-weld cleaning and preparation. Today I know better. Good welds depend on how clean the metal and filler rod are before you weld. Don't skip this step.

to wire brush clean mill scale off 4130 steel tubing. The wire brush will erode the steel and it will imbed wire bristle material into the chrome moly and contaminate it. Again, sandpaper is the best way to clean aluminum, stainless steel and 4130 steel.

Glass Bead Cleaning

Cleaning parts for welding by glass bead cleaning can be a mixed bag. Yes, the glass beads will clean the parts with little or no erosion of the metal. However, if you are cleaning tubing, the glass gets inside the tubing and is hard to remove. On general principles, you don't want your tubular structure/assembly to be full of little, tiny beads of glass.

Several weldors I have worked with would glass bead clean the weld scale off their parts between weld passes, but the glass dust must be rinsed off with solvent after the glass bead cleaning, then the solvent must be rinsed off with acetone to remove the solvent oils.

The glass bead cabinet is a very good way to clean parts after welding, in preparation for painting. But again, be sure to blow and rinse off all the glass bead residue so that it will not contaminate your paint.

Glass bead cabinets come in many sizes and price ranges. I bought a $99.00 special cabinet, built a wooden stand for it, and attached my shop vacuum cleaner to it to remove the dust and to evacuate the air pressure so the suction nozzle would work efficiently. It works great for occasional cleaning jobs.

If you use a glass bead cabinet on a daily basis, you seriously need a professional-sized cabinet with quality features. These come in various sizes to suit your needs. In the airplane repair business, I have even seen walk-in glass bead cabinets where the person cleaning parts had to wear a pressurized suit to prevent glass beads and dust from getting into his breathing and vision apparatus. He looked like a deep-sea diver in a sealed suit.

Chemical Cleaning

There are several metal-cleaning chemicals on the market, but you may have to go to a paint store or to a swimming pool supply outlet to find them.

• Rust can easily be removed from steel and cast iron by applying phosphoric acid (manganese phospolene) to the metal and then

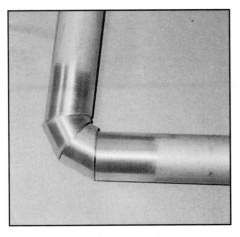

Here you see the clean, shiny areas on the tubing where the Alclad was removed to prevent it from contaminating the TIG weld.

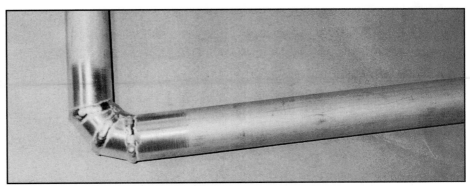

The end result: good, high-quality TIG welds on 6061-T6 tubing that was properly cleaned.

rinsing with clear water. Phosphoric acid diluted comes in many forms. Auto paint stores sell it as "metal conditioner" or "metal prep." House paint stores sell it as manganese phospolene—OSHPO. And it really works. If you have never tried liquid rust removal, you will be amazed at how well it works.

•Naval Jelly stops rust, but it will not float rust away like water-thinned phosphoric acid.

•Another liquid metal cleaner is swimming pool acid, actually muratic acid with a small amount of hydrochloric acid in it. You can buy this acid in hardware stores and, of course, in swimming pool supply stores.

•I keep a spray-bottle with a mixture of metal etch, phosphoric acid, or muratic acid to spray on parts to be welded. I also have a gallon-size glass jar with this rust remover so that I can dip small parts in it for pre-weld cleaning. Usually, 30 minutes in the acid will dissolve light rust, and a water rinse prepares the parts for welding.

Acetone, Denatured Alcohol— Although acetone is highly flammable, I keep a 1-quart can of it where I can clean the weld area and the welding rod just prior to welding. Of course, due care must be taken to prevent sparks and flame from contacting the acetone while I am welding.

The wisest thing to do is to clean the parts and the welding rod in an area outside of the welding shop and then immediately before welding, bring the parts into the weld shop.

Use Discretion

Of course, it is not necessary to clean the pieces of a utility trailer to the same level of cleanliness that you would clean your 4130 steel tubing for an airplane engine mount. But for any parts worth welding, it is wise to clean them as if you were preparing the parts for painting. Remove all rust, grease, mill scale and dirt, to the level of the intended use of the weld project.

Mill Scale

Most hot rolled, hot formed steel will have a thick layer of mill scale on it. Even the most expensive aircraft-quality 4130 chrome moly tubing and 4130 steel plate will have this thin

layer of mill scale on it. Mill scale will not fuse into a weld, and it must be removed to prevent it from weakening the weld. Use sandpaper, emery cloth or abrasive screen to remove mill scale from tubing. Use sheet sandpaper or a small air angle sander and a flexible pad to remove mill scale from 4130 and mild steel sheet, plate, and angle.

Aluminum Oxide

Aluminum tubing and sheet comes in two finishes, bare and *Alclad*. Bare is used when anodizing is planned. Clad is more common and the cladding is actually a thin film of oxide that is allowed to form on the aluminum to prevent further corrosion.

The cladding is not weldable, and it will contaminate the weld and weaken it. Mechanical removal is recommended, including sanding and burnishing with Scotchbrite® abrasive pads. Further cleaning by metal etch or phosphoric acid with a water rinse is recommended.

The best policy is: Clean the parts before you weld them as clean as you would if you were painting them. ∎

Gas Welding & Heat Forming

7

This chapter tells you how to set your shop up for gas welding, and then how to gas weld chrome moly 4130 steel, mild steel, and aluminum. You will also learn how to gas-weld stainless steel. The important thing about this chapter on gas welding is that when you learn the techniques for metal fusion in gas welding, you have acquired the techniques that are basic to most other forms of welding.

Learning to gas weld or becoming better at it will provide you with welding talents that will benefit you in most of the other types of welding that you may do in the future.

When you become good at gas welding, you will be able to control the heat and control the weld metal puddle. You will be able to fuse the metal parts together so that they are actually one permanent part.

CHOICE OF EQUIPMENT

In Chapter 2, we lightly covered oxyacetylene welding processes, and in Chapter 3, we explained the various kinds of welding equipment you might want to use to do your welding. Again, if I could only have one welder, it would be an oxyacetylene rig with several welding tips, a cutting torch with at least two sizes of cutting

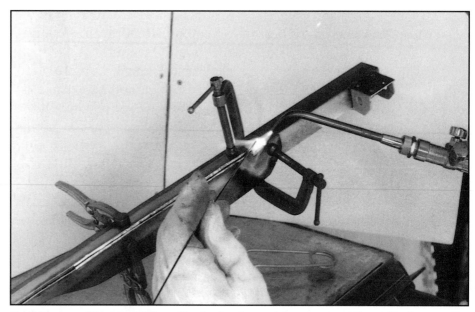

It all begins with gas welding. The basics learned from gas welding will give you the fundamentals to learn all other types of welding.

tips, and at least one heating or rosebud tip. Several companies make excellent starter kits that are light and portable but also effective in accomplishing some very professional welding projects. The only drawback to the small portable units is that the gas bottles don't hold very many cubic feet of gas. You may have to refill the bottles before you finish your project.

Gas Bottle Size

However, I find that the medium-sized oxygen and acetylene tanks that hold approximately 125 cubic feet each will last through building at least

three or four full-sized airplane fuselage projects and through an equal number of race car projects like the car frame shown on page 78. You may find that a portable plastic gas welding kit and a pair of 110-cubic feet tanks will last you at least for the completion of your project, and that refilling the tanks once a year is what you will expect.

If you find later that you are using your oxyacetylene cutting torch a lot, and that you use a lot of oxygen compared to the amount of acetylene you use, you might want to use a larger oxygen tank. If you do mostly

48

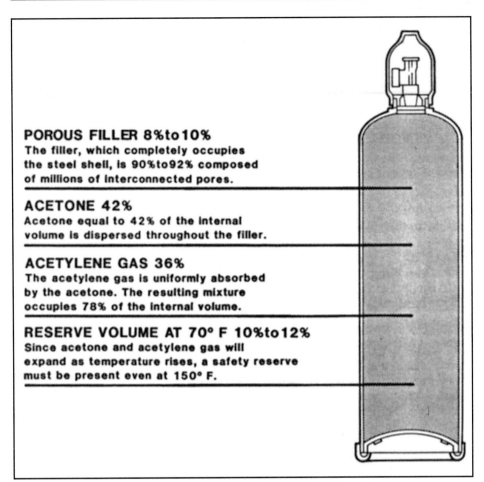

POROUS FILLER 8%to10%
The filler, which completely occupies the steel shell, is 90%to92% composed of millions of interconnected pores.

ACETONE 42%
Acetone equal to 42% of the internal volume is dispersed throughout the filler.

ACETYLENE GAS 36%
The acetylene gas is uniformly absorbed by the acetone. The resulting mixture occupies 78% of the internal volume.

RESERVE VOLUME AT 70° F 10%to12%
Since acetone and acetylene gas will expand as temperature rises, a safety reserve must be present even at 150° F.

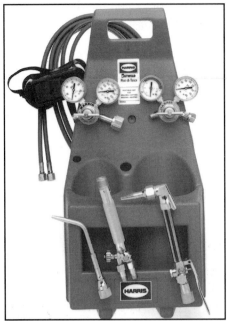

This portable gas welding torch set comes without bottles so you can rent or lease bottles from your local welding supply dealer, a less troublesome solution to the bottle problem. Photo: Harris Calorific/Lincoln Welding.

Acetylene is a compound of carbon and hydrogen (C2H2). It is a versatile industrial fuel gas used in cutting, heating, welding, brazing, soldering, flame hardening, metallizing, and stress relieving applications. It is produced when calcium carbide is submerged in water or from petrochemical processes. The gas from the acetylene generator is then compressed into cylinders or fed into piping systems. Acetylene becomes unstable when compressed in its gaseous state above 15 PSIG. Therefore, it cannot be stored in a hollow cylinder under high pressure the way oxygen, for example, is stored. Acetylene cylinders are filled with a porous material creating, in effect, a "solid" as opposed to a "hollow" cylinder. The porous filling is then saturated with liquid acetone. When acetylene is pumped into the cylinder, it is absorbed by the liquid acetone throughout the porous filling. It is held in a stable condition. Filling acetylene cylinders is a delicate process requiring special equipment and training. Acetylene cylinders must, therefore, be refilled only by authorized gas distributors. Acetylene cylinders must never be transfilled. Courtesy Victor Welding Co.

Rick Wuchner, Manager of the Ventura, CA, So-Cal Airgas store, shows some of the many oxygen bottle sizes his store has for rent.

neutral flame welding, brazing and soldering with your torch, you will empty both equal sized tanks at about the same rate.

Lease, Rent or Buy?

Probably the number one difficulty that I have experienced in all the years I have welded is in leasing, buying, or renting oxygen, acetylene and argon tanks. At the time this second revision is being written, I have two sets of tanks in my welding workshop because no local welding supply dealer will refill my two 99-year-lease tanks. They won't fill my tanks because they didn't get the business of leasing the tanks to me originally. So I had to rent a pair of oxyacetylene tanks from the nearest welding supply dealer in order to continue my projects. And I have a pair of tanks sitting empty in my workshop.

After surveying several welding supply dealers from the USA West Coast to the East Coast and back again, here is my recommendation on tanks:

First Choice—Lease a pair of tanks

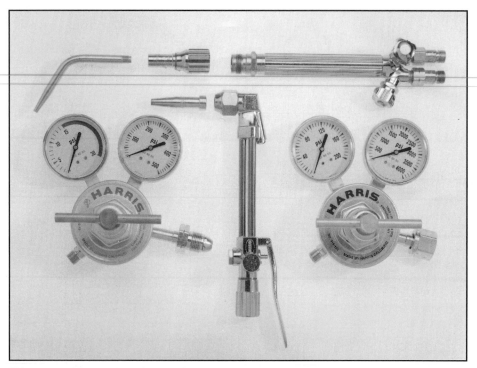

This gas welding and cutting set is a good setup for a beginner as well as a professional fabrication shop. You can buy your own gas hose in a length that suits your needs. Photo: Lincoln/Harris Calorific.

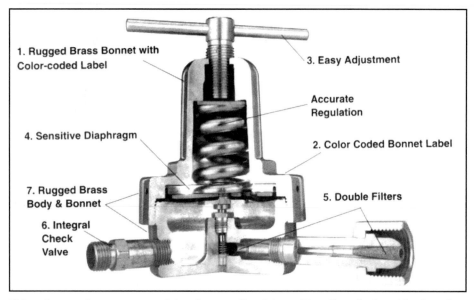

1. Rugged Brass Bonnet with Color-coded Label

3. Easy Adjustment

Accurate Regulation

4. Sensitive Diaphragm

2. Color Coded Bonnet Label

7. Rugged Brass Body & Bonnet

5. Double Filters

6. Integral Check Valve

This cut-away of an oxygen regulator gives you the picture of how the adjustment knob pushes on the spring to give pressure settings from zero psi to 200 psi. Photo: Smith's Equipment.

from the nearest welding dealer. Make sure that he will refill your tanks for you when they are empty. Try to get him to buy the tanks back from you if you move to another area or if you hear that he is moving. The price of the lease should be about the same as

2 years of month-to-month rent on the tanks.

Second Choice—Buy a pair of welding tanks from the local welding equipment dealer, but make sure he will agree to refill your tanks when they are empty. If you move, or if the

dealer moves, try to sell the tanks back to him.

Third Choice—Simply rent the tanks on a month-to-month basis. This is the easiest way to do it, but the most expensive. Two years of rent will equal purchasing the tanks outright.

Regulators

A *single-stage* regulator drops cylinder pressures from up to 2200 psi to 2–3 psi in one stage. Most small gas-torch kits come with single-stage regulators. The biggest problem with single-stage regulators is that they allow outlet pressure to drop as inlet pressure drops. Also, regulated pressure changes with temperature. Higher temperature raises pressure and vice versa. Therefore, you must keep your eye on the regulator gauges to maintain the desired outlet pressure.

A *two-stage regulator* automatically reduces cylinder pressure of 2200 psi down to about 50 psi. Pressure is adjustable down to 1–15 psi. Many large gas torches come with two-stage regulators because they use gas much faster—cylinder pressures are likely to drop rapidly due to a high gas-flow rate.

Although it would be nice if a small torch were available with two-stage regulators, many race cars or certified airplanes have been welded with small torches fed by single-stage regulators. You can always buy a set of two-stage regulators later if you absolutely must have the best equipment.

Regulators represent about 75% of the cost of a complete gas-welding starter kit. This doesn't include the cost of the cylinders.

I keep these tools in my gas-welding toolbox. From left to right: wire brush, stainless-steel wire brush, acetylene cylinder adapter, pliers, gas cylinder wrench, machinist hammer, flint striker, safety glasses, lightweight leather gloves, single-lens goggles, soapstone, temperature-indicating crayon, tip cleaners, rattail file, half-round file and flat file.

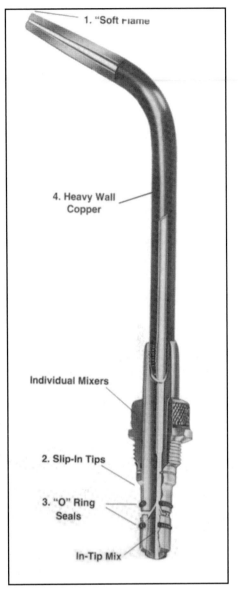

Cut-away of a Smiths torch tip, showing the intricate parts necessary to mix a good welding flame. Photo: Smith's Equipment.

Accessories

Even though you may plan to buy a gas-welding kit with several welding tips, you'll still need a few extra things to make welding easier.

Torch Lighter—Because you must *never light a torch with an open flame.* Get a torch lighter, or striker. Flint-type lighters make sparks similar to the way you would strike a match, and will work until the flint striker wears down. More expensive, electrical-discharge torch lighters will last many years.

Long-Handle Wire Brush—Use this brush to clean rust and welding scale from parts before welding. Weld seams must be clean and rust-free for complete fusion of base metal and filler metal.

Stainless-Steel Wire Brush—Use this small brush to clean welding scale while welding. I keep such a brush in my back pocket so it will be handy for cleaning off scale that develops during rest periods.

Welding Cylinder Wrench—This wrench is handy for removing and replacing the pressure regulators while changing cylinders. Also, this wrench can be used for opening and closing the acetylene- or hydrogen-cylinder valve if it doesn't have its own knob.

Pliers—Use pliers to pick up hot pieces of metal just welded. You can also use them to hold pieces in place while you tack-weld them.

Small Machinist Hammer—Often, you'll need a small hammer to bend hot metal into place, or tap a part into place before continuing the weld.

Safety Glasses—These clear-lens glasses are an absolute must for a weldor's toolbox. Don't confuse them with welding goggles. Wear safety glasses to protect your eyes whenever chipping, filing, grinding or sawing metal. Welding goggles reduce available light too much.

Leather Gloves—Wear leather gloves to shield your hands from welding heat. They allow you to weld for longer periods of time without the need to stop to cool your hands. Because they'll burn, never pick up hot metal with leather gloves. Use pliers instead.

Soapstone Marker—This chalk-like marker doesn't burn off until the metal melts. Use it for making reference marks on metal or to mark lines for a cutting torch.

Temperature Indicators—Temperature-indicating crayons and paint are convenient for determining the temperature of metal for heating or forming. Read about these on page 7.

Weld-Tip Cleaner—As with paint brushes, gas-welding and gas-cutting tips must be cleaned. Cleaning the outside of a tip is easy, but cleaning

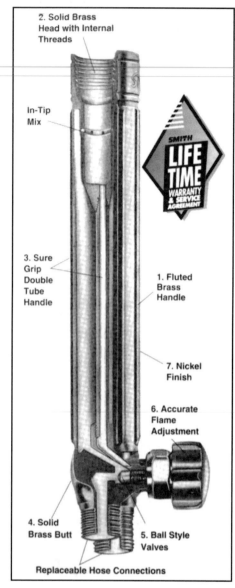

This cut-away of a gas welding/cutting torch body gives you an idea of why we say that your gas welding rig is considered delicate equipment. Handle it carefully and it will give you long, dependable service. Photo: Smith's Equipment.

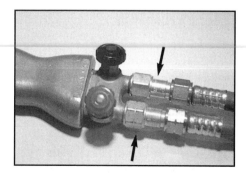

If your torch doesn't have safety check valves, or flashback arrestors, install one between torch and ease hose (arrows).

CARE OF GAS-WELDING EQUIPMENT

Treat your gas-welding torch and regulators as you would a quality camera, target pistol or any piece of precision equipment. Never let gas-welding equipment get wet or oily. Never leave the torch or hoses lying on the floor where they can be stepped on or driven over. Torch hoses last a long time, but not when subjected to that kind of abuse. And never, ever, lay the torch on the shop floor. Always coil the hoses and hang them on a hook, off the floor. And torch hoses are relatively inexpensive. Buy new hoses when the old hoses get stiff and frayed.

Do not bump or hit the regulators or gauges. These precision-calibrated instruments could be damaged. In most cases, return gas-welding equipment to a dealer for cleaning and repairs—even for torch-tip O-ring replacement. A quality gas-welding outfit should last many years when properly cared for.

CYLINDER SIZES

	Oxygen				Acetylene		
Size	Cu. Ft.	Height	Wt. Full	Size	Cu. Ft.	Height	Wt. Full
R	20	14"	14 lb	MC	8	14"	8 lb
AXL	58	41"	54 lb	B	33	23"	26 lb
Q	92	35"	70 lb	2AWQ	55	31"	61 lb
D	125	48"	124 lb	#4	90-150	36"	113 lb
S	155	51"	92 lb	#4	151-230	37"	150 lb
K	251	56"	153 lb	#5 WK	250-380	43"	220 lb
H	281	56"	162 lb				
T	337	60"	172 lb				

the inside requires a spiral tip-cleaning rod. The cleaner comes with a variety of rod sizes, each matched to a specific tip size. Each rod has a precision fit in its tip hole. Use the tip cleaner as you would a rifle-bore rod.

Metal Files—I use three different files in my welding toolbox: a coarse round file, a coarse half-round file, and a flat mill file. Files are used to fit parts before welding.

Acetylene Regulator Adapter—You might get an exchange acetylene bottle with male threads. Consequently, the male-thread (standard) regulator won't screw on.

To avoid this problem, get a male-to-male thread adapter for acetylene.

Welding Cart—Instead of buying a welding cart, build your own. It's relatively easy and you'll gain valuable experience doing it. I built mine many years ago, and it's still in use. You need training and practice before you start a complicated welding project. So, what better way to gain experience and proficiency than building something simple, but useful? Welding cart plans are provided on page 148.

To make the cart, you'll need to use many of the procedures described in

GAS WELDING TIPS, SIZES, GAS FLOW, DATA CHART

Decimal Metal Thickness	Fractions Metal Thickness	Victor Tip Size	Smith's Tip Size	Henrob Dillon Tip Size	Drill Size	Oxygen Pressure (PSIG)	Acetylene Pressure (PSIG)
.015"	1/64" to 1/32"	000	AW200	No Band .017	.020	3	3
.030"	1/32" to 3/64"	00	AW201	—	.025	3	3
.070"	1/32" to 5/64"	0	AW203	1 Band	.035	3	3
.090"	3/64" to 3/32"	1	AW204	—	.040	3	3
.125"	1/16" to 1/8"	2	AW205	2 Bands	.046	4	4
.190"	1/8" to 3/16"	3	AW207	—	.060	4	4
.250"	3/16" to 1/4"	4	AW209	—	.073	4	4
.375"	1/4" to 1/2"	5	AW210	—	.090*	5	5

* For metal thickness over .375", 3/8", 9.5 mm, use arc welding.

SAFETY TIPS FOR GAS WELDING

• Never tilt acetylene cylinders on the side when in use. The acetone stabilizer will flow into the regulator and damage it.
• Mark full cylinders FULL with a marking pen. Mark empty cylinders EMPTY with a marking pen.
• Store cylinders at less than 125 °F (52C). Make sure valves are closed and caps are on stored cylinders.
• Chain or otherwise restrain all cylinders in an upright position.
• Always crack—slightly open—the valve to blow out dust before attaching a regulator. This helps prevent contamination of the regulator, which may cause erroneous gauge readings.
• Never haul cylinders in the closed trunk of a car because of the explosion hazard from escaping gases.
• Wear goggles with the correct filter lens. See the chart, page 26.
• Don't wear oily or greasy clothes when welding.
• Wear leather or denim clothes when welding; leather is best. Blue jeans are great.
• Don't cut or weld material coated with zinc, lead, cadmium or galvanized coating. Poisonous fumes are generated as the coating burns off. Coated sheet steel is used in an increasing percentage of late-model cars and trucks, especially in rust-prone areas such as rocker panels, fenders, rear quarters, door outers, cowl plenums and trunk floors.
• Never use oil or grease on gas-welding equipment. It's extremely flammable at high temperatures, particularly in the presence of oxygen.
• Leak-test hoses, regulator and torch before lighting. Use soap solution or non-oily leak-detector solution on the connectors and look for bubbles. Never use an open flame to check for leaks.
• Oxygen fittings have right-hand threads. Hose is green or black.
• Acetylene fittings have left-hand threads. Grooves around flats on nuts identify left-hand threads. Hose is red.
• Never adjust an acetylene regulator to more than 14 psi. More pressure and flow will draw acetone from the cylinder. Without acetone, acetylene is very unstable and could explode.
• Keep pliers handy to pick up parts you've just welded or cut so you don't burn your gloves or fingers.
• To minimize soot when lighting an acetylene flame, add more acetylene. Don't use equal amounts of acetylene and oxygen. The resulting loud pop may cause an accident. Be prepared to add oxygen when the flame ignites. Never try to light the torch with a small amount of acetylene. This causes tremendous amounts of soot.

Adjust oxygen regulator to about 3 psi on low-pressure gauge, left. Low-pressure acetylene regulator should also be adjusted to 3 psi.

this book: fitting, cutting, butt welding, corner welding, T-welding and brazing. Read the sections that apply to each.

GETTING STARTED

It is a very good idea to gather up a lot of scrap pieces of the kind of metal you plan to use in your projects. If you plan to build an experimental airplane, buy several scrap pieces of 4130 steel tubing from one of the aircraft supply houses. All of the tubing supply companies sell weldors practice kits for just a few dollars.

If you plan to build race cars or even trailers, go to your local steel and aluminum supply dealer and buy several pounds of cut-offs. You can also go to local welding shops and ask to go through their scrap bins for metal to do your practice welding on. Then, of course, the welding projects in Chapter 16 of this book will provide you with some very good practice welding on things that do not have to be perfect at first.

My first serious practical gas-welding project was welding a go-kart. After several dozen go-karts, I progressed to welding bird-cage

frames for sports racing cars. Long after that, I decided to attend welding classes at the local community college.

Prepare Your "Coupons"

The welding trade calls sample welds and practice welds "coupons." This chapter shows you drawings and photos of several kinds of welding coupons for gas welding. Do a good job of fitting and preparation (cleaning) on your coupons. You might as well have good metal to practice on: dirty, poorly fitted metal will not teach you much.

Get Organized First—Make sure your weld area is well-organized and that there are no flammables in the shop. This includes paint cans, oil cans, rags, wood, and for sure, no gasoline cans. Remember that one gallon of gasoline has the explosive power to move a car 20 miles or more when the gasoline is "exploded" inside the engine.

Make sure that your clothes are not readily flammable. Wear a denim shirt and pants and cotton socks, and good, tight-fitting shoes. Wearing a weldor's cap is advised also. Soft leather gloves that allow good hand movement should be worn. But don't wear heavy leather gloves, because you will not be able to handle the torch and welding rod with heavy gloves.

Welding Table

Before you build your portable welding table, you can simply place the 3/8"-thick steel plate on top of a portable stool or other suitable platform, and put at least two firebricks on the steel plate to insulate your coupons on the torch flame from

the steel plate. After you learn to weld, and when you build your portable welding table from the plans on page 146, you will not have to search for a suitable place to do your welding.

Open the Cylinder Valve

There really is a correct procedure for opening the oxygen and the acetylene cylinder valves. Each valve is to be opened slowly and carefully to avoid damaging the diaphragms in the regulator. Open the acetylene tank valve about 3/4 to one turn and no more. This is to prevent too much accidental flow of acetylene as might happen if the valve were fully open and if you cranked in the adjustment past 15 psi. If the acetylene tank flow exceeds 1/7 of the tank volume per hour, the acetylene could become unstable and cause the tank to explode.

Also, if you had an emergency and needed to shut off the (fuel) acetylene in a hurry, 3/4 to one turn open makes it easier to shut off the gas.

Next, slowly open the oxygen valve, but open it all the way, several turns to make the valve seal set properly. The oxygen tank valve could leak if the valve is in the mid-open position. Again, if you open the high-pressure oxygen tank valve too quickly, the pressure surge could damage the oxygen regulator diaphragm. Now you are ready to light the torch, to adjust the flame, and to practice gas welding.

Safe Procedures for Lighting & Shutting Off Oxyacetylene

For many years, weldors were taught to light the torch with both

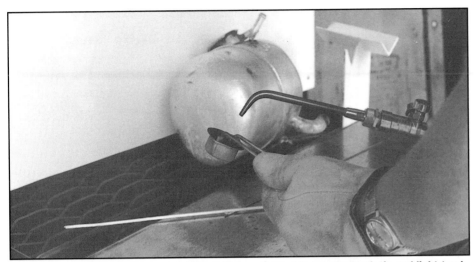

Open torch acetylene valve about 1/2 to 3/4 turn. Hold striker under torch tip and light torch. Remember, NEVER LIGHT AN ACETYLENE TORCH WITH AN OPEN FLAME!

Acetylene-only flame should look like this. Quickly add oxygen to eliminate soot.

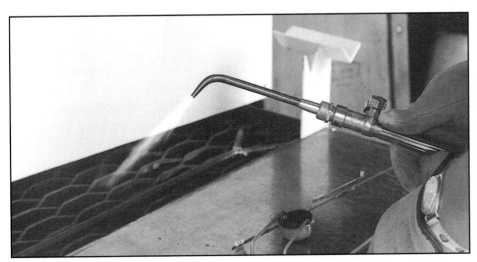

As you add oxygen, three distinct flame cones will appear. Continue adding oxygen until you have a neutral flame. It looks like this with #3 Victor tip, 3 psi oxygen pressure and 3 psi acetylene pressure. One distinct cone is at center with light blue outer flame.

oxygen and acetylene flowing, and to shut off the acetylene fuel first. But this procedure often caused popping and even flashback accidents, so a new procedure has been initiated to prevent gas welding explosions.

Lighting the Torch

1. Adjust the acetylene regulator to hold a predetermined pressure with the torch acetylene valve opened 1/2 turn. Then shut off the valve.

2. Adjust the oxygen regulator to hold a predetermined pressure with the torch oxygen valve opened 1/2 turn. Then shut off the valve.

3. Open the acetylene valve 1/8 turn and light the torch with the striker. Some black soot will come out, so increase the acetylene flow by opening the torch valve until the flame gets bigger and the smoking stops.

4. Next, open the torch oxygen valve about 1/8 to 1/4 turn and adjust for a neutral flame.

Shutting Off the Torch

1. Close the oxygen valve first. Oxygen supports combustion and aids explosions, so get rid of the oxygen first.

2. Next, shut off the acetylene valve.

3. If you intend to light the torch again within a few minutes, you need not bleed the lines and regulators.

4. When you are through welding or heating or cutting for even a few hours, close the oxygen and acetylene tank valves and bleed the pressure from the regulators, hoses and torch by opening each valve separately.

5. When the pressure is bled off, close the torch valves and back out the regulator adjustment nut until no drag is felt on the screw. Backing out the adjustment screws relieves

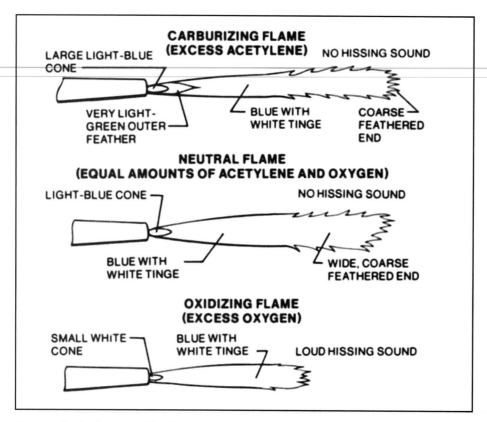

Here are the basic gas-welding flames. Each has a distinctive shape, color and sound. Sounds vary from soft (low gas flow) to medium (medium gas flow) to loud (full gas flow) at specific pressures. Neutral flame is most used.

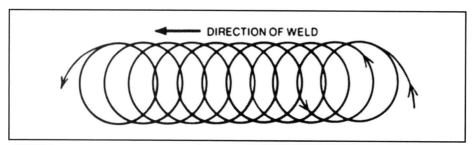

Use circular motion to preheat base metal before forming puddle. Circular motion spreads heat evenly. When steel turns dull red, you are ready to make molten puddle.

Because of the unstable nature of acetylene, the bottle could overheat and explode.

If more heat is required, for instance when preheating or maintaining heat for brazing a large cast iron structure, it would be advisable to use 2 or more acetylene tanks to provide the 1/7 volume per hour of heat flow. Also, consider 2 or 3 extra helpers, each operating a separate, independent oxyacetylene heating torch to provide the required heat. Think of the heating requirements in units, BTU's. If one torch only provides 1/2 the necessary heat in BTU's, then use 1 or 2 extra torches to bring up the BTU's.

pressure on the adjustment springs and reduces strain on the regulator.

Every time your oxygen-acetylene torch pops, carbon and sometimes flame will flow back into the torch body and will eventually cause torch burn-out and possibly a torch explosion.

However, I have welded almost constantly for 40 years or more and have never personally experienced a torch explosion.

WELDING TECHNIQUES

Now it is time to light up the torch and to begin practice welding projects. Adjust the oxygen and acetylene regulators for 4 psi each and put a .040" (#1) tip on the torch handle. You are ready to light up and to adjust for a neutral flame. Next you can start practicing on a piece of steel.

Making a Puddle

The first thing to do is make a molten puddle on the steel plate. With your welding goggles on and over your eyes, direct the neutral flame at the steel. Oscillate the torch tip in a half-moon, zig-zag or circular pattern as shown. The exact pattern is not important. The idea is to keep moving the torch in a rhythmic, repeatable pattern as you move the weld puddle along, but without overheating it. Use the pattern that works best for you.

Working Distance

Hold the torch about 1" from the work. The steel should start to turn red within 5–10 seconds. If it doesn't you're not holding the torch close enough, or the tip you selected isn't

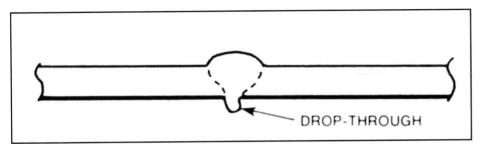

Slight drop-through is OK as it ensures 100% penetration. Reduce heat to reduce or eliminate drop-through.

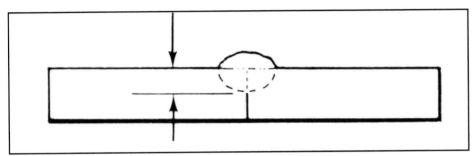

Good weld must have sufficient penetration, or depth of fusion. Weld has 50% penetration. Increase heat or slow down to increase penetration.

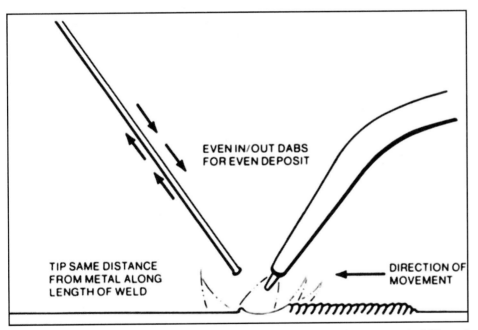

EVEN IN/OUT DABS FOR EVEN DEPOSIT

TIP SAME DISTANCE FROM METAL ALONG LENGTH OF WELD

DIRECTION OF MOVEMENT

Forehand welding: It takes coordination to use filler rod. Deposit an even amount of filler rod with each dab as you move molten puddle along weld seam. Practice will develop rhythm. Drawing by Ron Fournier.

large enough for the metal thickness.

I usually tell first-time welding students to overheat the metal first and melt holes in it. This is because the normal tendency is to weld too cold, resulting in poor penetration. So go ahead and burn a few holes until you can control the heat and maintain a puddle.

Learn to manipulate the torch by moving it closer to the metal, then backing up quickly if a hole starts to form. After experimenting with the puddle for about 10 minutes, you're ready to advance to the next step.

Running a Weld Bead

The next step is to practice running a bead on your piece of scrap steel. Start by making another molten puddle. With your left hand—if right-handed—momentarily dip the welding rod into the puddle, then withdraw it. If you're left-handed the torch goes in the left hand and the rod in the right. Always dip the rod into the molten puddle. Never try to heat the rod and puddle together. If you do, the flame will melt or even vaporize the small-diameter rod before the base metal gets hot enough to puddle. Remember, form a puddle, then intermittently dip the rod into the puddle to add filler material as you run the bead.

You've seen those beautiful welds that look like a row of fish scales? Well, they resulted from a weldor doing the dip, dip, dip thing. If the welding rod sticks in the puddle, point the flame at it, melt it off and try again. Sticking is caused by not dipping fast enough or not keeping the puddle molten. Just keep practicing.

Forehand Welding—Forehand or backhand welding refers to the direction you point the torch tip in relation to the direction you're running the weld bead. If you're forehand welding, the torch is angled so it points in the direction of the weld. This is to preheat the base metal so it puddles easily as you move along with the weld bead.

Backhand Welding—Like walking backward, backhand welding is similar to welding backward. The technique is to point the torch at the

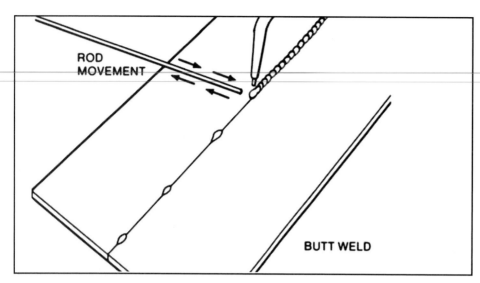

Sheet metal needs plenty of tacks–about every 1"–to reduce warpage. Tack welds are melted into weld bead as final bead is made. Drawing by Ron Fournier.

already-welded seam, away from the unwelded seam. This prevents the base metal from being preheated—usually an undesirable feature. Backhand welding is rarely used, except to avoid burning through very thin metal. The added mass of weld bead may help absorb the extra heat. However, so does pulling the torch away from the work.

Tack Weld—As previously discussed, a tack weld is nothing more than a very short weld that's used for holding two pieces in place prior to final welding. You'll make a lot of tack welds.

Stitch Weld—A stitch weld is used where a continuous weld bead would be too costly and time-consuming, and where maximum strength is not required. Although it can vary with the application, a stitch weld typically is made up of short weld beads about 3/4" long, spaced by equal gaps.

Butt Weld—Once you've mastered the art of running a bead, you're ready to try welding two pieces of metal together. Let's start with a basic joint. A butt weld is a weld made between two pieces laying alongside and butted against one another, edge to edge or end to end.

Place two pieces of metal side-by-side and butt them together. There should be no gap. The seam will be welded into one solid bead.

PRACTICE, PRACTICE

By now, you should have collected some scraps of steel to practice welding on. Ideal for welding practice would be several 2 X 5-in. pieces of 0.032–0.060-in. thick mild steel. They don't have to be exactly this size, but your practice work will look better if the pieces are uniform.

To avoid contaminating the weld bead with firebrick or the welding table, raise the metal pieces off the table or brick by inserting extra pieces of scrap underneath each workpiece, but not under the weld seam.

Next, tack-weld the two pieces together, first at each end, then about 1 in. apart, along the length of the seam. This keeps the metal aligned during the welding. The trick here is to keep both edges at the same temperature by manipulating the torch. Add a little heat until the puddle forms, then dip the rod in two or three times until you have a good tack weld.

After tack-welding, use your pliers to hold the work so you can check for warpage at the weld seam. Straighten the pieces by tapping on one piece with a hammer when you hold the other with the pliers. You're now ready to run a solid bead. If you're right-handed, do a forehand weld by starting at the right end of the seam,

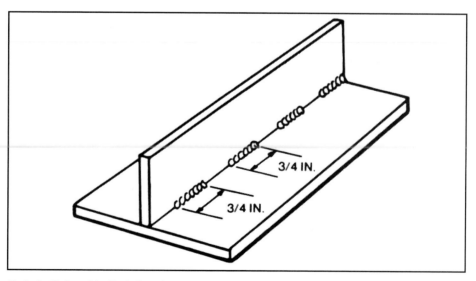

Typical stitch welds: Technique is used where continuous weld is not required for strength, and it is also used to avoid warpage on long sections. The stitch weld procedure can also be used for Arc, TIG, and MIG welds.

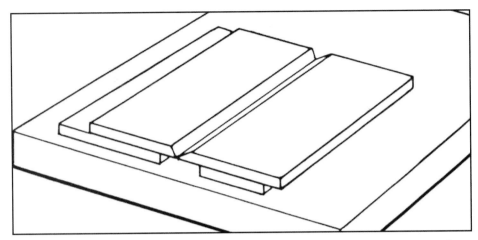

Space pieces to be butt welded off table on scrap pieces or firebrick so you won't contaminate weld, waste heat or damage tabletop.

make a puddle—torch in right hand—dip the rod and keep going until you get to the left end. If you're left-handed, reverse hands and start welding at the left end of the seam.

As you come to each tack weld, remelt it into the puddle. When complete, you won't be able to see the tack welds; they'll be part of the weld bead.

Test Weld

Because the appearance of a weld can fool the beginner, test each weld for soundness. "Pretty" welds can literally break in two if there is insufficient penetration—not enough filler fused with the base metal. The weld bead may only be "laying" on the base metal. Ideal penetration can be from as low as 15%—weld bead is fused into the base metal by 15% of overall thickness—to over 100%—it's fused the full thickness of the base metal and sagging through to the back side. Clamp the piece to be tested in a vise, just below the weld seam. With a big hammer, bend the top piece toward the top of the weld bead. Chances are that the weld will break through the back side, perhaps

completely if penetration is poor. A common cause of broken welds made by beginning weldors is crystallization. Crystallization is caused by excessive gas pressure—usually, too much oxygen pressure.

It is much better to weld with 1–2 psi gas pressure for any size tip and avoid overheating the weld. If your weld breaks, don't give up. Look for lack of penetration. With your next test weld, try to get a good puddle going before you dip in the filler rod. Practice makes better welds.

Outside-Corner Weld

A fillet, or outside-corner, weld is a weld performed on two pieces of metal joined in a V-type configuration. The weld bead is run on the outside rather than in the V's inside, or "crotch." An outside-corner weld is easier to make than a butt weld because it takes less heat to maintain a puddle and run the weld bead. It's easier to get the edges hot because they're up in the air and only the edges are being heated. Less heat is lost to supporting members, such as the steel work table.

Block the two pieces of metal up

like the peaked roof of a house. You can use a heavy piece of metal at the sides to hold the two pieces in place for tack welding. When you do this, you have just created your first welding jig!

Tack-weld the two ends, then make some tack welds in between. Now, run a continuous weld bead. A near-perfect corner weld is one with slight penetration through to the

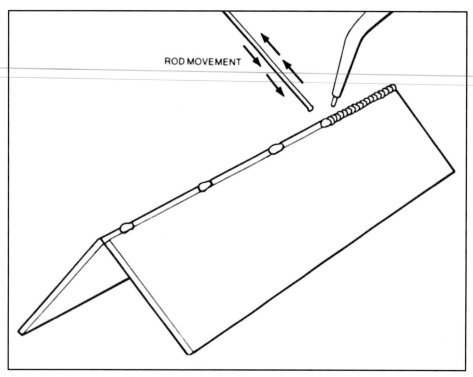

Outside-corner weld requires less heat than inside-corner weld. For maximum penetration, metal edges should not overlap, but form a V. Drawing by Ron Fournier.

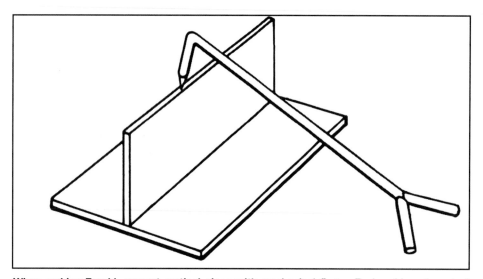

When making T-weld, support vertical piece with mechanical finger. Tack-weld each end of joint. Point flame at horizontal piece 70% of welding time, but manipulate torch to put equal heat on both pieces as you run bead.

underside—100% penetration of the base metal.

T-Weld

T-welds are the most difficult of all the practice welds. That's why we waited until last to try it. When making a T-weld, you're making another type of fillet weld—welding in a corner where two pieces join at 90° or so.

Block up two pieces of scrap metal with a metal finger as shown in the accompanying illustration. Viewed from the end, the two pieces form an upside-down T. Block up the flat piece in the area of the weld so the welding table won't absorb the heat of the weld. Firebricks or short sections of angle stock are great for this.

Next, tack-weld the two ends as before. Then make more tack welds about 1 in. apart along the weld seam. Again, if you are right-handed, start welding from the right end—vice versa for you southpaws. Direct most of the heat to the horizontal piece and less to the vertical—at about a 2-to-1 ratio.

The reason for directing more heat at the horizontal piece is that the weld is being made in its center, so there's more volume of metal to absorb heat. There's only half as much volume in the vertical piece because you're welding its edge. Remember, manipulate the torch to keep the puddle going on both pieces, and feed the rod into the puddle by intermittent dipping.

A common problem with T-welds is ending up with an otherwise decent weld that has an undercut or gouge in the vertical piece. This is caused by failing to adequately manipulate the torch to keep equal puddles on both pieces and overheating the vertical piece. The molten base metal from the vertical piece runs down and solidifies into the weld bead.

The solution is to tilt the torch away from the vertical piece when you see the undercut start, and at the same time, dip the rod into the undercut part of the puddle. You'll have to do a little torch twisting, but that's what it takes—that and a lot of practice.

Remember, control the temperature and the puddle, and the weld will take care of itself.

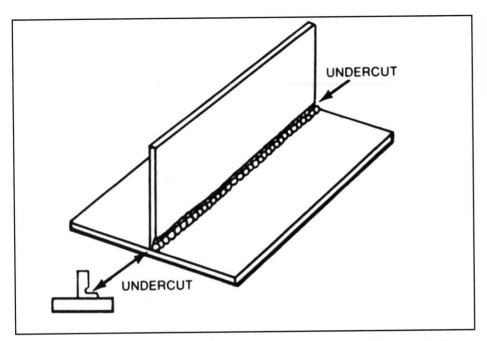

Problem with making T-weld is undercutting vertical workpiece and insufficient penetration on horizontal workpiece. Undercutting can be eliminated by correct manipulation of torch and rod. If your weld looks like this, point flame at undercut and add filler rod as puddle forms there.

GAS WELDING TORCH SOUNDS

Different volumes of oxygen and acetylene can be fed through a specific-sized welding tip to produce varying amounts of heat. For instance, a Victor or Smiths welding tip size no. 1, or AW204, can be used to solder (250°F), braze (800°F) or fusion-weld steel (3000°F). I have even used a heavy-duty cutting torch to solder 1/2" copper tubing to a cold drink aluminum can.

The amount of oxygen and acetylene flowing through the welding tip will provide varying BTU's of heat to the weld. Here is how I adjust the gas flows for 3 different heat requirements with a #1 welding tip:

1. Soft flame, for soldering, 4 psi oxygen and 4 psi acetylene pressures: Adjust for a very quiet flame sound, just barely a whisper of flame voice from the flame. Solder should flow in 2 to 3 seconds after applying the flame.

2. Medium flame, for brazing, 4 psi oxygen, 4 psi acetylene pressures: Adjust for a stronger flame sound. The sound will be like the sound of water running from your aerated bath lavatory faucet while you are washing your hands.

3. Maximum flame, for fusion welding, should have a very strong sound, obviously putting out a good amount of heat. It could also be described as a hissing sound. Use this flame for welding thicker metals, over .080" thick, and for heating parts for bending.

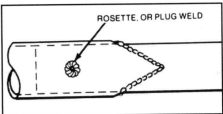

Plug weld is made to help join tubes, one slipped inside another. Note angle-cut–scarfed–end of larger tube to increase weld-bead length. This weld is also called a rosette weld.

COMMON PROBLEMS

Torch Pop

This will scare the dickens out of you and blow sparks all over the place! Torch pop is ignition of the gases inside the torch. It is more common when welding in corners such as when making a T-weld. Torch pop is caused by an overheated tip too close to the metal. This causes a small explosion inside the tip. After the torch pops several times, the tip gets dirty from the metal splattered on it, causing it to pop even when you aren't too close.

The solution for preventing torch pop caused by a dirty tip is to shut off the torch and clean the tip with your tip cleaner. Light the torch and try welding again. If cleaning the tip didn't cure the problem, increase oxygen and acetylene pressures 1 psi from the previous setting and adjust for a hotter flame. If this doesn't cure the popping problem, try the next larger tip.

Flaky Welds/Poor Penetration

Such welds break apart when you bend them. They're caused by not making the puddle hot enough before dipping the rod. You just can't melt rod and drop it onto the base metal,

Using the weldor's mechanical finger to hold two parts in place for welding, use the fire brick to insulate the parts from the cold welding table.

hoping it will stick. Instead, it must become a homogeneous part of the base metal by mixing, or fusing, while in the molten stage.

The solution is to get the weld puddle hotter. Do this by using a larger tip or by holding the torch closer to the work, but not so close that torch pop results. Remember, if you can't make a molten puddle in 5–10 seconds, use a larger tip. This is why I tell new weldors to melt holes in the metal, if necessary, but get it hot! Usually, it takes no more than 15 minutes of extra practice and you can be making good welds with this technique.

Rod Sticks to Base Metal

Every beginner experiences rod-sticking problems because the weld puddle isn't hot enough. The solution is simply to heat and maintain a molten puddle. The puddle melts the rod, not the torch. If you keep a good puddle going, then merely dip the rod where you want filler material. You'll get a good weld bead. Concentrate on the puddle, and the weld will take care of itself.

Flashback

This is a potentially dangerous condition where the gas burns back through the torch and hose to the regulator and cylinders, damaging the torch, hose and regulator. The cylinders are next, and an explosion is possible!

Flashback is usually accompanied by a loud hiss or squeal. If it occurs, flashback must not be allowed to continue! Immediately shut off oxygen at the tank if flashback occurs, then shut off acetylene. The oxygen is first because it supports combustion. Flashback is usually caused by a clogged torch barrel or mixture passage. Don't relight the torch until you cure the problem.

For these reasons, every oxyacetylene torch should be equipped with flashback safety arrestors. Basically, these are one-way valves that install in the torch gas lines.

Solutions

Don't be afraid to move the torch as necessary. You have 6300° F (3482C) available at the tip of a gas torch. Position the tip close enough to a piece of steel that melts at 2750° F

CORRECT GAS PRESSURES

As an EAA Technical Counselor, I am often asked to observe gas-welding and cutting practices by experimental aircraft builders. The most common mistake I see is too much gas pressure, especially too much oxygen pressure. About 3 to 4 pounds oxygen and acetylene pressure is about right for most fusion welding of thin-wall tubing and other steel parts that are less than .080" thick.

It seems that most self-taught weldors found that 10 pounds acetylene and 20 pounds or more oxygen pressure worked fairly well for cutting torch work, so they just use the same pressures for gas welding. (The proper pressures for cutting torch are 3 to 4 psi acetylene and 20 to 25 psi oxygen to cut 1/4" to 3/8" steel.)

Too much gas pressure, especially too much oxygen pressure, will burn up and crystallize your welds. After all, in cutting torch work, you use 20 to 25 pounds oxygen pressure to OXIDIZE and VAPORIZE the steel you are cutting.

Check the chart on page 53 for correct gas pressures to use in gas-welding.

(1510C), and the metal will melt!

If the puddle gets too big, pull the torch away for a second or two to give the metal a chance to cool and solidify. If you burn holes on one side of the weld seam and are not getting the other side red, direct the torch

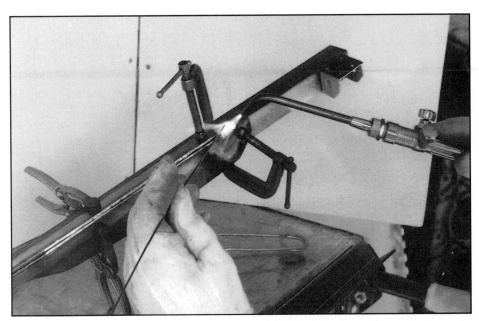

Tack-welding outside corner weld of 4130 steel engine bracket: Note how torch is held. As soon as puddle forms, welding rod is dipped into it to complete tack weld.

away from the hot side and toward the cold side to get more even heat.

You are probably getting tired of hearing this, but temperature control is the key. Control the temperature and you control the weld puddle. After selecting the correct tip size, temperature is controlled by the direction of the torch, the distance the tip is from the work and by gas-pressure settings. Follow these simple rules:

• Point the torch where you want the heat.
• Aim the torch away from where you don't want the heat.
• Back away if the puddle is too hot.
• Move closer if you are not getting the puddle hot enough.
• Increase heat by opening the torch valves if you can't get enough heat.
• Decrease heat by closing the torch valves if the puddle is too hot.
• Move the torch in an oscillating pattern.

GAS-WELDING SAE 4130 STEEL

After you've built your gas-welding table, and maybe another useful project or two, you should be ready to try welding chrome moly, or SAE 4130, steel. SAE 4130 welds about the same as mild steel, but it's more likely to become air-hardened and brittle from improper welding. Nevertheless, don't be afraid of 4130. If you can weld a nice bead with mild steel, you can learn to do the same with 4130.

Don't use copper-coated rod for welding 4130 steel. It may cause cracks and bubbles in the weld. Use only bare mild-steel or bare 4130 rod. For most jobs, 1/16-in. diameter rod is the best size to use. It comes in 36-in. lengths. I cut them in half for better control—did you ever try writing with a yard-long pencil?

Never braze 4130 steel. Its woodlike grain will open up and let brass flow into it. When the brass solidifies, the steel will then have thousands of little wedges that cause cracks between the grains. Sometimes the cracks will propagate as you watch!

Cleaning—Keep 4130 tubing or sheet clean of all oil, rust and dust. Clean it before you weld it. Don't even touch the weld area with your fingers after cleaning. Use methylethyl ketone (MEK), acetone or alcohol to clean both the base metal and welding rod. You can't get it too clean!

File, sand or sandblast all scale from previous welds before welding over them. The scale could contaminate your weld if not removed.

Shop Area

Keep the welding area clean, well-lit and draft-free—especially for welding 4130. A bright, clean shop area helps you make clean welds. A dark, dirty welding shop will contribute to cracks, pinholes and generally poor welds. Before you actually start building a long-term project, go out to the welding shop and write someone a letter in the position you'll be welding in! If you're not comfortable writing the letter, you certainly won't be comfortable welding. And never allow any drafts of air, cold or hot. One welding instructor once advised me to not even let my dog wag his tail in the welding shop!

Weld Technique

Evenly preheat the weld area to about 375 °F (190C). Although preheating to the precise temperature is not critical, a temperature-indicating crayon or paint can be used for getting the feel for how hot this is.

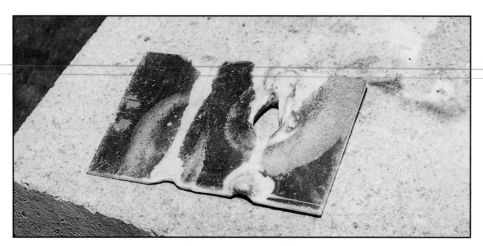

Two seams are shown here in aluminum that I am gas welding. The seam on the left is 100% penetration, but I have overheated and melted a big hole in the seam on the right. Too much heat is bad, too little is bad, too. Learn to CONTROL the heat.

Typical aircraft fuselage gas welding. Next step is to wipe the tubes down with metal prep liquid to remove the rust film, and then prime with rust preventative primer.

Play the flame—move it back and forth—over the entire weld seam, holding the torch tip about 4 in. from the metal.

Start welding where a minimum of preheating is required to form a puddle—such as on the edges. After running a bead for a fraction of an inch or so, the metal is automatically preheated, particularly if you're using the forehand method. This saves preheating time and reduces the chance of overheating the weld.

If you tack-welded the seam prior to running the final weld, be sure to remelt the tack welds along with the base metal and include them in the weld as you come to each.

Never jerk the torch away as you complete a weld. Hydrogen and oxygen in the air will contaminate the weld and it will cool too rapidly, possibly cracking it. After finishing a weld, pull the torch back slowly. Let the weld cool to a dull red before removing the torch completely. Not only does this reduce the chance of cracking the weld, pulling the torch back slowly also allows the molecules to relax gradually and stress-relieve (page 68) somewhat. Even when

stopping for more welding rod, hold the torch 4 in. from the work so the flame "bathes" the weld in heat.

Never weld the back side of a 4130 weld unless the designer specified it. If welded properly, the joint will be strong enough without doing so. Besides, the back side of a weld probably has scale that should be sandblasted prior to welding.

When redoing cracked welds on 4130 steel, file or saw out any bad welds and start over. You might even have to put a patch plate over the joint if excess metal was removed. A patch plate is usually made of the same material and thickness as that being patched. Extend the patch plate 200% past the damaged area and weld the plate all the way around.

Drill a relief hole in tubing that's being welded closed. If you don't, air pressure building up from the heat inside the tube will blow out the last of your weld as you finish sealing the tube. Therefore, drill a #40 or 3/32" hole in a non-stressed area about 1" from the end of every tube to be welded shut. If you want, squirt spray preservative such as LPS-or WD-40 into this hole. Or you can leave the

tube dry as I recommend, and either weld the hole shut or seal it with a Pop rivet. If you rivet the hole shut, coat the rivet with sealer to keep out moisture.

Rust Prevention

Rarely is it necessary to add oil to preserve the inside of a 4130-steel tubular structure. If moisture can't get inside the tubes, they won't rust. Most rust occurs from outside. Paint will protect it there. Oil is heavy, messy, and may contain chemicals harmful to 4130. I've repaired rusty fuselages from airplanes built in the '30's, and the rust was on the outside, not inside. There was no oil preservative inside the tubing.

GAS WELDING ALUMINUM

Most people equate oxyacetylene welding with gas welding. That's because acetylene is by far the most popular fuel used for gas welding. But when it comes to gas-welding aluminum, hydrogen is often recommended instead of acetylene. If

ALUMINUM WELDING Q&A

A handy question-and-answer pamphlet for welding aluminum was prepared for the annual Experimental Aircraft Association (EAA) Fly-in at Oshkosh, Wisconsin. This pamphlet was based on the most commonly asked questions concerning aluminum welding by the 200,000 EAA members. With their permission, portions of the pamphlet follow.

Question: What aluminum alloys are weldable?
Answer: 1100, 3003, 3004, 5050, 5052, 6061 and 6063 are weldable. Specifically, 1100 is dead soft and not good for structures; 5052 is medium hard and good for fuel tanks; and 6061-0 is soft, but can be heat-treated after welding to make it very hard and strong. 3003, 3004, 5050 and 6063 are weldable, but seldom used. You can weld 3003, 3004, 5050 and 6063, but don't use these for your project. Stick with 5052 and 6061 for better results.

Question: What kind of rod should be used?
Answer: 1100 rod for 1100 material and 4043 rod for all other alloys. Flux-cored rods work well also.

Question: What flux should be used?
Answer: Antiborax #5 for cast aluminum and #8 for sheet aluminum.
Note: There are many other fine aluminum-welding fluxes.

Question: At what temperature does aluminum melt?
Answer: Pure aluminum melts at 1217 °F (658C)—less than half that of steel—but alloys melt at lower temperatures. Aluminum oxide, a corrosive film that forms on aluminum immediately after cleaning, melts at a much higher temperature than aluminum. This oxide must be removed before welding and inhibited during welding. Remove oxide with a stainless-steel wire brush, Scotchbrite abrasive pad or acid. Use flux before and during welding to prevent oxide formation.

Question: What equipment is needed to weld aluminum with oxyhydrogen?
Answer: You need a standard gas-welding torch, one oxygen regulator and cylinder, another oxygen regulator converted for use on a hydrogen cylinder and a cylinder of hydrogen. You should also use cobalt-blue lenses in your welding goggles. With the conventional green lenses, all you would see of the flame and weld puddle would be a large yellow spot. The blue lens filters out the yellow light blocking your view.

Question: What size welding tip should be used for welding aluminum with oxyhydrogen?
Answer: Use a tip three times larger than the one used for welding 4130 steel of the same thickness. For example, if you would use a #1 tip for welding 4130 steel, use a #4 tip for welding aluminum of the same thickness.

you do use hydrogen in place of acetylene for welding aluminum, be sure to switch back to acetylene to weld steel. Welding steel with oxyhydrogen will cause hydrogen embrittlement—hydrogen contamination of the weld joint—causing it to be brittle.

Oxyhydrogen

For many years, aluminum welding was a mystery to me. I thought TIG was the only way to weld aluminum. Then I bought some aluminum welding flux, bare aluminum rod and tried gas welding aluminum. It surely is different from gas-welding steel, but it works!

Oxyhydrogen is the preferred method of welding aluminum. This is because its 4000° F (2204C) neutral-flame temperature is closer to the 1271° F (658C) melting temperature of aluminum than is the 6300 °F (3482C) flame temperature of oxyacetylene.

Aluminum vs. Steel

When using oxyhydrogen to weld aluminum, remember that an oxyhydrogen flame has little color. It doesn't look like an oxyacetylene flame. Therefore, don't try to adjust it to the same color. It will have two distinct cones—inner and outer—as described for oxyacetylene welding. When properly adjusted, oxyhydrogen flame has an almost clear outer cone and pale-blue inner cone. However, the adjusting procedure is the same.

Use hydrogen just as you would acetylene. Almost all the procedures for lighting the torch, choosing tip size, tip-to-work distance and other techniques for welding aluminum are essentially the same as for welding

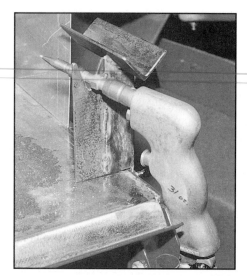

Make a weld bench hangar for your gas torch so you won't have to shut it off and relight each time you stop to adjust or setup your weld. A Dillon torch is shown here.

The V.H.T. ceramic paint-coated exhaust headers on my aircraft conversion Buick V6 engine were gas welded. The engine mount was TIG welded.

steel. Temperature control through practice is still the secret, regardless of the material or method.

One major difference in welding aluminum compared to steel is that the puddle is much cooler than steel and will not melt the welding rod as easily. So you must preheat the welding rod slightly by holding it near the puddle and partly in the flame as you move the puddle along. The dip, dip, dip process is the same.

Another difference with welding aluminum vs. steel is that there's little change in base-metal color as it melts to form a puddle. Aluminum doesn't change color and glow red like steel when it's heated. Instead, it stays the same color until a few degrees before reaching its molten state. At this instant, it becomes shiny where the puddle forms. If you continue heating the aluminum puddle past this state, looking for a color change, the puddle will drop out. You'll have a hole rather than a puddle.

Always use a welding rod with a diameter closest to base-metal thickness. If you have a metal shear

and you're welding sheet stock, filler rod can be made by shearing square strips from the sheet. This will automatically give you the precise filler-rod size and alloy.

For best results, the aluminum pieces must be absolutely clean. You should even wipe the welding rod with a clean, white cloth before welding.

Aluminum welding rod must be used with flux. Otherwise, you won't be able to weld aluminum successfully. Flux must be used to remove and inhibit the harmful oxides that form on aluminum. Mix aluminum welding flux powder with water or alcohol. While welding, frequently dip the welding rod into the mixture to keep the rod coated with fresh flux. Brush on the flux every two or three minutes. The base metal must have fresh flux on it while welding.

Keep a large bucket of fresh, clean water nearby. After cooling, dip the parts in the water to rinse off the flux.

Flux left on aluminum causes corrosion.

HEATING & FORMING

When you heat steel to dark blood red (1050°F or 566C), it bends easier than when cold. Heat steel to bright cherry red (1735°F or 746C) and you won't believe how easily it bends! Steel molecules become more "plastic," or pliable, the hotter they get.

You can use heat to assist in bending large, thick pieces of steel with a simple bench vise and a small pry bar. I modify new trailer hitches to fit old cars simply by heating and bending them. I learned that trick when I was about seven years old, helping my uncle make new plow tips for his farm tractor. Red-hot steel bends like warm taffy candy.

Practice bending hot steel and pretty soon you'll be doing things you never thought possible. Just use the appropriate-sized welding tip, cutting

Here is a 1930's aircraft engine mount that was gas welded. Sixty years later it is still good welding, and it is working.

ROSEBUD TIPS—HEATING WITH ACETYLENE

Victor Torch Number	Smith's Torch Number	Acetylene Pressure	Oxygen Pressure
4	AT603	6–10	8–12
6	AT605	8–12	10–15
8	MT603	10–14	20–30
10	MT605	12–14	30–40
12	MT610	12–14	50–60
15	—	12–14	50–60

Other fuel gases that can be used for heating include propane, natural gas, propylene, and MAPP gas.

tip or rosebud tip to apply the amount of heat needed. I try to use a tip that heats the metal to cherry red in less than one minute. If it takes longer, the tip is too small. This wastes both time and gas.

Rosebud Tip

A rosebud tip is so called because its flame configuration looks like a rosebud. Strictly meant for heating, this tip is actually useful for both heating and forming metal. A rosebud tip is no good for welding, but it does generate gobs of heat. It also uses a lot of oxygen and acetylene, so make sure your tanks aren't low before starting a project.

The first time I lit a rosebud tip, it produced a pop that sounded like a shotgun! The next six times I did it, I got the same loud noise! Then it occurred to me that 4 psi oxygen and acetylene pressure were insufficient to operate that big tip! I then increased oxygen pressure to 25 psi and acetylene pressure to 10 psi; the torch lit with only a soft pop. It takes a lot of pressure to operate a rosebud tip!

Rosebud tips are good for freeing stuck parts. However, because of their high-heat output, use a temperature-indicating crayon or paint to check the temperature so you don't overheat the part.

Cutting-Torch Tip

If necessary, you can use a cutting-torch tip for heating. It puts out more heat than a welding tip, but less than a rosebud tip. Adjust the torch for the maximum neutral flame without the cutting lever while you're heating a part, or you'll oxidize or cut it!

AUTOBODY GAS WELDING

From the first steel-bodied cars until about 1980, gas welding was the only way to repair severely damaged car bodies. After 1980, an increasing number of car bodies use high-strength steel (HSS) panels. Depending on the alloy, many high-strength steels have a crystalline grain structure that can be destroyed if heated above a certain temperature. The limit for martensitic steel, for example, is 700 °F (371C). This severely weakens the metal and can cause cracking.

For this reason, auto manufacturers recommend that HSS panels be welded with other techniques—MIG welding is one of them. But for all those older cars, you can make good use of your 6300 °F (3482C) gas welder for body-and-fender repair.

To determine whether the body panel on a late-model automobile or light-duty truck is HSS, consult the shop manual. Because each manufacturer uses different types of HSS and may have different repair procedures for even the same alloy, follow its procedures to the letter. Many HSS panels are critical structural members and the repair procedures are difficult, so don't be surprised if the manufacturer recommends replacing a damaged HSS panel rather than repairing it. ∎

STRESS-RELIEVING MYTHS

For at least 50 years, old-time weldors have advised new weldors to reheat tubular structures after the structure has been completely welded. They tell us that the purpose of reheating the welded cluster to dull red is to "stress-relieve" the weld area.

The truth is that open-air, manual application of oxyacetylene flame to a weld cluster is likely to do more damage to the structure than to simply leave it alone after welding. I will describe the correct procedure for stress-relieving a 4130 steel tube aircraft engine mount, and you be the judge of whether it is possible to accomplish the safe, correct procedure in your own workshop situation:

1. To prevent the stresses from distorting the mount, it should be firmly bolted to a very heavy fixture, and the fixture should be a heat-resistant stainless alloy. To prevent stresses in the fixture, it should be bolted and not welded together. The fixture to hold a 15-pound engine mount would weigh more than 100 pounds.

2. The engine mount and fixture would be placed in an air-tight oven and brought up to 1,150 °F to 1,175 °F temperature and held at that temperature for 2 hours. The color of the ENTIRE engine mount will be DARK CHERRY RED, and it will hold that color for 2 hours. Two hours are required to allow the entire mount to undergo a complete residual stress relief process.

3. After holding the mount and stress-relief fixture in the oven for 2 hours, the mount is removed from the oven and allowed to cool in still air (not wind, and positively not in water or oil). The complete cooling process generally takes 4 hours.

4. After the mount and fixture cools to 100°F or less, the mount is removed from the jig. Comments: It should be obvious that this correct stress-relieving process could not be accomplished in a welding shop.

• Heating the mount to 1,575°F, bright red and scaling, will anneal the mount, soften it, and cause it to lose strength, possibly back to less tensile strength than mild steel.

• Expert heat treaters, who work to mil-specs, agree that 4130 tubular steel weldments, as found in engine mounts, when properly fitted and cleaned before welding, and properly welded, will never stress crack in normal use. Furthermore, if no cracks exist immediately after welding, the likelihood of cracking long after welding is extremely remote.

• Magnaflux or dye penetrant checks for cracks and porosity after welding is really sufficient. You probably should check the engine mount for residual magnetism that might exist after MIG or TIG welding. Demagnetize it to prevent compass distractions.

Torch Cutting

8

Old-timers in the welding trade like to call torch cutting "burning it off," and they are partly correct in using this term because oxygen-fuel cutting actually works by oxidizing the metal in the cut.

GAS CUTTING & OXIDIZING

Now that you've "mastered" the art of gas welding, the next item on the list is learning to flame-cut with acetylene–sometimes referred to as a *blue wrench* or *gas wrench*. These terms came from mechanics who used cutting torches for loosening or removing stubborn bolts, nuts or other seized parts, usually due to rust. Our main purpose is to use the acetylene cutting torch for fabricating steel parts.

The primary chemical reaction in flame-cutting steel is oxidation. Because cutting is really oxidizing, you cannot cut metals that do not oxidize (rust) easily, such as aluminum and stainless steel.

So primarily, you can flame-cut mild steel and cold-rolled steel. These steels make up a large part of things we use, such as automobiles, trailers, farm equipment and more. Flame-cutting is useful for cutting elaborate shapes not suitable for cutting with a bandsaw. Remember that flame-

A sales representative of Victor Equipment Co. demonstrates with a motor drive oxyacetylene cutting torch, self-contained in one unit.

cutting leaves rough edges with slag that must be final-trimmed later. This is useful if you're doing arts and crafts, but a problem when doing precision fitting.

USING A CUTTING TORCH

After learning to gas weld, you will discover that a cutting torch is relatively easy to operate. Simply light the torch, adjust the flame to neutral, make a little puddle, push the oxygen lever and you're cutting steel! Here are the steps for cutting 1/4" thick steel plate:

Select a piece of 1/4" scrap steel plate. Also find a piece of angle steel similar to the one shown. Use it to guide your torch hand and help you cut a straight or smooth curved line. Secure it with C-clamps or locking pliers, if necessary. Mark the cut-line with soapstone.

Position the plate so the cut-line hangs over the edge of the welding table. Or, lay two short sections of angle iron face down on the bench and lay the plate on top. This will prevent a nice cut being made across your workbench top.

Before you light the torch, check your clothing. You should be wearing

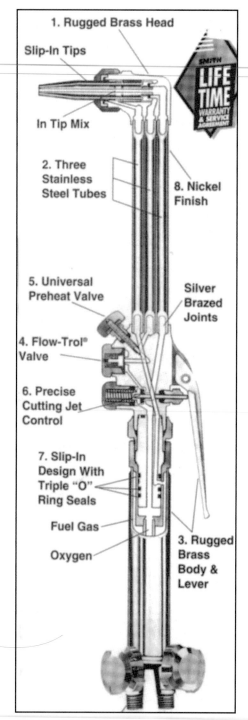

1. Rugged Brass Head

Slip-In Tips

In Tip Mix

2. Three Stainless Steel Tubes

8. Nickel Finish

5. Universal Preheat Valve

Silver Brazed Joints

4. Flow-Trol® Valve

6. Precise Cutting Jet Control

7. Slip-In Design With Triple "O" Ring Seals

Fuel Gas

Oxygen

3. Rugged Brass Body & Lever

The cut-away schematic of a typical Smiths cutting torch shows its complex design. Study the cut-away to see which valves adjust the acetylene, the preheat oxygen and the cutting oxygen flows. Photo: Smiths Equipment.

cuffless pants, high-top shoes, a long-sleeve shirt, gloves and welding goggles. If all is OK, proceed.

Just-lighted, acetylene-rich cutting torch looks like this. Acetylene valve is open about one turn to minimize sooty flame.

Opened cutting-tip oxygen valve neutralizes flame. Remember: Oxygen valve on torch body is fully opened.

Light Cutting Torch

Adjust the regulators for about 1–2 psi acetylene and 10–15 psi oxygen. Oxygen pressure is much higher than acetylene pressure because the oxygen does most of the work.

Next, preadjust the cutting tip. Shut off the oxygen valve on the cutting tip and fully open the oxygen valve on the torch handle. This is important. Open the acetylene valve about one turn and light the torch using a flint or electric striker. Add oxygen by opening the oxygen at the cutting tip until you have a neutral flame.

Start Cutting

If you're right-handed, support the torch with your left hand while resting it on the plate. This will allow you to guide the torch along the cut-line for a more even cut. Grip the torch with your right hand and have your thumb ready to press the oxygen-cutting lever. Start your cut by heating the edge of the steel plate at the right end of the cut-line. Reverse everything if you're left-handed.

Note that when cutting with acetylene, material is removed. The resulting void is the kerf. Consequently, you must cut on the outside of the cut-line—scrap side.

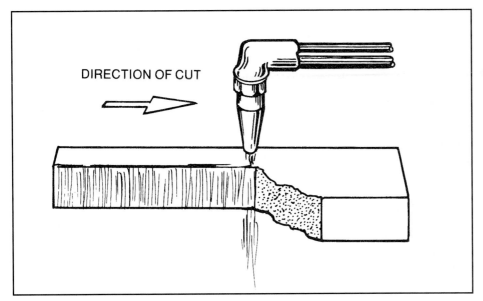

Drop cut occurs when cutting speed and oxygen flow are correctly matched so there is no drag. Flame exits kerf immediately below where it enters.

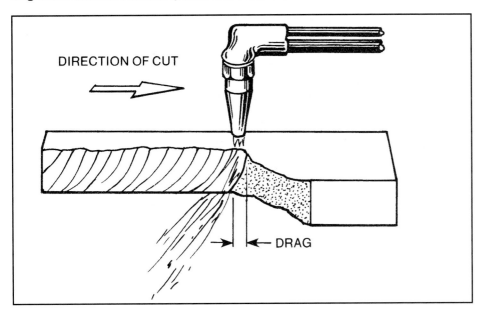

If speed is too high or oxygen flow not enough, drag occurs when cutting thick material. Reduce or eliminate drag by reducing speed or increasing oxygen flow–pressure.

For safe cutting, so the sparks don't fly everywhere, a 55-gallon oil drum makes a very good cutting table. Just make sure the drum has been professionally cleaned out if it has ever been used to store flammable liquids. Photo: Lincoln Weld Products.

Otherwise, the part you're making will be short by the kerf width.

When a puddle develops, press on the oxygen lever. The flame should immediately begin to blow away—oxidize—the metal. If it doesn't, you didn't have a good puddle. Continue heating the metal and try again. When the sparks fly as the cut begins, carefully guide the torch along the cut-line. With a 1/4" steel plate, you move along at about 1" every three seconds, following the cut-line. Move too fast and you'll get a shower of sparks back in your face because the metal wasn't heated enough. Move too slow and you'll overheat the metal, resulting in excessive slag.

Be careful when you reach the end of your cut. A chunk of hot metal may fall to the floor if you've made a clean cut. But, chances are the slag will hold the pieces together. A light tap with a hammer should break them apart.

Tips for Better Flame Cutting

Excess slag at bottom of cut indicates the preheat flame is too hot. Correct by reducing acetylene pressure or using a smaller tip.
• Metal doesn't have to be super clean for cutting.
• Most beginners force the cut by moving too fast. Slow down. Refer to the chart for ideal cutting speeds for different metal thicknesses.
• Clean cutting tip periodically. The cutting process tends to splatter molten metal back on the cutting tip, reducing cutting efficiency.

Flame-Cutting Aids

There are a few items you can use that'll make a cutting project easier. They range from a piece of angle iron used as a guide to cut a straight line to a sophisticated flame-cutting

71

The cut is started by heating a corner of the steel to RED hot BEFORE squeezing the cutting trigger. If you squeeze the cutting trigger before the steel gets red, the cut will not start.

A circle cutting guide attachment can be adapted to most cutting torches for cutting circles and accurate radiuses up to 24".

Once the cut starts, move the torch smoothly and steadily in the direction of the desired cut, holding the trigger down. To make neat, clean cuts, practice, practice, etc.

The proper way to hold a circle cutter attachment and cutting torch.

machine. Chances are you won't need the cutting machine unless you'll be duplicating several pieces from heavy steel plate.

Marking Tools—I already discussed soapstone, center punches and scribes for laying out and marking cut-lines, page 33. To assist with marking, you should make patterns. A pattern duplicates the shape you want to cut out. It can also be used for checking the fit of the final part without actually having the part.

Once you've developed the pattern, lay it on the material to be used for the final part. Then trace around the pattern with soapstone or whatever marker you desire.

Pattern material is inexpensive. All you need is plenty of thin, flexible sheets of cardboard. Such material is available at office-supply stores. Corrugated cardboard from boxes is cumbersome but OK to use in a pinch. To mark patterns, you need soft-lead

pencils and a pencil compass. Additional drawing equipment such as straightedges, 30°/60° and 45° triangles and felt-tip markers should also be on your list of pattern-making items. Finally, you'll need a pair of heavy-duty scissors for cutting out patterns.

Cutting Table—If you'll be doing a lot of flame cutting, it would be helpful to have a cutting table. See the drawing, page 146. A cutting torch can't distinguish a metal-top welding table from a workpiece. Consequently, you shouldn't flame-cut anything that's laying directly on top of the table. Otherwise, you'll end up

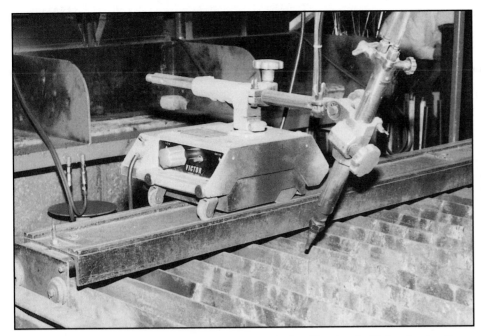

Typical welding school motor drive straight cut oxy-fuel cutting torch. Obviously, steadiness and consistent movement of the torch makes for better cuts.

pattern, the multiple cutting torches cut out several duplicate parts simultaneously.

Chances are you won't need such equipment. For the hobbyist or small welding shop, inexpensive flame-cutting machines for cutting out small parts are available, such as the one from Williams' Low-Buck Tools. Manually operated with a tracer, this machine is designed to install on a 55-gallon drum. The drum contains all the sparks and scrap that would otherwise fall to the floor.

PISTOL-GRIP CUTTING TORCH

A recent development in oxyacetylene cutting and welding equipment is a torch that is held like a pistol rather than like a regular torch. This torch was originally called the Dillon MK III after its inventor, but recently is marketed as the Henrob torch. This torch does a very good job at cutting steel plate and thin sheet steel.

making a cut through the table top as well.

To avoid this, either raise the workpiece off the table by supporting it with scrap angle iron, hang it over the edge of the table or set it somewhere else. Whatever you do, don't support it with anything you don't want to be cut or damaged. Ideally, it is best to support the work with a cutting table.

A flame-cutting surface doesn't have to be exotic. It can be as simple as several sections of angle iron, positioned corners up, bridging two metal saw-type horses. Or, it can be an honest-to-goodness table. The table would consist of an angle-iron frame with four legs. Instead of using a solid steel top, string several sections of angle iron loosely between the frame. This will allow you to adjust the angle iron to support the work as desired. You can also replace them easily when they are cut.

Cutting Machines—Cutting machines used in the welding industry

can flame-cut parts to exacting dimensions, duplicate parts or cut out many parts at one time. Some of this equipment is very expensive, such as an electric-powered machine that runs on its own track. More expensive machines have multiple cutting heads. While a tracer follows a single

A new, never-used portable motor drive cutting torch can be carried to a field job and set up for accurate cuts directly on the steel plate.

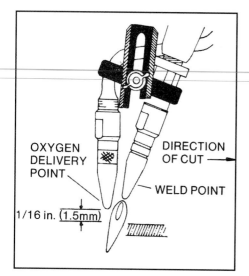

Preheat flame is separate from oxygen-delivery tip on Dillon MK III cutting torch. Consequently, cut must be made with oxygen tip trailing preheat tip as shown. Oxygen tip is adjusted to other side of preheat tip for cutting sheet metal. Drawing courtesy Shannon Marketing, Inc.

Cutting

Traditional gas torches use an oxyacetylene cutting tip with six small flames in a small circle to preheat the steel, and a large orifice in the middle of the six smaller holes. The larger orifice is where the stream of oxygen comes out when you depress the lever on the torch body for cutting. This means that no matter which direction you travel with the cutting torch, you are following the nice, clean-cut kerf, with at least one or two more oxyacetylene flames that tend to melt the metal back together.

The Dillon cutting torch attachment is different. It uses a single oxyacetylene flame to preheat the steel and a single oxygen-only stream of higher-pressure gas to make the cut. No acetylene flame follows, so the cut stays clean. But you don't get something for nothing. The Dillon cutting attachment works only in one direction. If you want to cut a circle, you have to move the cutting

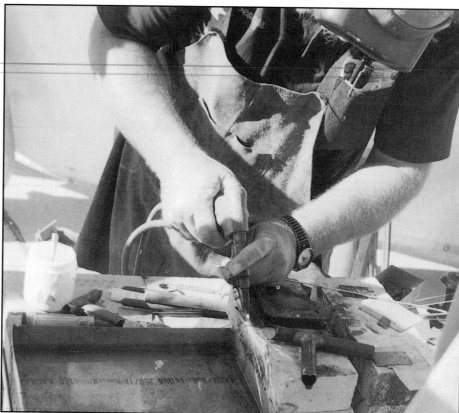

Dillon cutting torch has guide wheel assembly that clamps to torch to assist in making precise cuts. Torch must be turned to follow cut-line. Photo courtesy of K. Woods, Inc.

CUTTING STEEL WITH OXYGEN ONLY, NO FUEL!

Here's a trick I show new welding students to illustrate that cutting with oxyacetylene is primarily an oxidizing process. Try it after you learn to use the cutting torch.

Start your cut in a piece of steel plate about 1/4" thick. After establishing the cutting speed and travel, shut off the acetylene torch valve with your left hand–right hand, if you're left-handed–while you continue to cut. You'll still be able to cut the steel with oxygen only! Of course, if you don't move steadily, you'll lose the molten puddle and the oxidizing or cutting process will stop. You'll then have to relight the torch to resume the cut.

attachment in a circle to keep the preheated oxyacetylene flame and oxygen in a leading-trailing relationship.

Cutting thin sheet metal, such as car fenders, is really where this torch excels. For sheet-metal cutting, use a different attachment that places the oxygen-only tip at the rear of the flame so you make the cut going away from you. This allows very thin sheet metal to be cut quickly and with almost no heat-affected zone adjacent to the kerf. This feature is ideal for the new high-strength steels that tend to crack in the heat-affected zone. ∎

SETUPS FOR CUTTING STEEL

Material Thickness (in.)	1/8	1/4	1/2	3/4	1	1-1/2	2	4	5	6
Suggested Tip Number	00	0	1	1	2	2	2	3	3	4
Oxygen Pressure (psi)	5 to 10	10 to 15	10 to 20	15 to 25	20 to 30	25 to 35	30 to 40	35 to 45	40 to 45	45 to 50
Acetylene Pressure (psi)	1 to 2	1 to 2	2 to 3	2 to 3	2 to 4	2 to 4	2 to 4	3 to 5	3 to 5	4 to 6
Cutting Speed Per Minute (in.)	20 to 23	18 to 20	14 to 16	12 to 16	10 to 14	8 to 12	6 to 10	5 to 8	4 to 6	3 to 5
Oxygen Used Per Hour (cu ft)	50	80	120	140	150	170	220	350	420	500
Acetylene Used Per Hour (cu ft)	8	10	12	15	16	17	19	25	29	30

VICTOR ROSEBUD #6 = 6 holes x .043"
Dillon Cutting Copper 1 hole x .050" 8 bands

OXYACETYLENE CUTTING TIP SIZES, GAS FLOW DATA

Decimal Metal	Fraction Metal	Victor Tip Size No.	Smiths Size No.	Henrob Tip Size	Thousandths of inch	Oxygen psi	Acetylene psi
.125"	1/8"	000	–	Copper 8	.020"	20	3
.250"	1/4"	00	MC-12-00	–	.025"	20	4
.375"	3/8"	0	MC-12-0	–	.035"	25	4
.500"	1/2"	1	MC-12-1	–	.040"	30	4
1.000"	1"	2	MC-12-2	–	.046"	35	4
2.000"	2"	3	–	–		40	5

Gas Brazing & Soldering

9

Brazing and soldering are metal-joining methods that do not rely on melting the base metal to join two or more pieces. Instead of fusing the filler and base metals, they depend on surface adhesion of the solder or braze filler. This is made possible by capillary action—surface of the molten filler is attracted to fixed molecules nearby. When the filler metal cools, it bonds to the base-metal surface.

Brazing and soldering are more akin to adhesive bonding than welding. A major advantage of these processes over welding is that brazing and soldering are done at lower temperatures. Soldering is done below 800° F (427C); brazing is done below the melting point of the base metal, usually below 1500° F (816C). Therefore, warpage and temperature-induced stress in the base metal are lower. For example, to fusion-weld a bicycle frame, you must heat the metal to the melting point of steel—more than 2700° F (1482C). To braze that same frame, 1000° F (538C) is all that's necessary.

Another big advantage of brazing and soldering is that field repairs are simple. All you need is a torch. To arc- or TIG-weld in the field, such as repairing farm equipment, you'd have to perform the operation near an

I decided to braze the two-legged brace to the weldor's mechanical finger, see page 148, because it was more convenient than starting up my TIG or Arc welders. Brazing is completely suitable for many metal joining tasks.

electrical outlet or a portable welder. That's not always easy to do.

In case you were wondering, brazing does not mean brass any more than soldering means lead or silver. For instance, there is brass, aluminum and silver brazing. Brazing refers to the temperature, not the metal.

BRAZING

Most people think that brazing isn't as strong as arc welding because they've seen or heard of brazed joints breaking. To disprove this, compare the tensile strengths of welding and

brazing rods. E-7018 is considered to be the best arc-welding rod and it has a 70,000 psi tensile strength. Now, look at the tensile strength of several brazing rods on page 139— 80,000 psi or more!

To demonstrate the tensile strength of a brazed joint, I often ask my welding students to guess how much their cars weigh. The answers usually range from 2500 to 5000 lbs. You probably know what's coming next. I then take two 1" x 5" long, 0.060" thick mild-steel strips, overlap the ends 1" and braze them together.

Next, I put the brazed strips in the

I built the frame and steering on this IKF World Championship go-kart just with brazing—no fusion welds were used. I went on to race it to a third place finish in an IKF World Karting Championship race.

heat also allows you to braze near rubber parts because of the reduced chance of burning the rubber.

The lower heat and induced stresses also mean that you don't have to be as careful about avoiding air currents when brazing as you do when fusion-welding sensitive metals such as 4130 steel tubing. And materials of different thicknesses can be joined easily. You don't burn up the thin part trying to heat the thick part.

Brazed joints look good and usually require only flux removal to maximize their final appearance. Brazing, by its nature, will flow into a smooth fillet, giving the joint a finished look without filing or machining.

Remember: Never fusion-weld a joint that was previously brazed. If you have a project that requires both fusion welding and brazing, fusion-weld first, then do the braze joint. Never braze first. Otherwise, you'll boil away the braze filler with the higher temperature of the fusion weld.

You can maximize the strength of a brazed joint by giving the joint more

welding shop pull-test machine and let the class watch as the machine pulls over 3000 lbs before the base metal stretches and breaks! The brazed joint never breaks. Instead, the metal at one end of the joint fails. By this time, the whole class is saying, "That little 1 sq. in. of brazing could lift that entire 3000-lb car!" More importantly, the brazed joint is stronger than the metal itself.

Seam or Joint Design

The design of a seam or joint is very important in deciding whether to use solder or braze. For instance, avoid brazing or soldering butt joints. The main reason for this is the lack of sufficient wetted base-metal surface for filler to adhere to. Consequently, the base metal pulls away from the braze or solder under tension—pulling—or bending loads. Lap joints are another matter. The wetted surface area can be increased by simply increasing the overlap of the two pieces. And the joint is in shear—the

best type of joint for brazing or soldering. See the illustrations of types of joints and loads, below.

Low Heat Required

Because brazing requires less heat, and therefore results in less warpage than fusion-welding, brazing is used extensively in autobody repair. Less

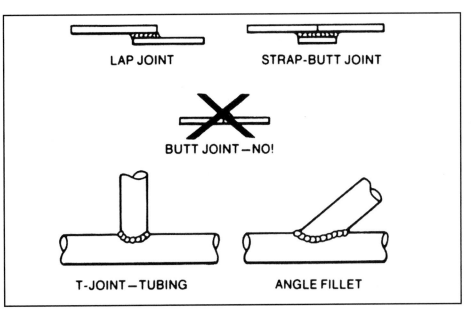

Best and worst joints for brazing, silver brazing or soldering: Never braze butt joints.

Most of the parts you see in this picture were brazed together, even the coil-over-spring shocks and independent suspension. Brazing has a tensile strength of 80,000 to 120,000 psi, stronger than the steel in the frame.

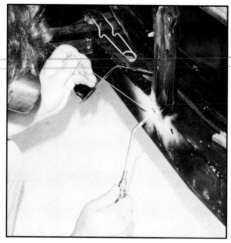

To braze sheet-metal brace to Corvair rear frame, I'm holding torch about 6 in. from metal for a softer and cooler braze joint. After braze is completed, smoke and flux residue can be brushed or washed off with water.

surface area. Depending on base-metal thickness, roughen it by grinding, sanding, sandblasting or coarse filing. All those little surface scratches increase surface area and provide a tooth the filler can cling to.

Identify the base metal and use the correct filler metal. For instance, if you tried to braze shiny stainless steel with shiny aluminum filler rod, the metals just would not mix. You'd end up with a hot mess that will fall apart.

Always heat the base metal sufficiently, especially cast iron. But don't overheat small areas. This may cause the braze filler to fume and boil.

Metals That Can Be Brazed

Metallurgy and metal joining methods are changing as fast as our computers and electronic things are changing. Be constantly aware of new materials and new joining methods. Read Chapter 15 for new methods and materials for joining metals.

A wide variety of metals can be joined by brazing. These include stainless steel, cast iron, brass, copper, bronze, aluminum (with aluminum brazing rod), mild steel, cold-rolled steel, chrome-plated metal, cast metal, galvanized steel and other zinc-coated steels.

Dissimilar metals can be joined, such as copper to steel, or copper to brass. This usually is not possible in fusion welding.

Gas Brazing Procedure

Brazing Cautions—You must never breathe brazing vapors. Use ventilation as necessary. When galvanized or zinc-coated metals are heated with a welding torch, fumes are given off that are extremely dangerous if inhaled. These fumes appear white, similar to cigarette smoke. It's safest to avoid welding, cutting, brazing or soldering galvanized or zinc-coated metals. Instead, let an experienced weldor do it for you. But, if you feel you must weld galvanized pipe or sheet metal, follow these precautions:

•Weld galvanized metal only outdoors in open air so fumes will not concentrate. Or ...
•Use a commercial air extractor to suck the fumes into a filter and away from humans and animals.
•Wear a high-quality breathing respirator while welding galvanized metal.

When practicing brazing, you'll be doing similar, but fewer, operations than you did when practicing gas-welding steel. Brazing involves three operations instead of five. You won't puddle brass or do butt welds. Instead, you'll concentrate on lap welding.

You'll need several 2" x 5" pieces of 0.030" thick mild-steel scrap. Your three projects are:

•Running a bead with brazing rod.
•Lap-brazing two pieces.
•T-brazing two pieces.

Required Equipment:
•Oxyacetylene welding outfit.
•1/16" brazing rod (36" long, cut in half). For the required filler material, see page 136.
•Powdered brazing flux or flux-coated rod.
•Bucket of water for removing flux from braze bead.

Common Mistake—Many inexperienced braze weldors overheat the base metal. After learning to fusion-weld mild steel at 2700° F (1482C), they must learn to braze at 1050-1075° F (566-580C). Remember that all metals start to vaporize at their boiling point. Brass brazing rod will boil, vaporize and generally ruin your project if heated to the melting point of steel!

For best results when brazing mild steel, heat the base metal to about

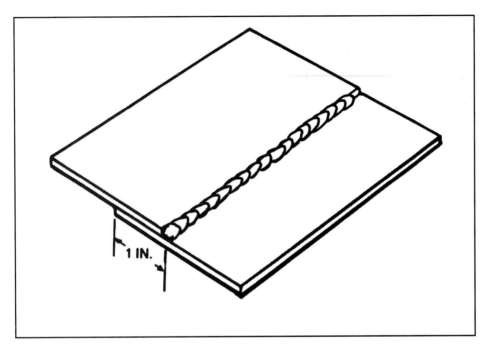

Overlap braze lap joint about 1". Heat joint evenly until both pieces are blood red. Apply brazing rod and watch it flow into joint.

blood red to dark cherry red—no hotter! To do this with a 6300° F (3482C) gas-welding torch, hold it farther away from the metal than if you were fusion-welding. You may think that a smaller torch tip will give less heat—true. But, for brazing, you should use a large soft flame rather than a small hot or harsh flame.

To get a soft or quiet flame, use a medium tip with very low gas pressures. You'll hear the difference between the soft flame compared to a neutral flame for welding steel. You'll like the way the soft flame sounds.

Flux the Rod—Open the can of brazing flux. Using a 0.040" tip, set both oxygen and acetylene regulators to 2–3 psi. Before you light your torch, double-check that you have on the proper apparel: long-sleeved shirt, long pants, gloves and welding goggles.

Light the torch and adjust for a soft flame; not loud and hot. Brush, or bathe, the working end of the brass

rod with the flame to get it warm—not molten. Quickly dip the rod into the powdered flux. When you pull out the rod, flux should have adhered to about 2" of the hot end of the rod, completely covering. Use plenty of flux. Don't worry about using too much. A can of flux goes a long way. One can will last me years, even when I do a lot of brazing.

Precoated Rods—You can buy brazing rod precoated with flux, eliminating the inconvenience of repeatedly dipping the rod into the flux. Sometimes, even an old pro like me buys some.

The problem with flux-coated rods is that they must be protected from moisture and rough handling. The flux coating can break and flake off. If this happens, there won't be enough flux on the rod to do a good braze job. I buy just enough rod to last 30 days or less, so the rods are always fresh.

Run a Braze Bead—Set your piece of scrap metal on firebricks. If you

haven't already done so, light the torch, adjust it to a soft flame and hold the tip 2–3" from the workpiece metal until a 1"-or-so round, blood-red spot develops. Now, just touch the hot spot with the brazing rod. The filler should melt and flow onto the steel. Continue this heating and touching process until the rod needs more flux. Dip the rod back into the flux and keep going.

Brazing Joints

Lap Joint—Now you're going to braze a lap joint and see capillary action at work. Make sure your two pieces of scrap steel are clean and rust-free. They should be flat so the edges will not bow up and look bad afterward. See the accompanying drawing for how to fit the two pieces. Remember to support the workpiece off the table so it doesn't soak up the heat.

Light the torch as before and coat the rod with flux. Play the torch along one end of the seam to heat both pieces to dull cherry red. When you're sure that both pieces are the same color and temperature, touch the edge of the seam with the flux-coated brazing rod. Watch the molten filler flow into the seam! That's caused by capillary action. Continue along the seam until you have it filled end-to-end with filler.

Shut off the torch and let the metal cool for about 3–4 minutes. Pick up the metal with your pliers and dip it into a bucket of water to cool and soften the flux. The flux turns into a glass-like substance after cooling from its molten state. Although you can chip it off, it's easier to let water soften it. Then, it's an easy job to remove the flux with a wire brush.

If you succeeded in getting the filler to flow completely into the seam, the

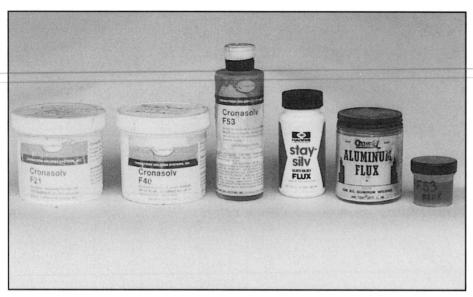

Several different kinds of flux are required for the many and varied kinds of brazing and soldering processes. Here are 6 kinds of flux I keep in my corrosives cabinet. Yes, flux is almost always corrosive, but it can be easily neutralized with warm water.

joint should be capable of lifting a car!

T-Joint—After you've mastered brazing the lap joint, practice brazing a T-Joint. Support the pieces so heat isn't absorbed into the work table. Hold the T in place with a "mechanical finger"—page 148. Tack-braze it in place. A tack should go at each end of the T.

As you start brazing the seam, remember to control the temperature of the pieces. Remember gas-welding a T-joint? You cannot heat just the bottom metal piece and expect the brass to flow onto the vertical piece. Manipulate the torch so both pieces are heated equally—dark cherry red. Remember, if you overheat the steel, the brazing rod will fume or boil! After you've finished the seam, shut off the torch.

The seam must be filled with filler to be strong. Adhered surface area must be maximized. The drawings nearby show the best seams for brass and silver brazing.

Brazing Aluminum

Brazing aluminum is similar to brazing steel or other materials. It also has the same problem as gas-welding aluminum: The base-metal color doesn't change as it's heated.

The clue to judging correct temperature for brazing aluminum is to watch the flux that's on the base metal. When it starts to melt and flow, the base metal is ready to be brazed.

Use a slightly carburizing flame to reduce aluminum oxidation.

Not only must aluminum-brazing rod have flux on it, you must also apply flux to both sides of the weld joint. I like to use a small metal-handled acid brush to apply the liquid flux, but you could even paint it on with a thick-bristled paint brush. Don't use a brush with plastic bristles—the bristles will melt. Acid brushes are relatively inexpensive—so inexpensive that you could throw them away after one use. For this reason, I keep a dozen or so brushes around at all times.

Brazing Copper, Cast Iron & Other Metals

By the time you're ready to braze other metals, you should be able to determine when the base-metal temperature is right for brazing. Copper turns red, stainless steel blue, and cast iron yellow when they've reached the right temperature. But, as with brazing aluminum, the best way to judge when the proper temperature has been reached is to watch the flux. When it melts, the base metal is ready to accept the filler metal.

Silver Brazing

You can silver-braze just about any metal that can be brass-brazed. Silver wets the metal better than brass. It also sticks to some metals where brass will not, such as carbide tool steel. This very hard steel is used for tipping saw blades and other cutting tools. Silver braze also gives a superior appearance to some projects such as costume jewelry.

Because silver is one of the metals in silver-brazing rod, it costs much more than brass filler. A silver-bearing alloy of low-tensile strength—about 20,000 psi—can be used to join dissimilar metals such as aluminum, steel, copper, stainless steel and monel. This particular low tensile-strength silver alloy melts at low temperature—about 500° F (260C).

Silver-Brazing Procedure—The first thing to do is thoroughly clean the joint surfaces. The joint clearance should be 0.002"–0.006". If you can't judge this spacing by eye, use a feeler gauge to set the spacing at exactly 0.004". After setting up parts with a feeler gauge a few times, you should be able to judge the 0.002" to 0.006" gap by eye with ease.

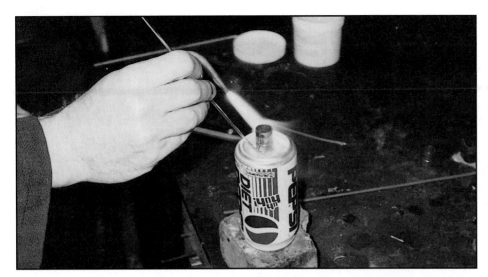

With the right kind of solder and flux, you can join any or all metals together with 7,000 psi tensile strength. Here, Cronatron #53 solder and flux is being used to solder copper to aluminum. See Chapter 15 for specifications.

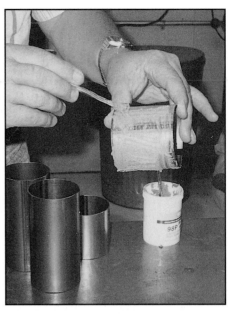

Here I am using a disposable brush to apply solder paste to a stainless steel tube. The tube will be soldered in a regular electric or gas kitchen oven set at 350° F.

Paint the joint area with flux thinned with water or alcohol. Coat both sides. Use a slightly carburizing flame and heat a broad area, keeping the torch in motion. When the flux turns clear and starts to run, add enough silver alloy to completely fill the joint. When finished, shut off the torch and allow the joint to cool for 3–4 minutes. Remove flux with hot water.

SOLDERING

Soldering is another welding operation done below 800° F (427C). Lead soldering and silver soldering are similar processes. The major difference is that lead soldering is more akin to brazing because of the lower temperatures used. Brazing, lead soldering and silver soldering all require the use of heat, flux and capillary action.

Metals that are easily soldered are platinum, gold, copper, silver, cadmium plate and tin. Less easy to solder are nickel plate, brass and bronze. Metals that are more difficult to solder because they don't wet easily are mild steel, galvanized plate,

and aluminum alloys 1100, 3003, 5005, 6061 and 7072.

Electric Soldering

Most people are familiar with soldering copper wire with an electric soldering iron or gun. I concentrate on flame soldering. The reason for this is that electric soldering generally is restricted to copper wiring and similar parts that have a small amount of mass. Although it's a low-temperature welding process, a conventional electric soldering iron, simply won't heat much more mass than a thumbtack to a temperature sufficient for soldering! The temperatures involved are the same, but the quantity of heat is different.

Soldering Procedure

Although welding goggles are not required to solder because intense light is not generated, you should wear safety glasses to prevent eye injury in case the solder pops or splatters. Once the goggles are on and the torch is in hand, follow these steps to flame solder:

•Clean the base metal with a Scotchbrite abrasive pad, steel wool or emery paper. Remove all oil, grease, paint and anything not part of the base metal. Otherwise, the solder will not adhere to the metal.
•Apply flux to the base metal. Choose the correct flux as suggested in Chapter 15. Use an acid brush to apply it to all surfaces you intend to solder.
•Heat the base metal to soldering temperature by playing a soft flame over the base metal until the flux melts and starts to run. Then touch the metal with solder. If heated correctly, it will flow into the joint by capillary action.
•Apply solder until the joint is filled. Apply heat as needed.
•If necessary, remove flux from the joint. Use a wire brush or water for this. ■

Arc Welding

10

Before reading this chapter, you should read Chapter 7 on *Gas Welding and Heat Forming*. As mentioned throughout this book, gas welding should be mastered, or at least understood, before you attempt any other welding techniques.

The first type of electric welding to be invented was the arc welder, and it continues to be a very useful welding tool. The selection of filler metal (welding rods) for the arc welder is extensive, and the rods are very easy to carry and to use. Small fabrication shops and the largest ship builders still find the arc welder to be a very handy fabricating tool.

The best way to become really good at arc welding is to read these instructions, then go practice, then read some more, and then practice, practice, practice. So, let's do it!

ARC WELDING BASICS

This section describes how to set up the arc welder you need and can afford. I say "afford" because arc-welder prices range from $99 to $50,000. You probably don't want or need the cheapest or most expensive welder.

Just as with oxyacetylene equipment, you must determine your welding needs, then choose the arc welder that meets those needs.

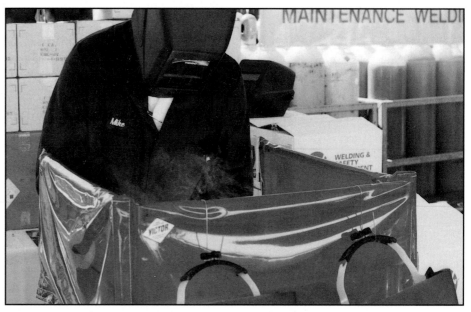

Small arc welding jobs can be safely done behind a small welding shield, as is being used here. You are responsible for eye protection when you are welding.

Remember, every welding project in Chapter 16 can be made with arc welders costing less than $300.

For most of you, a large-capacity arc welder is unnecessary. Even commercial welding shops often adjust their largest-capacity arc welders to about one-third of their capacity. For example, a 400-amp machine usually will be set at 100–125 amps.

Duty Cycle

More important than a welding machine's amperage is its duty cycle—the percentage of time the welder can be used at its rated output before it must "rest." This rest allows the machine to cool before resuming the weld. Exceed the duty cycle and the machine will gradually develop less than its rated output. A typical duty-cycle rating is 60% at 200 amps. This means the machine can be used for 6 out of 10 minutes while set at 200 amps. Duty cycle goes down when amperage is increased; it increases when amperage is reduced. A cost-efficient choice is a welder with a 90% or 100% duty cycle at 100–125 amps.

When considering duty cycle, keep

Here are 3 different 50-amp, 220-volt plug sets. If your 220-volt receptacle does not fit your welder, you can make a short adapter cord or change your welder plug.

It is easy and inexpensive to make a 50-amp extension cord for your welding machine. This one uses 12-3 Type S cord. 8-3 S cord would be better because it will carry more amps.

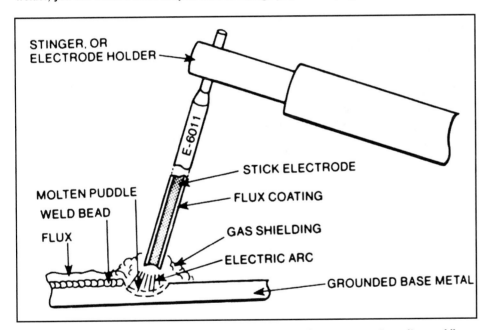

Arc welding generates high-temperature, gas-shielded metal spray to create molten puddle on base metal. Molten puddle solidifies as rod is moved along weld joint, leaving slag-coated weld bead.

in mind that you won't be able to weld 100% of the time regardless of what the welder is capable of. You must stop to change rods, take a rest, change positions, reset the welder, or stop to weld another part or weld seam.

Installing a 220-Volt or 440-Volt Welder

You don't have to pay much for a small AC 225-amp buzz-box welder. However, if your workshop doesn't have an outlet to plug it into, it will take you a day or two and more than the cost of the welder to have an electrician wire it in. And, no, you cannot plug a 220-volt welder into an outlet meant for a 220V electric clothes dryer. Even though the female plug will likely be 220V-50 amp, the plug style will likely be different.

You can either change the wall outlet, or you can buy 2 plugs, one male that fits into the wall plug, and one female plug that your welder will plug into. Then you buy 20 ft. or so of 8-3 "S" cord and make up a handy extension cord and 220V adapter. Problem solved!

If you buy a 440-volt welder, it's possible that you'll have to rewire the building's main service box to obtain the power needed for your welder.

Now that you're aware of the possible electrical problems, don't be bashful about buying an arc welder. Once you overcome any wiring problems, final setup is easy.

USEFUL EQUIPMENT

Arc-welder cart. Every welding shop I've worked in had the arc welder on wheels—on a cart or even a trailer. The reason for this is basic. It isn't possible to bring all welding jobs to the machine. Instead, you'll have to take the machine to the job now and then.

An arc welder should be supplemented with a gas-welding rig, including a cutting torch. Otherwise, you would have to do all your cutting with a hacksaw, metal shear or other metalworking tool. To give you an idea of how useful one can be, I use my gas-welding outfit 10 times more often than my arc welder.

•Hand grinder for dressing welds and beveling edges of metal plate prior to welding. A disc sander fitted with a stone or 4" grinder works well for this.

•Bench grinder. This type of grinder is handy for grinding small parts. Either mount the grinder on your workbench, fabricate a stand for your grinder or buy one.

•Chipping hammer. This special hammer is necessary to remove slag from arc welds.

•Several wire brushes. Use a wire brush to clean off the slag after chipping.

•C-clamps. These are like having several extra sets of hands. C-clamps are a must for holding parts together or in position for welding.

•Marking Tools. Turn to Chapter 5 for information about these tools. Regardless of the type of welding or cutting you'll be doing, you'll need to accurately indicate where cuts and welds are to be made.

•Set of hand tools for disassembly and assembly work.

•Spare weldor's helmet and lens for a helper to use or for a friend to watch with.

•First-aid kit with burn ointment. You'll need a first-aid kit eventually, even though you are super careful. It's almost impossible to work around hot metal without getting burned now and then. Eye burn fluid is recommended also.

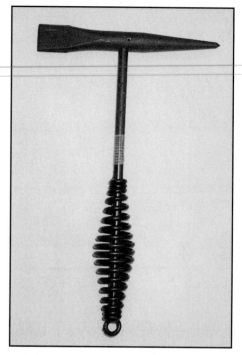

A chipping hammer like this one is a requirement for removing slag from arc welds.

Arc Welding Rod

Refer to Chapter 15 for specifics on which fillers to practice with. Pick two or three kinds of welding rod. I suggest starting out with 5 lbs of E-6011, E-6013 and E-7018.

Storing Arc-Welding Rod—All coated electrodes should be stored in a dry, warm atmosphere. I've seen it stored the following ways:

•In plastic pouches with the ends taped airtight and a couple of packages of desiccant crystals inside. The desiccant absorbs moisture. Although this method keeps the rod dry, the pouches are cumbersome to use when welding.

•In a tube with a tight-sealing lid and desiccant inside.

•In a 5-gallon metal can, or rod oven, with a seal on the lid and a 60-watt light bulb inside. The light burns 24 hours a day. This method works OK, but keeping the 60-watt bulb lit continuously gets expensive.

•In a commercially built rod oven. Because of the expense, this method is practical only for welding shops using over 50 lbs. of rod a month.

When doing certified welding with E-7018 low hydrogen welding rod, dry rod is very important. Weld quality is so important at the nuclear power plants that if the rod is exposed to the atmosphere more than eight hours, it must be thrown away. Although E-7018 rod can be reheated to drive out moisture absorbed from the air, it is never as easy to weld with as it was when fresh and dry.

Of course, all arc-welding rod should be kept dry. I've even seen E-6011 rod fail to maintain an arc because it was improperly stored for several weeks and, as a result, absorbed considerable moisture.

ARC WELDING TECHNIQUES

In this section, I describe how to use AC welding machines. If you want to

SUGGESTED ARC-WELDING ROD FOR PRACTICE

ROD	COMMENTS
E-6010	Easy to use, very little spatter.
E-6011	Easy to use, but spatters a lot.
E-6013	Easy to use, has little spatter and produces average bead.
E-7018	Harder to use, but process beautiful welds on clean metal. Use one of these rods when making your first practice bead on steel.

use a DC machine to practice, read the section on polarity, page 91.

Practice

If you've been collecting pieces of scrap metal, you should have some pieces of 3/16" or 1/8" mild steel. Use a cutting torch to make this 2" x 5" pieces on which to practice. This is a good size—large enough to work with, yet small enough to conserve your scrap pile.

Basic Practice Steps—As you did when learning to gas weld, learn the four basic types of welds before you begin a project. These are:

• Running a bead
• Butt weld
• T-weld
• Lap weld

For practice welding, you need a table with 1/4" to 1/2" thick, 2" x 3" steel top. In a pinch, you could simply lay a steel plate across two wooden stools or sawhorses. Later on, you could build an arc welding and cutting table. See page 146 for plans. Actually, I strongly recommend that your first welding project be this cutting and welding table. It can last you a lifetime, and it will streamline all your future welding projects.

Basic Arc Welding Principles

Before you strike the first arc, you should know what happens at the electrode tip. A 6,000–10,000 °F (3320–5540C) temperature is generated by an electric arc between the electrode tip and the workpiece. The flux coating on the welding rod is heated to a gas and liquid. This shields the molten puddle from the atmosphere; thus the name shielded metal-arc welding (SMAW). The shield prevents the molten puddle from chemically reacting with or being contaminated by atmospheric gases, which can cause hydrogen embrittlement, porosity and other bad effects.

As the weld puddle solidifies, the flux also solidifies, forming a coating on the weld bead and protecting it from the atmosphere as it cools. This resolidified flux—*slag*—is glass-like, and can then be chipped off to reveal the weld bead.

In the drawing nearby, you can see the arc welding process with stick electrode. The arc-welding rod actually sprays molten metal into the molten puddle on the base metal.

Remember: Where you point an arc-welding rod is where the weld metal goes! The heat and sprayed metal come off the end of the rod like a spray gun! Point the rod where you want the weld bead!

Striking an Arc

I teach my students how to strike an arc, run a bead, and actually weld something practical in less than two hours! For a complete novice, it only takes five minutes to learn to strike and maintain an arc!

Ready the Welder—The first thing to do when getting ready to arc weld is to ground the workpiece. You can't start an arc without a ground. Either connect the ground clamp directly to the work or to the metal table you'll be welding on. If you don't connect the ground clamp to a suitable ground, you could become the ground!

Once you've grounded the work, adjust the machine to 130 amps. Although this is a hot setting, you'll have an easier time learning how to strike and maintain an arc. Once you've learned to do these two things, you can readjust the machine to a lower amperage setting. Again, to make things easier, you'll need about five E-6011, 1/8" electrodes for practice.

The welding machine is now ready to be turned on. Make sure the weldor is. Regardless, don't do it while the working end of the welder—electrode holder, or stinger—is laying on the grounded table or workpiece. You may see some premature arcing.

Ready the Weldor—Don't turn on the welding machine until you've prepared yourself. You must be wearing the correct welding apparel: long-sleeve shirt, cuffless pants, high-top shoes, gloves and a welding hood.

INTERPASS TEMPERATURES

In all welding, gas or arc, if you need to run another pass on top of or beside the weld bead that you just completed, let the weld cool to below 500°F before you run the second and subsequent weld passes. If you don't let the weld cool, the molecular structure will be negatively affected and you could end up with crystallized, crack-prone, brittle welds. Metal is a lot more sensitive to cold and heat than you might think. Use Tempil sticks to check your interpass temperatures and don't resume welding until the metal is below 500 °F.

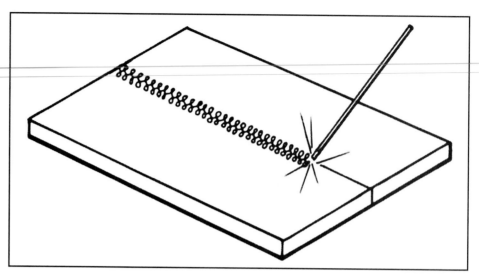

A common arc welding bead is this weave pattern. Use it only when making single-pass welds.

A leather apron is not necessary, but a good idea.

Turn on the Machine—Turn the machine to ON. With the electrode in hand—your right hand if you're right-handed, and vice versa if you're a lefty—squeeze it to open the jaws and insert the bare end of an electrode. Usually, there are grooves in an electrode-holder's jaws—for holding the rod at 90° to the holder 45° forward, 45° backward and in-line with the holder. For now, position the rod so it's 90° to the holder.

Strike an Arc—Weldors compare striking a welding arc to striking a match. However, a freshly lit match is immediately moved away from the striking surface. Not so with an arc welder. Once the arc is struck, you must keep the electrode tip near the work to maintain the arc. Instead of a match, the arc welder can now be compared to a spark plug. If a spark plug gap is excessive, it will not operate. The same holds true with the arc welder. Usually, you should maintain an electrode-tip-to-work gap of 1/8" to 1/4".

With these points in mind, let's get on with the business of arc-welding. With the stinger in both hands and welding hood or tinted shield flipped up—depending on the type of hood you have—hold the electrode tip about 1" from your workpiece. You are going to scratch or "tickle" the work with the rod to strike, or start, the arc.

Running a Bead

With a mental picture of where the electrode tip is, nod your head so the helmet will fall down, covering your face and eyes. The next light—let's hope—will be the arc. Like writing with chalk on a blackboard, scratch the work with the electrode to start the arc. You can't just touch the electrode tip to the work to start the arc—the tip must be moving. Once started, keep the small gap just suggested to maintain the arc, and move along slowly. But, chances are you won't get this far on the first few tries.

If the rod sticks to the work—it happens to everyone—swing the stinger from side to side to break it loose. Do this quickly. Otherwise, the rod will get red hot and soft. If this happens and you can't break it loose, squeeze the holder to release the rod. You can then use pliers to work the rod loose. Don't grab the hot electrode with your hands, even if you're wearing gloves!

After learning to strike an arc, you'll have to maintain it. Do this by moving along while maintaining the correct gap. To do this, you'll have to move the holder closer to the workpiece as you move along. The rod foreshortens as it melts to create the weld bead.

Once you've learned to strike an arc and maintain it, keep practicing while it's fresh in your mind. The object is to run a good weld bead.

Run the Bead—On a piece of scrap steel, practice running a bead until you're satisfied with its appearance. Use the pictures in this books as examples, or the welds on a car's trailer hitch or trailer. After welding a

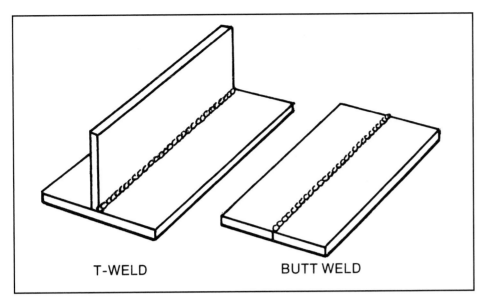

Different weld-joint configuration s require different heat. T-weld (left) requires more heat than butt weld (right). Also, majority of heat must be directed toward piece with most mass. For example, majority of heat must be directed toward horizontal piece when doing T-weld.

half-dozen or so acceptable beads on scrap steel, you should be able to try making a butt weld with two pieces of scrap metal.

Butt Weld

Items required for practicing a butt weld include two 2" x 5" steel plates, about 1/4" thick, and two E-6011 or E-6013 welding rods. The plates should be trimmed straight and even so they'll butt without any gaps. Place the two plates on your welding table. Butt them together, then tack-weld each end together.

As the first tack weld cools, the opposite end of the butt-weld seam will open up in a V-shape. Close the seam by supporting the back edge of one plate and tap on the edge of the other one with a small hammer. Once the gap is closed, tack-weld the other end of the seam. If the machine is still set at 130 amps, turn it down to 90 amps at this point in the practice. However, if the rod sticks in subsequent welding, turn up the machine to 130 amps again.

You're now ready to run your first butt-weld bead. Strike an arc at the right end of the seam, if right-handed, and slowly run a weld bead the length of the seam. The weld beam should be centered in the seam. After the weld is completed, let it cool for about five minutes in air. You can now pick up the metal with your pliers and stick it in water to complete the cooling.

Check for weld penetration. Look at the back side of the seam. If penetration is good, you should see signs of the weld puddle dropping out of the bottom of the seam or extreme discoloration of the metal. If penetration appears to be insufficient, turn the heat up 15 amps and try again on two fresh strips.

Practice making butt welds until you think they look pretty good. To remove doubt as to the quality of your welds—a good looking weld isn't necessarily a quality weld—take your best samples to a technical-school welding shop. Ask for the instructor's opinion.

Test Weld—An easy way to test a butt weld is to clamp one side of the weld sample in a large vise, just below the weld seam. Hit the top of the side opposite the weld bead with a hammer. This will bend the metal toward the top side of the weld bead. If the weld is weak due to poor penetration, it will break through the back side of the weld seam.

Common mistakes for beginners are moving the rod too fast, resulting in poor penetration. Slow down! It's better to have too much weld bead than not enough.

For beginners, the best weld bead is

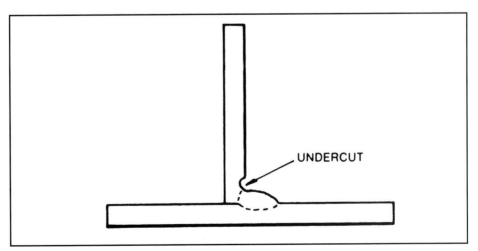

T-weld undercut is caused by a vertical piece being overheated. Base metal then flows onto horizontal piece. Manipulate electrode so horizontal piece is heated most, but aim electrode at vertical piece. Point the rod where you want the heat to go.

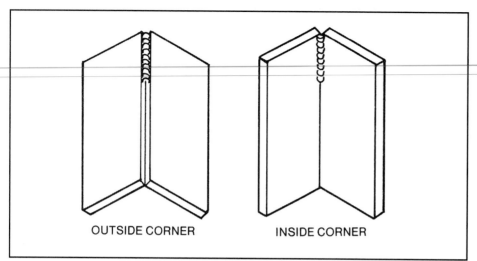

OUTSIDE CORNER   INSIDE CORNER

Different welds take different heat. Outside-corner weld (left) takes less heat than inside-corner weld (right).

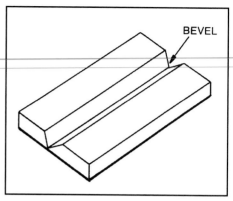

BEVEL

When butt-welding thick material, seam is beveled to obtain maximum penetration.

obtained by moving the arc-welding rod at the same speed as the second hand on a wrist watch—about 3" per minute. As your skill improves, adjust travel speed to maintain a molten puddle.

T-Weld

Doing a T-weld with an arc welder requires that you manipulate the rod to avoid undercutting the vertical piece and to get good penetration in both pieces of metal. The tendency is to burn through the vertical piece and get insufficient penetration on the horizontal piece. This is the same problem encountered when doing a T-weld with oxyacetylene. The difference between the two types of welds is that you must now manipulate an electrode rather than a gas-welding tip.

To practice doing a T-weld, you need two 1/4"-thick steel plates that measure about 2" x 5". Place the two plates on your welding table so they form an upside-down T. Use a mechanical finger to hold the vertical section in place while you tack-weld each end. If the first tack weld causes the vertical plate to raise at the seam, tap it back in place with your hammer. Tack weld the other end of the seam.

You'll need two sticks of E-6011 or E-6013 welding rod. If you're right-handed, strike an arc at the left side of the seam and start the bead. Apply about 70% of the heat to the flat part and 30% to the vertical part. This means you spend 70% of the time pointing the rod at the flat piece and 30% pointing the rod at the vertical piece. You'll have to swing the holder back-and-forth as you go while maintaining the arc gap and welding puddle. Have fun!

If you're left-handed, start at the right end of the seam and strike your arc on the lower piece. Starting on the upper piece would cause slag to be buried under the weld bead on the bottom piece. To time your stinger movement, count out loud, one-two-three, one-two-three, if it helps. You'll be constantly moving your rod-holding hand to accomplish this.

Again, when the weld is finished, let it cool. Chip off the slag and inspect the weld. If the vertical piece is undercut, suspect an excessive arc

gap or rate of travel. Do it again on more scrap pieces.

Keep practicing until you're satisfied with the looks of your welds. Once you've mastered the problem of undercutting the vertical piece and can get sufficient penetration in the horizontal piece with the same weld, you can do T-welding.

Lap Weld

A lap weld is the most difficult type of weld to do with an AC arc welder. You'll need two 3/8" thick, 2" x 5" steel plates. Lay them on your welding table so one overlaps the other, as shown. Tack-weld each end. E-6013 rod is best for lap welds, but E-6011 is OK in a pinch. As with a T-weld, the trick for doing successful lap welding is to prevent undercutting the top piece.

Again, strike the arc on the lower piece. Weave the rod in the motion shown on page 86. Be sure to pause slightly at the bottom each time you make a swing to prevent undercutting the top piece.

After you've completed running the lap-weld bead, cool the piece and chip off the slag. Inspect the weld for undercutting and penetration. Practice until you are making consistently good lap welds.

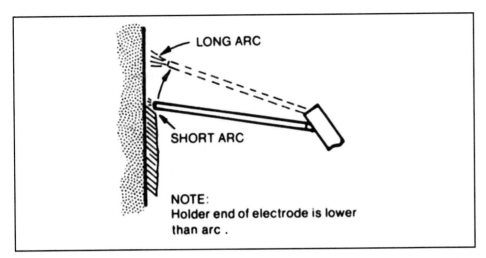

LONG ARC

SHORT ARC

NOTE:
Holder end of electrode is lower than arc .

Keep electrode horizontal or pointed slightly upward when welding vertically up. As molten metal begins to deposit, move electrode tip 0.5" to 0.75" upward to allow puddle to solidify. Bring tip back to deposit metal. Continue this movement while you only watch the molten puddle.

OUT OF POSITION ARC WELDING

Anything other than welding a seam that lays flat is out-of-position welding. This applies to all types of welds: butt, T or lap. And, out-of-position welding "will separate the men from the boys." Not only are they the most difficult to make, some are more difficult than others. Although I can give you tips on how to do each type of out-of-position weld, the only way to master each is with practice.

Regardless of your experience, always try to weld everything in the flat position, if possible. But many times it's impractical or impossible to reposition an item to make welding easier. So, you weld it where it is. Although there are "tricks" when doing each type of out-of-position weld, the number one "trick" is to point the rod where you want the puddle to go. You're fighting gravity when welding out of position.

I'll never forget some out-of-position welding I once did while on the top rung of a 30-foot ladder. I was

hanging upside down by my knees, welding the bottom side of a bracket on a light pole! While making that very uncomfortable weld, I said to myself, "Now this is really out-of-position welding." That was several years ago, and I'm sure the bracket is still holding tight to that light pole.

If you have to arc-weld upside down, here's how to do it:

Overhead Welding

Use E-6011 or E-6013 rod. Aim the rod so it points almost straight up, about 30° from vertical. Also, you should wear tight-fitting clothes. Sparks will fall on you.

After you start the arc, hold the electrode tip closer to the work than the 1/8" gap that's normal for 1/8" rod. I like to push the rod and puddle up against the base metal.

If you see a drip starting, push it back into the fresh weld bead, or if too big, fling it out of the puddle and start over. You may have to stop occasionally to let the weld bead cool. Usually, three or four seconds is enough cooling time. If you stop

longer, chip out the slag to prevent slag inclusions in the weld.

Vertical and Horizontal Welding

For practicing vertical and horizontal welds, I recommend a 1/8" E-6013 rod. Other rods can be used. For instance, E-6011 is OK, but E-7018 is more difficult to use. Even though you may be using E-6013 rod, don't think vertical or horizontal welding is easy. This type of welding is the most difficult to perform and get a good weld.

For vertical welding, point the rod about 15–25° upward from horizontal; 30–45° when welding down. If you weld at 90° to the work, the puddle will sag or fall out.

It is particularly important to watch the puddle, not the arc, when doing a vertical or horizontal weld. Otherwise, you won't be able to control the puddle. If it begins to sag, momentarily pull the rod back to lengthen the arc and stop depositing metal. This also cools the puddle slightly. Move back in to continue the weld. This in-and-out movement is indicative of how to do such a weld, particularly vertical up.

Vertical-Up—The puddle movement is up—which is difficult to make because the puddle is below the arc. You can't use the rod to "hold" the puddle. One way to keep the puddle in position is to momentarily interrupt the arc, "freezing the puddle so it stays in position. You can also momentarily bury the end of the rod in the puddle, also reducing heat.

Vertical-Down—Puddle movement is down—which makes it easier to control because you can use the rod to push or hold the puddle in position as it tries to fall or drop out. However,

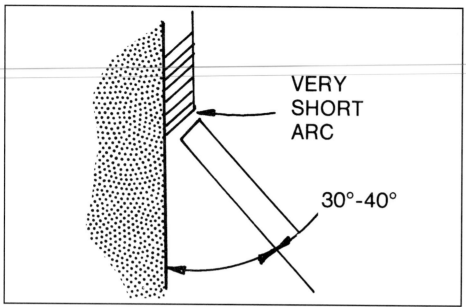

VERY
SHORT
ARC

30°-40°

Maintain 30–45° electrode-to-workpiece angle when welding vertically down. Move fast, otherwise slag will catch up with arc.

penetration on a vertical-down weld is not as good as that for a vertical-up weld. For this reason, vertical-down welds are not permitted when welding certain types of industrial pipe, such as high-pressure steam pipes in power plants.

Finally, don't forget to point the rod where you want the puddle to go. Although you've read it before, here it is again. All it takes is practice, practice, practice.

Stopping & Starting Arc Welds

One thing common to all types of joints or arc-welding positions is the need to stop, then continue the bead. You may have "burned" the electrode to a stub and need a new one, need to change position, or rest yourself or the machine. Whatever the reason, you must get a smooth transition without voids at the end of one bead and the beginning of the next.

To get a good transition between two overlapping weld beads, do this: Stop and allow the weld bead to cool

for about two minutes. With your chipping hammer, knock off the slag at the end of the bead. This is particularly important with deeply grooved weld joints. The slag will

flow into the groove at the end of the weld, causing an inclusion or void in the weld if the weld bead is continued over it. Therefore, all slag must be removed completely.

Chip out the slag with the pointed end of a chipping hammer. Finish up by wire-brushing the end of the weld bead. When the seam is perfectly clean, continue welding. Overlap the welds by striking the arc in the small crater at the end of the weld bead and immediately continue running the bead once a puddle forms. If done right, you should have difficulty seeing the transition between the two beads.

ARC WELDING SHEET METAL

Welding sheet metal with an arc welder is difficult to learn because sheet metal is thin and easy to burn

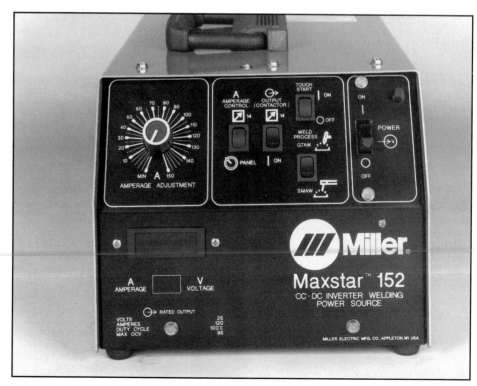

This portable constant-current–DIRECT current (DC)–inverter provides power for stick welding or DC only, scratch start TIG welding. Photo: Miller Electric Mfg. Co.

POLARITY

When welding with alternating current (AC), you don't have to set polarity—direction of current flow. In the U.S., AC constantly switches back and forth, 120 times a second, between positive and negative. A complete cycle occurs 60 times a second. In other countries, it may be 50 or 90 times per second. In direct-current (DC) welding, polarity makes a big difference. The above illustrations dramatize the effect of polarity in DC welding.

Almost all DC welding is done with reverse polarity—electrode is positive—because the welding rod gets hotter than the workpiece! Reverse polarity provides a steadier arc, and electrode-to-work metal transfer is smoother than with straight polarity—electrode is negative. It is easier to weld with a shorter arc and low amperage. Therefore, DC is better for making out-of-position welds. Some electrodes can be used with reverse or straight polarity; these are called AC/DC electrodes. And, there are some that can be used only with straight polarity.

The chart nearby gives recommended polarity settings for various metals. Make a copy and keep it near your welding machine so you can refer to it.

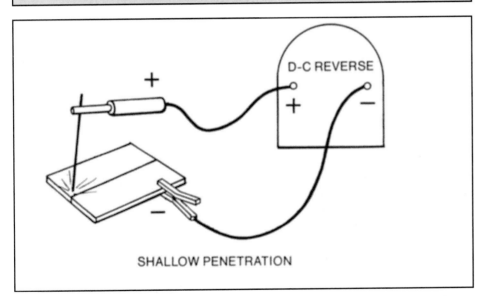

DC welder set to (+), REVERSE, or POSITIVE polarity, has this circuitry. Welding rod gets hotter when machine is adjusted this way.

through. Here are some tricks to make it easier:

•Weld at low-amperage settings. Try 60-75 amps with 1/8" rod or 40-60 amps with 3/32" rod.
•Hold a very close arc. This keeps excess heat down long enough for the puddle to stick to the base metal.
•Stitch-weld or spot-weld, then tack and fill the gaps. This prevents local heat buildup and burning holes in the metal.
•Use lap welds, if possible. This thickens the metal, creating more heat sink—mass to absorb heat.
•If all else fails, use copper strips of heat-sink compound to back up the weld seam. The weld will not stick to copper; you can remove the strips after the weld has cooled.

ARC WELDING 4130 STEEL

Often referred to as chrome moly, 4130 steel welds similar to mild steel. However, the resulting weld is more prone to cracking after it cools because of 4130's "graininess." Here are some tips for AC arc welding 4130 steel:

•The larger the piece, the more important it is to preheat before welding. Always try to weld 4130 steel in 70° F (21C) or higher temperatures, and preheat to 200-300° F (93-149C) in the expected heat-affected zone. Use a temperature-indicating crayon or paint to monitor preheat temperature.
•Preheat with an oxyacetylene torch—rosebud tip if it's a large part—or heat the part in a kitchen oven if it fits. For huge parts such as nuclear power-plant reactors, an electric blanket is used for preheating.
•Always use E-7018 rod for welding 4130 steel, or the recommended rod.
•Make sure the base metal is clean and free of rust, paint and grease. Otherwise, you'll end up with a defective weld.
•Bevel the weld joints to get maximum weld penetration.
•Before taking a chance on ruining an expensive piece of 4310 tubing or whatever, practice on scrap.

Welding 4130 steel is discussed in more detail in Chapter 7, because gas-welding is more suitable in most instances. For example, gas-welding is better for building a chrome-moly

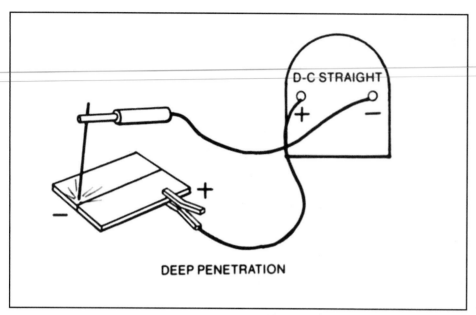

DC welder set to (—), STRAIGHT, or NEGATIVE polarity, has this circuitry. Workpiece gets hotter when machine is adjusted this way.

ARC-WELDING POLARITY

Metal	Polarity (AC/DC) in order of preference	Recommended Electrode
Stainless Steel	DC Reverse (Positive)	E-308-15, E-310-15
	AC	E-308-16, E-347-16
Bronze	DC Reverse (Positive)	E-CuSn-C
Aluminum	DC Reverse (Positive)	AL-43
Cast Iron	DC Reverse (Positive)	ESt
High-Tensile Steel	DC Reverse (Positive	E-7010-A1, E-8018-C3
	AC	E-7027-A1, E-8018-C1
Mild Steel	AC	E-6011, E-7014, E-7018
	DC Reverse (Positive)	E-6010, 5P, E-7018

airplane fuselage or race-car-suspension parts. A gas torch is easier to manipulate around the small parts and preheating is almost automatic.

ARC WELDING ALUMINUM

Although uncommon, it is possible to arc weld aluminum plate,

aluminum castings and aluminum sheet with a DC arc welder. The resulting weld bead will look rough compared to arc-welded steel. I've used it for welding large, thick aluminum plates, building up work edges on aluminum pieces and welding 1/4" aluminum for toolboxes, barbecue grilles and shelf brackets.

As with steel, you should preheat

1/4"-and-thicker aluminum to 300-400° F before arc welding. Expect a very bright arc, and a lot of noise and spatter when using aluminum arc-welding rod. The resulting arc-weld bead will be about 50% weaker than it would be if it was TIG welded.

ARC WELDING STAINLESS STEEL

Although there are no special tricks to arc-welding stainless steel, don't expect beautiful welds. Stainless weld beads are not pretty unless they are completely protected from the atmosphere. The back side of the weld usually will appear black and rough.

Appearance of a stainless-steel weld can be improved by coating the back side of the seam with flux paste. This protects the seam from oxygen in the atmosphere, minimizing crystallization of the weld.

The best welding processes for stainless steel are TIG and wire-feed (MIG). But if MIG or TIG are not available, you can do an acceptable job with an AC arc welder. Select the correct stainless-steel rod from the chart, page 136. Again, there are no special methods necessary for arc-welding stainless steel. Preheating is not necessary.

Do the same as you would when welding any material for the first time. Practice on scrap stainless before you try welding an actual part. If you don't have any stainless-steel scraps to practice on, you can use mild steel or 4130 steel scraps with a stainless-steel rod.

If you were wondering, yes, it is possible to arc-weld mild steel, 4130 steel and stainless steel together in one assembly.

Gasoline engine-powered constant current DC welding generator makes certified welds in pipe and structural steel. Photo: Miller Electric Mfg. Co.

ARC-WELDING CAST IRON

You may encounter the need to arc-weld cast iron. Usually, this involves the repair of a cast-iron machine base, farm equipment, transmission case or engine part. The last time I welded cast iron was when a local high school asked me to build a chariot for a football game halftime show. The students wanted it to look like Roman warriors coming into the arena. I was asked to weld late-model car axles to some cast-iron spoke wheels from a piece of old farm equipment.

I used NIROD—nickel-based welding rod—for doing this job. See Chapter 15 for other recommend-ations on cast iron rod.

It's hard to weld cast iron without it cracking. The reason is its rigidity. When one small area is heated, causing it to expand, the unheated area resists. Unfortunately, the cooler area loses the battle because cast iron is much stronger in compression than in tension. Thus, the cooler area—in tension—cracks. This is why it is extremely important to thoroughly preheat cast iron before welding it.

As you may suspect, welding cast iron requires a lot of patience. Start by heating the entire casting to 400-1200 °F (204-649C) before welding. Here's another application for temperature-indicating crayons or paint. And, only weld while the casting is hot. This is easy to do with small castings that fit into your kitchen oven, but large castings require a lot of heat. Because of this, it's standard practice for shops that weld cast iron to place a small natural-gas burner under the casting and heat it while it's being welded.

Another approach to welding cast iron—especially big castings that are impractical to preheat—is to arc-weld them at room temperature, but only 1/2"-long beads at a time, then stop. The 1/2" weld is chipped and allowed to cool for two or three minutes before another 1/2" weld is done. This allows the weld and heated area to "relax" as the heat is absorbed or dissipated into the casting. Some people recommend hammer-peening each short, fresh weld until it is cool.

Brazing Is Best—The best way to join cast iron is to braze it. And you can't do that with an arc welder. Get out your gas-welding torch. I've brazed a lot of small cast-iron pieces with success. The secret is to start by V-notching the crack or joint completely to the center from both sides of the part, or all the way through from one side. After you've done this, preheat the casting to about 350 °F (177C). The welding torch will help maintain heat as you braze. Read about brazing in Chapter 9. ∎

MIG Welding 11

MIG (or wire feed) welding machines are easy to use. Almost anyone can buy a new MIG welder and be welding just minutes after uncrating it. Because of this fact, MIG welding has become extremely popular in recent years.

When the first edition of this book was printed in 1980, MIG welders were rather simple machines. But recent electronic inventions and advancements have taken over the arc-welding business just the same as technology has revolutionized the computer industry. In 1980, computers were expensive, slow and not very powerful. In 1980, MIG welders were plain, heavy and not very accurate. Today you can buy a very powerful computer and today you can buy a very powerful, accurate but rather complicated MIG welder. The next few years will see even more improvements.

So MIG welding machines are available in many sizes and one size does not fit all needs. You really need to know what the welder will be used for during its long and useful life. Small, low-amp MIG welders can never be upgraded to do heavy work. Rather large MIG welders can be tuned and adjusted down to do light welding, as well as somewhat heavy work, however, many of the large

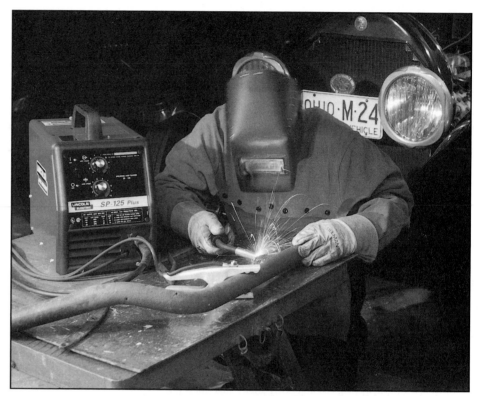

Home workshop antique auto restorers find the Lincoln 110-volt SP-125 MIG welder very useful in making exhaust system repairs. Note that this hobbyist is using a folding picnic table, covered with a sheet of plywood, as a welding table. Photo: Lincoln Electric Co.

units cannot be tuned down and slowed down to do thin-wall, small diameter tubing and autobody sheet metal.

The major advantage of MIG welding is its simplicity and speed. If tested beside any stick welding machine for 1 hour, the MIG machine would be able to weld 4 to 10 times more than the stick welding machine.

Several years ago I wanted to

increase production rates of aircraft seats for a six-passenger, twin-engine airplane. The first 400 ship-sets of seats (2,400 seats) had been welded with TIG, and it was really hard for the welding department to keep up with airplane production. It took one weldor 8 hours to build one seat with TIG welding.

So we purchased a high-quality MIG welding machine and had the

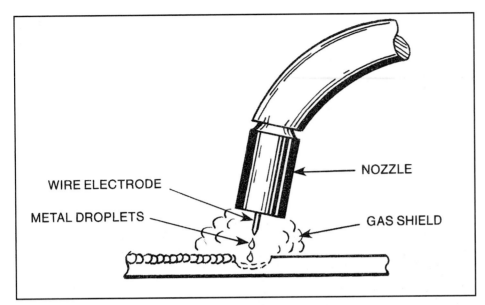

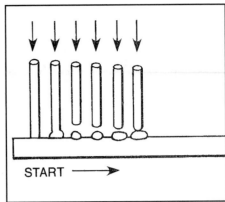

Arc starts at left and weld bead moves to right. Wire continuously advances and melts as it contacts base metal. With machine adjusted correctly, cycle occurs smoothly and quickly, sounding as if it were bacon frying.

Small-diameter, consumable electrode—wire—is fed into weld puddle at a high rate, or up to 700 in. per minute (ipm). Instead of argon shielding gas typically used with TIG welding, carbon dioxide (CO_2) is preferred for MIG welding. Shielding gas is not required when flux-cored wire is used.

weldors take certification tests with it. After a week, we were ready to build certified aircraft seats. Production of seats instantly went up. One weldor could build up to 4 seats in 8 hours with the MIG welder. That is a 400% increase in production.

The Downside

Yep, you guessed it: Nothing in life or in welding is all plusses. The MIG welds were not as pretty as the old TIG welds, but the seats all passed Magnaflux inspection, and 9 g's crash testing. However, when we later tried to weld tubular engine mounts with MIG rather than TIG, we found that the corrosion protection oil inside the engine mount tube leaked out at several places because of cold starts. A cold start is a lack of penetration where the weld bead appears to cover a seam but is only lapped over until the bead progresses. I'll explain this MIG term later in this chapter.

MIG welding is notably less accurate than TIG welding and somewhat less accurate than arc welding. At recent race car shows and welding trade shows, builders and buyers were heard commenting about the inherent lack of accuracy of MIG-welded tubular framework as found in race car frames and aircraft frames. The problem is that the weld bead will look good, when in fact 80% or more of the bead is on one member and not on the other member, making this weld only 40% as strong as it should be. We found the same problem when trying to MIG weld the engine mounts on those twin-engine General Aviation airplanes.

Why Less Accuracy?

The most common reason is that the MIG gun hides the weld puddle. The solution to that problem is to view the puddle from the side, rather than from behind the MIG gun.

The second reason is the bright light produced by the MIG weld process, and the smoke and spatter combine to make the weld puddle much harder to see than with other types of welding. A partial solution is to install a small, high intensity headlight on the gun, and to weld with gas rather than flux-cored wire. A third solution is to use high-quality welding wire rather than the cheapest wire you can find. Read Chapter 15, to learn more about the differences in welding rod and wires.

So, can the accuracy problem be overcome? Yes, it can—read on. But MIG is inherently less accurate than TIG. Read Chapter 12 to find out why.

WIRE-FEED OPERATION

Here's how a wire-feed welder operates: The gun is positioned over the weld seam at the same angle you would hold an arc-welding rod. Cup-to-work distance should be about equal to the distance across the cup opening. The gun trigger is then pulled, activating the DC current, positively-charged wire electrode—reverse polarity—and gas flow. (Straight polarity is rarely used because the arc is unstable and erratic. Also, penetration is lower.) The wire

So-Cal Airgas makes this Gold Gas™ mix of argon and oxygen for MIG welding. Ask your gas supplier to advise you on the best mix to use for MIG welding.

MIG welding makes barbecue grille assembly easy, fast work. These BBQ grilles are a welding school project at Allan Hancock College.

Nozzle Dip or Nozzle Spray is used in MIG welding to make the inevitable spatter easier to remove from the gun nozzle. Allan Hancock College.

is simultaneously fed through the gun nozzle and contacts the grounded base metal, causing a short circuit and resulting arc to start. Resistance heating melts both the base metal and the ends of the electrode. The wire then melts back faster than it is being fed to the base metal, momentarily breaking the arc and depositing metal. The arc force flattens the molten metal.

But the wire electrode is still advancing into the puddle, repeatedly arcing and melting off again and again. This on-and-off process occurs about 60 times per second, causing the characteristic buzz you hear. Some people describe a properly adjusted MIG welder as sounding like frying bacon.

Metal inert-gas (MIG) welding is so-called because of the types of electrode and shielding used. Unlike TIG welding, a MIG welder uses a consumable metal electrode. This electrode is a continuously fed wire that exits from the center of the welding torch, where a TIG welder tungsten would normally be; thus, the name wire-feed. Typical wire sizes are 0.024", 0.030", 0.035" and 0.045". Up

to 1/16" diameter can be used with special equipment. CO_2 shielding gas is used in place of argon because CO_2 is less expensive. Hollow flux-core electrodes are used frequently, eliminating the need for gas shielding, but with lots of smoke and spatter.

Fast & Clean—The major advantage of wire-feed welding is that it's fast. Unlike arc welding or TIG welding, you rarely have to stop for a new welding rod. Also, its weld rate—inches per hour of weld bead—is fast, especially when compared to TIG. Another advantage of the MIG welder is clean welds—much cleaner than those possible from an arc welder, when used with gas shielding.

Modes of MIG Welding

•Short-Circuit Transfer was the original method of MIG welding. It still exists today in the 110-volt machines and in the more basic types

of 220-volt machines. A drawing of this process is shown on page 95.

•Globular Transfer means that the weld wire only touches the metal when the weld begins. After that, globs of molten wire are expelled into the puddle. Globular transfer occurs especially with higher voltage, CO_2 gas and mild steel electrodes.

•Spray Arc Transfer occurs at higher amps and volts and wire feed speed, and with argon shielding gas. Higher metal-deposition rates occur, and the arc gives out a higher frequency humming sound. Spray arc transfer is desirable in the flat position.

•Pulsed Spray Transfer is possible when using a special welding machine that is designed to provide optional pulsed arc. Because the pulsed spray operates at two heat pulses, the weld puddle is allowed to freeze slightly between pulses. This feature provides for better control of thin sections, for welding aluminum, and for welding out-of-position with steel wire.

•Spot Welding is possible with a MIG welder by adding a spot-welding timer to the machine. And the weldor can also spot weld with short, quick trigger pulls of the MIG gun. But this is not as clean a method as a specific-

Notice the welds on this new Ducati motorcycle. The weld beads are thick and rounded, more for looks than for strength, although you do want strong frame welds when riding at over 150 mph! MIG welds have a certain look that TIG or oxyacetylene does not have.

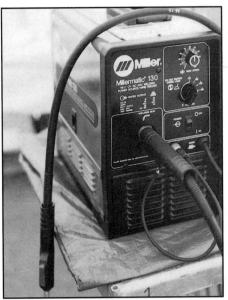

This lightweight, portable MIG welder by Miller Electric will weld between 30 and 130 amps and operates off 115-volts house current.

Close-up view of the driver's foot-well area on the tubular frame of a "stock car" shows how the tubes are fishmouth-fitted, then MIG welded.

purpose spot welder. See Chapter 14 for spot welder information.

WIRE-FEED MACHINES

The MIG welder is a simple, compact welding machine. It consists of the welding gun, power supply, wire-drive mechanism and control unit, shielding-gas supply and, for some heavy-duty units, a water-cooling system.

Gun

In place of the TIG torch or arc-welder stinger is a gun. The typical gun looks like a pistol and directs the filler metal and shielding gas to the weld seam. Service lines running to the gun include an electric-power cable, electrode conduit, and gas hose, if used. Heavy-duty industrial-type guns also have water lines for water-cooling. Otherwise, gun cooling is done with air. Electric power is transferred to the wire electrode via a sliding contact with the copper electrode guide tube in the gun.

The gun nozzle, which is usually interchangeable, determines the gas-shield coverage of the weld puddle. Nozzle-orifice size varies from about 3/8" to 7/8" (10 to 22mm). A larger orifice gives additional shielding, as does a larger TIG torch cup.

Power Supply

Almost all wire-feed welders supply DC current. This requires a transformer-rectifier when using an AC power source. Depending on the machine, output can range from 15 to 1200 amps. Required power supplies typically range from 110 to 200/230 volts, or all the way up to 575 volts, depending on machine output. Duty cycle is either 60% or 100%.

Wire-Drive Mechanism & Control Unit

The wire-drive mechanism is relatively simple. It consists of a wire spool and DC motor powered drive rolls—two wheels that run against each other with the wire in between. Sometimes, two sets of wheels are used. The drive-roll mechanism pulls the wire off the spool and pushes it through the conduit to the gun at a weldor-adjusted rate.

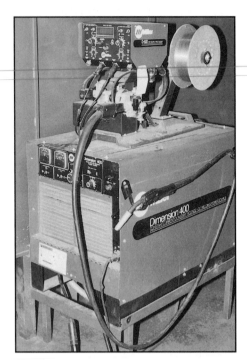

Heavy-duty, modular MIG machine is used for production MIG welding and here at the Allan Hancock College welding lab. The digital wire feed unit on top is programmable for repeat settings.

The MIG welder control unit regulates arc starting and stopping, as well as wire-feed, gas flow and, sometimes, water flow rates. It also synchronizes these functions. Usually, there's a jogging feature that feeds the wire to or through the gun while not welding.

Similar to the wire-jogging feature, the control unit also has a shielding-gas purge switch to manually control gas flow. In addition, timers control preweld and postweld gas flow automatically. The purge switch can override the automatic timers. Another timer controls water flow, if water cooling is used. Finally, a wire-feed brake stops the electrode the instant the gun switch is released. This prevents wire from being fed to the puddle when the arc is interrupted.

Shielding Gas

Except for CO_2 used in place of

argon, a MIG welding shielding-gas setup is similar to that used for TIG welding. Not only is CO_2 less expensive than argon, it also has superior heat conductivity.

Regardless of the gas used, constant pressure and flow must be maintained while welding. Also, you must be able to adjust pressure and flow for different applications. Therefore, the gas cylinder must be equipped just as if it were used for TIG welding. Although different gases use different flow meters or flow gauges, the CO_2 must have its pressure and flow regulated.

Types of Wire Feed Welders

When considering a wire-feed welder, remember the old adage: You can't drive a railroad spike with a tack hammer or a tack with a sledge hammer. You must match the machine to the job. If you have both heavy and light-duty welding to do, you need two welding machines, one heavy and one light.

Check out all the MIG machines that are available today. Take your time in making your decision about which MIG machine to buy. There are many options. Several MIG machines can have add-on modules installed so that you can also TIG weld with your MIG power supply.

110-Volt—There are many acceptable 110-volt wire-feed welders available. Hobart Brothers Company makes a light-duty portable wire-feed welder for sheet metal and body shop work that plugs into 110-volt service. This welder has some limitations, but it's effective at doing what it was designed for—welding 24 gauge (0.0239") sheet metal. It will not weld heavy-gauge stock, such as that used

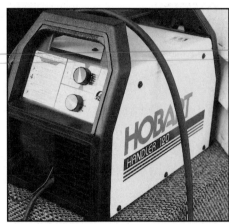

Hobart makes this 110-volt, 120-amp portable wire-feed MIG welder.

for trailer hitches or farm equipment. Some of the specs for this machine are:

- 100-amp maximum.
- 0.024" and 0.030" steel wire.
- 0.035" and smaller aluminum wire.
- Gas-flow timer allows making stitch welds—tack welds about 1/2" to 1" long—and tack welds. Timer guarantees faster gas shutoff after each weld is completed.
- Wire-feed speed control regulates arc heat.
- Unit is self-contained except for CO_2 shielding gas cylinder. CO_2 is used instead of argon because it's effective and inexpensive. Where high-quality welds are required in 4130 steel, a mixture of 75% CO_2 and 25% argon may be used. This comes premixed in one cylinder from welding supply stores.

200/230 Volt—Several companies make conventional 200/230-volt wire-feed welders. These are the most common machines. Most machines this size can be made portable, but rarely will they plug into a wall socket. Chances are you'll have to change the socket to fit the plug on

The right front wheel A-Arm on George Scott's dirt track stock car is MIG welded.

A major advantage of MIG welding is that it is fast and clean—making it ideal for commercial use. Shown here is a typical setup in a large commercial shop. Courtesy Lincoln Electric.

Properly dressed for MIG welding, Ron Chase of Western Welding Co. shows how to repair a folding table with a gas-MIG 220-volt machine. Note that Ron's left hand is guiding the gun for accuracy.

the machine. They can handle 0.024", 0.030," 0.035" and 0.045" wire.

No CO_2 Gas Required, Your Choice—If you use flux-cored wire—0.045" diameter—no gas shielding is required. It's similar to stick welding, except you don't have to stop for new rod. Flux is on the inside of the electrode, so it doesn't chip and flake off.

When welding any kind of sheet steel, plain or galvanized, this wire gives good weld performance with spatter. It can also weld thicker metals at higher volt and amp settings.

575-Volt Heavy-Duty Welder—The minimum these 800-amp machines will weld is 100 amps. They have a 100% duty cycle at 800 amps! You can operate the machine at 800 amps all day, without cool-down, if necessary. If you have some heavy-duty welding to do, this type of machine will handle the job. But it's definitely not for the average home weldor! If you think you absolutely must have one, try it first. Remember, you don't need a sledge hammer to drive tacks, and you don't need a 100-amp-minimum welder to build a utility trailer.

MIG WELDING STEEL & ALUMINUM

Most welding instructors give you about five minutes of discussion about a MIG-welding machine and then tell you to go run a few beads on a piece of steel or aluminum. And that will be the extent of the instruction. That's how easy it is to start wire-feed welding.

However, there's more to wire-feed

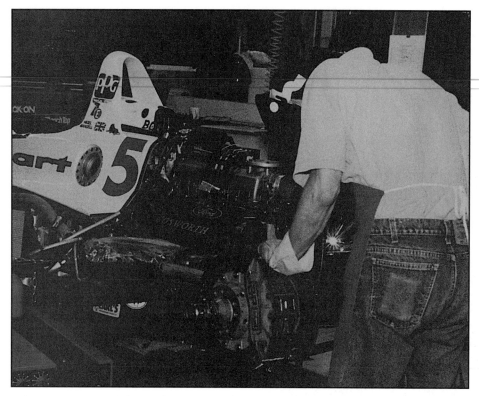

Jeff Noland of H.T.P. makes a quick MIG weld repair on former world champion Nigel Mansell's Indy race car, using an H.T.P. MIG 140 wire-feed machine, equipped with gas. Photo: H.T.P.

Preparation

Before you start welding, attach a small notepad or a sheet of paper to the MIG machine to record settings that work for you. Copy the following chart to make your notes. See page 102.

Settings vary from one machine to another. Even though two machines are the same model, they are not the same machine. Some examples of these variables are:

• How fast the wire filler rod is fed at a setting of 1, 2 or 3.
• How hot the wire gets when the volt and amp settings are adjusted to mid-range.
• How much shielding gas is supplied to the weld puddle when adjusted for 20 cfh.

Not only will one machine vary slightly from another, each project will vary in adjustment requirements. For example, if I were describing how

welding than pointing the gun at the weld seam and pulling the trigger. Problems can occur if you don't take precautions. The wire may get tangled in the drive mechanism or stick in the collet. The cup on the gun may fill with spatter, making weld quality erratic. Just call it Murphy's Law of MIG welding.

Adjustments

There are also several adjustments that have to be exactly right or the weld will end up looking worse than if it were done with an E-6011 arc-welding rod in the hands of a beginner. Here are some adjustments you'll have to make:

• Wire-feed speed.
• Power—amps.
• CO_2 or argon gas-flow rates.
• Wire size.
• Amps and volts while welding.

You'll need a helper to monitor these readings as you weld, and make adjustments as necessary.

George Scott's A.R.S. Chevy-powered race car features lots of MIG-welded tubing in the frame and even in the exhaust system.

75% argon and 25% CO_2 is used here to weld the mild steel tubular race car driver cage. The welder is a Miller Constant Voltage DC wire-feed unit. Photo: Miller Electric Equipment.

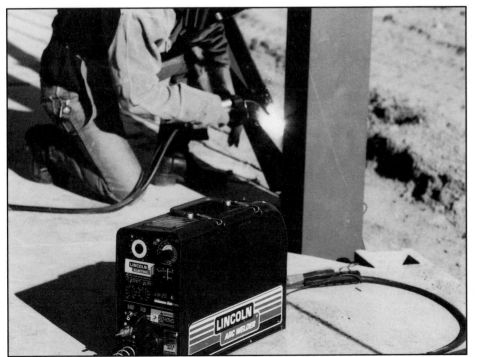

Lincoln Electric also makes a portable wire-feed welder that can be plugged into a motor-driven 400 amp or larger welder, then moved directly to the job, as is done here in this building construction site. Photo: Lincoln Electric Co.

Semi-automatic aluminum welding is made easy by the use of this foot-pedal speed-controlled positioner/turntable, with a Cobramatic MIG gun attached to the rotater/positioner.

to drive my car, I couldn't tell you how far to depress the accelerator to maintain a certain speed. It's the same with MIG welders. Although you can come close on your initial settings, trial and error are required to get them exactly right. So pick a reasonable starting setup such as described in the chart for flux-core wire on page 102. Fine-tune the machine after making a few test beads. Practice on some scrap metal pieces of the same metal you're welding until you can determine the right settings.

As you find the correct settings for each welding situation, record them on your "Wire-Feed Adjustment" chart. Then, next time you're welding 0.020" steel or whatever, you can set up your machine perfectly by just referring to your chart.

Adjust by Sound—I adjust my welder by listening to the arc. Seriously, when it's right, the arc sounds almost like bacon frying on a grille over an open fire. Of course, you should also check for proper weld deposit and penetration. Once you're

101

RECOMMENDED MIG PROCEDURES, GAS, SETTINGS

METAL & MODE	POLARITY	GAS	ELECTRODE	AMPS	VOLTS	WIRE FEED SPEED
Mild Steel Short Circuit	DCEP	75% Argon 25% CO_2	ER 70S-3 .023 .030 .035	30-90 40-145 50-180	14-19 15-21 16-22	100-400 90-340 80-380
Mild Steel Spray Mode	DCEP	Argon +1% 5% O_2	.023 .030 .035	— 135-230 165-300	— 24-28 24-30	— 330-650 340-625
4130 Steel Short Circuit	DCEP	75% Argon 25% CO_2	ER 70S-5 .023 .030 .035	30-90 40-145 50-180	14-19 15-21 16-22	100-400 90-340 80-380
4130 Steel Spray Mode	DCEP	Argon	.023 .030 .035	— 135-230 165-300	— 24-28 24-30	— 330-650 340-625
Stainless Steel Short Circuit Type 304-308	DCEP	90% He 7.5% Ar 2.5% CO_2	ER 308 .030 .035 .045	60-125 75-160 100-200	16-23 16-23 16-24	150-280 125-280 110-230
Stainless Steel Spray Mode	DCEP	Argon 2% O_2	.035 .045 1/16	180-300 200-450 220-500	24-33 24-35 24-36	290-600 250-475 180-300
Titanium Short Circuit Ti-O.15 Pd	DCEP	Argon	ERTi-0.2 Pd	TEST FIRST	TEST FIRST	TEST FIRST
Titanium Spray Mode	DCEP	Argon	ERTi-0.2 Pd	TEST FIRST	TEST FIRST	TEST FIRST

(continued next page)

RECOMMENDED MIG PROCEDURES, GAS, SETTINGS (cont.)

METAL & MODE	POLARITY	GAS	ELECTRODE	AMPS	VOLTS	WIRE FEED SPEED
Aluminum Short Circuit	DCEP	Argon	ER 4043			
		Argon	.030	50-120	16-19	250-550
		Argon	ER 4043			
			.035	65-140	17-20	240-425
			ER 4043			
			3/64	75-170	17-22	160-325
Aluminum Spray Mode	DCEP	Argon	ER 4043			
			.030	95-200	20-27	505-1200
		Argon	ER 4043			
			.035	110-220	20-27	425-850
		Argon	ER 4043			
			3/64	130-290	22-31	250-65

WIRE-FEED ADJUSTMENT

Metal Thickness (in.)	Amps	Volts	Gas Flow	Wire Size	Wire Speed
0.020					
0.030					
0.040					

Make copies of chart for recording MIG-welder setups. You'll then be able to make quick setups when doing similar jobs.

satisfied with the arc sound and weld quality, have someone read the volt and amp gauges while you are welding. Record these numbers in your chart, too.

Wire Cutters—Always keep a pair of small diagonal cutters handy. You'll soon learn why. When you pull the trigger to start the arc and it doesn't, you'll end up with a "pile" of wire.

Even though the arc doesn't start, shielding-gas flow and wire feed do. You can't release the trigger fast enough to prevent this, only minimize it. Before you can continue welding, you must cut off the excess wire. Simply snip it off with cutters and continue welding.

Maintenance—Lack of maintenance causes frequent wire-feed welding problems. The cup will get dirty—it must be kept clean. The cup and nozzle must be cleaned of spatter regularly. Special sprays and jellies are made for this purpose. To keep your MIG welder operating trouble-free, perform these simple maintenance operations:

• Keep the cup clean. Normal weld spatter will clog the cup quickly, blocking gas flow and therefore cause the welder to be unprotected from air. To prevent this, you should coat the inside of the cup with an anti-spatter spray or gel such as Nozzle-Kleen or Nozzle-Dip Gel from Weld-Aid Products. As a less-expensive alternative, many weldors use Pam, an aerosol cooking-oil substitute available at supermarkets. If the nozzle does get dirty, clean it. Special ream-like MIG welder nozzle cleaners are available.

HIGH-STRENGTH STEEL CAR BODIES

To reduce weight for improved fuel economy, many late-model cars use high-strength steel in the body structure. Because the high-strength steel has excellent tensile strength—about 40,000-120,000 psi compared to mild steel's 30,000 psi—panel thickness can be reduced, resulting in considerable weight savings.

But there's a hitch. High-strength steel cannot be gas-welded or brazed. It will harden and crack. Because of its unique grain structure, it must be arc-welded with a low-hydrogen stick electrode such as E-7014, or wire-feed welded. Ford Motor Company and Chrysler Corporation, for example, recommend MIG welding. Here are some of the additional reasons they give for recommending the wire-feed process:

- Welds are made quickly on all types of steel.
- Low current can be used, resulting in less distortion of sheet metal.
- No extensive training is necessary.
- Wire-feed equipment is no bulkier than a set of oxyacetylene cylinders.
- MIG spot welding is more tolerant of gaps and misfits in seams.
- Severe gaps can be spot welded by making several spots atop each other.
- Simple to weld vertically and overhead.
- Metals of different thicknesses can be welded easily with the same-diameter wire.
- Almost all autobody sheet metal can be welded with one wire type.

- Keep the drive gears and rollers clean. Copper-plated wire will clog the drive rolls in a hurry, so avoid copper-plated wire, if at all possible. Every 2-3 hours of welding time, check the drive rollers for metal filings, and brush them away before continuing. To reduce wire drag and clogging, lubricate the wire. Use a treated-felt applicator to lube steel wire; untreated applicator for aluminum wire.

MIG Welding Stainless Steel

The correct wire for MIG-welding stainless steel depends on its alloy. In most cases, 300-series MIG-welding wire will work with the more common 300-series stainless steel. If the alloy is unknown, try ER-308, a general purpose stainless wire. You can even weld mild steel with ER-308 wire. The weld on mild steel will be much less ductile than the base metal. Some common stainless steel wire numbers are ER-308, ER-309, ER-310, ER311, ER-348, ER-410, ER-420, ER-430 and ER-502.

Each number represents different carbon and alloy content. For instance, the chemical content of ER-308 is carbon 0.08%, chromium 20%, nickel 10%, manganese 2%, silicon 0.50%, phosphorus 0.03%, sulfur 0.03%.

NOTE: Most high-strength steel used in late-model cars is confined to body structures, reinforcements, gussets, brackets and supports. In most cases, the outer panels remain regular mild steel and can still be gas-welded or brazed. ■

Welding wire in MIG machines does not like to be pushed much more than 10 to 15 feet. A solution to MIG welding many feet away from the motor-driven power supply is this portable wire feeder that plugs into the main power supply. Photo: Miller Electric Equipment.

TIG Welding

12

Years ago, TIG welding, also known as "heli-arc" welding, was thought to be magic! It definitely had an aura about it. When I was asked if I would like to try my hand at TIG welding aluminum, I responded with a big "Yes!" The next day, I was welding patio screen-door frames for government housing. Nothing to it. It took less than one hour of help from a professional TIG weldor and about 5 hours of practice to get the hang of it. Afterward, I took the test for certified aircraft weldor—and passed!

Passing the certification test for TIG weldor was relatively easy. I was already an accomplished oxy-acetylene weldor. The point is, you should become proficient at gas-welding before trying TIG. Gas-welding requires the same basic skill—controlling the puddle, moving the torch, dipping the rod, running the bead, etc. And, just like riding a bicycle, these are skills you won't forget as you apply them to TIG and other advanced forms of welding.

Name Calling

Many people will say, "Don't call it Heli-arc, call it TIG." Some even say, "Call it GTAW."

Linde developed their trade name for TIG welding, Heli-arc, from the words helium arc welding. Helium is

TIG Welding is like magic! It is relatively simple to master. It is also very versatile. Miller Electric makes this AC/DC Econotig welder that welds steel, stainless steel and even titanium, and aluminum. It is a great machine for race car and aircraft projects. Photo: Miller Electric Co.

an inert shielding-gas that envelops the weld puddle, keeping it free from atmospheric contamination. Today, helium has been replaced (largely) by argon, argon/hydrogen or argon/helium mixtures. However, the name Heli-arc stuck in common usage, even when referring to TIG

welders manufactured by other companies. Consequently, most weldors have to stop and think about what TIG means—tungsten inert gas or, more specifically, GTAW for gas tungsten-arc weld. It's important for you to understand that these three terms refer to the same welding

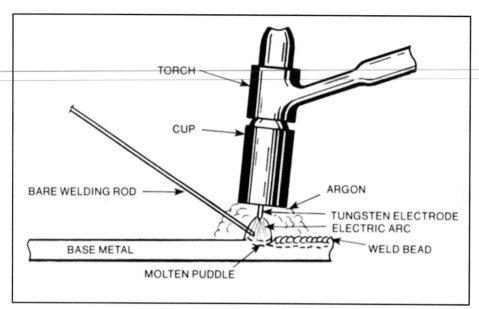

In TIG welding, like gas welding, the filler rod is dipped into the puddle while the tungsten electrode is maintaining the puddle. The tungsten must not touch the puddle and the filler rod must not touch the tungsten, or weld contamination will result. When this happens—and it will—stop, clean the weld, clean the tungsten, and start over again.

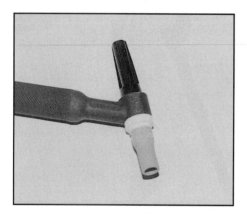

This WP-10 Weldcraft TIG torch is water-cooled and works great for all race car parts, airplane parts, and even ocean-going yacht railings. I like to use a short back cap as often as possible.

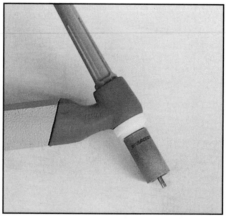

A larger air-cooled torch with a long back cap is useful for field welds on pipe and thicker steel and aluminum materials. Sliding amp control can be added to most torches.

process, not three different ones.

This section describes various TIG welding setups, from the expensive ones to setups that'll get you by. I start with the very best. As with other welding processes, you should know what you really need before you buy. Because a TIG outfit can be the most expensive of all welders, don't buy one until you have it explained and demonstrated to your satisfaction.

HOW TIG WORKS

TIG is the neatest, most precise and controllable of all hand-held welders. You could almost weld a razor blade to a boat anchor or shim stock to a crankshaft.

A small, pointed tungsten electrode—non-consumable—provides a concentrated high-temperature arc with pinpoint accuracy. You don't have to heat the whole area to start a puddle. Once the puddle starts, add filler just as you would with a gas welder.

Because of the TIG welder's high-heat concentration, but reduced heating of the workpiece, TIG is great for welding aluminum. Aluminum dissipates heat quickly, and the less heat absorbed by the aluminum, the better for the weld. If your workshop has a TIG machine, you could use it for most fusion-welding jobs except for rough ones such as building a race-car trailer. In fact, it can replace the gas-welding torch for all jobs except brazing or soldering. TIG welding has one major drawback—it's slow. So, for projects that don't require pretty welds but need to be done quickly, use an arc welder or MIG welder. MIG is covered in Chapter 11.

TIG Components

Torch—Although more complex than a gas welder, the working end of the TIG welder is also called a torch. Instead of a flame, an electric arc is directed at the work to make and maintain the weld puddle. The arc occurs between the tungsten—a high melting-point, non-consumable electrode in the torch—and the workpiece. A collet in the torch clamps the tungsten so it can be adjusted in and out of the torch and retained in place.

Surrounding the tungsten is an open-ended cup that directs this shielding gas to the immediate area of the weld bead. Cups are ceramic because of the intense heat. Speaking of heat, TIG welding torches must be cooled because of the close proximity to the weld puddle. It's not uncommon for a torch to get so hot that it's too

The world's smallest cup, smallest tungsten torch is nice to have in your tool kit when doing those hard-to-reach welds. The cup is heat-proof Pyrex™ glass.

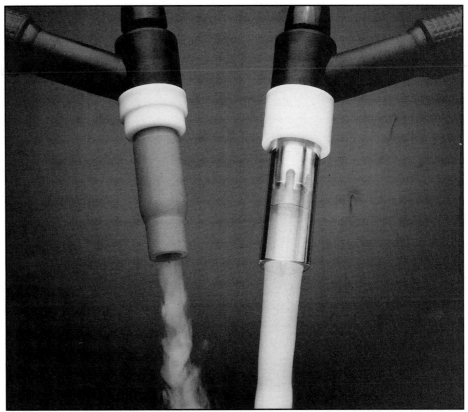

Gas flow, simulated here with smoke, is obviously better with the torch on the right, compared to gas flow on a conventional torch on the left. The glass cup also aids in seeing the TIG weld puddle better. Photo: CK/Conley Kleppen.

uncomfortable to handle.

Torch and cable cooling are done with air or liquid—usually water. Water cooling is preferred by the serious user. However, an air-cooled torch is suitable for doing small jobs. Air cooling is not sufficient when welding for long periods or for welding thick material. How do you know when the torch gets too hot? Simple. It will burn your hand. The cable can get hot enough to melt the insulation!

Water, inert gas and electric power must be fed to a water-cooled torch. Consequently, the torch has a cable for electric power and three hoses—one each for gas, water supply, and water return. Water is circulated through the torch and returned to the reservoir or dumped so the torch will receive a continuous supply of cool water.

Cups—The ceramic cup used on a TIG torch directs inert-gas flow over the weld puddle. A larger cup gives more gas coverage and improved weld quality. There are, however, times when a large cup such as a #10 will not fit into a corner or other tight area. Consequently, a smaller cup must be used. Don't go smaller than a #4 cup. Gas coverage will be inadequate.

I use a #10 cup for flat seams, #8 cup for welding 1" diameter tubing, engine mounts and race-car suspensions, and #6 or #4 for tight corners of aluminum air boxes and oil tanks.

Back Caps—These are used to clamp the collet and prevent the opposite end of the tungsten from arcing. Back caps are available in various lengths; short ones are used to weld in tight corners.

COMPLETE TIG SETUPS

If money is no problem, several companies sell complete, first-class TIG welding outfits. Such outfits will have built-in features to make welding easier. Here are some of these features:

High Frequency

High frequency is provided to start the arc by jumping a spark gap like a spark plug. This is done by superimposing high voltage on the welding circuit. Otherwise, you would have to touch the tungsten to the base metal to start the arc. Touching the base metal is not desirable because the tungsten tip usually breaks off and ends up in the weld puddle. If it breaks off, the result is a contaminated weld bead. A broken tip also shortens the life of the tungsten. Touching the base metal may also contaminate the tungsten.

Although it is possible to weld steel without high frequency, it is required

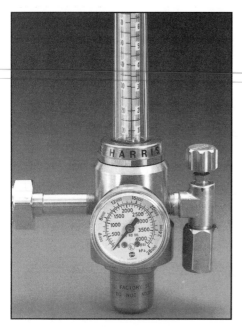

Keep a spare flow meter and a spare argon bottle in your shop for back-gassing stainless steel welds and titanium welds. Harris Co. Calorific Div. Lincoln Elect.

This Daytona MIG argon regulator and flow meter registers flow with a gauge and needle rather than with a floating ball. Compare it to the ball type flow meter in photo at left.

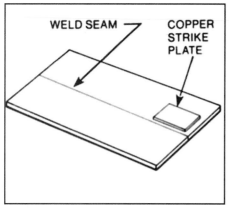

For TIG welding without high-frequency starting, use small copper strike plate for starting arc. Unlike steel, copper will not cause tungsten to break off when starting.

to weld aluminum or magnesium. Read on for more about this.

Flow Meter

A flow meter allows you to monitor and regulate gas flow to the torch. Gas flow should be 10-25 cubic feet per hour (cfh). Any less would not purge the weld—displace all atmospheric contaminants around the puddle—allowing the weld to become contaminated. Much more than 30 cfh would be wasteful and could cool the weld too fast. Flow meters have built-in pressure regulators set by the manufacturer. There are flow gauges available that allow you to adjust both gas flow and pressure.

Cylinder—A cylinder is for storing a long-time supply of inert gas, usually argon.

Recommended TIG Spare Parts List

To avoid making too many trips to the welding supply store, keep a supply of TIG welder parts in your toolbox. A shopping list is located nearby.

I recommend that you use a spare argon flow meter when you are welding stainless steel or titanium and need to back-gas the weld to prevent "sugar" on the back side of the weld. I also recommend a spare TIG torch with power and water hoses.

Power Supply

The TIG power supply is similar to an arc-welder power supply. A TIG power supply may also include the following:

•Foot-pedal amperage control adjusts arc temperature and intensity. Rocking the pedal with your foot also starts and stops arc voltage. Pushing with your toe starts the arc and increases amperage for increased heat and weld penetration; pushing with your heel reduces amperage and stops the arc. It can also be used for stick-electrode welding. When using a foot control, I set amperage about 20% higher than what I think is needed. This way, I get extra heat by pushing on the foot pedal. I don't have to stop the readjust the power-supply amperage control. Note: Stick welding with a foot pedal enables you to "back off" the amps at thin spots. Likewise, you can add amps where extra heat is needed. And stopping the weld with the foot pedal allows you to reduce heat at the edge of the

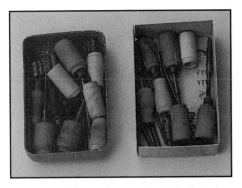

I keep two separate trays for tungsten: one for pure tungsten and one for 2% thoriated tungsten. The pure tungsten, cups and collets are kept in a green painted tray, the 2% thoriated tungsten, cups and collets are kept in a red painted tray so I won't mix them up, even when the color codes are ground off the tungsten.

workpiece and avoid a flat, thin bead common with conventional stick welding.

• Solenoids built into welder cabinet start and stop gas flow and cooling water. These solenoids self-time water flow to cool the torch and give timed post-purge of gas so the hot weld bead is not contaminated while cooling. Consequently, water and gas flow continue for a short time, even though the arc has been stopped. To take advantage of the post-purge gas, the torch must be held for a moment over the end of the weld bead after the arc has been stopped.

• A high-frequency power supply has a cleaning feature required for TIG welding aluminum and magnesium. Its continuous spark actually cleans the puddle area by providing an electric-field shield over the weld bead. This is in addition to the feature that allows the arc to be started without the need to touch the tungsten electrode to the base metal. High frequency is shut off once the arc is established when DC current is used; it continues with AC current. A separate switch provides for start high-frequency or continuous high-frequency.

•Range and polarity switches select AC, DC straight or DC reverse polarity for any kind of welding.

•115-volt AC receptacle accepts work light, extension cord or other accessories.

•Portable base or cart contains necessary items such as gas bottle, water reservoir and foot-control pedal.

•Water reservoir is extra nice because it allows continuous or high-amperage welding without danger of overheating the torch. A good water reservoir holds about 5 gallons and includes a motor-drive pump. The water should be mixed with 50% ethylene-glycol antifreeze if below-freezing temperatures are expected. If you use antifreeze, change it periodically. Antifreeze degrades and gives off a noxious odor when stale.

Necessary Extras

To supplement the deluxe TIG setup, you'll need a good arc-welding helmet, gloves and a small toolbox with trays for collets, chucks, cups, extra helmet lens and tungsten electrodes. I have two small trays for tungsten—one for dull electrodes and one for sharp electrodes. When the tray of sharp tungstens is depleted, I sharpen the dull ones and transfer them to the sharp tray. This saves time when I need to change electrodes.

In the toolbox, keep several pins, clamps and setup fixtures handy. An important tool is a 6" x 12" framing square. If you recall from the gas-welding chapter, the framing square is useful for setting up parts and checking for possible warpage. Most things we weld are square to something, so a square comes in handy.

TIG SPARE PARTS LIST

Number	Description
3	#10 ceramic cups
3	#8 ceramic cups
6	#6 ceramic cups
3	#4 ceramic cups
4	1/8" tungsten, 2% thoriated (for steel)
4	1/8" tungsten, pure (for aluminum)
4	3/32" tungsten, 2% *thoriated (for steel)
4	3/32" tungsten, pure (for aluminum)
4	1/16" tungsten, 2% thoriated (for steel)
4	1/16" tungsten, pure (for aluminum)
2	1/8" collets
2	1/8" chucks
2	3/32" collets
2	3/32" chucks
4	1/16" collets
4	1/16" chucks
1	stubby back cap
1	2" back cap
2	small stainless-steel wire brushes

*Thoriated means tungsten includes thorium alloy for making it easier to start the arc. Unfortunately, thorium also can contaminate aluminum and magnesium, so pure tungsten should be used for these applications.

Reduce trips to the welding supply store by keeping a supply of TIG parts in your toolbox.

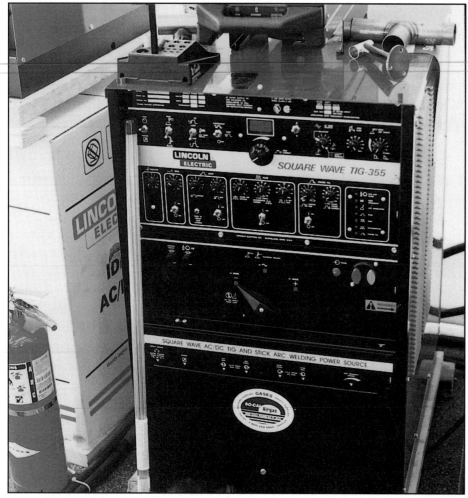

This 355-amp square wave TIG welding machine provides repeatable, certified-quality welds that are much easier to achieve than with less featured machines. If you think you are a good weldor, this machine will make you a great weldor.

Daytona MIG Co. sells this portable pulsed TIG welder that will nicely weld steel and stainless steel up to 3/16" thick. The welder and argon bottle weigh under 80 lbs.

• Files of various sizes and descriptions. As discussed in the gas-welding chapter, you should have a coarse round file, a coarse half-round file and a flat mill file.

• Bench grinder with fine- and medium-abrasive wheels. Use these for sharpening tungsten and fitting small parts.

• Belt sander and disc sander for fine-dressing tungsten and fitting parts to be welded.

• Other niceties include a comfortable chair, clean welding table and good lighting. Mount a fluorescent light over your welding table. Two or three portable flood lights for welding such things as engine mounts, race car frames, trailers or airplane fuselages may come in handy. A clean, well-lit welding area is safer and will result in better welds.

Maximum TIG Setups

This is as opposed to MINIMUM TIG welding setups.

If money is not a problem, or if you want to set up a first-class race car or airplane shop, you should investigate the top-of-the-line TIG machines. These come with computer, digital memory programs that you can repeat each time you do a specific welding job. And these machines can and do provide much better arc quality than the bottom-of-the-line welding machines.

Pulsed Arc—Years ago, weldors had to pump the foot pedal to keep from melting holes in their parts. Now, the welding machine can be adjusted to give pulses of full heat with low amp power alternated in between the full power pulses. Pulsed current welds allow the puddle to "freeze" between filler rod application so that better puddle control is possible. Ask your welding supply for a demonstration before you buy.

Minimum TIG Setups

DC Conversion—If you already have a DC arc welder, you can convert it to a TIG welder by adding an air-cooled torch, and an argon flow meter/flow gauge and tank. This will restrict your TIG welding to steel and stainless steel because of the lack of a

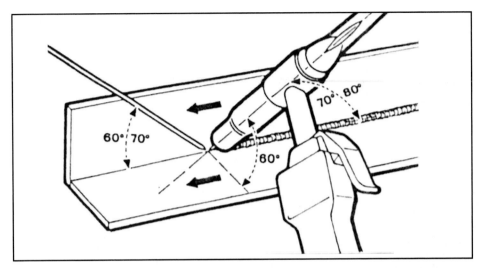

The correct angle for TIG welding in a corner is shown here. But always manipulate the torch angle to point the heat where required for good welds. Courtesy Daytona MIG Co.

Inverter Technology

In the 1970s and early 1980s, a good TIG machine weighed as much as 400 to 500 pounds. Today, because of inverter technology, a machine even more powerful and more precise can weigh less than 100 pounds. But this current (no pun intended) technology has allowed more features to be included in top-of-the-line welding machines with no penalty in size and weight.

Be aware of the newest welding machines that provide capacitor start, pulsed start, balanced and unbalanced wave forms, as well as asymmetric wave forms. All these features contribute to more precise machine settings and more precise welds. These machines will not be covered in this book because they deserve a complete book dedicated solely to their setup and operation. There is just too much to say about them to try to put it in this single chapter.

USING A TIG WELDER

To refresh your memory, an electric arc is generated between a non-

high-frequency unit. Because of this, you'll also have to start an arc on the base metal or a copper strike plate. A strike plate makes arc-starting easier and reduces tungsten contamination from the base metal. Also, copper won't break the tungsten as easily as steel or aluminum, but it's not as good as high-frequency arc-starting. Without a foot-operated amperage control, you must stop welding to make amperage adjustments at the machine.

AC Conversion—If you have an AC/DC arc welder, you can fit a high-frequency unit to it—for about the cost of your AC/DC machine—a torch, hoses and regulator, and argon gas cylinder. This will allow you to TIG weld aluminum. Again, you won't have a foot-operated amp control. You'll also have to touch the tungsten to the base metal or strike plate to start the arc.

Which ever TIG setup you choose, you'll be able to weld things you never thought possible, such as welding thin parts to thick parts or complicated assemblies without burning them up. But mainly, TIG is

so "squeaky" clean that you can weld in your Sunday clothes and not get dirty. Although I don't recommend this, I've actually welded airplane parts in a suit, white shirt and tie!

Although slow compared to arc and MIG welding, TIG welding is considered the most precise method of shop welding. Consequently, it's used to repair bad welds where accuracy counts.

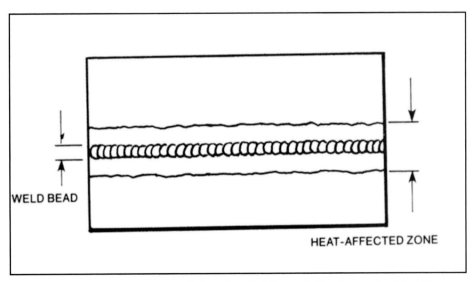

Heat-affected area of TIG weld seam is less than that for most other types of welding. Because of high temperature differential and small heat-affected area, avoid TIG welding in a draft.

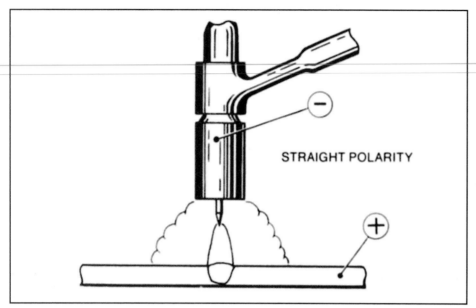

Set TIG welder to DC current, straight polarity for welding mild steel, 4130 steel, stainless steel and titanium.

consumable tungsten electrode and the base metal in TIG welding. Non-consumable means that the electrode is not intentionally melted into the weld puddle as in conventional arc welding. The tungsten will, however, erode and become contaminated in use. Consequently, it will also be ground away as you dress it to the desired point again and again.

The type of metal being welded determines the tungsten tip shape required. For instance, the sharp point needed to weld steel confines the arc to a smaller area, resulting in more concentrated heat at the weld seam. But a crayon-shaped point is used to weld aluminum because that metal dissipates heat more quickly and needs more area heated at the weld seam. This is done with the resulting broad arc. More on dressing tungsten tips later.

As the arc heats a molten puddle on the base metal, dip the filler rod into it as you would if gas welding. The inert gas from the torch shields the puddle from atmospheric contamination.

To ensure that shielding gas covers the weld while solidifying, keep the torch over the bead after completion until the purge gas stops. The timer on the TIG machine usually provides 5-6 seconds of gas flow after the torch is off. This keeps 4130 steel from cracking and helps prevent crystallization in stainless steel. Also,

titanium demands post-purge gas to prevent atmospheric contamination of the weld bead.

Polarity Settings

Most DC TIG welding is done on straight polarity—electrode is negative. There is no polarity with AC current. AC current constantly changes direction. There are 60 complete cycles a second in the U.S. AC current frequency is 50 cycles per second in Europe and 90 cycles per second in other parts of the world.

When welding most mild steel, stainless steel, 4130 steel and titanium, set your machine for DC current, straight polarity. This concentrates most of the heat at the work—about 70%. However, when welding aluminum or magnesium, use AC current. The reason for this is the oxide-cleaning feature of reversing current—very important when welding these non-ferrous metals.

As for TIG welding with DC current, reverse polarity—electrode is positive—you'll rarely have the

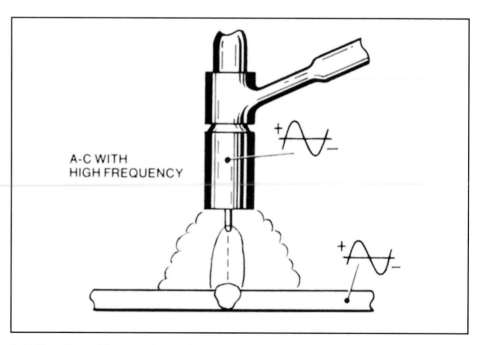

Set TIG welder to AC current for welding aluminum and magnesium.

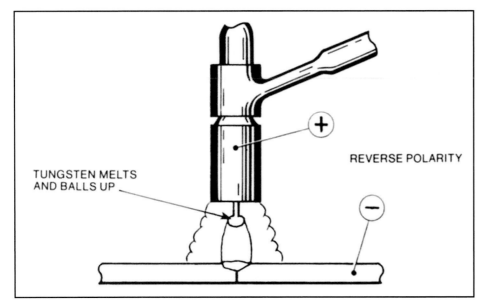

DC current, reverse polarity is a seldom-used TIG setting because tungsten melts before base metal. Higher heat is at electrode, not work. Therefore, its only application is for welding thin-gauge metal.

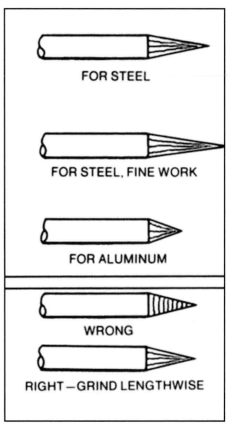

Shape of tungsten electrode is important. Sharpen tungsten to look like a pencil for welding steel. Sharpen to needle point for fine work. Sharpen tungsten to crayon-like point for welding aluminum. Always grind tungsten lengthwise.

occasion. The tungsten may melt before the base metal because about 70% of the heat is at the electrode, not at the work. Consequently, penetration of the base metal is poor. However, DC current, positive polarity does have a useful application. It is best used for welding very thin sheet metal—not aluminum or magnesium. Shallow penetration then becomes an advantage.

Tungsten Sharpening

I coarse-sharpen tungsten electrodes on a small bench grinder, then finish dressing them on a power sander with 120-grit paper. The grinder removes metal fast, and the sander does a good dressing job. Remember to grind tungsten slightly blunt for welding aluminum and sharp for welding steel, stainless steel and titanium.

A sharp tungsten tip is best for welding steel because it provides for better control and concentrated heat. But, because the heat for welding aluminum needs to be more evenly distributed and AC current melts the

sharp electrode point—tungsten/base-metal heat distribution is about 50/50 AC—a sharp tip would contaminate the weld. The blunt tip cures both of these problems. It scatters the arc, distributing the heat, and it doesn't melt.

Don't go to the trouble of balling, or rounding the tip of the tungsten for aluminum welding as some weldors do. It will ball itself in about two seconds of welding.

Sharpen tungsten lengthwise using a series of straight cuts toward the tip. Never sharpen tungsten by rotating it against a grinder. Sharpening a tungsten in this manner will result in a poorly controlled arc pattern. See drawings at right.

Install & Adjust Tungsten—Loosen the collet on the back of the torch to free the tungsten. This will allow it to slide in and out. Adjust the tungsten tip so it projects about 1/8" to 1/4" past the tip of the cup.

TIG Welding Tips

To make things easier, read the

following before you start TIG welding:

• I know you've heard this once before, but here it is again: Master gas welding before you attempt TIG welding.

• Clean the base metal as if you were going to eat off it. Seriously, you can't get the metal too clean for TIG welding. There is no flux to float off impurities.

• Cut the welding rod into 18"-long pieces. This usually means cutting the 36"-long rods in half. Shorter pieces are easier to use.

Practice, Practice, Practice. These practice welds in aluminum and stainless were made with a Lincoln square wave, pulsed TIG machine.

• Tungsten diameter should be about half base-metal thickness. For example, use 1/6" diameter tungsten to weld 1/8" thick metal.

• Cup size should be as large as possible without restricting access to the weld. For instance, you'll have to use a smaller cup to weld in tight corners. Use a #8, #10 or #12 cup for flat steel and where access is good; #4 or #6 cup for corners and where access is poor.

• Clean the welding rod before you start welding. Use MEK or alcohol on a clean, white cloth. Even dust contaminates the weld.

• Make sure the lighting is good because the light from the arc is less intense than in other welding. Use a clean, #10 helmet lens for most TIG welding.

• Do not allow even the slightest breeze or draft in the weld area. Cool air will crack a TIG weld because the heat-affected area is smaller and more sensitive to rapid cooling. A breeze can also blow away the shielding gas.

•Never touch the hot tungsten to the puddle or filler rod. Capillary action causes molten metal to wick up—flow up—the tungsten, contaminating it. The weld is also contaminated as the wicked metal oxidizes and boils off the tungsten, blowing oxidized metal into the puddle. If this happens, stop welding immediately and grind the wicked metal off the tungsten. This is why you'll use more tungsten now than after you have more experience.

• Use a gas lens cup if you have room. The gas flow will be more uniform and your welds will look better.

• When—not if—you burn a hole through the base metal, completely let up on the pedal or break the arc. Let the puddle cool before continuing.

• Always shield the TIG weld light so you won't burn someone's eyes. TIG appears to give off less light, but its ultraviolet radiation is just as dangerous.

• Sit down while welding, if possible. Be comfortable. Take advantage of

the fact that TIG welding generates no sparks to fly into your lap.

• Always tack-weld parts before running your final bead.

• Before making a critical TIG weld, try your procedure on a test specimen first.

• As in oxyacetylene welding, make a molten puddle, then dip filler rod into it. Don't try to melt the rod with the arc. A cold weld, with poor penetration, results if you try to drop molten rod into the seam.

Ready to Weld

The torch should be in your right hand if you're right-handed. Switch hands if you're left-handed. To begin, rest the corner of the cup on the work at about 45°. Don't allow the tungsten to touch the work. Later on, you should be able to judge a 1/4" to 1/8" arc gap without touching the cup to the work.

Get comfortable. Rest your arms—covered with long shirt sleeves—on the welding table or workpiece. Assume the most comfortable

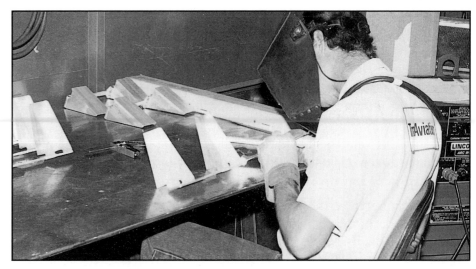

Typical production weld shop for making aircraft parts. The weldor should be wearing a long-sleeved shirt to prevent arc burns from the TIG ultraviolet light.

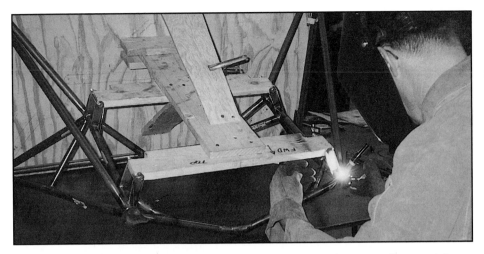

Dale Johnson is TIG welding a 4130 steel bracket on my auto engine conversion mount for my Grumman Traveler airplane. Notice the wood and plywood welding jig. It worked perfectly. The mount fit the airplane without any tweaking.

TIG welded 4130 steel tubing adds strength to Ken Duttweiler's Buick V-6 drag race car.

The Daytona 24-hour race winning Olds Aurora powered race car has a carbon fiber reinforced and TIG welded 4130 steel frame.

position that allows you to operate the foot pedal. Never weld with your arms unsupported. If possible, lean your shoulder against something steady.

Button up your collar to prevent neck burns from the ultraviolet light.

Keep the welding rod away from the arc until you have a molten puddle. After it has started, dip the rod in and out of the puddle as you move along the weld seam. Do not try to melt the rod with the arc. Let the

molten puddle do the melting. Dipping the rod into the puddle cools the puddle slightly. So the rhythmic in-and-out motion of the rod maintains a constant puddle temperature. If dipping cools the puddle too much, compensate by increasing heat slightly.

Tip: This one is worth repeating, particularly for doing TIG welding. Drill a #40 hole in tubing when welding it closed. This will keep the puddle from blowing out onto the

tungsten as you close the seam. Hot air expanding inside the tubing causes this.

The following are specifics on welding popular metals. Suggestions are given on how to set up the machine and provide for additional shielding, if needed. Except for magnesium, refer to Chapter 15, for information on specific alloys and the respective filler rods to use. Magnesium rods are covered in this chapter, page 136.

TIG-WELDING STEEL

Because mild steel and chrome moly are TIG welded with the same process—machine and filler material—you can weld mild steel to 4130 or vice versa. For welding either, set up the machine by using the following procedure:

• Use steel welding rod #7018 or equivalent. See chart, page 142, for details.

• Use DC straight polarity—negative electrode.

• Set high-frequency switch to

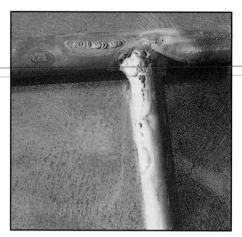

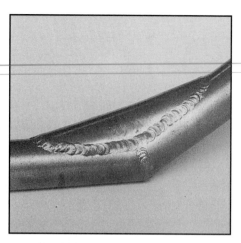

Troubleshooting defective welds is how you learn to be a better weldor. Rough welds and smoky deposit at the weld indicates not enough argon flow, or in certain cases, it could indicate contaminated argon, or even the wrong gas in the bottle.

Another defective weld. Note the small crater at the right side of the weld bead, caused by shutting the torch off abruptly while the puddle was still molten.

This otherwise excellent TIG weld on this aircraft 4130 steel tube mount is defective because the weld bead is concave and subsequently weak in one spot. Adding another weld bead on top of the low one would be acceptable.

I welded this boat radar antenna mount from 316 stainless steel. 316 grade stainless is least susceptible to salt water corrosion. The radar antenna manually tilts to match the tilt of the yacht when under full sail. Photo: "Vicky."

START, if so equipped.

•Adjust gas flow to 20 cfh.

•Set water-cooling switch to ON, if so equipped.

•Set coarse- and fine-amp controls—coarse to 75 amps, fine to 80% for 0.063" steel. This gives 75 X 0.80 = 60 amps with the pedal all the way down. For thick metal, try these settings: coarse to 125 amps and fine to 35%, or 44 amps with full pedal. If this doesn't provide enough heat, increase the fine-amp setting. Go 50%, 80% or whatever heat you need. The reason for the lower fine-amp setting for thicker materials is to increase duty cycle. Regardless of the fine-amp setting, final amperage at the electrode is controlled by you at the pedal.

•Turn on contactor and amp-control switches to remote foot pedal, if so equipped. Contactor switch turns "on" the foot pedal.

TIG WELDING STAINLESS STEEL

TIG welding stainless steel is similar to welding mild steel and 4130 steel. However, you'll need some extra items: an extra bottle of inert gas, extra flow meter and about 20 feet of argon hose. This assembly of items will be used to purge, or shield, the back side of the weld.

Back-Gas Purging

As mentioned earlier, purging is the process of displacing all atmospheric gases and replacing them with an inert shielding gas such as argon. I usually set purge-gas flow the same as or about 25% more than torch flow.

The reason for using back-gas purge is that molten stainless crystallizes if it's exposed to air. Sugar, or crystallization, on the back side of the weld would weaken the weld and base metal considerably.

It's easy to back-gas stainless tubing such as an engine exhaust pipe. You simply fill the tube with gas. Cap both ends of the tube with making tape, punch holes in the tape, stick the purge hose in one end and a short piece of open tubing in the other end to exhaust the argon. At 15-20 cfh gas flow, the average exhaust pipe can be purged in air in 4-5 minutes. Larger pieces, such as long, 12" ID tubing, take more time to purge.

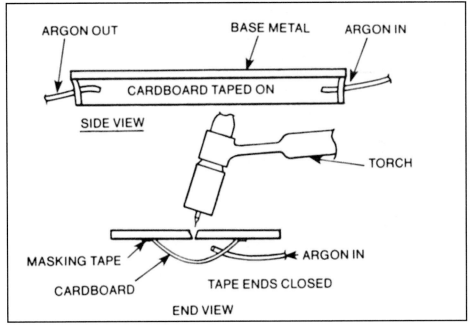

Simple back-gas shield for TIG welding titanium can be made from cardboard. Masking tape is used to hold it to workpiece. Use the same kind of shield for welding stainless steel and 4130 steel.

Flat stainless plate is harder to purge. But you can build a cardboard-and-masking-tape cover over most seams to act as a purge shield. Simply enclose the back side of the weld seam with craft-type cardboard and masking tape. Shape the cardboard so it straddles, but doesn't touch, the weld seam. Burn a large hole in the cardboard and the purge gas will be lost. Cap the ends with masking tape and insert the purge and exhaust hoses. The accompanying drawing illustrates how to make a back-gas shield for flat plate.

TIG WELDING TITANIUM

Although most alloys of this expensive, lightweight metal can be TIG welded with the same basic machine setup as mild steel and 4130 steel, it requires a more elaborate shielding-gas apparatus than stainless steel. As a result, some titanium alloys are not weldable at all. This is because

of titanium's extreme sensitivity to contamination. Hot titanium reacts with the atmosphere and dissimilar metals, causing weld embrittlement.

This contamination is serious if carbon, oxygen or nitrogen is present in sufficient quantities. In the solid state, as in a weld heated above 1200° F (649C), titanium absorbs oxygen and nitrogen from the air.

An argon-gas shield must cover both sides of a weld seam while titanium is at the 3263° F (1795C) molten stage and all the way down to 800° F (427C). Otherwise, embrittlement from contamination and resulting cracking will occur.

Make a device to provide a trailing shield to cover the weld until it cools to below 800° F (427C). Usually, after the trailing shield passes over the completed weld, the titanium has cooled sufficiently. You can also use temperature-indicating crayon or paint to be sure. But if you do, be careful not to contaminate the weld with the crayon or paint. See the drawings of trailing-shield device.

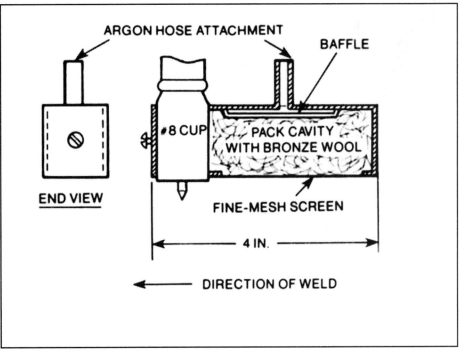

Trailing-gas shield for TIG welding titanium protects completed welds as it cools. Shield must fit tightly to cup. Practice on stainless steel before welding expensive titanium parts.

117

Gas Chamber

Because it is extremely reactive with nitrogen, oxygen and hydrogen, the best place to weld titanium is in a total inert atmosphere such as that in outer space. So, if you can get a ride on the Space Shuttle to do your welding, great. Otherwise, the next best thing is a gas chamber. Such a chamber looks similar to an incubator for newborn babies or a bead-blasting cabinet. Put the part to be welded inside the chamber, close it, then put your hands into built-in gloves to do the welding. The chamber is completely purged with argon so no air exists. Much purer titanium welds will result.

TIG WELDING ALUMINUM

TIG welding aluminum is slightly different than welding steel. Machine settings are different. And, as you may remember from the gas-welding chapter, aluminum doesn't change color as it forms a puddle—it gets shiny instead. However, unlike gas welding, flux isn't needed to TIG weld aluminum. In fact, flux would really mess up the weld and your TIG welder. The following tips will help you TIG weld aluminum:

• The weld seam area should be as clean as possible, and it should be free of aluminum oxide. Remove the oxide immediately before you weld—not a week or two before—by mechanical or chemical cleaning. Mechanical cleaning is done with a stainless steel wire brush, sandpaper or abrasive pads. Afterward, wash off the dust with soapy water and rinse with clear water. Then, wipe down the seam area with alcohol, MEK

KNOW BEFORE YOU START

Because welding titanium is so specialized and the technique so critical, you should discuss your titanium-welding project with a weldor who specializes in titanium before attempting it. If you can't find such a person, RMI Company, Niles, OH 44446 is a company that specializes in welding titanium. They would be glad to answer your questions. Or contact the American Society for Metals, Metals Park, OH 44073. Their educational division, the Metals Engineering Institute, can provide instructional material through their Home Study and Extension Courses for all types of welding techniques and metals. Finally, there's the American Welding Society (AWS), 2501 Northwest 7th Street, Miami, FL 33125. Their four-volume *Welding Handbook* will give you all you ever wanted to know about welding and more.

GAS SHIELDING FOR TITANIUM WELDING

Material Thickness	Torch Argon Flow (cfh)	Trailing Shield Argon Flow (cfh)	Back Gas Helium or Argon (cfh)
0.030	15	15	3
0.060	15	20	4
0.090	20	20	4
0.125	20	30	5

or acetone.

• To eliminate the need for cleaning newly fabricated aluminum parts, use the more expensive paper-covered sheet stock. Adhesive-backed, paper-covered sheet can be cut, formed and fitted with the covering in place. When you're ready to weld, simply peel back the covering from the weld seam to expose super-clean aluminum, and you're ready to weld.

• If TIG welding aluminum castings or forgings, V-groove the joint all the way through. Or V-groove the joint from both sides and weld on both sides, if possible.

• If welding aluminum plate that's more than 1/16" thick, V-groove the joint for better penetration.

• Don't attempt to weld 2024 aluminum or other non-weldable aluminum alloys. Read on for a handy test to check the weldability of unknown aluminum alloys. Weldable alloys include 1100, 5052, 6061 and all castings.

Machine Settings
• Polarity selector switched to AC.

• High-frequency unit set to CONTINUOUS—an absolute necessity.

GUIDELINES FOR TIG-WELDING ALUMINUM

Material Thickness (in.)	Current (amps)	Tungsten Diameter (in.)	Weld Rod Diameter (in.)	Argon Flow (cfh)
0.020	25	0.040	1/32	16
0.040	34	1/16	1/16	18
0.063	50	1/16	1/16	20
0.080	75	3/32	3/32	20
0.100	100	3/32	3/32	22
0.125	125	3/32	1/8	25
0.250	150	1/8	1/8	35

Chart gives approximate settings for welding aluminum. Adjust amps to suit the particular conditions.

ALUMINUM HEAT TREATMENT NOMENCLATURE

T0	Completely soft, no temper	T6	Solution bath heat-treated, then artificially aged
T2	Annealed by heat to soften (cast only)	T7	Solution heat-treated, then stabilized
T3	Solution heat-treated, then cold-worked	T8	Solution treated, cold-worked, artificially aged
T4	Solution (salt bath) heat-treated	T9	Solution treated, aged, then cold-worked
T5	Artificially aged		

• Argon flow meter set to 20 cfh, or check flow chart.

• Water cooling switched ON, if so equipped.

• Contactor and amp switches on REMOTE, if equipped with foot pedal.

• Set coarse amp adjustment to about 60 amps; fine amp adjustment to 70%. Reset amp setting(s) if heat is not right.

Also, you'll need to use the following items:

•Pure tungsten rather than 2% thoriated tungsten for TIG welding aluminum. As you may recall, thorium contaminates aluminum welds.

•Aluminum welding rod 4043 for most alloys. Check the chart to be sure.

Preheating

Not too many years ago, I toured one of the largest aircraft engine repair shops in the world. I was shown the welding shop where cracked and broken cast-aluminum cylinder heads were repaired. They used a natural gas oven to preheat three or four cylinder heads at a time before welding. The cylinder head was heated evenly to 350° F (177C).

Any aluminum more than 1/4" thick benefits from preheating before welding. A simple way to preheat is in a gas or electric oven with thermostatic control, such as a kitchen oven. Just make sure the aluminum you're welding is absolutely clean, or you'll stink up the house for a week!

It's usually not necessary to re-preheat during welding, because the weld heat keeps the part hot enough.

Weldability

Several aluminum alloys are not weldable, most notably 2024. Usually found in sheet form, this alloy is commonly used on airplane-wing and fuselage skins, wing ribs and fuselage bulkheads. If you're not sure about the alloy, look for previous welds. If it was welded before, it can be welded again. But if there is no sign of welding, try a sample weld to test it. Finding a part to do sample welds on might not be easy, but I would even go to the trouble of finding an unusable part to practice on to avoid ruining a repairable part. Race car shops and airplane shops usually have scrap parts in a corner somewhere. So, if an aluminum part is not weldable, repair it with rivets or nuts and bolts, or replace it.

Grounding Aluminum

When TIG welding aluminum on a grounded tabletop, it usually arcs between the table and the work. Consequently, you may make a beautiful weld on a part that took all day to cut, shape and fit, only to turn it over and find arc burns and craters where it contacted the table.

Arcing can be prevented by grounding the workpiece directly or providing a simple ground for the work. I usually lay a heavy mechanical finger on the part to help hold it against the weld table. If you use a separate ground cable, make sure it's welding-cable size. A small-diameter cable will overheat from high amperage.

Weld Craters

When running an aluminum weld bead, don't break the arc or rapidly shut off the arc with the foot pedal as you reach the end of the seam. This will cause a small depression or crater at the end of the bead. Instead, back off the foot pedal slowly, then lead—move—the arc back to the already solidified bead. This will "freeze" the puddle while it's still convex.

Fitting Parts

Fitting parts closely is crucial when welding aluminum. Refer to

ALUMINUM ALLOYS

Alloy Number	Tensile Strength (psi)	Heat Treatable	Weldable
1100	15,000	No	Yes
3003	26,000	No	Yes
3105	23,000	Yes	Yes
5005	26,000	No	Yes
5052	41,000	No	Yes
5086	47,000	No	Yes
2024	61,000	Yes	No
6061	42,000	Yes	Yes
7075	65,000-75,000	Yes	No

Note: If welding becomes necessary after part is heat-treated, heat-affected area is likely annealed or softened. If heat-affected area is small, it usually does not require heat-treating again. But if affected area is large, part should be heat-treated again.

Refer to chart to determine if aluminum alloy is weldable.

Griffin Aluminum Radiators exhibit excellent TIG welding. Try to make your aluminum TIG welds look this good.

Chapter 5. You can fill a gap, but the back side of the weld will look like sand or even gravel because the molten aluminum solidifies too quickly and forms a lumpy bead.

Back-Gas Purge

TIG welding aluminum does not require a back gas to purge oxygen and hydrogen from the weld as with stainless steel and titanium. Although it can improve the appearance of the

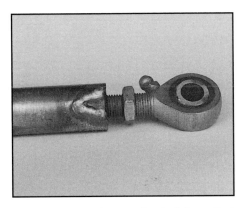

Race car suspension member is fishmouthed and TIG welded.

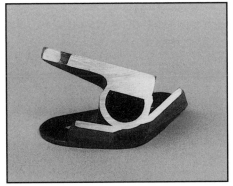

Don't try to TIG weld the back side of your aircraft or race car parts unless the welding engineer specifies welding on both sides. This TIG welded aircraft turbocharger bracket shows what cross section you should expect to see in your welds. The bracket has been sectioned to certify my production welds. The test passed aircraft certification.

back side of the weld bead, it will not noticeably improve weld integrity. Aluminum does not pick up atmospheric contamination as does stainless steel or titanium. And, aluminum will solidify in open air with nothing more than oxidation of the surface. Consequently, using back gas to weld aluminum is a waste of valuable time, money and equipment.

Heat-Treating After Welding

Aluminum is very easy to shape, form and TIG weld in the dead-soft—O—condition. After these operations are completed, it is then heat-treated to give the soft aluminum strength and rigidity.

The 6061-O alloy is commonly used to make fuel-tank bulkheads, wing ribs and other parts for airplanes and race cars. The complete assembly or subassembly is then heat-treated in an oven or brine—salt—solution. In the oven process, 6061-O is heated to 950° F (510C) for about 15-30 minutes and then air-cooled.

The brine solution process involves heating the solution to 1000° F (538C), a temperature at which it does not boil. The 6061-O alloy is immersed in the brine for 15-30 minutes. It is then immediately quenched—cooled—in 70° F (21C) water. At this point, the aluminum isn't completely hard, but after sitting 24 hours at room temperature, it age hardens to full strength and hardness. Age hardening is also called precipitation hardening.

Note: After 6061-O is heat-treated, it is transformed into either 6061-T4 (solution heat-treated) or 6061-T6 (oven heat-treated).

The chart on page 120 details heat-treatment nomenclature. For example, for aluminum-alloy 6061-TX, the T means temper, the following number indicates heat-treatment type.

Annealing—Annealing aluminum is the process of heating it to 750° F (399C) and allowing it to air-cool slowly to remove the effects of previous heat-treatment so it can be cold-formed without cracking. It can be reheated after annealing.

TIG-WELDING MAGNESIUM

CAUTION: Magnesium can burn and support its own combustion. Water or dry-powder fire extinguishers will not put out a magnesium fire. In practical terms, the only way a magnesium fire can be extinguished is to wait for all the magnesium to be consumed. I once saw a race car with a magnesium transmission case and wheels burn to the ground. All that remained of the transmission were the axle shafts, gears, nuts and bolts. The magnesium parts burned to ashes. The intense heat melted the aluminum engine, leaving crankshaft and connecting rods laying in the dirt. A fully-equipped fire truck was not able to extinguish the fire.

So, when welding magnesium, try to do so outside, away from flammables. Magnesium isn't likely to catch fire unless there are magnesium filings nearby—such as those created by machining or dressing the weld seam. If magnesium does catch fire, stand back and let it burn. You probably can't stop it.

Clean Before Welding

As with other metals, magnesium should be cleaned of all scale and corrosion in the weld-seam area before TIG welding. Use some aluminum wool, steel wool or a stainless-steel brush to remove the white, powder-like corrosion.

If the corrosion can't be removed by mechanical means, use chemicals. Mix 24 oz. chromic acid, 5 1/3 oz. ferric nitrate and 1/16 oz. potassium fluoride in 70-90° F (21-32C) water to make a 1-gal. chemical cleaning solution. Dip the part in this solution for three minutes, then remove it and rinse in hot water. Let it air dry before welding. Don't use compressed air for drying. Compressed air may be contaminated with dirt, water and oil.

GUIDELINES FOR TIG-WELDING MAGNESIUM

Material Thickness (in.)	Current (amps)	Tungsten Diameter (in.)	Weld Rod Diameter (in.)	Argon Flow (cfh)
0.040	35	1/16	1/16	12
0.063	50	1/16	1/16	12
0.080	75	3/32	3/32	12
0.100	100	3/32	3/32	15
0.125	125	1/8	3/32	15
0.250	175	1/8	1/8	20

Listed are approximate values for welding magnesium. Make adjustments for conditions. Follow above recommendations for setting up to weld magnesium.

STRESS-RELIEVING MAGNESIUM THROUGH HEAT TREATMENT

MAGNESIUM SHEET			MAGNESIUM CASTINGS		
Alloy	Temp (F)	Time (min)	Alloy	Temp (F)	Time (min)
AZ31B-0	500	15	AM100A	500	60
AZ31B-H24	300	60	AZ63A	500	60
HK31A-H24	600	30	AZ81A	500	60
HM21A-T8	700	30	AZ91C	500	60
ZE10A-0	750	30	AZ92A	500	60
ZE10A-0	450	30			
ZE10A-H24	275	60			

To prevent cracking, magnesium sheet and castings must be stress-relieved by heating to given temperature and held there for times given. Heating is best done in an oven.

Stress-Relieving Magnesium

Magnesium alloyed with aluminum is susceptible to a unique phenomenon called stress-corrosion cracking. For example, if you drill a hole in magnesium and put in a tight-fitting bolt, the area corrodes, then cracks as a result of the corrosion. Check with the manufacturer to determine the alloy content of a particular magnesium alloy.

Otherwise, you may need to have the magnesium analyzed metallurgically—an expensive process. These alloys must be heat-treated to remove the welding stresses, which would otherwise result in corrosion and cracking. See the chart above for stress-relieving by heat-treatment. ■

Plasma Cutting & Welding

13

In this chapter, we explain a relatively new shop tool, the plasma cutter, and we explain another type of welding equipment, the PAW (plasma arc welder). Plasma cutters are extremely handy for making quick, clean cuts in any metal, including brass, copper, stainless steel or aluminum.

Plasma is a gas heated to an extremely high temperature and ionized so it becomes electrically conductive. The plasma-arc welding (PAW) process uses such a gas to transfer an electric arc to the workpiece and to constrict, or contain, the arc for welding.

The Process

In the basic plasma-arc process, invented and developed by Linde, an electrode is located within a torch nozzle. This nozzle has an arc-constricting orifice. Inert gas, usually argon or nitrogen, is fed through the nozzle, where it is heated as high as 50,000° F (27,760C), the plasma temperature range. The plasma arc emerging from the orifice is hot enough to melt any metal. In welding operations, a shielding gas is simultaneously introduced through a separate concentric passage of the torch. This protects the weld puddle from atmospheric contamination in a

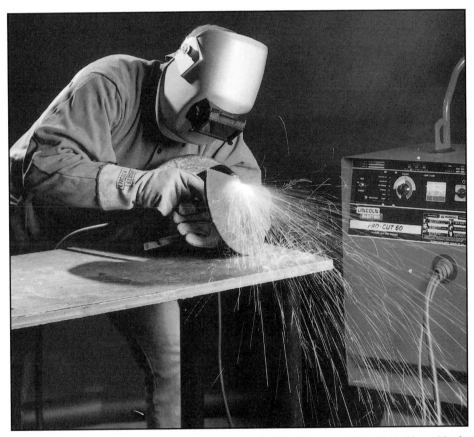

I have cut lots of stainless steel with a plasma cutter like this one. Notice that this weldor is properly clothed and protected to avoid burns. Photo: Lincoln Electric Co.

manner similar to a TIG welding torch.

Because of the modest amount of skill required, plasma-arc is widely accepted for welding and cutting of ferrous and non-ferrous metals. The plasma arc's straight, narrow, column-like shape and high-current density mean that it is not critical to maintain

a certain nozzle-to-workpiece distance to obtain weld or cut consistency. Also, greater nozzle-to-workpiece distance possible with the plasma techniques mean better visibility of the workpiece for controlling the puddle or cut. However, the best cuts occur when the nozzle is 1/8" to 1/4" from the work, and less torch

123

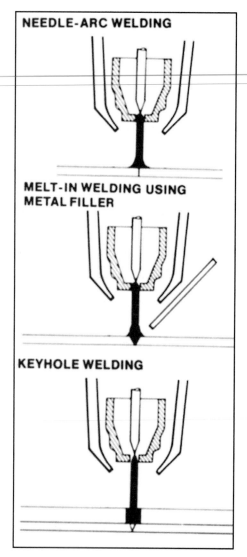

NEEDLE-ARC WELDING

MELT-IN WELDING USING METAL FILLER

KEYHOLE WELDING

Plasma-arc welding processes, from top to bottom: needle-arc welding, melt-in welding using filler metal, and keyhole welding. Drawings courtesy Linde Welding Products.

contamination occurs when the torch does not touch the work.

PLASMA-ARC WELDING

In plasma-arc welding, the plasma produces a much longer, hotter and easier-to-handle arc than does a TIG welder. At low currents—under 100 amps—so-called needle-arc welding can be done. This long, needle-shaped arc is used to join very thin metal— 0.001" to 0.125" thick. Results are

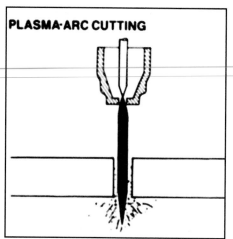

PLASMA-ARC CUTTING

Plasma-arc cutting requires only high-energy gas column for cutting. Shielding gas is not required. Constricted-gas jet produces narrow, smooth kerf. Drawing courtesy Linde Welding Products.

comparable to those of other mechanized fusion-welding processes that require sophisticated controls to maintain precise torch-to-workpiece distance.

Higher currents can also be used in plasma-arc welding. Although a wider arc is generated, high-quality welds can be made on workpieces up to 1" (25mm) thick using currents up to 400 amps.

Inert Atmosphere

Plasma-arc welding works best on metals that are sensitive to atmospheric contamination such as titanium and stainless steel. A vacuum-welding chamber setup is recommended. The vacuum chamber is like a skylight dome on the roof of a house, except that it has ports for welding gloves, like a sandblasting cabinet.

Two Modes

There are two modes of penetration in plasma-arc welding: melt-in and keyhole.

Melt-In utilizes the plasma arc for

Circle-in-a-circle cut by numerically controlled plasma cutter.

conventional, manual and mechanized fusion welding. The major advantages of TIG welding are better operator control of torch-to-work distance and the elimination of tungsten electrode contamination, because the electrode is protected inside the nozzle. High-quality, narrow butt welds or lap welds on joints up to 1/8" thick can be accomplished. Filler metal can be used.

Keyhole plasma-arc welding gives a long, narrow arc that completely penetrates the workpiece to form a keyhole at the center of the weld puddle. If a close-fitting butt-type weld seam is used, filler metal is not required. As the torch travels forward, molten metal forms at the leading side of the arc, flows around the arc and rises to form a small weld bead behind it. A complete weld on both the top and bottom surfaces is formed in one pass to give a 100% weld. The complete penetration of the workpiece thickness and movement of the molten metal purges impurities and gases from the weld prior to solidification. This gives the highest possible weld quality. Keyhole welding can be done with metals up to 1/4" thick.

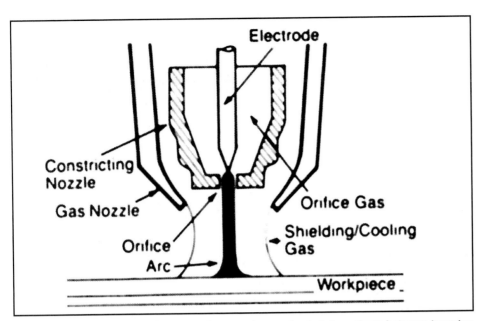

Unlike TIG welder, arc from plasma torch occurs in superheated gas column between tungsten tip and workpiece. Like TIG welding, shielding and cooling is done by flow of gas exiting torch between cup and constricting nozzle. Drawing courtesy Linde Welding Products.

Plasma Prof 150 with a pantograph cutting table makes cutting things such as stainless steel exhaust flanges for production header fabrication a snap to do. Photo: Daytona MIG.

Optical tracking plasma cutter will follow any line drawing to cut exact patterns in all metals.

Equipment

The equipment used for plasma welding is similar to that for TIG welding. At first glance, the torch looks the same. But concealed in the plasma-welding torch is a non-consumable tungsten electrode and a ceramic shielding-gas nozzle.

A power supply initiates the arc and contains the gas supply and water-cooling system. Capacity ratings range from 10 to 400 amps at 60% to 100% duty cycle. Direct current, straight polarity is usually used. Reverse polarity—positive electrode—is used with a water-cooled copper electrode. Although filler rod is usually added the same as in TIG welding, mechanized wire-feed systems can also be used for industrial applications.

The main difference between TIG and plasma-welding equipment is the gas supply. Although the same gas is used, two gas supplies are required with plasma welding; one for the orifice gas–plasma gas, and one for the shielding gas.

Which gas should be used depends largely on the type of welding technique—melt-in or keyhole—and the type of metal to be welded. For example, argon is used for welding steel–carbon, low-alloy or stainless and aluminum with keyhole or melt-in techniques. A 75% helium/25% argon mixture is used if the material is over 1.8" thick when using the melt-in technique. For reactive metals such as titanium, use argon if the material is less than 1/4" thick. In other applications, use an argon/helium

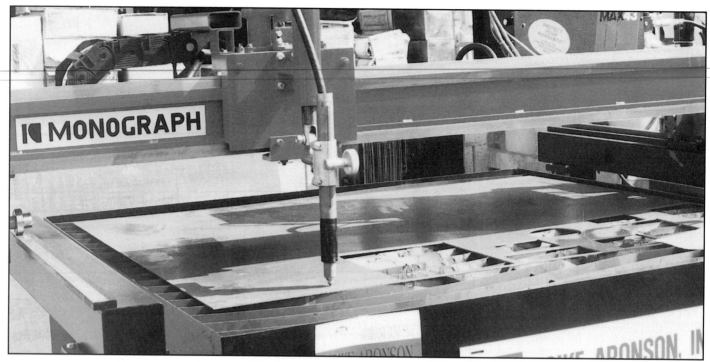

Numerically controlled pantograph plasma cutter is great for prototype work.

mixture; 50%-75% helium for keyhole and 75% helium for melt-in.

Using Plasma Arc

Because the electrode doesn't extend from the nozzle, you can't touch the electrode to the work to start an arc, as is possible with a TIG-welding torch. So an arc must be initiated at the torch, either mechanically or electrically. The mechanical method involves extending the electrode until it touches the nozzle. The electrical method is done with a high-frequency, AC power supply such as that used with TIG welding, or with high voltage superimposed on the welding current. Whichever method is used, the orifice gas is ionized, causing it to conduct the pilot arc current, initiating the torch-to-work arc. Once started, plasma-arc welding technique is virtually the same as in TIG welding.

PLASMA-ARC CUTTING

Plasma-arc cutting uses a highly constricted, high-velocity arc that penetrates the metal similar to keyhole plasma-arc welding. However, up to 50,000 volts is used to melt the metal. Either compressed air from a shop air compressor or a blended, inert shielding gas is used to blow the molten metal out of the kerf. Because it uses a narrow, straight, column-like arc, there is minimal kerf width. And, because of the clean cut, the cut surfaces do not generally require cleanup. Metal up to 6" thick can be cut with a plasma-arc setup, depending on the type of metal and the arc current used.

Plasma arc can be set up to cut with nitrogen as its shielding gas. Although shielding gas is preferred for clean, no-oxidation cuts, shop air is also used because of its low cost. Because high-pressure, high-velocity shop air

is extremely noisy, it can be distracting. The sound is similar to listening to a compressed-air blow gun at close range.

All front-mounted controls are very convenient, as well as the portability of this small, light plasma cutter. Photo: H.T.P.

Daytona MIG sells this 110-volt operated plasma cutting machine that weighs 55 pounds, will cut 100 inches per minute in any metal and up to 1/8" thick. Photo: Daytona MIG.

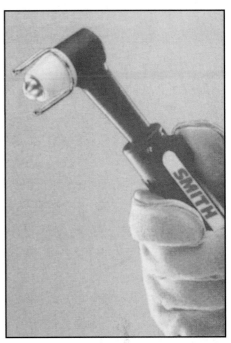

Plasma torch should be operated while wearing heavy leather gloves and face protection because of 50,000° F temperature generated by the plasma process. Photo: Smith's Equipment Co.

The plasma-arc cutting torch is excellent for autobody shop work because it will cut through paint, undercoating, body putty and dirty metals. No precleaning is necessary. Because it doesn't use the oxidizing process to cut metal, plasma arc is ideal for cutting high-strength, low-alloy steel used in new unibody cars. This non-oxidizing feature also makes it a natural for cutting stainless steel; and non-ferrous metals such as aluminum, copper and brass.

The big advantage of plasma-arc cutting is speed—up to 50 times faster than oxyacetylene. Available now are portable units complete with gas, torch and power supply, ready to plug into 220-volt, 60-cycle, single-phase power, and there are units that plug into 110-volt, 60-cycle power.

110-VOLT PLASMA TORCHES

These extremely lightweight (60-lb) machines are very handy and useful in small shop operations, as well as for portable units to take with you for field repairs. There is one minor drawback. Because of simplicity and cost, the smaller units do not have high frequency arc initiation. The 110-volt torch must be pressed against the metal that is to be cut, then lifted slightly as the cut/arc is started. On very light-gauge metal, this can be a problem, but a start block could be used, say to cut .016" aluminum sheet.

Torch Guide—To prolong the life of the PAC consumable tips, the torch should be held about 1/4" to 3/8" off the cut kerf. A guide wire or stand-off is recommended for all PAC torch operations. The guide for 110-volt PAC torches is spring-loaded to allow for contact start cuts.

The torch cut guide can be removed for irregular cuts. Check the photos of torches in this chapter for how to install the guides. ■

Special Welding Processes 14

O ther special welding processes used by hobbyists, craftsmen and production shops are: steel spot welding, aluminum spot welding, metal spraying, jewelry brazing and welding, metal shaping, abrasive cutting and—believe it or not—epoxy welding. Yes, there are times when welding heat is not appropriate, or when disassembly of the cracked block or oil pan would take too much time, and in certain selected cases, epoxy repairs are the answer. But first, we will discuss spot welding.

SPOT WELDING

Used strictly for joining one sheet-metal panel to another, spot welding is one of the oldest production-welding methods. The primary function of a spot welder is to make many welds with little effort and in a short time. It is also clean and requires no filler. Spot welding has been used extensively since the introduction of unibody—unitized body/frame—cars. It is also the standard welding procedure used by most sheet-metal fabricating industries.

Steel flange spot-welding. Note the previous spot welds on either side of the spot welder tongs. Photo: H.T.P.

This is one welding technique that can be understood and done without first learning how to gas weld. All that's necessary to produce a spot-weldable part is to have two clean pieces of sheet metal that will lay flat against each other at the weld joint. You cannot spot-weld a butt joint or T-joint because these joints lack

sufficient surface area to clamp the two together. Therefore, all spot welds use *lap joints*.

In many cases, a lap joint is made by forming what is termed a weld flange. For example, if a panel butts into the side of another panel, as is the case of T-joint, a flange about 1/2" wide is formed at the butting edge of

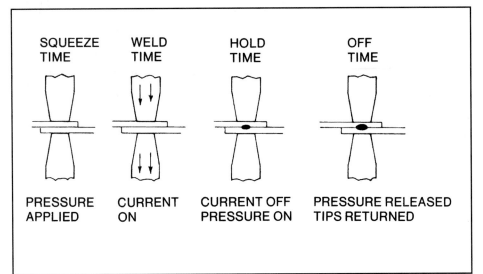

Four primary cycles of a spot weld are: squeeze, weld, hold and release. Pressure and current must be precisely controlled to assure good welds. Drawing courtesy The James F. Lincoln Arc Welding Foundation.

the panel to provide an overlap—lap joint—with the other panel. This overlapping flange, or weld flange, can then be spot-welded. You see such flanges every time you look under a car. It's the flange that runs lengthwise of the rocker panel, joining it to the floorpan and inner rocker panel.

Types of Spot Welders

Spot welding, more appropriately called resistance welding, uses pressure and electrical resistance through two metal pieces as the main ingredients of the weld. Heat is produced by high amperage routed through two mating workpieces clamped between two electrical contact points. No filler, flux or shielding is used. Because the pieces must be clamped together, access to both sides of the joint normally is required. Heating is local, usually a 1/4"-diameter spot.

Spot welding can be done to almost all steel and aluminum alloys, but equipment cost varies widely. Whereas a spot welder capable of

welding steel costs less than a buzz-box arc welder, a spot welder for welding aluminum usually costs more than $50,000!

Although spot welding is simple by itself, complete spot-welding setups can get extremely complicated. The degree of complexity depends on how the welding unit is mounted. For

instance, robot-operated spot welders are in wide use, particularly in the auto industry. However, for the small fabrication of a piece in a home shop, a pedestal-mounted spot welder requires that you take the work to the machine.

Regardless of the mounting, most spot welders operate on single-phase AC power. A step-down transformer converts the power-line voltage to about 250 volts.

Single-Phase Resistance—The standard, sheet metal shop spot welder is a single-phase resistance-type machine. With this type, AC current is passed to two copper electrodes that clamp two thin pieces of metal together. The resulting short circuit causes the metal to heat to the melting point where the two electrodes meet. This melts and forces the panels together, locally, or at a spot.

Three Phase Rectifier—This welder consists of a three-phase step-down transformer with diodes

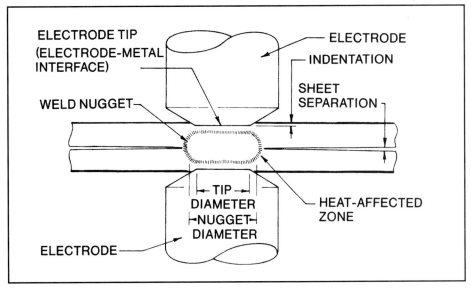

In spot-welding, heat is produced by electrical resistance between copper electrodes. Pressure is simultaneously applied to electrode tips to force metal together to complete fusing process. Spot-weld nugget size is directly related to tip size. Drawing courtesy The James F. Lincoln Arc Welding Foundation.

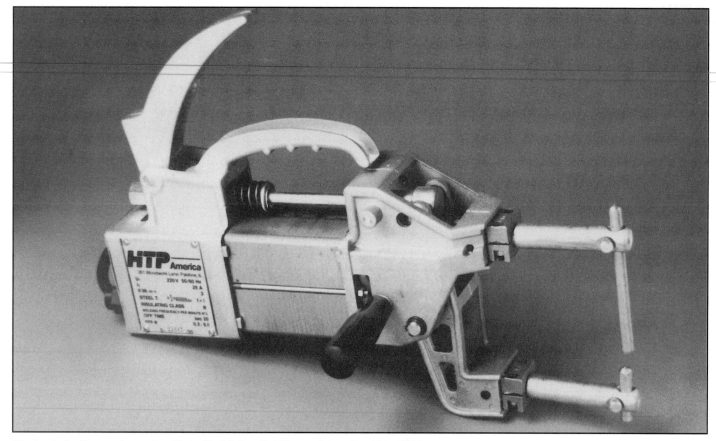

Steel auto bodies can be spot-welded with this handheld heavy-duty unit sold by H.T.P.
Photo: H.T.P.

connected to the secondary circuit. These water-cooled, silicone diodes are connected in parallel. Current is also passed through two clamping electrodes.

Capacitor Discharge—With the capacitor-discharge welder, which also uses clamping electrodes, a bank of capacitors is charged from a three-phase rectifier and then discharged into an inductive transformer. It can be used to spot-weld dissimilar metals or delicate electronic parts.

Single Electrode—This is a simple unit that looks like a pistol and relies on a portable arc welder for its power source.

Unlike other spot-welder designs, the sheet-metal panels are not clamped together. This design features make it suitable for "skinning" sheet-

metal panels on automobile bodies where access to the inner panel is blocked. The primary appeal of this light-duty spot welder is for the auto hobbyist.

Aluminum Spot Welder—If you try welding aluminum with a spot welder meant for welding steel, it will only burn the metal. And it won't fuse the pieces together. Aluminum spot welders are sophisticated devices programmed to slowly apply pressure to the spot weld while gradually increasing voltage and current. They develop up to 76,000 amps to spot-weld 1/8" thick aluminum.

Other Spot Welders—There are several other types of resistance welders, but most are high-production factory units, unsuitable for small workshops. For instance, projection

welding utilizes small dimples, or projections, in the sheet metal. These projections arc as they touch another panel to complete the welding circuit, flash-butt welding the two together. A high electrical resistance at the projections is created, causing the panels to fuse together.

Seam Welding—Another type of resistance weld is called *seam welding*. Two pieces of sheet metal are drawn between two rollers while pressure and electrical current are applied to the knurled rollers. This action provides for a continuous resistance weld which is leak-free. A common application is in the manufacture of automotive gas tanks. The top and bottom halves are continuously welded at a flanged seam.

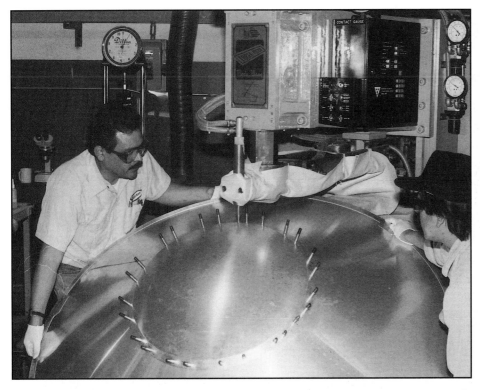

Spot welding an aluminum bulkhead for large jet airliner fuselage. Clecos hold the center in place while spot welds are made, then rivets fill the Cleco holes.

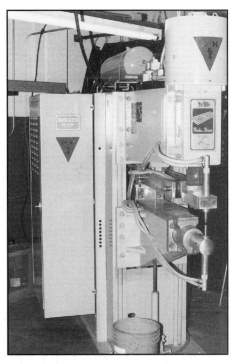

Aluminum spot welders costs thousands of times more than the steel spot welder, and are much more complex.

Spot Welders Are Complex

Most spot-welder designs are complex even though using them is relatively easy. Spot-welder settings for current applied, on-and-off cycle timing, electrode pressure, and the shape and condition of the electrodes all affect weld quality. Once the proper setup is achieved, the operator can make numerous welds of equal quality.

Aluminum Spot Welding

When spot-welding low-carbon steel, a good weld can be made with a generalized setup using a wide range of current adjustments and an even wider range of clamp, current-applied, current-off, pressure-held, and pressure-released cycles or steps. However, aluminum is much harder

to spot-weld. It requires a precise clamping cycle and slow application of an initially low current, building up to a spike of high current, then tapering off to low current again. The clamping force is initially high to prevent arcing of this high-resistance metal, then backed off to prevent thinning of the aluminum, and finally increased as the weld cools. These steps must be carefully done in an accurate timing mechanism. And, as you may recall from the TIG and MIG chapters, aluminum is highly susceptible to surface oxidation that adversely affects weld quality. Consequently, aluminum must be absolutely clean to spot-weld.

Also, the spot-welder contacts must be dressed—filed—often to keep them clean and to maintain electrode shape and size. Tip size is important. For example, when 31,960-psi pressure is applied to a 1/4" diameter

surface, reduction of this contact-patch size to 3/16"-diameter will reduce the applied pressure to 17,978 psi—a 44% reduction! This will ruin the weld.

Because of the critical factors and complex machinery involved, aluminum spot welders can cost more than an entire welding shop equipped with MIG, TIG, arc and gas welders!

Not to worry, though. There are other ways to join sheet aluminum. For instance, at Aerostar Airplanes I used an aluminum spot-welding machine to weld two-piece landing gear doors. When the spot welder broke down, we substituted rivets for the spot welds—rivets and spot welds are similar in strength. Although spot welding is usually faster and much less prone to working, or moving, of the two sheets of metal joined, solid rivets will work nearly as well.

METALS THAT CAN BE SPOT-WELDED

	Aluminum	Stainless Steel	Brass	Copper	Galvanized Steel	Steel	Monel	Tin	Zinc
Aluminum	X						X	X	
Stainless		X	X	X	X	X	X	X	
Brass		X	X	X	X	X	X	X	X
Copper		X	X	X	X	X	X	X	X
Galv Steel		X	X	X	X	X	X	X	
Steel		X	X	X	X	X	X	X	
Monel		X	X	X	X	X	X	X	
Tin	X	X	X	X	X	X	X	X	
Zinc	X		X	X					X

X indicates combinations that can be spot-welded.

Weldability of Metals by Spot Welding

Almost any metal that can be fusion-welded can be spot-welded. However, oil, grease, dirt or paint on either or both of the parts to be spot-welded can have a negative effect on weldability. For this reason, it is imperative that you clean the sheet metal thoroughly prior to attempting spot welding.

Galvanized steel is a natural for spot welding. Galvanized coating was once used extensively in heating and air-conditioning ducting. It is now used in automobile bodies for rust prevention. Galvanized steel spot-welds easily with no special preparation to the metal. However, you should use a good breathing respirator because it generates poisonous zinc-based gases.

Stainless steel can be spot-welded, but it should be silver-brazed or TIG-welded instead.

Nickel-plated steel, such as that used in car-body trim, can be spot-welded. Unfortunately, heat will discolor the plating.

Refer to the weldability chart above. It can be helpful for determining which metals can be spot-welded.

Metal Coatings—Special weld-through sealers are used in the auto industry for rust protection. They prevent water from entering blind areas in the body through weld seams. The sealer is applied to the facing side of a weld-seam half, then the panels are put in place and the seam is spot-welded right through the sealer. This makes a watertight spot-weld seam.

How to Use a Basic Spot Welder

Most simple spot welders have four controls:

•On-off power switch.

•Foot pedal or hand lever to cause the two electrodes to come together and make the spot weld.

•Electrode contact-pressure adjustment, usually air pressure.

•Amperage and timer adjustment. As metal thickness increases, so does arm-contact time and amperage.

Before making final spot welds, cut out some test scraps of the same material, about 2" square. Hold them together with locking pliers while spot-welding.

Start off with low pressure and amperage. Metals that have high electrical resistance, such as aluminum, require less overall current, but more exact control of the cycle. Steel and other metals with low resistance require greater amounts of current. Keep raising the pressure and amperage until the spot is about 3/16" to 1/4" diameter.

For instance, a spot welder with 1/4" electrodes should be set to the following values to assure good welds in steel sheet: 9800 amps, 32,000-psi ram pressure. Clamp time and weld dwell time are programmed into the welder. Make few spot welds using the suggestions just given and adjust

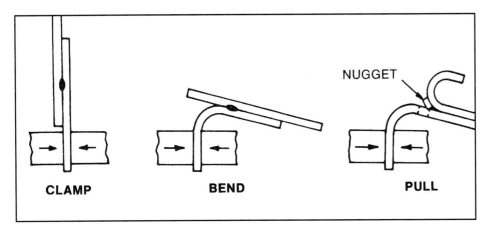

Test spot weld by clamping one piece in vise and bending two pieces apart. If spot weld pulls apart, penetration is poor. If spot weld pulls a nugget–spot-welded metal–from one piece, penetration should be good. Drawing courtesy The James F. Lincoln Arc Welding Foundation.

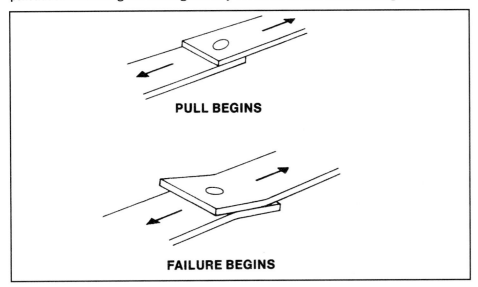

You can also test spot weld by pulling apart spot-welded strips in test rig. Drawing courtesy The James F. Lincoln Arc Welding Foundation.

This pull test machine tests spot-weld coupons to assure the spot welds are structurally as strong as the design engineer specifies.

the welder as required for good welds.

The distance between each weld is pitch. For most applications, weld pitch should be about 1" to 1 1/2". Judge the spacing as you move along a weld seam, making one weld after another. After making a few spot welds, perform the following tests.

Spot Weld Testing

The ultimate method of checking spot-weld quality is destructive testing whereby the two spot-welded pieces are pulled apart. However, you can tell a lot about a spot weld by its appearance.

Good spot welds are essentially round. The depression matches the size of the spot-welder electrode-contact points, with only a slight amount of heat-affected metal around the spot. The amount of indentation in the weld should be slight.

Bad spot welds look much different. Severe discoloration around a spot weld indicates overheating. Either the current was too high or the dwell time—duration—of the current was too long.

Little or no contact spot indicates a poorly fused spot weld. The current was too low or current dwell time too short. Spitting and sparking around the spot weld indicate one or more problems. Either the metal was dirty with paint, oil or grease, or the current was much too high for the metal thickness. It is also possible that clamping pressure was insufficient.

As with anything, it is easier to make corrections to spot-welding machine settings when you know what caused the problem. Be sure to analyze the first two or three spot welds before continuing to make welds.

Destructive Testing—The most common way to check the operation of a spot welder is to perform a pull test. Spot-weld two metal test strips together. Start by visually inspecting the spot weld; then tear the two strips

133

Cast-iron repairs can be easily made by metal spraying, as is demonstrated here.

This weldor is metal-spraying a chisel point with hard face material. Flame spraying can also be used to build up worn shafts without melting the shaft.

apart. Look at the torn-apart spot weld to determine whether it is good or bad. The amount of force it takes to tear the strips apart is the best indicator. If you have to really wreck the metal to tear the weld apart, the weld is probably good.

To tear the strips apart, clamp one strip in a vise. With pliers or vise grips, peel the other strip away from the first strip. After you have torn the strips apart, look at the weld area. Usually, there will be a hole in one piece of metal and a weld nugget—fused metal spot—stuck to the other piece. Pulling a nugget indicates the weld was stronger than the base metal.

Check the size of the nugget. It should be almost the size of the face diameter of the electrode. Check the shape of the nugget. It should have some of the metal from the other test strip attached to it. Look for brittle, crystalline-looking metal in the nugget. Brittleness is usually caused by excess heat, not enough pressure, or too much heat time. Look at the nearby drawings for how to do a spot-weld pull test.

Using a Single Spot Welder

If you are restoring an antique car, or building a new car and don't want to use rivets, a hand-held spot welder such as the Braze'N'Spot Welder from The Eastwood Company, Malvern, PA, may give you the desired results. The advantage of this type of spot welder is that you can weld panels where access to the back side is restricted. Eastwood's spot welder operates off a 50-amp buzz-box/transformer and plugs into 110-volt house current.

To use it, first ground the welder to the work. Next, push the welding gun against the outer panel, forcing it against the inner. Do this while you retract the carbon electrode by pulling the gun trigger. To make the spot weld, apply current and heat to the outer panel by releasing the electrode trigger. A spring will force the electrode against the outer panel, allowing amperage to heat the metal to a molten puddle. This should fuse the outer panel to the inner.

When the metal glows red to white hot, retract the electrode, but continue holding pressure on the gun and panel

for 5 or 6 seconds until the hot spot cools. If everything was clean and fitted properly, you should have a good spot weld.

METAL SPRAYING

This process has been around for at least 75 years, but very little is commonly known about it. The first two uses of the metal-spraying processes were to apply hard-facing alloy steel layers to saw blades, mining equipment and to lathe tools, and the second use for metal spraying was to build up worn pulleys, oil seal surfaces and shafts. Almost any metal powder can be sprayed with the oxyacetylene heated torch.

More recent uses for metal spray are to fill cracks in cast-iron cylinder heads and cracks in cast-iron exhaust manifolds. The metal-spray equipment is easy to set up and to use. The photos in this chapter show two different setups and how they are used.

JEWELRY MAKING

Look at the chart on page 3 that

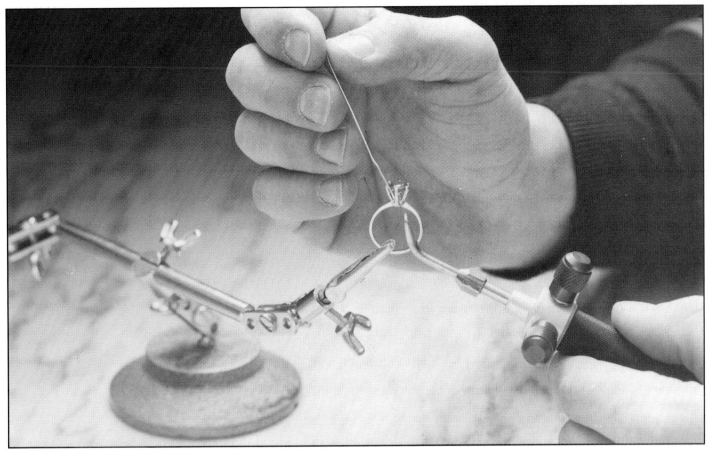

Oxyacetylene flame is used here to make gold and silver jewelry. But the torch is very small, jewelry making size. Photo: Smith's Equipment Co.

gives the melting points of metals. You can see that gold melts at 1,946° F (1,063C), and silver melts at 1,761° F (960C). That is a higher melting point than the temperature that aluminum melts at—1,217° F (658C)—so it shows that oxyacetylene will be a good heat source for welding rings, bracelets and other jewelry.

But don't use a full-sized gas welder to weld your wedding rings together! The amount of heat would quickly turn them both into a shapeless blob of gold! You need a tiny, little flame to accurately control the very small amount of heat necessary to do a super-delicate job like that. Conversely, the small jeweler's oxyacetylene torch could never weld the tubular framework on a race car or

even a bicycle. Just remember the example of tack hammers and 10-pound sledge hammers, related to railroad spikes and picture nails. It takes the right tool to do each job. Also, remember the lessons learned in chapter 1 about heat control.

Self-teaching yourself to do jewelry repair and jewelry-making could be the start of a whole new career or a retirement income. But practice first before you try the wedding and engagement ring fusion.

COLD REPAIRS

Several years ago a motorcycle rider came into the welding shop with a pencil-diameter-sized hole punched in the transmission case of his expensive new motorcycle. He wanted me to

weld it "at low temperature," he said. After I explained to him that his aluminum transmission case melted at 1,300° F, and that a welding temperature of at least that heat was required to fusion-weld the hole in the case, he asked what could be done to avoid taking the transmission apart for welding.

I suggested that he buy a package of epoxy-and-steel powder and repair the leak with J.B. Weld. He did as I suggested, and the "cold weld" is still holding. Try steel-based epoxy for similar metal repairs. Clear epoxy does not have the heat resistance or the flexibility that metal base epoxy does. You could save yourself a lot of work with the "cold weld" method. Try it sometime. ■

Welding Rods, Wires & Fluxes

15

Welding rod comes in grades of quality and strength, just like nuts and bolts do. Even though a rod is marketed and sold as 4130 rod, there are at least 6 grades of what the vendors call 4130.

You can buy at least 6 different grades of 1/4" fine-thread bolts, ranging from hardware store, no markings, no source, to aircraft-quality 1/4" fine-thread bolts with markings and even certification papers if you need them. The hardware store 1/4" bolt might get the job done if you are using it to repair lawn furniture, but you would not want to use hardware store bolts to hold the tail on your aerobatic airplane.

GRADES OF ROD

Let's look at the grades of welding rod, and then you be the judge of which ones you should stock in your welding shop. We will look at bare steel 36"-long rod as an example.

#1. Commercial grade, copper-coated mild steel. No brand, no source, probably made from remelt scrap steel, no known chemical content. Good for use as shop tie wire, nonstructural. Okay for use in welding rusty garden equipment and

5-pound boxes of arc welding rods and brazing rods are available from local welding supply dealers. Know what rod is best for your welder and your project before you go shopping. Photo: SoCal Airgas.

rusty tail pipes. Cost: $1.50 per pound.

#2. Commercial grade, copper-coated mild steel. Sold by a brand-name supplier. Probably the same metal as #1. Brittle welds, cost: $2.00 per pound.

#3. Commercial grade, copper-coated 4130 steel. Brand name, but made from remelt, scrap steel, not certified.

Cost: $5.00 per pound.

#4. Commercial grade, bare (no copper) 4130 steel. Brand name, certification papers say that the metal has 4130 material, specific carbon amounts. Cost: $10.00 per pound, minimum of 10- to 20-lb boxes.

#5. MC grade, bare 4130 rod. Brand name, certification papers specify new material, not scrap. Cost is $20.00 per

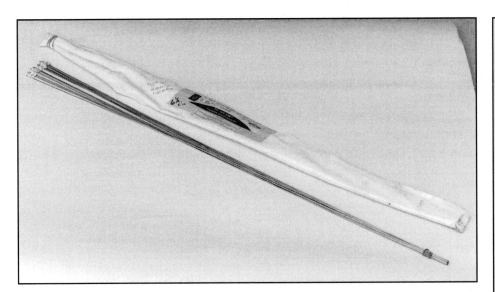

Most packages containing high-quality steel welding wire warn: DO NOT TOUCH WIRE WITH BARE HANDS OR UNCLEAN GLOVES. Welding rod must be kept clean and free of rust, so always keep it boxed to keep it free from contamination and dust. Photo courtesy United States Welding Corp.

CAUTION!

Copper fumes generated in welding are hazardous to your health. If you are welding with copper-coated welding rod or wire, be sure to evacuate the welding fumes with air suction equipment to prevent breathing copper fumes. Copper fumes accumulate in the body, which means that they can add up over a period of time. To be safe, just don't breathe copper fumes, or—better yet—don't use copper-coated welding products at all if you can help it.

pound in 10- to 20-lb boxes.

#6. MC grade, vacuum melt, rolled, not die formed, certified 4130 and certified clean, no imbedded impurities, no copper, no oil, no rust, sealed containers, with desiccant dryers. Sells for $40.00 per pound in 6-lb minimum orders.

COPPER-COATED WELDING ROD

Weldors use strips of copper to back-up their welds in their sheet metal because copper will insulate the weld and because it won't stick to the steel, stainless steel or aluminum that is being welded.

Copper strike strips are used to scratch-start TIG torches because copper will not wick up (flow) to the tungsten electrode.

Copper is never fusion-welded to steel because copper and steel won't mix.

Copper is used to coat welding wire to make it easier to draw the wire to 1/8", 1/16" or whatever diameter is desired. Copper is applied to the rod or wire to lubricate the dies in the drawing or sizing equipment.

Copper is not applied to base welding rod to prevent rust. Copper-coated welding rod will rust if not protected from moisture in sealed containers. Even MIG wire in rolls should be kept in sealed containers to prevent moisture and dust contamination. Copper-coated MIG wire will flake off and sometimes clog the rollers, guides and cables and prevent proper operation of the MIG welders.

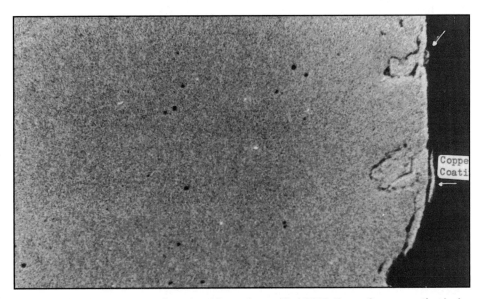

Cross section of copper-coated steel welding rod magnified 3000 times. Copper coating looks foreign to rest of metal. It can do the same thing to your weld. Although it's OK to use for welding exhaust systems and metal furniture, don't use copper-coated rod to gas-weld expensive parts. Photo courtesy United States Welding Corporation.

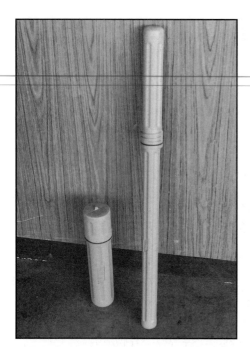

Plastic containers like the two shown here will protect bare welding or brazing rod (the long container) and flux-coated arc welding rod (the short container) from dust and moisture. Ask your welding supply dealer for these containers.

Copper-coated welding rod often causes cracks and bubbles in 4130 steel welds. Since copper does not fuse with steel, it flows into the grain of 4130 steel like brass does, and wedges the steel apart, causing cracks, and because copper-coated welding rod is usually made from inferior steel, it contributes contamination to the weld bead. If it is worth welding, take a little more time to do the job right, and use better-quality welding rod.

METALLURGICAL ADVANCES

In recent years, welding rod and consumable metallurgy has advanced almost as quickly as the electronics industry has advanced. High-quality filler metals are available today that did not even exist 20 years ago. When you are selecting a stock of welding rods, fluxes and wires, take a little extra time to search for the good stuff. In one specific example, I can spend $90 for the 4130 chrome-moly tubing to build an aircraft engine mount. The new 200-horsepower aircraft engine sells for $30,000. The primer and paint to paint the completed mount will cost $40. I need 1/2 pound of welding rod to weld the tubes together. I can either pay $1 for cheap rod or $25 for vacuum-melted, metallurgically controlled, certified welding rod that incidentally does a much better job. It makes more sense to use the $50-per-pound certified welding rod.

Similarly, there are solders that allow all metals to be joined, but these new high-tech solders cost 10 to 20 times as much per pound as the old 90-year-old technology solders do. But they get the job done, whereas the 90-year-old stuff doesn't.

AWS (American Welding Society) classifications for welding generally indicate the correct rod for the material and welding process you are looking for. Consult the AWS charts whenever possible.

CHOOSING THE RIGHT ROD

Choosing the correct filler rod for welding a specific material is as important as knowing how to weld. You can buy scores of different types of welding rods at your local welding-supply outlet. There are hundreds of different types. So it's obvious that a weldor needs some help in choosing the correct welding rod.

First, find out what kind(s) of metal you are welding. Then, decide which welding process is best for that metal. In many cases, you have several choices.

For instance, you may want to weld or braze a broken office chair. If the break occurred at other than a previously welded joint, you'd have at least four choices of how to fix it: gas-weld with oxyacetylene, arc-weld with small-diameter welding rod, TIG-weld it, or MIG-weld it.

In the case of repairing a break at an old weld, you'd first have to determine how it was welded, and then probably reweld it using the same method. Why? Some welds are not compatible with others. To determine the previous metal-joining method, make a visual inspection with a magnifying glass. An arc weld usually has some slag and spatter; a gas weld has flaking. TIG and wire-feed weld beads are clean. Brass braze is a gold or bronze color.

The easiest way to weld the broken chair, assuming it's a new break, would be to gas-weld it with either bare steel wire or copper-coated steel wire—if you have access to a gas welder.

Any of the other three choices would work.

To choose the correct welding rod, first determine the type of welding equipment available, then pick the correct welding rod. Welding rod choices are grouped according to the specific welding equipment.

In my home workshop, I have about ten different kinds of gas-welding and brazing rod, six kinds of arc-welding rod, and four or five kinds of solder. A commercial welding shop would have a similar assortment of filler materials. For TIG welding, you want at least ten different kinds and sizes of welding rod for welding common steel and aluminum alloys of various

Moisture is a very detrimental thing to flux-coated stick welding rod. This Phoenix Dry Rod portable rod oven can be plugged into 110-volt power to keep your stick electrode dry and warm, right at the job.

thicknesses.

In addition to being used as a filler material, welding rod has numerous other uses. For instance, you can use an arc welder and welding rod to build up a worn surface, such as a crankshaft journal. Or, welding rod can be used to add a hardened surface.

Lead Soldering w/Gas Welder

50/50 Wire Solder—This typically comes in spools and is used with brush-on flux to solder copper tubing with an oxyacetylene torch. Melting temperature is extremely low: 250-400° F.

Wire solder is available in several different alloys: 10/980, 15/85, 20/80, 40/60 and 63/37. In the recommended example, 50/50 indicates 50% tin and 50% lead composition by weight. High lead-content solder such as 10/90, 15/85 and 20/80 are used for sealing brass auto radiators and filling seams in steel automobile bodies. These solders are used primarily with acid-base fluxes.

Inorganic Fluxes—These are the strongest, yet most corrosive. Because inorganic fluxes are corrosive, they should be used only where it's easy to remove them after soldering. They are usually made of salts such as ammonium chloride or zinc chloride dissolved in water. Use inorganic fluxes only when soldering copper or steel buckets, or small pieces that are easy to dip or wash clean.

Medium-Strength Organic Fluxes—These are used for soldering copper tubing and brass, as well as steel. Usually they are called acid fluxes because they are made from glutamic or stearic acid. Otay brand flux is called No. 5 soldering paste, and is for cleaning and fluxing all metals except aluminum, magnesium and stainless steel.

Rosin Flux Solders—These are the weakest type and must not be used for flame soldering. The base for rosin flux comes from pine-tree resin. It activates when heated and deactivates as it cools. Use rosin flux only for soldering electrical and electronic components.

Aluminum Soldering w/Gas Welder

Welco 1509—This is a low-temperature (500° F/260C) solder for joining aluminum, zinc, die-cast metal, copper, brass, stainless steel

and other metals to each other, or to themselves. It requires a flux for cleaning and soldering. I use Welco 380 flux. This solder has a tensile strength—stress at which it breaks while under tension—of 29,000 psi. It comes in 1-lb wire spools and is available in 1/16", 3/32" and 1/8" sizes. Welco 1509 can be used on aluminum car radiators, A/C evaporators and condensers, and other similar metals.

This is a low-temperature self-fluxing solder alloy for aluminum window frames, zinc-based carburetors and outboard-motor housings. It comes in 1/8" diameter rods and melts at 700° F (371C).

Silver Brazing w/Gas Welder

All State No. 101 and 101FC Trucote Braze—These are general-purpose silver-brazing alloy rods. They have a high tensile strength of 52,00 psi with a working temperature of 1145° F (618C). Either rod may be used to join almost any metal with a melting temperature above 1150° F (621C). Number 101 or 101FC Trucote braze rod is especially good for soldering copper, brass, steel, stainless steel or aluminum. These rods may be used with a separate flux or can be bought with a blue-colored flux coating. They come in diameters of 1/16", 3/32" and 1/8".

Welco 200 Braze—This is a high-strength, 56% silver alloy for ferrous and non-ferrous metals. Its bonding temperature is 1155° F (624C); tensile strength is 85,000 psi! By contrast, the tensile strength of chrome-moly steel is only 70,000 psi! This silver brazing rod can be used for race car suspensions and airplane wing ribs. It requires liquid flux.

Large fabrication shops and welding schools keep the arc welding rod dry by keeping it in large ovens like those pictured here. Photo: Allan Hancock College.

Welco 200 Flux—Ideal for brazing, it comes in a 12-ounce plastic jar in paste form. It can be removed with warm water after brazing.

Brass, Bronze & Aluminum Brazing w/Gas Welder

All State Nickel/Bronze Braze Rod—This rod has a high tensile strength of 85,000 psi. It melts at 1200-1750° F (649-954C), so it's a little harder to work with than lower-temperature brazing rods. Regardless, braze rod does a good job when used properly, for assembling bicycle frames and ornamental railings. It should not be used on chrome-moly steel because it penetrates the grain of the base metal and cracks it.

All State No. 41FC Braze—This is a high-quality rod with a tensile strength of 60,000 psi. It is flux-coated, eliminating the inconvenience of using separate flu. No. 41FC brazing rod is widely used to make bicycle frames, race car frames, and for repairing auto bodies. If exposed to the atmosphere over a period of time, the flux coating will flake off and the rod becomes unusable. Therefore, I buy flux-covered rods for jobs I expect to complete within a few days.

All State No. 31 Aluminum Braze—This is primarily meant for use with thin sheet aluminum such as fuel tanks, oil tanks and truck, race car and trailer bodies. It can also be used to repair aluminum irrigation pipe. No. 31 braze rod should not be used on 2024- or 7075-series aluminum. It comes in 18" lengths and 1/16", 3/32" and 1/8" diameters. It has a high melting temperature of 1075° F (579C). A good flux such as All State No. 31 should be used.

All State No. 33 Aluminum Braze—This is for brazing aluminum castings. This filler will repair cracked or broken castings and will fill holes or build up areas that have been worn away or broken off. Use All State No. 31 flux with this rod.

Welco 10 Aluminum Braze—This rod is a 30,000-psi tensile strength alloy used on aluminum sheet. Use Welco No. 10 flux.

Anti-Borax Flux—This flux comes in 1-lb cans and is used for brass brazing of brass, bronze, steel and cast iron. It can be used in paste form by mixing with water. Or, it can be used in its dry powder form by simply dipping the brazing rod into the can after heating the rod with the torch. I like to use the heat-the-rod-and-dip method for best results.

Gas Welding Aluminum or Steel

Welco 120 Welding Rod—This welding rod is a high-quality alloy for fabrication and repair of most weldable aluminum alloys. It is easy to control, and the bead solidifies rapidly, producing a nice-looking weld. Use oxyhydrogen if you plan to do much aluminum welding. Use Welco No. 10 flux when welding with this rod.

Never use coat hanger wire for gas welding. It is always coated with paint or varnish, and it is made of the cheapest steel, sources unknown. Welds made with coat hanger wire usually crack.

No. 1100 Aluminum Welding Rod—No. 1100 is the most common rod to use for all-purpose gas welding with a separate flux. Use Oxweld Aluminum flux No. 725FOO. This flux comes in 1/4-lb jars, and the label states "for all aluminum welding."

All State Sealcore—This s a unique tubular aluminum welding rod with the flux contained inside its hollow core. This welding rod is claimed to be the most versatile of all aluminum welding rod for torch welding. It comes in 1/8" and 3/16" diameter rods, 20" and 32" long. It is supposed to be good for field work such as repair of irrigation pipe and farm equipment.

Coat Hangers—For many years, weldors have used coat hangers to oxyacetylene-weld car fenders or anything else they could think of. Don't do it! Paint on coat-hanger wire contaminates the weld. And the alloy in the wire is unknown. Usually, coat hangers are so brittle that they break when you try to straighten them. I suspect they are made of the cheapest steel available. You wouldn't want a weld to crack because of poor-quality

filler material.

Welco W-1060 Mild-Steel Rod—This gas-welding rod is copper-coated, but it is a good choice for gas-welding mild steel. It is available in 1/16", 5/64" and 1/8" diameters, 36" long. My last invoice from an Airco dealer called this #7 Steel. It works well for automobile exhaust systems and, if used carefully, OK for gas-welding non-structural 4130 or 4340 aviation steel. But the copper coating contaminates the weld if the rod is not used carefully. See the above photo of a scanning electron micrograph of the copper coating.

Airco or Oxweld Bare Mild-Steel Rod—I use this rod for torch welding, gas welding or oxyacetylene welding 4130 steel. And it's what the Experimental Aircraft Association recommends in their welding schools for welding airplane parts. It comes in 1/16", 5/64" and 1/8" diameter rods, 36" long.

Arc Welding

The charts on pages 142 indicate how to identify arc-welding rod. Welding-rod diameter refers to the wire size, not the diameter of the coating. Wire diameter is measured easily at the holder-end of the rod.

E-6011—This is the easiest to use all-purpose welding rod for arc welding mild steel with a 220-volt AC buzz-box arc welder in all positions. This is a good welding rod when the work is dirty or oily, and you don't have time to make the job pretty. The E-6011 rod produces a considerable amount of unsightly spatter—small globules of molten metal that stick to the base metal—in the area of the weld. I recommend E-6011 arc welding rod for making repairs on farm equipment.

To weld with E-6011 rod, hold a 1/8" or shorter arc. Move at a steady pace that is just fast enough to stay ahead of the molten slag. For welding overhead or vertical-up, reduce the current setting by one notch on the buzz-box.

E-6011 comes in 1/16", 5/64", 1/8" and larger sizes.

E-6013—This is an excellent general-purpose welding rod for use with 220-volt AC buzz-box welders when both easy operation and outstanding weld appearance are important. This rod can be used in all positions. The work must be cleaner than when welding with E-6011 rod. I recommend this 60,000 psi tensile-strength rod for projects such as building trailers.

To weld mild steel with E-6013 welding rod, drag the tip of the rod lightly against the work. Do not hold a gap as you would with E-6011 rod. Some companies call this a contact rod because you always keep it in contact with the base metal. Move steadily and just fast enough to stay ahead of the molten puddle. When welding sheet metal with E-6013, weld downhill. E-6013 welding rod comes in 1/16", 5/64", 1/8" and larger diameters.

E-7014—This welding rod is also designed to be used with 220-volt AC buzz-box welders. Compared to E-6011 and E-6013 rods, E-7014 is a slightly higher-strength welding rod with good appearance characteristics. The first two numbers in the identification indicate 70,000 psi tensile strength. This rod is commonly used for sheet-metal welding.

You should lightly drag the tip of the rod on the base metal when laying a bead. Therefore, you don't have to maintain a gap with the arc. This is a

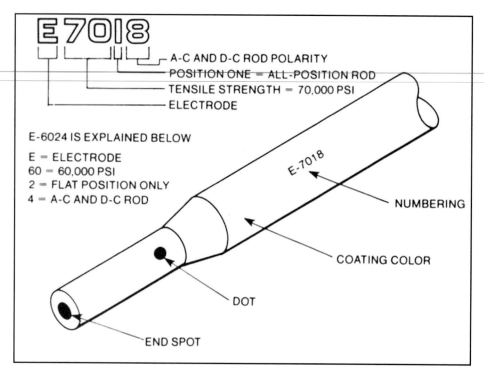

E 7018
- A-C AND D-C ROD POLARITY
- POSITION ONE = ALL-POSITION ROD
- TENSILE STRENGTH = 70,000 PSI
- ELECTRODE

E-6024 IS EXPLAINED BELOW

E = ELECTRODE
60 = 60,000 PSI
2 = FLAT POSITION ONLY
4 = A-C AND D-C ROD

E-7018 — NUMBERING — COATING COLOR — DOT — END SPOT

Color of spot or dot on stick electrode is for quick identification. End spot can be seen when electrodes are in container.

good welding rod for beginners, but so is E-6011. E-7014 comes in 3/32", 1/8" and larger diameters.

E-6010—Sometimes called 5P by oil-field pipe weldors, it is an all-purpose, deep-penetrating welding rod for use with 220/440-volt DC welders. It works similar to all purpose AC/DC E-6011 rod, but leaves a much smoother weld and produces almost no spatter. It will work on dirty, oily or rusty pipe and other steel. E-6010 welding rod should not be used with an AC welding machine. Set the machine for DC, positive or reverse polarity when welding with this rod. It comes in 1/8" diameter and larger sizes.

E-7018 LO-HI—Sometimes called low hydrogen, this rod should be used with arc welders. This welding rod was developed originally for 70,000 psi tensile-strength, X-ray quality welds in the nuclear power industry. Regardless, the cost of E-7018 rod is

similar to other welding rod, so it is one of the most commonly used welding rods. It produces high-quality, good-looking welds suitable for pipe welding that must be certified. E-7018 rod also works well for trailer frames, race car frames and mild steel.

The work must be thoroughly cleaned and prepared. If you are sloppy, it is easy to bury slag pockets with E-7018 rod. If the metal is clean and properly prepared, the slag will actually peel off as the weld cools. This rod is highly susceptible to moisture damage, so it must be kept dry and clean at all times. Once you get the hang of using E-7108 rod, you'll like it.

EST & Lincoln Ferroweld—Excellent steel welding rods for making high-strength welds in cast iron when no machining is required afterward. These welding rods are used with 220-volt AC welders or

220-volt DC welders set to reverse—positive—polarity.

Maintain a short arc as you would with an E-6011 rod—don't touch the work with the rod tip. It's best to preheat the entire part to 400° F (204C) prior to welding to minimize stresses. Cast iron should be welded in short, 1"-long beads and allowed to cool in between. This reduces heat buildup and the likelihood of cracking.

E-308-16 Stainless Steel Rod—This rod may be used with a 220-volt welder for certain stainless steel welds where weld appearance is not critical, such as pipe welding. It leaves a finish similar to E-6011 rod—a lot of spatter. Use AC current or DC current set to positive polarity.

Other specialty welding rods designed for use in electric arc-welding machines can be found at your local welding shop:

•Aluminum
•Bronze
•Hard surfacing—a very hard coating of steel applied over mild steel surfaces where abrasive wear occurs, such as bulldozer blades, plow blade and steam-shovel scoops.

TIG Welding

4043—This is an uncoated aluminum welding rod for TIG welding aluminum. I use this welding-rod alloy for most TIG welding jobs on aluminum. It is available in several diameters: 0.020", 0.040", 1/16", 3/32", 1/8" and 5/32". You can weld the following aluminum alloy with 4043 welding rod: 1100, 5052, 6061, and 356 (casting).

Welco W-1200, AWS A5.2-69, Class RG60—This is the most common TIG welding rod for welding

Commercial-grade TIG wire for 4130 steel magnified 1300 times is full of voids and surface imperfections. Compare to next micrograph that shows metallurgically controlled wire; obviously a better product. Photo courtesy United States Welding Corporation.

aircraft structural components and rocket motor cases.

These welding wires/rods are rolled to size, not drawn through dies, and the rolling process produces very smooth wire with no imbedded impurities. Drawing wire through dies usually imbeds lubricants from the dies, which causes defects in the weld. See the 1600 power micrograph pictures for comparisons.

308 Stainless Steel Rod—This rod is what's needed for TIG welding most stainless steel alloys. There are more than 80 different stainless steel alloys; 90% of them are welded with 308 rod! Contact your dealer for special cases, or if you have any questions about which rod to use.

Stainless rod is non-magnetic, of course, and is identified with a small white tag taped to each rod with the identification number on it. This 308 stainless steel rod is used for welding such things as race car and airplane

steel. If the box of welding rod has a heat number—a reference number from the manufacturer's quality-control department—the particular batch of rod has been checked for quality. If a question comes up about the rod for one reason or another, the manufacturer can trace its history through this number. A heat number usually means the welding rod is for low-carbon, high-strength steel such as 4130.

Don't let anybody sell you copper-coated rod for TIG welding. The copper coating can cause blowholes in a weld and generates fumes that can be hazardous to your lungs.

Vacuum Melt 4130—When you are welding 4130 steel parts that will later be heated-treated for additional strength, the very best rod to use is United States Welding Corporation 4130-6457V. This rod is certified to content, source of elements, and is the highest-grade welding rod available.

Other welding rods available from

United States Welding Corporation are D6AC, vacuum melted, controlled chemistry, primarily used with gas tungsten arc process for critical

Vacuum melt, metallurgically pure, roller-formed (not die-formed) wire, magnified 1300 times, is much smoother than commercial grade 4130 in previous photo. Photo courtesy United States Welding Corporation.

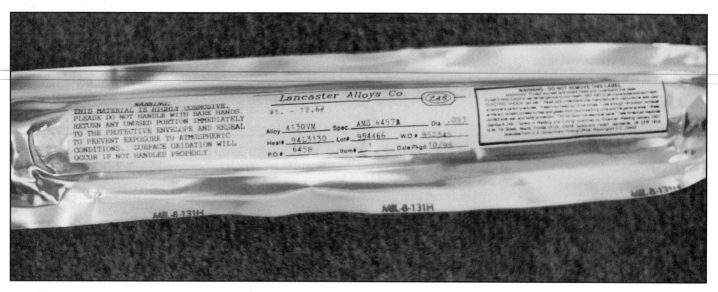

5- and 10-pound quantities of vacuum melt, MC grade bare welding rod comes packed in hermetically-sealed plastic bags like this 4130 VM rod from Lancaster Alloys Co.

exhaust systems, kitchen cookware and missile parts.

316 stainless steel rod is used for TIG welding where salt water corrosion would be a problem, as in ocean-going yachts and power boats. It is also a good rod for welding race car exhaust systems.

SFA 5.16 ERTI-2 Titanium Rod—This rod is for TIG welding one titanium alloy. The rod must exactly match the alloy. Another titanium alloy, Ti-6A1-4V—commonly referred to as 6-4, is the one most commonly used because it's considered the "4130" of titanium. To weld 6-4 titanium, a 6-4 titanium rod must be used.

Titanium rods come in the same sizes as stainless rods. They are also identified with paper tags and usually must be special-ordered from most welding supply shops.

Magnesium Rod—This rod is not carried by most welding supply shops. Because of its limited use, it's a special-order item. Use the same rod as the magnesium alloy being welded, such as AZ92A, AZ101Z or AZ61A

rod. The best rod for TIG welding wrought magnesium is AZ61A rod.

MIG Welding

Electrodes used for MIG welding are smaller than those used in other types of welding. This is because of the high current and speed at which the filler metal is introduced into the weld puddle. For instance, wire sizes start as low as 0.020" diameter and increase in steps of about 0.005". Average size is 0.045" with a normal maximum of 0.090". A maximum diameter of 0.125" has been used in heavy industrial applications. Note: A smaller electrode will achieve more penetration at the same amperage.

E4043, E4146 & #5183 Aluminum Wire Alloys—These are available for MIG welding aluminum. E4043 is most common for shop-welding projects, and for building aluminum trailers for heavy-duty, freight-hauling trucks.

E70S-1, -2, -3, -4, -5 & -6 Steel Alloy Wire—This wire is for welding mild steel. Prefix E indicates electrode. The second digit (7) refers

to tensile strength in 10,000 psi, or 70,000 psi. The third digit (0) refers to position—horizontal in this instance. S means solid electrode, versus hollow core. A T instead of an S indicates hollow-core electrode. Dash numbers indicate the chemical composition of the wire as specified by the American Welding Society.

All numbers indicate varying percentages of carbon (C) and silicon (Si) except for -2. The E70S-2 wire also contains titanium (Ti), zirconium (Zr) and aluminum (Al). The higher the dash number, the higher the silicon content. ■

Welding Projects

16

This chapter should be the most rewarding chapter in the whole book. You spent all that money on welding equipment and the partner in the house wants to know what you plan to do with it. Now is the time when you can prove that the time-and-money investment was worth it. The projects in this chapter are entirely within your skill level if you have studied and practiced, practiced, practiced the techniques in this book.

I start the project off with the welding and cutting table. It is the single most useful thing I have ever built for my shop, except for my 8-foot workbench. I have no doubt that all my go-karts, race cars and airplane projects would have been a lot easier and faster to build if I had built this table for my shop way back in my teenage years. So, build this table first, and it will help all your other projects along. As you can see in the construction and assembly pictures, you can build this table out on the driveway or on the shop floor. Then, when it is finished, you can cut, fit, bend and weld all your future projects on the welding and cutting table.

In each of the projects, I list the welding and cutting processes

Welding supply dealers or steel supply dealers can sell you as much or as little angle, strap, U-channel or round stock as you need. If you're going to be serious about this, then you need to develop a good relatiohship with your local dealer. Courtesy Western Welding Inc., Goleta, CA.

recommended for fabricating the project. For instance, in Project #1, the welding and cutting table, I recommend the cutting torch and arc welding, although MIG or TIG welding could also be used. And remember, use good-quality welding rod. The projects deserve good-quality fabrication materials.

PROJECT #1
WELDING & CUTTING TABLE

It will probably take 40 to 50 hours for a beginner to build this welding and cutting table, but it is worth the effort.

Welding Process—Cutting torch and arc welding.

Materials—Look in the phone book yellow pages under "Steel."

Cut the Parts to Fit—This is a more advanced project, so do your own trimming to make the lower frame and secondary parts. Cut the legs to length. Make the top frame of 1-1/4" x 1-1/4" angle. Make it square. Weld on the 3/8"-thick tabletop. Trim all other parts to fit. Leave the sides open or fill them in with 16-gauge (0.060") sheet metal. You can make doors and a catch tray for sparks from the cutting torch. Tack-weld the 1" x 1/4" straps in place so you can replace them easily when they get "ratty" from cutting-torch flame. Paint everything but the tabletop.

(continued next page)

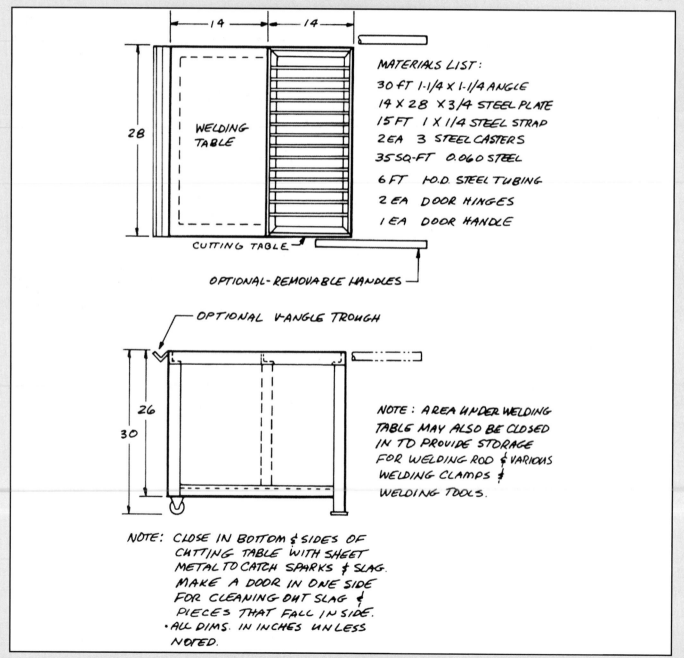

MATERIALS LIST:

30 FT 1-1/4 X 1-1/4 ANGLE
14 X 28 X 3/4 STEEL PLATE
15 FT 1 X 1/4 STEEL STRAP
2 EA 3 STEEL CASTERS
35 SQ-FT 0.060 STEEL
6 FT 1-O.D. STEEL TUBING
2 EA DOOR HINGES
1 EA DOOR HANDLE

WELDING TABLE

CUTTING TABLE

OPTIONAL-REMOVABLE HANDLES

OPTIONAL V-ANGLE TROUGH

NOTE: AREA UNDER WELDING TABLE MAY ALSO BE CLOSED IN TO PROVIDE STORAGE FOR WELDING ROD & VARIOUS WELDING CLAMPS & WELDING TOOLS.

NOTE: CLOSE IN BOTTOM & SIDES OF CUTTING TABLE WITH SHEET METAL TO CATCH SPARKS & SLAG. MAKE A DOOR IN ONE SIDE FOR CLEANING OUT SLAG & PIECES THAT FALL INSIDE.
• ALL DIMS. IN INCHES UNLESS NOTED.

(Welding & Cutting Table, continued)

The welding table project started with a trip to Western Welding, Inc., where they cut the metal to correct lengths for me, for about 1% of the total cost of material.

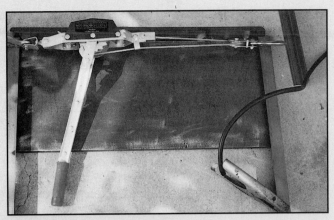

A fence stretcher come-along holds the table sides in place while I tack-weld them. This project is laid out on concrete.

Welding/cutting table legs, sides and top were arc-welded with a 225-amp buzz-box welder, using E-6010 arc-welding rod 1/8" diameter.

1/8" holes were drilled in framework and the aluminum sides were Cleco'ed in place.

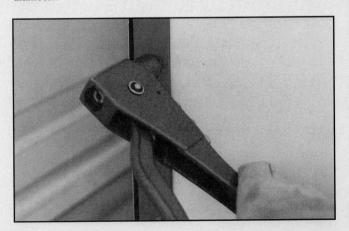

When all the 1/8" rivet holes were drilled and deburred, 1/8" steel pop rivets were installed to hold the aluminum sides on the table. Aluminum cannot be welded to steel, so I used rivets to fasten the aluminum sides, bottom and front to the steel framework.

This front quarter view of the completed table shows the class A B C dry chemical fire extinguisher mounted where it is easy to reach, and it shows the 4" bench vise mounted on the cutting table side. A $5.00 used lawn chair makes welding a comfortable job.

PROJECT #2 MECHANICAL FINGER

You can build one of these weldor's mechanical fingers in about one hour. It will be a handy tool as long as you weld, braze or solder metal.

Welding Process—Gas welding, brazing, TIG or wire-feed welding. Arc welding is OK, too.

Materials—Look in the phone book yellow pages under "Steel." Get 2-foot lengths of 1/4", 3/8" and 1/2" round rod.

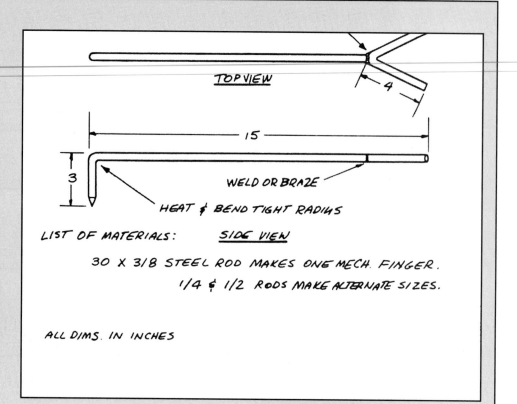

TOP VIEW

WELD OR BRAZE

HEAT & BEND TIGHT RADIUS

LIST OF MATERIALS: SIDE VIEW

30 X 3/8 STEEL ROD MAKES ONE MECH. FINGER.

1/4 & 1/2 RODS MAKE ALTERNATE SIZES.

ALL DIMS. IN INCHES

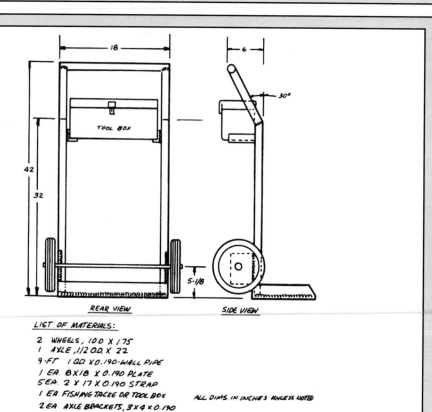

TOOL BOX

REAR VIEW   SIDE VIEW

LIST OF MATERIALS:

2 WHEELS, 10.0 X 1.75
1 AXLE, 1/2 O.D. X 22
9-FT 1 O.D. X 0.190-WALL PIPE
1 EA. 8 X 18 X 0.190 PLATE
5 EA. 2 X 17 X 0.190 STRAP
1 EA. FISHING TACKLE OR TOOL BOX
2 EA. AXLE BRACKETS, 3 X 4 X 0.190
1 EA. CHAIN OR STRAP TO HOLD BOTTLES
8 EA. 28 LONG X 1.00 TUBING FOR ROD HOLDERS
ASSORTED SCRAP METAL ROD & STRAP FOR MAKING
BRACKETS & TABS.

ALL DIMS. IN INCHES UNLESS NOTED

PROJECT #3 GAS-WELDING CART

Once I built my first gas welding cart, my friends all wanted plans so they could build their own carts.

Welding Process—Oxyacetylene welding, brazing and cutting with a cutting torch. Arc, TIG and wire-feed welding are also acceptable.

Materials—Purchase the materials at a business that sells steel. Look in the phone book yellow pages. Buy the wheels from a hardware or building-supply store.

Cut the Parts to Fit—With a cutting torch, cut the sheet metal to fit; cut the tubing and rod with a hacksaw. Heat the two long side tubes with a welding tip and bend to a 30° angle to form handles.

(continued next page)

(Gas Cart, continued)

Tack-Weld—Start with the flat base and tack-weld the handles to the base. Keep everything square. Now, tack-weld all other parts into place.

Weld—You can gas-weld most of the cart. But, if you can't get a good puddle because of too-little flame and too-thick metal, braze or arc-weld it. Add hold-down straps or chain as shown in the photograph. Add some accessory hooks as desired. Paint to suit.

Completed gas-welding cart has been put to good use.

PROJECT #4
SANDER & GRINDER STAND

Build this grinder stand in one to two hours, then count the many hours of use it will see afterwards.

Welding Process—Weld with an arc welder and cut with a cutting torch.

Materials—Buy the materials at a business that sells steel. You may also be able to pick up some of the materials at a scrap-metal yard.

Cut Parts to Fit—With a cutting torch, cut the pipe and plates to size.

Weld—Arc-weld the plates to the ends of the pipe. Paint to suit.

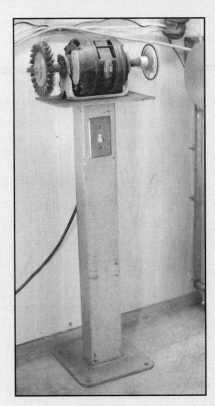

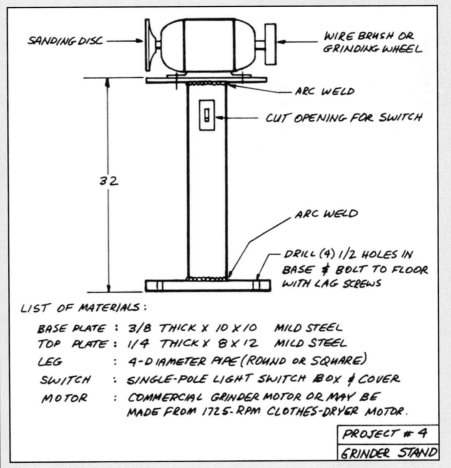

SANDING DISC

WIRE BRUSH OR GRINDING WHEEL

ARC WELD

CUT OPENING FOR SWITCH

32

ARC WELD

DRILL (4) 1/2 HOLES IN BASE & BOLT TO FLOOR WITH LAG SCREWS

LIST OF MATERIALS:

BASE PLATE : 3/8 THICK X 10 X 10   MILD STEEL
TOP PLATE : 1/4 THICK X 8 X 12   MILD STEEL
LEG   : 4-DIAMETER PIPE (ROUND OR SQUARE)
SWITCH   : SINGLE-POLE LIGHT SWITCH BOX & COVER
MOTOR   : COMMERCIAL GRINDER MOTOR OR MAY BE MADE FROM 1725-RPM CLOTHES-DRYER MOTOR.

PROJECT # 4
GRINDER STAND

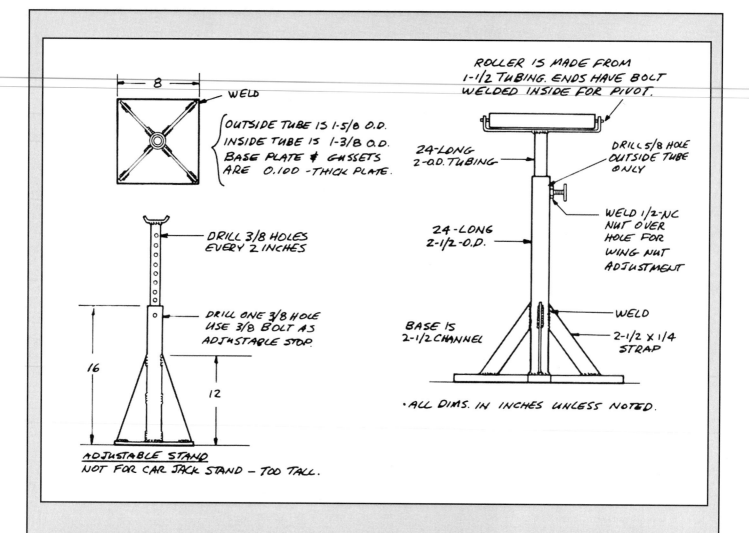

ROLLER IS MADE FROM
1-1/2 TUBING. ENDS HAVE BOLT
WELDED INSIDE FOR PIVOT.

WELD

OUTSIDE TUBE IS 1-5/8 O.D.
INSIDE TUBE IS 1-3/8 O.D.
BASE PLATE & GUSSETS
ARE 0.100-THICK PLATE.

24-LONG
2-O.D. TUBING

DRILL 5/8 HOLE
OUTSIDE TUBE
ONLY

DRILL 3/8 HOLES
EVERY 2 INCHES

24-LONG
2-1/2-O.D.

WELD 1/2-NC
NUT OVER
HOLE FOR
WING NUT
ADJUSTMENT

DRILL ONE 3/8 HOLE
USE 3/8 BOLT AS
ADJUSTABLE STOP.

BASE IS
2-1/2 CHANNEL

WELD

2-1/2 X 1/4
STRAP

16

12

ADJUSTABLE STAND
NOT FOR CAR JACK STAND — TOO TALL.

• ALL DIMS. IN INCHES UNLESS NOTED.

PROJECT #5
JACK STANDS & WORK STANDS

These shop-made work stands will come in handy when you start building the trailers shown in this chapter.

Welding Process—Arc welding, cutting torch.

Materials—1-1/2" to 2" pipe for center shaft. Metal plate is 0.190" to 0.375." thick, depending on which of the pictured stands you make.

Plan Before Cutting—When you decide which work stand to build, make a simple drawing like the ones in this chapter. don't just start cutting and welding. Other people may want you to make work stands for them once they see yours.

Sandblast & Paint—This prevents corrosion and improves the project's appearance.

NOTE: All the projects listed in this chapter will benefit from pre-cleaning before welding. Sandblast or glass-bead clean the raw steel. Power sand the joints to be welded before welding and then sandblast or wire brush and power sand the welds after welding. Make it real pretty!

PROJECT #6 TOW BAR FOR CARS & JEEPS

Welding Process—Cutting torch and arc, TIG or wire-feed welding.

Materials—Buy only good-quality new materials. Two tow-bar capacities are described. The light tow bar will safely tow cars up to 3500 lbs total weight; heavy bar up to 6,000 lbs. My light tow bar has pulled my Corvairs across the U.S. and a Scirocco race car up and down the West Coast. If the tow bar is bolted snugly to the towed vehicle and careful towing habits are observed, cars will track well at all legal speeds.

Cutting and Fitting—A tow bar must be a perfect triangle to tow a car straight. Consequently, it's best to make a jig. Use a 1-7/8" hitch ball as the center point to line up the hitch and lay out a triangle on a piece of 1/2" plywood. Use a 1/2" steel bar as the base of the triangle. Tack-weld the tow bar in the jig, then remove it for final welding.

Attaching to Tow Vehicle—On my Scirocco race car, I removed the front bumper guards, exposing a bolt hole under each guard. In each hole, I put a 3/8" Grade-5 bolt. I also drilled two more 3/8" holes on the bottom flange of the bumper for two more bolts—and the tow-bar angle. These four bolts and the two 1/2" hinge bolts were all double-nutted and checked every 100-200 miles of towing. The heavy-duty tow bar uses larger bolts, of course.

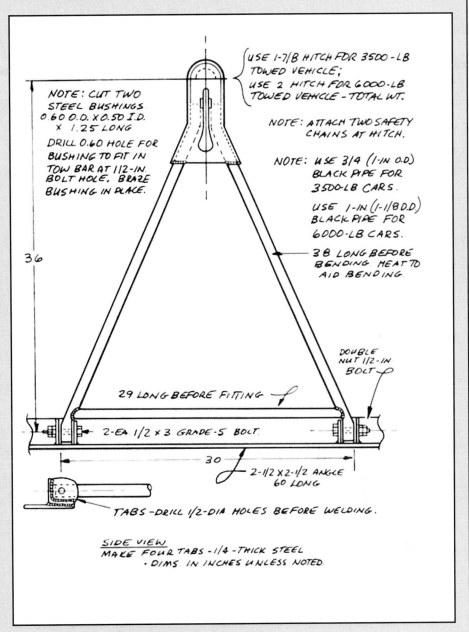

NOTE: CUT TWO STEEL BUSHINGS. 0.60 O.D. X 0.50 I.D. X 1.25 LONG

DRILL 0.60 HOLE FOR BUSHING TO FIT IN TOW BAR AT 1/2-IN. BOLT HOLE. BRAZE BUSHING IN PLACE.

USE 1-7/8 HITCH FOR 3500-LB TOWED VEHICLE; USE 2 HITCH FOR 6000-LB TOWED VEHICLE - TOTAL WT.

NOTE: ATTACH TWO SAFETY CHAINS AT HITCH.

NOTE: USE 3/4 (1-IN. O.D.) BLACK PIPE FOR 3500-LB CARS.

USE 1-IN (1-1/8 O.D.) BLACK PIPE FOR 6000-LB CARS.

3 8 LONG BEFORE BENDING. HEAT TO AID BENDING.

36

DOUBLE NUT 1/2-IN BOLT

29 LONG BEFORE FITTING

2-EA. 1/2 X 3 GRADE-5 BOLT.

30

2-1/2 X 2-1/2 ANGLE 60 LONG

TABS-DRILL 1/2-DIA HOLES BEFORE WELDING.

SIDE VIEW
MAKE FOUR TABS -1/4 -THICK STEEL
· DIMS. IN INCHES UNLESS NOTED.

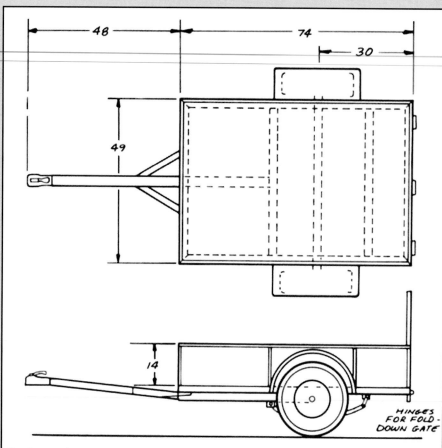

MATERIALS:
FRAME: 2-1/2 MILD-STEEL ANGLE
SIDE RAILS 1-IN. MILD STEEL ANGLE
FOLDING TAIL GATE: 1-IN MILD-STEEL ANGLE
          EXPANDED-STEEL MESH
FLOOR: 3/8-THICK EMBOSSED STEEL
WHEELS: 13 IN.
SPRINGS: BOAT-TRAILER LEAF SPRINGS
TONGUE: 2-1/2-SQ TUBE OR
          2-1/2 C-CHANNEL

DIMS. IN INCHES UNLESS NOTED.

PROJECT #7
UTILITY TRAILER

These plans can be slightly modified to include a plywood floor or a full mesh floor and sides.

I purposely chose a Chevy Citation rear-axle assembly as the basis for this project because the Citation wheel bolt pattern is the same size as my Chevy Cavalier convertible that will tow the trailer. That way, my spare tire in the Cavalier convertible will fit the utility trailer if I ever have a flat tire on the trailer. Conversely, the trailer wheels and tires will also fit my car if I need an extra spare for it. It makes sense to standardize. Also, the trailer weighed 300 pounds when it was finished, very light and no problem for towing behind a compact car.

Welding Process—Cutting torch, arc welding on the frame, TIG welding on the axle. Plasma arc cutting for the smaller angle brackets and tail light holders.

Materials—The axle shaft was replaced with 2-1/2" x .125" wall square tubing. The tongue was made from 2" OD schedule 40 pipe. Other materials are given on the plans.

Variations—As shown in the photos, this trailer can be varied to carry 3 motorcycles for a steel bed lawn equipment or snowmobile trailer, or a plywood bed utility trailer. Be sure to add tail and stop lights. Get your trailers registered with your state Department of Motor Vehicles.

(continued next page)

(Utility Trailer, continued)

The trailer, Project #7, was laid out on a concrete driveway and welded with a Lincoln 225 stick welder and E-6011 and E-6010 welding rod.

Trailer frame and tongue is welded and ready for fitting the springs, axle and wheels.

To save weight and make a prettier trailer, I made an axle from 2" x 2" x .125" wall square tubing.

I TIG-welded the axle to wheel spindle adapters, using Cronatron #222T high-strength rod. Note the really pretty welds!

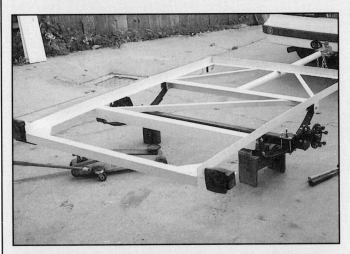

The frame is now painted white with black trim to match my tow cars, a Chrysler LeBaron and a Chevrolet Cavalier. The tail lights are wired and the bed is ready for the 5/8" outdoor plywood floor.

Late-night fabricating of wood fenders to be ready for Department of Motor Vehicles inspection and registration the next day. The trailer weighed 300 pounds at the DMV inspection. Cost with new tires was $300, or $1 per pound.

PROJECT #8
TRAILERS FOR RACE CARS & AIRPLANES

Welding Process—Cutting torch, arc welding.

Materials—For the axle, I use the rear-axle assembly from a front-wheel-drive car. I just pick a wheel I like, and cut and splice or cut and narrow the axle to suit the desired dimensions. Springs can be purchased at a boat-trailer supply house or wheel manufacturer. Never build a trailer without springs. That $100 you might save by not buying springs could result in $1000 worth of damage to whatever you're hauling due to the rough ride. For a light trailer, try the rear axle from a 1980-or-later GM X-Body (Chevy Citation, Buick Skylark, etc.). For a heavier-duty axle, use Cadillac Eldorado, Olds Toronado or 1979-and-later Buick Riviera rear axle.

Dimensions—These plans are for a trailer that will haul a 2000-lb car. Scale up the dimensions for a slightly larger car or down for hauling smaller items such as a home-built airplane. The trailer tongue can be lengthened at least 10 feet for airplane towing. Plans for a tandem-axle car trailer can be ordered from J.C. Whitney.

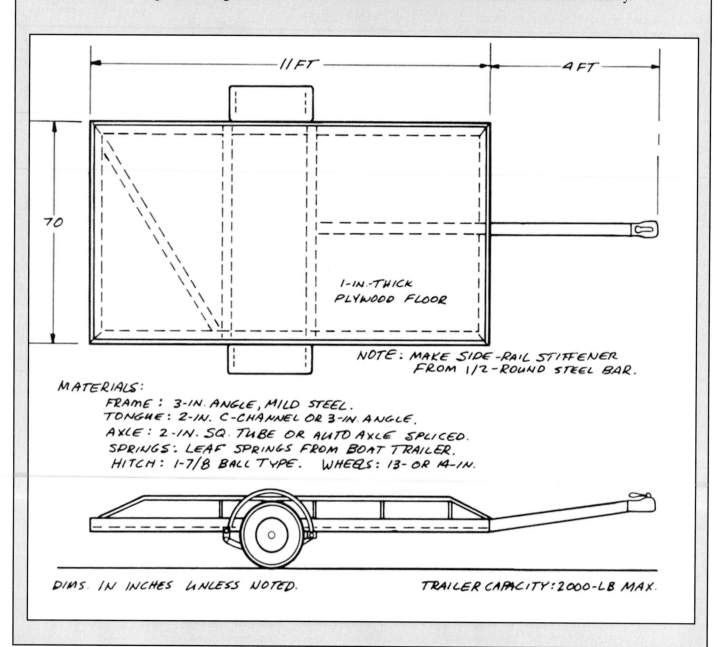

11 FT   4 FT

70

1-IN.-THICK PLYWOOD FLOOR

NOTE: MAKE SIDE-RAIL STIFFENER FROM 1/2-ROUND STEEL BAR.

MATERIALS:
    FRAME: 3-IN. ANGLE, MILD STEEL.
    TONGUE: 2-IN. C-CHANNEL OR 3-IN. ANGLE.
    AXLE: 2-IN. SQ. TUBE OR AUTO AXLE SPLICED.
    SPRINGS: LEAF SPRINGS FROM BOAT TRAILER.
    HITCH: 1-7/8 BALL TYPE.   WHEELS: 13- OR 14-IN.

DIMS. IN INCHES UNLESS NOTED.   TRAILER CAPACITY: 2000-LB MAX.

Welding Certification & Training

17

The reason for including this chapter in *Welder's Handbook* is to give you a taste of what high-quality welding should look like. Even if you are only building utility trailers and metal work benches, why not be an expert craftsman and make your welding projects as "pretty" as you possibly can?

Not long ago I was photographing new products at a racer's trade show, and I saw a booth where several first-class, beautiful work benches were displayed. The work benches were fabricated from steel and aluminum sheet, and every corner and every seam was fitted and welded perfectly. The work benches were well thought out and looked strong enough to last 100 to 200 years or more. Why not? If it is worth building, why not make it strong and make it look good?

A new truck, lockable enclosed tool boxes and a diesel generator-powered welder in the truck bed are features of a professional, certified weldor and his portable welding business.

Study Good Welds

You can learn a lot about welding by looking closely at welds that are obviously very good, aircraft or race car quality. Look and take note of "pretty" welds, then practice, practice, practice until your welds look as good. That is one very good way to improve your welding skills. It is also a good idea to try to pass the following certification tests, even if you don't plan to become a certified weldor. Try to be the best weldor you can be.

CERTIFICATION

Certification, or qualification as it's sometimes called, is very helpful if you plan to make a living as a weldor. However, you can weld professionally without being certified. Certification is something like having a college degree in sales. You can earn a living selling without it, but being certified may help you get a job by proving you are qualified. Regardless of certification, most large businesses and agencies conduct their own certification tests, or they will have an outside testing lab validate your certification tests.

155

Several weldors are busy in this picture, erecting the steel (earthquake) seismic structure for Santa Barbara, California's newest Chevy dealership.

The same Chevy dealership construction project pictured in the previous photo, but several months later. Welding exists in many buildings where it does not show.

Documentation

What do you have to show once you're certified? You'll be given a wallet-sized card listing the metals you are certified to weld and by what process(es). Also indicated will be whether you are a Class A (excellent), Class B (good) or Class C (OK) weldor. Typically, the card and certification expire in 3, 6 or 12 months. You must be retested to stay certified.

You can be certified in different specialized areas: spot-welding contacts on home-heater thermostatic controls, TIG-welding stainless-steel pipe in the horizontal position in nuclear power plants, and so on. Two well-known certifications are for petroleum pipe-welding and aircraft TIG-welding. Certification testing is always conducted under strict supervision.

Pipe Welding Certification

The exam for this certification process is welding a section of 5/16" wall, 6"-OD pipe called a coupon. The ends of two 6"-long sections of pipe are beveled in a lathe and placed end-to-end for tack-welding. Usually, a backing ring is placed at the pipe joint with 1/8" spacers sticking out to obtain the proper root opening— distance between the pipe ends at the weld seam—to ensure that 100% penetration is obtained.

The pipe is placed at the weldor's eye level. This positions the weld seam so all welding positions— horizontal, vertical and overhead— and combinations of each must be used to complete one pass. The weldor starts with a root pass—the first bead. More than one pass is required to fill the weld seam. Except

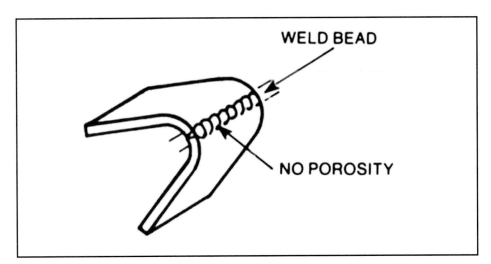

Cracks, tears or porosity in weld are not allowed. Failures due to flux or slag inclusions are most common defects.

for the last one, following passes are appropriately called fill passes—usually one or two are required. The last pass is called the cap, or weld-out pass.

Slag must be removed after each pass to eliminate possible slag inclusions.

After welding is completed, a coupon from the weld seam is removed and scrutinized. Coupons are cut from the pipe with a saw or cutting torch. The weld bead is then ground flat and even with the pipe surface. Each coupon is placed in a test machine and bent backward to a horseshoe shape. Any porosity or cracking of the weld or base metal will fail the weldor.

When certifying for arc welding, 5P welding rod is used for oil-field mild-steel pipe; E-7018 rod for 4130 steel pipe in steam and nuclear power-plant welding. TIG and wire-feed techniques are also used. As mentioned, the employer usually gives the certification test. However, there are also individual testing laboratories. One such laboratory is:

Advanced Testing Laboratories
4345 E. Imperial Hwy.
Lynwood, CA 92063

Aircraft Welding Certification

Several coupons—both pipe and sheet—must be welded successfully before you can become a Class A certified aircraft weldor. Exact procedures may vary from country to country, depending on that government's regulatory laws. But basically, there are four groups of metals to be certified on:

*Group I, 4130 Steel—One cluster, cross-sectioned, polished and then etched for examination.
*Group II, Stainless Steel—Butt-weld 0.032" and 0.063" plates together. Complete fusion is required. Weld is sectioned, ground and bent. Visual inspection follows.
*Group IV, Aluminum—Butt-weld 0.032" and 0.063" plates together. Complete fusion is required. Weld is sectioned, ground and bent. Visual inspection follows.
*Group VI, Titanium—Special request; test production part by sectioning, polishing and 10X magnification inspection.

Group I, 4130 Steel Cluster Test Procedure—Test consists of TIG welding three tubes to a flat plate with specified angles, dimensions and weld seams. The plate is 1/8" thick x 8" x

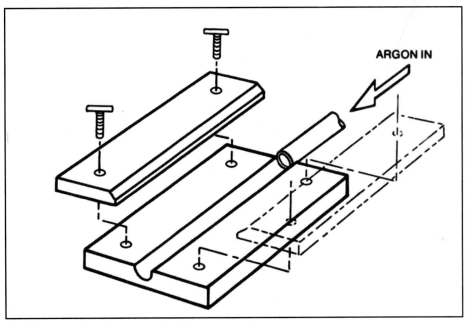

To pass certification on stainless steel, back-gas and clamping fixture must be used.

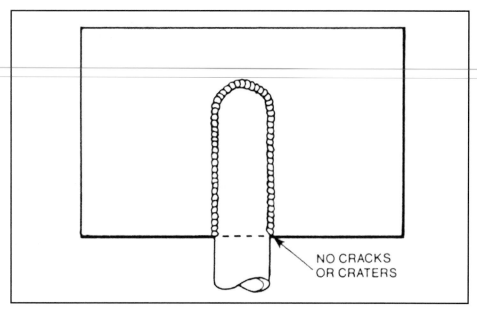

After first tube is welded 100% on both sides, inspect for craters and cracks where tube and plate meet.

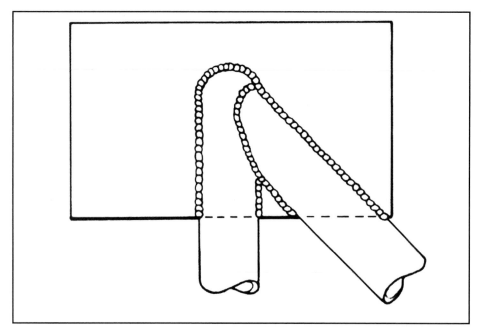

Second and third tubes must be fitted closely. Any gaps could result in failure of the certification test. Weld 100% on both sides.

6" 4130 condition N—normalized, or stress relieved. The three 1" OD x 0.063"-wall tubes are 6" long.

The sequence is to TIG-weld the first tube 100% all around on one side of the plate. After each weld pass, the torch should be held over the weld until argon-gas flow stops—usually 5-6 seconds. It is then allowed to cool in draft-free, still air for 20 minutes. The weld seam must fit tight because fitting is also graded.

After the first weld cools, scale is carefully removed from the other side of the plate and tube. Sandblasting should not be done to remove scale on the back side of the weld because sand would contaminate the tubing. Instead, scale should be removed with a stainless-steel wire brush. Be sure to clean the weld seam with MEK or alcohol prior to fitting and welding. Otherwise, the weld will be contaminated. Now, the back side of the first tube can be welded 100%.

Again, all scale is removed from the coupon and the second tube is fitted. The second tube should be filed so it fits tightly around the weld bead on the first tube. If too much filing is done, causing a gap in the fit, you could fail. The plate and third tube should be cleaned as before. Weld the third tube like the second.

After completing the final weld, the coupon is cleaned and submitted to the lab for inspection. The testing laboratory will not grant certification if cracks, porosity, cratering, incomplete fusion or a poor fit is found.

Aluminum Butt Plates Test Procedure—Minutes prior to actually making the TIG weld, a Scotchbrite abrasive pad should be used to slightly roughen each piece of aluminum about 1/2" and a 1/4" air space should be left under the weld seam. This ensures that 100% penetration is achieved. If the bead is not thicker than the base metal on both sides, you fail the test. Even though I don't use back gas, I use my stainless-steel back-gas clamping fixture to clamp the aluminum.

When the testing lab checks the aluminum weld coupon, they sand the weld bead to the thickness of the base metal, cut the coupon into strips and pull-test them. The aluminum strips must not break in the heat-affected area. Also, craters and porosity are not allowed.

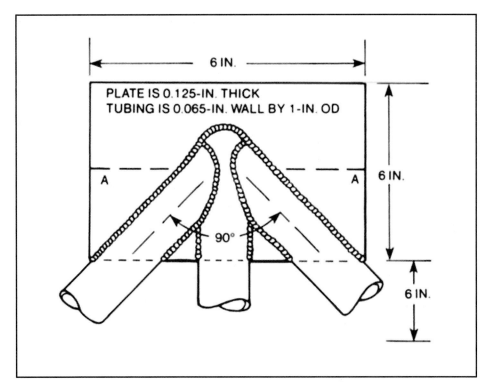

PLATE IS 0.125-IN. THICK
TUBING IS 0.065-IN. WALL BY 1-IN. OD

Penetration on aircraft-weldor's certification test cluster must be at least 15% into 1/8" plate. Weld is not the only thing that's graded. Fitting tubes to specifications is also graded.

Stainless Steel Butt Plates Test Procedure—Stainless steel coupons are always TIG welded with back-gas purging to prevent crystallization on the back side of the bead. Clamping and back-gas purging are done with the fixture pictured above.

Prior to welding, clean the stainless-steel plates with MEK or alcohol. Use clean, high-quality welding rod. To ensure that argon gas covers the weld while solidifying, keep the torch over the bead after completion until the purge-gas stops. The timer on the TIG machine usually provides 5-6 seconds of gas flow after the torch is off.

The testing laboratory sands the stainless steel weld down to base-metal thickness, then cuts it into test strips. The strips are then bend-tested at the weld seam. Without 100% penetration, the bend test will break through at the bead. If so, the weld will fail.

Welding Titanium—This is an on-the-job test that uses an actual production part. The part is then destruction-tested. It is sectioned through, pulled and bent, and subsequently polished and inspected with a 10X magnifying glass.

WELDING FOR A LIVING

I once saw a bumper sticker that read, "Welding holds the world together." Welding actually does hold more things together than most people realize. For instance, your automobile has hundreds of spot welds, seam welds and weld beads. Take a drive and you'll see steel bridges with thousands of welds. You'll pass by skyscrapers that are welded-steel structures underneath all that brick, mortar and glass. People welded those things together. Even robot welders have to be set up and adjusted by a weldor—er, uh, welding technician—so the welds are as good as those welded by human hands.

Ordinarily, welding pays as well as most skilled trades. In many cases, a weldor's salary matches that of a degreed engineer. Likewise, a welding

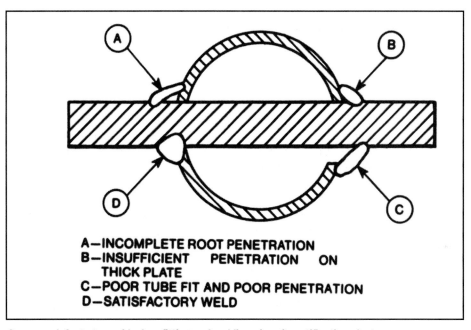

A—INCOMPLETE ROOT PENETRATION
B—INSUFFICIENT PENETRATION ON THICK PLATE
C—POOR TUBE FIT AND POOR PENETRATION
D—SATISFACTORY WELD

Common defects to avoid when fitting and welding aircraft certification cluster.

My proudest achievement in my welding career is this Kennedy Space Center, Space Shuttle Launch Pad modification to Pad B, just prior to Return to Flight after the Challenger disaster. I was promoted to Senior Project Engineer for Space Shuttle Astronaut Emergency Escape Systems. That is a really big accomplishment for a Texas farm boy who learned to weld in Buddy Bobbitt's tractor repair shop! Things like this are still possible for any beginning weldor.

their factories. Lincoln's school, the James P. Lincoln Arc Welding Foundation, provides scholarship awards for students of other colleges and universities. For more information contact:

Aluminum Company of America
(ALCOA)
Technology Marketing Division
303-C ALCOA Building
Pittsburgh, PA 15219

Hobart Brothers
Box HW-34
Troy, OH 45373

The James P. Lincoln Arc Welding
Foundation
Box 17035
Cleveland, OH 44147 ■

engineer is one of the highest-paid engineers. The highest-paying jobs for welding engineers are in offshore structures and structural-steel industries. Why? The forces of supply and demand govern here. Welding is a skilled trade and there is an element of danger involved.

Welding is like riding a bicycle; once you learn how, you never forget. But you do need to practice from time to time. Once your friends and neighbors discover that you can weld, you'll be asked to fix many things— now that will keep you in practice. You could even make a little extra money.

WELDING SCHOOLS

Some colleges and industrial arts schools provide evening and weekend classes for weldors desiring to become certified. Trade schools also provide classes for welding certification. In these classes, a student is first taught the basics on pieces of scrap metal, and then given several weeks of nothing but practice, practice, practice to improve skills.

If you are serious about wanting to become a professional weldor, start by taking a welding class. Most high schools teach welding as an elective. And most two-year colleges provide a degree in welding technology. It's interesting to note, though, that many four-year colleges and universities do not provide welding courses or degrees. Those that do usually are engineering schools.

Most welding manufacturers, such as Lincoln Electric Co. and Hobart Brothers, operate welding schools at

Welding on projects such as this bridge may not require certification, but it won't hurt. On the other hand, some welding jobs do require certification, such as in nuclear power plants, aircraft industry and pipeline construction.

Glossary

Air-Arc Gouging—An electric-arc process that cuts metal by melting it with a carbon or copper electrode, and simultaneously blows away the molten metal with a 100 psi air blast through the center of the electrode. It's a very noisy and messy process, but gets the job done cheaply where a lot of metal has to be removed.

Alloy—Basic metal modified by chemical compositions to improve its hardness or corrosion-resistant characteristics.

Aluminum Heat-Treating—Process by which aluminum is heated to 960°-980° F (516-527C), then quickly cooled, or quenched. The temperature is reached by placing the part in an oven or in a bath of liquid salts called solution heat treatment. Quenching is accomplished by using cool air or water. Quick cooling is the secret to heat-treating.

Annealing—Opposite of hardening. It's done to remove hardness in certain metals where drilling or other machining is desired. The metal is usually heated to about the same temperature as for heat treating, but then allowed to cool slowly. Aluminum and steel may be annealed. Usually, the part can be heat-treated again after annealing.

Arc Blow—Deflection of the arc from its normal path by magnetic forces, usually associated with d-c welding.

Arc Welding--A welding process that fuses metal by heating it with an electric arc and simultaneously depositing the electrode in the molten puddle.

Backfire—Momentary, loud pop at the oxyacetylene torch tip. It is caused by the flame backing up into, or combustion occurring inside, the tip. It's usually a result of the weldor trying to get more heat from a torch with low gas pressure by holding it too close to the work and overheating the tip. This occurs more readily in a large torch with a rosebud tip because of low gas pressure. Backfire can be dangerous and should not be allowed to continue.

Backhand Welding—Like walking backward, it is welding backward. The weldor points the torch at the already welded seam, away from the unwelded seam. I doubt the need for this procedure, but it could be useful to avoid burning through very thin metal. The added mass of weld bead could help absorb the extra heat.

Backing Ring—Metal ring placed inside the seam of pipe being butt welded—welded end to end. The ring provides for full weld penetration and 100% strength in butt welded pipe seams. It is usually tapered for smooth flow of liquids, steam or gas once the weld is completed. The ring is then left inside the pipe

Backing Strip—Metal strip that serves the same purpose as a backing ring—to ensure 100% strength and weld penetration.

Bare Electrode—A consumable, bare electrode used in arc welding with no flux coating.

Base Metal—Prime metal to be welded, brazed or cut. In auto bodywork, the car's steel fender is the base metal and the welding rod the filler metal.

Bead or Weld Bead—Result of fusing together a seam in two or more pieces of metal with welding rod. Usually, the bead is thicker than the base metal.

Bevel—Preparatory step prior to welding, whereby the edges of the base metal are angle-filed or ground to better accept the filler metal. When welding thick metal the bevel forms a V-groove that promotes better weld penetration.

Braze—Non-fusion weld produced by heating a base metal above 800° F (427C) and using a non ferrous filler metal. The liquid (above 800° F) filler metal flows between closely fitted surfaces of the metal joint by capillary action. The base metal is not melted in braze welding.

Butt Joint—Joint between two pieces of metal lying flat, end to end.

Capillary Action—Action whereby the surface of a liquid—including metal heated to a liquid state— is raised, lowered or otherwise attracted to fixed molecules nearby. A plumber sweats together copper pipe by using liquid brass and other liquid solder to flow into tight-fitting places by capillary action. Dictionary defines it as "the force of adhesion between a solid and a liquid."

Carburizing—1. A heat-treating process that hardens iron-based alloys by diffusing carbon into the metal. The metal is heated for several hours while in contact with carbon, then quenched; 2. In gas welding, an acetylene-rich flame that coats the metal with black soot.

Corrosion—Gradual chemical attack on metal by moisture, the atmosphere or other agents. This includes rust on iron or steel, oxidation on aluminum, and acid pitting and etching on stainless steel. Corrosion is the biggest long-term problem for fabricated metal construction. Corrosion prevention can be accomplished by painting, plating, oiling or any other coating that keeps oxygen away from the base metal.

Cover Glass—Clear glass used in goggles and welding helmets to protect the more expensive colored lens from weld spatter.

Covered Electrode—Arc-welding electrode, used as a filler metal, that is covered with flux to protect the molten weld puddle from the atmosphere until the puddle solidifies. Commonly called slick electrode.

Deposit—Filler metal added during the welding operation.

Depth of Fusion—Depth that fused filler material extends into the base metal.

Duty Cycle—Seldom-understood term that applies to electric-arc welders, not gas welders. The duty cycle is a ten-minute period. If an arc welder has a 100% duty cycle, it can be used 10 minutes out of every 10 minutes or 100% of the time. If a welder has a 20% duty cycle, it can be used two minutes and must cool off for eight minutes out of each 10 minutes. Usually, a small arc welder will have a

90% duty cycle at the lower amp settings, tapering off to 5% at the highest settings.

Dye-Penetrant Testing—An inexpensive process to check welds for cracks and other defects. The process consists of three chemicals in solution: a cleaner, a spray-on or brush-on penetrating red dye, and a white developer solution. After the area is cleaned and the red dye allowed to soak in a few minutes, the developer is sprayed on. Defects show red and smooth areas appear white. After inspection, the cleaning solution can be used to remove the dye.

Field Weld—Weld done at the site or in the field rather than in a welding shop.

Filler Metal—Welding rod or other metal added to the seam to assure a maximum thickness weld bead.

Fillet Weld—Weld deposit of filler metal approximately triangular in shape. Usually made when welding a T-joint or 90° intersection.

Flash Burn—Burn caused by ultraviolet-light radiation from the arc in arc welding. Usually painful and more severe than sunburn, especially when the eyes are flash burned.

Flashback—Burning of mixed gases inside the torch body or hoses. Usually accompanied by a loud hiss or squeal. Must not be allowed to continue! Shut off immediately if flashback occurs, then shut off acetylene. Use in-line arrestors to prevent flashback. If flashback occurs, do not light up the torch again until you find the cause and eliminate it.

Flux—Chemical powder or paste that cleans the base metal and protects it from atmospheric contamination during soldering or brazing. Flux consists of chemicals and minerals that properly clean and protect each type of metal. Therefore, each type of metal joining requires a specific formulation of flux. Flux is not used in fusion welding, except as a coating over arc-welding rod or in submerged arc welding.

Fusion Welding—The only true kind of welding. The metal pieces to be welded are heated to a liquid state along the weld seam, and usually filler metal of the same or similar type is added to the molten puddle and allowed to cool, forming one continuous piece of metal. Thus, the weld should be stronger than the added filler metal.

Gas Welding—Also known as oxyacetylene welding. Common term to describe welding accomplished by burning oxygen and acetylene to make a 6300° F (3482C) flame.

Gas Metal-Arc Welding—Or GMAW, a process in which an inert gas such as argon, helium or carbon dioxide is fed into the weld to shield the molten filler metal. This displaces atmospheric air and inhibits oxygen from combining with the molten metal and forming oxides and other impurities that would weaken the weld. Also known as metal inert gas—MIG—welding, but commonly called wire-feed welding.

Gas Tungsten-Arc Welding—Or GTAW, another type of inert-gas arc welding with tungsten as the electrode material. Tungsten is used because it will not melt at welding temperatures. The arc is similar to the heat from an oxyacetylene torch except that it can be concentrated in a much smaller space. Helium or argon gas is used to shield the weld puddle—argon is preferred. The filler metal is uncoated. This process can be used to weld steel, stainless steel titanium, aluminum, magnesium and several other metals. Also known as tungsten inert gas—TIG—welding and commonly known as Heliarc welding. Heliarc is a registered trademark of Linde Corp.

Hard Facing—Applying a very hard metal face to a softer metal to improve wear characteristics, such as on a bulldozer blade. The welding rod itself is the hard metal.

Hardness Testing—There are three types of tests: Rockwell, Brinell and Shore. Rockwell Hardness is a two-stage ball-impression test. Brinell Hardness is a one-stage ball-indentation test. Shore Hardness is a drop test for indenting metal. All three hardness tests use the principle that the harder the metal is, the less likely it is to be dented by a given force. Obviously, all three methods require calibrated special equipment, and a trained operator to analyze the results.

Heat-Affected Zone—Portion of the base metal that has not melted, but that has become discolored or blued by the heat from welding or cutting. Usually, metal strength is changed in the heat affected zone. The heat-affected zone is usually detectable by eye.

Heat Sink—A mass of metal, water-soaked rag or other heat absorbing material, placed so it absorbs heat, thus preventing overheating of a component or area. The use of a heat sink in welding can prevent or limit burn-through or warpage.

Heat-Treating—A process that adds strength and brittleness to metal. Almost all metals have a critical temperature at which their grain structure changes. This involves controlled heating and cooling of the metal to achieve the desired change in crystalline structure. Not all metals can be heat-treated.

Heliarc—Trademark of Linde. See TIG and Gas Tungsten Arc Welding.

Horizontal Position—Weld seam is horizontal.

Interpass Temperature—When several passes or beads are made in welding a joint, the lowest temperature of the weld bead before the next pass is started. Keep it as low as possible when welding cast iron to prevent cracking of the weld.

Joint—Junction where two or more metal pieces are joined by welding, brazing or soldering.

Kerf—Width of the cut, in oxyacetylene or plasma-arc cutting.

Keyholing—Usually occurs when butt-welding two very thin pieces of aluminum. A small hole melts all the way through but is filled with filler rod.

Magnaflux—A magic word in the welding or metallurgy business. An inspection process used with magnetic (ferrous) materials to detect cracks or other flaws. A fine iron powder is sprayed over the inspection area, then a strong magnetic field is induced electrically to cause any crack or defect to show as a separation of the iron particles. In most cases, this process causes the part to become magnetized, requiring demagnetization after the inspection is complete. Industrial Magnaflux equipment is similar in size and cost to a large arc welder.

Melting Point—Temperature at which a metal melts and becomes liquid. The melting point of frozen water (ice) is 32° F (0C). Mild steel's melting point is 2700° F (1482C).

MIG Welding—Metal inert-gas, or wire-feed, welding. The "M" for metal is a spool of wire fed through a shield of inert gas, usually 25% helium and 75% argon. It produces a result similar to tungsten inert-gas—TIG—welding. In the case of

MIG welding, the filler rod or wire becomes the electrode and melts. Therefore, it must be continuously fed to the weld puddle. Stainless steel, steel, aluminum and other metals that can be fusion-welded can also be MIG-welded.

Neutral Flame—An oxyacetylene flame with equal amounts of oxygen and acetylene.

Nitriding—A surface-hardening process for certain steels, whereby nitrogen is introduced in contact with anhydrous ammonia gas in the 935–1000° F (502–538C) range. Quenching is not required. It's accomplished in similar manner to carburizing.

Normalizing—This process is usually used on high-strength steels such as 4130 to remove strains in fabricated parts or in material intended for bending or machining. The metal is heated to a point above the critical transformation temperature and then allowed to cool in still air at room temperature—with no drafts.

Overhead Position—Welding position in which the seam is welded from the underside of the joint.

Oxidize—1. Effect of applying excess oxygen, causing metal to vaporize during welding, as in an oxidizing flame; 2. Slow chemical process whereby oxygen and water combine to attack ferrous metals, resulting in rust and corrosion.

Oxidizing Flame—Gas-welding flame with excess oxygen.

Oxygen Cutting—Special cutting process that cuts metals by the chemical reaction of oxygen and the base metal at elevated temperatures. No acetylene is used except to start the oxidizing process, then only oxygen is used.

Peening—Working metal with a small pointed hammer or spraying with small steel shot. Peening is usually done to improve the surface strength of the metal by putting it in compression. This prevents cracks from starting at the surface.

Penetration—Depth of weld metal from molten puddle into the base metal. Ideal penetration is 15% to 110%. Less penetration makes a weak weld. At more than 100%, the weld bead is thicker than the base metal.

Pickling—Just like putting cucumbers in solution, you put metal in a diluted acid or other chemical to clean oil, scale and other unwanted matter from its surface. Usually, a corrosion inhibitor, such as wax, is applied to the surface after pickling to prevent corrosion until the metal is painted or plated.

Plating—Outer coating of chromium, copper, nickel, zinc, cadmium or other heavy metal to enhance appearance or inhibit corrosion of parent metal. Usually, plating is accomplished by immersion in an acid solution with cathode and anode electric current, causing the plating material to deposit on the parent metal. Most ferrous metals can be plated.

Plug Weld—Also called rosette weld. A circular weld made through a hole in one piece of tubing or blind channel, connecting another piece slipped inside.

Polarity—In direct-current arc welding, TIG welding and wire feed welding, how current flows either positive to negative, or negative to positive. AC welding has no polarity because it switches between positive and negative polarity and back again 60 times per second. Heat is about equal at the electrode and workpiece. DC straight polarity (DCSP) is also called negative polarity. In DCSP, the workpiece is positive and the electrode negative. Heat is greater at the workpiece than at the electrode. DC, reverse polarity (DCRP) is also called positive polarity. In DCRP, the workpiece is negative and the electrode positive. Heat is greater at the electrode than at the workpiece and penetration is shallow.

Postheating—Heating a weld after it is completed, usually for stress relief.

Preheating—Heating the weld area beforehand to avoid thermal shock and thermal stresses. Metal is more sensitive to temperature than you would think. For instance, a steel pipe welded at 32° F (0C) will be more brittle than one heated to 90° F (32C), or even to 300° F (149C). The larger the mass to be welded, the more it needs preheating.

Puddle—Liquid area of the weld where heat is being applied either by flame or electric arc. This is the most important part of welding! If the puddle is properly controlled, the weld will automatically be good.

Reducing Flame—In oxyacetylene welding, a flame with excess acetylene, imparting excess carbon to the weld. Also called a carburizing flame.

Residual Stress—Stress remaining in a structure after the weld joint cools. Metal wants to warp when it is heated. If a very strong jig or a triangulated structure prevents warping, stresses remain in the area near the weld.

Root Opening—Distance between two pieces of metal to be joined.

Root of Weld—Point of weld farthest from heat source. Intersection point between bottom of weld and base-metal surface.

Rosette Weld—See Plug Weld

Sandblasting—Fast, easy method of cleaning certain metals before welding and painting. A high velocity air blast, carrying sand, is directed at the metal and the particles of sand abrade its surface. Obviously, this cleaning process should be used with caution to protect eyes and lungs. It leaves a rough surface and cleanup is messy.

Seam Welding—A form of spot welding in which two pieces of sheet metal are resistance-welded in a continuous seam.

Shielded-Metal Arc Welding— SMAW, also commonly known as stick welding or as the layman knows it—simply arc welding. Shielding is the flux coating on the metal rod.

Shop Weld—To prefabricate or weld subassemblies in a shop or controlled environment before taking them on-site for final assembly. Often used in large welding projects such as oil-drilling sites and nuclear power plants.

Slag—Impurities resulting from heating metal and boiling off dirt and scale present in most open-air welding. Slag is found at the kerf from oxyacetylene torch cutting. Slag will also be found in the hardened flux on top of an arc-welded bead.

Soldering—Metal joining process similar to brazing. Metal pieces are joined with molten solder, without melting the base metal. Solder is drawn into the joint by capillary action. As it cools, it sticks to the base metal. If two pieces of lead solder were joined by melting them together, technically that could be called welding. But I would just call it melting lead!

Spatter—Small, unsightly droplets of metal that deposit alongside the weld bead. Especially common in arc welding with E-6011 rod.

Spot Welding—A production welding method to join sheet metal. Electrical resistance heating and clamping pressure are used to fuse panels together with a series of small "spots." Filler metal is not used. Also called resistance welding.

Steel Heat-Treating—Process of heating and rapidly cooling steel in the solid state to obtain certain desired properties: workability, microstructure, corrosion resistance, and so on. Depending on its mass and alloy type, the steel is oven-heated to 1475–1650° F (802–900C), then quenched by dipping it in water or oil.

Stickout—Length of electrode (tungsten or wire) that sticks out past the gas lens, cup or gun.

Stitch Weld—Tack-welding technique with short weld beads about 3/4" long, spaced by equally long gaps with no welding. Used where a solid weld bead would be too costly and time-consuming, and where maximum strength is not required.

Stress Cracking—Metal cracking at the weld due to temperature changes or molecular changes. Overheated welds are more prone to stress cracking than underheated welds.

Stress-Relief Heat-Treating—When a complicated, rigidly braced structure such as an airplane engine mount or race-car suspension member is welded, stresses remaining in the metal will cause premature fatigue cracking unless they are relieved. Stress relieving is accomplished by heating part or all of the structure to about two-thirds of the melting point and then cooling it slowly. This allows the molecules in the structure to relax and stay relaxed.

Stringer Bead—A straight weld bead made without oscillation.

Submerged-Arc Welding—SAW or sub-arc welding. A process in which the electric arc is submerged in powder flux, thereby protecting the weld from atmospheric contamination. The system is usually automatic feed and travel and a base-metal rod is used. It's used where high accuracy and weld quality are desired.

Sugar—Crystallization in a weld. It usually occurs when welding stainless steel if the back side of the weld seam is not protected by an inert gas such as argon. Sugar has no strength and should not be allowed in a weld.

Tempering—When metal has been hardened by heat-treating, it usually becomes brittle. In order to relieve the internal strains, the metal is usually reheated to about one-quarter or one-half the temperature originally used in heat-treating.

Thoriated Tungsten—Tungsten electrode with 1–2% thorium added to provide a more stable arc. Thoriated tungsten is used to weld steel. However, pure tungsten must be used to weld aluminum and magnesium.

TIG—Tungsten inert-gas welding. Often called Heliarc, because helium was first used as an inert gas for this welding process. Argon and other gas mixtures are now used. Electrode is tungsten because it doesn't melt at welding temperatures. The inert gas shields the weld from atmospheric impurities, providing a high-quality weld. Filler rod is hand fed to the weld.

Tungsten Electrode—A non-consumable electrode used in TIG welding. The melting point of tungsten is 5432° F (3000C).

Ultrasonic Testing—Process to test metal parts for defects. High frequency sound waves are directed at the part, and their reflections are picked up by a receiver. Cracks and flaws inside the metal are detected by discontinuities in the return sound. Ultrasonic test equipment is expensive and requires trained personnel to operate and analyze the results.

Vertical Position—Type of weld in which the metal to be welded is vertical and the weld bead progresses upward or downward.

Weave Bead—Weld bead made with a transverse oscillation such as a figure-8, or a "Z" motion while moving forward along the seam. This bead deposits more filler metal and ties the two pieces together more effectively than a straight bead, but also provides the possibility of including slag or flux in the weld, thereby contaminating it.

Weld—Local melting together and fusing of metal produced by heating the base metal and, in most cases, applying filler rod to the molten puddle. The filler rod usually has a melting point approximately the same as the base metal, but above 800° F (427C).

Weldability—Capacity of specific metals to be welded and to perform satisfactorily for the intended service. Not all metals are weldable. See Chapter 1 for what can and cannot be welded.

Welder—Piece of equipment used for welding.

Weldor—Person who performs the welding operation.

X-Ray—Inspection for stress cracks, internal corrosion or other defects. An actual X-ray picture is taken of the part. The operator must be specially trained because of the radioactive materials involved. This process is used to inspect welded seams on high pressure nuclear powerplant piping, airplane parts and other situations in which any defect would cause an expensive or potentially dangerous problem.

TERMS FOR WELDING DEFECTS

Arc Strike—Unintentional arc start outside of the weld bead. Usually more of a problem in TIG welding in which strict quality control standards are observed.

Cold Weld—Poor penetration of the weld bead, usually less than 5% of the bead thickness.

Crater—In arc welding and TIG welding, a depression at the end of the weld bead caused by stopping the weld with too much heat applied.

Crater Crack—Crack in the crater at the end of the TIG-weld bead caused by stopping the weld with too much heat applied and with drawing the shielding gas before the weld solidifies.

Drop Through—Filler material that sags through on the underside of the weld, caused by either too much heat or poor joint fit.

Discontinuity—An interruption in the basic weld bead, usually excess filler material, but not necessarily a defect.

Inadequate Penetration—Depth of filler metal is less than 15% of the weld-bead thickness.

Porosity—Usually, gas pockets caused by the wrong weld temperature or a dirty, contaminated weld. Most porosity is caused by getting the weld bead too hot.

Slag Inclusion—Dirty weld due to flux trapped in the weld bead or scale or dirt from the base metal or welding rod.

Undercut—Cutting away of the base metal by improper application of temperature. The weldor just pointed the heat at the weld bead and capillary action pulled the molten puddle away from the cooler base metal to the hotter molten puddle. Undercut can be avoided by paying more attention to temperature control.

Index

A

AC (alternating current) polarities, 91
Acetylene, safety using 29; See also Oxyacetylene
Aircraft welding certification, 157-159
Airplanes, trailers for, 154
Aluminum
  alloys, 120
  arc welding, 92
  cast, 4
  gas brazing, 80
  gas welding, 64-66, 140-141
MIG welding, 99-104
  sheet, 4-5
  spot welding, 131
TIG welding, 118-121
American Welding Society Journal, 144
Arc cutting, 11
Arc rods, magnesium, 11
Arc welding, 10-11, 141-142
  aluminum, 92
  basics, 82-84
  butt welds, 87-88
  cast iron, 93
  duty cycles, 82-83
  equipment, 17-18, 84
  eye burns, 28
  4130 steel, 91-92
  installing 220 volt welders, 83
  installing 440 volt welders, 83
  interpass temperatures, 86
  lap welds, 88-89
  lenses, 26-27
  horizontal welding, 89-90
  overhead welding, 89
  vertical welding, 89-90
  polarities, 92
  practice, 85
  principles, 85
  rods, 84
  rods for practice, 85
  running beads, 86-87
  sheet metal, 90-91
  stainless steel, 92-93
  striking arcs, 85-86
  T welds, 88
  techniques, 84-89
Arc welds, starting & stopping, 90
Autobody gas welding, 68

AWS (American Welding Society), 118, 138

B

Back-gas purging, 116-117, 121
Backhand welding, 57-58
Bandsaw
  converting to cut steel and aluminum, 42-43
    speeds, 43
Brass
  bronze & aluminum brazing w/gas welders, 140
    cast, 4
Brazing, 8-15
  oxyacetylene, 9-10
Brazing, gas, 76-81
  aluminum, 80
  copper, cast iron & other metals, 86
  joint designs, 77
  joints, 79-80
    lap joint, 79-80
    T joints, 80
  low heat required, 77-78
  metals that can be brazed, 78
  procedures, 78-79
  seam designs, 77
  silver, 80-81
Burns
  eye, 28
  radiation, 25
Butt welds, 58, 87-88

C

Car bodies, high-strength steel, 104
Careers, welding for a living, 159-160
Carts, gas welding, 148-149
Cast aluminum, 4
Cast brass, 4
Cast iron, 3, 93
Certification, 155-159
  aircraft welding, 157-159
  documentation, 156
  pipe welding, 156-157
Chart temperatures, 2
Chrome moly steel, 4
Cleaning
  before welding, 44-47

  methods, 44
  aluminum oxide, 47
  chemical cleaning, 46-47
  glass bead cleaning, 46
  mill scale, 47
  power sanding, 45
  sandblasting, 45
  sandpaper cleaning, 44-45
  using discretion, 47
  wire brushes, 45-46
Colors
  changes of metals, 1-2
  of steel at various temperatures, 6
  for tempering carbon steel, 6
Containers, flammable, 25
Copper fumes, 137
Coupons
  defined, 54
  preparing one's, 54
Cutting, 8-15
  arc, 11
  flame, 71
  aids, 71-73
  laser, 13-14
  machines, 73
  oxy-flame, 10
  plasma, 13
  processes, 128-135
  tables, 72-73
  torches, 69-73
Cylinder sizes, 52

D

DC (direct-current) polarities, 91

E

EBW (electron beam welding), 13-14
Electric soldering, 81
Equipment
  convenience, 22-23
  gas welding, 48-54
  metal markers, 23
  plasma cutting, 21
  sandblasters, 22-23
  spot welding, 21-22
  temperature indicators, 23
  welding, 16-23
Eye burns, arc welding, 28
Eyeglasses, 29-30

F

Ferrous & non-ferrous metals, 2-5
thin-wall tubing, 31-32
tools, 33
    hand, 33-34
    marking, 33
    measuring, 33
    power, 34-35
    safety, 35
Fitting, 31-35
    defined, 31
    fishmouth joints, 32-33
    angle & plate steel, 33
    heavy-gauge material, 33
Flame cutting
    aids, 71-73
    cutting machines, 73
    cutting tables, 72-73
    marking tools, 72
    tips for better, 71
Flammable containers & welding, 25
Forehand welding, 57
Forged steel, 3-4

G

Gas
    brazing, 76-81, 86
    correct pressures, 62
    flow data, 75
Gas cutting oxidizing, 69
Gas welders
    aluminum soldering with, 139
    brass, bronze & aluminum brazing
with, 140
    lead soldering with, 138
    silver brazing with, 139-140
Gas welding, 48-68
    aluminum, 64-66, 140-141
    oxyhydrogen, 65
    vs. steel, 65-66
    autobody, 68
    carts, 148-149
    common problems, 61-64
        flaky welds/poor penetration, 61-62
        flashbacks, 62
        rod sticks to base metals, 62
        solutions, 62-63
        torch pop, 61
    equipment, 16-17, 48-54
    accessories, 51-54
    care of, 52
    gas bottle sizes, 48-49
    lease, rent or buy, 49-50
    regulators, 50

getting started, 54-56
heating & forming, 66-68
    cutting-torch tips, 68
    rosebud tips, 67
lighting & shutting off oxyacetylene,
54-56
    opening the cylinder valve, 54
    practice, 58-60
        outside-corner welds, 59-60
        T welds, 60
        test welds, 59
    SAE 4130 steel, 63-64
    rust prevention, 64
    weld techniques, 63-64
    safety, 28-29, 53
    steel, 140-141
    stress relieving myths, 68
    techniques, 56-58
        making puddles, 56
        running weld beads, 57-58
        working distances, 56-57
    tips, sizes, gas flow, data chart, 53
    torch sounds, 61
    welding tables, 54
Gases, 1
Glass bead cleaning, 46
Glasses, safety, 27
Gloves, 25
Goggles, 28

H

Heat control, 1-7
Heating
    and forming, 66-68
    and gas welding, 9
Heli-Arc welding, 8
Helmets, choice of, 27-28
Horizontal welding, 89-90

I

Identification markings, 5-6
    ferrous or non-ferrous, 5-6
    permanent metal markers, 5
Interpass temperatures, 86
Iron, cast, 3

J

Jack stands, 150
Jeeps, tow bars for, 151
Jewelry making, 134-135
Jigging, 35-43
Jigs
    and fixtures, 38-42

particle board and blocks, 40
permanent steel, 36-37
plywood and bolt, 40-41
plywood and nails, 39-40
surface plates, 41
tubular and angle, 41-42
welding, 35-38
wooden, 36
Joints, fishmouth, 32-33

K

Keyhole plasma-arc welding, 124

L

Lap welds, 88-89
Laser cutting, 13-14
LBC (laser beam cutting), 14
LBW (laser beam welding), 14
Leak testing, 59
Lenses, welding, 26-27
Liquids, 1

M

Magnesium
    arc rods, 11
    TIG welding, 121-122
Markers, metal, 23
Markings, identification, 5-6
Mechanical fingers, 148
MEK (methylethyl ketone), 63
Melt-in plasma-arc welding, 124
Metal Fabricator's Handbook, 5
Metals
    basics, 1-7
    color changes of, 1-2
    ferrous & non-ferrous, 2-5
    cast aluminum, 4
    cast brass, 4
    cast iron, 3
    chrome moly steel, 4
    forged steel, 3-4
    mild steel, 4
    sheet aluminum, 4-5
    stainless steel, 4
    titanium, 5
    markers, 23
    melting points of, 3
    spraying, 134
    weights, boiling points of, 3
    weights, melting points of, 3
    and welding processes, 15
MIG welding, 11-12, 94-104, 144
    equipment, 18-19
    high-strength steel car bodies, 104
    modes of, 96-97

procedures, gas, settings, 102-103
steel & aluminum, 99-104
adjustments, 100
MIG welding stainless steel, 104
preparations, 100-104

O

110-volt plasma torches, 127
Outside-corner welds, 59-60
Overhead welding, 89
Oxyacetylene, 8
brazing & soldering, 9-10
cutting tip sizes, 75
lighting & shutting off, 54-56
Oxy-flame cutting, 10

P

PAC (plasma-arc cutting) 126-127
PAW (plasma-arc welding); *See Plasma-arc welding*
Permanent steel jigs, 36-37
Pipe welding certification, 156-157
Plasma cutting, 13
equipment, 21
and welding, 123-127
Plasma torches, 110-volt, 127
Plasma-arc cutting, 126-127
Plasma-arc welding, 123-126
equipment, 125-126
inert atmosphere, 124
process, 123-124
two modes, 124
keyhole, 124
melt-in, 124
using, 126
Plywood and bolt jigs, 40-41
Plywood and nails jigs, 39-40
Polarities
AC (alternating current), 91
arc welding, 92
DC (direct-current), 91
settings, 112-113
Power sanding, 45
Power supplies, 108-109
Projects, welding, 145-154
Propane, welding with, 7
Purging, back-gas, 116-117

R

Radiation burns, 25
Regulators, 50
Repairs, cold, 135
Rods
choosing the right, 138-144
aluminum soldering w/gas welders,

139
lead soldering w/gas welders, 139
silver brazing w/gas welders, 139-140
copper-coated welding, 137-138
grades of, 136-137
welding metallurgical advances, 138

S

Safety
acetylene, 29
for bystanders, 27-28
glasses, 27
soldering, 29
eyeglasses, 29-30
for gas welding, 53
grinding sparks, 29
work areas, 29
Sandblasters, 22-23
Sandblasting, 45
Sander & grinder stands, 149
Sanding, power, 45
Sandpaper cleaning, 44-45
Schools, welding, 160
Sheet aluminum, 4-5
Sheet metal, arc welding, 90-91
Silver, gas brazing, 80-81
SMAW (shield metal arc welding), 10-11
Soldering, 81
electric, 81
procedures, 81
Soldering safety, 29
Solids, 1
Sparks, grinding, 29
Spot weld testing, 133-134
Spot welders
complex, 131
types of, 129-130
using, 132-134
Spot welding, 13-14, 128-134
aluminum, 131
equipment, 21-22
weldability of metals by, 132
Stainless steel, 4; See also Steel
arc welding, 92-93
MIG welding, 104
TIG welding, 116-117
Steel; See also Stainless Steel
angle & plate, 33
chrome moly, 4
colors for tempering carbon, 6
cutting with oxygen only, 74
forged, 3-4
gas welding, 140-141
high-strength car bodies, 104
MIG welding aluminum and, 99-104

mild, 4
setups for cutting, 75
TIG welding, 115-116
Stitch welds, 58
Stress relieving myths, 68

T

T welds, 60, 88
Tables, welding & cutting, 54, 146-147
Tack welds, 38, 58
Temperatures
chart, 2
color of steel at various, 6
conversions, 43
indicators, 23
interpass, 86
of soldering, brazing & welding processes, 2
welding, 6-7
overheat at first, 7
temperature control, 7
of welding fuels, 6
Test welds, 59, 87-88
Testing, leak, 59
TIG setups
inverter technology, 111
maximum, 110
minimum, 110-111
TIG welders
using, 111-115
polarity settings, 112-113
tungsten sharpening, 113
welding tips, 113-114
TIG welding, 12-13, 105-122, 142-144
aluminum, 118-121
alloys, 120
back-gas purge, 121
fitting parts, 120-121
grounding aluminum, 120
guidelines for, 119
heat treatment nomenclature, 119
heat-treating after welding, 121
machine settings, 118-119
preheating, 119
weld craters, 120
weldability, 119-120
components, 105-106
equipment, 19-21
gas shielding for titanium welding, 118
magnesium, 121-122
cleaning before welding, 121-122
guidelines, 122
stress-relieving, 122
setups, 107-111
flow meters, 108

high frequency, 107-108
necessary extras, 109-110
power supplies, 108-109
spare parts list, 108
spare parts list, 109
stainless steel, 116-117
steel, 115-116
titanium, 117-118
gas chambers, 118
know before you start, 118
Titanium, 5
gas shielding for welding, 118
TIG welding, 117-118
Torches
cutting, 69-75
gas cutting oxidizing, 69
pistol grip, 73-74
flame cutting aids, 71-73
light cutting, 70
110-volt plasma, 127
tips for better flame cutting, 71
using cutting, 69-73
Tow bars for cars & jeeps, 151
Trailers
for race cars & airplanes, 154
utility, 152-153
Training, 160
Tubular and angle jigs, 41-42
Tungsten sharpening, 113

U

Utility trailers, 152-153

V

Vertical welding, 89-90

W

Welder's Handbook, 155
Welding
aluminum, 140-141
arc, 10-11, 82-93, 141-142
areas, 22
backhand, 57-58
brazing & cutting, 8-15
cleaning before, 44-47
electron beam welding (EBW), 13-14
equipment, 16-23, 17-18
gas, 16-17
MIG, 18-19
spot, 21-22
TIG, 19-21
where to buy, 16
forehand, 57
gas welding & heating, 9
gases (bottled), 28-29

Heli-Arc, 8
horizontal, 89-90
jigs, 35-38
laser beam welding (LBW), 14
lenses, 26-27
MIG, 11-12, 94-104, 144
overhead, 89
overview of processes, 8-14
oxyacetylene brazing & soldering, 9-10
projects, 145-154
gas welding carts, 148-149
jack stands, 150
mechanical fingers, 148
sander & grinder stands, 149
tow bars for cars & jeeps, 151
trailers for race cars & airplanes, 154
utility trailers, 152-153
welding & cutting tables, 146-147
work stands, 150
with propane, 7
rods, wires & fluxes, 136-144
safety
arc-welding lenses, 26-27
choice of helmets, 27-28
comfort, 30
general safety tips, 29-30
schools, 160
special, 128-135
spot, 13-14, 128-134
steel, 140-141
stress-relieving assemblies, 37-38
avoiding warpage, 37-38
gas-welded structures, 38
tack-weld first, 38
TIG-welded structures, 38
tables, 54
temperatures, 6-7
TIG, 8, 12-13, 105-122, 142-144
vertical, 89-90
wire feed, 94-104
Welding Handbook (American Welding Society), 118
Welding safety, 24-30
arc, 25-27
clothing, 25-26
radiation burns, 25
gas, 28-29
bottled welding gases, 28-29
goggles, 28
Welds
butt, 58, 87-88
lap, 88-89
outside-corner, 59-60
stitch, 58

T, 60, 88
tack, 38, 58
test, 59, 87-88
Wire feed
adjustments, 103
machines, 97-99
control units, 97-98
guns, 97
power supplies, 97
shielding gas, 98
wire-drive mechanisms, 97-98
operations, 95-97
welders
110-volt, 98
200/230 volt, 98-99
575-volt heavy-duty welder, 99
no CO_2 gas required, 99
types of, 98-99
welding, 11-12, 94-104, 144; *See also MIG welding*
Wooden jigs, 36
Work stands, 150

About the Author

Richard Finch, shown above driving a race car that he designed and welded, has been fortunate enough to see his self-taught welding skills take him around the world many times.

Richard learned welding the hard way, by trial and error. He began welding on drag race cars, trailers and farm equipment while in high school and college. His first significant welding projects were go-karts and trailers for race cars. This experience progressed to building road racing cars and then parts for experimental airplanes. He worked full-time as a Civil Service weldor at Vandenberg AFB, then at the Ted Smith Aerostar Aircraft Factory, and taught welding at Allan Hancock College in Santa Maria, CA. He worked as a welding trouble-shooter for Piper Aircraft, Fairchild Aircraft, and as a Service Engineering Manager for SAAB-SCANIA Aircraft.

But Richard's most elite welding project was being selected as Emergency Egress Project Engineer at Kennedy Space Center, FL, to oversee the launch pad modifications necessary to ensure Space Shuttle Astronaut safety for "return to flight" operations after the Challenger disaster. For his success in that job, Richard was given a commendation plaque signed by ten space shuttle astronauts: Carl Meade, Bill Shepherd, Kathy Thornton, P. Wilhoit Jr., Dr. Jay Apt, Ken Cameron, Steve Sewell, Prince Tuhat, Sam Gemor, and G. David Low. For that project, Richard was also nominated for the prestigious NASA Excellence Award. Prior to the Challenger explosion, astronauts Dick Scobee, S. Christa McAuliffe, school teacher, and Judy Resnik, signed Richard's commercial pilot bi-annual review.

Twenty years after starting to college, Richard earned an AA Degree in Business. Nine years later, an AS Degree in Welding, and 4 years later, BS and MS Degrees in Engineering. Richard still knows that welding is an ever-changing science, and it is a very satisfying accomplishment—he highly recommends welding as a skill that can be learned by almost anyone. ■

OTHER BOOKS FROM HPBOOKS AUTOMOTIVE

HANDBOOKS

Auto Electrical Handbook: 0-89586-238-7
Auto Upholstery & Interiors: 1-55788-265-7
Brake Handbook: 0-89586-232-8
Car Builder's Handbook: 1-55788-278-9
Street Rodder's Handbook: 0-89586-369-3
Turbo Hydra-matic 350 Handbook: 0-89586-051-1
Welder's Handbook: 1-55788-264-9

BODYWORK & PAINTING

Automotive Detailing: 1-55788-288-6
Automotive Paint Handbook: 1-55788-291-6
Fiberglass & Composite Materials: 1-55788-239-8
Metal Fabricator's Handbook: 0-89586-870-9
Paint & Body Handbook: 1-55788-082-4
Sheet Metal Handbook: 0-89586-757-5

INDUCTION

Holley 4150: 0-89586-047-3
Holley Carburetors, Manifolds & Fuel Injection: 1-55788-052-2
Rochester Carburetors: 0-89586-301-4
Turbochargers: 0-89586-135-6
Weber Carburetors: 0-89586-377-4

PERFORMANCE

Aerodynamics For Racing & Performance Cars: 1-55788-267-3
Baja Bugs & Buggies: 0-89586-186-0
Big-Block Chevy Performance: 1-55788-216-9
Big Block Mopar Performance: 1-55788-302-5
Bracket Racing: 1-55788-266-5
Brake Systems: 1-55788-281-9
Camaro Performance: 1-55788-057-3
Chassis Engineering: 1-55788-055-7
Chevrolet Power: 1-55788-087-5
Ford Windsor Small-Block Performance: 1-55788-323-8
Honda/Acura Performance: 1-55788-324-6
High Performance Hardware: 1-55788-304-1
How to Build Tri-Five Chevy Trucks ('55-'57): 1-55788-285-1
How to Hot Rod Big-Block Chevys:0-912656-04-2
How to Hot Rod Small-Block Chevys:0-912656-06-9
How to Hot Rod Small-Block Mopar Engines: 0-89586-479-7
How to Hot Rod VW Engines:0-912656-03-4
How to Make Your Car Handle:0-912656-46-8
John Lingenfelter: Modifying Small-Block Chevy: 1-55788-238-X
Mustang 5.0 Projects: 1-55788-275-4

Mustang Performance ('79-'93): 1-55788-193-6
Mustang Performance 2 ('79-'93): 1-55788-202-9
1001 High Performance Tech Tips: 1-55788-199-5
Performance Ignition Systems: 1-55788-306-8
Performance Wheels & Tires: 1-55788-286-X
Race Car Engineering & Mechanics: 1-55788-064-6
Small-Block Chevy Performance: 1-55788-253-3

ENGINE REBUILDING

Engine Builder's Handbook: 1-55788-245-2
Rebuild Air-Cooled VW Engines: 0-89586-225-5
Rebuild Big-Block Chevy Engines: 0-89586-175-5
Rebuild Big-Block Ford Engines: 0-89586-070-8
Rebuild Big-Block Mopar Engines: 1-55788-190-1
Rebuild Ford V-8 Engines: 0-89586-036-8
Rebuild Small-Block Chevy Engines: 1-55788-029-8
Rebuild Small-Block Ford Engines:0-912656-89-1
Rebuild Small-Block Mopar Engines: 0-89586-128-3

RESTORATION, MAINTENANCE, REPAIR

Camaro Owner's Handbook ('67-'81): 1-55788-301-7
Camaro Restoration Handbook ('67-'81): 0-89586-375-8
Classic Car Restorer's Handbook: 1-55788-194-4
Corvette Weekend Projects ('68-'82): 1-55788-218-5
Mustang Restoration Handbook('64 1/2-'70): 0-89586-402-9
Mustang Weekend Projects ('64-'67): 1-55788-230-4
Mustang Weekend Projects 2 ('68-'70): 1-55788-256-8
Tri-Five Chevy Owner's ('55-'57): 1-55788-285-1

GENERAL REFERENCE

Auto Math:1-55788-020-4
Fabulous Funny Cars: 1-55788-069-7
Guide to GM Muscle Cars: 1-55788-003-4
Stock Cars!: 1-55788-308-4

MARINE

Big-Block Chevy Marine Performance: 1-55788-297-5

HPBOOKS ARE AVAILABLE AT BOOK AND SPECIALTY RETAILERS OR TO
ORDER CALL: 1-800-788-6262, ext. 1

HPBooks
A division of Penguin Putnam Inc.
375 Hudson Street
New York, NY 10014

ISSN 1534-5407

NATIONAL SECURITY

Thomas Wiloch

INFORMATION PLUS® REFERENCE SERIES
Formerly published by Information Plus, Wylie, Texas

THOMSON

GALE

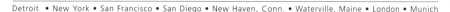

Detroit • New York • San Francisco • San Diego • New Haven, Conn. • Waterville, Maine • London • Munich

National Security

Thomas Wiloch

Paula Kepos, Series Editor

Project Editor
John McCoy

Permissions
Margaret Abendroth, Edna Hedblad, Emma Hull

Composition and Electronic Prepress
Evi Seoud

Manufacturing
Drew Kalasky

LIBRARY OF CONGRESS CATALOGING-IN-PUBLICATION DATA

ISBN 0-7876-5103-6 (set)
ISBN 0-7876-9079-1
ISSN 1543-5407

Printed in the United States of America
10 9 8 7 6 5 4 3 2 1

TABLE OF CONTENTS

PREFACE vii

CHAPTER 1
An Introduction to National Security. 1

National security may be defined as protection from threats to a country's territory, people, or values and policies. During the cold war from the 1940s through the 1980s, America's national security policy relied on containment and a shifting balance between conventional and nuclear forces. Although it has been an uncontested global leader since the Soviet Union's fall in 1991, the United States currently faces many threats: international terrorism, weapons of mass destruction, reliance on imported oil and gas, the Arab-Israeli conflict, and the war on drugs, among them. The so-called "transition states" (Russia, China, and India) and "states of concern" (North Korea, Iran, Libya, Syria, Sudan, and Cuba) also influence U.S. security planning. Entrusted with formulating national security are the president, the National Security Council, the intelligence infrastructure, the military, and Congress.

CHAPTER 2
The Threat of Conventional Weapons 11

Although in decline since 1987, transfers of major conventional weapons (any armaments that do not fall into the "weapons of mass destruction" category) remain sizable. U.S. arms sales are regulated by legislation and the State Department's Office of Defense Trade Controls. Foreign nations, particularly European countries, compete with the United States as arms suppliers; and the demand for weapons is especially strong in the Persian Gulf area. On the other hand, many international and nongovernmental organizations work to limit the small arms, light weapons, and land mines that have killed so many civilians around the world. Also, many nations have signed international agreements (such as the Convention on Anti-Personnel Mines, the CFE and CFE 1A treaties, and the Wassenaar Arrangement) intended to restrain the development and use of conventional weapons.

CHAPTER 3
Proliferation of Weapons of Mass Destruction (WMD) . . . 21

Weapons of mass destruction include biological, chemical, and nuclear agents, as well as conventional weapons capable of inflicting widespread casualties. Throughout history the inventory of such weapons has grown. Currently they are of great concern in China, Egypt, India, Iran, Israel, Libya, North Korea, Pakistan, and Syria. Iraq's alleged development of WMD, despite the 1991 Gulf War and United Nations inspections, and the 2003 U.S. invasion of that country partly because of its alleged WMD programs, provide a case study. During the early and mid-1990s, illicit trafficking in nuclear and fis-

sile materials posed another serious threat, which was addressed by a U.S.-Russian accord in 1995. At least twelve other nonproliferation regimes and treaties currently cover WMD.

CHAPTER 4
Countries of Proliferation Concern 33

The end of the cold war in the early 1990s failed to diminish the threat posed by countries whose political objectives or national security concerns are not in concert with those of the United States. In addition to Russia, there are several other nations whose chemical, biological, nuclear, and conventional weapons programs may imperil global security, among them China, Egypt, India, Iran, Israel, Libya, North Korea, Pakistan, and Syria. Some of these nations have shared scientific developments with each other—particularly with regard to nuclear-weapons technology—according to monitoring efforts conducted by the Nuclear Threat Initiative (NTI). Others are thought to have secretly violated the terms of the Biological and Toxin Weapons Convention (BTWC) or the Nuclear Nonproliferation Treaty (NPT). Iraq under Saddam Hussein proved an especially ominous threat, and its weapons-development program was the subject of United Nations Security Council resolutions, as well as efforts to bring it into international compliance accords, for several years preceding the 2003 U.S.-led invasion.

CHAPTER 5
Preparing for Biological and Chemical Attacks 47

The federal government funds both research and preparedness efforts to battle bioterrorism, spending much of its budget in this area on the National Guard and the Centers for Disease Control and Prevention (CDC). To assist state and local health departments, the CDC has prioritized various biological and chemical weapons by threat level and created a laboratory response network. The Animal and Plant Health Inspection Service (APHIS), another federal agency important in guarding against bioattack, monitors potential threats to U.S. agricultural products and livestock.

CHAPTER 6
International Terrorism 57

Terrorism is generally motivated by religious and/or political conflicts and, though tactics vary, is almost always violent. Because of tensions in the Middle East, many states currently sponsoring anti-American terrorism are Islamic, including Iran, Syria, and Libya. The United States also accuses Sudan, North Korea, and Cuba of state-sponsored violence, while some of its own international actions have been labeled terrorist by other nations. Though anti-American terrorism by Middle Eastern extremists dates back to the 1980s, the September 11, 2001, attacks of Osama bin Laden and al Qaeda were the most devastating. The federal government reacted by trying to cut off financing for terrorist operations, reorganizing its bureaucracy, improving transportation security, passing the Patriot Act of 2001, and declaring war on terrorism. Several opinion polls captured the public's reactions to 9/11 and its aftermath.

CHAPTER 7

Domestic Terrorism 79

Terrorist attacks carried out by groups or individuals operating exclusively in the United States have involved guns, bombs, and germs; have targeted Congressmen, churchgoers, federal workers, spectators at Olympic games, and employees of abortion clinics; and have occurred at locations ranging from Washington, D.C., to Oklahoma City, Oklahoma. Some have been attributed to left-wing organizations espousing socialist ideologies, others to antigovernment right-wing extremists. Special-interest and single-issue terrorism often falls into the categories of ecoterrorism and anti-abortion activism. In response to hate crimes, such watchdog groups as the Anti-Defamation League and Southern Poverty Law Center monitor domestic extremists.

CHAPTER 8

Civilian National Security Infrastructure 93

The president works with the White House staff, National Security Council, U.S. Department of State, Office of Management and Budget, and intelligence community to propose national security or foreign policy initiatives that Congress then votes for or against. Reflecting the importance of strategic information to national security, the sprawling intelligence community encompasses fourteen executive-branch units and agencies, the most important of which are the CIA and FBI. The Department of Homeland Security, established after the terrorist attacks on September 11, 2001, incorporates Citizenship and Immigration Services, a government agency dedicated to protecting America's borders.

CHAPTER 9

The Military, Peacekeeping, and National Security 113

The Department of Defense (DOD) places an active-duty force of approximately 1.4 million men and women at the president's disposal. Presided over by the secretary of defense (also the commander in chief's top military advisor), the secretary's staff, and the Joint Chiefs of Staff (who function as civilian-military go-betweens), the DOD comprises the army, navy, and air force, plus the Marines and Coast Guard. There are also nine unified combatant commands (multiservice groups that control combat troops in various geographic locales) and a range of defense agencies that provide logistical support and related services. Since the early 1990s, peacekeeping has become a more important military function, but some people question U.S. involvement in peacekeeping missions based on their cost, risks, and effect on combat readiness.

CHAPTER 10

Global Dynamics of National Security: Alliances and Resources 123

The United States pursues its national security objectives through participation in international organizations and strategic alliances. Some of the most important current U.S. commitments are to the United Nations, the North Atlantic Treaty Organization, Israel, and the Gulf Cooperation Council. Because adequate supplies of oil and gas are so vital, energy security has become a component of national security. The Persian Gulf region, which provides much of the world's supply, is an area of increasing concern, along with more than half a dozen world oil transit chokepoints. Like oil and gas, water is a critical natural resource that may give rise to armed conflict in the near future, especially among countries bordering the Nile River. Conflicts over natural resources on the African continent may foreshadow future problems in resource-stressed areas around the world.

CHAPTER 11

New Arenas: Organized Crime and Emerging Technologies. 131

Both the domestic national security infrastructure and the threats to American security are changing. Policy makers now recognize organized crime as a significant danger both in the United States and abroad. Often associated with drug trafficking, financial fraud, environmental degradation, and contraband smuggling, it can destabilize governments through political corruption. Computer technology, with its constant advances, also poses new risks. Besides defending against cyber warfare in the form of computer attacks on American businesses, government, critical infrastructure, and military operations, the United States increasingly employs digital information in its battlefield systems technology and devises new techniques for disrupting enemy computer networks.

IMPORTANT NAMES AND ADDRESSES 147

RESOURCES 151

INDEX 153

PREFACE

National Security is part of the *Information Plus Reference Series*. The purpose of each volume of the series is to present the latest facts on a topic of pressing concern in modern American life. These topics include today's most controversial and most studied social issues: abortion, capital punishment, care for the elderly, crime, the environment, health care, immigration, minorities, national security, social welfare, women, youth, and many more. Although written especially for the high school and undergraduate student, this series is an excellent resource for anyone in need of factual information on current affairs.

By presenting the facts, it is Thomson Gale's intention to provide its readers with everything they need to reach an informed opinion on current issues. To that end, there is a particular emphasis in this series on the presentation of scientific studies, surveys, and statistics. These data are generally presented in the form of tables, charts, and other graphics placed within the text of each book. Every graphic is directly referred to and carefully explained in the text. The source of each graphic is presented within the graphic itself. The data used in these graphics are drawn from the most reputable and reliable sources, in particular from the various branches of the U.S. government and from major independent polling organizations. Every effort was made to secure the most recent information available. The reader should bear in mind that many major studies take years to conduct, and that additional years often pass before the data from these studies are made available to the public. Therefore, in many cases the most recent information available in 2005 dated from 2002 or 2003. Older statistics are sometimes presented as well, if they are of particular interest and no more-recent information exists.

Although statistics are a major focus of the *Information Plus Reference Series*, they are by no means its only content. Each book also presents the widely held positions and important ideas that shape how the book's subject is discussed in the United States. These positions are explained in detail and, where possible, in the words of their proponents. Some of the other material to be found in these books includes: historical background; descriptions of major events related to the subject; relevant laws and court cases; and examples of how these issues play out in American life. Some books also feature primary documents, or have pro and con debate sections giving the words and opinions of prominent Americans on both sides of a controversial topic. All material is presented in an even-handed and unbiased manner; the reader will never be encouraged to accept one view of an issue over another.

HOW TO USE THIS BOOK

National security has been foremost on the minds of many Americans since the September 11, 2001, terrorist attacks. The United States has taken many measures as a result, including the formation of the Department of Homeland Security, increased security measures at airports, and the enactment of the Patriot Act of 2001. This book covers all of these topics and more, providing the history of national security in the United States; descriptions of the various conventional and nonproliferation treaties and regimes; and information on countries of proliferation concern to the United States. Nuclear, chemical, and biological weapons are discussed in detail, as are domestic and international terrorism and Americans' feelings regarding national security after the September 11, 2001, terrorist attacks.

National Security consists of eleven chapters and three appendices. Each of the chapters is devoted to a particular aspect of U.S. national security. For a summary of the information covered in each chapter, please see the synopses provided in the Table of Contents at the front of the book. Chapters generally begin with an overview of the basic facts and background information on the chapter's topic, then proceed to examine sub-topics of particular interest. For example, Chapter 5: Preparing for Biological and Chemical Attacks begins with a description of the federal role in preparing for chemical and/or biological attacks, including funding for research and preparedness. The chapter then moves into a discussion of various chemical and biological agents, including the possible threat and preparedness activities for each agent. One such biological attack in the United States, the Dalles Incident, is covered in detail. Finally the chapter examines the threat of chemical or biological attacks on animals and plants in the United States, their potential economic impact, and how the U.S. government defends against this threat. Readers can find their way through a chapter by looking for the section and sub-section headings, which are clearly set off from the text. Or, they can refer to the book's extensive index, if they already know what they are looking for.

Statistical Information

The tables and figures featured throughout *National Security* will be of particular use to the reader in learning about this topic. These tables and figures represent an extensive collection of the most recent and valuable statistics on national security, as well as related issues—for example, graphics in the book cover common chemical warfare agents, U.S. State Department-designated foreign terrorist organizations, the organization of the U.S. Department of Defense, the numbers and locations of domestic hate groups, and world oil transit "chokepoints." Thomson Gale believes that making this information available to the reader is the most important way in which we fulfill the goal of this book: to help readers understand the issues and controversies surrounding national security in the United States and reach their own conclusions.

Each table or figure has a unique identifier appearing above it, for ease of identification and reference. Titles for the tables and figures explain their purpose. At the end of each table or figure, the original source of the data is provided.

In order to help readers understand these often complicated statistics, all tables and figures are explained in the text. References in the text direct the reader to the relevant statistics. Furthermore, the contents of all tables and figures are fully indexed. Please see the opening section of the index at the back of this volume for a description of how to find tables and figures within it.

Appendices

In addition to the main body text and images, *National Security* has three appendices. The first is the Important Names and Addresses directory. Here the reader will find contact information for a number of government and private organizations that can provide further information on aspects of national security. The second appendix is the Resources section, which can also assist the reader in conducting his or her own research. In this section, the author and editors of *National Security* describe some of the sources that were most useful during the compilation of this book. The final appendix is the index.

ADVISORY BOARD CONTRIBUTIONS

The staff of Information Plus would like to extend their heartfelt appreciation to the Information Plus Advisory Board. This dedicated group of media professionals provides feedback on the series on an ongoing basis. Their comments allow the editorial staff who work on the project to continually make the series better and more user-friendly. Our top priorities are to produce the highest-quality and most useful books possible, and the Advisory Board's contributions to this process are invaluable.

The members of the Information Plus Advisory Board are:

- Kathleen R. Bonn, Librarian, Newbury Park High School, Newbury Park, California

- Madelyn Garner, Librarian, San Jacinto College—North Campus, Houston, Texas

- Anne Oxenrider, Media Specialist, Dundee High School, Dundee, Michigan

- Charles R. Rodgers, Director of Libraries, Pasco-Hernando Community College, Dade City, Florida

- James N. Zitzelsberger, Library Media Department Chairman, Oshkosh West High School, Oshkosh, Wisconsin

In addition, Information Plus staff owe special thanks to Dr. Harold Molineu, Professor of Political Science at Ohio University, for his particular assistance as an acting advisor on *National Security*. Dr. Molineu's substantial background in the field allowed him to provide expert advice and indispensable recommendations on content and organization.

COMMENTS AND SUGGESTIONS

The editors of the *Information Plus Reference Series* welcome your feedback on *National Security*. Please direct all correspondence to:

Editors
Information Plus Reference Series
27500 Drake Rd.
Farmington Hills, MI 48331-3535

CHAPTER 1

AN INTRODUCTION TO NATIONAL SECURITY

Throughout its history the United States has acted to protect its territories, citizenry, and interests at home and abroad. The terrorist events of September 11, 2001, heightened public interest in national security matters, and the U.S. government has asked all Americans to be watchful, suspicious, and alert to signs of danger or potential security threats. Since September 11, 2001, rarely does a day pass without media attention focused on national security issues.

As new conflicts emerge and power shifts occur worldwide, America's role and responsibilities in terms of ensuring its own and other nations' security is evolving. As a global leader, the United States is challenged to continually revisit its fundamental values and to reset its national security agenda accordingly.

DEFINING NATIONAL SECURITY

At the beginning of the twenty-first century, the term national security is used in the United States to describe both the concept and philosophy of protecting and defending the nation and as well as the specific programs and actions undertaken to achieve this important goal. The concept of national security has been defined in different ways throughout the years. However, most definitions of national security center not only on building and supporting the capacity to safeguard U.S. citizens but also on maintaining public confidence in the government's ability to defend against threats to national values, integrity, and property. In 1962 Arnold Wolfers wrote in *Discord and Collaboration* (Baltimore, MD: Johns Hopkins University Press) that "Security, in an objective sense, measures the absence of threats to acquired values, in a subjective sense, the absence of fear that such values will be attacked." More than three decades later, Sam Sarkesian observed in *U.S. National Security: Policy Makers, Processes, and Politics* (Boulder, CO: Lynne Reiner Publishers, 2nd ed., 1995) that "National security is the confidence held by the

great majority of the nation's people that the nation has the military capability and effective policy to prevent its adversaries from effectively using force in preventing the nation's pursuit of its national interests."

In 2002 President George W. Bush described the philosophical underpinning of the U.S. national security strategy as efforts aimed at protecting the fundamental values of freedom and human dignity.

> The U.S. national security strategy will be based on a distinctly American internationalism that reflects the union of our values and our national interests. The aim of this strategy is to help make the world not just safer but better. Our goals on the path to progress are clear: political and economic freedom, peaceful relations with other states, and respect for human dignity. — President George W. Bush, *The National Security Strategy of the United States of America,* 2002

FEDERAL AGENCIES AND ORGANIZATIONS THAT FOCUS ON NATIONAL SECURITY ISSUES

Providing for the national security of the United States is one of the primary duties of the federal government. The Executive Office of the President, U.S. Senate, and the House of Representatives act to establish policy and allocate resources to ensure national security. These policies are then carried out by the wide variety of federal agencies and offices that focus on diplomacy, trade, arms control, military intervention, espionage, money laundering, the environment, and immigration.

Executive Office of the President

The president is the highest executive leader of the United States and the commander in chief of all of the armed forces. As such, the president and his deputies are ultimately responsible for most decisions about national security.

Established by the National Security Act of 1947 (and later amended by the National Security Act Amendments of 1949), the National Security Council (NSC) is a part of the Executive Office of the President. Its primary function is to assist the president on all foreign policy and national security matters. The NSC is headed by the president, and its regular participants include the vice president, secretary of state, secretary of the treasury, secretary of defense, assistant to the president for national security affairs (also called the national security adviser), the chairman of the Joint Chiefs of Staff, and the director of central intelligence. Other key personnel, such as the president's chief of staff, counsel to the president, attorney general, and the assistant to the president for economic policy, are also invited to attend NSC meetings when necessary. Other branches of the Executive Office of the President that are involved in national security matters include:

- Office of Management and Budget
- National Security and International Affairs Division
- Office of National Drug Control Policy
- Office of Science and Technology Policy
- Office of the U.S. Trade Representative

The United States Congress

The U.S. Congress, comprising the House of Representatives and the Senate, is crucial to any national security policy. The U.S. Constitution does not assign the executive branch supreme authority—it must have the support of its legislature. All laws and resolutions must be passed through both houses of Congress before the president can act on them. Treaties negotiated by the president must be ratified before the Senate before they are official. Select committees within Congress (such as the Committee on International Relations and the Subcommittee on Intelligence) also serve to oversee policies related to national security and international affairs. Congressional committees and offices related to national security include:

- Congress Budget Office, Defense and International Affairs
- Congressional Research Service
- U.S. Senate Committees: Appropriations; Armed Services; Commerce, Science, and Transportation; Energy and Natural Resources; Finance; Committee on Foreign Relations; Committee on the Judiciary; Select Committee on Intelligence
- U.S. House of Representatives Committees: Appropriations; Banking and Financial Services; Budget; Commerce; Economic and Educational Opportunities; International Relations; Judiciary; National Security; Resources; Science; Transportation and Infrastructure; Permanent Select Committee on Intelligence

Government Agencies and Organizations

Key agencies and organizations of the U.S. government are involved in the implementation of national secu-rity initiatives and programs that gather tactical military intelligence; respond to transnational threats, terrorism, or cyber warfare; provide border security; and aim to thwart narcotics trafficking, international organized crime, and the covert employment of weapons of mass destruction. Perhaps chief among government organizations is the Department of Defense, which includes the U.S. Air Force, Army, Marine Corps, and Navy. The armed forces all play vital roles in safeguarding the country's national security and are called in to enforce and defend U.S. security policies when needed. Figure 1.1 shows the involvement of these forces in the intelligence community. Additional agencies within the Department of Defense support military and security operations. Other federal entities involved in national security include:

- The Department of Commerce—Bureau of Industry and Security, International Trade Administration
- The Department of Energy—National Nuclear Security Administration, National Laboratories and Technology Centers
- The Department of Homeland Security—the U.S. Coast Guard, the Citizenship and Immigration Service, the Customs and Border Protection Service, the Office for Domestic Preparedness, the Transportation Security Administration
- The Department of Justice—the Bureau of Alcohol, Tobacco, Firearms and Explosives, the Drug Enforcement Administration, the Federal Bureau of Investigation, the National Drug Intelligence Center
- The Department of State—Foreign embassies, regional and functional bureaus, the National Foreign Affairs Training Center
- The Department of the Treasury—Office of International Affairs
- Independent Agencies—the Peace Corps, U.S. Arms Control and Disarmament Agency, U.S. International Trade Commission, Export-Import Bank, Overseas Private Investment Corporation, U.S. Agency for International Development
- Federally Funded Research and Development Centers—the Center for Naval Analyses, RAND Corporation, the Institute for Defense Analyses

THE INTELLIGENCE COMMUNITY. Headed by the director of central intelligence (DCI), the U.S. intelligence community includes the Central Intelligence Agency, Defense Intelligence Agency, National Security Agency, Army Intelligence, Navy Intelligence, Air Force Intelligence, Marine Corps Intelligence, National Geospatial Intelligence Agency, National Reconnaissance Office, and aspects of the Federal Bureau of Investigation, Department of Homeland Security, Department of the

FIGURE 1.1

The intelligence community

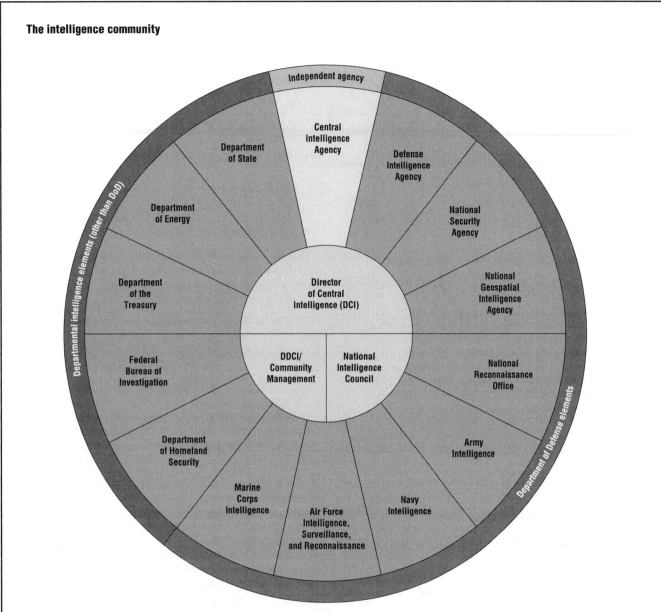

SOURCE: "The Intelligence Community," in *Factbook on Intelligence, 2004,* Central Intelligence Agency, 2004, http://www.cia.gov/cia/publications/facttell/intelligence_community.html (accessed September 22, 2004)

Treasury, Department of Energy, and Department of State. The main goal of the intelligence community is to prevent, detect, and defend against immediate and long-term threats to national security. The DCI is aided by another group, called the National Intelligence Council, which is composed of government intelligence officers as well as members of the public (academic or nonprofit organizations) and private sectors.

The intelligence community also includes national centers to address specific security threats: the Counterterrorist Center, the Crime and Narcotics Center, the Directorate for MASINT (Measurement and Signature Intelligence) and Technical Collection, the El Paso Intelligence Center, Information Analysis Infrastructure Protec-

tion, the National Drug Intelligence Center, the National Virtual Translation Center, the Terrorist Screening Center, the Terrorist Threat Integration Center, and the Weapons Intelligence, Nonproliferation and Arms Control Center.

DEFINING THREATS TO NATIONAL SECURITY

In general, threats to national security are distinguished from other common crimes by their scale and scope of action and because they:

• Have the potential to take more lives or cause much more damage and disruption

• Arise from a source in the international arena, usually as a reaction against a state's foreign policy

- Aim to make a political, ideological, or religious statement

- Are larger, more menacing, and perceived as a greater threats than those normally managed by local law enforcement authorities

National security threats may come from governments (also called "states"), subnational entities (such as terrorist groups, organized-crime networks, or companies practicing industrial espionage), external intelligence-gathering agencies (private individuals or agencies hired to gather information), or even U.S. citizens. Prime targets for threats against U.S. national security include:

- U.S. citizens outside the country, including tourists traveling abroad, soldiers on active duty, and the diplomatic community

- U.S. property outside the country, including U.S. embassies, military facilities, naval ships, factories, and offices

- U.S. citizens, transportation centers, landmarks, ports, edifices and other physical structures within American borders

Throughout its history the U.S. government has not hesitated to take action to defend national security. Usually such actions, whether preemptive (to prevent or avert a potential threat) or defensive (in response to a real and present threat), are taken when there are significant risks to national interests. Risks to national interests are classified in response to the urgency and immediacy of the threat and its potential targets. The three broad classes of national security risks are:

1. Vital interests, or issues directly concerning the survival and safety of the country. These threats may involve physical territory, safety of citizens, or even danger to a close ally. A threat to a vital interest may well be met with military force.

2. Important national interests or risks in which national survival is not at stake. Such risks may imperil the world at large and could potentially escalate to threaten vital interests. Military mobilization and resource commitment vary depending on the situation.

3. Humanitarian or other interests, or those where the primary focus is on containing the problem and averting escalation rather than defending national security interests.

POWER AND NATIONAL SECURITY

A state's power, measured in terms of its resources, largely determines its ability to fulfill its national security agenda. Since definitions of the term "power" and the metrics applied to it vary, quantifying a state's power is challenging. In *Foundations of National Power* (New York: Van Nostrand, 1951), Harold and Margaret Sprout posited that a nation's power can be measured roughly by using the following equation:

> National Power = human resources + physical habitat + foodstuffs and raw materials + tools and skills + organization + morale and political behavior + external conditions and circumstances.

Nation-states (a term used to describe a relatively homogenous population or persons of common nationality who inhabit a state) apply power in order to bring about the changes in the political, social economic, or military arenas.

MILESTONES IN THE DEVELOPMENT OF U.S. NATIONAL SECURITY POLICY

The Early Twentieth Century

Modern U.S. national security policy and activities had their origins in intelligence gathering, code-breaking, and monitoring of international cables conducted during World War I (1914–18). In the following decade mechanical cipher machines automated encryption (the process of translating a message into an encoded message), and aerial photography enhanced surveillance capabilities.

During World War II (1939–45) U.S. military intelligence activities intensified. Cryptographers triumphed when they deciphered Japanese codes, and radar emerged as yet another means of surveillance. At the close of World War II, the United States emerged as the most powerful nation in the world. Still much of the world was ideologically and politically divided between the capitalist camp, headed by the United States, and the communist bloc (a group of nations), led by the Soviet Union. The fear and distrust that existed between the United States and the Soviet Union prompted the cold war (1946–90), a period of suspicion and rivalry that rapidly progressed to overt hostility between the United States and the Soviet Union. Cold war policies, particularly the emphasis on military readiness, laid the foundation for national security policy for decades to come.

The cold war standoff led to a massive arms expansion on both sides. Although the Soviet Union had a stronger military land force, America's naval superiority and its sole possession of the atomic bomb until 1949 gave it unrivaled power. An evaluation conducted by a joint Department of State/Department of Defense committee, which delivered its findings to the National Security Council in 1950 in a document labeled NSC 68, also spurred the U.S. military buildup. The document called for a massive buildup and increased funding for the armed forces in an effort to prevent the expansion of Soviet influence. This "containment" policy shaped U.S. national security thinking for decades.

U.S. policies for containing the Soviet Union were further supported by the Truman Doctrine and the Marshall Plan. The 1947 Truman Doctrine was a response to a growing communist threat in Greece and Turkey and marked the beginning of U.S. intervention around the world in the name of anticommunism. President Truman stressed the duty of the United States to combat totalitarian regimes (governments that seek total control of economic and political matters as well as the attitudes, values, and beliefs of the populace) worldwide. In an address before a joint session of Congress on March 12, 1947, he stated: "I believe that it must be the policy of the United States to support free peoples who are resisting attempted subjugation by armed minorities or by outside pressures. I believe that we must assist free peoples to work out their own destinies in their own way. I believe that our help should be primarily through economic and financial aid which is essential to economic stability and orderly political processes."

The Marshall Plan, named for its author, Secretary of State George C. Marshall, was an economic aid package intended to help restore the economies of Europe (including West Germany) following the end of World War II. The plan served to re-establish stability in Europe and promote American interests.

In July 1947 the National Security Act created the National Military Establishment (Department of Defense), the Central Intelligence Agency (CIA), an independent Air Force (the Air Force had formerly been a division of the Army), and a cabinet-level Secretary of Defense. In 1948 the Armed Forces Security Agency was created to oversee the military intelligence agencies—Army Security Agency, Naval Security Group, and Air Force Security Service. One year later, in 1949, the Soviet Union tested its first atomic bomb.

The 1950s

During the 1950s the Korean War (1950–52) and advances in Soviet nuclear technology led the United States to reassess its strategic national security policies. Enormous military buildup and nuclear weapons made both the Soviets and Americans virtually impenetrable. In the mid-1950s President Dwight Eisenhower sought to achieve a balance between military spending and the health of the domestic economy. The Joint Chiefs of Staff (the chairman, vice chairman, Chief of Staff of the Army, Chief of Naval Operations, Chief of Staff of the Air Force, and the Commandant of the Marine Corps) examined U.S. strategic forces and made several recommendations:

- Withdrawal of some U.S. troops from abroad

- Creation of a mobile strategic reserve (a well-prepared, well-trained, and well-equipped military unit on reserve—as opposed to active duty—that could be mobilized quickly in the event of a threat to national security)

- Strengthening alliances while allowing allied forces to rely primarily on their own defenses

- Further investment in U.S. air defenses

In 1954 President Eisenhower introduced his "New Look" policy, which implemented these recommendations of the Joint Chiefs of Staff. It also articulated the threat of massive retaliation as a deterrent against Soviet aggression. "Massive retaliation" involved instantaneous reaction to any Soviet threat using any means necessary, including nuclear weapons. U.S. officials reasoned that the Soviet Union would not attack the United States, even if it knew it could do major damage, if it knew that the inevitable result of such an attack would be severe damage to itself.

Also during the 1950s science and technology began to play an increasing role in American intelligence gathering and strategy. In March 1950 the United States Communications Intelligence Board was created, and in 1952 President Truman created the National Security Agency (NSA) under the authority of the Secretary of Defense. The NSA replaced the Armed Forces Security Agency and initially focused on code-breaking and other intelligence activities.

"Vulnerability of U.S. Strategic Air Power," a RAND report issued in April 1953, asserted that the United States was at risk of attack by Russian bombers because the country lacked a warning system. The report also assailed U.S. intelligence gathering capabilities and along with the perceived intelligence failures during the Korean War, it prompted the NSA to establish the first worldwide network of listening posts to intercept and monitor communications. Initially equipped with radio receivers, the stations quickly employed more sophisticated technology, including radar, computers, and satellites to obtain information.

The 1960s

The 1962 Cuban Missile Crisis demonstrated the utility of the U.S. intelligence community's increasing ability to collect and analyze information. NSA listening posts detected communications between Russian ground controllers and Cuban pilots that revealed the Soviet Union's intention to move missiles into Cuba. U.S. surveillance flights over Cuba detected the Cuban missile sites, and enhanced intelligence capabilities proved to be pivotal in assisting the U.S. strategic management of the crisis.

The rapid nuclear arms buildup by both the Soviets and the Americans prompted the United States to reconsider its heavy reliance on nuclear weapons and to explore other policy options. Under the administration of President John F. Kennedy, the 1960s witnessed the beginning of yet another national security doctrine, called "flexible response." It was developed as the United States began to reexamine its diminishing reliance on conventional weapons and the increasing role played by tactical nuclear weapons in Europe. Many U.S. policy makers and military leaders believed that if tensions heightened considerably, little would prevent an escalation from use of tactical nuclear weapons (small-scale nuclear weapons for use on the battlefield) to strategic nuclear weapons (more powerful weapons used to strike an enemy's military, economic, or political power sources). Policy makers speculated not only that the doctrine of massive retaliation would be ineffective when dealing with lower-level conflicts, such as clashes with Soviet proxies in smaller countries, but also that a real possibility existed that the world might become engaged in an all-out nuclear war.

Flexible response policy gave the president the ability to choose the appropriate level of force needed to deal with a wide range of challenges. The president could opt for a massive nuclear retaliation or a limited counterforce (attacking only the opponent's force structure) or countervalue (attacking the opponent's cities and populace) nuclear strike. Conventional forces were also strengthened and improved under this doctrine in order to shift away from heavy reliance on nuclear weapons. The military reserves and National Guard were expanded, the number of navy warships and army divisions was increased, and counterinsurgency (antirevolution or anti-revolt) forces were enlarged. These conventional military forces were put to use in fighting the Soviet-backed North Vietnamese during the Vietnam War (1959–75).

Flexible response did not mean the end of the buildup of U.S. and Soviet nuclear weapons. By the late 1960s both the Soviet Union and the United States had acquired second-strike nuclear capability, meaning that each country was able to mount a serious retaliation against a nuclear strike on its territory.

The close of the decade saw the intensified efforts to gather domestic and foreign intelligence. The Intelligence Evaluation Committee (IEC) was established, the CIA ordered installation of monitoring devices in foreign embassies, and the NSA initiated a program that monitored the communications of some six hundred American citizens.

The 1970s

By the 1970s serious questions about U.S. national security policy arising from the unsuccessful American intervention in the Vietnam conflict prompted another reevaluation. The need for U.S. involvement to counter communism and contain the Soviet threat was coming under increasing scrutiny by both U.S. critics and the world at large. During this period U.S. policy makers developed the idea of "strategic sufficiency" (maintaining enough military might to deter the enemy from coercing a country or its allies) in order to preserve the doctrine of flexible response while simultaneously ensuring enough retaliatory power to guarantee mutually assured destruction (MAD) in case of a war. To reduce tensions and the possibility of a conflict between the United States and the Soviet Union, the administration of President Richard M. Nixon introduced the con-

cept of "détente." As part of this approach, representatives of the United States and the Soviet Union met in 1969 for the Strategic Arms Limitations Talks (SALT I), in hopes of limiting the number of missiles each country deployed. In May 1972 Richard Nixon became the first American president to visit the Soviet Union, and he signed the SALT I Treaty, freezing the number of missiles at current levels for five years. President Nixon also pledged to increase trade and scientific cooperation between the two countries in such areas as space exploration.

In 1978 the Foreign Intelligence Surveillance Act authorized the creation of a Foreign Intelligence Surveillance Court—eleven judges with the authority to issue warrants to the FBI and NSA for domestic surveillance. The following year the Special Collection Service (SCS)—a joint organization of CIA and NSA—began deploying trained agents on intelligence gathering missions.

The End of the Cold War

The cooperation of the détente period ended with the election of President Ronald W. Reagan in 1980. Reagan believed the Soviet Union had benefited disproportionately under détente and that the United States was held to fewer nuclear missiles than the Soviet Union. In addition Reagan believed that a renewed arms race would bankrupt the economically unsound communist regime and hasten its collapse. During the Reagan administration, spending on the U.S. military increased sharply. While some observers claim the collapse of the Soviet Union was the result in part of the heavy losses the Soviet army suffered in its invasion of Afghanistan (1979–88), others credit Reagan's defense spending as driving the weak Soviet economy to the breaking point. What is undeniable is that in the late 1980s and early 1990s, the Soviet Union fell apart. Eastern European states previously dominated by the Soviets overthrew their communist governments, and in 1991 the Soviet Union itself broke down into over a dozen different nations, of which Russia was by far the largest.

THE GLOBAL CONSEQUENCES OF THE BREAKUP OF THE SOVIET UNION

The 1991 dissolution of the Soviet Union into independent states and its gradual conversion to a capitalist economy brought an end to the bipolar global structure in which most world power rested in the hands of the U.S. and Soviet blocs. What has happened to the balance of power in the world since then is a matter of debate. One theory is that the world has primarily followed an unrivaled U.S. leadership, or "unipolar hegemony." This notion of unipolarity was strongly validated when the United States led a forceful alliance against Iraq during the Gulf War in 1991.

Other international analysts contend that this theory grossly overstates American dominance and that, in fact, the world is much more "multipolar" in nature. A multipolar global order relies on international interdependence, in which each region finds its own optimal power structure. Power, in a multipolar world—whether military, economic, or political—varies from nation to nation, and each uses its strength to fight for survival and dominance.

Joseph Nye, Dean of the Kennedy School of Government at Harvard University and a prominent theorist on international relations, developed yet another model—that of "multilateral interdependence," in which the world power structure can be compared to a three-layer cake. The top military layer is mainly unipolar, since not many states can rival the military might of the United States. The middle economic layer is tripolar, consisting primarily of the strong U.S., western European, and Japanese economies. The bottom layer is made up of transnational interdependence among a number of states. As the idea of interdependence grows, states realize that cooperation is essential to carrying out their policies, especially in the realm of national security.

For the United States this notion is evidenced by the country's participation in various international organizations and regional alliances under the umbrella of "collective security." Collective security is the idea that a group of states sharing common interests should join together against any potential aggression or opposition. A threat to one means a threat to all. This is not a recent phenomenon and can be traced back centuries to ancient times. In the early twenty-first century, the United Nations (UN) and the North Atlantic Treaty Organization (NATO) are the primary bodies through which the United States engages in collective security. Bilateral military alliances (between the United States and one other country) or multilateral alliances (involving several countries) are other avenues for collective security.

CHANGING GLOBAL DYNAMICS: THE UNITED STATES AS A GLOBAL LEADER

To understand national security, it is necessary to consider a country's policies within the context of its history, ideology, and existing political governance. As societies and states continue to evolve, so do the relationships between them. Each generation sees a new set of global conflicts and political alliances, as well as shifts in interstate dynamics.

The twentieth century witnessed a host of political phenomena, such as the fall of colonialism, the rise of capitalist economies, the growth of industry, growing concern for human rights, and the spread of nuclear weapons. All countries, including the United States, were challenged to adapt and modify their policies in response to shifts in global power. Until 1991 much of the globe was involved in the cold war, which divided the world along ideological lines. Many countries found themselves allying with either the Soviet or U.S.-led blocs, though others remained nonaligned.

The world in the early twenty-first century is very different from what it was even just a few decades ago. The United States is a leading superpower both in economic and military might. The economic, social, cultural, and military trends that accompany globalization (the increasing integration of world markets for goods, services, and capital that transcend national borders) have influenced the United States to adopt policies that strengthen its position as a global leader. This global presence and preeminence is a double-edged sword—it creates opportunities for the U.S. domestically and abroad and also incites conflicts, threatening national security by rendering the United States and its allies targets. It also pressures the United States to stretch, and possibly even overextend, its military forces.

UNDERSTANDING THREATS TO NATIONAL SECURITY

All states seek self-preservation, and ultimately it is up to each country to craft a security framework to ensure its own survival. Although the strategies have varied, the goals of American national security programs have always been to protect the sovereignty of the United States and to protect U.S. interests. The country's political leaders, military, and intelligence community establish and execute a national security policy. To enact policy, they employ such actions and strategies as diplomacy, military intervention, trade agreements, and alliances to advance the U.S. national security agenda.

Since the events of September 11, 2001, preventing and defending against international terrorism and strengthening U.S. homeland security have dominated the national security agenda; however, other international concerns continue to influence America's national security policies:

• The United States depends on oil and gas imports from the Persian Gulf. The unpredictable prices of these imports, combined with declining U.S. oil production, present problems for the United States. In addition, U.S. and UN sanctions against Iran, first imposed in October 1987 because of that country's continuing support of international terrorism and its aggressive actions against shipping in the Persian Gulf (*Iran: What You Need to Know about U.S. Economic Sanctions,* Washington, DC: U.S. Department of the Treasury, Office of Foreign Assets Control, 2003), do not allow it to increase its oil output and have generated hostility toward the nations and organizations imposing the sanctions.

• Several nations, including North Korea and Iran, have aggressive programs to develop weapons of mass destruction (WMD).

- The United States and other countries, including U.S. allies, have significant political and ideological differences about policies in the Middle East and Southeast Asia.

- The United States is concerned about the ongoing Arab-Israeli conflict.

- There is rising anti-U.S. sentiment around the world, ranging from sharp disapproval of U.S. foreign policy to extreme anti-Americanism.

- WMD and small arms threaten to proliferate worldwide.

- Combating the illegal drug trade drains millions of dollars each year and commits national resources and personnel to a fight that has no foreseeable conclusion.

- Various "transition states" and "states of concern," two types of geopolitical entities, may be either potential allies or may work overtly or covertly to undermine U.S. national security. Transition states and states of concern are explained in greater detail in the next two sections.

Transition States

A transition state is a country that is slowly becoming more like a traditional Western capitalist society. Examples of transition states include Eastern European countries, China, and India. To encourage these countries to make the transition to capitalism, the United States promotes (a) the growth of market democracy in the country and (b) increases in the country's per capita gross domestic product (GDP), or the total value of the goods and services it produces. This policy often involves the United States in interventions in unstable parts of the world. Often, American involvement and intervention in these countries is viewed as evidence of increasing Western interference with indigenous cultures and ways of life. For example, U.S. support for the dismantling of state-owned industries, water, and electrical power often causes sharp price increases. As a result, poor people cannot afford these vital resources. However, the private utility companies that provide and purchase these resources benefit.

RUSSIA, CHINA, AND INDIA. The three large transitional states on the Eurasian landmass—Russia, China, and India—present the United States with opportunities in terms of economics and strategic alliances. However, the United States also prefers to maintain a balance of power and prevent the regional dominance, or hegemony, of any one of these countries. Following established national security policy guidelines, America strives to counter the spread of conventional weapons and WMD in and from these countries.

Russia, China, and India are not strong U.S. allies like some western European nations, and their transition to free-market economies is far from complete. Among these transition states, China may have the best chance to fulfill the promise of its potential. Between 1978 and 2003, China's gross domestic product (GDP) quadrupled, according to the *The World Factbook* (Washington, DC: Central Intelligence Agency, 2004), rising to $6.45 trillion. India's 2003 GDP was $3.03 trillion, while Russia's was $1.28 trillion. India has averaged an annual growth in GDP of 6% since 1990. China was, from 1979 through 1999, the world's fastest-growing economy.

With a population of nearly 1.3 billion people (according to the CIA's *World Factbook*) and an expanding role in Asia, China is a rising power. The Soviet Union's collapse permitted China to reduce its military forces in the north and devote more military resources to the south and southeast. China has also expanded its participation in the Association of Southeast Asian Nations (ASEAN) and the ASEAN Regional Forum. In the late 1990s its economic success allowed it to provide a $1 billion loan to Thailand through the International Monetary Fund. It is one of the five powers (along with Russia, North Korea, South Korea, and the United States) negotiating the future of the Korean Peninsula.

China's defense spending has risen steadily, from $14.6 billion in 2000 to $20.4 billion in 2002. (See Table 1.1.) The estimate for 2003 is $60 billion, according to the *World Factbook*. In addition, the amount China spends on defense as a percentage of its GDP has also risen, from 1.2% in 1998 to 1.5% in 2001. (See Figure 1.2.)

China has also become more assertive regionally. It claims the Senakaku Islands, which Japan also claims, and the Spratly Islands in the South China Sea, which are also claimed by the Philippines, Vietnam, Brunei, Malaysia, and Taiwan. To reinforce the latter claim, China seized Mischief Reef in the Spratlys in 1995. In May 1996 China formally expanded its claimed sea area from 370,000 to 3,000,000 square kilometers (from 142,858 to 1,158,307 square miles).

Perhaps most important, China claims that the island of Taiwan is part of China because it was considered Chinese territory until the end of World War II, when military leader Chiang Kai-shek fled to the island to form a separate noncommunist state. Taiwan, which is richer, more confident, and more democratic than China, but much smaller, challenges the assertion that it should be considered part of mainland China.

The world got a glimpse of China's threatening ambitions in 1996, when it fired ballistic missiles into the waters around Taiwan. The Taiwan issue is a flashpoint in Chinese-American relations. Officially, the United States calls for a "one-China" policy, which recognizes the rights of the mainland government, based in the capital of Beijing. However, the United States also pledges to defend Taiwan at all costs if an attempt is made to militarily force the island to unify with mainland China. For

TABLE 1.1

China's annual defense expenditures, 2000–02

(in billions of U.S. dollars. 1 U.S. dollar=8.3 ren min bi)

Year	Personnel	Maintenance and operations	Equipment	Total
2000	4.89	4.97	4.69	14.55
2001	5.56	5.85	5.96	17.37
2002	6.51	7.00	6.90	20.41

SOURCE: Howard M. Krawitz, "China's Annual Defense Expenditures, 2000–02," in "Modernizing China's Military: A High-Stakes Gamble?" *Strategic Forum,* Institute for National Strategic Studies, National Defense University, December 2003, http://www.ndu.edu/inss/strforum/SF204/ SF204.pdf (accessed September 22, 2004)

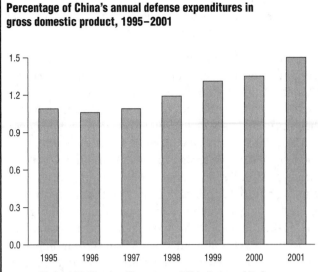

FIGURE 1.2

Percentage of China's annual defense expenditures in gross domestic product, 1995–2001

SOURCE: Howard M. Krawitz, "Percentage of China's Annual Defense Expenditures in Gross Domestic Product, 1995–2001," in "Modernizing China's Military: A High-Stakes Gamble?" *Strategic Forum,* Institute for National Strategic Studies, National Defense University, December 2003, http://www.ndu.edu/inss/strforum/SF204/SF204.pdf (accessed September 22, 2004)

American policy makers, the ultimate goal is to seek a peaceful reunification between the two countries.

In the late 1990s the administration of President Bill Clinton called for a strategy of "engagement" in U.S.-China relations that would reflect the principles of "taming the dragon." In a speech presented at the National Geographic Society in June 1998, President Clinton stated that "bringing China into the community of nations, rather than trying to shut it out, is plainly the best way to advance both our interests and our values."

Russia, China, and India are major military powers in Asia in terms of both conventional and nuclear weapons. Russia is India's leading arms supplier, although Britain, France, Germany, and Israel have contributed to the development of its weapons arsenal as well. In the decades since India and Pakistan were partitioned in 1948, the two countries have continued to dispute ownership of the region of Kashmir. Each nation blames the other for the consequent arms buildup, which includes nuclear weapons. Both countries have tested nuclear weapons, creating further tensions and animosity. Pakistan continues to engage in significant weapons transactions with China and North Korea.

India exercises regional influence over such surrounding states as Bangladesh, Sri Lanka, Nepal, and Bhutan. China, however, like Pakistan, presents a serious obstacle to India and is a serious strategic rival. In late 1996 Chinese president Jiang Zemin's visit to India led to a significant thaw in India-China relations. The two sides agreed to set aside border disputes, but the relationship between the two countries continues to be one of serious rivalry over regional influence, global status, energy access, foreign investment, and trade. Senior Indian officials identified China, not Pakistan, as the key reason for their May 1998 nuclear tests.

Relations between India and China are of strategic concern to the United States because both countries have nuclear weapons and both possess ambitions to regional domination. Good relations between India and China are perceived as necessary by the United States in order to maintain regional stability for both economic and political purposes.

All three transition states—Russia, India, and China—have the military potential to greatly upset their regions of the world. According to Global Security.org (http://www.globalsecurity.org), Russia has 1.2 million troops in Eurasia and plans to reduce that number to 800,000 by 2006. Its offensive forces no longer menace Europe, but it is still far stronger than its immediate neighbors. In South Asia, India has twice as many troops as its rival, Pakistan, but both are equipped with nuclear weapons, and in recent years the two countries' animosity over disputed Kashmir has intensified.

Cooperation with and among these states greatly improves U.S. national security in and around Eurasia. Although the United States cannot dictate policies to sovereign nations, it can offer diplomatic channels and mediation opportunities to try to enhance the countries' transitions to free-market economies and democratic societies.

States of Concern

States of concern (also known as "rogue states") are countries perceived as dangerous to the United States and its allies. Such states as Iran, North Korea, Libya, Syria, Sudan, and Cuba, which have demonstrated hostility toward the United States, are considered states of concern. Iraq was considered a rogue state until the U.S. invasion in 2003 overthrew that country's leader, Saddam Hussein. Libya, too, changed its policies toward the United States. In 2003 Libya took responsibility for its role in the terrorist bombing of Pan American World Airways (Pan Am) Flight 103 over Lockerbie, Scotland, in December 1988, and agreed to pay restitution. In December 2003 it agreed to end its programs to develop WMD.

The U.S. Department of State claims that rogue countries have established links to terrorist networks. The Clinton administration defined such states as "recalcitrant and outlaw states that not only choose to remain outside the family of democracies, but also that assault democratic values" (*Strategic Assessment 1999: Priorities for a Turbulent World,* Washington, DC: Institute for National Strategic Studies, National Defense University, 1999). This definition applies to countries that threaten U.S. interests by unconventional and violent means. It is generally agreed that such states can swiftly destabilize their respective regions and other regimes in surrounding areas, and also threaten vital U.S. interests and national security.

As of 2004 many states of concern were ruled by long-standing leaders who were not democratically elected.

- Cuba's leader, Fidel Castro, seized power in 1959 and has led the communist-controlled government ever since.

- Communist leader Kim Jong-il appeared to have consolidated power in North Korea after the death of his father, Kim Il-sung (who led the country from 1948 until his death in 1994).

- Libya's Muammar Qadhafi, who took control in a military coup in 1969, did not face any organized opposition, nor was there a more moderate leadership capable of replacing him.

- Syria's leader, Bashar al-Assad, inherited control of the country after his father, President Hafez al-Assad, who had ruled for thirty years, died in 2000.

- Iran's Muslim cleric leadership, which came to power in the Islamic Revolution of 1979, remained in power, with its chief of state (supreme leader), Ayatollah Ali Khamenei, at the helm since 1989. Though the more liberal government of Iran's elected president, Mohammad Khatami, had made overtures toward reform and had been seeking more democracy in the Islamic republic, Khamenei and his allies opposed many of the changes.

These states seek to gain military strength to support their regimes by buying weapons and materials from sympathetic states manufacturing conventional weapons or WMD. This makes U.S. efforts to isolate such countries by imposing sanctions on weapons and other imports and exports ineffective. Some transition states, as well as some of America's closest allies, do not support U.S. policies such as sanctions to isolate states of concern. For example, most of America's European allies reject U.S. efforts to punish companies doing business with Cuba, Libya, or Iran. France's oil consortium openly challenged such U.S. sanctions by investing in Iranian oil fields, and Canadian companies regularly invest in Cuban businesses. Other countries continue to sell weapons to states considered "rogues" by the United States.

States of concern to the United States are capable of using violence to alter the regional status quo. Both Iran and Iraq competed to control the Persian Gulf region, with Iraq having attacked Iran in 1980 and Kuwait in 1990. North Korea seeks control of the entire Korean Peninsula, which is also claimed by South Korea. Syria seeks to intimidate Israel and to control its part of the Middle East. States of concern may threaten neighboring nations—including U.S. allies—or control local resources, such as petroleum, that are of vital interest to the West.

CHAPTER 2

THE THREAT OF CONVENTIONAL WEAPONS

Weapons are an integral part of any military. Conventional weapons of the early twenty-first century are accurate and deadly enough to destroy almost all types of military targets, including buried command centers, hardened aircraft shelters, and tanks and other armored vehicles. Challenging and combating the proliferation, or spread, of weapons that can be used against the United States and its allies is a top priority on the national security agenda.

Weapons proliferation involves both the spread of arms across national borders and the buildup of states' arsenals. An increase in weapons sales and production, however, is not always associated with wars or other conflicts. Such factors as power, prestige, ideology, and perceived threats can influence greatly a state's decision to buy, sell, or produce weapons. Often companies and states increase their weapons production and export levels because of political and economic factors or the perceived need to stay ahead in technological innovation.

Both conventional arms and weapons of mass destruction (WMD) pose their own set of unique problems. Conventional weapons include all types of armaments that do not fall under the WMD category, including small arms, machine guns, grenades, land mines, armored vehicles, radar equipment, aircraft, submarines, and ships. Key players in arms trafficking are individuals, transnational groups (which may include terrorists, organized crime, religious groups, drug traffickers, multinational businesses, or others), defense contractors, and governments themselves. Trade in surplus weapons is a significant source of revenue for several countries, including the United States.

Key factors in the proliferation of conventional weapons since 1990 have included Iraqi aggression against Kuwait, which made some Persian Gulf states increasingly insecure about their own national security; the collapse of the Soviet Union in 1991, which released weapons systems into the international market that were no longer needed by the newly independent states; and technological innovation that has reduced the time between a weapon's development and its replacement by a newer system, creating a surplus of outdated, but still powerful, weapons that find their way onto the world market.

TRADE IN CONVENTIONAL WEAPONS

Although international transfers of major conventional weapons have generally declined since their peak in 1997, vigorous trade in such weapons continues. (See Figure 2.1.)

Most arms sales go to developing countries, particularly those involved in regional conflicts or disputes. Among the biggest importers of arms are China, Taiwan, Israel, India, and Turkey. As a region, the Near East is the largest buyer of arms, followed by Asia, Africa, and Latin America. African countries have shown the largest increase in arms purchases in the period from 1995 through 2002. (See Table 2.1.)

Trends and Players

In previous decades, the cold war saw the increased export of arms from the U.S. and Soviet blocs to countries around the world to help maintain regional strongholds and, thereby, the balance of power. Both sides were also eager to ship weapons to their Third World allies to support various "national movements" or to suppress insurgent (rebel) activities. Such regional hotspots as Iran/Iraq, Pakistan/India, and Afghanistan became lucrative markets for arms sales. Unfortunately, these weapons were often used by corrupt regimes against ethnic minorities or to commit other human rights violations. Developing countries that spent millions of dollars acquiring weapons and weapons systems diverted resources from critical social programs.

The end of the cold war disrupted the worldwide arms flow but did not bring an end to global hostilities. Con-

FIGURE 2.1

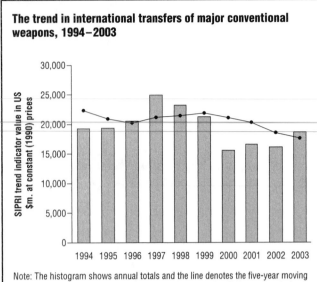

The trend in international transfers of major conventional weapons, 1994–2003

Note: The histogram shows annual totals and the line denotes the five-year moving average. Five-year averages are plotted at the last year of each five-year period.

SOURCE: "Figure 12.1.1.The Trend in International Transfers of Major Conventional Weapons, 1994–2003," in *SIPRI Yearbook 2004*, Stockholm International Peace Research Institute, August 2003, http://web.sipri.org/contents/armstrad/atglobal_trend_94-03.pdf (accessed September 23, 2004)

flicts in the Balkans and Africa, as well as intrastate and interstate tensions in the former Soviet bloc, continued to create markets for weapons. Although the disintegration of the Soviet Union temporarily halted arms sales from that region, the immediate need for hard currency in the newly independent states and the abundance of Soviet-made weapons and weapons systems drove them onto the arms market.

The first years of the twenty-first century have seen an interesting pattern of buyers and sellers that is different from the sales patterns established during the cold war. Key Western allies, such as Turkey and the United Arab Emirates (UAE), have turned to the former Soviet states for certain weapons systems, as have China and Iran. Russia aggressively marketed its defense equipment in order to raise funds to help stabilize its fledgling market economy and finance its oil sector. Despite Russia's internal political upheavals and problems within the military, countries including Kuwait, India, the UAE, and Malaysia still view it as a strong military supplier. Russia sold T-72 tanks to Syria and Su-30 fighter jets and SA-20 surface-to-air missiles to China, much to the displeasure of the American government.

The U.S. Role

U.S. ARMS EXPORTS. Western arms sales go not only to developing countries but also to such key U.S. allies as Israel, Taiwan, certain Persian Gulf states, and NATO countries. Frequently, countries such as Pakistan and Israel use U.S. military assistance programs to finance the pur-

chase of American military equipment. These high-quality, American-made weapons, though sold to U.S. allies, can eventually find their way into various states of concern and into the hands of dangerous transnational groups (terrorists and organized crime) after the initial exchange. U.S. armed forces are accustomed to using these weapons, not facing them in the hands of opposing armed forces. This is a source of concern to defense planners.

Despite U.S. concern about American-made weapons being used against U.S. troops, the United States was the world's largest weapons supplier between 1995 and 2002. (See Table 2.2.)

"U.S. Policy on Small/Light Arms Export," a report prepared by Lora Lumpe of the Federation of American Scientists for the Academy of Arts and Sciences Conference on Controlling Small Arms, December 1997, identified five primary channels through which arms leave the United States:

1. Foreign military sales (FMS), which are negotiated at a state level between governments

2. Excess defense weapons under the Foreign Assistance Act

3. Direct industry sales

4. Covert government operations

5. Illegal markets

According to the Congressional Research Service's report *Conventional Arms Transfers to Developing Nations, 1995–2002*, some of the major weapons sales by the United States in 2002 included:

• Sixteen AH-64 Apache helicopters and related equipment to Kuwait (valued at $870 million)

• Ten F-16 C/D combat fighter aircraft to Chile (valued at $500 million)

• Three Aegis combat systems to South Korea for its KDX-3 destroyers (valued at over $960 million)

• Twelve F-16 C/D fighter aircraft, munitions, and support to Oman (valued at over $700 million)

THE REGULATION AND CONTROL OF U.S. ARMS SALES. U.S. arms sales are technically governed by regulations that forbid sales to certain countries of concern and regimes known to be oppressive. The 1976 Arms Export Control Act (AECA), overseen by the Office of Defense Trade Controls (ODTC) within the Bureau of Political Military Affairs of the U.S. Department of State, is the primary law regulating U.S. arms sales. AECA mandates that American weapons must be exported only for United Nations (UN) operations, self-defense purposes, and as responses to internal security threats. It also requires the Departments of Defense and State to produce regular

TABLE 2.1

Regional arms transfer agreements, by supplier, 1995–2002

(in millions of current U.S. dollars)

	Asia		Near East		Latin America		Africa	
	1995–98	1999–02	1995–98	1999–02	1995–98	1999–02	1995–98	1999–02
United States	5,426	6,462	13,314	27,207	1,245	1,877	89	109
Russia	13,100	18,000	1,800	2,200	400	200	600	1,500
France	1,100	3,400	8,600	500	500	0	100	600
United Kingdom	3,800	500	1,200	600	0	0	200	700
China	1,200	2,600	1,400	600	100	100	500	800
Germany	1,600	1,000	100	400	200	100	0	1,600
Italy	800	100	100	400	400	200	100	300
All other European	1,900	1,200	3,100	2,100	1,900	600	700	3,800
All others	2,700	2,000	1,200	1,900	1,000	600	800	800
Major West European*	7,300	5,000	10,000	1,900	1,100	300	400	3,200
Total	**31,626**	**35,262**	**30,814**	**35,907**	**5,745**	**3,677**	**3,089**	**10,209**

Note: All foreign data are rounded to the nearest $100 million. The United States total for Near East in 1999–2002 includes a $6.432 billion licensed commercial agreement with the United Arab Emirates in 2000 for 80 F-16 aircraft.
*Major West European category included France, United Kingdom, Germany, Italy.

SOURCE: "Table 1C. Regional Arms Transfer Agreements, by Supplier, 1995–2002," in *Conventional Arms Transfers to Developing Nations, 1995–2002,* Congressional Research Service, September 22, 2003, http://www.fas.org/man/crs/RL32084.pdf (accessed September 23, 2004)

TABLE 2.2

Arms deliveries to the world, by supplier, 1995–2002

	1995	1996	1997	1998	1999	2000	2001	2002	Total 1995–2002
United States	19,382	17,633	19,044	19,097	19,877	13,871	9,987	10,241	129,133
Russia	4,482	3,909	3,018	2,490	3,429	4,312	4,402	3,100	29,141
France	3,755	4,620	7,777	8,261	4,646	2,695	1,886	1,800	35,440
United Kingdom	6,420	7,701	7,893	4,300	5,530	7,007	4,716	4,700	48,267
China	848	829	1,277	792	442	862	734	800	6,585
Germany	2,665	2,251	1,393	1,697	2,433	1,401	629	500	12,970
Italy	242	118	464	226	664	323	210	400	2,648
All other European	4,239	4,028	5,107	3,734	3,208	3,126	2,096	1,800	27,339
All others	2,422	2,369	2,902	2,037	2,544	2,156	2,306	2,100	18,836
Total	**44,455**	**43,459**	**48,875**	**42,634**	**42,773**	**35,756**	**26,965**	**25,441**	**310,358**

SOURCE: "Table 9A. Arms Deliveries to the World, by Supplier, 1995–2002," in *Conventional Arms Transfers to Developing Nations, 1995–2002,* Congressional Research Service, September 22, 2003, http://www.fas.org/man/crs/RL32084.pdf (accessed September 23, 2004)

reports and notify Congress of any significant arms sales. The ODTC is responsible for the International Traffic in Arms Regulations, which contains a list of all munitions acceptable for export. All companies wishing to sell such equipment need to register with the ODTC and obtain the appropriate export license before concluding any sales.

Another important law regulating U.S. arms sales is the 1961 Foreign Assistance Act, which calls for military and other developmental aid to friendly governments. This law bars any sales to oppressive governments and countries adhering to policies contrary to American values and beliefs.

These regulations have not always proved effective, as demonstrated by arms sales to oppressive regimes such as Iraq (during the Iran-Iraq war of 1980–88) and Indonesia. Perhaps the biggest blow to U.S. credibility on its arms export policies was the Iran-contra affair. In November 1986 officials of the administration of President Ronald Reagan announced that some of the money earned from arms sales to Iran had been redirected to aid the Nicaraguan contras (rebels), an act that violated an existing ban on aid to Nicaraguan military and paramilitary activities. The arms sales to Iran had been conducted primarily to secure the release of Americans held hostage in the Middle East. High-ranking members of the Reagan administration, including President Reagan himself, Vice President George H. W. Bush, National Security Advisor John Poindexter, Secretary of State George Schultz, Secretary of Defense Caspar Weinberger, and Director of Central Intelligence William Casey were aware of the arms sales, but how many members of the administration were directly aware that the money was being rerouted to the contras is not clear. The independent investigation that followed embroiled the Reagan administration in a major scandal; National Security Advisor John Poindexter, former National Security Advisor Robert McFarlane, and their assistant Lt. Col. Oliver North resigned as a result.

The U.S. Department of State began the Blue Lantern program in 1990 to strengthen U.S. export controls to make sure U.S. arms exports do not end up in the wrong hands or abet any illicit supply networks. The primary function of the program is to perform end-use checks on all U.S. arms exports. Generally conducted by U.S. personnel abroad, their purpose is to ensure the material is not acquired using fraudulent documentation. The Blue Lantern program is designed to promote the responsible sale of weapons to allies and to prevent the transfer of dual-use technology to adversaries. (Dual-use items are controlled materials that have either military or civilian applications.)

According to the State Department, in 2003 the ODTC conducted 413 checks and discovered 76 transactions that were in violation of established policies. Of these, the majority (47%) of recipients were in the Western Hemisphere, followed by Asia (22%), Europe (18%), the Near East and South Asia (at 5% each), and Africa (3%). (See Figure 2.2.) Most of the sales in the Western Hemisphere involved firearms, ammunition, and explosives. Sales in Asia were often aircraft and helicopter spare parts. In "End-Use Monitoring of Defense Articles and Defense Services Commercial Exports," the State Department reported that in fiscal year (FY) 2003 cooperation between U.S. Customs and the State Department led to more than 665 commercial arms seizures, worth more than $106 million.

Other Countries

A number of countries acquire weapons to either expand or replace military equipment, and there is intense and increasing competition among weapons suppliers. In addition to the United States, countries including France, Russia, China, the United Kingdom, and Germany are also significant players on the world arms market. According to the Stocklholm International Peace Research Institute in 2004, Russia's weapon sales were about $7 billion in 2003. For the period 1999 through 2003, Russia sold $26.2 billion worth of weapons. France, Germany, and the United Kingdom followed with $6.4, $5.2, and $4.2 billion worth of sales, respectively, over the same period.

Although weapons sales by European countries declined in the early twenty-first century, especially sales to developing countries, Western European contractors heavily compete with American companies for sales. Additionally, they have historically served as significant suppliers in the conventional arms market to those countries that are not traditional clients of the United States.

Russia's increased arms trade is a direct result of the economic and political turmoil Russia faced after the

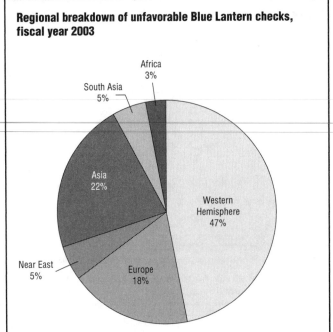

FIGURE 2.2

Regional breakdown of unfavorable Blue Lantern checks, fiscal year 2003

SOURCE: "Regional Breakdown of Blue Lantern Cases Initiated in FY 2003," in *End-Use Monitoring Report for FY 2003*, U.S. Department of State, http://www.pmdtc.org/docs/End_Use_FY2003.pdf (accessed September 23, 2004)

Soviet breakup. It continues to supply its past allies and those already familiar with Russian technology, including India, China, and Iran. In March 2003 the U.S. charged (*Daily Telegraph,* March 25, 2003) that Russia had violated UN trade sanctions by supplying Iraq with advanced weaponry, including anti-tank missiles and night-vision goggles. Illicit arms sales are also a continuing problem. According to a 2001 UN report on illicit arms traffic (*Report of the Panel of Experts Pursuant to Security Council Resolution 1343 (2001), paragraph 19, concerning Liberia,* United Nations, New York, NY, October 26, 2001), former KGB officer Victor Bout is a leading arms dealer who sells millions of dollars worth of Russian and Eastern European weapons to embargoed African countries engaged in civil wars.

In comparison with other states, China entered the arms market somewhat late, in the 1980s. With sales of approximately $1.5 billion from 1999 to 2003, China supplies arms to countries unable to afford newer, more sophisticated technology. It sold antiship missiles to Iran and remains the primary supplier for Pakistan. China is rumored to have provided missile technology to Iran and North Korea, raising concerns within the arms control community.

The Persian Gulf remains one of the most active theaters when it comes to arms sales. Five states surrounding the Gulf (Iran, Iraq, Saudi Arabia, the UAE, and Yemen) and five more in its near vicinity (Egypt, Israel, Libya, Syria, and Turkey) possess ballistic missile capabilities and are prime buyers of Western weapons equipment. Along with these missile capabilities, several other factors make this region of particular importance to the West. These include energy resources (there is an abundance of oil and natural gas fields in the region), the Iranian threat (the country has defiantly continued to develop its nuclear weapons capability), and strategic military alliances (Israel, Egypt, and Jordan all serve as important Western allies in the area). Extensive talks have taken place among U.S. and regional policy makers about the implications of deploying a theater missile defense in the Gulf. The main purpose of deploying this system would be to protect U.S. vital strategic interests and those of its main regional allies: member states of the Gulf Cooperation Council (GCC), including Bahrain, Kuwait, Oman, Qatar, Saudi Arabia, and the UAE. Contributing to a heightened perception of threats for most states in this theater are past use of WMD, revelations about Iranian and Iraqi WMD programs, and the instability in other Gulf states and the former Soviet Union. Accordingly, purchasing top-of-the-line equipment is a strong priority for the defense ministries of these countries.

An ongoing Cooperative Defense Initiative between the United States and its GCC allies, focuses, among other things, on improved interoperability (coordination between two separate entities, such as by holding joint exercises and simulations) among these countries. Also, through the Hizam al-Taawun ("belt of cooperation") project, these states are coordinating their shared early warning and air defense systems. This should theoretically allow the GCC states to pool their defense resources, spending less money individually but sharing in collective security. The project was initially undertaken in 1997, when GCC defense ministers collectively agreed to purchase "a $500 million ground-based early warning system that would link the GCC states' radars and communication systems" (Robert Shuey, "Theater Missile Defense: Issues for Congress," *Congressional Research Service*, May 22, 2001). Nevertheless, besides a few Patriot units deployed in Kuwait and Saudi Arabia, the region still lacks a cohesive joint infrastructure to protect against missile attacks. (Patriot units consist of the Patriot missile, its launcher, high-tech radar, and an engagement control station [a station where operators control the missiles and identify targets]. The Patriot missile is a high-velocity missile that focuses on neutralizing incoming missiles before they hit the ground. It can achieve supersonic speed within twenty feet of leaving the launcher and has a range of at least 100 kilometers.)

Arms Importers

On the receiving end of arms sales, the United States obtained less than $1.3 billion worth of arms from 1999 to 2003, ranking number twenty-seven in the world in arms imports, according to the Stockholm International Peace Research Institute in 2004. China was the largest arms importer, receiving $11.8 billion worth of major conventional weapons from 1999 to 2003. India ranked second with $7.8 billion worth of imports over the same period, followed by Greece ($4.4 billion), Turkey ($3.5 billion), and the United Kingdom ($3.3 billion). Two countries showed particularly dramatic increases in their arms imports during this period: Greece (from $556 million annually to $2 billion annually) and India (from $1 billion annually to $3.6 billion annually).

SMALL ARMS SALES AND THE ROLE OF IGOS AND NGOS

Small arms are highly desirable to guerilla and other subnational groups because they tend to be cheap, easy to conceal, and easy to transport. Though definitions vary, examples of small arms include revolvers and self-loading pistols, rifles and carbines, light machine guns, submachine guns, and assault rifles. Examples of light weapons include heavy machine guns, portable anti-aircraft guns, portable antitank guns and recoilless rifles, portable launchers of antitank missile and rocket systems, portable launchers of anti-aircraft missile systems, and mortars of calibers less than 100 millimeters. Such weapons are of grave concern to individual states and the international community, but the issue has not been addressed multilaterally (by more than two nations) until quite recently. The definition of terms and concerns about sovereignty, among other things, have impeded a global consensus on small arms proliferation.

Organizations Battle Small Arms Proliferation

International government organizations (IGOs)—organizations made up of various countries, typically represented by government officials of these individual countries—and nongovernmental organizations (NGOs)—organizations set up to study particular areas of focus and are typically not made up of any government officials—have been very strong proponents of halting small arms proliferation.

THE UNITED NATIONS (UN). The UN organized a platform for debate on conventional small arms sales. After its initial proposition in 1997 and subsequent reiteration in 1999 by a panel of experts, the UN Conference on the Illicit Trade in Small Arms and Light Weapons was held July 9–20, 2001. Representatives from the governments of more than 150 countries attended. The conference adopted a program of action, described in the *DDA 2001 Update* (UN Department for Disarmament Affairs, June-July, 2001) as "a comprehensive document, containing unprecedented political commitments and concrete measures at the national, regional and global levels to tackle the illegal trade in small arms."

The United States, in particular, had reservations regarding rules and regulations proposed at the conference because of domestic considerations. The National Rifle Association (NRA), a powerful American gun owners' lobbying group, among others, maintained that the UN's recommendations violate the Second Amendment of the U.S. Constitution. The Second Amendment deals with individual citizens' "right to bear arms" and therefore affects measures that might constrain the legal manufacture of small arms and prohibit civilians from possessing them. As a matter of sovereignty (a nation's right to make its own laws and policies), the United States would not accept any language that would infringe on Americans' national right to bear arms.

The end result of the conference was an eighty-six-paragraph report (the "Programme of Action") dealing with illegal trade in small arms and assistance to affected states. The Programme of Action called for implementing an arms embargo, improving interstate cooperation, and encouraging cooperation among civil society organizations (another label for NGOs).

The future of the UN conference and the Programme of Action is questionable. Although various international groups strongly support it, states remain hesitant about endorsing the agreement. Furthermore, ambiguous language and the conference's inability to agree on definitions leave it susceptible to violations. Verification of states' compliance would also be difficult. However, the conference is still viewed by many gun-control activists as a step in the right direction. A follow-up series of meetings was held in New York in July 2003 to discuss progress toward full implementation of the Programme of Action. According to a summary report issued at the conclusion of the meetings by Kuniko Inoguchi of Japan, who served as the meetings' chairperson, "Barely two years after the adoption of the Programme of Action, progress has been made across the world in public disclosures about the origins, destinations, modus operandi and profiling of groups engaged in illicit arms trade." A review conference will be held to assess the broader success of the program no later than 2006.

THE WORLD HEALTH ORGANIZATION (WHO). The WHO is another IGO that confronts the problem of arms sales. The WHO believes it has a direct stake in the issue, since thousands of people are killed or injured by violent armed attacks annually. In October 2002 it issued the *World Report on Violence and Health,* a study that included seeking "practical, internationally agreed responses to the global drug trade and the global arms trade" among its key recommendations.

THE INTERNATIONAL CRIMINAL POLICE ORGANIZATION (INTERPOL). The International Criminal Police Organization (Interpol) tackles arms trafficking by creating a platform where countries agree to share information to help eliminate it. Interpol has 181 member nations and runs a database to collect information on illegal firearms and track stolen and recovered weapons.

THE ORGANIZATION FOR SECURITY AND COOPERATION IN EUROPE (OSCE). Small arms trafficking is one of many issues related to international security and stability dealt with by the Organization for Security and Cooperation in Europe (OSCE). In April 2000 the OSCE Forum for Security and Cooperation held a seminar on small arms proliferation and conventional weapons trafficking. A product of this seminar was the OSCE Document on Small Arms and Light Weapons, which aimed to help the UN combat illicit weapons trafficking. The OSCE document pledges that member states will take extreme precautions when it comes to arms transfers (which are to be undertaken only for legitimate purposes) as well as develop confidence-building and transparency measures. (Confidence building implies unilateral or joint efforts to boost trust between member parties, while transparency implies openness. Generally, these are achieved by holding joint exercises as well as on-site inspections and data exchanges.) It is important to note that the document does not call for the creation of a new authority to combat small arms trafficking. Rather, it relies on the voluntary declaration of participating states to stand by the principles of the document.

THE INTERNATIONAL ACTION NETWORK ON SMALL ARMS (IANSA). The International Action Network on Small Arms (IANSA) is an umbrella organization representing more than 340 NGOs dedicated to combating arms proliferation. IANSA encourages coordination among human rights groups, development agencies, gun control lobby groups, public health organizations, and religious groups. These groups also focus on ways to reincorporate former combatants, including child soldiers, into everyday life.

AFRICAN AND LATIN AMERICAN NGOS. Not all NGOs believe that loose coordination among international groups is the best method for eliminating the arms trade. Some NGOs in Africa and Latin America, such as the Vivario—a Brazilian organization that fights to abolish arms and educate the public about their dangers—are proponents of establishing a strong, centrally organized, international body to fight conventional arms proliferation.

LAND MINES

Land mines are another type of conventional weapon of concern to the United States. Once a land mine has been placed in an area it can remain active for years, even decades. Since they are usually planted in large numbers they can render entire regions unsafe for anyone, combatant or civilian, to enter even long after a war is over. Over 100 countries possess more than 250 million antipersonnel land mines (APMs), and a worldwide grassroots movement has emerged to rid the globe of these deadly weapons. Land mines and other unexploded ordnance affect hundreds of thousands of people. According to the U.S. State Department, more than sixty countries have unexploded land mines on their territory. The U.S. Humanitarian Demining Program estimated on its Web site (http://www.humanitarian-demining.org/) in 2004 that some fifty-five million land mines deployed worldwide cause some ten thousand casualties each year. The State Department considers these types of weapons especially dangerous because they:

- Kill and maim thousands of people annually.

- Create millions of refugees.

- Prevent productive land use, especially for agricultural and industrial purposes.

- Deny road/travel access.

- Deny access to water and create food scarcities, leading to starvation and malnutrition.

- Inflict long-term psychological effects on victims.

- Undermine political and economic stability.

The global anti-land mine campaign is spearheaded by the International Campaign to Ban Landmines (ICBL), the recipient of the 1997 Nobel Peace Prize. The efforts of this group culminated in the 1997 Convention on the Prohibition of the Use, Stockpiling, Production and Transfer of Anti-Personnel Mines and on Their Destruction. The treaty entered into force on March 1, 1999.

The convention has 143 countries as parties, while nine countries have signed but not ratified the treaty. As of October 2004, forty-two countries had not signed the treaty, including China, India, Iran, Iraq, Israel, Pakistan, Russia, and the United States. One of the criticisms of the treaty is the lack of a strong monitoring and verification authority. Most countries that refuse to sign the treaty rely heavily on APMs in their areas of conflict. Increased mine laying by India and Pakistan along their disputed border has caused serious concern among antimine activists. Countries such as Sri Lanka and Russia claim they have not signed in order to defend themselves against mines laid by internal insurgents and terrorists.

The United States meanwhile believes that mines serve an important purpose in the Demilitarized Zone (DMZ) between North and South Korea, where U.S. military personnel are deployed. American officials believe the United States does not have a strong enough military presence along the DMZ to defeat potential North Korean aggression without the help of APMs. The United States had been a leading proponent of the anti-land mine treaty as long as it gained an exception for the Korean issue. When such an exception was refused, the U.S. delegation withdrew from treaty negotiations and refused to sign the treaty.

Though the United States has not signed the ICBL treaty, it has had a moratorium on producing mines since 1996. It has also unilaterally destroyed 3.3 million mines from its arsenal. In February 2004 a new land mine policy was announced by Lincoln P. Bloomfield. Jr., the Assistant Secretary of State for Political-Military Affairs. The new policy commits the United States to eliminate persistent land mines (those without an automatic self-destruct or self-deactivation mechanism) from its arsenal, to destroy all of its non-detectable land mines within one year, and to use land mines only for the defense of the South Korean border (*Fact Sheet: New United States Policy on Landmines: Reducing Humanitarian Risk and Saving Lives of American Soldiers,* Washington, DC: Bureau of Political-Military Affairs, February 27, 2004).

The United States has also provided assistance with fighting the land mine problem through the U.S. Humanitarian Demining Program. Operating under the Department of State, demining programs assist countries around the world, both financially and through direct training. According to a fact sheet issued by the State Department's Bureau of Political-Military Affairs, the United States contributed more than $800 million toward demining activities from 1993 through fiscal year 2003. In 2001 the United States and the government of Mozambique partnered in the creation of a Quick Reaction Demining Force (QRDF), which could be immediately deployed to mine crises around the world. A QRDF was sent to Sri Lanka to assess the mine threat there and perform short-term clearance tasks, primarily in the heavily mined Jaffna peninsula, as well as the Vanni and Killinochchi regions. The team previously worked in stabilization efforts in Afghanistan and postconflict Kosovo. Following the success of Operation Iraqi Freedom in liberating Iraq, the U.S. Humanitarian Demining Program worked in Iraq to find and destroy an estimated ten to fifteen million land mines deployed in some 2,500 minefields. Along with this effort to destroy land mines, the program also seeks to eliminate unexploded ordnance, much of it left over from the Iraq-Iran War, and abandoned munitions depots.

TREATIES AND REGIMES

International treaties are intended to restrain the development and potential use of conventional weapons. They are sometimes limited, either by the types of weapons included in the treaty (for example, a treaty dealing only with land mines) or their application (they may not be signed by all parties with the type or types of weapons concerned). In efforts to reduce the threat to its own security, the United States has proposed, signed, and ratified many such treaties over the years, particularly with the Soviet Union during the cold war era. Many of these treaties are bilateral (involving two parties), but some have been ratified or signed by other countries, especially the Warsaw Pact Organization countries, which were allied with the Soviet Union, and the countries of NATO, which are allied with the United States.

Treaty on Conventional Armed Forces in Europe (CFE)

The Treaty on Conventional Armed Forces in Europe (CFE) was originally drafted primarily for NATO and Warsaw Pact countries and was signed on November 19, 1990. Because of the breakup of the Soviet Union and other changes in Europe, thirty states now are parties to this treaty, which aims to restrict the overwhelming number of conventional forces in Europe. As of 2004, members of the CFE include Armenia, Azerbaijan, Belarus, Belgium, Bulgaria, Canada, the Czech Republic, Denmark, France, Georgia, Germany, Greece, Hungary, Iceland, Italy, Kazakhstan, Luxembourg, Moldova, the Netherlands, Norway, Poland, Portugal, Romania, Russia, Slovakia, Spain, Turkey, Ukraine, the United Kingdom, and the United States. As new members are added to NATO, each will be asked to comply with CFE requirements.

In November 2002, at a NATO conference in Prague, Secretary General George Robertson stated that NATO: "reiterated the goals, principles and commitments contained in the Founding Act on Mutual Relations, Cooperation and Security, and in the Rome Declaration. Reaffirming adherence to the CFE Treaty as a cornerstone of European security, [NATO] agreed to continue to work cooperatively toward ratification by all the States Parties and entry into force of the Agreement on Adaptation of the CFE Treaty, which would permit accession by non-CFE states." He also "welcomed the approach of those non-CFE countries who have stated their intention to request accession to the adapted CFE Treaty upon its entry into force, and agreed that their accession would provide an important additional contribution to European stability and security."

"The Atlantic to the Urals" is the so-called area of application (AOA) for the CFE. This means that parties to the treaty can deploy only a limited number and certain types of weaponry within the AOA. Some signatory countries actually lie outside the AOA or their territories extend beyond it; for them, limits apply only to any of their forces stationed in the zones of Europe established by the treaty. The first negotiations addressed only equipment levels in the AOA; follow-up negotiations addressed troop limits, resulting in the CFE 1A document (discussed below).

The CFE limits the use of five categories of arms within its AOA: tanks, artillery, armored combat vehicles (ACVs), combat aircraft, and attack helicopters. Limits on the equipment levels allotted to each country are stated in the treaty according to each country's boundaries and according to the zones beyond its sovereign territory in which it may operate. These four zones lie in a concentric fashion that calls for fewer troops and fewer weapons

deployed the closer one moves to the center of Europe. The smallest of the zones is the central zone; the largest is the flank zone, on the northern and southern flanks of Europe.

Further limits on the number of combat vehicles a single country could possess altogether are also stated in the treaty. A restriction to 20,000 tanks, 30,000 ACVs, 20,000 heavy artillery pieces, 2,000 attack helicopters, and 6,800 combat aircraft in the AOA for all NATO and former Warsaw Pact members is also imposed. Each alliance divides its "bloc" limit among member parties. One state cannot possess more than one-third of the treaty's allowed maximums.

As of 2004, Russia's continued presence in Chechnya stood starkly in violation of the CFE, but Russia continued to unilaterally destroy military items it inherited from the Soviet Union. On the tenth anniversary of the treaty in November 2000, Russian President Vladimir Putin reiterated his support of the treaty and said Russia would comply with the CFE as soon as its military operations in Chechnya ended.

The CFE embraces several methods of compliance, including onsite inspections, information exchanges, and national/multinational technical means. All of these are overseen by the Joint Consultative Group of Vienna, Austria. The types of inspection permitted by the treaty are several: announced inspections of declared sites, challenge inspections within a specified area (for which, if a country refuses the inspection, it is required to issue a reasonable assurance of compliance), and inspections to verify the destruction or redeployment of equipment.

The Concluding Act of the Negotiation on Personnel Strength of Conventional Armed Forces in Europe

The Concluding Act of the Negotiation on Personnel Strength of Conventional Armed Forces in Europe, also called the Conventional Armed Forces in Europe 1A (CFE 1A) Treaty, was intended as a politically, not legally, binding document that does not have to be ratified. Signed on July 17, 1992, its point was to limit or reduce personnel levels in the AOA of the CFE treaty. Parties to the treaty are generally members of NATO and the former Warsaw Pact, plus those countries that were part of the former Soviet Union located in the AOA of the CFE. Each country sets its own limits on personnel levels. Once set, these limits are open to discussion but not negotiation. Personnel counted within the limits can be (1) active-duty land or air forces, including land-based air defense; (2) command and staff of those units; (3) land-based naval aircraft and naval infantry, coastal defense units, and other forces holding equipment under the CFE treaty; or (4) reserve personnel called up for active duty for more than ninety days. Sea-based naval personnel, internal security units, and forces under UN command are exempt. The treaty sets forth additional measures to stabilize personnel, such as forty-two-day advance notification required to: increase personnel strength by more than 1,000; increase an air force unit by more than 500; or call up more than 35,000 reservists (except if the reservists are called up for natural disasters or other emergencies). At a meeting held in Vienna, Austria, in 2001, the thirty states who are parties to the treaty reaffirmed their continuing commitment to the CFE.

The Wassenaar Arrangement on Export Controls for Conventional Arms and Dual-Use Goods and Technologies

Signed by thirty-three countries, the Wassenaar Arrangement on Export Controls for Conventional Arms and Dual-Use Goods and Technologies (often referred to as simply the Wassenaar Arrangement) was developed in 1998 to promote transparency and set arms sales limitations on certain weapons and dual-use goods and technologies. Its signatories include Argentina, Australia, Austria, Belgium, Bulgaria, Canada, the Czech Republic, Denmark, Finland, France, Germany, Greece, Hungary, Ireland, Italy, Japan, Luxembourg, the Netherlands, New Zealand, Norway, Poland, Portugal, the Republic of Korea, Romania, the Russian Federation, Slovakia, Spain, Sweden, Switzerland, Turkey, Ukraine, the United Kingdom, and the United States. The countries have agreed to a control list in terms of their weapons exports.

These countries actively participate in preventative enforcement, follow-up investigations, and information exchanges to control dual-use equipment. Dual-use items are equipment and materials considered to be controlled commodities that either cannot be exported at all or that require an export license because of the potential for misuse. The items, such as certain telecommunications equipment, chemicals, or even microorganisms and toxins, have potential civilian as well as military uses.

To ensure that surplus military equipment does not get into the wrong hands, parties to the Wassenaar Arrangement agreed to increase safeguards on military equipment. They agreed that surplus military equipment should fall under the same export controls as new materials and that the physical security of, and inventory controls on, these materials should be increased. To ensure that its goals are met, the Wassenaar Arrangement requires data exchanges biannually (in April and October) to report on transactions from the previous six months. The arrangement continued to undergo refinement in 2003 and 2004 as additional categories of small arms and light weapons were added to control lists.

CHAPTER 3

PROLIFERATION OF
WEAPONS OF MASS DESTRUCTION (WMD)

HISTORY OF USAGE AND PROLIFERATION

The use of gases, poisons, and toxins by states at war can be traced back centuries. As Table 3.1 shows, chemical and biological weapons have a long history dating back to the fifth century B.C.

One of the first people to contemplate the use of biological weapons in North America was Lord Jeffrey Amherst. Amherst was the commanding general of British forces in North America during the final battles of the French and Indian War (1754–63). Carl Waldman's *Atlas of the North American Indian* (New York: Facts on File, 1985) describes a siege at Fort Pitt (Pittsburgh) by the forces of Native American leader Chief Pontiac during the summer of 1763. Amherst sent a letter to another British officer, encouraging him to send smallpox-infected blankets and handkerchiefs to the Indians surrounding the fort in an effort to start an epidemic. Although there were epidemics of smallpox among some of the Indian tribes in the area, it is uncertain if such a plan was executed or if the smallpox was related to this early proposal of "germ warfare."

Still, the transformation of biological, chemical, and nuclear agents into weapons of mass destruction (WMD) is a relatively recent phenomenon in the history of warfare. What exactly is a weapon of mass destruction? Several definitions exist. Some analysts include only nonconventional chemical, biological, radiological, and nuclear (CBRN) weapons in this category. According to the U.S. Code, Title 5, "War and National Defense," a WMD is "any weapon or device that is intended, or has the capability, to cause death or serious bodily injury to a significant number of people through the release, dissemination, or impact of (A) toxic or poisonous chemicals or their precursors; (B) a disease organism; or (C) radiation or radioactivity."

However, several other policy analysts and experts look at WMDs more broadly. Conventional weapons

capable of creating widespread casualties or "mass destruction" are also classified as WMD. The Federal Bureau of Investigation, for instance, in "The FBI and Weapons of Mass Destruction," August 1999, stated that a "weapon of mass destruction (WMD), though typically associated with nuclear/radiological, chemical, or biological agents, may also take the form of explosives, such as in the bombing of the Alfred P. Murrah Federal Building in Oklahoma City, Oklahoma, in 1995. A weapon crosses the WMD threshold when the consequences of its release overwhelm local responders." For the purpose of this text, WMD shall hereafter solely refer to CBRN weapons and their delivery systems.

It was not until World War I (1914–18) that WMD were first used strategically in a battlefield environment to inflict massive casualties. On April 22, 1915, the German army released chlorine gas from cylinders in Ypres, Belgium, causing at least 2,800 casualties. The British retaliated later that year, using the same gas against German troops. In total, about 124,000 tons of chemical weapons were used by all sides during World War I.

Japan made use of chemical and biological weapons while fighting in China and Manchuria before and during World War II (1939–45). World War II also saw the introduction of nuclear weapons, when the United States dropped atomic bombs on Hiroshima and Nagasaki, Japan. After World War II, the use and stockpiling of WMD continued. The ensuing cold war between the United States and the Soviet Union witnessed an alarming buildup of WMD arsenals and the spread of WMD capabilities to such nations as the United Kingdom (1952) and France (1960).

Though the timeline in Table 3.1 ends in 1998, several additional deployments of WMD have occurred since. In 1998 the United States bombed sites in Iraq that allegedly contained WMD. Between 1998 and 2001 several anthrax hoaxes and actual attacks were launched by various indi-

TABLE 3.1

Chronology of biological and chemical weapons use and control, 429 B.C.–1998

- 429 B.C. - Spartans ignite pitch and sulphur to create toxic fumes in the Peloponnesian War (CW)
- 424 B.C. - Toxic fumes used in siege of Delium during the Peloponnesian War (CW)
- 960-1279 A.D. - Arsenical smoke used in battle during China's Sung Dynasty (CW)
- 1346-1347 - Mongols catapult corpses contaminated with plague over the walls into Kaffa (in Crimea), forcing besieged Genoans to flee (BW)
- 1456 - City of Belgrade defeats invading Turks by igniting rags dipped in poison to create a toxic cloud (CW)
- 1710 - Russian troops allegedly use plague-infected corpses against Swedes (BW)
- 1767 - During the French and Indian Wars, the British give blankets used to wrap British smallpox victims to hostile Indian tribes (BW)
- April 24, 1863 - The U.S. War Department issues General Order 100, proclaiming "The use of poison in any manner, be it to poison wells, or foods, or arms, is wholly excluded from modern warfare"
- July 29, 1899 - "Hague Convention (II) with Respect to the Laws and Customs of War on Land" is signed. The Convention declares "it is especially prohibited... To employ poison or poisoned arms"
- 1914 - French begin using tear gas in grenades and Germans retaliate with tear gas in artillery shells (CW)
- April 22, 1915 - Germans attack the French with chlorine gas at Ypres, France. This was the first significant use of chemical warfare in WWI (CW)
- September 25, 1915 - First British chemical weapons attack; chlorine gas is used against Germans at the Battle of Loos (CW)
- 1916-1918 - German agents use anthrax and the equine disease glanders to infect livestock and feed for export to Allied forces Incidents include the infection of Romanian sheep with anthrax and glanders for export to Russia, Argentinian mules with anthrax for export to Allied troops, and American horses and feed with glanders for export to France (BW)
- February 26, 1918 - Germans launch the first projectile attack against U.S. troops with phosgene and chloropicrin shells. The first major use of gas against American forces (CW)
- June 1918 - First U.S. use of gas in warfare (CW)
- June 28, 1918 - The United States begins its formal chemical weapons program with the establishment of the Chemical Warfare Service (CW)
- 1919 - British use Adamsite against the Bolsheviks during the Russian Civil War (CW)
- 1922-1927 - The Spanish use chemical weapons against the Rif rebels in Spanish Morocco (CW)
- June 17, 1925 - "Geneva Protocol for the Prohibition of the Use in War of Asphyxiating, Poisonous or Other Gases, and of Bacteriological Methods of Warfare" is signed - not ratified by U.S. and not signed by Japan
- 1936 - Italy uses mustard gas against Ethiopians during its invasion of Abyssinia (CW)
- 1937 - Japan begins its offensive biological weapons program. Unit 731, the biological weapons research and development unit, is located in Harbin, Manchuria. Over the course of the program, at least 10,000 prisoners are killed in Japanese experiments (BW)
- 1939 - Nomonhan Incident - Japanese poison Soviet water supply with intestinal typhoid bacteria at former Mongolian border. First use of biological weapons by Japanese (BW)
- 1940 - The Japanese drop rice and wheat mixed with plague-carrying fleas over China and Manchuria (BW)
- 1942 - U.S. begins its offensive biological weapons program and chooses Camp Detrick, Frederick, Maryland as its research and development site (BW)
- 1942 - Nazis begin using Zyklon B (hydrocyanic acid) in gas chambers for the mass murder of concentration camp prisoners (CW)
- December 1943 - A U.S. ship loaded with mustard bombs is attacked in the port of Bari, Italy by Germans; 83 U.S. troops die in poisoned waters (CW)
- April 1945 - Germans manufacture and stockpile large amounts of tabun and sarin nerve gases but do not use them (CW)
- May, 1945 - Only known tactical use of biological weapons by Germany. A large reservoir in Bohemia is poisoned with sewage (BW)
- September, 1950-February, 1951 - In a test of biological weapons dispersal methods, biological simulants are sprayed over San Francisco (BW)
- 1962-1970 - U.S. uses tear gas and four types of defoliant, including Agent Orange, in Vietnam (CW)
- 1963-1967 - Egypt uses chemical weapons (phosgene, mustard) against Yemen (CW)
- June, 1966 - The United States conducts a test of vulnerability to covert biological weapons attack by releasing a harmless biological simulant into the New York City subway system (BW)
- November 25, 1969 - President Nixon announces unilateral dismantlement of the U.S. offensive biological weapons program (BW)
- February 14, 1970 - President Nixon extends the dismantlement efforts to toxins, closing a loophole which might have allowed for their production (BW)
- April 10, 1972 - "Convention on the Prohibition of the Development, Production and Stockpiling of Bacteriological (Biological) and Toxin Weapons and on Their Destruction" (BWC) is opened for signature
- 1975 - U.S. ratifies Geneva Protocol (1925) and BWC
- 1975-1983 - Alleged use of Yellow Rain (trichothecene mycotoxins) by Soviet-backed forces in Laos and Kampuchea. There is evidence to suggest use of T-2 toxin, but an alternative hypothesis suggests that the yellow spots labeled Yellow Rain were caused by swarms of defecating bees (CW)
- 1978 - In a case of Soviet state-sponsored assassination, Bulgarian exile Georgi Markov, living in London, is stabbed with an umbrella that injects him with a tiny pellet containing ricin (BW)
- 1979 - The U.S. government alleges Soviets use of chemical weapons in Afghanistan, including Yellow Rain (CW)
- April 2, 1979 - Outbreak of pulmonary anthrax in Sverdlovsk, Soviet Union. In 1992, Russian president Boris Yeltsin acknowledges that the outbreak was caused by an accidental release of anthrax spores from a Soviet military microbiological facility (BW)
- August, 1983 - Iraq begins using chemical weapons (mustard gas), in Iran-Iraq War (CW)
- 1984 - First ever use of nerve agent tabun on the battlefield, by Iraq during Iran-Iraq War (CW)
- 1985-1991 - Iraq develops an offensive biological weapons capability including anthrax, botulium toxin, and aflatoxin (BW)
- 1987-1988 - Iraq uses chemical weapons (hydrogen cyanide, mustard gas) in its Anfal Campaign against the Kurds, most notably in the Halabja Massacre of 1988 (CW)
- September 3, 1992 - "Convention on the Prohibition of the Development, Production, Stockpiling and Use of Chemical Weapons and on their Destruction" (CWC) approved by United Nations
- April 29, 1997 - Entry into force of CWC
- 1998 - Iraq is suspected of maintaining an active CBW program in violation of the ceasefire agreement it signed with the UN Security Council. Baghdad refuses to allow UNSCOM inspectors to visit undeclared sites (CW/BW)

CW: Chemical Weapons Use
BW: Biological Weapons Use

SOURCE: Adapted from "Chronology of State Use and Biological and Chemical Weapons Control," in *Chemical and Biological Weapons Resource Page*, Center for Nonproliferation Studies, October 24, 2001, http://cns.miis.edu/research/cbw/pastuse.htm, (accessed September 23, 2004)

viduals and organizations. Media organizations including NBC and the *Washington Post,* government offices in the U.S. State Department, the White House, congressional offices, U.S. post offices, and abortion clinics across the country were targeted. Anthrax exposure, infection, and even deaths resulted from some of the attacks on media organizations and in post offices where anthrax-laced mail was handled. Intelligence and homeland security planners must assume that this type of weapon will continue to be used intermittently in the future.

WHY NATIONS DEVELOP WEAPONS OF MASS DESTRUCTION

The dangers of modern WMD are significant enough to warrant increasing global concern, especially given the large number of countries that possess some sort of WMD capabilities. While it is difficult to posit a single explanation of various nations' rationales for developing WMD capabilities, in many instances the reasons include one or more of the following:

- National security/lack of conventional weapons capability
- Perception of an imminent threat
- Deterrence/balance of power
- Regional stability
- Leadership personalities
- Pride/prestige
- Politically powerful commercial defense industries
- Technological imperatives (the race to acquire the best technological capabilities in order to maintain positions of leadership)

The enormous and widespread damage and costs of WMD, especially nuclear weapons, have discouraged states from using them during conflict. Instead, most stockpile them primarily for purposes of deterrence. Most countries that possess nuclear weapons have a "no first strike" policy that calls on WMD for defensive purposes only (i.e., they will not launch nuclear weapons at another country first but only in reaction to an attack against their own nation). Many countries and international organizations have spearheaded arms control plans because they recognize the potential disastrous effects of WMD proliferation, and they cite such reasons to control them as fear of retaliation, moral considerations, and difficulty controlling their effects. These initiatives designed to curb WMD buildup generally take the form of treaties and agreements, although compliance with, and adherence to, such treaties has varied.

CHEMICAL WEAPONS

Chemical warfare agents are poisonous chemical materials used to kill or incapacitate. These agents can be delivered in a variety of ways, including canisters,

TABLE 3.2

Dual use chemicals

Dual-Use Chemical	Chemical Warfare Agent	Other Uses
Thiodiglycol	Sulfur Mustard	Plastics, dyes, inks
Thionyl chloride	Sulfur Mustard	Pesticides
Sodium sulfide	Sulfur Mustard	Paper
Phosphorus trichloride	Sulfur Mustard	Insecticides
Phosphorus Oxychloride	Tabun	Insecticides
Dimethylamine	Tabun	Detergents
Sodium Cyanide	Tabun	Dyes, Pigments
Dimethyl methylphosphonate	G (nerve) Agents	Fire retardants
Dimethyl hydrochloride	G (nerve) Agents	Pharmaceuticals
Potassium bifluoride	G (nerve) Agents	Ceramics
Diethyl phosphite	G (nerve) Agents	Paint Solvent
Methylphosphonic difluoride	G (nerve) Agents and VX	Organic Chemical Synthesis
Phosphorus pentasulfide	VX	Lubricants, pesticides

SOURCE: Compiled by Information Plus staff based on "United States Efforts in Curbing Chemical Weapons Proliferation," in *World Military Expenditures and Arms Transfers,* U.S. Arms Control and Disarmament Agency, 1990, and Eric Croddy et al., *Chemical and Biological Warfare: A Comprehensive Survey for the Concerned Citizen,* Copernicus Books, 2002

artillery shells, artillery rockets, aerial bombs, mines, missile warheads, grenades, sprayers, and even released by individuals. Although the international community generally condemns the use of such weapons, several states have had or do have full-fledged chemical weapons programs. Figure 3.1 shows which countries have abandoned chemical weapons programs and which may have or were suspected to have chemical weapons as of 2004.

Besides their use in World War I, chemical weapons saw limited application during the twentieth century. Weaponized chemical releases after World War I include those during the 1930s by the Italians in Ethiopia and the Japanese in China, as well as Iraq's use against Iran during the 1980s. In addition, the United States, Russia (the former Soviet Union), United Kingdom, North Korea, Libya, Iran, Israel, and Syria have all been accused of having developed chemical weapons.

One factor that makes chemical weapons programs hard to detect is the dual-use nature of many of the chemicals used—these same chemicals have legitimate uses in medicine, scientific research, farming, pest control, and other applications. Table 3.2 provides examples of chemicals used both in weapons and in other fields.

Characteristics of Chemical Weapons

The weaponization of a chemical agent, or the process that turns an ordinary chemical into a weapon, is a laborious process. The compound must be stabilized, a delivery method must be created and implemented, and the weapon has to be stored and transported. The potency of a chemical agent greatly depends on environmental factors, the quality of the agent, and its means of delivery.

FIGURE 3.1

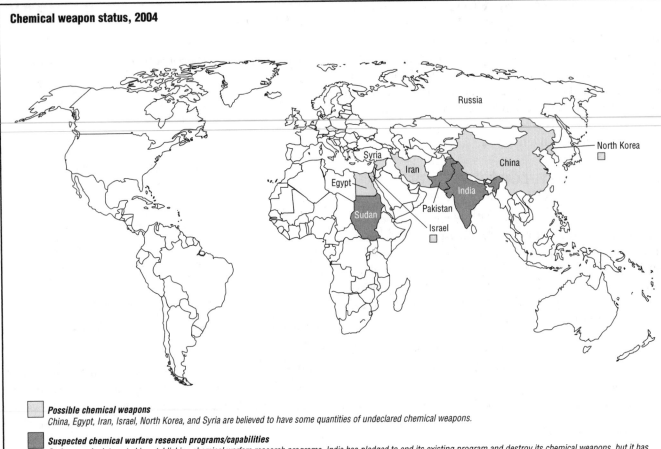

Chemical weapon status, 2004

Possible chemical weapons
China, Egypt, Iran, Israel, North Korea, and Syria are believed to have some quantities of undeclared chemical weapons.

Suspected chemical warfare research programs/capabilities
Sudan may be interested in establishing chemical warfare research programs. India has pledged to end its existing program and destroy its chemical weapons, but it has not yet done so completely. Pakistan may have the capability to produce chemical warfare agents but has not done so.

Countries that have abandoned chemical weapon programs
Until the Chemical Weapons Convention, many nations had chemical warfare programs, but a growing number have since ended their programs and are destroying their weapons, including the United States, the United Kingdom, France, Canada, Germany, Italy, Japan, Russia* and other states of the former Soviet Union, South Africa, South Korea, the Federal Republic of Yugoslavia, and now Libya and Iraq.

Significant countries that have not joined the CWC
Egypt, Iraq, Israel, North Korea, and Syria.

Other possible programs
Some reports indicate that Cuba, Ethiopia, Myanmar, Saudi Arabia, South Korea, Taiwan, and Vietnam may be interested in developing or may operate chemical warfare programs but the evidence is inconclusive.

*Russia had declared 44,000 tons of Soviet era mustard gas, lewisite, satrin, VX, and other chemicals, which it promised to destroy in joining the Chemical Weapons Convention, but it may have additional undeclared capabilities.

SOURCE: "Chemical Weapon Status 2004," in *Proliferation News and Resources*, Carnegie Endowment for International Peace, http://www.ceip.org/files/projects/npp/resources/DeadlyArsenals/maps/chem.jpg (accessed September 23, 2004)

The usefulness of chemical agents vary depending on their intended purpose. Chemical weapons are usually categorized according to the following factors:

1. Toxicity, or the lethality of an agent. Less-toxic substances can be used to incapacitate (as in riot control) rather than kill.

2. Physical state. Whether an agent is in solid, liquid, or gaseous form plays a significant role in its weaponization and delivery.

3. Mode of action, or whether the agent is delivered via inhalation (breathing), ingestion (eating or drinking), or through the skin.

4. Speed, or the amount of time between exposure to an agent and the appearance of symptoms.

5. Half-life or ability to persevere. This refers to how long an agent can retain its characteristics in the environment in which it is used and continue to pose a threat. The ability to persevere is relevant for response and decontamination following a chemical attack.

Classes of Chemical Weapons

Chemical weapons fall under one of the following classes: blister agents, blood agents, choking agents, nerve agents, or nonlethal agents. Table 3.3 provides a summary of common examples of each type and their effects.

TABLE 3.3

Common chemical warfare agents

Types	Agents	Effects
Blister	Mustard Nitrogen Mustard Lewisite	Causes large skin blisters; respiratory damage; long-term debilitating injuries, including blindness
Choking	Phosgene	Death from lack of oxygen
Blood	Hydrogen Cyanide Cyanogen Chloride	Interferes with body's oxygen supply, causing death
Nerve	Tabun Sarin Soman Cyclosarin VX Fourth generation	Loss of muscular control, respiratory failure, and death
Other	TFNM[1] BZ[2]	Penetrates air filters; Incapacitation

[1]Trifluoronitrosomethane
[2]3-Quinuclidinyl Benzilate

SOURCE: "Common Chemical Warfare Agents," in *Proliferation: Threat and Response,* U.S. Department of Defense, Office of the Secretary of Defense, January 2001, http://www.defenselink.mil/pubs/ptr20010110.pdf (accessed September 23, 2004)

BLISTER AGENTS. Classified as first-generation chemical agents (World War I–era agents—among the first chemical agents used on the battlefield), blister agents are also known as vesicants. Their primary physiological effects include burning sensations to the eyes, skin, and mucous membranes. As the name implies, large, watery blisters can form, along with severe damage to the upper respiratory tract. Lewisite, nitrogen mustard, sulfur mustard, and phosgene oxime are all different types of vesicants. They can be dispersed in aerosol, liquid, or vapor form, and (except for lewisite) cause no immediate pain at the time of exposure. Blister agents are used primarily to incapacitate rather than kill the enemy, but large doses can result in death.

BLOOD AGENTS. Blood agents enter the body primarily via inhalation and incapacitate the blood tissues' ability to use oxygen properly, causing the target to asphyxiate. Hydrogen cyanide, cyanogen chloride, and arsine are blood agents that are highly volatile and disperse quickly under normal conditions.

CHOKING AGENTS. Also known as lung irritants, choking agents usually come in the form of heavier gases that tend to settle at ground level or in depressions such as trenches and foxholes. Chlorine, chloropicirin, phosgene, and diphosgene are choking agents that, when inhaled, cause a fluid buildup in the lungs so that victims drown and die of oxygen deficiency.

NERVE AGENTS. Nerve agents come in various forms, including VX and the G-series agents (so called because their U.S. Army codes begin with the letter "G") tabun,

sarin, soman, and cyclosarin. Highly deadly, these agents block the flow of acetylcholinestrase, an enzyme crucial to the functioning of the nervous system. Effects of exposure to nerve agents include seizures and a loss of body control as the agents exhaust their victims' muscles, including the heart.

NONLETHAL AGENTS. These incapacitants (which may be used separately or in conjunction with other chemical agents) include less potent chemicals that can be further subdivided into psychochemicals, tear gas agents, and vomiting agents. Psychochemicals are mainly hallucinogenic compounds such as lysergic acid diethylamide (LSD) and 3-quinuclidinyl benzilate that cause delusions and can incapacitate victims for a period of time. Tear gas agents, often used in riot control, are highly irritating to the eyes and respiratory tract. They include orthochlorobenzylidene malononitrile, chloroacetophenonoe, and brombenzyl cyanide. Vomiting agents, such as adamsite, are arsenic-based. They not only cause vomiting and but may also irritate the eyes and respiratory system.

Chemical Weapons Attack in Tokyo, 1995

In 1995 a sarin gas attack was perpetrated on civilians in the Tokyo subway by the Japanese cult Aum Shinrikyo ("Supreme Truth"), led by Shoko Asahara. The Aum cult, a religiously motivated apocalyptic terrorist group, spent the late 1980s and early 1990s experimenting with various warfare agents, seeking out chemical weapon components and other WMD from various states. In one instance they attempted to buy a MIG-29, one of the Soviet Union's most advanced fighter aircraft, and a nuclear warhead from Russia. They succeeded in buying a large Russian military helicopter. They also tested anthrax on sheep in Australia.

Sarin, in terms of its symptoms, is in a class with two other deadly nerve agents, soman and tabun. Symptoms of exposure to these chemicals include reduced vision, diarrhea, vomiting, paralysis, and respiratory failure (asphyxiation). Those sufficiently exposed can lapse into a coma and die. With sarin, doses that are potentially life threatening may be only slightly larger than those producing the least effects. Symptoms of overexposure occur within minutes or hours and include constriction of pupils (miosis), visual effects, headaches, runny nose and nasal congestion, salivation, chest tightness, nausea, vomiting, giddiness, anxiety, difficulty in thinking, difficulty sleeping, nightmares, muscle twitches, tremors, weakness, abdominal and thoracic cramps, diarrhea, and involuntary elimination. Severe exposure can cause convulsions, asphyxiation, and death.

During the morning rush hour on March 20, 1995, Aum Shinrikyo cult members carried bags of sarin onto five separate trains in the Tokyo subway system. They punctured the sarin-filled bags with the tip of a specially sharpened umbrella, then disembarked at the next stop. The five packages leaked onto the floor of the trains, and the sarin fumes began spreading almost immediately. Soon, many passengers were coughing and feeling nauseated. As the trains reached their next stops, some passengers collapsed on the platforms and others ran for the station exits. Within a few hours, twelve commuters were dead and 5,500 others were injured to varying degrees, some permanently.

In 2000 Toru Toyoda and Kenichi Hirose, two of the perpetrators of the attack, were sentenced to death by hanging. Another, Shigeo Sugimoto, was sentenced to life imprisonment. All three claimed that they had been brainwashed by the cult. By 2004 a total of thirteen cult members had been sentenced to death for their roles in the subway attack, though none had yet been executed.

BIOLOGICAL WEAPONS

Certain biological organisms and toxins have been developed as weapons that can be used against humans, livestock, and crops. Biological weapons are different from their chemical counterparts because they use living organisms or their products—viruses, bacteria, or toxins—such as ricin (which is derived from the castor bean) or mycotoxin (which is produced by fungi). Biological weapons attack a target by causing a deadly disease via inhalation, injection, ingestion, or entry through the skin into the body. They can be delivered through a variety of means, including bombs, warheads, sprayers, and individual delivery. Figure 3.2 shows the nations possessing or suspected of developing biological weapons as of 2004.

Depending on the agent, the incubation period for biological agents, or the time span between exposure and the first appearance of symptoms, can vary from a few hours to weeks. When it comes to weaponizing biological agents, certain characteristics make some organisms more ideal than others. These include the agent's ability to reliably infect, its contagiousness (whether or not it will spread easily from one person to another), stability, incubation time, ease of transportation, resistance to common antibiotics, and virulence (lethality). Weaponizing biological agents can be difficult because it is important to keep the pathogen alive and virulent through the delivery process and to make sure that the size of the agent is just right for optimum delivery.

Common Classes of Biological Weapons Agents

BACTERIA. Bacteria are single-celled organisms that can vary in lethality. Common bacteria used in biological weapons include *Bacillus anthracis* (causes anthrax), *Vibrio cholerae* (causes cholera), *Yersina pestis* (plague bacteria), and *Francisella tularensis* (causes tularemia). The bacterial incubation period is usually a few days.

RICKETTSIAE. Named after Howard Taylor Ricketts, the American pathologist who first identified them, rickettsial organisms are similar to bacteria, except that they exist within the intracellular environment and reproduce only in animal tissue. Rocky Mountain fever, Q fever, and typhus are all diseases caused by rickettsiae.

VIRUSES. Small in comparison with bacteria, viruses are also intracellular parasites and can affect plants and animals alike. Some diseases caused by viruses include smallpox, encephalitis, Ebola, yellow fever, lassa fever, and Venezuelan equine encephalitis.

TOXINS. Toxins differ from the other classes in that they are poisons produced by living organisms rather than living organisms themselves. Toxins may be proteins or nonproteinacious in nature, and they act by disrupting nerve impulse transmissions or blocking protein synthesis. Examples of toxins include *Clostridium botulinum* (causes botulism), found on poorly preserved food; *Ricinus communis* (ricin), found in the castor bean seed; and saxitoxin, found in certain shellfish.

NUCLEAR WEAPONS

Nuclear and other radiological (radioactive) weapons are some of humankind's deadliest creations. In August 1945 two atomic bombs were detonated in Japan by the United States. These bombs were a product of the top-secret Manhattan Project, which cost the U.S. government approximately $2 billion. The attack devastated the cities of Hiroshima and Nagasaki, causing about seventy thousand and forty thousand fatalities, respectively. The bombs destroyed everything except concrete-reinforced buildings in a mile-wide area below the blast point (ground zero), which was 1,800 feet in the air. Casualties from the immediate blasts were relatively small. Most victims were killed in subsequent fires caused by the tremendous heat from the blast, estimated at several million degrees. Blast winds from the explosions destroyed buildings several miles away. More than 150,000 people were injured by the atomic explosions at Hiroshima and Nagasaki. The after-effects of the nuclear radiation, such as radiation-induced cancer, persisted for decades. Today nuclear technology, which is much cheaper and easier to acquire than it was in 1945, poses one of the gravest international threats.

A nuclear reaction results from atomic fission or fusion, with the former being easier to accomplish in a weapons production process. When the atom of a material is bombarded by neutrons it is broken into two roughly equal fragments which releases energy. This process is called "fission," and the substances that are manipulated to release energy through fission are called "fissile materials." In a nuclear chain reaction, the minimum amount of fissile material required to sustain the reaction is called "critical mass." When additional material is added to the reaction, it results in the creation of a supercritical state

FIGURE 3.2

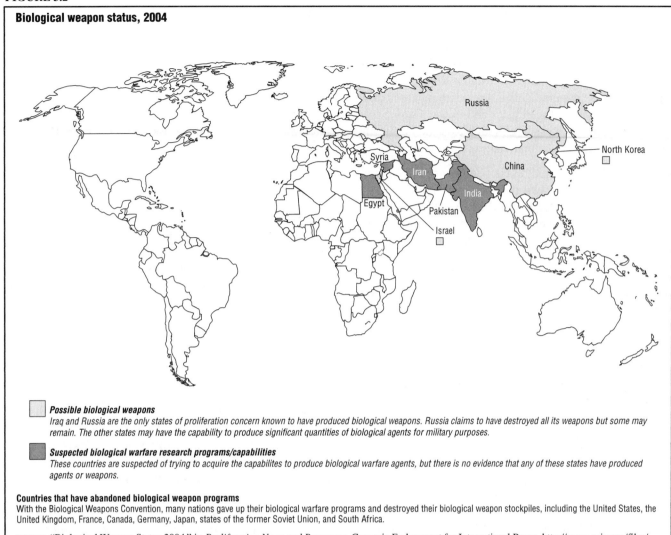

Biological weapon status, 2004

Possible biological weapons
Iraq and Russia are the only states of proliferation concern known to have produced biological weapons. Russia claims to have destroyed all its weapons but some may remain. The other states may have the capability to produce significant quantities of biological agents for military purposes.

Suspected biological warfare research programs/capabilities
These countries are suspected of trying to acquire the capabilites to produce biological warfare agents, but there is no evidence that any of these states have produced agents or weapons.

Countries that have abandoned biological weapon programs
With the Biological Weapons Convention, many nations gave up their biological warfare programs and destroyed their biological weapon stockpiles, including the United States, the United Kingdom, France, Canada, Germany, Japan, states of the former Soviet Union, and South Africa.

SOURCE: "Biological Weapon Status 2004," in *Proliferation News and Resources*, Carnegie Endowment for International Peace, http://www.ceip.org/files/ projects/npp/resources/DeadlyArsenals/maps/bio.jpg (accessed September 23, 2004)

(one in which the rate of the reaction constantly increases), where the mass rapidly expands because of the intense heat and pressure. The critical mass of a particular nuclear device depends on the type of fissile material used, its density, and the design of the weapon.

Fission weapons can be delivered in a number of ways, including bombs, several different types of missiles, and other dispersal devices. Most nuclear devices are either implosion-type weapons (those that burst inward) or gun-type assemblies (those that propel outward), and the most typical fissile materials used are plutonium and highly enriched uranium. Other fissile isotopes, such as cesium and cobalt, can also create significant damage as radiological weapons. The blast effects of a nuclear bomb are similar to those of a conventional bomb, but the results are far more deadly because of the high temperatures caused by an atomic explosion and because of the nuclear radiation that follows.

THE TRAFFICKING OF NUCLEAR AND FISSILE MATERIAL

The illicit trafficking of nuclear and fissile material was a grave U.S. national security threat in the early and mid-1990s. In the aftermath of the Soviet breakup, the amount of this trafficking increased dramatically. Major factors contributing to this problem were political and economic crises in the newly independent states and inadequate security and inventory systems at former Soviet nuclear facilities, especially the civilian ones used for energy production. Budget cuts severely reduced salaries for many scientific and security personnel, which increased incentives for such "insiders" to trade matériel for money.

From 1992 to 1994 seven cases were reported involving weapons-usable fissile materials, which are often referred to as "the seven significant cases." These cases are considered significant because of the quantity and/or quality of the materials involved. The last of the cases involved the seizure of 2.7 kilograms (six pounds) of highly enriched uranium in Prague on December 14, 1994. The uranium confiscated in Prague can be traced to a large criminal network, members of which are also linked to a seizure of highly enriched uranium in Landshut, Germany, in July 1994 and a seizure of plutonium-239 in Munich, Germany, in August 1994. Evidence suggests that the materials in all three of these cases originated at the Institute for Physics and Power Engineering (IPPE) in Obninsk, Russia, although Russian authorities continue to deny that the material came from Obninsk.

Although many gaps and uncertainties remain, certain conclusions may be drawn from the events that followed the seizure of the uranium in Prague. All the suspects indicted were middlemen—no buyer or supplier was ever arrested or even identified. The IPPE is an important research facility in the Russian Federation, and it is hard to imagine that such a sizable amount of uranium was diverted or stolen by a novice, or even by a single person. If the uranium did come from IPPE, it suggests the presence of an elaborate network of smugglers.

Cases such as this focused the world's attention on the urgent need to improve security inventory procedures in Russia and other republics of the former Soviet Union. According to the Russian-language newspaper *Moskovskiye Novosti,* at the time of the Prague incident, IPPE desperately lacked adequate security and systems of registering, controlling, and physically protecting nuclear materials.

The U.S. and Russian governments jointly initiated the Materials Protection, Control, and Accounting Program (MPCA) with nuclear facilities of the former Soviet Union in 1995. The IPPE was one of the first pilot facilities for the program. Under the auspices of the MPCA, U.S. laboratories cooperated with Russian facilities to introduce advanced material protection and accounting systems. These included developing computerized materials inventory and accounting databases, training Russian specialists, and implementing video monitoring systems and portal monitors. To guard against nuclear theft, portal monitors are placed in such facilities as uranium enrichment plants, weapons manufacturing and storage plants, nuclear laboratories, and nuclear waste disposal sites. They scan vehicular or human traffic and sound an alarm if they detect radioactivity. The project was successful in securing tons of fissile materials, and by March 1996 the program had received more funding from the U.S. Department of Energy and a new computerized MPCA system was established at IPPE. Following the attacks of September 11, 2001, the Bush administration pushed to accelerate the MCPA timeline, moving forward the date for securing all weapons-usable nuclear material to 2008 from a projected 2010 (Nuclear Threat Initiative Web site, www.nti.org). The 2004 budget for the MCPA was $227 million.

Since 1995 no thefts or diversions of significant quantities of weapons-grade fissile material have been confirmed. Less clear, however, is whether this change is permanent, or merely a hiatus. Some nonproliferation experts believe that lack of demand, increased awareness, and international assistance to the newly independent states have all contributed to this ebb in the flow of smuggled nuclear substances. On the other hand, the international intelligence community's failure to share information and the increasing sophistication of nuclear smugglers may have merely created the false impression that nuclear thefts and diversions no longer occur.

NONPROLIFERATION REGIMES AND TREATIES

An international treaty is usually negotiated between two (bilateral) or more (multilateral) states and typically enumerates the rights and duties each party has in reference to the issue being addressed. They are usually signed by the legitimate ruling administration of a sovereign state, but signing the treaty is not usually the last step taken in its approval; it also needs to be ratified in order for it to take effect. In the United States, treaty ratification requires the approval of two-thirds of the U.S. Senate. When the parties have achieved ratification, a treaty comes into effect.

International biological, chemical, nuclear, and missile capabilities are covered under several treaties and agreements, including the Biological and Toxin Weapons Convention (BTWC), Chemical Weapons Convention (CWC), Nuclear Nonproliferation Treaty (NPT), and the Missile Technology Control Regime (MTCR), among others. In addition, treaties declaring certain territories to be either entirely weapons-free—such as the Antarctic— or nuclear weapons-free—such as the Caribbean and Africa—are also in effect.

The Biological and Toxin Weapons Convention (BTWC)

The BTWC, which prohibits the development, production, and stockpiling of biological and toxin weapons, was opened for signature in 1972. As of 2004 it had 167 signatories, 151 of which had ratified the treaty. Its biggest drawback is the lack of an overarching monitoring or verification body to ensure that parties abide by the treaty. This shortcoming was especially evident when President Boris Yeltsin of Russia announced in 1992 that the former Soviet Union had aggressively pursued an offensive biological capability despite having signed the BTWC. So far, all negotiations for the proposed verification system have proved fruitless, especially since it would apply only to those states that ratify the protocol. The result would be two tiers of states, those subject to verification procedures and those that would be exempt. Also, there is no single way to distinguish which biological facilities fall under the BTWC. It must be noted, however, that unlike the NPT or the MTCR, the BTWC is nondiscriminatory—all parties involved are subject to the same procedures.

The Chemical Weapons Convention (CWC)

The CWC was developed and opened for signature in 1993; as of 2004 it had 164 member states. The CWC has intrusive and strict measures. It calls for the prohibition of the development, production, stockpiling, and retention of chemical weapons. It also discourages states from assisting or inducing other parties to develop such capabilities. The organization responsible for overseeing the CWC is the Organization for the Prohibition of Chemical Weapons (OPCW), based in The Hague, Netherlands. The OPCW conducts routine inspections as well as challenges inspections of facilities believed to be violating the CWC. Its September 2004 status report indicated that a total of 1,800 inspections in sixty-five countries had been conducted from 1997 to 2004.

The Australia Group

In addition to the BTWC and the CWC, an informal group of thirty-eight countries and the European Commission came together in 1985 in response to the use of chemical weapons during the Iran-Iraq war and formed the Australia Group. These states created a list of materials that could potentially be used to develop biological and/or chemical weapons and restricted exports of these materials to known or suspected states of proliferation concern.

The Nuclear Nonproliferation Treaty (NPT)

The Nuclear Nonproliferation Treaty entered into force in 1970 and as of 2004 had 189 member states. The NPT is a two-tiered agreement, which means there are separate obligations for nuclear weapons states (including the United States, United Kingdom, Russia, China, and France) and nonnuclear weapons states. The nuclear weapons states are required not to transfer any weapons capabilities to nonnuclear weapons states and to work on eventual disarmament. Meanwhile, the nonnuclear weapons states resolve not to develop nuclear capabilities, in return for technological assistance (in the energy sector) from nuclear weapons states. The NPT has an escape clause that allows a country to withdraw upon three months' notice. To date, the only states of concern that have refused to sign the NPT are India, Pakistan, and Israel; North Korea, which signed the NPT in 1985, announced in January 2003 that it intended to withdraw from the treaty.

The Nuclear Suppliers Group (NSG)

The Nuclear Suppliers Group (NSG) is a group of forty nuclear supplier countries that seeks to ensure that the international trade in nuclear items for peaceful purposes does not contribute to the proliferation of nuclear weapons. The group's guidelines for trade cover a range of nuclear materials and technology relating to nuclear reactors and the equipment needed to operate them. The NSG helps member countries engage in peaceful nuclear cooperation while meeting existing international nuclear nonproliferation agreements.

The Missile Technology Control Regime (MTCR)

Unlike the NPT, the MTCR is not a formal treaty; however, it is by far the most effective regime that addresses the proliferation of missile technology. Formed in 1987, the MTCR is a group of thirty-four states that aim to limit or prohibit transfer of missile and dual-use nuclear/missile capabilities. In 1993 the MTCR extended its guidelines on export controls and restrictions to cover missiles with biological and chemical capabilities.

The Treaty on the Principles Governing the Activities of States in the Exploration and Use of Outer Space, Including the Moon and Other Celestial Bodies

The Treaty on the Principles Governing the Activities of States in the Exploration and Use of Outer Space, Including the Moon and Other Celestial Bodies, also called the Outer Space Treaty, was signed January 27, 1967, and has 127 parties, although it was negotiated predominantly by the United States and Soviet Union. Its intent is to limit the militarization of the moon, outer space, and celestial bodies. The treaty requires that countries use celestial bodies for peaceful purposes only—not for any military bases, fortifications, or weapons testing.

The Treaty between the United States and the Soviet Union on the Limitation of Antiballistic Missile Systems (ABM Treaty)

The Treaty between the United States and the Soviet Union on the Limitation of Antiballistic Missile Systems, also called the Antiballistic Missile (ABM) treaty, was signed May 26, 1972, and had only two parties: the United States and the Soviet Union (whose obligations were later assumed by the Russian Federation). It prohibited deployment of an antiballistic missile (ABM) system, or one designed to counter missiles in flight, for the defense of territory or the building of bases for such a defense. In December 2001 President George W. Bush formally notified Russia of his intention to pull out of the ABM treaty by the following June. On June 13, 2002, he explained, "With the Treaty now behind us, our task is to develop and deploy effective defenses against limited missile attacks. As the events of September 11 made clear, we no longer live in the Cold War world for which the ABM Treaty was designed. We now face new threats from terrorists who seek to destroy our civilization by any means available to rogue states armed with weapons of mass destruction and long-range missiles. Defending the American people against these threats is my highest priority as Commander-in-Chief."

The Interim Agreement between the United States and the Union of Soviet Socialist Republics on Certain Measures with Respect to the Limitation of Strategic Offensive Arms (SALT I)

The Interim Agreement between the United States and the Union of Soviet Socialist Republics on Certain Measures with Respect to the Limitation of Strategic Offensive Arms, also called the Strategic Arms Limitation Treaty I (SALT I), was signed May 26, 1972, and included as parties the United States and the Soviet Union. This agreement, like the ABM treaty, arose from the first series of Strategic Arms Limitation Talks (SALT), which ran from November 1969 through May 1972. The United States and the Soviet Union agreed in this treaty not to build any more intercontinental ballistic missiles (ICBMs), the largest and most powerful type of nuclear missiles. An increase in the number of submarine-launched ballistic missiles (SLBMs) was allowed to each side if an equal number of land-based launchers were destroyed.

The Treaty between the United States and the Union of Soviet Socialist Republics on the Limitation of Strategic Offensive Arms (SALT II)

The Treaty between the United States and the Union of Soviet Socialist Republics on the Limitation of Strategic Offensive Arms, also called the Strategic Arms Limitation Treaty II (SALT II), was signed June 18, 1979, by the United States and the Soviet Union. It was negotiated as a result of the SALT II talks, which ran from 1972 to 1979, and set limits on the number of ballistic missiles and their launchers. Each country was limited to 2,250 launchers plus 1,320 launchers for multiple independently targetable reentry vehicles (MIRVed) missiles. A MIRVed missile is a nuclear delivery vehicle capable of carrying more than one warhead, where each warhead can be independently targeted toward different objectives. Newer ICBMs and air-to-surface ballistic missiles (ASBMs) were limited to ten warheads per missile, while SLBMs were allowed fourteen warheads per missile. Also under this treaty, space-based weapons were prohibited.

A protocol (a less formal agreement than a treaty) lasting two years was also signed at the same time as the treaty, and called for a prohibition on the deployment of air-launched ballistic missiles, mobile ICBMs, ground-launched cruise missiles, and sea-launched cruise missiles with a range of over six hundred kilometers (373 miles).

Although President Jimmy Carter submitted the treaty for ratification to the U.S. Senate immediately after signing, congressional concerns and the Soviet invasion of Afghanistan in 1979 caused the treaty to be removed from consideration. For that reason, the treaty was never signed and never became a binding legal agreement, although for approximately seven years it had the force of a politically binding agreement. In May 1986 President Ronald Reagan,

citing Soviet violations, declared that the United States would no longer honor the SALT II limits. The United States then exceeded those limits in November of that year.

The Treaty between the United States of America and the Union of Soviet Socialist Republics on the Reduction and Limitation of Strategic Offensive Arms (START I)

The Treaty between the United States of America and the Union of Soviet Socialist Republics on the Reduction and Limitation of Strategic Offensive Arms, also called the Strategic Arms Reduction Treaty I (START I), was signed July 31, 1991, by the United States and the Soviet Union. As of 2004 the following states of the former Soviet Union have agreed to the treaty limitations: Belarus, Kazakhstan, the Russian Federation, and Ukraine. The point of START I was to reduce the numbers of U.S. and Soviet strategic offensive arms, including ICBMs, SLBMs, and heavy bombers, and to limit the number of nuclear warheads for each party to six thousand. The parties agreed to limits of 4,900 warheads on deployed ballistic missiles and 1,100 on deployed mobile ICBMs. The treaty also limited the former Soviet Union to only 154 deployed heavy ICBMs, versus the 308 that were in place before the treaty (with each allegedly carrying ten warheads). The parties agreed to exchange telemetric information, or data radioed from the missiles themselves, from all test flights of ICBMs and SLBMs and to exchange the equipment necessary to interpret these data.

Compliance with the START limits is achieved through verification measures. Verification is an obstacle to the ratification of many arms control agreements because it is inherently intrusive and is often used as an excuse by politicians not to reach an arms control agreement. Verification includes both verification itself and monitoring. Monitoring involves intelligence gathering, analyses, and data exchanges. Verification is more of a legal formality and a policy process that either supports or questions the conclusions reached through monitoring. For START signatories, the chief body assigned to monitor compliance is the Joint Compliance and Inspection Commission, which has met in Geneva, Switzerland, since 1991.

The Russian Federation succeeded the Soviet Union as a party to the treaty after the latter's breakup, but many strategic offensive arms had been located in the former Soviet states of Belarus, Kazakhstan, and Ukraine. Consequently, a protocol was signed in Lisbon, Portugal, on May 23, 1992, making START I a multiparty treaty of five nations (the United States, the Russian Federation, Belarus, Kazakhstan, and Ukraine) instead of a bilateral treaty exclusively with the United States and the Russian Federation.

The Treaty on Open Skies

The Treaty on Open Skies, signed March 24, 1992, in Helsinki, Finland, is composed of members of NATO and the former Warsaw Pact. The thirty participating states have the right to conduct, and the obligation to receive, overhead flights by unarmed observation aircraft, excluding helicopters. These aircraft are authorized to carry certain accessories such as cameras—panoramic, still-frame, and video—and infrared scanning devices. Host nations may require that a host aircraft be used during the flight; this is known as the "taxi option." Otherwise, the inspecting party provides the aircraft used in the flight. All aircraft and sensor suites, prior to use, must undergo certification inspections. Negotiated annual quotas limit the number of flights each country can conduct and must receive. Each country must accept as many flights as it is allowed to conduct. Countries with larger landmasses are allotted larger quotas. For example, in 2004 the U.S. and Russian quota was forty-two flights per year whereas Portugal had only two. Data from any such flight may be acquired by any state.

The Treaty between the United States and the Russian Federation on Further Reduction and Limitation of Strategic Offensive Arms (START II)

The Treaty between the United States and the Russian Federation on Further Reduction and Limitation of Strategic Offensive Arms, also called the Strategic Arms Reduction Treaty II (START II), was signed January 3, 1993, by the United States President George H. W. Bush and the Russian Federation President Boris Yeltsin. The objectives of the START II treaty were the elimination of all MIRVed ICBMs (missiles carrying multiple warheads) and a significant reduction in SLBMs. START II took nearly all of its definitions, procedures, and compliance schemes from START I. Each party agreed to decrease its deployed strategic weapons to 3,000–3,500 warheads by 2003. Bombers, such as the B-2, were to be held up to more scrutiny. B-2s must be exhibited and inspection-ready and could no longer test with long-range nuclear air launched cruise missiles (ALCMs).

The Proliferation Security Initiative (PSI) and Other Measures

Proposed in May 2003 by President George W. Bush, the Proliferation Security Initiative (PSI) is an agreement concerning specific steps to follow to prevent shipments of WMD, their delivery systems, and related equipment and materials. The PSI is not a formal treaty but rather a partnership between participating countries. It calls for joint training exercises and the development of common activities designed to stop, search, and seize WMD shipments, especially in international waters. In February 2004 President Bush called for the PSI to be expanded beyond seizing WMD shipments in transit. He called for partner countries to work together to identify and break up the criminal networks that traffic in such weapons.

In September 2003 President Bush called upon the UN to pass a resolution requiring all states to criminalize proliferation, enact strict export controls, and secure sensitive materials within their borders. In April 2004 the UN Security Council did just that, adopting Resolution 1540, a measure concerning the spread of WMD. It decided that "all States shall refrain from providing any form of support to non-State actors that attempt to develop, acquire, manufacture, possess, transport, transfer or use nuclear, chemical or biological weapons and their means of delivery." The resolution also calls for states to criminalize the creation or possession of WMD by nonstate actors and to "take and enforce effective measures to establish domestic controls" of WMD.

The Future of WMD Arms Control

On May 24, 2002, President Bush and Russian president Vladimir Putin signed the Strategic Offensive Reductions Treaty (SORT), which calls for each country to deploy no more than 1,700 to 2,200 strategic warheads by December 31, 2012. At the time, the U.S. reduction plan specified the retirement of all fifty of its ten-warhead Peacekeeper ICBMs and the conversion of four Trident submarines from strategic (carrying nuclear warheads) to conventional service. The U.S. Senate ratified the treaty in March 2003, and the State Duma (the lower house of Russia's legislature), approved it the following May.

In November 2002 the International Code of Conduct against Ballistic Missile Proliferation went into effect. By January 2004, 111 countries had signed the agreement. It is intended to supplement, not supplant, the earlier Missile Technology Control Regime (MTCR) and focuses in particular on efforts to prevent the spread of ballistic missiles.

CHAPTER 4

COUNTRIES OF PROLIFERATION CONCERN

Since the end of the cold war and the breakup of the Soviet Union in the early 1990s, the weapons of mass destruction (WMD) threat no longer focuses solely on two superpowers but includes a host of nations, among them China, Egypt, India, Iran, Israel, Libya, North Korea, Pakistan, Russia, and Syria. Figure 3.2 and Figure 3.1 (both in Chapter 3) show the countries actively involved in developing biological and chemical weapons. Although several countries including the United States, Russia, the United Kingdom, France, and China reduced stockpiles from 1986 to 2002, significant stores of these weapons remain in these countries. North Korea's declaration of an active nuclear weapons program in late 2002 is an illustration of the gravity of this transnational threat.

Many nations are also working to develop missile systems capable of carrying nuclear weapons. (See Figure 4.1.) Figure 4.1 and Table 4.1 list the differing ranges of such missiles. In a June 2000 address to the Asia Society, U.S. Assistant Secretary for Nonproliferation Robert J. Einhorn called the proliferation of WMD and their missile-delivery systems the gravest threat to world security.

This chapter provides an overview of the nuclear, chemical, biological and conventional weapons programs and capabilities of selected nations. It also presents a more detailed examination of the history of Iraq's weapons development, program, and capabilities as well as the events in Iraq that led to its invasion by the United States in March 2003.

CHINA

One of the five nuclear weapons states of the Nuclear Nonproliferation Treaty (NPT), China has been developing WMD since 1955. It conducted its first nuclear test in 1964 and maintains a nuclear arsenal that consists of missiles and various other munitions. China's missile collection includes intercontinental ballistic missiles (ICBMs),

TABLE 4.1

Missiles, by range and country of possession, 2004

Range	Country
Intercontinental and/or submarine-launched ballistic missiles (>5,500 km)	China, France, Russia, United Kingdom, United States, North Korea (Taepo Dong 2 or Taepo Dong ICBM)
Intermediate-range ballistic missiles (3,000–5,500 km)	India, Iran, possibly North Korea
Medium-range ballistic missiles (1,000–3,000 km)	Israel, North Korea, Saudi Arabia, China, India, Pakistan, Iran
Short-range ballistic missiles (70–1,000 km)	Afghanistan, Algeria, Argentina, Armenia, Belarus, Bulgaria, China, Czech Republic, Egypt, Greece, India, Iran, Iraq, Israel, Kazakhstan, Libya, Netherlands, North Korea, Pakistan, Romania, Russia, Serbia, Slovakia, South Korea, Syria, Taiwan, Turkey, Turkmenistan, Ukraine, United Arab Emirates, Vietnam, and Yemen.

SOURCE: "Table 1. Missiles by Categories of Range," in *Missile Survey: Ballistic and Cruise Missiles of Foreign Countries,* Congressional Research Service, March 5, 2004, http://fpc.state.gov/documents/organization/31999.pdf (accessed September 23, 2004)

submarine-launched ballistic missiles (SLBMs), and theater missiles, according to inventories compiled by the Arms Control Association (a nonpartisan organization dedicated to promoting public understanding of and support for effective arms control policies). The country is estimated to possess around four hundred nuclear warheads, and maintains missiles targeted at the United States and Taiwan. Currently, China and Russia are the only two potential U.S. adversaries with the capability of deploying missiles that can target and reach U.S. cities. China has, however, repeatedly pledged a "no first use" policy with its nuclear forces, meaning it would only use nuclear weapons to retaliate for an offensive nuclear attack.

Development of improved missile systems is a high priority for China. In December 2002 it successfully tested a DF-21 medium-range missile (1,800 kilometers or approximately 1,100 miles) with multiple warheads. Test-

FIGURE 4.1

Ballistic missile proliferation status, 2004

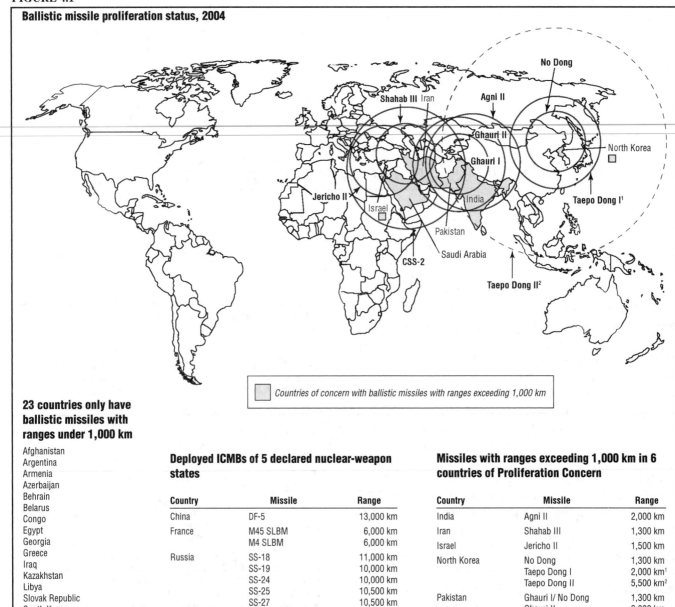

Countries of concern with ballistic missiles with ranges exceeding 1,000 km

23 countries only have ballistic missiles with ranges under 1,000 km

Afghanistan
Argentina
Armenia
Azerbaijan
Behrain
Belarus
Congo
Egypt
Georgia
Greece
Iraq
Kazakhstan
Libya
Slovak Republic
South Korea
Syria
Taiwan
Turkey
Turkmenistan
Ukraine
United Arab Emirates
Vietnam
Yeman

Deployed ICMBs of 5 declared nuclear-weapon states

Country	Missile	Range
China	DF-5	13,000 km
France	M45 SLBM	6,000 km
	M4 SLBM	6,000 km
Russia	SS-18	11,000 km
	SS-19	10,000 km
	SS-24	10,000 km
	SS-25	10,500 km
	SS-27	10,500 km
	SS-N-18 SLBM	6,500/8,000 km
	SS-N-20 SLBM	8,300 km
	SS-N-23 SLBM	8,300 km
United Kingdom	Trident II/D-5 SLBM	7,400 km
United States	Minuteman II	9,650 km
	MX Peacekeeper	9,650 km
	Trident I/C-4 SLBM	7,400 km
	Trident II/D-5 SLBM	7,400 km

Missiles with ranges exceeding 1,000 km in 6 countries of Proliferation Concern

Country	Missile	Range
India	Agni II	2,000 km
Iran	Shahab III	1,300 km
Israel	Jericho II	1,500 km
North Korea	No Dong	1,300 km
	Taepo Dong I	2,000 km[1]
	Taepo Dong II	5,500 km[2]
Pakistan	Ghauri I/ No Dong	1,300 km
	Ghauri II	2,000 km
Saudi Arabia	CSS-2	2,600 km[3]

[1]The sole test of the Taepo Dong I flew 1,320 km. Some experts speculate that an operational third stage and reentry vehicle would allow the Taepo Dong I to deliver a light payload over 5,500 km.
[2]The Taepo Dong II has not been flight-tested. The 2001 National Intelligence Estimate of the Ballistic Missile Threat speculates that, with a lighter payload, it could have a 10,000-km range.
[3]Saudi Arabia puchased CSS-2 missiles from China in 1987 and has never tested them. Experts question whether these missiles are operational.

SOURCE: "Ballistic Missile Proliferation Status 2004," in *Proliferation News and Resources*, Carnegie Endowment for International Peace, http://www.ceip .org/files/projects/npp/resources/DeadlyArsenals/maps/missile.jpg (accessed September 23, 2004)

ing of multiple warheads on the DF-31, a missile under development with a range of eight thousand kilometers (nearly five thousand miles), is expected.

China agreed to the NPT in 1992 and pledged not to export assembled ground-to-ground missiles (missiles originating from a land-based launcher and directed toward another land-based target) two years later. In 1996 it signed the Comprehensive Test Ban Treaty and the Chemical Weapons Convention (CWC), stating at the time that former chemical weapons production facilities in China had the ability to produce warfare agents such as mustard and lewisite, but that all chemical weapons stockpiles had been destroyed. The U.S. government remains skeptical about the veracity of these claims. U.S. defense officials also question China's claim that it does not possess biological weapons agents, despite the fact that China signed the Biological and Toxin Weapons Convention (BTWC) in 1984.

According to *China and Proliferation of Weapons of Mass Destruction and Missiles: Policy Issues* (Washington, DC: Congressional Research Service, August 8, 2003), China has repeatedly ignored promises it has made about not selling WMD, particularly missile technology, to other countries. At the close of 2004, the Bush administration had imposed sanctions against Chinese companies on at least eight occasions for transfers related to ballistic missiles, chemical weapons, and cruise missiles to Pakistan and Iran. It is widely accepted among analysts, and documented in Arms Control Association (ACA) fact sheets, that China directly assisted Pakistan's short-range ballistic missile and medium-range ballistic missile programs with raw materials and technical expertise, and also provided short-range ballistic missiles to Iran and intermediate-range ballistic missiles to Saudi Arabia.

The Nuclear Threat Initiative (NTI is a nonprofit organization founded in 2001 by CNN founder Ted Turner and former Senator Sam Nunn that seeks to strengthen global security by reducing the risk of use and preventing the spread of nuclear, biological, and chemical weapons) reports that China has also entered into governmental nuclear cooperation agreements with about twenty countries. These agreements are exclusively limited to the peaceful uses of nuclear energy and contain clauses that guarantee against the re-transfer of material or equipment by either country without prior consent by the other country and require adequate physical protection on all imported material and equipment in the territory of either country.

EGYPT

Although Egypt has historically been an ally of the United States in the Middle East, it remains on the American list of countries to monitor for WMD proliferation. In its *Unclassified Report to Congress on the Acquisition of Technology Relating to Weapons of Mass Destruction and Advanced Conventional Munitions, 1 January through 30 June 2001* (Washington, DC: Central Intelligence Agency, 2001), the Central Intelligence Agency (CIA) warned the U.S. Congress about Egypt's continued purchase of missiles and technology from North Korea and its ongoing acquisition of various weapon systems. Egypt began its nuclear program in the 1950s with assistance from, first, the Soviet Union and, later, the United States. It also produced its own Scud-B and Scud-C missiles, along with a host of rockets, to deliver WMD. Egypt agreed to the NPT in 1981 and has also called for the creation of a Middle Eastern nuclear weapon-free zone.

Egypt is one of the first countries to have trained its own military in chemical weapons defense, and it reportedly used mustard gas against northern Yemen during the mid-1960s. Its chemical weapons arsenal is believed to include mustard and phosgene, which are deliverable through missile warheads, rockets, mines, and artillery shells. Egypt's development of chemical weapons and its refusal to sign the CWC are considered a direct response to the development of an Israeli nuclear program. It has nevertheless officially pledged not to acquire and produce chemical warfare agents.

Information on Egypt's biological weapons program is limited. It signed the BTWC on April 10, 1972, and declares that it does not have biological weapons capabilities. On the other hand, Egypt has a strong technological base and the necessary resources for developing a significant biological weapons program, and its past efforts have been linked to developing biological agents such as plague and the encephalitis virus. According to the NTI, Israel has charged Egypt with conducting research to weaponize anthrax, plague bacteria, botulinum toxin, and Rift Valley fever virus. The Egyptian government vehemently denies these allegations.

INDIA

India is one of the most recent entrants into the nuclear weapons arena. The NTI reports that India tested its first nuclear device in 1974 and performed five additional underground nuclear weapons tests in May 1998. India, one of a handful of states that refuses to sign the Nuclear Nonproliferation Treaty, has a strong nuclear power program, for which it receives assistance from a host of countries. It has developed ballistic missiles and advanced conventional weapons to serve as delivery modes for its nuclear warheads. Prithvi series missiles have ranges between 150 and 250 kilometers (ninety-three to 155 miles). The Danush is a naval version of the Prithvi with a range of 250 kilometers (155 miles). The Agni I is said to have a range of 700 to 750 kilometers (435–466 miles), while the Agni II is reported to have a range of 1,500 kilometers (932 miles). In September 2003 India announced plans to develop an Agni III missile with a range of three thousand kilometers (1,864 miles).

India ratified the CWC in 1996 and declared an existing stockpile of chemical weapons. Under terms of the CWC, it must destroy this weapons stockpile by 2007. There is little information on whether India has offensive biological weapons capabilities, although it possesses a strong civilian biotechnology infrastructure. India has been a signatory of the BTWC since 1974.

IRAN

A threat to the United States since the overthrow of the Shah of Iran in 1979, Iran is believed to have developed an active WMD program to counter the Israeli threat, as well as to discourage opponents and establish regional dominance in the Persian Gulf and Caspian Sea. Iran agreed to the NPT in 1970, but concerns remain that Iran is covertly developing a nuclear weapons capability under the guise of building and running nuclear power plants to generate electricity. The NTI reported that in late 2002 American intelligence established via satellite photographs that Iran was secretly building and operating two nuclear facilities—a uranium enrichment facility at Natanz and a heavy water production plant near Arak. The International Atomic Energy Agency (IAEA), which conducts inspections under the NPT, admitted that Iran had delayed IAEA inspections of those two plants. In February 2003 an IAEA delegation visited the plant at Natanz and IAEA director Mohamed ElBaradei confirmed that the Natanz facility was enriching uranium, a key component in nuclear weapons development.

In the summer of 2004 concern about Iran's nuclear intentions renewed when it was learned that the country had resumed building centrifuges and restarted equipment used to make uranium hexaflouride gas, both of which are necessary to build nuclear weapons. Iran had promised Britain, France, and Germany to suspend building centrifuges to demonstrate its intent to cooperate with the IAEA. In April 2004 Iran resolved to cooperate fully with the IAEA, claiming that it had suspended enrichment programs and agreeing to an IAEA inspection. A June 2004 IAEA resolution rebuked Iran for failing to be forthcoming about its nuclear program. The United States had been pushing the IAEA to bring its concerns before the UN Security Council. A war of words ensued, with Israeli officials threatening a preemptive strike against Iran's nuclear sites, and Iran announcing that any preemptive strike would be met with an attack on Israeli nuclear plants. Figure 4.2 shows where Iran's nuclear facilities are located.

In August 2004, amid escalating tension over its nuclear program, Iran announced that it had carried out a field test of its new Shahab-3 missile, with an estimated range of eight hundred miles. The missile is capable of hitting targets in Israel and U.S. bases in the Persian Gulf region. With North Korea's aid, Iran has been developing missiles for many years. Based on old Russian designs, and updated by both North Korea and Iran, the missiles already developed and stockpiled are capable of reaching at least five hundred kilometers (311 miles), with the potential to go up to four thousand kilometers (2,485 miles). (See Figure 4.3.)

Iran is one of the few countries that has had chemical weapons used against it (by Iraq, in the 1980–88 war). The United States claims that Iran has been working on developing a chemical weapons program since the war with Iraq. According to the NTI, Iran's chemical weapon arsenal may include sarin, mustard, phosgene, and hydrocyanic acid. According to U.S. government estimates, Iran has the capacity to produce 1,000 metric tons of chemical agents per year and may have a stockpile of at least several thousand metric tons of weaponized and bulk chemical agents. Iran ratified the CWC in 1997 and strongly denies the existence of a chemical weapons program. In 2003 the CIA reported (*Unclassified Report to Congress on the Acquisition of Technology Relating to Weapons of Mass Destruction and Advanced Conventional Munitions, 1 January through 30 June 2003,* Washington, DC: Central Intelligence Agency, November, 2003) that Iran was actively pursuing contacts with Chinese companies to acquire the technology and expertise to produce its own nerve agents. Iran also ratified the BTWC (in 1973) but is believed to retain the resources and expertise to conduct an offensive biological weapons program. The United States asserts that Iran may have produced small quantities of biological weapons including mycotoxins, ricin, and the smallpox virus.

ISRAEL

Israel has the most sophisticated conventional and nuclear weapons program in the Middle East, largely because it has received considerable financial assistance from the West. Nevertheless, Israel has not signed the Nuclear Nonproliferation Treaty (NPT) and has chosen to pursue a nuclear option because it does not believe that the United States would effectively protect it in the case of a first-strike WMD attack from its immediate neighbors. Even though Israel considers the United States a strong ally, it is firmly independent and believes that it can rely only on itself for protection. Israel's geographic location makes it vulnerable to attacks by Arab neighbors. Israel has not overtly declared its nuclear capability, but there is little disagreement among experts that Israel has a well-developed nuclear program, based outside the town of Dimona. In the late 1990s, U.S. intelligence estimated that Israel could possess as many as seventy-five to 130 nuclear weapons. Estimates from other observers ranged as high as four hundred. (See Figure 4.4.) In July 2004 IAEA director Mohamed ElBaradei visited Israel; however, Israeli officials contend that they will not consider disarmament until a comprehensive Middle Eastern peace is obtained, and they will not permit IAEA inspection of the Dimona nuclear complex.

Little has been published about Israel's biological and chemical weapons capabilities. However, Israel has not signed the BWC, and the NTI reports that neighboring states allege that Israel has an active biological weapons program. According to the Federation of American Scientists, Israel's offensive biological and chemical warfare program is located at Ness-Ziona. Although Israel signed the CWC, as of 2004 it had not ratified the convention. Several reports published by scientists working at the Department of Pharmacology in the Israel Institute for Biological Research at Ness-Ziona described nerve agents, and in 1992 a plane crashed en route to Ness-

FIGURE 4.2

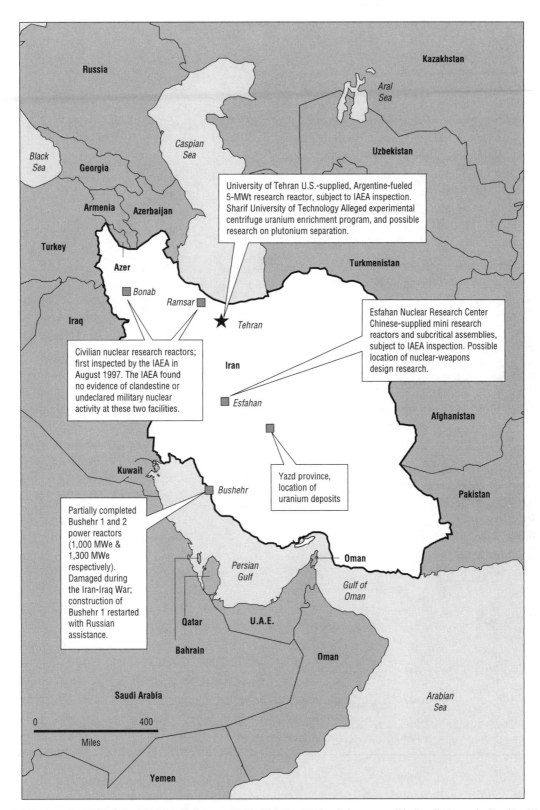

University of Tehran U.S.-supplied, Argentine-fueled 5-MWt research reactor, subject to IAEA inspection. Sharif University of Technology Alleged experimental centrifuge uranium enrichment program, and possible research on plutonium separation.

Esfahan Nuclear Research Center Chinese-supplied mini research reactors and subcritical assemblies, subject to IAEA inspection. Possible location of nuclear-weapons design research.

Civilian nuclear research reactors; first inspected by the IAEA in August 1997. The IAEA found no evidence of clandestine or undeclared military nuclear activity at these two facilities.

Yazd province, location of uranium deposits

Partially completed Bushehr 1 and 2 power reactors (1,000 MWe & 1,300 MWe respectively). Damaged during the Iran-Iraq War; construction of Bushehr 1 restarted with Russian assistance.

SOURCE: Joseph Cirincione, Jon Wolfsthal, and Miriam Rajkumar, "Table 15.2. Iran Nuclear Infrastructure," in *Deadly Arsenals: Tracking Weapons of Mass Destruction,* Carnegie Endowment for International Peace, 2002, http://www.carnegieendowment.org/pdf/npp/15-Iran.pdf (accessed September 23, 2004)

FIGURE 4.3

Ranges of Iran's missiles

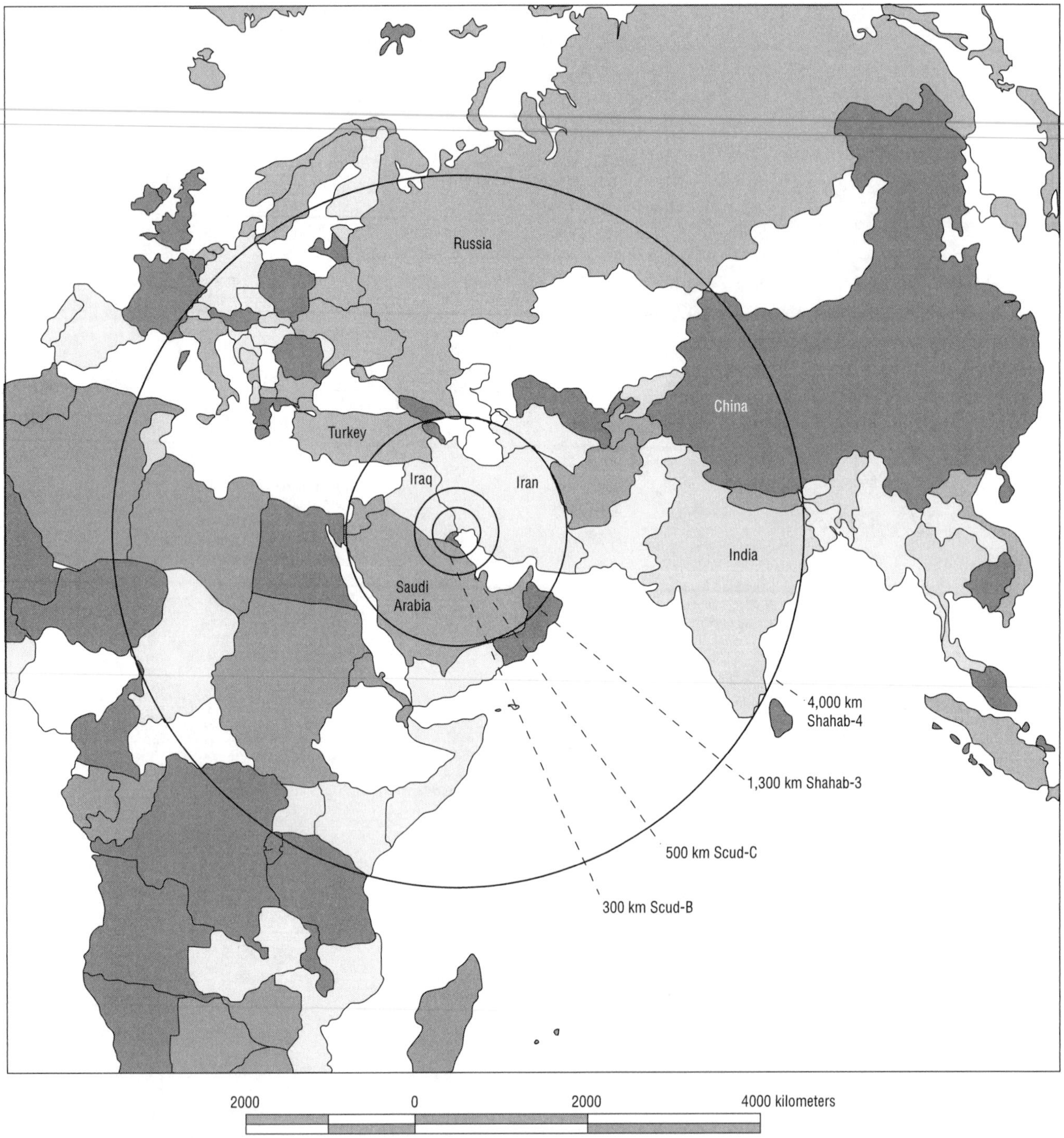
Russia
China
Turkey
Iraq
Iran
India
Saudi Arabia
4,000 km Shahab-4
1,300 km Shahab-3
500 km Scud-C
300 km Scud-B
2000
0
2000
4000 kilometers

SOURCE: "Ranges of Iran's Missiles," in *Missile Survey: Ballistic and Cruise Missiles of Foreign Countries*, Congressional Research Service, March 5, 2004, http://fpc.state.gov/documents/organization/31999.pdf (accessed September 23, 2004)

Ziona that contained fifty gallons of a chemical precursor of a sarin nerve agent.

LIBYA

For many years Libya was motivated to engage in aggressive pursuit of WMD in response to Israel's nuclear program and the nation's desire to assume a more prominent role in regional politics. Libya used chemical weapons in the 1987 conflict in Chad, and stockpiled nuclear weapons technology for many years. Once considered a serious danger to the Middle East because of its support of terrorist organizations and pursuit of nuclear, chemical, and

FIGURE 4.4

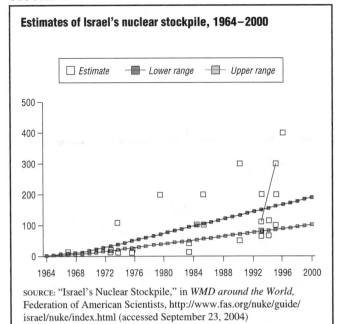

Estimates of Israel's nuclear stockpile, 1964–2000

SOURCE: "Israel's Nuclear Stockpile," in *WMD around the World*, Federation of American Scientists, http://www.fas.org/nuke/guide/israel/nuke/index.html (accessed September 23, 2004)

biological weapons, Libya took a dramatic turn in 2003. Following the fall of Saddam Hussein's Iraqi regime, Libya began to quietly negotiate with the United States and Britain to end its pursuit of nuclear weapons in exchange for normalized relations with the West. By December it had agreed to end all of its WMD programs and allow international inspectors into the country. As part of the agreement, Libya also agreed to eliminate ballistic missiles with a range over three hundred kilometers (186 miles), abide by the Missile Technology Control Regime (MTCR) guidelines, and sign the Comprehensive Test Ban Treaty.

According to *Disarming Libya: Weapons of Mass Destruction* (Washington, DC: Congressional Research Service, April 22, 2004), Libya's development of nuclear weapons was farther along than the United States or Britain had believed. Libya possessed a large number of centrifuges required to enrich uranium for weapons use as well as weapons design information. "The design closely resembles a 1960s vintage Chinese nuclear warhead," the report stated. In January 2004 it was revealed that both Libya and Iran had received nuclear development assistance from Abdul Qadeer Khan, Pakistan's leading nuclear scientist. Destruction or removal of Libya's WMD capabilities began in January 2004 with the removal of 55,000 pounds of documents and components to the United States. In March 2004 more than a thousand tons of missile and centrifuge parts, missile launchers, and related equipment were shipped out.

NORTH KOREA

North Korea poses a serious threat to the peace of Asia. An isolated nation with a large military, it has a

TABLE 4.2

North Korea's nuclear, biological, and chemical weapons and missile programs, January 2001

Nuclear	Plutonium production at Yongbyon and Taechon facilities frozen by the 1994 Agreed Framework; freeze verified by IAEA.
	Believed to have produced and diverted sufficient plutonium prior to 1992 for at least one nuclear weapon.
	Concerns remain over possible covert nuclear weapons effort.
	Ratified the NPT; later declared it has a special status. This status is not recognized by the United States or the United Nations. Has not signed the CTBT.
Biological	Pursued biological warfare capabilities since 1960s.
	Possesses infrastructure that can be used to produce biological warfare agents; may have biological weapons available for use.
	Acceded to the Biological and Toxin Weapons Convention.
Chemical	Believed to possess large stockpile of chemical precursors and chemical warfare agents.
	Probably would employ chemical agents against U.S. and allied forces under certain scenarios.
	Has not signed the CWC.
Ballistic missiles	Produces and capable of using SCUD B and SCUD C SRBMs, and No Dong MRBM.
	Successfully launched variant of Taepo Dong 1 MRBM in failed attempt to orbit satellite. (August 1998)
	Developing Taepo Dong 2 ICBM-range missile; agreed to flight test moratorium on long-range missiles in September 1999; reaffirmed in June 2000.
	Remains capable of conducting test.
	Not a member of the MTCR.
Other means of delivery available	Land- and sea-launched anti-ship cruise missiles; none have NBC warheads.
	Aircraft: fighters, bombers, helicopters.
	Ground systems: artillery, rocket launchers, mortars, sprayers.
	Special Operations Forces.

IAEA = International Atomic Energy Agency
CTBT = Comprehensive Test Ban Treaty
CWC = Chemical Weapons Convention
SRBM = Short Range Ballistic Missile (Range: 1000 kilometers or less)
MRBM = Medium Range Ballistic Missile
ICBM = Intercontinental Ballistic Missile (Range: greater than 5,500 kilometers)
MTCR = Military Technology Control Regime
NPT = Nonproliferation Treaty
NBC = Nuclear, Biological, or Chemical

SOURCE: "North Korea: NBC Weapons and Missile Programs," in *Proliferation: Threat and Response*, U.S. Department of Defense, Office of the Secretary of Defense, January 2001, http://www.defenselink.mil/pubs/ptr20010110.pdf (accessed September 23, 2004)

highly developed nuclear weapons program and has constructed missiles capable of hitting targets in neighboring countries, including the United States. Table 4.2 demonstrates North Korea's pursuit of various NBC (nuclear, biological, and chemical) weapons and missile programs. North Korean nuclear research harks back to the 1960s, when the country established a research reactor with the help of the Soviet Union. It signed the NPT in 1985, but there were discrepancies between its nuclear declarations and the results of IAEA inspections. North Korea agreed to halt its nuclear program in 1994 when it entered into an agreement with the United States, which pledged to help it develop civilian nuclear energy. North Korea nullified this agreement in 2002, when it revealed its uranium-enrichment program for nuclear weapons.

Although North Korea's deputy foreign minister admitted to a U.S. official that his country did, in fact, possess a nuclear weapon, the question has still not been answered officially. The U.S. National Intelligence Council estimated in December 2001 that North Korea had already produced one, possibly two, nuclear weapons. In January 2003 North Korea announced its intention to withdraw from the Nuclear Nonproliferation Treaty. In June 2003 it openly announced its intention to build a nuclear deterrent force. Since late 2003 North Korea has engaged in talks with the United States, China, Russia, Japan, and South Korea in an effort to end its nuclear weapons program. However, by the close of 2004 no progress had been made despite U.S. insistence on dismantling the North Korean nuclear program before addressing its economic and security concerns.

North Korea also has a full-scale missile program. It successfully developed a series of Scud missiles and flight-tested an ICBM in 1998. Figure 4.5 and Figure 4.6 show the ranges of North Korea's existing short- and medium-range missiles, as well as the potential range of long-range missiles it is developing. In 2004 *Jane's Defense Weekly* reported that North Korea was deploying a new land-based ballistic missile with a range of 2,500–4,000 kilometers (1,553–2,485 miles) and a sea-based missile with a range of at least 2,500 kilometers (1,553 miles). Both missiles can carry nuclear weapons and are capable of striking the United States. North Korea's stockpile of approximately six hundred ballistic missiles reportedly includes about a hundred medium-range No-Dong missiles. These missiles could reach an estimated 1,300 kilometers (808 miles).

North Korea has refused to sign the CWC and is believed to maintain a significant chemical weapons capability. The NTI reports that North Korea has twelve chemical weapons facilities—where raw chemicals, precursors, and actual agents are produced—and six major storage depots. Its stockpile of chemical weapons is said to include sarin, phosgene, and mustard, as well as several types of delivery munitions. The NTI estimates North Korea's chemical weapon production capacity as about 4,500 tons per year with the potential to triple in wartime.

North Korea signed the BTWC in 1987, but U.S. officials suspect that it is secretly developing biological weapons agents, including the bacterias that cause anthrax and plague.

PAKISTAN

Pakistan has pursued various NBC weapons and missile programs. In 1998 it became the world's seventh nuclear power. Pakistan officially developed its nuclear weapons program in response to a perceived threat from India. It received significant scientific and technical assistance from China and North Korea. As part of an aid package involving

FIGURE 4.5

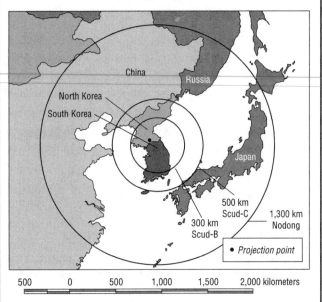

North Korean short- and medium-range missile capabilities, 1999

SOURCE: "Figure 1. North Korean Short- and Medium-Range Missile Capabilities," in *Missile Survey: Ballistic and Cruise Missiles of Foreign Countries*, Congressional Research Service, March 5, 2004, http://fpc.state.gov/documents/organization/31999.pdf (accessed September 23, 2004)

some $3 billion over five years, Pakistan agreed in 2003 to cease nuclear proliferation and assist the United States in its war on terrorism. Abdul Qadeer Khan, the scientist who was instrumental in making Pakistan a nuclear power, was revealed in early 2004 to have been selling nuclear equipment and expertise to Iran and Libya. His black market operations extended worldwide and made him a wealthy man. Authorities in Pakistan declined to imprison Khan, who is considered a national hero, but he will be under virtual house arrest for the remainder of his life. Pakistan has a small number of ballistic missiles capable of reaching India, most of which are said to be reverse-engineered—copied from a functional device by breaking it down to its components—from Chinese and North Korean missiles. Among the short-range missiles in Pakistan's arsenal are the solid-fuel Hatf-2 and Hatf-3 and the Shaheen-1. Figure 4.7 shows the ranges for Pakistan's ballistic missiles.

Pakistan signed the CWC in 1993 and ratified it in 1997, but there is little information, if any, on declared chemical weapons agents. While its biotechnology infrastructure is not as developed as that of India, Pakistan has well-established laboratories capable of carrying out biological weapons research. It signed the BTWC in April 1972 and ratified it in 1974.

SYRIA

Syria hosts a nuclear research center that operates under IAEA safeguards at Dyr al-Jajar and has been a sig-

FIGURE 4.6

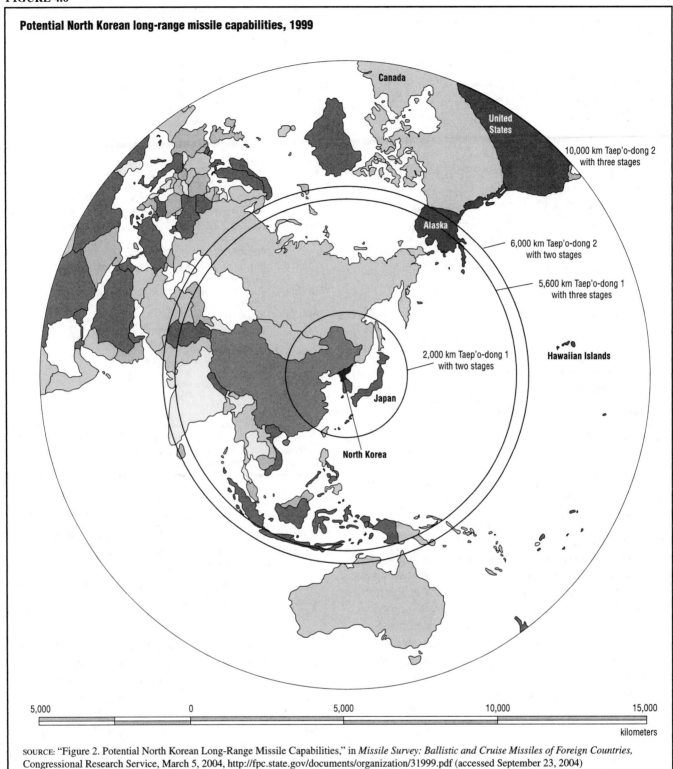

Potential North Korean long-range missile capabilities, 1999

10,000 km Taep'o-dong 2
with three stages

6,000 km Taep'o-dong 2
with two stages

5,600 km Taep'o-dong 1
with three stages

2,000 km Taep'o-dong 1
with two stages

Canada

United States

Alaska

Hawaiian Islands

Japan

North Korea

5,000   0   5,000   10,000   15,000

kilometers

SOURCE: "Figure 2. Potential North Korean Long-Range Missile Capabilities," in *Missile Survey: Ballistic and Cruise Missiles of Foreign Countries,* Congressional Research Service, March 5, 2004, http://fpc.state.gov/documents/organization/31999.pdf (accessed September 23, 2004)

natory to the NPT since 1968. Like other countries in the region, Syria's primary impetus for acquiring WMD is to counter Israel's nuclear and much greater conventional military strength. Table 4.3 shows Syria's NBC weapons and missile programs. Its missile program can be traced back to the early 1970s. In the early years of the twenty-first century it possessed one of the largest collections of ballistic

missiles in the Middle East, although the range of most of the missiles is estimated to be limited to five hundred kilometers (310 miles). The newer Scud D missile, first test-fired in September 2000, has an estimated range of seven hundred kilometers (441 miles). Over the years it has relied on the Soviet Union, Iran, and North Korea to help develop its missile program. Syria has several hundred Scud C,

FIGURE 4.7

Ranges of Pakistan's current and future ballistic missiles, 2004

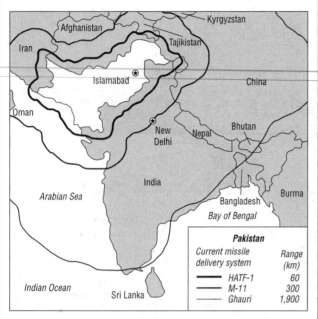

| Pakistan | |
Current missile delivery system	Range (km)
HATF-1	60
M-11	300
Ghauri	1,900

SOURCE: "Ranges of Current and Future Ballistic Missiles," in *WMD around the World*, Federation of American Scientists, http://www.fas.org/nuke/guide/pakistan/missile/index.html (accessed September 23, 2004)

TABLE 4.3

Syria's nuclear, biological, and chemical weapons and missile programs, January 2001

Nuclear	Is not pursuing the development of nuclear weapons.
	Ratified the NPT; has not signed the CTBT.
Biological	Possesses adequate biotechnical infrastructure to support limited biological warfare program.
	Believed to be pursuing biological agent development, but no major agent production effort likely is underway.
	Signed but not ratified the BWC.
Chemical	Possesses and is capable of delivering nerve agents; may be developing more advanced VX nerve agent.
	Making improvements to chemical infrastructure.
	Has not signed the CWC.
Ballistic missiles	Maintains and is capable of using force of SCUD B, SCUD C, and SS-21 missiles.
	Producing SCUD Cs with North Korean assistance.
	Making improvements to missile production infrastructure.
	Not a member of the MTCR.
Other means of delivery available	Land- and sea-launched anti-ship cruise missiles; none have NBC warheads.
	Aircraft: fighters, helicopters.
	Ground systems: artillery, rockets.

NPT = Nuclear Nonproliferation Treaty
BWC = Biological and Toxin Weapons Convention
CWC = Chemical Weapons Convention
MTCR = Military Technology Control Regime
CTBT = Comprehensive Test Ban Treaty
NBC = Nuclear, Biological, or Chemical

SOURCE: "Syria: NBC Weapons and Missile Programs," in *Proliferation: Threat and Response*, U.S. Department of Defense, Office of the Secretary of Defense, January 2001, http://www.defenselink.mil/pubs/ptr20010110.pdf (accessed September 23, 2004)

Scud D, and SS-21 SRBM missiles. At least some of this arsenal is fitted with chemical-weapon warheads.

According to NTI reports, Syria is also believed to harbor an extensive collection of chemical weapons, including a stockpile of the nerve agent sarin. It allegedly received assistance from Egypt in the chemical weapons arena before the 1973 war against Israel. Syria is not a signatory to the CWC. The country did, however, sign the BTWC in 1972. There is no definitive evidence of a Syrian biological weapons program, but the country has an extensive biotechnology and pharmaceutical infrastructure that employs many dual-use items that could be diverted to a biological weapons program.

IRAQ

During the 1980s Iraq invested heavily in nuclear, biological, chemical, and missile programs in an effort to deter enemies and gain preeminence in the Persian Gulf. By the start of the Gulf War in 1991, Iraq had developed and refined plans for nuclear weapons. NTI reports speculate that in 1991 Iraq was just three years away from constructing a nuclear weapon. At the end of the Gulf War, the United Nations Special Commission (UNSCOM) was mandated by UN Security Council Resolution 687 to assist the IAEA to verify and dismantle all of Iraq's non-nuclear WMD capabilities.

In the seven years from 1991 to 1998, UNSCOM accounted for 817 of 819 ballistic missiles and uncovered and destroyed a vast undeclared WMD arsenal, including forty-eight Scud missiles, three thousand tons of precursor chemicals, 690 tons of chemical weapons agents, 38,537 munitions, and a biological weapons facility at al-Hakam.

Although it was established as a verification unit dealing with a sovereign state's most sensitive security matters, evidence that UNSCOM was being used by the CIA to gather military intelligence eventually led to its demise following the Desert Fox air strikes of December 1998. Desert Fox was a joint air attack by U.S. and British forces against Iraq's "nuclear, chemical and biological weapons programs and its military capacity to threaten its neighbors," as President Bill Clinton explained in a televised address to the nation on December 16, 1998.

On December 17, 1999, UN Security Council Resolution 1284 replaced UNSCOM with a newer verification unit called the UN Monitoring, Verification, and Inspections Commission (UNMOVIC), led by former IAEA director Hans Blix. Iraq did not accept the new inspectors, and demanded that sanctions against the country be lifted before inspections could resume. The United States refused to lift the sanctions until Iraq demonstrated its complete destruction of all WMD.

In 2002 the United States once again focused its attention on Iraq, a longstanding concern because of its consistent attempts to establish NBC and missile programs. (See Table 4.4.) In November 2002 the UN Security Council passed Resolution 1441, which called for the return of UN inspectors to Iraq in order to determine if Iraq had renewed its secret WMD programs since 1998. That year members of the UN Special Commission on Iraq (UNSCOM) and the IAEA Action Team had been prohibited from entering Iraq and conducting regular monitoring and verification tasks.

Even as the inspectors were allowed to begin their new mission in late 2002, unanswered questions remained: During the four years that UN arms inspectors were not allowed into the country had Iraq resumed development of WMD capability or had Iraq more or less remained benign because of its economic plight, the result of extensive multilateral sanctions? The answers vary: Some observers contended that the ongoing sanctions had damaged the Iraqi people and the domestic economy, that UNSCOM had already uncovered all of Iraq's WMD capability, and that the UN should have lifted the embargo before any inspection teams were allowed in. Supporters of arms inspection, on the other hand, maintained that substantial Iraqi WMD capability still remained unaccounted for and that there needed to be further verification that Iraq had met all disarmament conditions before sanctions were eased.

Background: The Sources of U.S. Information on Iraqi WMD Capability

Iraq was extremely careful to conceal its WMD procurement activities, and had it not been for the defection and disclosures of two high-ranking Iraqi officials, Dr. Khidir Hamza (in 1994) and General Hussein Kamel (in 1995), much would still remain concealed.

Hamza obtained his training in the United States, receiving degrees from the Massachusetts Institute of Technology and Florida State University, and began working for the Iraqi nuclear program in 1970. He eventually became the director of Iraq's nuclear weapons program, making him the highest-ranking scientist to defect.

General Hussein Kamel was in charge of the Ministry of Industry and Military Industrialization, the primary agency responsible for secretly developing Iraq's WMD, and was married to one of Saddam Hussein's daughters. In 1995 Kamel defected to Jordan, and panicked Iraqi officials, fearing Kamel's disclosures, hastily provided IAEA inspectors with more than 140 boxes of documents detailing matters related to the Iraqi nuclear program. Kamel left Jordan and returned to Iraq after Saddam Hussein promised him asylum. However, within days of

TABLE 4.4

Iraq's nuclear, biological, and chemical weapons and missile programs, January 2001

Nuclear	Had comprehensive nuclear weapons development program prior to Operation Desert Storm. Infrastructure suffered considerable damage from Coalition bombing and IAEA dismantlement.
	Retains scientists, engineers, and nuclear weapons design information; without fissile material, would need five or more years and significant foreign assistance to rebuild program and produce nuclear devices; less time would be needed if sufficient fissile material were acquired illicitly.
	Ratified the NPT; has not signed the CTBT.
Biological	Produced and weaponized significant quantities of biological warfare agents prior to Desert Storm.
	Admitted biological warfare effort in 1995, after four years of denial; claimed to have destroyed all agents, but offered no credible proof.
	May have begun program reconstitution in absence of UN inspections and monitoring.
	Acceded to the BWC.
Chemical	Rebuilt some of its chemical production infrastructure allegedly for commercial use.
	UNSCOM discovered evidence of VX persistent nerve agent in missile warheads in 1998, despite Iraqi denials for seven years that it had not weaponized VX.
	May have begun program reconstitution in absence of UN inspections and monitoring.
	Has not signed the CWC.
Ballistic missiles	Probably retains limited number of SCUD-variant missiles, launchers, and warheads capable of delivering biological and chemical agents. Retains significant missile production capability.
	Continues work on liquid- and solid-propellant SRBMs (150 kilometers) allowed by UNSCR 687; likely will use technical experience gained for future longer range missile development effort.
	Not a member of the MTCR.
Other means of delivery available	Land-launched anti-ship cruise missiles; air-launched tactical missiles; none have NBC warheads; stockpile likely is very limited.
	Air systems: fighters, helicopters, UAVs.
	Ground systems: artillery, rockets.

IAEA = International Atomic Energy Agency
NPT = Nuclear Nonproliferation Treaty
CTBT = Comprehensive Test Ban Treaty
UN = United Nations
BWC = Biological and Toxin Weapons Convention
UNSCOM = United Nations Special Commission
CWC = Chemical Weapons Convention
SRBM = Short Range Ballistic Missile (Range: 1000 kilometers or less)
MTCR = Military Technology Control Regime
NBC = Nuclear, Biological, or Chemical
UAV = Unmanned Aerial Vehicle

SOURCE: "Iraq: NBC Weapons and Missile Programs," in *Proliferation: Threat and Response*, U.S. Department of Defense, Office of the Secretary of Defense, January 2001, http://www.defenselink.mil/pubs/ptr20010110.pdf (accessed September 23, 2004)

returning, he was executed. Much of the information the United States obtained about possible Iraqi WMD programs came from these two men.

The Nuclear Weapons Program

Under the leadership of then vice-president Saddam Hussein, Iraq began developing a nuclear program in the early 1970s, concentrating its efforts on acquiring nuclear technology abroad. Because Iraq was a signatory to the NPT, it was legally prohibited from developing nuclear weapons, so its efforts had to occur secretly. As a result, Iraq had some trouble acquiring the materials it needed to develop a successful weapons program, particularly fissile materials. It bought the Tamuz-1 (or Osirak) 40-megawatt test reactor from France in 1975, which used weapons-grade uranium and could produce weapons-grade plutonium. However, in 1981, just before Osirak was ready to produce enough plutonium to test the IAEA safeguards, Israel destroyed it. By 1990 Iraq had begun a program to chemically process unirradiated and irradiated research reactor fuel to recover a significant quantity of highly enriched uranium for a low-yield nuclear device. However, the project did not come to fruition because the research center at Tuwaitha—the site of Iraq's covert fuel development activities—was destroyed in the January 1991 Desert Storm bombings.

There has been much debate as to how far along Iraq was in developing a nuclear weapon at the time of the Persian Gulf War. One thing is clear, though, from the documents obtained after General Kamel fled Iraq: The Iraqi nuclear program was plagued by bitter infighting, mismanagement, and a lack of infrastructure. At the same time, the documents made it clear that Iraq would go to any lengths to pursue nuclear technology. If Iraq had not invaded Kuwait and in turn been attacked by a UN coalition force led by the United States in the early 1990s, it is highly probable that it could have developed a small nuclear armory by the late 1990s.

The Chemical Weapons Program

Unlike its nuclear and biological programs, Iraq could not easily conceal its chemical weapons capability from the rest of the world because it had used such weapons in the past, against Iran and against Kurdish rebels during the Iran-Iraq war (1980–88). It first used riot-control agents during the early part of the war. By 1984 Iraq progressed to the use of mustard gas and tabun, and later still it added nerve agents such as sarin and cyclosarin. The infamous attack against the Kurdish town of Halabja remains one of the deadliest chemical weapons attacks on a civilian population. On March 16, 1988, Iraq used an amalgam of chemical weapons, including sarin, tabun, and VX, against the Iraqi Kurds seeking their independence, killing approximately five thousand civilians. According to the Federation of American Scientists, Iraqi use of chemical weapons during the Iran-Iraq war occurred in three distinct phases:

1. 1983–86: Chemical weapons played a defensive role, deflecting Iranian human-wave assaults (volunteers seeking martyrdom deployed in masses before the Iranian troops). Around 5,500 Iranians were killed by tabun-filled aerial bombs, and approximately sixteen thousand were killed by the blister agent mustard gas.

2. 1986–early 1988: Iraq used chemical weapons to disrupt offensive Iranian maneuvers.

3. Early 1988–conclusion of war: Iraq integrated its nerve agent strikes into its overall offensive, which later that year led to a cease-fire.

Iraq started producing blister agents in 1981 and disclosed in 1995 that it had a stock of 2,850 tons of mustard gas. The nerve gases sarin and tabun were not produced until 1984, and the program itself faced several problems with stabilization and storage. In 1995 Iraq declared that it had produced more than 210 tons of tabun and 790 tons of sarin. However, it is believed that the quality of these agents was relatively poor. In addition, UNSCOM destroyed about thirty tons of tabun, thirty tons of sarin, and six hundred tons of mustard gas during 1992 and 1994.

Iraq also focused on developing the deadly nerve agent VX, importing about five hundred tons of precursor chemicals (chemicals used in the production of a CW agent) between 1987 and 1988. Iraq admitted to filling aerial bombs with VX, but the program itself was unsuccessful and was eventually abandoned in late 1988. Iraq was accused by the British of developing Agent-15, an incapacitating gas.

Because of the deceitful nature of many declarations Iraq submitted to the UN, the entire scope of Iraq's chemical weapons program was never fully determined. Based on information provided by Iraq, the program did not heavily rely on domestic resources—munitions for the program were procured from abroad through legal or illegal means. Through 1997 UNSCOM helped destroy approximately 38,000 filled and unfilled munitions. Citing UNSCOM reports, the Federation of American Scientists claimed, in the article "UNSCOM and Iraqi Chemical Weapons" posted on their Web site (www.fas.org), that the Iraqi chemical weapons program included the use of "binary artillery munitions and aerial bombs, chemical warheads for short-range missiles, cluster aerial bombs, and spray tanks." Dr. Hamza stated that Iraq was fully capable of rebuilding its chemical weapons facilities.

Western intelligence reports corroborated Dr. Hamza's statements, strongly indicating that Saddam Hussein was rebuilding and stockpiling Iraq's chemical weapons arsenal. Satellite images revealed that the Republican Guard (a unit within the Iraqi military) had shifted weapons, including tons of precursor chemicals for VX, to new hiding places, which included schools and hospitals. Evidence also suggested that Iraq was working on rebuilding chemical facilities that were destroyed during the bombings. Furthermore, the U.S. State Department asserted that Iraq had failed to account for 1.5 tons of VX, one thousand tons of mustard gas, and fifty-five munitions containing mustard gas during the UN inspections.

The Biological Weapons Program

The Iraqi biological warfare program, begun in 1974, was comprehensive and included a range of agents, such as botulinum toxin, anthrax, ricin, and others. Iraq legitimately acquired much of its seed stock from U.S. and European suppliers under the guise of laboratory research. The American Type Culture Collection in Rockville, Maryland, provided Iraq with most of its anthrax strains. Bacteria came from Bedford, England, as well as Fluka Chimie, a Swiss firm. The NTI reports that by 1990 the biological weapons program had produced twenty-five missile warheads and 166 aerial bombs filled with anthrax, botulinum toxin, or aflatoxin.

By 1997 UNSCOM determined that seventy-nine sites were providing active support to the Iraqi biological weapons infrastructure. The main facilities were located at al-Salman and al-Hakam. Al-Salman conducted experiments on the effects of agents and toxins on larger animals, such as sheep, monkeys, and dogs, in the laboratory and in the field. There were rumors that human subjects were used, but this has never been confirmed. Al-Hakam was initially overlooked because its plain and insecure appearance belied the activities that occurred in the facility. It took UN inspectors four years to discover that Al-Hakam was an integral part of the weapons program, where researchers carried out work on anthrax, botulinum toxin, and aflatoxin.

Iraq also had several medical, university, and veterinary facilities that conducted biological research and may have been involved in a covert biological weapons program. Iraq retained the laboratory equipment, know-how, and means for delivering agents after the Gulf War. Speculation about Iraqi capabilities and programs abounded. In 2000 the United Kingdom theorized that Iraq could resurrect its biological weapons program in a matter of months and that at its peak production, could have produced 350 liters of weapons-grade anthrax per week.

Missiles

Although missiles are not direct constituents of WMD capability, they play a significant role in the delivery of these deadly weapons. UN Security Council Resolution 687 called for the destruction of all Iraqi long-range missiles, forbidding Iraq to have any missiles with a range of more than 150 kilometers (93 miles). Yet a 1999 intelligence report from the White House to Congress indicated that Iraq might have retained seven or more complete missile systems and their components. In 1995 Iraq was caught trying to smuggle gyroscopes (part of the missile's onboard guidance system) in from Russia to bolster its missile capabilities. The Center for Nonproliferation Studies concluded that Iraq still retained certain components of its missile systems and might have started work on al-Hussein and Scud missiles within a year after inspections stopped.

Operation Iraqi Freedom

By 2002 the administration of President George W. Bush had become concerned that Iraq was developing WMD and that it might supply such weapons to terrorist organizations. U.S. and other intelligence agencies believed that Iraq had chemical and biological weapons and was actively trying to develop nuclear weapons. President Bush called for stronger action to force Iraq to disarm and to allow the resumption of inspections.

Given its past history of deception and concealment, it was difficult to simply accept declarations of innocence from Iraq at face value. Its refusal to allow inspectors back into the country had also created significant doubt about the credibility of its claims. Compliance with UN inspection requirements would have likely resulted in a lifting of sanctions, but Iraq failed to do so consistently, despite the economic cost. Estimates are that Iraq lost over $120 billion because of the sanctions. Nevertheless, many argued that there was no solid evidence that Iraq was lying about having ended its WMD programs.

In October 2002 the U.S. Congress issued the Iraq Liberation Act (Public Law 105-338) authorizing the use of force against Iraq should it become necessary. In November 2002, the UN Security Council unanimously adopted Security Council Resolution 1441, which found Iraq to be in violation of several earlier UN resolutions and demanded that Iraq allow inspections to resume.

Iraq agreed to the resolution, and later that month UNMOVIC inspectors entered the country for the first time in four years. Four months of inspections failed to uncover any evidence that Iraq had resumed its WMD programs, but the United States, the United Kingdom, and a number of other countries remained convinced that Iraq was a threat. On March 17, 2003, when the United States issued an ultimatum to Iraq, giving Saddam Hussein and his two sons forty-eight hours to leave the country peacefully, the UN inspectors were withdrawn. On March 20, 2003, the United States and its allies attacked.

The invasion of Iraq met with light military resistance and took only a few weeks to secure the capital city of Baghdad. In July 2003 Hussein's two sons, Uday and Qusay, were killed in a gun battle with allied forces in the city of Mosul. Saddam Hussein was captured in December 2003 and is slated to stand trial in 2005. In June 2004 Iraqi sovereignty was handed over to a civilian interim government that scheduled democratic elections for 2005. A U.S.-led weapons inspection effort, the Iraq Survey Group (ISG), began searching Iraq for WMD beginning shortly after the invasion. Their final report, *Comprehensive Report of the Special Advisor to the DCI on Iraq's WMD* (Washington, DC: Central Intelligence Agency, September 2004), concluded that there were no WMD in Iraq.

Aftermath

The decision to invade Iraq was based on four assumptions:

1. Saddam Hussein's regime was on the verge of acquiring nuclear weapons and had already amassed stockpiles of chemical and biological weapons.

2. Hussein's regime purportedly had significant links with al Qaeda and may have been involved with the attacks of September 11, 2001.

3. Following the fall of Hussein's regime, there would be celebration and rapid, peaceful democratization.

4. The democratization of Iraq would catalyze similar transformations in the region and there would be a new willingness among Arab governments and people to make peace between Israel and a Palestinian state.

Although the first two assumptions were stated by U.S. Vice President Dick Cheney and repeated by other members of the Bush administration, both proved to be false. The fall of Saddam Hussein was well received by some Iraqis but angered others, and was followed by postwar chaos, violence, unemployment, and power shortages. At the end of 2004 it was too early to tell if the fourth assumption would prove true. Critics have suggested that the U.S. invasion of Iraq may actually hinder progress toward enduring peace in the Middle East.

The invasion purchased freedoms for the people of Iraq that they could not have had without Hussein's fall, but costs for both the United States and Iraq have been steep. Lawrence Lindsey, former head of the President's Council of Economic Advisers, projected a financial toll of about $200 billion by the end of fiscal year 2004. The loss of human lives has also exceeded expectations. By December 2004 there were more than 1,100 American military deaths and more than eight thousand wounded. Iraqi losses were much greater—according to a study by researchers from the Johns Hopkins Bloomberg School of Public Health, Columbia University School of Nursing, and Al-Mustansiriya University in Baghdad conducted in September 2004, more than 100,000 Iraqi civilians had been killed than would have been expected had the invasion not occurred. The American military was expected to remain in Iraq for an extended period to oversee the country's transition to stability. However, the new government's reign appeared tenuous. Weekly and sometimes daily terrorist actions, including car bombings and kidnappings, were taking place as of December 2004.

Even before the invasion, some political observers discounted the fear of Iraq's WMD and purported links to terrorists as the rationales for the U.S. invasion. They contend that the United States' prime motivation was to establish dominance over the main oil-producing region of the world—the Mideast. Mark Dunlea, chair of the New York State Green Party articulated this opinion in February 24, 2003 (www.gp.org), when he said, "Oil isn't the only motivation, but it's a major reason President Bush plans to launch an invasion of Iraq. While the White House and its apologists deny it regularly, the evidence is clear that the Bush Administration and American oil companies are trying to win control over the world's second largest source of oil." The failure to find any WMD in Iraq after the invasion strengthened such criticism.

Conversely, some observers persist in believing that Iraq may have had WMD capabilities. Although large stockpiles of weapons materials were never found, they observe that smaller items were found, including a centrifuge used in enriching uranium for use in nuclear weapons. The centrifuge was buried in the backyard of an Iraqi scientist, confirming how easy it is to hide the components of a weapons program. In August 2004 the ISG announced that before the invasion the Iraqi Intelligence Service had replaced guards at the Iraq-Syria border with their own agents to allow trucks to pass through unhindered. The cargo of these trucks was not known, but the ISG suspects that they may have been carrying WMD materials to Syria. In its final report, the ISG stated that Saddam Hussein wanted to re-create his country's WMD programs, especially chemical weapons and ballistic missiles, as soon as sanctions against Iraq could be ended.

It is unlikely that lingering doubts about Iraq's WMD capabilities will be resolved to the satisfaction of all parties; however, it is important to observe that to a significant extent, doubts and speculation prompted the U.S. invasion of Iraq. This observation underscores the vital role of intelligence in national security programs and also challenges governments to exercise caution when gathering, communicating, and confirming international intelligence.

CHAPTER 5

PREPARING FOR BIOLOGICAL AND CHEMICAL ATTACKS

While conventional weapons, such as explosives and firearms, remain the most likely means by which terrorists might attempt to harm U.S. civilians, the possibility of an attack involving biological or chemical weapons has increased. Many nations and terrorist groups have explored the use of such weapons on small and large scales, and many countries, including the United States, have chemical and/or biological weapons programs or materials used in these types of weapons.

Biological warfare–related technology, materials, information, and expertise—including information on potential U.S. vulnerabilities—have become more readily available. Genetic engineering is only one of several technologies that might allow countries or groups to develop agents, such as modified viruses, that would be difficult to detect and diagnose or that could defeat current procedures for protection and treatment. Furthermore, all the materials needed to produce such agents are dual-use in nature, meaning they have both military and civilian applications, so they are readily available. Any country with political will and competent scientists can produce agents. The threat from chemical warfare may also grow in coming years. Many states have chemical warfare programs, and these capabilities will likely spread to additional states and terrorist groups. Government officials consider smaller-scale bioterrorist events to be more likely than large-scale ones because they are less difficult to engineer. However, federal public health agencies, such as the Centers for Disease Control and Prevention (CDC), have little choice but to prepare for a variety of attacks.

THE FEDERAL ROLE

The U.S. General Accounting Office (GAO) defines biological and chemical terrorism as the threatened or intentional release of viruses, bacteria, poisonous gases, liquids, or other toxic substances for the purpose of influencing the conduct of government, intimidating or coercing a population, or simply intending to cause widespread harm. Any such act that has the potential for, or the intention of, infecting, injuring, or killing hundreds, thousands, or even millions of people is considered terrorism.

In 2001 the GAO identified more than forty federal departments and agencies with some role in combating terrorism, with twenty-nine of those having some role in preparing for, or responding to, the public health and medical consequences of a biological or chemical attack. The cabinet-level departments involved include the U.S. Departments of Agriculture (USDA), Commerce, Defense (DOD), Energy (DOE), Health and Human Services, Justice, Transportation, Treasury, and Veterans' Affairs, along with two independent agencies, the Environmental Protection Agency (EPA) and the Federal Emergency Management Agency (FEMA). Within these larger divisions, departmental agencies take on various roles. The Department of Homeland Security, established in 2002, also plays an important role.

These departments and agencies may work alone or with other agencies in emergency planning for averting or responding to attacks. These units participate in activities that include, but are not limited to: (1) detecting biological agents; (2) developing a national stockpile of drugs with which to treat victims of disasters; and (3) developing vaccines, such as the anthrax vaccine, for the widespread inoculation of U.S. citizens and residents.

Funding for Research

In fiscal year 2001 federal departments and agencies reported total funding for research on biological and chemical terrorism in the amount of $156.8 million. Some of the activities funded include the development of technologies to detect bioterrorist attacks. Funding for bioterrorism research increased dramatically in response to the terrorist attacks of September 11, 2001. In September 2003 Health and Human Services Secretary Tommy G.

Thompson announced $350 million in grants over five years to establish eight Regional Centers of Excellence for Biodefense and Emerging Infectious Diseases Research. The eight institutions receiving grants include the Harvard Medical School, Duke University, the University of Chicago, and the University of Texas Medical Branch. According to a statement by the National Institute of Allergy and Infectious Diseases, the research conducted by the Regional Centers will include:

- Developing new approaches to blocking the action of anthrax, botulinum, and cholera toxins

- Developing new vaccines against anthrax, plague, tularemia, smallpox, and Ebola

- Developing new antibiotics and other therapeutic strategies

- Studying bacterial and viral disease processes

- Designing new advanced diagnostic approaches for biodefense and for emerging diseases

- Conducting immunological studies of diseases caused by potential agents of bioterrorism

- Developing computational and genomic approaches to combating disease agents

- Creating new immunization strategies and delivery systems

On July 21, 2004, President George W. Bush signed into law the Project BioShield Act. Project BioShield authorizes $5.6 billion over ten years for the government to purchase and stockpile vaccines and drugs to fight anthrax, smallpox, and other diseases that may be used in a bioterror attack. The Department of Health and Human Services will purchase seventy-five million doses of an improved anthrax vaccine for the Strategic National Stockpile. In addition, grants for bioterror research will be expedited under Project BioShield, as will the delivery of newly developed drugs to victims who may require assistance.

Funding for Preparedness

In 1999 a National Pharmaceutical Stockpile—a reserve of antibiotics, chemical antidotes, antitoxins, life-support medications, IV administration, airway maintenance supplies, and medical/surgical items for use in an emergency—was first authorized by Congress. In March 2003 it was renamed the Strategic National Stockpile. Initially run by the Department of Health and Human Services and the Centers for Disease Control and Prevention, it is now managed by the Department of Homeland Security. Funding for the stockpile has increased fivefold since its inception. Among its accomplishments is the purchase and storage of enough smallpox vaccine to protect every American citizen.

Federal departments and agencies spent almost $650 million on bioterrorism- and terrorism-preparedness activities for fiscal years 2000 and 2001: $296 million in 2000 and $347 million in 2001. In January 2002 the U.S. Department of Health and Human Services announced $1 billion in bioterrorism preparedness grants to be given to states and cities to increase their ability to respond to bioterror emergencies. The CDC announced in June 2004 that grants of some $840 million were available to assist public health facilities to prepare for and respond to bioterrorism.

The Centers for Disease Control and Prevention (CDC)

Most investments in national defense increase national security, especially by acting as a deterrent against hostile acts. Similarly, investments in the public health system, most experts believe, will provide the best civil defense against biological and chemical terrorism. In the lead among federal agencies preparing for future homeland biological and chemical terrorism incidents is the CDC, headquartered in Atlanta, Georgia. The CDC's programs to fight terrorism—particularly bioterrorism—integrate planning and training to develop public health preparedness and include surveillance (monitoring trends), epidemiology (studying the incidence, distribution, and control of disease), rapid laboratory diagnosis, emergency response, and information systems (computers and telecommunications).

The CDC assists state and local public health departments by:

- Identifying the biological agents likely to be involved in a terrorist attack

- Developing case definitions to assist in detecting and managing infection with these agents

- Establishing a Rapid Response and Advanced Technology laboratory, which can provide fast identification of biological and chemical agents rarely seen in the United States

- Developing a nationwide integrated information, communications, and training network with the Health Alert Network, the National Electronic Data Surveillance System, and Epidemic Information Exchange. The Health Alert Network (HAN) is, according to the CDC Web site (http://www.cdc.gov), meant to "ensure that each community has rapid and timely access to emergent health information; a cadre of highly-trained professional personnel; and evidence-based practices and procedures for effective public health preparedness, response, and service on a 24/7 basis." Considered an improvement on the existing system of tracking diseases, the National Electronic Data Surveillance System (NEDSS) will create a nationwide standard for the collection and analysis of all health related data. According to the CDC, the Epidemic

Information Exchange (Epi-X) is "the nation's secure, Internet-based communications network for public health investigation and response. Epi-X provides public health officials throughout the United States with up-to-the-minute information, reports, alerts, and discussions regarding terrorist events, toxic exposures, disease outbreaks, and other public health events."

In other programs, the CDC seeks to enhance the public health system and to expand response capacity, provides training in preparedness and response for public health employees, and continues to support and grow its networked information systems.

BIOATTACK: THE DALLES INCIDENT

Civilians are vulnerable to foodborne or waterborne bioterrorism, as demonstrated by the intentional salmonella contamination of restaurant salad bars in and around The Dalles, Oregon, in September and October 1984. A total of 751 people developed salmonella gastroenteritis from eating or working at those salad bars.

The outbreak occurred in two waves: September 9–September 18 and September 19–October 10. Most cases occurred in ten restaurants. Epidemiological studies of customers at four restaurants and of employees at all ten restaurants indicated that eating from salad bars was the major risk factor for infection. Eight of the ten affected restaurants (80%) operated a salad bar, compared with only three of the twenty-eight nonaffected restaurants (11%) in The Dalles.

The investigation did not identify any water supply, food item, supplier, or distributor common to all affected restaurants, nor were the employees exposed to any single common source. Infected employees may have contributed to the spread of the illness, but they did not initiate the outbreak, nor did food-rotation errors or inadequate refrigeration of the salad bars (although they may have promoted bacteria growth).

A criminal investigation revealed that members of a religious commune, the Rajneeshees, had deliberately contaminated the salad bars with salmonella bacteria as part of a plan to influence a county election in favor of candidates they endorsed. By making residents of The Dalles too ill to go to the polls on election day, the group hoped to seize control of the county government. According to most accounts, commune members planned to contaminate The Dalles's municipal water system just before election day. About a month before the election, they began experimenting with the bacteria by poisoning the refreshments they served to two county commissioners who were visiting the Rajneeshees' compound. Later, commune members sprinkled salmonella on produce in grocery stores. Finally, the Rajneeshees sprinkled salmonella bacteria in and around the salad bars of the town's ten most popular restaurants. Within a few weeks, more than seven hundred people had become ill.

CRITICAL BIOLOGICAL AND CHEMICAL AGENTS

Biological Agents

THREAT DELINEATION. The first step in preparing for biological or chemical attacks is to detect threats. The CDC has gone to great lengths to identify and prioritize biological and chemical weapons agents. Priorities are based less on the likelihood of an agent's use than on its potential to cause widespread catastrophe. Agents have traditionally been evaluated based on military concerns and troop protection, but civilian populations differ in many ways from military populations, having a wider age range and a wider range of health conditions. In general, civilian populations are more vulnerable, and consequences of an attack would be more severe. This means that military priority lists cannot simply be carried over and applied to civilian threats.

In 1999 Congress began to upgrade public health capabilities to respond to potential biological and chemical attacks, making the CDC the lead agency for overall public health planning. The CDC, in turn, formed a Bioterrorism Preparedness and Response Office to focus on several areas of preparedness, including planning, improved surveillance and epidemiological capabilities, rapid laboratory diagnostics, enhanced communications, and medical therapeutics stockpiling.

To focus the preparedness efforts properly, the first step was to identify and prioritize critical biological and chemical agents. Many biological agents affect human beings, but relatively few, authorities reasoned, have the potential to create public health catastrophes that would severely strain U.S. public health and medical systems, so the CDC sought a new threat-assessment method that could be reviewed, reproduced, and standardized.

On June 3–4, 1999, the CDC convened a meeting of national experts to review the threat potential of various biological and chemical agents to civilian populations. The experts included academic infectious disease experts, national public health experts, CDC personnel, civilian and military intelligence experts, and law enforcement officials. They identified agents they believed had the potential for great public health impact based on subjective assessments in four general categories: overall public health impact (the death or disease rates), dissemination potential (how much the disease could spread), public perception of its impact, and the special preparedness needed for each agent. These criteria were weighted on a scale from zero to three for each agent in order to evaluate the potential threat from each. A factor given the most weight received a three (+++), and the factor given the least weight received a zero (0). Final category assignments—A, B, or C threat status—were based on the ratings the agents received in each of the four areas. (See Table 5.1.)

Category A agents have the greatest potential for causing disruption, disease, and mass casualties, and require the broadest public health preparedness, including improved surveillance, laboratory diagnosis, and medication stockpiling. Examples of Category A agents are those that cause smallpox, anthrax, plague, botulism, and tularaemia.

Category B agents have the potential for large-scale catastrophe but generally would cause fewer cases of severe illness and death than Category A agents. They would have a smaller public health and medical impact, have lower public awareness, and require fewer special preparedness measures. Although these, too, should receive heightened awareness from the medical and emergency communities, along with more surveillance and improved laboratory diagnostic capabilities, these are not needed for Category B agents on the order suggested for Category A agents. In Category B are some agents that the CDC and its experts know have undergone development as weapons but that otherwise do not meet Category A criteria, as well as some agents of concern for food and water safety. Examples of Category B agents include organisms that cause Q fever, brucellosis, and glanders, as well as food- or waterborne agents such as salmonella and *E. coli* pathogens.

Category C agents do not currently appear to present a high bioterrorism threat but may emerge as future threats as scientific knowledge about them improves. These agents are addressed by the CDC's overall preparedness efforts—efforts intended to improve detection and treatment of unexplained illnesses and emerging infectious diseases. Category C agents include the Nipah virus, hantaviruses, yellow fever, and multidrug-resistant tuberculosis.

The agents were categorized based on the evaluation criteria applied to them, especially in Categories A and B. For example, the public health impact of smallpox (Category A) ranks higher than that of brucellosis (Category B) because the mortality for those untreated is higher for the former (about 30%) than the latter (about 2%). In addition, smallpox has a higher dissemination potential because it can be transmitted from person to person. It ranks higher for special public health preparedness, as well, because additional vaccine must be made and stockpiled, and improved surveillance, educational, and diagnostic efforts are necessary. Other Category A threats, such as inhalation anthrax and plague, also have higher public impact ratings than brucellosis because of their higher morbidity (illness) and mortality (death) rates. Although mass production of Category B agents *Vibrio cholerae* (the organism causing cholera) and *Shigella spp* (the cause of shigellosis) would be easier than that of anthrax spores, these agents produce lower morbidity and

TABLE 5.1

Critical biological agents that pose a risk to national security

Category A

The U.S. public health system and primary health-care providers must be prepared to address varied biological agents, including pathogens that are rarely seen in the United States. High-priority agents include organisms that pose a risk to national security because they

- can be easily disseminated or transmitted person-to-person;
- cause high mortality, with potential for major public health impact;
- might cause public panic and social disruption; and
- require special action for public health preparedness (Box 2).

Category A agents include

- variola major (smallpox);
- *Bacillus anthracis* (anthrax);
- *Yersinia pestis* (plague);
- *Clostridium botulinum* toxin (botulism);
- *Francisella tularensis* (tularaemia);
- filoviruses,
  — Ebola hemorrhagic fever,
  — Marburg hemorrhagic fever; and
- arenaviruses,
  — Lassa (Lassa fever),
  — Junin (Argentine hemorrhagic fever) and related viruses.

Category B

Second highest priority agents include those that

- are moderately easy to disseminate;
- cause moderate morbidity and low mortality; and
- require specific enhancements of CDC's diagnostic capacity and enhanced disease surveillance.

Category B agents include

- *Coxiella burnetti* (Q fever);
- *Brucella* species (brucellosis);
- *Burkholderia mallei* (glanders);
- alphaviruses,
  — Venezuelan encephalomyelitis,
  — eastern and western equine encephalomyelitis;
- ricin toxin from *Ricinus communis* (castor beans);
- epsilon toxin of *Clostridium perfringens;* and
- *Staphylococcus* enterotoxin B.
  A subset of List B agents includes pathogens that are food- or waterborne. These pathogens include but are not limited to
- *Salmonella* species,
- *Shigella dysenteriae*,
- *Escherichia coli* O157:H7,
- *Vibrio cholerae*, and
- *Cryptosporidium parvum*.

Category C

Third highest priority agents include emerging pathogens that could be engineered for mass dissemination in the future because of

- availability;
- ease of production and dissemination; and
- potential for high morbidity and mortality and major health impact.

Category C agents include

- Nipah virus,
- hantaviruses,
- tickborne hemorrhagic fever viruses,
- tickborne encephalitis viruses,
- yellow fever, and
- multidrug-resistant tuberculosis.

Preparedness for List C agents requires ongoing research to improve disease detection, diagnosis, treatment, and prevention. Knowing in advance which newly emergent pathogens might be employed by terrorists is not possible; therefore, linking bioterrorism preparedness efforts with ongoing disease surveillance and outbreak response activities as defined in CDC's emerging infectious disease strategy is imperative.

SOURCE: Ali S. Kahn, Alexandra M. Levitt, and Michael J. Sage, "BOX 3. Critical Biological Agents," in "Biological and Chemical Terrorism: Strategic Plan for Preparedness and Response," *Morbidity and Mortality Weekly Report*, vol. 49, no. RR-4, April 21, 2000, http://www.cdc.gov/mmwr/preview/mmwrhtml/rr4904a1.htm (accessed September 23, 2004)

mortality, so their public health impact, or dissemination threat, would be less. Although infectious doses of these bacteria are very low, it would be very difficult for a terrorist to use them effectively. The total amount of bacteria required, as well as the advanced state of current water purification and food-processing techniques, would limit these agents' effectiveness for intentional, large-scale water or food contamination.

PREPAREDNESS ACTIVITIES. In addition to identifying major biological agents and threats, the CDC (in its publication *Morbidity and Mortality Weekly,* vol. 49, no. RR-4, April 21, 2000) provided nine basic steps to prepare public health agencies for biological attacks:

1. Enhance epidemiologic capacity to detect and respond to biological attacks

2. Supply diagnostic reagents to state and local public health agencies

3. Establish communication programs to ensure delivery of accurate information

4. Enhance bioterrorism-related education and training for health-care professionals

5. Prepare educational materials that will inform and reassure the public during and after a biological attack

6. Stockpile appropriate vaccines and drugs

7. Establish molecular surveillance for microbial strains, including unusual or drug-resistant strains

8. Support the development of diagnostic tests

9. Encourage research on antiviral drugs and vaccines

Enhancing epidemiologic capacity means adding additional resources to trace the source and spread of disease. Supplying diagnostic reagents means providing chemical compounds to state and local public health agencies for a variety of medical purposes ranging from detection to prevention.

Chemical Agents

THREAT DELINEATION. Chemical agents can range from warfare agents to toxic substances in common commercial use. The fact that chemical warfare technologies are increasingly available, coupled with the relative ease with which chemical agents can be produced, increases the U.S. government's concern that terrorist states or groups may use them in the future.

The CDC takes a similar approach to combating chemical threats as it does to biological ones, providing many of the same resources to state and local public health agencies and emergency services teams. However, the CDC's identification and prioritization of critical chemical agents differs from that of biological agents. Because hundreds of new chemicals are introduced inter-

nationally each month, the categories of chemical agents are necessarily more generic than for biological agents.

The CDC identifies and prioritizes chemical agents according to criteria including the following:

• Are the agents already known to be used as weapons?

• Are they readily available to hostile states and terrorists?

• Are they likely to cause morbidity or mortality?

• Are they likely to cause panic or disruption?

• Do they require special actions for public health preparedness?

Table 5.2 lists the CDC's chemical agent categories, along with notable examples of each. The chemical agents most likely to be used are nerve agents (tabun, sarin, soman, GF, and VX), blood agents (hydrogen cyanide and cyanogen chloride), blister agents (lewisite, mustards, and phosgene oxime), and heavy metals (arsenic, lead, and mercury).

PREPAREDNESS ACTIVITIES. The CDC provides recommendations to help public health agencies prepare for potential chemical attacks. First, agencies should take a generic approach to the treatment of chemical agent injuries, treating those exposed according to clinical syndrome, or the group of symptoms they have, rather than the specific agent. These syndromes include burns and trauma, cardio-respiratory failure, neurological damage, and shock. Those who respond and treat affected individuals must also communicate with the authorities responsible for environmental sampling for, and decontamination of areas affected by, such chemical agents.

The CDC's five steps in preparing public health agencies for chemical attacks are listed in their publication *Morbidity and Mortality Weekly* (vol. 49, no. RR-4, April 21, 2000):

• Enhance epidemiologic capacity for detecting and responding to chemical attacks

• Enhance awareness of chemical terrorism among emergency medical service personnel, police officers, firefighters, physicians, and nurses

• Stockpile chemical antidotes

• Develop and provide bioassays for detection and diagnosis of chemical injuries

• Prepare educational materials to inform the public during and after a chemical attack

Enhancing epidemiological capacity refers to mapping the origin and spread of disease symptoms. Bioassays are intended to determine the relative strength of a chemical agent by comparing its effect on a test organism with that of a standard-strength preparation.

Laboratory Response Network

The CDC has described five key focus areas of state and local efforts to prepare for biological and chemical attacks:

TABLE 5.2

Chemical agents that might be used by terrorists

Chemical agents that might be used by terrorists range from warfare agents to toxic chemicals commonly used in industry. Criteria for determining priority chemical agents include

- chemical agents already known to be used as weaponry;
- availability of chemical agents to potential terrorists;
- chemical agents likely to cause major morbidity or mortality;
- potential of agents for causing public panic and social disruption; and
- agents that require special action for public health preparedness (Box 4).

Categories of chemical agents include

- nerve agents,
  — tabun (ethyl N,N-dimethylphosphoramidocyanidate),
  — sarin (isopropyl methylphosphanofluoridate),
  — soman (pinacolyl methyl phosphonofluoridate),
  — GF (cyclohexylmethylphosphonofluoridate),
  — VX (o-ethyl-[S]-[2-diisopropylaminoethyl]-methylphosphonothiolate);
- blood agents,
  — hydrogen cyanide,
  — cyanogen chloride;
- blister agents,
  — lewisite (an aliphatic arsenic compound, 2-chlorovinyldichloroarsine),
  — nitrogen and sulfur mustards,
  — phosgene oxime;
- heavy metals,
  — arsenic,
  — lead,
  — mercury;
- Volatile toxins,
  — benzene,
  — chloroform,
  — trihalomethanes;
- pulmonary agents,
  — phosgene,
  — chlorine,
  — vinyl chloride;
- incapacitating agents,
  — BZ (3-quinuclidinyl benzilate);
- pesticides, persistent and nonpersistent;
- dioxins, furans, and polychlorinated biphenyls (PCBs);
- explosive nitro compounds and oxidizers,
  — ammonium nitrate combined with fuel oil;
- flammable industrial gases and liquids,
  — gasoline,
  — propane;
- poison industrial gases, liquids, and solids,
  — cyanides,
  — nitriles; and
- corrosive industrial acids and bases,
  — nitric acid,
  — sulfuric acid.

SOURCE: Ali S. Kahn, Alexandra M. Levitt, and Michael J. Sage, "BOX 5. Chemical Agents," in "Biological and Chemical Terrorism: Strategic Plan for Preparedness and Response," in *Morbidity and Mortality Weekly Report*, vol. 49, no. RR-4, April 21, 2000, http://www.cdc.gov/mmwr/preview/mmwrhtml/rr4904a1.htm (accessed September 23, 2004)

1. Preparedness and prevention
2. Detection and surveillance
3. Diagnosis and characterization of biological and chemical agents
4. Response
5. Communication

Perhaps the most technically challenging of the five focus areas is the third: diagnosis and characterization of biological and chemical agents. For that reason, the CDC and its partners created two multilevel laboratory response networks, one for biological terrorism and one for chemical terrorism. These networks link state-of-the-art clinical labs to state and local public health agencies in all states, districts, territories, and selected cities and counties. Each network is a three-level hierarchy of laboratories, according to their respective capabilities. The Laboratory Network for Biological Terrorism consists of three types of laboratories:

1. National Laboratories are responsible for specialized strain characterizations, bioforensics, select agent activity, and handling highly infectious biological agents.

2. Reference Laboratories are responsible for investigation and/or referral of specimens.

3. Sentinel Laboratories play a key role in the early detection of biological agents.

National Laboratories are those run by the CDC or by the U.S. Army Medical Research Institute of Infectious Diseases. There are over a hundred Reference Laboratories across the United States and in Australia and Canada, and some 25,000 designated Sentinel Laboratories.

The Laboratory Network for Chemical Terrorism consists of three levels as well, designated as Levels 1, 2, and 3. All sixty-two laboratories in the network perform Level 1 duties, which are:

• Working with hospitals in their jurisdiction

• Knowing how to properly collect and ship clinical specimens

• Ensuring that specimens, which can be used as evidence in a criminal investigation, are handled properly and chain-of-custody procedures are followed

• Being familiar with chemical agents and their health effects

• Training on anticipated clinical sample flow and shipping regulations

• Working to develop a coordinated response plan for their respective state and jurisdiction

Of the sixty-two chemical terrorism network laboratories, forty-one also participate in Level 2 activities in which laboratory personnel are trained to detect exposure to a limited number of toxic chemical agents in human blood or urine. In Level 3 laboratories, which comprise five of the total sixty-two, personnel are trained to detect exposure to an expanded number of chemicals in human blood or urine, plus analyses for mustard agents, nerve agents, and other toxic chemicals.

CITIES READINESS INITIATIVE

In June 2004 the CDC announced the Cities Readiness Initiative pilot program. In this eight-month program the federal government provided direct assistance to cities to help

TABLE 5.3

Why agriterrorism may be an attractive tool for terrorists

Factor	Description
Lower physical risk	Disseminating a plant or livestock disease pathogen presents less physical risk to the perpetrator than releasing human disease pathogens or lethal chemicals.
Smaller chance of outrage and backlash	Agriterrorism is not likely to create the same kind of backlash as using a method of terrorism that kills people.
Similarity to natural outbreaks	Livestock and crops can be attacked in a way that the disease outbreak mimics a natural disease occurrence, complicating epidemiological investigation and reducing risk of detection.
Lower technical barriers	Agriterrorism can be carried out fairly easily, by comparatively low-tech means. The cost and the technical/scientific skills and education required to collect, produce, and deliver biological agents against animal agriculture are modest. Pathogens could be isolated from infected animals or diseased crops, and small quantities could easily be carried across a Customs checkpoint or unregulated border area, or sent through the mail. Then, infection with some pathogens would be simple. (For example, a terrible epidemic could be caused by dropping Newcastle disease-contaminated bird droppings into a feeding trough, or placing tongue scrapings from foot-and-mouth disease-infected animals into the ventilation system of a large hog operation.)

SOURCE: "Why Agriterrorism May Be an Attractive Tool for Terrorists," in "Appendix E: Agriterrorism," *Tool Kit for Managing the Emergency Consequences of Terrorist Incidents,* Federal Emergency Management Agency, July 2002, http://www.fema.gov/doc/reg-viii/tkapp-e.doc (accessed September 23, 2004)

them successfully receive and dispense medicine and medical supplies from the Strategic National Stockpile. Local, state, and national plans to distribute emergency medicines were combined to create a consistent nationwide approach that will ensure deliveries of needed supplies during such emergencies as a bioterrorist attack, a nuclear accident, or an outbreak of disease. Twenty cities and the District of Columbia participated in the $25 million pilot program.

PROTECTING ANIMALS AND PLANTS

The Animal and Plant Health Inspection Service (APHIS)

Animals and plants, especially those that human beings depend on for food, are also subject to attack. On the front line of this problem is the Department of Homeland Security's Animal and Plant Health Inspection Service (APHIS). APHIS is a largely unknown agency but an important one. It monitors the nation's borders for foreign agricultural diseases and pests. It protects farm animals from disease and pests and provides a host of services to cattle ranchers, milk producers, turkey farmers, and other agrarian groups.

According to the Center for Nonproliferation Studies report *Chronology of CBW Attacks Targeting Crops and Livestock, 1915–2000* (October 2001), bioterror attacks on a nation's animals and agriculture are not uncommon in times of war. The German secret service deployed anthrax against horses and mules slated for use by the Allied powers in World War I. In Afghanistan in the 1980s, the Russian military used toxins against the horses of the rebel mujaheddin. England suspected that Nazi Germany had dropped Colorado potato beetles on rural areas during World War II to destroy potato crops; in 1950, communist East Germany accused the United States of a similar attack.

One in eight American jobs and 13% of the U.S. gross national product (the value of all the goods and services produced in the country) are dependent on agriculture. The country's economic stability depends on a safe and readily available food supply. U.S. crops and livestock could be tempting targets to terrorists, especially because of the perceived ease of attacking such targets. (See Table 5.3.) Livestock and plant pathogens could threaten U.S. agricultural productivity and cause economic damage. Such factors as the resilience of the agent used, the density of the targeted animal population, and the susceptibility of targeted plants and animals to disease determine the level of vulnerability. (See Table 5.4.)

Potential threats to U.S. agricultural products and livestock come from a number of pathogens and agents. Animals could contract many types of disease, including anthrax, Q fever, brucellosis, foot-and-mouth disease (FMD), Venezuelan equine encephalitis, hog cholera, African swine fever, avian influenza, Newcastle disease, Rift Valley fever, rinderpest, and others. The Office International des Epizooties (OIE) is a 155-member organization that sets the animal health standards for international trade. They maintain two lists of diseases: List A includes diseases with the potential for very rapid spread across national borders and serious socio-economic or public health consequences. List B includes diseases that are considered dangerous within the individual country where an outbreak occurs. Table 5.5 shows the List A and many of the List B animal diseases. Many staple plants, such as corn, wheat, rice, and soybeans, are susceptible to disease. Soybean rust, for example, can be easily introduced and spreads quickly, which could cause U.S. soybean producers, processors, livestock producers (who feed soy products to their animals), and consumers to lose up to $8 billion annually, according to USDA estimates. Table 5.6 lists plant diseases of particular agriterrorism concern.

Some of these plant and animal agents can be found outside U.S. borders, and many can be readily transported, inadvertently or intentionally, into the United States, some with low risk of detection. APHIS is the agency

TABLE 5.4

Factors that affect vulnerability to agriterrorism

Factors	Description
Number of agents	There are many agents (at least 22) that are lethal and highly contagious to animals, many of which are not vaccinated against.
Resilience	Most of these agents are environmentally resilient. They can live for a long time in organic matter (e.g., soil).
Susceptibility	Antibiotic and steroid programs, and husbandry programs designed to improve quality and quantity of meat, have made U.S. livestock more disease prone. U.S. livestock and poultry are especially susceptible to exotic diseases because most serious diseases that affect them have been eradicated or brought under control with U.S. borders, so the animals lack antibodies to fight these agents. In crops, widespread use of commercial hybrids has limited their genetic diversity, making them more vulnerable to a killer pathogen.
Concentrated populations	Animal populations are highly concentrated, and large herds make ideal targets for infection and contagion. For example: • About 75% of the swine industries concentrated in nine Midwestern States; the most successful swine farms each have 10,000 hogs or more. • Beef cattle are fattened in large feedlots—some containing 150,000 to 300,000 animals at a time. • Dairies usually have as many as 1,500 lactating cows at one time. • Poultry has a heavy concentration in the Delaware/Maryland/Virginia peninsula. Chickens are usually grown in floor pens with 10,000 to 20,000 birds per pen.
Mobility	Animal populations are highly mobile. The animals are typically born in one location, moved halfway across the country to a feedlot for final fattening, then moved again for slaughter. Chicken breeding stocks and eggs are shipped great distances for the purpose of genetic improvements. Animals that are incubating disease during these movements can greatly increase the spread of the disease.
Inadequate security	Agricultural facilities are not highly secure. Food processors lacking sufficient security and safety preparedness methods have proliferated over the years.
Limited detection capabilities	The United States is even more vulnerable because it is unprepared to prevent such an attack or even quickly detect an outbreak. (Veterinary students receive minimal education in foreign animal diseases). Our primary recourse would be response, after an attack has occurred.

SOURCE: "Factors That Affect Vulnerability," in "Appendix E: Agriterrorism," *Tool Kit for Managing the Emergency Consequences of Terrorist Incidents*, Federal Emergency Management Agency, July 2002, http://www.fema.gov/doc/reg-viii/tkapp-e.doc (accessed September 23, 2004)

TABLE 5.5

Transmissible animal diseases identified as threats by the International Office of Epizootics, 2002

List A diseases[1]	Selected List B diseases[2]	
• African horse sickness	**Multiple species:**	**Cattle:**
• African swine fever	• Anthrax	• Bovine anaplasmosis
• Bluetongue	• Aujeszky's disease	• Bovine babesiosis
• Classical swine fever	• Echinococcosis/hydatidosis	• Bovine brucellosis
• Contagious bovine pleuropneumonia	• Heartwater	• Bovine cysticercosis
• Foot-and-mouth disease	• Leptospirosis	• Bovine genital campylobacteriosis
• Highly pathogenic avian influenza	• New World screwworm Cochliomyia	• Bovine spongiform encephalopathy
• Lumpy skin disease	hominivorax)	(BSE)
• Newcastle disease	• Old World screwworm (Chrysomya	• Bovine tuberculosis
• Peste des petits ruminants	bezziana)	• Dermatophilosis
• Rift Valley fever	• Paratuberculosis	• Enzootic bovine leukosis
• Rinderpest	• Q fever	• Haemorrhagic septicaemia
• Sheep pox and goat pox	• Rabies	• Infectious bovine rhinotracheitis/
• Swine vesicular disease	**Avian:**	infectious pustular vulvovaginitis
• Vesicular stomatitis	• Avian infectious bronchitis	• Malignant catarrhal fever
	• Avian infectious laryngotracheitis	• Theileriosis
	• Avian mycoplasmosis (M. avian	• Trichomonosis
	chlamydiosis gallisepticum)	• Trypanosomosis (tsetse-borne)
	• Avian tuberculosis	**Swine:**
	• Duck virus hepatitis	• Atrophic rhinitis of swine enterovirus
	• Duck virus enteritis	encephalomyelitis
	• Fowl cholera	• Porcine brucellosis
	• Fowl pox	• Porcine cysticercosis
	• Fowl typhoid	• Porcine reproductive and respiratory
	• Infectious bursal disease (Gumboro	syndrome
	disease)	• Transmissible gastroenteritis
	• Marek's disease	• Trichinellosis
	• Pullorum disease	

[1]List A: Transmissible diseases which have the potential for very serious and rapid spread, irrespective of national borders, which are of serious socio-economic or public health consequence and which are of major importance in the international trade of animals and animal products.
[2]List B: Transmissible diseases that are considered to be of socioeconomic and/or public health importance within countries and which are significant in the international trade of animals and animal products. Other categories of List B diseases include equine, sheep, goat, fish, crustacean, bee, Lagomorph, mollusc, and other.

SOURCE: "Animal Diseases: List A Diseases and Selected List B Diseases," in "Appendix E. Agriterrorism," *Tool Kit for Managing the Emergency Consequences of Terrorist Incidents*, Federal Emergency Management Agency, July 2002, http://www.fema.gov/doc/reg-viii/tkapp-e.doc (accessed September 23, 2004)

TABLE 5.6

Crop diseases of particular agriterrorism concern

Crop affected	Disease	Pathogen	Pathogen type	Primary mode of transmission
Cereals (wheat, barley, rye)	Stem rust of wheat	*Puccinia graminis*	Fungus	Airborne spores
	Stem rust of cereals	*Puccinia glumarum*	Fungus	Airborne spores
	Powdery mildew of cereals	*Erysiphe graminis*	Fungus	Airborne spores
Corn	Corn blight	*Pseudomonas alboprecipitans*	Bacteria	Waterborne cells
Rice	Rice blast	*Pyricularia oryzae*	Fungus	Airborne spores
	Rice blight	*Xanthomonas oryzae*	Bacteria	Waterborne cells
	Rice brown-spot disease	*Helminthosporium oryzae*	Fungus	Airborne spores
Potato	Late blight of potato	*Phytophthora infestans*	Fungus	Airborne spores

SOURCE: "Crop Diseases of Particular Agriterrorism Concern," in "Appendix E. Agriterrorism," *Tool Kit for Managing the Emergency Consequences of Terrorist Incidents,* Federal Emergency Management Agency, July 2002, http://www.fema.gov/doc/reg-viii/tkapp-e.doc (accessed September 23, 2004)

responsible for diagnosing and managing all suspicious agricultural disease outbreaks. APHIS's authority, depending on the pathogen involved, extends as far as confiscating property and eradicating all plant and animal hosts within quarantine zones. Binding international agreements force countries to immediately disclose select plant and animal disease outbreaks, regardless of severity. Such disclosures can have an instant impact on export trade, as other countries prohibit potentially contaminated items from entering their borders. National security and public trust can both be threatened in such cases, depending on the extent of disease transmission, the success of the government's response, and the amount of time it takes to bring conditions back to normal.

Foot-and-Mouth Disease (FMD)

One example of how countries' livestock industries can be affected by disease has been the various outbreaks of foot-and-mouth (also called hoof-and-mouth) disease. During an outbreak in the United Kingdom in 1967 and 1968, for example, more than 430,000 animals were destroyed; an outbreak in 2000–01 in the United Kingdom and Ireland forced the destruction of more than eight million animals at over ten thousand locations. The outbreak caused severe economic hardship throughout the United Kingdom and parts of Europe.

A member of the picornavirus family, FMD is endemic in many parts of the world, but the United States has not seen cases since the 1920s. Thus, few American vet-

erinarians are familiar with the early stages of FMD infection. An animal becomes infected shortly after exposure but well before the onset of clinical symptoms. When symptoms do occur, they may include a sudden rise in temperature, followed by an eruption of blisters in the mouth, in the nostrils, on areas of tender skin, and on the feet. The blisters expand, then break, exposing raw, eroded skin surfaces. Eating becomes difficult and painful. Because the soft tissue under the hooves is swollen, the animal limps. Livestock raised for meat lose weight, and dairy cattle and goats give far less milk. FMD kills very young animals and causes pregnant females to abort.

Merely transporting infected tissue can start an epidemic—a single infected cow or pig can generate enough viral particles to communicate the disease over vast geographic areas in weeks. An outbreak of this disease could be easily introduced to the United States and might debilitate the U.S. livestock industry. According to the USDA, an outbreak could cost as much as $20 billion over fifteen years in increased consumer costs, reduced livestock productivity, and restricted trade.

APHIS does not permit imports of FMD-positive animals. While scientists appear close to developing an effective vaccine, vaccinating all susceptible animals would cost about $1 billion annually. In addition, the vaccine would not eradicate the disease. Currently the only effective countermeasure against FMD is slaughtering and incinerating all exposed and infected animals.

CHAPTER 6
INTERNATIONAL TERRORISM

On September 11, 2001, nineteen members of the al Qaeda terrorist group hijacked four U.S. commercial airliners and flew two of them into the twin towers of the World Trade Center in New York City and one into the Pentagon in Washington, D.C. The fourth plane crashed in Pennsylvania. More than three thousand people were killed and thousands more injured as a result of these devastating attacks, which caught the United States and the rest of the world by surprise. After spending years on the back burner, the term "terrorism" captured the world's attention. It caused a media frenzy and spread fear and insecurity among the American public at a rate unparalleled since the early days of the cold war. Before this attack on U.S. territory, Osama bin Laden and the organization he led, "al Qaeda," were little known outside of terrorism experts. Afterward these names, and many other terms associated with terrorism, were omnipresent in the media and in politics.

DEFINING TERRORISM

As with "national security," "terrorism" is a difficult term to define, particularly because it involves subjective social issues and relies on the unique perceptions of the definer. The saying "one man's terrorist is another man's freedom fighter" illustrates this difficulty. Most analysts find it much easier to define whether a specific act is terrorist in nature than to find a broad definition of terrorism itself.

Although many experts believe that there is generally an intuitive understanding of what constitutes terrorism, it is important for governments to define the term for the purposes of prevention and retaliation. Some definitions of "terrorism" offered by academic sources and government-policy analysts include:

- "The political use of violence or intimidation." —*American Heritage Dictionary*

- "The threat of violence, individual acts of violence, or a campaign of violence designed primarily to instill fear." —Brian Michael Jenkins, "International Terrorism: A New Mode of Conflict," in David Carlton and Carlo Schaerf (eds.), *International Terrorism and World Security* (London: Croom Helm, 1975)

- "The illegitimate use of force to achieve a political objective when innocent people are targeted." —Walter Laqueur, *The Age of Terrorism* (Boston: Little, Brown, 1987)

- "Coercive intimidation, premeditated acts or threats of violence systemically aimed at instilling such fear in the target that it will force the target to alter its behavior in the way desired by the terrorists." —Paul Wilkinson, *Technology and Terrorism* (London: Frank Cass, 1993)

- "The recurrent use or threatened use of politically motivated and clandestinely organized violence, by a group whose aim is to influence a psychological target in order to make it behave in a way which the group desires." —C. J. M. Drake, *Terrorist Target Selection* (New York: St. Martin's Press, 1998)

- "The unlawful use of force or violence against persons or property to intimidate or coerce a government, the civilian population, or any segment thereof, in furtherance of political or social objectives." —Federal Bureau of Investigation (FBI)

- "The unlawful use of—or threatened use of—force or violence against individuals or property to coerce or intimidate governments or societies, often to achieve political, religious, or ideological objectives." —U.S. Department of Defense

- "Premeditated, politically motivated violence perpetrated against noncombatant targets by subnational groups or clandestine agents, usually intended to influence an audience." —U.S. Department of State

MOTIVATIONS AND TRENDS

Intelligence specialists call terrorism a "transnational" threat because the people who make up terrorist groups may not come from, represent, or be sponsored by a particular country. Instead, they can operate across international boundaries and against any number of countries to further their cause or objectives. Though sometimes state sponsored, it is rare for terrorists to be what American officials call "state actors," such as foreign governments; rather, they are generally doing their own work and pursuing their own goals.

Terrorism generally involves some political or religious message and is almost always violent. Some of the actions that the U.S. State Department defines as terrorist activities include:

- The hijacking or sabotage of any conveyance (including an aircraft, vessel, or vehicle)

- Seizing or detaining, and threatening to kill, injure, or continue to detain, another individual in order to compel a third person (or governmental organization) to do or abstain from doing any act as an explicit or implicit condition for the release of the individual seized or detained

- A violent attack upon an internationally protected person (defined as (1) a chief of state, head of government, or foreign minister in a country other than his or her own and any accompanying family member; or (2) any other representative, officer, employee, or agent of the U.S. government, a foreign government, or international organization and any member of his or her family/household) or upon the liberty of such a person

- An assassination

- Using any biological agent, chemical agent, nuclear weapon or device, explosive, firearm, or other weapon or dangerous device with intent to endanger, directly or indirectly, the safety of one or more individuals or to cause substantial damage to property (other than for mere personal monetary gain)

- A threat, attempt, or conspiracy to do any of these activities

Violence and terrorism usually occur together, as the message delivered by the terrorists is intended to reach an audience beyond simply the targets or victims of an attack. As a result, most terrorist attacks are also symbolic. In the September 11 attacks, the World Trade Center towers and the Pentagon represented the might of the U.S. economy and military. Other symbolic terrorist targets have included places of worship, government offices, military bases and barracks, and police personnel.

Terrorist tactics have varied over the years as well. They have ranged from single-person attacks to those involving mass destruction and casualties. Kidnappings, sabotage, assassinations, knifing campaigns, hijackings, murders, bombings, bank robberies, and cyber attacks are all tactics employed by terrorists.

Most terrorist attacks are motivated by political or religious conflict. Terror groups motivated by religious obligations have been on the rise since the 1980s. Throughout history, religious believers have often felt justified in perpetrating violence on behalf of their causes, reasoning that their actions are sanctioned by God. Secular (nonreligious) groups must maintain the support of their constituencies and so cannot commit heinous acts without the risk of alienating such support or facing widespread condemnation. Religious terrorists may believe that God is their main audience or that they are fighting for downtrodden people everywhere, whom they seek to defend against nonbelievers in a holy war. For these reasons, it may be easier for a religiously motivated terrorist group to perpetrate an attack that causes massive casualties.

Religion and politics are not always easy to separate. In the case of cults or millenarian groups (those believing in the thousand-year period of Christian triumph on Earth as predicted in the Bible) that shut themselves away from society, the single-minded religious agenda is not hard to identify. However, both religion and politics motivate certain Islamic groups. Hizballah, for example, which has a defining Islamic philosophy, also serves as a strong political faction in Lebanon. This is also the case with Hamas, which uses religion to recruit followers but is also a formidable political alternative to the Palestinian Authority in the West Bank and Gaza in Israel.

Another trend in the world of terrorism is the emergence of "cells," or ad hoc groups that form to conduct a particular attack without a strong organizational structure, base, or leader. The danger in such small, short-lived groups is that they are hard to track because of their lack of established patterns. Single-issue terrorist groups, such as those fighting against abortion or to protect the environment, can also fall into the "cell" pattern.

Terrorism and the United States

Understanding the motivations of terrorists who strike at the United States is very important in helping policy makers counter the problem in the long run. These motivations vary between different groups and individuals within these groups. But there are common themes. Over the years, disaffected individuals from repressed societies have grown increasingly irate with American foreign policies. This is especially true of U.S. involvement in the Middle East.

Some Muslims consider U.S. actions in the region as evidence of collaboration with regimes that compromise pure Islamic values and consequently lead to global oppression of Muslims. Governments of some of the predominantly Islamic countries in the Middle East, includ-

FIGURE 6.1

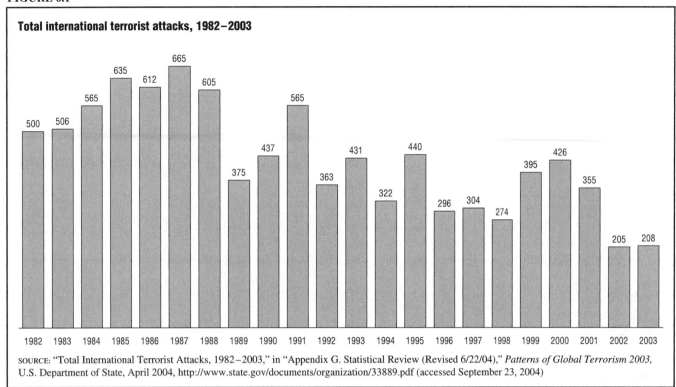

Total international terrorist attacks, 1982–2003

SOURCE: "Total International Terrorist Attacks, 1982–2003," in "Appendix G. Statistical Review (Revised 6/22/04)," *Patterns of Global Terrorism 2003*, U.S. Department of State, April 2004, http://www.state.gov/documents/organization/33889.pdf (accessed September 23, 2004)

ing Kuwait, Jordan, and Saudi Arabia, have supported the United States in the past. Others, especially Pakistan, Saudi Arabia, and Egypt, were particularly helpful after the events of September 11, 2001. Still, many of these countries faced some degree of opposition from domestic groups who believe that the West exploits the region and supports oppressive regimes. Such perceptions of American support for illegitimate and tyrannical governments are one of the primary motivations for groups such as al Qaeda, in which charismatic leaders exploit individual frustrations and channel them into a hatred for the West, particularly the United States.

Besides collaboration, other U.S. policies in the Middle East also contributed to animosities that may lead to hostile acts such as terrorism. Immediately following the 1991 Persian Gulf War, the United States established military bases in Saudi Arabia to have a forward-deployed force (a military force positioned in a region to reach a tactical advantage) in case tensions resumed. For many Muslims around the world, the notion of a foreign military "occupying" the soil that hosts two of the most sacred sites in Islam (Mecca and Medina) is a sacrilege. Strong American support, both political and military, for the Jewish state of Israel in a region that is predominantly Arab and Muslim also inflames anti-American sentiment. Because the region is a major supplier of oil and natural gas and holds large deposits of these vital resources, Western dependence on oil makes American strategic involvement in the region crucial to ensure that oil continues to flow out of the region at reasonable prices.

INTERNATIONAL TERRORISM STATISTICS

The average number of international terrorist attacks from 1982 to 2003 was about 430 per year. (See Figure 6.1.) The mid-1980s saw a peak in terrorist attacks, averaging more than six hundred per year from 1985 to 1988. The number of terrorist attacks was generally lower after 1988, with the exception of 1991, and was especially low between 1996 and 1998. They rose slightly in 1999 through 2001. Following the September 11, 2001 attacks, the numbers again dropped dramatically in 2002 and 2003, down to just over two hundred attacks per year.

Of the terror attacks against U.S. targets in 2003, those against businesses (sixty-one) far outnumbered those against other types of targets (sixteen attacks against government targets, fifteen against diplomatic targets, and three against military targets). (See Figure 6.2.) Terrorist attacks against businesses fell dramatically from a high in 2000 of 408 attacks. In 2000 there were 152 bombings of the Colombian pipeline, a multinational oil pipeline in Colombia, that accounted for 40% of international terrorist acts in that year. Of the 346 international terrorist attacks in 2001, 178 (51%) were bombings of the Colombian pipeline. By 2003, thanks to increased government surveillance powers and a new program designed to demobilize terrorists who turned themselves in peacefully, attacks on the Colombian oil pipeline were negligible.

The number of U.S. fatalities from international terrorism from 1998 to 2000 was low, with the highest number, twenty-three, occurring in 2000. Because nearly three

FIGURE 6.2

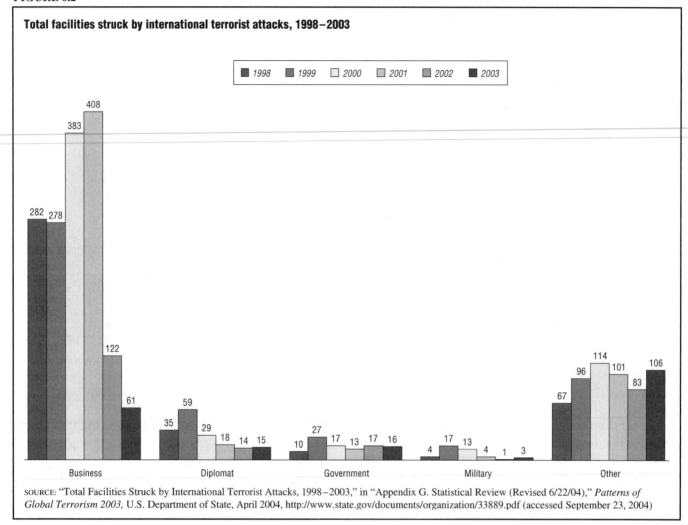

Total facilities struck by international terrorist attacks, 1998–2003

Legend: ■ 1998 ■ 1999 □ 2000 □ 2001 ▨ 2002 ■ 2003

SOURCE: "Total Facilities Struck by International Terrorist Attacks, 1998–2003," in "Appendix G. Statistical Review (Revised 6/22/04)," *Patterns of Global Terrorism 2003,* U.S. Department of State, April 2004, http://www.state.gov/documents/organization/33889.pdf (accessed September 23, 2004)

thousand people were confirmed dead in the September 11, 2001, attacks by al Qaeda, casualties in 2001 far outnumbered those of any previous year. (See Figure 6.3.) In addition to those killed or injured in the September 11 attacks, eight U.S. citizens were killed and fifteen were wounded in separate international terrorist attacks in 2001.

STATE-SPONSORED TERRORISM

The Nature of State-Sponsored Terrorism

Characterizing state-sponsored terrorism briefly is almost as difficult as coming up with a succinct definition for the term "terrorism" itself. Nevertheless, a crucial part of the concept is a sovereign government's involvement (at varying levels) with individual actors or organizations that carry out acts of terror. Boaz Ganor, an expert at the International Policy Institute for Counterterrorism in Herzliya, Israel, claims in his article "Countering State-Sponsored Terrorism," posted at the Milnet Web site (http://www.milnet.com/ict/counter.htm), that states choose to sponsor terrorism for a variety of reasons. According to Ray S. Cline and Yonah Alexander, who are quoted by Gonar, one of the most obvious reasons is the ability to "achieve

strategic ends in circumstances where the use of conventional armed forces is deemed inappropriate, ineffective, too risky or too difficult."

State sponsorship of terrorism dates back centuries, as rulers and governments throughout history have aided subnational organizations to wreak havoc on their enemies. The cold war of the twentieth century witnessed abundant support for revolutionary and guerilla organizations by the Soviet and U.S. superpowers. (Guerilla groups are generally irregular forces that fight for revolutionary causes and employ tactics not usually employed by states' regular military forces, like hit-and-run operations and illicit fundraising.) The United States and the Soviet Union saw these groups as conduits to establish more potential allies and fewer potential enemies without the larger threat of a global war. Such efforts occasionally led to the establishment of "puppet" regimes—essentially, governments put into power through the efforts of a larger, more powerful sponsoring state that would preach the sponsor's doctrine and promote its interests and philosophies in a given region.

One distinct characteristic of state-sponsored terrorism is its covert nature, including plausible deniability, or

FIGURE 6.3

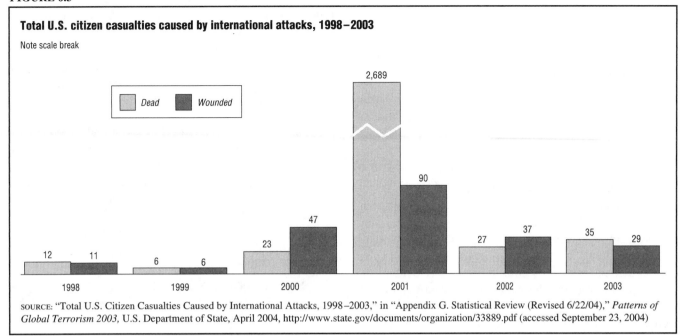

Total U.S. citizen casualties caused by international attacks, 1998–2003

Note scale break

Dead Wounded

2,689

90

47

23

12 11

6 6

27 37

35 29

1998   1999   2000   2001   2002   2003

SOURCE: "Total U.S. Citizen Casualties Caused by International Attacks, 1998–2003," in "Appendix G. Statistical Review (Revised 6/22/04)," *Patterns of Global Terrorism 2003,* U.S. Department of State, April 2004, http://www.state.gov/documents/organization/33889.pdf (accessed September 23, 2004)

the ability of a state to credibly claim it has no knowledge of the terrorist activities occurring. State support for terrorist groups allows the violent expression of the state's unfulfilled goals without overtly pressuring other governments toward certain political ends. State sponsorship of terrorism aspires to promote domestic and foreign policy through clandestine, or secret, means.

State-sponsored terrorism is also frequently employed to silence dissidents of a regime. Iran and Syria are two countries the U.S. Department of State identifies as sponsoring such crimes.

Terrorist groups often see state sponsors as desirable because of the resources they can provide that would otherwise be difficult and costly to acquire. According to *Patterns of Global Terrorism, 1993* (Washington, DC: U.S. Department of State, April, 1994), "International terrorism would not have flourished as it has during the past few decades without the funding, training, safe haven, weapons, and logistic support provided to terrorists by sovereign states."

One of the most blatant acts of state-sponsored terrorism, and one that put the phenomenon on the center stage of international relations, was the 1979 taking of hostages by Iranian "revolutionists." For more than a year, Americans waited anxiously as a group of fundamentalist "students" held fifty-two Americans hostage in Iran for 444 days. Many of the so-called students were actually agents of the incoming radical Islamic regime headed by Ayatollah Khomeini. This fundamentalist regime was later implicated as a sponsor of several acts of terrorism in the region over the years.

Hossein Sheikhosleslam, a leading figure among the revolutionary students, went on to hold the prestigious post of assistant for political affairs in the Iranian Foreign Ministry after the coup. The Iranian hostage crisis was followed by more than a decade of violence by fundamentalist Islamic and nationalist groups in the Middle East, fueled by anti-Western sentiments and supported by a handful of states sharing similar ideologies.

Levels of Support

In "Countering State-Sponsored Terrorism" (http://www.milnet.com/ict.counter.htm), Ganor has outlined the levels of culpability (blame or guilt) a state incurs based on its depth of involvement in the sponsorship of terrorist activities. According to Ganor, the escalating scale of state involvement begins with mere ideological support and ends with direct terrorist attacks by government agents. The six levels of state involvement Ganor outlined are:

1. Ideological support. Communism, democracy, and Islamic fundamentalism are a few of the doctrines that a state might choose to support through a puppet regime or terrorist organization.

2. Financial support. This requires funding terrorist organizations with cash in order to carry out their operations.

3. Military support. Training and the supplying of weapons fall under this category. For example, many experts agree that the attack on the USS *Cole* in Aden, Yemen, would not have been possible without some form of military assistance from a state.

4. Operational support. This category involves the logistical support that is involved in carrying out any ter-

rorist operation, such as providing falsified documents and safe havens.

5. Initiating/orchestrating terrorist attacks. It is at this level that states move from passive and indirect complicity to orchestrating terrorist attacks themselves.

6. Direct attacks by government agents. In such instances, agents of the state carry out the attacks themselves to further state agendas and interests.

State Sponsors of International Terrorism

Each year the U.S. State Department identifies countries it considers terror sponsors, or countries that repeatedly support international terrorism, on a watch list. Countries on the list suffer four sets of U.S. government sanctions. First, a ban on arms-related exports and sales goes into effect. Second, a thirty-day congressional notification is required for the export of dual-use items, or those with both civilian and military uses, that could increase military capability or the ability to support terrorism. Third, prohibitions are implemented on economic assistance. Finally, there are various financial and other restrictions, such as U.S. opposition of loans by international financial institutions to terror list countries and the lifting of diplomatic immunity to allow families of terrorist victims to file civil lawsuits in U.S. courts.

The State Department's 2003 list appeared in its annual publication *Patterns of Global Terrorism*. It featured Iran, Iraq, Syria, Libya, Sudan, North Korea, and Cuba. Libya and Sudan have taken positive steps toward cooperating in the war against terrorism. Since Iraq's liberation and the overthrow of Saddam Hussein, that country is no longer officially a state sponsor of terrorism. However, because terrorist incidents still occur within its borders, Iraq has become "a central front in the global war on terrorism," according to the State Department.

The State Department's list of state sponsors of terrorism is often debated among terrorism experts, many of whom believe it is outdated and merely a political tool the United States uses in order to impose sanctions. Critics of the list also claim that it intentionally does not include some countries, such as Saudi Arabia, that are known to have terrorist links but are important to the United States for economic or other reasons.

After the attacks of September 11, 2001, President George W. Bush stated, "Every nation, in every region, now has a decision to make. Either you are with us, or you are with the terrorists. From this day forward, any nation that continues to harbor or support terrorism will be regarded by the United States as a hostile regime." Some state sponsors on the list, such as Libya, Sudan, Iran, and Syria, made initial limited moves to cooperate with the international community's campaign against terrorism. Despite hostile relations with the United States, all four

states offered deep condolences after the September 11 attacks. Iran publicly stated that the United States had every right to seek retaliation for these attacks and even closed its boundaries with Afghanistan to prevent terrorists from escaping U.S. forces. The Sudan clamped down on extremists within its own boundaries and arrested several alleged members of al Qaeda.

Despite these moves, the State Department emphasizes that none of its seven state sponsors of terrorism has yet taken all necessary actions to divest itself fully of terrorist ties. For example, the State Department claims Iran and Syria have sought to have it both ways. They clamped down on certain terrorist groups, such as al Qaeda, but maintained their support for others, like Hamas and Hizballah, insisting they are national liberation movements.

MIDDLE EAST TENSIONS: THE ISRAELI/PALESTINIAN CONFLICT AND OTHER ISSUES. Tensions with many of the countries in the Middle East have been a problem for the United States for many years. There is much resentment of U.S. support for Israel, a Jewish nation located in the midst of mainly Muslim states. The dispute between Israel and its neighbors, particularly the Palestinians, goes back to 1948, when European Jews gained independence from Western powers occupying the Middle East (mainly Great Britain) and formed an independent Jewish state. The Jews felt the need for their own state in the area of the Middle East that three religions, Judaism, Islam, and Christianity, consider sacred ground. During World War II (1939–45), the Nazis had exterminated more than six million European Jews during the Holocaust. Many of those who survived had lost numerous relatives and loved ones. They not only felt uncomfortable remaining in Europe but yearned for a homeland in the Holy Land—the biblical "Promised Land" of their ancestors—and a place in which to arrange for their own protection.

However, the indigenous Palestinians and other Arabs in the area were pushed out of some regions by the creation of the new state. They resented that lands they believed were theirs had been taken over by Israel in the vacuum created by the departure of the British. Since that time, the Israelis have fought several wars with their Arab neighbors, including the Six-Day War of June 1967. In that conflict, Israel seized some of the most contested areas in the region: the Sinai Peninsula, the Golan Heights, the West Bank, the Gaza Strip, and East Jerusalem, while fighting Egypt, Jordan, and Syria. Iraq, Kuwait, Saudi Arabia, Sudan, and Algeria also offered aid against the Israeli attacks, but within six days Israel had taken control of the contested areas. Egypt lost the Sinai Peninsula and the Gaza Strip, Jordan lost East Jerusalem and the West Bank, and Syria lost the Golan Heights. Since that time, various peace agreements have forced Israel out of parts of those territories.

Many Palestinians, especially those in groups advocating violence (in order to reclaim land) like the Palestine Islamic Jihad and Hamas, feel Israel has taken a rigid and militant stance in favor of taking, holding, and defending Israel, a concept known as Zionism. To them, Zionism, the motivating force for Israel's original independence, continues more than half a century later in the occupied West Bank. There, a new generation of orthodox Jewish settlers has been residing and building homes and schools throughout the occupied territories. This perceived Zionism has bred Palestinian nationalist groups, which to Arab nations seem to be legitimate movements for national liberation, not terrorist groups. These Arab factions are also largely anti-American because American policy has generally favored Israel, which it sees as an outpost of Western-style democracy in the Middle East and a reliable ally.

The actions of the Israelis have increased tensions, and the Palestinians and Arab nations in the region, including Lebanon, Syria, Iran, and Iraq, have pursued terrorism to try to force Israel out of certain areas. Some terrorist groups and nations in the Middle East advocate the elimination of Israel and the creation of a Palestinian state at any cost. Increasingly, suicide bombing is being used as a tool against Israeli civilians as well as the military. Such bombers give their lives for what they perceive as martyrdom. These methods have drawn criticism from various outsiders, more retaliatory measures from Israel, and praise from groups of Arabs who feel more and more desperate. The United States has tried to serve as a broker of peace, but this peace brokering is viewed with skepticism by countries that believe U.S. solutions have a bias in Israel's favor. Consequently, a more even-handed American approach to resolving the region's rivalries is the U.S. foreign-policy goal in the volatile region.

In general, the Middle East has seen a resurgence of terrorism in the early years of the twenty-first century, mostly because of intensifying Arab anger at Israel's refusal to remove settlers and forces from the West Bank and the Gaza Strip and the U.S. and allied war on terrorism in Afghanistan and Iraq. Though Yassir Arafat's Palestinian Authority had subdued terrorist groups such as Hamas and the Islamic Jihad from their customary violence for about a year, the violence returned in the Palestinian Intifada (uprising) of 2000–02 and Israel's subsequent invasion of the occupied territories.

More than ever, previously peaceful Islamic charitable organizations are supporting harder-line political groups. Stronger, more organized, and more spontaneous Islamic political movements, many in the West believe, are providing both loyal support and conspiratorial cover for terrorist activities. Terrorist sponsors and "informal" sponsors, though they might be located in the Middle East, can support terrorist organizations around the world.

Neither the state sponsors nor the terrorist groups themselves respect international boundaries.

IRAN. Iran is a state of concern for the United States. The country underwent a conversion from a society favorable to U.S. interests to one unfavorable to the United States in 1979, when a fundamentalist Islamic revolution deposed the monarchy of Iran. In the early 2000s, more moderate forces began to have some influence in the Iranian government. Iran's hostility toward the West, particularly in the form of aggressive comments about harming Israel or the United States, lessened. It renounced the fatwa (a formal legal statement researched by an Islamic cleric in Islamic holy texts and through consultation with other Islamic scholars and clerics) death sentence that its clerics had called for earlier against expatriate Indian author Salman Rushdie. They had originally claimed Rushdie deserved death because of what they considered his sacrilegious writings, and the author had gone into hiding. Reformist President Mohammad Khatami claimed that Iran would no longer support terrorism. In addition, Iran began to offer assistance to the United States by actively investigating members of al Qaeda.

Despite these changes, according the State Department report *Patterns of Global Terrorism, 2003,* released in April 2004, Iran remained the most dangerous "rogue" state. The State Department justifies this designation by stating that even though Iran is seemingly assisting the United States in battling some terrorist groups, it is supporting others at the same time (such as Hizballah, Hamas, and the Palestine Islamic Jihad). By August 2004 Iran was also defying UN inspectors seeking to visit its nuclear plants.

IRAQ. Under the regime of Saddam Hussein, Iraq was linked to radical Palestinian terrorist organizations such as the Arab Liberation Front and the Abu Nidal Organization. In addition, Iraqi Intelligence Service (IIS) agents were believed to be involved in a plot to assassinate former President George H. W. Bush during a trip to Kuwait in April 1993. Abu Ibrahim, the former head of the now-disbanded 15 May Organization, which masterminded several bombings, sought refuge in Iraq, as did Abdul Yassin, a suspect in the 1993 World Trade Center bomb plot. Iraq also provided financial, military, and operational support to the Mujahedin-e Khalq Organization, an Iranian terrorist faction that opposes the existing Iranian government and has carried out several attacks and assassinations. Several terrorist attacks were carried out against UN relief workers and others attempting to remove land mines from territory in Iraq during the mid-1990s.

Iraqi agents are also believed to have been involved in several attacks against dissidents and "enemies of the state." An Iraqi scientist was assassinated in December 1992 as he was about to defect to Jordan. In early 1999, reports emerged from eastern Europe of a deadly plot to bomb the Prague, Czech Republic, headquarters of Radio Free Europe, which houses Radio Liberty. Radio Liberty had begun broadcasting its services to Iraq in 1998, much to the displeasure of Iraqi authorities. Although specific details of the case remain classified, it is believed that the alleged bomb plot was being orchestrated by the IIS.

Iraq was the only Arab Muslim country that did not condemn the September 11, 2001, attacks on the United States, and it even expressed sympathy for Osama bin Laden following U.S. retaliatory strikes. In 2001 Iraq continued to provide training, bases, and political encouragement to numerous terrorist groups. According to the article "Saddam Hussein's Philanthropy of Terror" by Derek Murdock (*American Outlook,* fall, 2003), Hussein had been rewarding Palestinian suicide bombers by granting their families bonuses of up to $25,000. Since the removal of Hussein in 2003, Iraq is no longer a state sponsor of terrorism.

SYRIA. Although Syria has made a point of clamping down on certain terrorist groups, such as al Qaeda, it continues to support other terrorist groups, including Hamas and Hizballah, insisting that these are purely national liberation movements. In 2004 Syria continued to harbor and provide logistics support to a number of terrorist groups, allowing Hamas, the Palestine Islamic Jihad, and other Palestinian groups to maintain offices in Damascus. There also is continuing suspicion that Syria has played a role in hiding Iraqi WMD materials. In January 2004 David Kay, former head of the U.S. Iraq Survey Group (ISG), was quoted in the *Daily Telegraph* (http://www.telegraph.co.uk/) as saying that interrogations with Iraqi officials showed that "a lot of material went to Syria before the war, including some components of Saddam's WMD programme." According to a *Washington Times* (http://www.washingtontimes.com/national/20040816-011235-4438r.htm) report in August 2004, the ISG had discovered that just before the U.S. invasion in 2003, Iraqi intelligence officials took over border crossings between Syria and Iraq, removed the regular guards, and allowed a number of eighteen-wheeled trucks to pass through into Syria unchecked. Pentagon and CIA officials have made no public comment about the cargo in those trucks.

LIBYA. In the 1980s Libya fired missiles at American aircraft doing maneuvers off the Libyan coast. Libyan agents were also involved in the April 1986 bombing of La Belle Discotheque, a popular Berlin, Germany, nightclub frequented by U.S. military personnel. A soldier and a civilian were killed in the attack and approximately two hundred people were injured. In retaliation, President Ronald Reagan ordered U.S. aircraft to strike targets within Libya in an operation code-named "El Dorado Canyon." Ten days after the bombing at La Belle, U.S. planes simultaneously struck five military targets within Libyan territory. Libyan leader Muammar Qaddafi's daughter was killed in the attack.

Two years later, on December 21, 1988, a bomb exploded on board Pan Am Flight 103 over Lockerbie, Scotland, and killed all 259 of the plane's passengers, plus eleven people on the ground. Two Libyans were connected to the incident but could not be charged, as they took refuge in Libyan territory. Nearly twelve years after the attack, Libya agreed to hand over the two suspects, al-Amin Khalifa Fhimah and Abdel Basset Ali al-Megrahi, to a special Scottish court convened in the Netherlands. After a two-year trial, the court delivered its verdict in January 2002. Abdel Basset Ali al-Megrahi was convicted of murder and sentenced to life in prison; al-Amin Khalifa Fhimah was found not guilty of the crimes. Al-Megrahi appealed his conviction, but the verdict was upheld in March 2002. In October 2002 Libya took responsibility for the Lockerbie bombing and settled with the families for $2.7 billion. In January 2004 Libya also settled with France over a 1989 plane bombing in which 170 people were killed.

The U.S. State Department noted in *Patterns of Global Terrorism, 2003* that Libya is one of two states (the other being Sudan) that seem closest to understanding what they must do to get out of the terrorism business, and each has taken steps in that direction. For example, following the September 11, 2001, terrorist attacks, Qaddafi issued a statement condemning the attacks as horrific and gruesome. He urged Libyans to donate blood for the U.S. victims. On September 16, 2001, he declared that the United States was justified in retaliating for the attacks. After September 11, Qaddafi repeatedly denounced terrorism. In December 2003 Libya stated that it would abandon its WMD program and allow UN inspectors unhindered access to all of its WMD research plants.

In general, Libya appears to have curtailed its support for international terrorism, although it may maintain residual contact with a few groups. Qaddafi's government in later years sought to recast itself as a peacemaker, offering to mediate a number of conflicts, such as the military standoff between India and Pakistan that began in December 2001. Still, Libya's past record of terrorist activity hinders Qaddafi's efforts to shed Libya's rogue state image. In April 2004 the United States lifted sanctions against Libya.

SUDAN. The United States and Sudan have continued and enhanced the counterterrorism dialogue they began in mid-2000. Like Libya, Sudan condemned the September 11, 2001, attacks. It pledged itself to combating terrorism and cooperating with the United States. The Sudanese government has stepped up its counterterrorist cooperation with various U.S. agencies, and Sudanese authorities have investigated and apprehended extremists suspected of involvement in terrorist activities. In late September 2001, the UN recognized these positive steps by removing UN sanctions on the country.

However, the U.S. State Department still designates Sudan as a state sponsor of terrorism, and unilateral U.S. sanctions remain in place. The United States contends that a number of international terrorist groups, including al Qaeda, the Egyptian Islamic Jihad, the Egyptian group al-Gama'a al-Islamiyya, the Palestine Islamic Jihad, and Hamas, continue to use Sudan as a safe haven, primarily for conducting logistical and other support activities. Still, the State Department concedes that press coverage about Sudan's cooperation with the United States may have led some terrorist elements to leave the country.

NORTH KOREA. North Korea, also known as the Democratic People's Republic of Korea, has been on the list of designated state sponsors of terrorism since 1988, after it allegedly shot down a South Korean plane carrying civilians in 1987. Another reason North Korea is on the list is that the country offered sanctuary to four Japanese Communist League/Red Army Faction members after they hijacked a Japan Airlines flight to North Korea in 1970. In addition, the U.S. State Department reports evidence of recent sales of small arms to terrorist groups.

In a statement released shortly after the September 11, 2001, attacks, North Korea repeated its official policy of opposing and not supporting terrorism. It also signed the UN Convention for the Suppression of the Financing of Terrorism, agreed to be bound by the Convention Against the Taking of Hostages, and indicated its willingness to sign five related agreements.

CUBA AND TIES TO LATIN AMERICA. Cuban leader Fidel Castro continues to be ambiguous about the U.S.-declared "war on terrorism." In October 2001 he called the war "worse than the original attacks—militaristic and fascist." When that statement did not gain him the support he anticipated, Castro instead declared Cuba's support for the war on terrorism. He eventually signed all twelve UN counterterrorism conventions, as well as the Ibero-American declaration on terrorism at the 2001 Ibero-American summit. Cuba did not protest the detention of approximately six hundred al Qaeda members captured in Afghanistan at the U.S. Naval Base at Guantanamo Bay, Cuba.

However, American officials believe that Castro still accepts terror as a political tactic. Twenty members of the Basque Homeland and Freedom (Euskadi Ta Askatasuna) guerrilla group, whose cause is the separation and independence of the Basque region from Spain, continue to reside in Cuba with the knowledge of the Cuban government. Cuba also provides a degree of safe haven and support to members of the Revolutionary Armed Forces of Colombia and National Liberation Army groups, both Colombian organizations that commit acts of terrorism against Colombian political, military, and economic targets, and kidnap and kill citizens of other countries, including the United States. In August 2001 a Cuban

spokesman revealed that Sinn Fein's official representative for Cuba and Latin America, Niall Connolly, one of three Irish Republican Army members arrested in Colombia on suspicion of providing explosives training to the Revolutionary Armed Forces of Colombia, had been based in Cuba for five years.

The triborder area of Latin America, where the boundaries of Argentina, Brazil, and Paraguay meet, is also considered to be a hotbed for terrorist and other illicit activities. The U.S. State Department has claimed that groups such as Hamas, Hizballah, and al Qaeda use the region for various logistical and financial (mostly money-laundering) purposes. The three governments have pledged to fight terrorism and, after September 11, 2001, made several arrests of individuals linked to terror groups.

SUBSTATE TERROR GROUPS

Beyond the countries named on the U.S. State Department's list of state sponsors of terrorism, a variety of terrorist groups are essentially transnational (beyond national boundaries) and substate (under a state level) in nature. The U.S. Secretary of State designates thirty-seven groups as "foreign terrorist organizations." These designations are made pursuant to section 219 of the Immigration and Nationality Act, as amended by the Antiterrorism and Effective Death Penalty Act of 1996, and carry legal consequences. Such a group must fit three specific criteria: (1) it must be a foreign entity; (2) it must engage in terrorist activity; and (3) the terrorist activities undertaken by the group must pose a threat to American nationals and U.S. national security.

In addition to the groups named as foreign terrorist organizations, various other terrorist groups are not a part of the list, because they do not fit the State Department's strict legal criteria. However, authorities closely monitor these groups, as their activities are considered essentially terrorist in nature.

Officially, it is the policy of the U.S. government not to negotiate with terrorists. In practice, however, the government has departed from that policy to negotiate at times with such terrorist groups as the Palestinian Liberation Organization and the Irish Republican Army as a way of getting those groups to the bargaining table with their regional enemies.

TERRORISM AROUND THE WORLD

Terrorist attacks sometimes stem from specific regional conflicts. One example of this type of conflict is in Northern Ireland, where the Irish Republican Army has launched terrorist acts against the British government and Irish Protestants in an attempt to gain independence for Northern Ireland. Another is in the Basque region of northern Spain and southern France, where the Basque separatist group Euskadi Ta Askatasuna seeks an independent Basque homeland. In addition, one of the most destructive and deadly regional conflicts is between the Palestinians and Israelis over disputed territory in the Middle East.

Chechen rebels allied with al Qaeda continue to strike inside Russia. On August 24, 2004, two Chechen female suicide bombers are believed to have brought down two Russian jetliners and killed eighty-nine people. On September 1, 2004, Chechen terrorists seized a school in Beslan, Russia, and some 1,100 parents and children were taken hostage. Although authorities negotiated with the hostage-takers, by September 4 the group had begun to kill those inside. When the terrorists exploded a bomb in the school's gymnasium and shot children in the back as they ran from the scene, Russian security forces moved in. An estimated 331 people, including 172 children, were killed in what Russian president Vladimir Putin declared "a massacre."

Although most terrorism across the globe is internal, this does not mean it is of less importance to the United States. Turmoil within even one nation can destabilize entire regions. In addition, there may be consequences to the United States from terrorist groups mainly interested in other countries. In Uzbekistan, the Islamic Movement of Uzbekistan kidnapped four U.S. citizens in August 2000 while they were mountain climbing (the four Americans later escaped). The Revolutionary Armed Forces of Colombia kidnapped three Americans in March 1999, later executing them in Venezuela. In February 2002 the Movement of Holy Warriors in Pakistan kidnapped and beheaded American journalist Daniel Pearl.

Also, since many terrorist groups use methods that indiscriminately kill civilians, Americans visiting certain areas that harbor internal terrorist groups may be at risk from random violence. The Tamil Tigers in Sri Lanka, the Revolutionary Armed Forces and National Liberation Army of Colombia, and the Irish Republican Army in Ireland and Great Britain have all used bombings that have killed scores of people. American citizens could become unintentional victims if they happened to be in the wrong place at the wrong time.

Southeast Asia is increasingly becoming of major concern to the United States because of terrorist organizations operating out of the region. Groups like the Abu Sayyaf Group in the Philippines choose methods of terrorism that include kidnapping foreign hostages to receive ransom money. An October 2002 nightclub bombing in Bali, Indonesia, by members of the Jemaah Islamiya militant Muslim network, which is linked to al Qaeda and seeks to set up an Islamic state in Southeast Asia, killed almost two hundred people, mostly foreign tourists including several Americans. More than three hundred were injured in the blast.

Also, because of the highly subjective nature of the term "terrorism," U.S. actions abroad have sometimes been called terrorist in character by other nations. Cuba is one of the most vociferous of the state and substate groups that accuse America of carrying out terrorist acts abroad. Some of the U.S. actions (mostly undertaken unilaterally) cited in support of such theories include:

• The U.S. blockade on Cuba since 1963

• Varied interventions in Latin America during the cold war, including support for the Pinochet government in Chile, the contras in Nicaragua, and various death squads in Honduras and El Salvador

• Support for the mujahideen, who fought the Soviets in Afghanistan during the 1980s

• The invasion of Grenada in 1983

• Support of Saddam Hussein's Iraq during the Iran-Iraq war

• The "accidental" shooting of an Iranian passenger airliner in 1988 that killed 290 civilians

• The 1989 invasion of Panama

• Intervention in the Middle East on behalf of Kuwait in 1991 and the ensuing bombing of Iraq

• The naval blockade of Serbia and Montenegro in 1993

• The 1994 intervention in Haiti

• A bombing of the Chinese embassy in Belgrade that killed three Chinese citizens

• Alleged support for numerous assassinations and attempted assassinations over the years of individuals such as François Duvalier (Haiti), Patrice Lumumba (Congo), Fidel Castro (Cuba), Raúl Castro (Cuba), Ernesto Che Guevara (Cuba), Salvador Allende (Chile), Mobutu Sese Seko (Zaire), Muammar Qaddafi (Libya), Ayatollah Khomeini (Iran), and Saddam Hussein (Iraq)

INTERNATIONAL TERRORISM DIRECTED AGAINST THE UNITED STATES

Introduction: September 11, 2001, and Anti-American Terrorism

In 1993 the World Trade Center in New York City, a symbol of American financial wealth and power, was the target of international terrorists, who detonated a bomb in the underground parking garage, killing six people and injuring a thousand. On September 11, 2001, the World Trade Center once again became the target of a Muslim extremist terrorist group, along with other symbolic American targets such as the Pentagon in Washington, D.C. During the attacks, nineteen Middle Eastern men, fifteen of whom were Saudi Arabian and all of whom were members of al Qaeda, hijacked and then crashed four commercial jetliners.

FIGURE 6.4

Route of American Airlines Flight 11, September 11, 2001

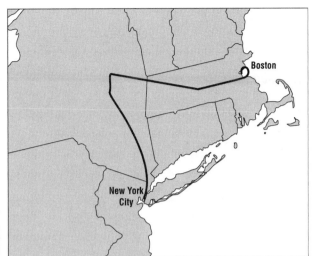

7:59	Takeoff
8:14	Last routine radio communication; likely takeover
8:19	Flight attendant notifies AA of hijacking
8:21	Transponder is turned off
8:23	AA attempts to contact the cockpit
8:25	Boston Center aware of hijacking
8:38	Boston Center notifies NEADS of hijacking
8:46	NEADS scrambles Otis fighter jets in search of AA 11
8:46:40	AA 11 crashes into 1 WTC (North Tower)
8:53	Otis fighter jets airborne
9:16	AA headquarters aware that Flight 11 has crashed into WTC
9:21	Boston Center advises NEADS that AA 11 is airborne headng for Washington
9:24	NEADS scrambles Langley fighter jets in search of AA 11

SOURCE: "American Airlines Flight 11 (AA 11) Boston to Los Angeles," in *9/11 Commission Report*, National Commission on Terrorist Attacks upon the United States, August 21, 2004, http://www.9-11commission .gov/report/911Report_Ch1.pdf (accessed September 23, 2004)

FIGURE 6.5

Route of United Airlines Flight 175, September 11, 2001

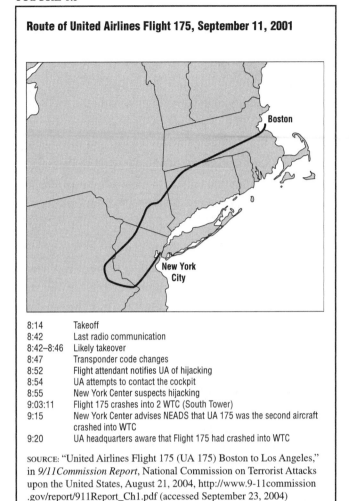

8:14	Takeoff
8:42	Last radio communication
8:42–8:46	Likely takeover
8:47	Transponder code changes
8:52	Flight attendant notifies UA of hijacking
8:54	UA attempts to contact the cockpit
8:55	New York Center suspects hijacking
9:03:11	Flight 175 crashes into 2 WTC (South Tower)
9:15	New York Center advises NEADS that UA 175 was the second aircraft crashed into WTC
9:20	UA headquarters aware that Flight 175 had crashed into WTC

SOURCE: "United Airlines Flight 175 (UA 175) Boston to Los Angeles," in *9/11 Commission Report*, National Commission on Terrorist Attacks upon the United States, August 21, 2004, http://www.9-11commission .gov/report/911Report_Ch1.pdf (accessed September 23, 2004)

Five of the terrorists hijacked American Airlines Flight 11, departing Boston, Massachusetts, for Los Angeles, California, at 7:45 A.M. At 8:45 A.M., they intentionally piloted the aircraft into the North Tower of the World Trade Center. (See Figure 6.4.) Another five terrorists hijacked United Airlines Flight 175, which departed Boston for Los Angeles at 7:58 A.M. At 9:05 A.M. they flew the plane into the South Tower of the World Trade Center. (See Figure 6.5.) The crashed planes, carrying tons of jet fuel in their full tanks for the long journey across the country, ignited upon impact, causing a fire with four-thousand-degree temperatures that melted the internal structure of the 104-story World Trade Center office towers. Both towers eventually completely collapsed, destroying other buildings and property as they fell. More than 255 firefighters and seventy police officers died inside the towers as they tried to rescue the thousands of office workers and facility personnel trapped inside.

When the official cleanup and recovery efforts in New York City ended with a final ceremony on May 30,

2002, the New York City Office of Emergency Management gave final totals for the destruction caused by the attacks. Of the 2,823 people killed in New York, 1,102 victims had been identified. An estimated 3.1 million hours of labor were spent on cleanup, and more than 1.8 million tons of debris had been removed in 108,342 truckloads. The leveling of the World Trade Center towers also caused property damage in the billions of dollars. Tens of thousands of people had to be evacuated from their homes in Manhattan. Air pollution initially increased, and authorities suspected there might be lasting health effects from the shattered debris, air pollution, and rubble.

Terrorists using knives and box cutters also hijacked American Airlines Flight 77, a Boeing 757 commercial airliner with sixty-four persons aboard. The plane had departed at 8:10 A.M. from Dulles International Airport, in suburban Herndon, Virginia, outside Washington, D.C. At 9:39 A.M., the terrorists directed the plane into the west side of the Pentagon in Washington, D.C. The left side of the building was destroyed. The number of those killed included sixty-four passengers and crew members aboard the plane and 125 military and civilian personnel on the ground. Another eighty were injured. (See Figure 6.6.)

FIGURE 6.6

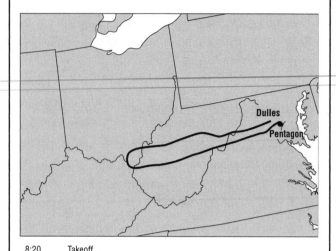

Route of American Airlines Flight 77, September 11, 2001

8:20	Takeoff
8:51	Last routine radio communication
8:51–8:54	Likely takeover
8:54	Flight 77 makes unauthorized turn to south
9:05	AA headquarters aware that Flight 77 is hijacked
9:25	Herndon Command Center orders nationwide ground stop
9:32	Dulles tower observes radar of fast-moving aircraft (later identified as AA 77)
9:34	FAA advises NEADS that AA 77 is missing
9:37:46	AA 77 crashes into the Pentagon
10:30	AA headquarters confirms Flight 77 crash into Pentagon

SOURCE: "American Airlines Flight 77 (AA 77) Washington, D.C. to Los Angeles," in *9/11 Commission Report*, National Commission on Terrorist Attacks upon the United States, August 21, 2004 http://www.9-11commission.gov/report/911Report_Ch1.pdf (accessed September 23, 2004)

FIGURE 6.7

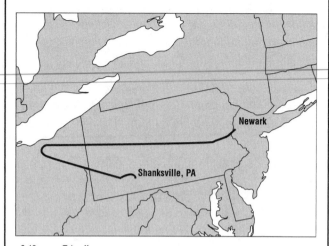

Route of United Airlines Flight 93, September 11, 2001

8:42	Takeoff
9:24	Flight 93 receives warning from UA about possible cockpit intrusion
9:27	Last routine radio communication
9:28	Likely takeover
9:34	Herndon Command Center advises FAA headquarters that UA 93 is hijacked
9:36	Flight attendant notifies UA of hijacking; UA attempts to contact the cockpit
9:41	Transponder is turned off
9:57	Passenger revolt begins
10:03:11	Flight 93 crashes in field in Shanksville, PA
10:07	Cleveland Center advises NEADS of UA 93 hijacking
10:15	UA headquaters aware that Flight 93 had crashed in PA; Washington Center advises NEADS that Flight 93 has crashed in PA

SOURCE: "United Airlines Flight 93 (UA 93) Newark to San Francisco," in *9/11 Commission Report*, National Commission on Terrorist Attacks upon the United States, August 21, 2004, http://www.9-11commission.gov/report/911Report_Ch1.pdf (accessed September 23, 2004)

Terrorists hijacked United Airlines Flight 93, also a Boeing 757, carrying forty-four passengers and crew from Newark International Airport in New Jersey to San Francisco International Airport in California. The hijackers took over the plane's controls and headed the aircraft toward Washington, D.C. It is believed the intended goal of the plane was the White House. But the passengers, having heard about the World Trade Center attacks during their flight, attempted to retake control of the plane and stormed the cockpit. The plane crashed in the countryside near Shanksville, Pennsylvania, killing all aboard. (See Figure 6.7.)

The September 11 attacks were the most destructive acts of war or terrorist violence against Americans on U.S. soil since the Japanese attack on the naval base at Pearl Harbor, Hawaii, in 1941. The total estimated death toll from the September 11 attacks was nearly three thousand people, including citizens of seventy-eight countries.

The U.S. Justice Department quickly determined that the attacks were conducted by al Qaeda, under the leadership of Osama bin Laden. From 1998 to 2001 al Qaeda was suspected of being responsible for the majority of U.S. deaths from international terrorism. The attacks of September 11, 2001, followed the al Qaeda–attributed

bombing of the USS *Cole* in the port of Aden, Yemen, on October 12, 2000, which killed seventeen U.S. sailors and injured thirty-nine others. The U.S. government also found al Qaeda responsible for the August 1998 bombings of U.S. embassies in Nairobi, Kenya, and Dar Es Salaam, Tanzania, in which twelve U.S. citizens were killed.

There are also other bombings that may not have been the work of al Qaeda but of related Muslim extremist terrorists. These would include the first bombing of the World Trade Center in 1993. In addition, in June 1996 the Khobar Towers military barracks near Dahran, Saudi Arabia, were bombed, and in 1994 a U.S. military assistance headquarters in Jiddah, Saudi Arabia, was bombed.

These acts, occurring roughly over the decade 1992–2002, represent a resurgence of anti-American terrorism by Middle Eastern extremists. The first round of such attacks in the 1980s killed hundreds of U.S. military and diplomatic personnel. In 1983 an Arab terrorist organization bombed the American embassy in Beirut, Lebanon, beginning a sustained period of violence from Middle Eastern terrorist organizations against U.S. targets overseas—mainly embassies, barracks, and other facilities.

Attacks against diplomatic, military, and government personnel or facilities are significant because they are symbols of U.S. strength. They are usually better protected than most businesses but make more attractive targets. Military targets, in terms of U.S. troops, are found worldwide, with 257,692 U.S. military personnel stationed abroad as of March 31, 2004. (See Table 6.1.) More than 116,000 were stationed in Europe, and 98,000 were stationed in East Asia and the Pacific. About 211,000 troops were stationed in Iraq.

Al Qaeda—Understanding the Phenomenon

Many Americans were not familiar with the name Osama bin Laden and the group al Qaeda prior to September 11, 2001, but the rise of this enigmatic terrorist leader and his organization can be traced back to the early 1980s. Many Muslim leaders around the world were outraged by the Soviet invasion of Afghanistan in December 1979 and rallied to declare a jihad ("holy war") against the invading superpower, which had an official platform of atheism. Many individuals, mainly Arabic, heeded the call and arrived in Afghanistan to fight as defenders of Islam. They came to be known as the "mujahideen," holy warriors who strove to protect their religion at all costs. One of these holy warriors was Osama bin Laden.

Bin Laden came from a wealthy Yemeni family in Saudi Arabia. Driven by the religious obligation he felt, he arrived in Afghanistan to defend his faith. Many scholars claim that bin Laden was more of a financier for the mujahideen than an actual fighter on the frontlines. Still, it was in Afghanistan that bin Laden met prominent militants, such as Muhammed Atef and Ayman al-Zawahiri, who would later become a vital part of the al Qaeda network.

After the Soviets left Afghanistan, bin Laden called for a worldwide jihad. He preached radical views of Islam and endorsed violent tactics, which led the Saudi government to strip him of his citizenship. He then moved to Sudan, where he set up a network of organizations and businesses to raise money for his cause. After being pressured by the United States, the Sudanese government asked bin Laden to leave. He relocated to war-torn Afghanistan, established a special relationship with the ruling Taliban authorities, and eventually was considered above the law in the country. Bin Laden set up various military camps to train young men from around the world in such skills as assassination and espionage.

The name al Qaeda ("the base") is not a term used by bin Laden himself. Western experts coined the phrase in order to label the unique loose-knit structure of the organization. There is not one cohesive group known as al Qaeda. Instead, it is primarily a network of various individuals, cells, and other organizations that come together for a main common cause, the defense of Islam. They receive military training and financial support from top al Qaeda leaders such as bin Laden.

Although the defense of Islam can be interpreted very broadly, bin Laden holds specific grievances against the United States. He specifically cites the U.S. military presence in Saudi Arabia, U.S.-led sanctions against Iraq, U.S. support for Israel, and other historical U.S. "terrorist" acts, such as the dropping of atomic bombs on Japan during World War II. Bin Laden calls for the creation of an Islamic nation, something along the lines of the Ottoman Empire. (The Ottoman Empire was a powerful Islamic kingdom that spread across Europe and parts of the Middle East from the early fourteenth century to the end of World War I.)

After the attacks of September 11, 2001, the U.S. military attacked Afghanistan, the one central location that could be associated with al Qaeda. The group's training camps were captured and destroyed. After that, the organization became even more decentralized. Although bin Laden eluded capture as of 2004, he remains a high-priority target for counterterrorism agencies everywhere and continues to boost the morale of his followers through video- and audiotaped recordings he has secretly delivered to news agencies. The destruction of the al Qaeda network has become a top priority for the United States in its efforts to combat terrorism.

FINANCING TERRORISM

Any terrorist organization, no matter its size or type, requires substantial amounts of money and resources to be able to carry out attacks and maintain some form of cohesion. Funding for such organizations can come from state sponsors, individual contributors, seemingly legitimate "front" organizations, and criminal activities:

- State sponsorship was common during the cold war, when both the United States and the Soviet Union supported various groups whose ideologies matched theirs or challenged the ideology of the other side. Afghanistan, Angola, South Africa, and parts of Latin America all served as battlegrounds in the war fought between the two major blocs. Iran and Libya have often been accused of supporting fundamental Islamic groups to export the 1979 Islamic revolution and encourage anti-Western sentiments.

- Individual contributors come from a wide spectrum of society. Fund-raisers target individuals' emotions to elicit money and other resources. Millionaires, expatriate nationals, and members of wealthy families are frequent fund-raising targets of terrorist groups.

- Laundering money through front organizations provides a way for groups to transfer cash from legitimate causes to terrorist ones.

TABLE 6.1

Active duty military personnel strengths, by regional area and country, March 31, 2004

Regional area/country	Total	Army	Navy	Marine Corps	Air Force
United States and Territories					
Continental United States (CONUS)	958,215	361,001	186,536	126,832	283,846
Alaska	17,989	7,843	91	20	10,035
Hawaii	35,810	16,978	8,260	5,819	4,753
Guam	3,315	36	1,419	6	1,854
Johnston Atoll	2	0	0	0	2
Puerto Rico	769	219	490	20	40
Trust Territory of the Pacific Islands	26	26	0	0	0
U.S. Virgin Islands	6	1	3	0	2
Transients	31,397	5,022	9,932	2,498	13,945
Afloat	120,666	0	120,537	129	0
Total—United States and Territories	**1,168,195**	**391,126**	**327,268**	**135,324**	**314,477**
Europe					
Albania	8	1	1	5	1
Austria	23	4	0	13	6
Belgium	1,534	883	92	36	523
Bosnia and Herzegovina	2,931	2,903	0	20	8
Bulgaria	12	4	1	5	2
Cyprus	26	3	0	14	9
Czech Republic	15	3	0	9	3
Denmark	23	2	4	6	11
Estonia	7	0	1	6	0
Finland	15	2	2	8	3
France	82	11	10	25	36
Germany*	75,603	58,598	296	213	16,496
Gibraltar	6	0	6	0	0
Greece	562	69	418	11	64
Greenland	138	0	0	0	138
Hungary	16	4	0	7	5
Iceland	1,754	2	1,031	1	720
Ireland	8	2	0	6	0
Italy*	13,354	3,196	5,218	61	4,879
Latvia	6	0	0	6	0
Lithuania	7	0	1	6	0
Luxembourg	15	10	0	5	0
Macedonia, The Former Yugoslav Republic of	104	76	1	21	6
Malta	7	0	0	7	0
Netherlands	722	358	27	14	323
Norway	85	12	12	12	49
Poland	20	5	1	10	4
Portugal	1,077	17	48	6	1,006
Romania	12	4	1	5	2
Serbia (includes Kosovo)	128	118	0	5	5
Slovenia	11	0	0	6	5
Spain	1,968	55	1,383	219	311
Sweden	10	1	1	5	3
Switzerland	20	1	2	12	5
Turkey	1,863	143	28	15	1,677
United Kingdom	11,801	432	1,188	97	10,084
Afloat	2,534	0	2,534	0	0
Total—Europe	**116,507**	**66,919**	**12,307**	**897**	**36,384**
Former Soviet Union					
Azerbaijan	5	0	0	5	0
Georgia	38	0	0	38	0
Kazakhstan	10	3	0	6	1
Kyrgyzstan	8	0	0	8	0
Russia	79	19	3	46	1
Turkmenistan	7	0	0	7	0
Ukraine	15	10	1	0	4
Total—Former Soviet Union	**162**	**32**	**4**	**110**	**16**
East Asia and Pacific					
Australia	205	16	77	36	76
Burma	10	3	0	6	1
Cambodia	5	5	0	0	0
China (includes Hong Kong)	60	9	14	30	7
Fiji	1	0	0	1	0
Indonesia (includes Timor)	24	7	4	10	3
Japan	40,045	1,864	5,396	18,112	14,673

Regional area/country	Total	Army	Navy	Marine Corps	Air Force
Korea, Democratic Peoples Republic of	14	0	0	14	0
Korea, Republic of	40,258	30,190	362	266	9,440
Laos	5	1	0	2	2
Malaysia	15	2	3	5	5
Mongolia	1	1	0	0	0
New Zealand	8	2	2	0	4
Philippines	144	10	6	120	8
Singapore	196	5	125	19	47
Thailand	113	43	11	29	30
Vietnam	19	6	0	11	2
Afloat	16,601	0	12,422	4,179	0
Total—East Asia and Pacific	**97,724**	**32,164**	**18,422**	**22,840**	**24,298**
North Africa, Near East, and South Asia					
Afghanistan (not available)	0	0	0	0	0
Algeria	9	1	0	7	1
Bahrain	1,496	20	1,321	131	24
Bangladesh	8	2	0	6	0
Diego Garcia	491	2	437	0	52
Egypt	350	277	5	20	48
India	28	7	4	11	6
Iraq (see OIF table)	0	0	0	0	0
Israel	38	6	3	16	13
Jordan	22	9	0	6	7
Kuwait (see OIF table)	6	0	6	0	0
Lebanon	4	4	0	0	0
Morocco	11	2	1	5	3
Nepal	9	3	0	6	0
Oman	31	3	1	7	20
Pakistan	26	4	2	15	5
Qatar	3,432	143	4	58	3,227
Saudi Arabia	291	136	24	25	106
Sri Lanka	6	0	1	5	0
Syria	9	3	0	6	0
Tunisia	15	6	2	6	1
United Arab Emirates	18	3	6	9	0
Yemen	15	4	0	11	0
Afloat	592	0	592	0	0
Total—North Africa, Near East, and South Asia	**6,907**	**635**	**2,409**	**350**	**3,513**
Sub-Saharan Africa					
Botswana	6	1	0	5	0
Burundi	5	0	0	5	0
Cameroon	6	1	0	4	1
Chad	9	3	0	6	0
Congo (Kinshasa)	8	2	0	5	1
Cote D'Ivoire	12	4	0	8	8
Djibouti	539	1	0	538	0
Eritrea	5	3	0	0	2
Ethiopia	12	4	0	8	0
Ghana	11	4	0	7	0
Guinea	6	1	0	5	0
Kenya	29	12	1	12	4
Liberia	9	1	0	7	1
Madagascar	1	0	1	0	0
Mali	6	0	0	6	0
Mozambique	6	0	0	6	0
Niger	7	1	0	6	0
Nigeria	16	4	0	9	3
Senegal	9	1	1	7	0
Sierra Leone	2	0	0	2	0
South Africa	29	4	1	20	4
St. Helena (includes Ascension Island)	2	0	0	0	2
Tanzania, United Republic of	6	0	1	5	0
Togo	6	0	0	6	0
Uganda	9	1	0	8	0
Zambia	6	1	0	5	0
Zimbabwe	8	4	0	4	0
Total—Sub-Saharan Africa	**770**	**53**	**5**	**694**	**18**
Western Hemisphere					
Antigua	2	0	0	0	2
Argentina	28	4	3	10	11

TABLE 6.1

Active duty military personnel strengths, by regional area and country, March 31, 2004 [CONTINUED]

Regional area/country	Total	Army	Navy	Marine Corps	Air Force
Bahamas, The	32	0	26	6	0
Barbados	7	1	0	6	0
Belize	2	1	1	0	0
Bolivia	21	7	1	5	5
Brazil	37	8	5	19	5
Canada	147	9	41	9	88
Chile	25	5	3	10	7
Colombia	55	17	4	27	7
Costa Rica	7	1	0	6	0
Cuba (Guantanamo)	700	6	526	167	1
Dominican Republic	14	2	1	8	3
Ecuador	35	6	2	6	21
El Salvador	21	8	0	11	2
Guatemala	16	8	0	7	1
Guyana	1	1	0	0	0
Haiti	455	8	0	447	0
Honduras	413	197	2	9	205
Jamaica	12	1	4	7	0
Mexico	21	9	3	5	4
Nicaragua	11	4	0	7	0
Panama	16	6	4	6	0
Paraguay	11	4	0	6	1
Peru	35	6	11	13	5
Suriname	2	2	0	0	0
Trinidad and Tobago	5	0	0	5	0
Uruguay	13	3	3	6	1
Venezuela	32	10	3	8	11
Afloat	25	0	25	0	0
Total—Western Hemisphere	**2,201**	**334**	**668**	**819**	**380**
Undistributed					
Ashore	29,025	2,553	11,890	14,582	0
Afloat	4,396	0	4,396	0	0
Total—undistributed	**33,421**	**2,553**	**16,286**	**14,582**	**0**
Total—foreign countries	**257,692**	**102,690**	**50,101**	**40,292**	**64,609**
Ashore	233,544	102,690	30,132	36,113	64,609
Afloat	24,148	0	19,969	4,179	0
NATO countries	110,494	63,800	9,756	747	36,191
Forward deployment Pacific Theater	101,610	32,218	20,286	22,889	26,217
Total— worldwide	**1,425,887**	**493,816**	**377,369**	**175,616**	**379,086**
Ashore	1,281,073	493,816	236,863	171,308	379,086
Afloat	144,814	0	140,506	4,308	0
Operation Iraqi Freedom (OIF data subject to change)					
Total (in/around Iraq as of March 31, 2004)	**211,028**	**155,291**	**14,838**	**25,568**	**15,331**

SOURCE: "Active Duty Military Personnel Strengths by Regional Area and Country (309A), March 31, 2004," U.S. Department of Defense, Washington Headquarters Services, Directorate for Information Operations and Reports, http://www.dior.whs.mil/mmid/M05/hst0403.pdf (accessed September 23, 2004)

• Criminal activities, such as narcotics smuggling, bank robberies, and kidnappings, can also raise a great deal of money. For example, a right-wing group called the Order stole about $3.6 million from an armored truck in California in 1984. The Turkish Kurdistan Workers' Party, the Revolutionary Armed Forces of Colombia, Peru's Shining Path, al Qaeda, and Lebanon's Hizballah have all been linked to the drug trade.

Domestic monitoring of possible money laundering on behalf of terrorist organizations has been stepped up under the provisions of the USA Patriot Act. While American banks have issued fewer suspicious activity reports (SARs) related to terrorism since late 2001, they are reporting more SARs based not on names appearing on government watch lists but on their own initiative. This trend seems to show that banking institutions are becoming more aware of possible terrorist financial activity and transactions. (See Figure 6.8.)

International efforts to curb the financing of terrorism were weak and underdeveloped for many years. The September 11 terrorist attacks jump-started domestic and international efforts to destroy the financial infrastructure of various terrorist groups, but such a goal is far from complete. Immediately following the attacks, President George W. Bush signed the Executive Order on Terrorist Financing, giving the U.S. Treasury Department the authority to block the assets of individuals and organizations associated with terrorist organizations. In April 2002 UN Resolution 1373 called for the suppression of all terrorism financing. Whether the cooperation and communication between governments needed to make these efforts effective will be forthcoming remains to be seen.

FIGURE 6.8

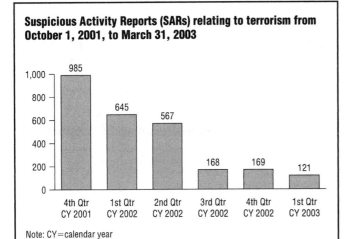

Suspicious Activity Reports (SARs) relating to terrorism from October 1, 2001, to March 31, 2003

Note: CY=calendar year

SOURCE: "Chart 1. SARs Filed Relating to Terrorism for the 18-Month Period (by Calendar Year Quarters), October 1, 2001 through March 31, 2003," in *International Narcotics Control Strategy Report 2003,* Bureau for International Narcotics and Law Enforcement Affairs, March 2004, http://www.state.gov/g/inl/rls/nrcrpt/2003/vol2/html/29910.htm (accessed September 23, 2004)

U.S. REACTION TO SEPTEMBER 11, 2001

Homeland Security

As authors Ashton B. Carter and William J. Perry observe in their book *Preventive Defense: A New Security Strategy for America* (Washington, DC: Brookings Institution Press, 1999), "Catastrophic terrorism is a military-scale threat divorced from the traditional context of foreign military conflict. This is entirely new in the American experience. Catastrophic terrorism challenges the U.S. government to reinvent a new national security structure from the ground up."

After the September 11, 2001, attacks, the U.S. government and public placed a new emphasis on "homeland security" and efforts to protect against homeland terrorism. Anthrax attacks in late 2001 encouraged a new examination of bioterrorism as well. The country was nervous and wanted to put new structures in place to prevent against future attacks.

In the wake of 9/11, the U.S. Congress, the president, and the intelligence community all knew a new terrorist attack could come at any time. Consequently, they believed that government efforts to protect against terrorism should increase and that these efforts should be directed by the White House. The federal government also bolstered the intelligence community—those parts of the government, including federal law enforcement, that cooperate with the U.S. Department of Defense to maintain national security.

President George W. Bush set up a new Office of Homeland Security (OHS) in the White House and appointed then–Pennsylvania governor Tom Ridge to head

it. The OHS coordinated the work of law enforcement officials, the military, and the intelligence community. Its major responsibilities included (1) supporting "first responders," those first on the scene of a homeland terrorist incident or catastrophe; (2) defending against bioterrorism; (3) securing America's borders; and (4) using up-to-date technology to secure the United States in the future.

The initial role of the OHS was to coordinate first responders to a terrorist or bioterrorist attack. First responders consist of the country's more than one million firefighters (approximately 750,000 of whom are volunteers); 556,000 full-time local police personnel, including approximately 436,000 sworn law enforcement officers; and 291,000 sheriff's office personnel, including 186,000 sworn officers. Another group of first responders comprises the country's 155,000 emergency medical technicians.

On March 12, 2002, the OHS implemented a system of threat conditions as a way of providing uniform advisories of possible terrorist threats. The five threat conditions range from "low" to "severe." Severe risk may require the closing of government offices and the deployment of emergency personnel. Intermediate threat conditions are "guarded" (a general risk of terrorist attacks), "elevated" (a significant risk of terrorist attacks), and "high" (a high risk of terrorist attacks). Colors were assigned to each threat level: low is indicated by green; guarded by blue; elevated by yellow; high by orange; and severe by red.

On June 6, 2002, President Bush proposed a major reorganization of the federal government, creating a permanent cabinet-level Department of Homeland Security. Drawing on various ideas put forward by Ridge, Congress, and outside studies and commissions, President Bush's plan sought to unify responsibility for protecting against terrorist attacks on American soil. Prominent among the twenty-two federal agencies included in the new department are the Coast Guard, Immigration and Naturalization Service, Border Patrol, Customs Service, Transportation Security Administration, and the Federal Emergency Management Agency. Four divisions within the new department, reflecting its four major responsibilities, are:

- Border and Transportation Security

- Emergency Preparedness and Response

- Science and Technology

- Information Analysis and Infrastructure Protection

H.R. 5710, which was approved by the U.S. Senate and signed into law by President Bush on November 26, 2002, officially established the Department of Homeland Security. Besides the establishment of the new department, the primary highlights of this 484-page document included: reorganization and tighter control of immigration within the United States; a shift of the Bureau of Alcohol, Tobacco, and Firearms from the Treasury Department to the Department of Justice; a call for greater research and development into possible increases in the Homeland Security infrastructure; and a provision of separate funds for the Homeland Security Advanced Research Projects Agency, which would help identify cutting-edge technology to aid the department. The law also called for greater coordination between the government and private sector to increase various critical infrastructures (such as power grids and telecommunication lines) across the country. Security measures strengthening the Coast Guard and airport security, along with the allocation of greater funds for domestic preparedness, were also written into the bill.

Some critics of the new department feared that, with 180,000 employees and a $36 billion budget, it would lack simplicity and flexibility. Others contended that the work of assembling this huge superagency might take away from more urgent actions needed to combat terrorism. That the intended new agency might not pay enough attention to the roles of state and local governments and the private sector caused some concern, as did its potentially insufficient enforcement powers and limited access to raw data from the FBI, Central Intelligence Agency, and National Security Agency, none of which were incorporated into the department. Counterproposals included appointing high-level liaisons to force cooperation among units of the federal government and changing existing agencies to improve their effectiveness at fighting terrorism (for example, establishing a special domestic security group within the FBI like Great Britain's MI5).

Airport/Port Security

After September 11, 2001, the security of air travel became a top concern. Airline travelers soon became accustomed to waiting in longer lines at airport ticket counters, baggage check-ins, and other preflight security checkpoints as more strict attention was paid to checking passenger identification and to other security measures. Congress took the responsibility for airport preflight and security screening away from the airlines and placed it with a new Transportation Security Administration (TSA). The TSA, with an expected workforce of 35,000 to 40,000 employees (including 28,000 passenger and baggage screeners), was predicted to become the largest U.S. government agency created since the 1960s. However, Congress also provided for a process through which airlines might be permitted to go back to previous methods of contracting out screening/security services within three years after the new TSA security inspectors began their duties.

Beginning in the mid-1960s, the United States had experienced a rash of airplane hijackings to Cuba. Conse-

quently, in 1968 the FAA launched a highly secret federal "sky marshal" program. Sky marshals are certified law enforcement officers who ride anonymously on certain air flights. They are allowed to carry firearms. Their primary responsibility is to maintain law and order during the flight. At the time of the September 11 terrorist attacks, thirty-three sky marshals were working within the air transport system, mostly on international flights to and from the United States. After September 11, the Federal Air Marshall Service was made part of the Department of Homeland Security's Immigration and Customs Enforcement agency, and the number of air marshals was increased.

In addition, security changes to airplanes themselves were implemented after the 9/11 hijackings. The FAA ordered temporary reinforcement of cockpit doors. This strengthened doors not only against intrusion but also against penetration by small-arms fire and grenades. The FAA required airlines to install permanent cockpit door improvements by 2003. Airlines have also offered special personal defense training to their pilots and flight attendants.

Much debate ensued over whether pilots should be allowed to carry guns. Many people, including many of the pilots themselves, supported the idea of armed pilots, while others preferred nonlethal weapons such as stun guns, tazer guns, or mace. Still others believed reinforced cockpit doors and specially trained air marshals should be enough to stave off any attack. After weighing the pros and cons of the issue, H.R. 4635, the Arming Pilots Against Terrorism Act, was passed on July 10, 2002. It allows airline pilots to undergo weapons training and carry arms in the cockpit.

Other measures were also implemented to enhance overall transportation security. One of these was the scanning of a small percentage of the thousands of cargo containers that arrive at U.S. seaports each day. The scanning is done to search for explosives, radioactive materials, and biohazards.

The Patriot Act of 2001

The September 11 attacks caused the government to round up suspects vigorously, attempting to increase security at the possible expense of civil liberties. The U.S. Department of Justice, the U.S. Department of the Treasury, the FBI, and other law enforcement agencies proceeded to detain, hold, or deport approximately a thousand people on immigration and other violations. Law enforcement agencies obtained more leeway to wiretap and detain suspects as well.

These authorizations came primarily through the Patriot Act, which was passed by Congress, signed by the president, and enacted on October 26, 2001. The act, which is 342 pages long, made changes, some large and some small, to more than fifteen statutes. The government was given the authority to monitor the online search engine requests of almost any American, obtain a wiretap of a suspected individual's cell or regular phone via one request to a judge, and add DNA samples to a federal DNA database of almost anyone convicted of "any crime of violence."

The Patriot Act also gave the FBI more access to the medical, financial, mental health, and educational records of individuals without having to show evidence of a crime and without a court order. The bill expanded the government's ability to conduct secret searches and permitted the attorney general to detain and incarcerate noncitizens based on suspicion of any act or behavior that might be seen as a threat to national security and to deny readmission to the United States of noncitizens (including lawful permanent residents) under certain conditions. Yet other steps were designed to tighten U.S. immigration practices and keep terrorists out of the country in the first place. "As of May 5, 2004," according to *Report from the Field: The USA Patriot Act at Work* (Washington, DC: Department of Justice, July 2004), "the Department has charged 310 defendants with criminal offenses as a result of terrorism investigations since the attacks of September 11, 2001, and 179 of those defendants have already been convicted."

Some Americans were alarmed that as the federal government moved to deal decisively with terrorist threats, civil liberties were seemingly restricted. Determining the appropriate balance between security precautions and personal freedom is a continuing matter of debate. As the National Commission on Terrorism wrote in its 2001 report to Congress, *Countering the Changing Threat of International Terrorism* (Washington, DC: U.S. Government Printing Office, 2001), "U.S. leaders must find the appropriate balance by adopting counterterrorism policies which are effective but also respect the democratic traditions which are the bedrock of America's strength."

War on Terrorism

After the terrorist attacks of September 11, 2001, President George W. Bush declared a "war on terrorism" with four basic principles. First, no concessions or deals would be made to individuals or groups holding any U.S. citizen hostage. Second, terrorists would be tracked down and brought to justice for their crimes, no matter how long it took. Third, any state that sponsors terrorism would be forced to change its behavior through isolation and applied pressure. Finally, training would be provided under the Antiterrorism Assistance program to strengthen the counterterrorist capabilities of countries working with the United States.

The war on terrorism has been unlike any other in U.S. history. In this war, the U.S. government and its citizens ceased to look upon the Atlantic and Pacific Oceans as shields from attack. For the first time, an American war was being conducted not against a foreign nation but

against transnational enemies—al Qaeda and other terrorist groups operating across international boundaries. These enemies had managed to bring major destruction and devastation, if not conventional war, to America's doorstep.

Even with knowledge of existing terrorist groups and cells, the government had to pin down who, what, or where the enemy may be. Abroad, the president and Congress targeted al Qaeda, the radical Muslim extremist group linked with the 9/11 attacks. Al Qaeda's leader, exiled Saudi Arabian Osama bin Laden, was known to be operating terrorist training camps in Afghanistan. In the aftermath of the September 11 attacks, the president sent thousands of U.S. troops, hundreds of ships and planes, and many bombs and other weapons to Afghanistan. U.S. forces attacked the ruling Taliban party that had harbored bin Laden and other al Qaeda terrorists.

Within a few months, the United States captured taped evidence that they believed proved bin Laden was the mastermind of the September 11 attacks. The tapes showed him gloating over the unexpected degree of his "success"—that is, getting the World Trade Center's twin towers to collapse. In earlier videos released by al Qaeda shortly after the attacks, bin Laden praised the attacks and taunted the American people. Intelligence officials believe he had hoped to foster rebellions in Muslim countries to become a folk hero.

The military actions ordered by President Bush in Afghanistan were generally considered successful. The war, code-named Operation Enduring Freedom, included U.S. use of twelve thousand bombs and missiles, the killing of at least three thousand enemy troops, and the capture of seven thousand or more hostile combatants. Although most of the forces killed or captured in the military effort were only indirectly related to al Qaeda's global terrorist activities, the action succeeded in ending the fundamentalist Taliban's rule over Afghanistan and in eliminating Afghanistan as an official safe haven for al Qaeda.

In 2003 the United States resolved to depose Iraq's leader Saddam Hussein because of his continuing refusal to cooperate with UN weapons inspectors. There was widespread fear that Iraq might pass along deadly WMD materials to terrorist organizations, including al Qaeda, for use against Americans. Following a series of failed diplomatic moves, the United States and a coalition of some thirty allied nations invaded Iraq and removed Hussein from power in Operation Iraqi Freedom. The former dictator was located and arrested in December 2003. While controversy still surrounds the U.S. decision to invade Iraq, little doubt remains that Iraqi support for terrorists is no longer a threat.

AMERICA'S NEW RESOLVE: HOW AMERICANS FELT AFTER SEPTEMBER 11, 2001

Until the September 11, 2001, events in New York City, Washington, D.C., and Pennsylvania, America had

FIGURE 6.9

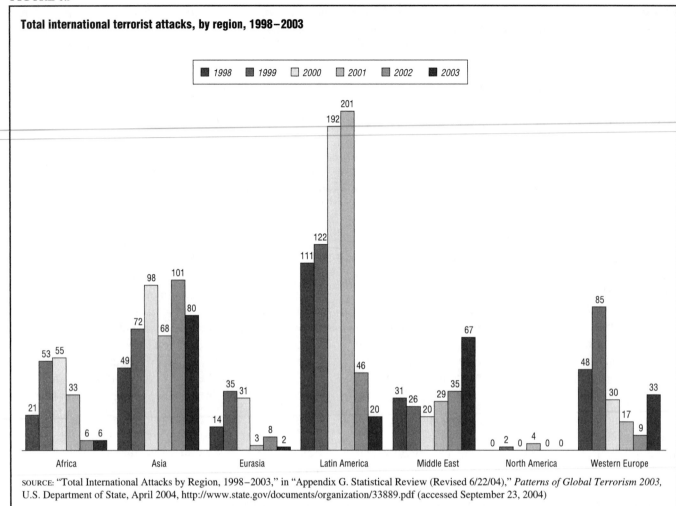

Total international terrorist attacks, by region, 1998–2003

Legend: ■ 1998 ■ 1999 □ 2000 ■ 2001 ■ 2002 ■ 2003

SOURCE: "Total International Attacks by Region, 1998–2003," in "Appendix G. Statistical Review (Revised 6/22/04)," *Patterns of Global Terrorism 2003*, U.S. Department of State, April 2004, http://www.state.gov/documents/organization/33889.pdf (accessed September 23, 2004)

suffered relatively few total terrorist acts, compared with all those committed against both developed and developing nations. Indeed, in the preceding decades, terrorism had become a weapon of choice in domestic, regional, and international disputes. Despite their prevalence in other areas of the world, the number of international terrorist attacks in 2003, or those involving the citizens or territory of more than one country, was lower for North America (0) than any other global region. (See Figure 6.9.) In that year, Asia suffered the most attacks (eighty). Even in 2001, although casualties were high, the total number of terrorist attacks on North America was still relatively low.

In the early 1990s, after the fall of the Soviet Union, U.S. defense planners believed that the principal threats to U.S. personnel and interests would still occur remotely—mostly in Europe, the Persian Gulf, or the Korean Peninsula. After September 11, 2001, however, planners recognized that major threats can enter the country almost as easily as people, goods, and money.

The breakup of the Soviet Union in 1991 had not weakened terrorism. Although it is likely that the Soviet

bloc had provided much aid to terrorist organizations and nations supporting them, after the end of the cold war terrorists simply found other sources of funding. Those sources reportedly include underground banking systems, money acquired through drug trafficking, and laundered money. Osama bin Laden himself, as a former Saudi prince, is thought by intelligence specialists to have great personal wealth (hundreds of millions of dollars). The fall of the Soviet Union also did not limit the breeding grounds of terrorism to the Middle East—there have been increasingly frequent and violent acts of terrorism in many other parts of the world, such as Sri Lanka, South Asia, and the Pacific Rim.

Since the September 2001 attacks, Americans have concluded that threats to the homeland are larger, more complex, more difficult, and more urgent than ever. They understand that it is much harder than they thought to confront terrorist threats and at the same time maintain their values, civil liberties, economic pursuits, and way of life. Still, the citizens and the government of the United States are resolved to do so. Although public opinion has changed over time since September 11, 2001, one thing that has remained constant is the U.S. determination to protect itself from other such catastrophes.

Public Reaction to the September 11, 2001, Terrorist Attacks

IMMEDIATE REACTION. As might be expected, the initial response of the American public to the terrorist attacks was fear, shock, and outrage. A Wirthlin Worldwide National Quorum telephone survey was conducted September 15–17, and the results were reported in *The Wirthlin Report: Current Trends in Public Opinion from Wirthlin Worldwide* (vol. 11, no. 8, September 2001) and its research supplement. The survey found that about three of five Americans (63%) felt the terrorist attacks had "shaken [their] own personal sense of safety and security" either "a great deal" or "a good amount." Only 10% responded that the attacks had not shaken them at all. When asked "What do you believe is the greatest fear of most Americans?," the item most frequently mentioned was terrorist attack. Also mentioned as fears were death/dying (6%), losing freedom (5%), a nuclear strike (3%), and the unknown (2%). Three out of five people surveyed felt they would have to implement changes in their everyday lives in the next five years because of the terrorist threat.

The "greatest fear of most Americans" mentioned second most frequently in the September 15–17 Wirthlin Worldwide survey was involvement in war, but more than 80% of Americans supported the use of military force against those responsible for the attacks. A Harris Poll from September 19–24, 2001, gave similar results: 66% of Americans felt that it would be worse to "fail to take very strong action against those who planned and supported these attacks" than to "take action that kills many innocent [sic] and loses the support of many of those who support us." Many (48%) thought it was impossible to take strong action against the terrorists without killing innocent people (*The Harris Poll #47,* September 26, 2001). In the same survey, 88% of respondents expressed the belief that "many other countries" would provide support to the United States in military action against the terrorists.

In Wirthlin Worldwide's September 15–17 poll, an overwhelming 96% of Americans felt that America would go to war against Afghanistan, but nearly half (45%) believed that military action would lead to more terrorist acts. The support of military action also remained strong in a survey conducted a week later (September 21–26) by Wirthlin Worldwide (*The Wirthlin Report,* vol. 11, no. 9, October 2001, and its research supplement), with 59% of respondents feeling that the death of American troops would not be too high a price to pay in order to respond to the attacks.

Fifty-seven percent of respondents felt President George W. Bush was dealing with the aftermath of Sep-

tember 11 well, according to the Wirthlin Worldwide poll of September 15–17; this figure jumped to 87% in the Wirthlin Worldwide survey conducted September 21–26. The September 19–24 Harris Poll (*The Harris Poll #48,* September 27, 2001) provided slightly higher numbers: 90% of the poll respondents said President Bush was doing either an excellent or a pretty good job.

Asked in a September 19–24 Harris Poll why America was attacked and why terrorists hate the United States, about one-quarter of respondents (26%) felt the most important reason was American democracy and freedom; only slightly smaller percentages thought the main reasons were U.S. support for Israel (22%) and "our values and way of life" (20%).

ONE TO THREE MONTHS LATER. Wirthlin Worldwide conducted telephone surveys over the periods October 19–22, 2001, and November 2–5, 2001, and asked respondents about some of the same issues they had covered in their September surveys (*The Wirthlin Report,* vol. 12, no. 2, March-April 2002). In these later surveys, support for military action to help stem terrorist activity remained strong. About three-quarters (74%) of Americans felt that the war on terrorism would be long and tough, with ramifications extending beyond Afghanistan. Respondents also provided answers to the question, "What must be accomplished for you to conclude that the war on terrorism has been won?" The most frequent responses to that question, given by 31% of those polled, fell into the general category of capturing/eliminating terrorists. Other respondents felt that the war on terrorism would never end (17%) or that it would not be over until there was peace or no more terrorism (14%).

The Harris Poll conducted its own research on Americans' feelings about, and understanding of, the war on terrorism in a November 14–20, 2001, poll and reported its findings in *The Harris Poll #58* (November 24, 2001) and *The Harris Poll #60* (December 5, 2001). According to the mid-November Harris survey, 86% of Americans felt very confident or somewhat confident that the United States had "a clear plan for winning the war on terrorism." Only 14% of respondents felt "not very" or "not at all" confident about the U.S. strategy. About three of five respondents (61%) felt that the American government had clearly defined what it would mean to win the war on terrorism.

The same November Harris Poll found that while nearly half of all Americans (47%) remained at least somewhat anxious about their personal safety, slightly more than half (52%) were either not very anxious or not at all anxious. Two of five respondents (42%) said they were more concerned about their safety than before September 11.

LATER POLLS. American feelings remained strong about the national security ramifications of the September 11 attacks through mid-2002. In a Harris Poll from January 24–30, an overwhelming majority of Americans (93%)

supported continuing the war on terrorism (*The Harris Poll #7*, February 6, 2002). A February 4–6, 2002, Gallup Poll found that 44% of Americans believed terrorism or international issues was America's largest problem. This number had dropped since October 2001 but remained relatively constant in December 2001, January 2002, and February 2002. Only 35% of respondents to the same Gallup Poll said they were at least somewhat worried that they or a family member would fall prey to a terrorist act.

Data from 2002 also defined who America felt its enemies were and how prepared the country was to deal with them. Gallup Polls from March 1–3, 2002, demonstrated that a large majority of Americans (78%) believed Muslim countries were generally unfavorable to the United States, as opposed to only 29% of respondents who felt the non-Muslim world was unfavorable to the United States. A February 4–6 Gallup Poll revealed that half of Americans (50%) believed that the strength of "the national defense is . . . about right at the present time," while a similar percentage (48%) felt the right amount of money was being spent for military purposes.

Data from *The Harris Poll #16* (April 3, 2002) compared the views of Americans on various increases in law enforcement powers from September 2001 and March 2002. While support for all of the proposals mentioned dropped slightly over that period, support for items such as expanded undercover activities and stronger security checks for travelers remained high, at 88% and 89% respectively. The least popular option, opposed by 51% of the March 2002 respondents, was expanded cell phone and e-mail monitoring by the U.S. government.

A Harris Poll conducted August 20, 2004, showed that many Americans felt "misled by the government's statements about Iraq's weapons of mass destruction and Iraq's links to Al Qaeda." Sixty percent of respondents believed that what the government had said about Iraq's WMD was misleading. However, most (51%) believed that the statements made about WMD had tried to be accurate. Asked whether they believed the war against Iraq had strengthened the war on terror, 50% believed it had while 40% believed it had not.

CHAPTER 7
DOMESTIC TERRORISM

The Federal Bureau of Investigation (FBI) divides terrorism into two distinct types: international terrorism and domestic terrorism. The FBI defines international terrorism as "the unlawful use of force or violence committed by a group or individual, who has some connection to a foreign power or whose activities transcend national boundaries, against persons or property to intimidate or coerce a government, the civilian population, or any segment thereof, in furtherance of political or social objectives." These incidents may take place within the United States or may involve U.S. citizens or interests overseas. Domestic terrorism, however, is defined by the FBI as terrorism that "involves groups or individuals who are based and operate entirely within the United States and Puerto Rico without foreign direction and whose acts are directed at elements of the U.S. government or population."

NOTABLE INCIDENTS OF DOMESTIC TERRORISM

Domestic terrorism is not new to the United States. In 1920 the financial district of New York City was a terrorist target—a massive bomb killed thirty people. An investigation centered on Sicilian, Romanian, and Russian terror groups, but the case was never solved. More than eighty years later, scars from the bombs can still be seen on buildings in New York's financial district.

In 1954 four armed, pro-independence Puerto Rican terrorists started shooting guns from the visitors' gallery of the U.S. House of Representatives. Five Congressmen were wounded.

Bombings

16th STREET BAPTIST CHURCH, BIRMINGHAM, AL, SEPTEMBER 1963. In 1963 the bombing of the 16th Street Baptist Church in Birmingham, Alabama, killed four female African-American children. Almost thirty years later, the case was finally closed when, on May 22, 2002, former Ku Klux Klan member Bobby Frank Cherry, age

seventy-one, was convicted of four counts of murder. Cherry, who was trained in demolitions in the U.S. Army, claimed during the trial that he could not have planted the bomb the night before the attack because he was at home watching wrestling on television with his cancer-stricken wife. Prosecutors were able to show not only that no wrestling was on television that night but also that Cherry's wife was not diagnosed with cancer until two years after the bombing. Thomas E. Blanton, an accomplice in the bombing, was convicted in 2001 and sentenced to life in prison. A third accomplice, Robert Chambliss, was convicted in 1977 and later died in prison.

THE WEATHER UNDERGROUND, 1960s AND 1970s. In the late 1960s and early 1970s, the radical left-wing Weather Underground, a splinter group of the Students for a Democratic Society (SDS), carried out some twenty-five bombings across the country. Among their targets were the New York City Police Headquarters in June 1970, the U.S. Capitol Building in March 1971, the Pentagon in May 1972, and the U.S. State Department in January 1975. In March 1970 an explosion ripped through a Manhattan townhouse where members of the group were making bombs, killing Theodore Gold, Diana Oughton, and Terry Robbins, all members of the Weather Underground. The bombs being made were antipersonnel weapons loaded with shrapnel. By the late 1970s the Weather Underground had turned to bank robberies to finance its operations. Along with members of the separatist Black Liberation Army (BLA), it was involved in the October 1981 robbery of a Brinks armored car in Nyack, New York, in which two policemen and a Brinks security guard were shot dead. Among those convicted of the robbery were Weather Underground members Kathy Boudin, who served twenty-two years in prison; Judith Clark, sentenced to seventy-five years in prison; and BLA member Donald Weems, who was sentenced to life in prison, where he died of AIDS in 1986. In 2003 the Nyack post office was renamed in honor of those killed in the robbery.

ALFRED P. MURRAH FEDERAL BUILDING, OKLAHOMA CITY, OK, APRIL 1995. On April 19, 1995, a two-ton truck bomb exploded just outside the Alfred P. Murrah federal building in Oklahoma City, Oklahoma, killing 168 people and injuring 518. Because a day care center was in the building very near the site of the explosion, many of the victims were children. Rescue workers searched for bodies in the rubble for almost two weeks after the blast. There was an enormous outpouring of grief for, and assistance to, bombing victims and their families. Oklahoma City residents and others aided the rescue workers and made monetary donations to assist the victims and their families. Several years later, a huge memorial was erected at the site of the bombing in honor of the victims.

Federal authorities arrested Timothy McVeigh for the crime. McVeigh, a disgruntled former army member who was rumored to be associated with an antigovernment militia group, evidently set the bomb in retaliation for the FBI's handling of the Branch Davidian cult standoff in Waco, Texas, in 1994, which resulted in the deaths of over eighty men, women, and children. McVeigh's bombing of the Murrah building on April 19 coincided with the date in 1993 that the Branch Davidian compound had been destroyed in a federal raid. He was convicted of the bombing and then executed by lethal injection on June 9, 2001. The government allowed the families of the victims to watch McVeigh's execution on closed-circuit television in the federal prison in which he died. Also arrested was McVeigh's accomplice, Terry L. Nichols. In 1998 he was convicted and sentenced to life for the deaths of eight law enforcement officers killed in the blast. In August 2004 Nichols received 161 life terms for the deaths of other victims.

CENTENNIAL PARK, OLYMPIC GAMES, ATLANTA, GA, JULY 1996. While the toll in lives and property damage was much lower than in the Oklahoma City bombing, a bombing in July at the 1996 Olympic Games in Atlanta, Georgia, created international alarm. When a nail-packed pipe bomb exploded in a large common area, one person was killed and more than a hundred were injured. Authorities believed the perpetrator might have been affiliated with a Christian Identity group, a militant white supremacist organization.

Shortly after the attack, suspicion centered on a security guard at Centennial Park, where the blast occurred, but he was later cleared and given an official apology. In May 1998 the FBI added Eric Robert Rudolph to its Top Ten Most Wanted list, seeking him for questioning about the Olympics bombing and two later incidents. Rudolph was also charged with bombing an abortion clinic in Birmingham, Alabama, in January 1998. In that blast, an off-duty police officer was killed and a nurse was seriously injured. The FBI, the Bureau of Alcohol, Tobacco, and Firearms, and the Birmingham Police Department offered a $1 million reward. Rudolph was captured in December 2003 in North Carolina, where he had been hiding in the rugged Nantahala National Forest.

THE UNABOMBER. Over a seventeen-year period, an individual nicknamed the "Unabomber" committed sixteen bombings in several states. Three people were killed and twenty-three injured in the attacks. After reading a fifty-six-page manuscript supposedly written by the Unabomber and published in the *New York Times* and the *Washington Post* newspapers in 1995, David Kaczynski contacted the FBI and shared his fears that his brother, Theodore, might be the Unabomber. The manhunt for Theodore Kaczynski was one of the longest and most difficult in U.S. history, involving hundreds of federal and state law enforcement agents. Kaczynski was later captured and pled guilty at his trial. Although he claimed the bombings (usually letter bombs) were directed against the U.S. federal government, the victims were generally not directly related to the government. In January 1998 Kaczynski was sentenced to life imprisonment, with no possibility of parole, for his actions as the Unabomber.

MAILBOX BOMBER, LUKE J. HELDER. Beginning on Friday, May 3, 2002, eighteen pipe bombs were placed in rural mailboxes in Illinois, Iowa, Nebraska, Colorado, and Texas, injuring five people. On Tuesday, May 7, 2002, the FBI arrested twenty-one-year-old college student Luke J. Helder in connection with the bombings. Helder was charged by federal prosecutors in Iowa with the use of an explosive device to maliciously destroy property affecting interstate commerce and with the use of a destructive device to commit a crime of violence, punishable by up to life imprisonment. The pipe bombs, some of which did not detonate, were accompanied by letters warning of excessive government control over individual behavior. In April 2004 Helder was deemed to be incompetent to stand trial.

Anthrax Attacks

Anthrax, classified by the U.S. government as a potential weapon of mass destruction, is a bacterial disease spread through spores. The spores can live in soil or the wool or hair of diseased animals. Humans acquire the disease when the spores are inhaled or ingested. Ulcerous sores on the skin and lesions on the lungs are symptomatic of the disease. While potentially deadly if it spreads to the lungs, anthrax is treatable if identified early.

Terrorist attacks using anthrax occurred in the autumn of 2001. Anthrax-tainted letters were sent through the U.S. postal system in the first major bioterrorist attack against the U.S. homeland. No link was demonstrated between the terrorist attacks of September 11, 2001, and the anthrax attacks, but the anthrax attacks did prove that terrorists could use the U.S. Postal Service (USPS) to unleash germ warfare against American citizens, news organizations, and congressional representatives.

Lethal anthrax bacilli infected the skin or the lungs of personnel at various offices, all of which had received letters containing a suspicious white powder: the *Sun* tabloid newspaper in Boca Raton, Florida; the headquarters of NBC News in New York's Rockefeller Center; the New York headquarters of CBS News; the offices of the *New York Post* in New York; the congressional offices in Washington, D.C., of Senator Tom Daschle of South Dakota and Senator Patrick Leahy of Vermont; and facilities of the Microsoft Corporation in Nevada. The anthrax also appeared at several USPS processing facilities and at several outlying mail-sorting centers for federal government agencies such as the State Department and the Department of Defense.

Twenty-two persons developed anthrax. Five died from it: two postal workers in Washington, D.C., a Florida newspaper editor, an elderly Connecticut woman, and a New York hospital worker. The government ordered thousands more people, mostly postal workers, to take the antibiotic Cipro as a precautionary measure. Nine months later, health specialists estimated that the tainted letters may have cross-infected as many as five thousand other pieces of mail.

The anthrax arrived in letters that contained a message referring to Allah (the name for God in Islam), and the message seemed to imply an association with Islamic terrorism. However, it became increasingly clear that a single person located within the United States could have packed the letters with anthrax spores. The attacks were not necessarily the work of a group, much less an Islamic terrorist group. The spores used had been highly "weaponized," or finely milled to diameters of between one and three microns. This technical feat ensured their maximum dispersal when the envelopes were opened or even as they shuttled from post office to post office. The level of sophistication in this refinement of the anthrax implied that a highly skilled scientist or technician within the U.S. military's own bioweapons research and testing program could have been responsible. As a result, although the perpetrators were still unknown as of late 2004, the anthrax attacks are generally considered domestic terrorism.

Another effect of the anthrax attacks was the slew of threats and hoaxes that followed. In late October 2001 a USPS employee, Sharon Ann Watson of Stafford, Virginia, was arrested on charges of perpetrating an anthrax hoax at the Falmouth, Virginia, post office where she worked. She was charged with knowingly mailing threatening communications and unlawful delay or destruction of mail. Each offense carried penalties of up to twenty years in prison.

By November 2001 a total of 353 postal facilities had been evacuated for varying amounts of time as a result of

8,674 hoaxes, threats, and suspicious mailing incidents, which averaged 578 per day. Postal inspectors had arrested twenty people for anthrax-related hoaxes, threats, and suspicious mailing incidents and continued to investigate eighteen additional incidents. A reward of up to $2.5 million for information leading to the arrest and conviction of anyone mailing anthrax resulted in 165 investigative leads. The attacks caused an expensive, difficult logjam in mail delivery that forced the U.S. government to buy multimillion-dollar machines to irradiate all mail in an attempt to kill any dangerous bacteria it might contain.

THE INCIDENCE OF DOMESTIC TERRORIST ATTACKS AND CASUALTIES IN THE UNITED STATES

From 1980 to 2001, the FBI recorded 482 incidents or planned incidents of terrorism within the United States. (See Figure 7.1.) According to *Terrorism: 2000–2001*, (Washington, DC: Federal Bureau of Investigation, 2002), these incidents killed 2,993 people and injured 14,047. Of the 482 incidents committed in the United States, 164 were committed by international terrorist groups, 130 by domestic left-wing groups, eighty-five by domestic right-wing groups, and eighty-one by domestic special-interest groups. (See Figure 7.2.) By region, terrorist attacks were most common in the Northeast (144), Puerto Rico (103), and the West (97). (See Figure 7.3.) Table 7.1 is a chronological summary of all domestic terror incidents from 1990 to 2001.

MOTIVATIONS AND TRENDS

Domestic terrorism has been driven by various motivations ranging across the political spectrum. Special-interest groups have undertaken terrorist attacks on U.S. soil as well.

Left-Wing Organizations

According to the FBI's *Terrorism in the United States 1999,* terrorist groups on the extreme left tend to "profess a revolutionary socialist doctrine and view themselves as protectors of the people against dehumanizing effects of capitalism and imperialism." These groups were more prominent during the days of the cold war between the United States and the Soviet Union, and they carried out a number of bombings and robberies from the 1960s to the 1980s. From 1980 to 1985, the FBI attributed eighty-six of the 184 recorded terrorist attacks to left-wing extremists. The fall of the Soviet Union and a global shift away from communist ideologies greatly affected the motivations and capabilities of such groups.

Some left-wing groups have been fighting for the independence of Puerto Rico. Groups such as the Popular Puerto Rican Army often employ violent means in their attempts to secure full Puerto Rican independence from

FIGURE 7.1

Terrorism in the United States, 1980–2001

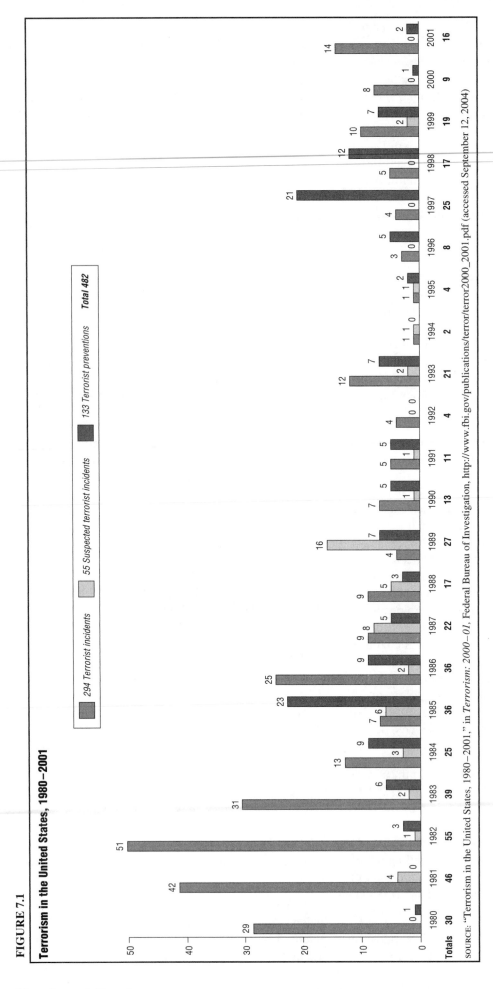

SOURCE: "Terrorism in the United States, 1980–2001," in *Terrorism: 2000–01*, Federal Bureau of Investigation, http://www.fbi.gov/publications/terror/terror2000_2001.pdf (accessed September 12, 2004)

FIGURE 7.2

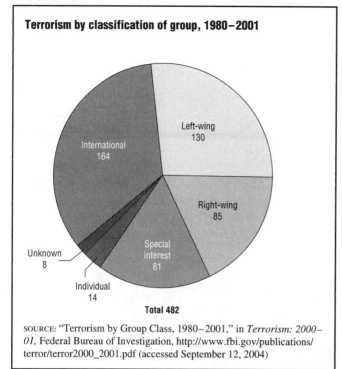

Terrorism by classification of group, 1980–2001

- International 164
- Left-wing 130
- Right-wing 85
- Special interest 81
- Unknown 8
- Individual 14

Total 482

SOURCE: "Terrorism by Group Class, 1980–2001," in *Terrorism: 2000–01,* Federal Bureau of Investigation, http://www.fbi.gov/publications/terror/terror2000_2001.pdf (accessed September 12, 2004)

the United States. In 1998 three of the five recorded acts of terrorism within the United States and its territories occurred in Puerto Rico and were attributed by the FBI to the Popular Puerto Rican Army. Groups fighting for Puerto Rican independence were more active during the 1980s and carried out several bombings and violent attacks.

Other types of left-wing groups include anarchists and social extremists, whose causes vary but remain political and anti-establishment. They operate in groups or as individuals. Such groups were responsible for extensive damage during riots in Seattle, Washington, in 1999, during demonstrations against the World Trade Organization ministerial meeting.

Right-Wing Organizations

Right-wing groups tend to regard the U.S. government as oppressive or unjust. Often, such groups believe in racial supremacy and refuse to follow any rules and regulations set forth by the government. The origins of some of these groups can be traced back to the nineteenth century. The widespread poverty and destitution in the Southern states after their defeat in the Civil War (1861–65), combined with attitudes of racial superiority, created an atmosphere that gave birth to such organizations as the Ku Klux Klan.

Contemporary right-wing extremists have toned down their rhetoric in order to attract a larger audience. Members of the extreme right often adhere to one or more of the following beliefs:

- Christian Identity adherents believe that Americans of white European descent are descendants of the ten lost

tribes of Israel, that the Aryan race is God's chosen race, and that whites will defeat Jews and nonwhites during the Second Coming of Christ.

- White supremacists call for the supremacy of the white race above all others; extreme members of such organizations also believe that a special homeland should be established to maintain the purity of the white race.

- Militias are armed paramilitary groups that strongly believe the U.S. government is out to destroy them. They preach elaborate conspiracy theories—for example, that the U.S. government is merely a cog in a "new world order" run by the United Nations (UN).

- Patriot Movement members consider themselves to be true patriots who disagree with how the government currently functions and refuse to adhere to any federal, state, or local laws. Many have racist ideologies. According to the Intelligence Project of the Southern Poverty Law Center (SPLC), 171 Patriot groups operate within the United States, forty-five of them militia groups (although not all of these groups advocate violence). (See Table 7.2.)

- Tax Protest Movement members believe that tax laws are incorrectly interpreted and that paying federal income tax should be voluntary.

The FBI claims that since about the mid-1990s there has been a rise in grass-roots patriot and militia movements that profess antigovernment sentiments and global conspiracy theories. This rise is the result of the increasing prominence of the UN, growing U.S. involvement around the world, the passage of increased gun-control legislation, and recent confrontations between militias and the law enforcement community. These groups present a unique threat to the federal government because they often stockpile weapons and refuse to acknowledge any law enforcement above the level of the county sheriff. Many also lack a cohesive organizational structure and an overall leader or headquarters, making these small, tightly knit groups hard to infiltrate or monitor.

The increase in activity by right-wing groups beginning in the 1990s was partly caused by a shift in tactics away from hierarchical organizations to what is termed "leaderless resistance." Using small cells of only a few members who commit acts of resistance, this strategy makes such groups more difficult for law enforcement to infiltrate. First popularized by "The Order," a right-wing group involved in armored-car robberies and the murder of a Jewish radio personality, "leaderless resistance" was also promoted in the 1978 novel *The Turner Diaries,* written by National Alliance founder William L. Pierce and an underground best-seller in far-right circles.

FIGURE 7.3

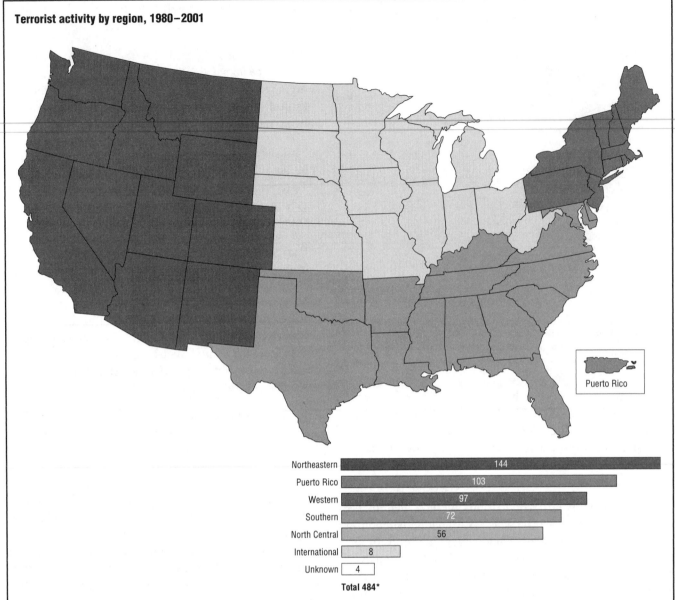

Terrorist activity by region, 1980–2001

Region	Count
Northeastern	144
Puerto Rico	103
Western	97
Southern	72
North Central	56
International	8
Unknown	4

Total 484*

*Although designated as a single act of international terrorism, the aircraft attacks of September 11, 2001, have been designated as one terrorist incident in the Northeastern region and one terrorist incident in the Southern region for the purposes of this graph. Similarly, although the anthrax mailings that occurred from September through November 2001 have been categorized as a single act of terrorism, the incidents have been designated as one terrorist incident in the Northeastern region and one terrorist incident in the Southern region for the purposes of this graph.

SOURCE: "Terrorist Activity by Region, 1980–2001," in *Terrorism: 2000–01,* Federal Bureau of Investigation, http://www.fbi.gov/publications/terror/terror2000_2001.pdf (accessed September 12, 2004)

Special-Interest and Single-Issue Terrorism

In addition to left-wing and right-wing groups, a variety of special-interest groups, such as animal liberation groups, environmentalists, antiabortion activists, and black separatists, have committed acts of terror to draw attention to one specific cause. They have carried out such destructive acts as arson, bombings, and even anthrax hoaxes in the past. These groups frequently use media outlets such as the Internet to disseminate their ideologies and recruit members.

ECOTERRORISM. The underground Earth Liberation Front (ELF) and Animal Liberation Front (ALF) are the leading groups that have engaged in ecoterrorism since the mid-1990s. In testimony before the House Resources Committee, Subcommittee on Forests and Forest Health, in February 2002, James F. Jarboe, the FBI domestic terrorism section chief, defined ecoterrorism as "the use or threatened use of violence of a criminal nature against innocent victims or property by an environmentally oriented, subnational group for environmental-political reasons, or aimed at an audience beyond the target, often of a symbolic nature." These groups are composed of radical environmentalists and are primarily active across North America. ELF, for example, describes itself in a

TABLE 7.1

Chronological summary of terrorist incidents in the United States, 1990–2001

Date	Location	Incident type	Group
1/12/90	Santurce, PR	Pipe bombing	*Brigodo Internocionolisto Eugenio Moria de Hostos de los Fuerzos Revolucionories Pedro Albizu Compos* (Eugenio Maria de Hostos International Brigade of the Pedro Albizu Campos Revolutionary Forces)
1/12/90	Carolina, PR	Pipe bombing	*Brigodo Internocionolisto Eugenio Moria de Hostos de los Fuerzos Revolucionories Pedro Albizu Compos* (Eugenio Maria de Hostos International Brigade of the Pedro Albizu Campos Revolutionary Forces)
2/22/90	Los Angeles, CA	Bombing	Up the IRS, Inc.
4/22/90	Santa Cruz County, CA	Malicious destruction of property	Earth Night Action Group
5/27/90	Mayaguez, PR	Arson	Unknown Puerto Rican group
9/17/90	Arecibo, PR	Bombing	Pedro Albizu Group Revolutionary Forces
9/17/90	Vega Baja, PR	Bombing	Pedro Albizu Group Revolutionary Forces
2/3/91	Mayaguez, PR	Arson	Popular Liberation Army (PLA)
2/18/91	Sabana Grande, PR	Arson	Popular Liberation Army (PLA)
3/17/91	Carolina, PR	Arson	Unknown Puerto Rican group
4/1/91	Fresno, CA	Bombing	Popular Liberation Army (PLA)
7/6/91	Punta Borinquen, PR	Bombing	Popular Liberation Army (PLA)
4/5/92	New York, NY	Hostile takeover	*Mujohedin-E-Kholq* (MEK)
11/19/92	Urbana, IL	Attempted firebombing	Mexican Revolutionary Movement
12/10/92	Chicago, IL	Car fire and attempted firebombing	Boricua Revolutionary Front (two incidents)
2/26/93	New York, NY	Car bombing	International Radical Terrorists
7/20/93	Tacoma, WA	Pipe bombing	American Front Skinheads
7/22/93	Tacoma, WA	Bombing	American Front Skinheads
11/27–28/93	Chicago, IL	Firebombing	Animal Liberation Front (nine incidents)
3/1/94	New York, NY	Shooting	Rashid Najib Baz convicted on November 30, 1994
4/19/95	Oklahoma City, OK	Truck bombing	Timothy McVeigh and Terry Nichols convicted. (Michael Fortier found guilty of failing to alert authorities of plot)
4/1/96	Spokane, WA	Pipe bombing/bank robbery	Spokane Bank Robbers
7/12/96	Spokane, WA	Pipe bombing/bank robbery	Spokane Bank Robbers
7/27/96	Atlanta, GA	Pipe bombing	Eric Robert Rudolph charged on October 14, 1998
1/2/97	Washington, DC	Letter bomb (counted as one incident)	Pending investigation No claim of responsibility
1/2/97	Leavenworth, KS	Letter bomb (counted as one incident)	Pending investigation No claim of responsibility
1/16/97	Atlanta, GA	Bombing of abortion clinic	Eric Robert Rudolph charged on October 14, 1998
2/21/97	Atlanta, GA	Bombing of alternative lifestyle nightclub	Eric Robert Rudolph charged on October 14, 1998
1/29/98	Birmingham, AL	Bombing of reproductive services clinic	Eric Robert Rudolph charged with the bombing on February 14, 1998
3/31/98	Arecibo, PR	Bombing of superaqueduct construction project	Claim of responsibility issued by *Los Macheteros*
6/9/98	Rio Piedras, PR	Bombing of bank branch office	Claim of responsibility issued by *Los Macheteros*
6/25/98	Santa Isabel, PR	Bombing of bank branch office	*Los Macheteros* suspected
10/19/98	Vail, CO	Arson fire at ski resort	Claim of responsibility issued by Earth Liberation Front
3/27/99	Franklin Township, NJ	Bombing of circus vehicles	Claim of responsibility issued by Animal Liberation Front
4/5/99	Minneapolis, St. Paul, MN	Malicious destruction and theft	Animal Liberation Front
5/9/99	Eugene, OR	Bombing	Animal Liberation Front
7/2–4/99	Chicago Skokie, IL Northbrook, Bloomington, IN	Multiple shootings	Benjamin Nathaniel Smith
8/10/99	Granada Hills, CA	Multiple shootings	Buford O'Neal Furrow
8/28–29/99	Orange, CA	Malicious destruction and theft	Claim of responsibility issued by Animal Liberation Front
10/24/99	Bellingham, WA	Malicious destruction and theft	Claim of responsibility issued by Animal Liberation Front
11/20/99	Puyallup, WA	Malicious destruction	Animal Liberation Front
12/25/99	Monmouth, OR	Arson	Claim of responsibility issued by Earth Liberation Front
12/31/99	East Lansing, MI	Arson	Claim of responsibility issued by Earth Liberation Front
1/3/00	Petaluma, CA	Incendiary attack	Animal Liberation Front
1/15/00	Petaluma, CA	Incendiary attack	Animal Liberation Front
1/22/00	Bloomington, IN	Arson	Earth Liberation Front
5/7/00	Olympia, WA	Arson	Revenge of the Trees
7/2/00	North Vernon, IN	Arson	Animal Liberation Front
7/20/00	Rhinelander, WI	Vandalism	Earth Liberation Front
12/1/00	Phoenix, AZ	Multiple arsons	Mark Warren Sands
12/9–30/00	Suffolk County, Long Island, NY	Multiple arsons	Earth Liberation Front
1/2/01	Glendale, OR	Arson	Earth Liberation Front
2/20/01	Visalia, CA	Arson	Earth Liberation Front
3/9/01	Culpeper, VA	Tree spiking	Earth Liberation Front
3/30/01	Eugene, OR	Arson	Earth Liberation Front
4/15/01	Portland, OR	Arson	Earth Liberation Front

statement posted on their Web site (http://www.earthliberationfront.com/) as "an international underground organization that uses direct action in the form of economic sabotage to stop the exploitation and destruction of the natural environment." Most ELF members believe in a form of deep ecology, or the theory that all nonhuman life has an intrinsic value and must be protected from humanity.

TABLE 7.1

Chronological summary of terrorist incidents in the United States, 1990–2001 [CONTINUED]

Date	Location	Incident type	Group
5/17/01	Harrisburg, PA	Bank robbery	Clayton Lee Waagner
5/21/01	Seattle, WA	Arson	Earth Liberation Front
5/21/01	Clatskanie, OR	Arson	Earth Liberation Front
7/24/01	Stateline, NV	Destruction of property	Earth Liberation Front
9/11/01	New York, NY	Aircraft Attack	*Al-Qaeda*
	Arlington, VA		
	Stony Creek, PA		
Fall 2001	New York, NY	*Bacillus anthrocis* mailings	Pending investigation
	Washington, DC		No claim of responsibility
	Lantana, FL		
10/14/01	Litchfield, CA	Arson	Earth Liberation Front
11/9/01	Morgantown, WV	Bank robbery	Clayton Lee Waagner
11/12/01	San Diego, CA	Burglary and vandalism	Animal Liberation Front

SOURCE: "Chronological Summary of Incidents in the United States, 1990–2001," in *Terrorism: 2000–01,* Federal Bureau of Investigation, http://www.fbi.gov/publications/terror/terror2000_2001.pdf (accessed September 12, 2004)

Both ALF and ELF were created without a hierarchical and centralized structure so that various subgroups and individuals are able to carry out actions under the umbrella of a larger group. In a statement made before the U.S. Senate in May 2001, FBI director Louis J. Freeh labeled ALF "one of the most active extremist elements in the United States." ALF and ELF have committed some six hundred criminal acts since 1996, according to the FBI. Their actions—including arson, vandalism, and bombings—resulted in some $43 million in damages between 1996 and 2002, while in 2003 alone, ecoterrorist damage estimates attributed to ELF and ALF surpassed $50 million. The FBI reports that there has been over $200 million in damages from all ecoterrorist incidents since the late 1980s.

Ecoterrorists have taken action against various targets they believe endanger the earth's environment in some way, including country clubs, ski resorts, oil companies, multinational corporations, research institutes involved in genetic modification, animal laboratories, lumber yards, and various U.S. government agencies. Their tactics have ranged from tree-spiking (inserting spikes in trees to damage saws) and sabotage to arson and firebombing. In October 1998, during a single attack on a ski resort in Vail, Colorado, members of ELF caused approximately $12 million worth of damage. On October 30, 2001, several members of ELF firebombed a wild-horse corral in California that belonged to the U.S. Bureau of Land Management. In August 2003 ELF took responsibility for burning a five-story apartment building under construction in San Diego, California, causing some $50 million in damages. In August 2003 a group calling itself the Animal Liberation Brigade–Revolutionary Cells bombed the Chiron Corp., a biotechnology firm in Emeryville, California. On September 26, 2003, the same group set off a bomb at Shaklee Corp. in Pleasanton, California.

ANTIABORTION ACTIVISM. Another cause that falls in the special-interest category is the antiabortion movement in the United States. Acts of violence against, and murders of,

health care professionals involved in providing abortions rose rapidly in the 1980s and 1990s. Individuals and groups pursuing such activities belong to a larger pro-life movement in the United States that believes the rights of unborn children must be protected. Though most members of the pro-life movement do not support killing medical professionals, a fundamentalist segment of the group strongly believes that killing abortion providers is the only way to protect the unborn.

These groups have no overall structural organization. Individuals sharing similar beliefs network primarily through pamphlets and the Internet. Some Web sites even list names of abortion providers in the United States. Law enforcement officials believe these lists provide "hit lists" for individuals who wish to kill abortion providers. Besides targeting medical professionals, antiabortion groups also commit arson, bombings, blockades (so that workers and patients cannot get into clinics), and anthrax hoaxes.

In response to increasing acts of violence committed against abortion providers and their clinics, Congress enacted the Freedom of Access to Clinical Entrances Act (FACE) in 1994. The legislation specified federal criminal penalties against any individual obstructing, harassing, or acting violently against abortion providers or recipients. Furthermore, in response to the 1998 murder of Dr. Barnett Slepian, a reproductive health care provider in New York, then–Attorney General Janet Reno established a Task Force on Violence against Health Care Providers. Falling under the auspices of the U.S. Department of Justice, the task force is headed by the assistant attorney general for the Civil Rights Division. It is staffed by lawyers and other personnel from the Civil Rights and Criminal Divisions of the Department of Justice, as well as investigators from the FBI, Bureau of Alcohol, Tobacco, and Firearms, U.S. Postal Inspection Service, and U.S. Marshals Service.

Black Separatism

The two largest black separatist groups in America are the Nation of Islam (NOI) and the New Black Panther Party for

TABLE 7.2

Active patriot groups, 2003

Alabama

Alabama Committee
Birmingham

Constitution Party
Holly Pond

Alaska

Constitution Party
Anchorage

Jefferson Party
Anchorage

Arizona

Constitution Party
Chandler

Ranch Rescue
Douglas

American Patriot Friends Network
Glendale

Civil Homeland Defense
Tombstone

Arkansas

Militia of Washington County
Feyetteville

Constitution Party
Little Rock

California

California Militia
Brea

John Birch Society
Brea

State Citizens Service Center
Research Headquarters
Canoga Park

Truth Radio
Delano

John Birch Society
Fountain Valley

Free Enterprise Society
Fresno

Second Amendment
Committee
Hanford

John Birch Society
Irvine

John Birch Society
Laguna Hills

John Birch Society
Mission Viejo

John Birch Society
Newport Beach

John Birch Society
Oceanside

Southern California High
Desert Militia
Oceanside

John Birch Society
Orange

Freedom Law School
Phelan

John Birch Society
Santa Ana

Truth in Taxation
Studio City

American Independent Party
Torrance

Colorado

Ranch Rescue
Boulder

American Freedom Network
Johnstown

District of Columbia

American Free Press
Washington

Florida

Citizens for Better Government
Gainesville

Constitution Party
Jupiter

Georgia

Militia of Georgia
Lawrenceville

Constitution Party
Woodstock

Idaho

Constitution Party
Boise

Sons of Liberty
Boise

Police & Military Against
the New World Order
Kamiah

Constitution Party
Post Falls

Illinois

Southern Illinois Patriots League
Benton

Constitution Party
Springfield

Indiana

Indiana Citizens Volunteer
Militia 2nd Brigade
Allen County

Old Paths Baptist Church
Campbellsburg

Indiana Citizens Volunteer
Militia 6th Brigade
Columbus

NORFED
Evansville

Indiana Citizens Volunteer
Militia 4th Brigade
Indianapolis

Indianapolis Baptist Temple
Indianapolis

Indiana Citizens Volunteer
Militia 1st Brigade
Lake County

Indiana Militia Corps 2nd Brigade
Northeastern Indiana

Indiana Militia Corps 1st Brigade
Northeastern Indiana

Indiana State Militia
14th Regiment
Owen County

Indiana Militia Corps
5th Brigade
Pendleton

Indiana Citizens Volunteer
Militia 7th Brigade
Perry County

Indiana Citizens Volunteer
Militia 5th Brigade
Putnam County

Indiana Militia Corps
4th Brigade
Southeastern Indiana

Indiana Militia Corps
3rd Brigade
Soutwestern Indiana

Indiana Citizens Volunteer
Militia 3rd Brigade
Tippeacanoe County

Iowa

Constitution Party
Randall

Kansas

Constitution Party
Wichita

Kentucky

Kentucky State Militia 5th
Battalion
Central Kentucky

Take Back Kentucky
Clarkson

Free Kentucky
Lebanon

Constitution Party
Nicholasville

Louisiana

John Birch Society
New Orleans

Maryland

Southern Sons of Liberty

Constitution Party
Pasadena

Save a Patriot Fellowship
Westminster

Michigan

Michigan Militia Corps
Wolverines
Big Rapids

Michigan Militia
Detroit

Patriot Broadcasting Network
Dexter

Michigan Militia Corps
Wolverines
Kalamazoo

U.S. Taxpayers Party
Lansing

Michigan Militia Corps
Wolverines
Livingston County

Michigan Militia Corps
Wolverines
Macomb County

Citizens Militia of St. Clair
County
Memphis

Michigan Militia Corps
Wolverines
Oakland County

Michigan Militia, Inc.
Redford

Southern Michigan
Regional Militia
St.Clair

Lawful Path
Tustin

Minnesota

Constitution Party
St. Paul

Missouri

Missouri 51st Militia
Grain Valley

7th Missouri Militia
Granby

Montana

Militia of Montana
Noxon

Nevada

Center for Action
Sandy Valley

Independent American Party
Sparks

New Jersey

Constitution Party
Palmyra

New Jersey Committee
of Safety
Shamong

New Jersey Militia
Trenton

New York

Constitution Party
Albany

We The People
Queensbury

North Carolina

Constitution Party
Rocky Point

North Dakota

Constitution Party
Casselton

Ohio

Right Way L.A.W.
Akron

Central Ohio
Unorganized Militia
Columbus County

Constitution Party
Columbus

E Pluribus Unum
Grove City

Unorganized Militia
of Champaign County
St. Paris

Oklahoma

Ranch Rescue
Marietta

Present Truth Ministry
Panama

Oregon

Emissary Publications
Clackamas

Southern Oregon Militia
Eagle Point

TABLE 7.2

Active patriot groups, 2003 [CONTINUED]

Freedom Bound
  International
Klamath Falls

Constitution Party
Scappoose

Embassy of Heaven
Stayton

Pennsylvania

American Nationalist Union
Allison Park

Constitution Party
Lancaster

John Birch Society
Pittsburgh

Northern Voice Bookstore
Wildwood

South Carolina

Aware Group
Greenville

Constitution Party
Greenville

Tennessee

Constitution Party
Chattanooga

Constitution Party
Cookeville

Take Back Tennessee
Maynardville

Constitution Party
Memphis

Constitution Party
Nashville

Constitution Party
Winchester

Texas

Constitution Party
Abilene

Ranch Rescue
Abilene

Constitution Party
Alice

Constitution Party
Arlington

Constitution Society
Austin

John Birch Society
Austin

Constitution Party
Beaumont

Constitution Party
Belton

Constitution Party
Brenham

Constitution Party
Bryan

Buffalo Creek Press
Cleburne

Constitution Party
Cleburne

Constitution Party
Cleveland

Constitution Party
Conroe

Constitution Party
Corpus Christi

Republic of Texas
Dallas

Constitution Party
Danbury

Constitution Party
Early

Constitution Party
El Paso

Constitution Party
Elkhart

Republic of Texas
Fort Worth

Constitution Party
Franklin

Constitution Party
Guthrie

Constitution Party
Hondo

Constitution Party
Houston

Republic of Texas
Houston

Constitution Party
Huntsville

Constitution Party
Iredell

Constitution Party
Jefferson

Constitution Party
Kopperl

Constitution Party
Marshall

Constitution Party
Mcqueeney

Constitution Party
Midland

God Said Ministries
Mount Enterprise

Constitution Party
Navasota

Constitution Party
Odessa

Republic of Texas
Overton

Constitution Party
Plano

Constitution Party
San Antonio

Constitution Party
Texarkana

Constitution Party
The Woodlands

Church of God
  Evangelistic Association
Waxahachie

Constitution Party
Waxahachie

Constitution Party
Weatherford

Utah

Constitution Party
Bountiful

Vermont

Constitution Party
Quechee

Virginia

Ranch Rescue
Ashburn

Kenton's Rangers Virginia
  Line Militia
Front Royal

Virginia Citizens Militia
Roanoke

Constitution Party
Vienna

Washington

Washington State
  Jural Society
Ellensburg

Ranch Rescue
Vancouver

Wisconsin

American Opinion
  Book Services
Appleton

Constitution Party
Appleton

John Birch Society
Appleton

State	Count
Alabama	2
Alaska	2
Arizona	4
Arkansas	2
California	18
Colorado	2
Connecticut	0
Delaware	0
District of Columbia	1
Florida	2
Georgia	2
Hawaii	0
Idaho	4
Illinois	2
Indiana	16
Iowa	1
Kansas	1
Kentucky	4
Louisiana	1
Maine	0
Maryland	3
Massachusetts	0
Michigan	12
Minnesota	1
Mississippi	0
Missouri	2
Montana	1
Nebraska	0
Nevada	2
New Hampshire	0
New Jersey	3
New Mexico	0
New York	2
North Carolina	1
North Dakota	1
Ohio	5
Oklahoma	2
Oregon	5
Pennsylvania	4
Rhode Island	0
South Carolina	2
South Dakota	0
Tennessee	6
Texas	44
Utah	1
Vermont	1
Virginia	4
Washington	2
West Virginia	0
Wisconsin	3
Wyoming	0
Total	**171**

SOURCE: "Active Patriot Groups in the United States in the Year 2003," in *Intelligence Report,* Southern Poverty Law Center, Spring 2004, http://www .splcenter.org/images/dynamic/intel/report/23/ir113_patriot_groups.pdf (accessed August 24, 2004)

Self Defense (NBPP). These groups promote a strongly anti-white, anti-Semitic position and call for a separation between the races. The NBPP also encourages members to arm themselves. For a list of black separatist hate groups, see Table 7.3.

NATION OF ISLAM. The NOI was founded in the 1930s by Elijah Muhammad, who taught that whites were "the devil race" and blacks were the "makers of the universe." Probably its most prominent member was Malcolm X, who eventually left the group and was murdered by three NOI members in February 1965. Following Muhammad's own death in 1974, Louis Farrakhan took over the organization. In addition to its hatred of whites, the NOI is also anti-Semitic. In its book *The Secret Relationship between Blacks and Jews,* the group alleges that the slave trade was organized and run by Jews. During the 1990s Farrakhan caused controversy by visiting with the heads of such countries as

TABLE 7.3

Black separatist hate groups, 2003

City	Chapter	Group
Alabama		
Birmingham	Nation of Islam	Black Separatist
	New Black Panther Party	Black Separatist
Huntsville	Nation of Islam	Black Separatist
Mobile	Nation of Islam	Black Separatist
Montgomery	Nation of Islam	Black Separatist
Arizona		
Phoenix	Nation of Islam	Black Separatist
	New Black Panther Party	Black Separatist
Arkansas		
Little Rock	Nation of Islam	Black Separatist
California		
Adelanto	Nation of Islam	Black Separatist
Bakersfield	New Black Panther Party	Black Separatist
Long Beach	Nation of Islam	Black Separatist
Los Angeles	Nation of Islam	Black Separatist
	New Black Panther Party	Black Separatist
Montclair	Nation of Islam	Black Separatist
Oakland	Nation of Islam	Black Separatist
Rialto	Nation of Islam	Black Separatist
Richmond	Nation of Islam	Black Separatist
Sacramento	Nation of Islam	Black Separatist
San Diego	Nation of Islam	Black Separatist
San Francisco	Nation of Islam	Black Separatist
Colorado		
Denver	Nation of Islam	Black Separatist
Connecticut		
Bridgeport	Nation of Islam	Black Separatist
Hartford	Nation of Islam	Black Separatist
	United Nuwaubian Nation of Moors	Black Separatist
New Haven	Nation of Islam	Black Separatist
Delaware		
Wilmington	Nation of Islam	Black Separatist
District of Columbia		
Washington	Nation of Islam	Black Separatist
	New Black Panther Party	Black Separatist
	United Nuwaubian Nation of Moors	Black Separatist
Florida		
Fort Lauderdale	Nation of Islam	Black Separatist
Jacksonville	Nation of Islam	Black Separatist
Miami	Nation of Islam	Black Separatist
Pensacola	Nation of Islam	Black Separatist
Tallahassee	Nation of Islam	Black Separatist
Tampa	Nation of Islam	Black Separatist
West Palm Beach	Nation of Islam	Black Separatist
Georgia		
Albany	Nation of Islam	Black Separatist
Athens	Nation of Islam	Black Separatist
	United Nuwaubian Nation of Moors	Black Separatist
Atlanta	Nation of Islam	Black Separatist
	New Black Panther Party	Black Separatist
	United Nuwaubian Nation of Moors	Black Separatist
Augusta	Nation of Islam	Black Separatist
	New Black Panther Party	Black Separatist
	United Nuwaubian Nation of Moors	Black Separatist
Columbus	Nation of Islam	Black Separatist
Decatur	United Nuwaubian Nation of Moors	Black Separatist
Eatonton	United Nuwaubian Nation of Moors	Black Separatist
Macon	Nation of Islam	Black Separatist
	United Nuwaubian Nation of Moors	Black Separatist
Savannah	Nation of Islam	Black Separatist
	New Black Panther Party	Black Separatist
Hawaii		
Hawaii County	Nation of Islam	Black Separatist

TABLE 7.3

Black separatist hate groups, 2003 [CONTINUED]

City	Chapter	Group
Illinois		
Chicago	Nation of Islam	Black Separatist
	New Black Panther Party	Black Separatist
East St. Louis	Nation of Islam	Black Separatist
Rockford	New Black Panther Party	Black Separatist
Indiana		
Indianapolis	Nation of Islam	Black Separatist
Iowa		
Waterloo	Nation of Islam	Black Separatist
Kentucky		
Louisville	Nation of Islam	Black Separatist
Louisiana		
Baton Rouge	Nation of Islam	Black Separatist
	New Black Panther Party	Black Separatist
New Orleans	Nation of Islam	Black Separatist
	New Black Panther Party	Black Separatist
Maryland		
Baltimore	Nation of Islam	Black Separatist
	New Black Panther Party	Black Separatist
Massachusetts		
Boston	Nation of Islam	Black Separatist
	New Black Panther Party	Black Separatist
Springfield	Nation of Islam	Black Separatist
Michigan		
Detroit	Nation of Islam	Black Separatist
	New Black Panther Party	Black Separatist
Minnesota		
St. Paul	Nation of Islam	Black Separatist
Mississippi		
Greenville	Nation of Islam	Black Separatist
Holly Springs	Nation of Islam	Black Separatist
Jackson	Nation of Islam	Black Separatist
McComb	New Black Panther Party	Black Separatist
Missouri		
Kansas City	Nation of Islam	Black Separatist
St. Louis	Nation of Islam	Black Separatist
	New Black Panther Party	Black Separatist
Nebraska		
Omaha	Nation of Islam	Black Separatist
Nevada		
Las Vegas	Nation of Islam	Black Separatist
	New Black Panther Party	Black Separatist
New Jersey		
Camden	Nation of Islam	Black Separatist
New Brunswick	Nation of Islam	Black Separatist
Newark	Nation of Islam	Black Separatist
	New Black Panther Party	Black Separatist
Paterson	New Black Panther Party	Black Separatist
Plainfield	Nation of Islam	Black Separatist
Trenton	New Black Panther Party	Black Separatist
New York		
Albany	Nation of Islam	Black Separatist
Brooklyn	United Nuwaubian Nation of Moors	Black Separatist
Buffalo	Nation of Islam	Black Separatist
Harlem	New Black Panther Party	Black Separatist
New York	Nation of Islam	Black Separatist
Rochester	New Black Panther Party	Black Separatist
Syracuse	Nation of Islam	Black Separatist
North Carolina		
Charlotte	Nation of Islam	Black Separatist
Durham	Nation of Islam	Black Separatist
Greensboro	Nation of Islam	Black Separatist
	New Black Panther Party	Black Separatist
Raleigh	Nation of Islam	Black Separatist
Reidsville	Nation of Islam	Black Separatist

Iran and Libya, even winning a promise from Libya's dictator Muammar Qaddafi of a $1 billion donation. Qaddafi had already given Farrakhan an interest-free $5 million loan in 1985, according to an *Intelligence Report* by Martin A. Lee of in the Southern Poverty Law Center.

NEW BLACK PANTHER PARTY FOR SELF DEFENSE. The NBPP was founded by Michael McGee as the Black Panther Militia. McGee, who was involved in Milwaukee, Wiscon-

TABLE 7.3

Black separatist hate groups, 2003 [CONTINUED]

City	Chapter	Group
Ohio		
Cincinnati	Nation of Islam	Black Separatist
	New Black Panther Party	Black Separatist
Cleveland	Nation of Islam	Black Separatist
	New Black Panther Party	Black Separatist
Columbus	Nation of Islam	Black Separatist
Dayton	Nation of Islam	Black Separatist
Toledo	Nation of Islam	Black Separatist
Oklahoma		
Tulsa	Nation of Islam	Black Separatist
Pennsylvania		
Harrisburg	Nation of Islam	Black Separatist
Philadelphia	Nation of Islam	Black Separatist
	New Black Panther Party	Black Separatist
	United Nuwaubian Nation of Moors	Black Separatist
Pittsburgh	Nation of Islam	Black Separatist
Rhode Island		
Providence	New Black Panther Party	Black Separatist
South Carolina		
Columbia	Nation of Islam	Black Separatist
	New Black Panther Party	Black Separatist
Tennessee		
Chattanooga	Nation of Islam	Black Separatist
Knoxville	Nation of Islam	Black Separatist
Memphis	Nation of Islam	Black Separatist
	New Black Panther Party	Black Separatist
Nashville	Nation of Islam	Black Separatist
Texas		
Austin	New Black Panther Party	Black Separatist
Dallas	Nation of Islam	Black Separatist
	New Black Panther Party	Black Separatist
El Paso	Nation of Islam	Black Separatist
Houston	Nation of Islam	Black Separatist
	New Black Panther Party	Black Separatist
Jasper	New Black Panther Party	Black Separatist
San Antonio	Nation of Islam	Black Separatist
Virginia		
Danville	Nation of Islam	Black Separatist
Petersburg	Nation of Islam	Black Separatist
Richmond	Nation of Islam	Black Separatist
Washington		
Seattle	Nation of Islam	Black Separatist
Wisconsin		
Milwaukee	Nation of Islam	Black Separatist

SOURCE: "Active U.S. Hate Groups in 2003: Black Separatist," in *Intelligence Report,* Southern Poverty Law Center, http://www.splcenter.org/intel/map/hate.jsp?T=10&m=2 (accessed September 23, 2004)

sin, politics as a city council member and alderman before turning to extremist politics, announced the creation of the Black Panther Militia in 1990 and sought to enlist local street gangs as members. "Our militia will be about violence. I'm talking actual fighting, bloodshed and urban guerilla warfare," McGee explained, according to the Anti-Defamation League's Web site (http://www.adl.org/). Renamed the NBPP, the group began appearing at rallies and demonstrations throughout the country, often armed with shotguns and automatic rifles. A May 1996 Dallas school board meeting was canceled after the NBPP threatened to come with loaded weapons. A high school in Wedowee, Alabama, was burned down hours after a speech was given there by NBPP leader Mmoja Ajabu. A local member was later acquitted of arson. The NBPP is strongly anti-white and anti-Semitic. It also calls for a separation of the races and the overthrow of the government. Following the terrorist attacks of September 11, 2001, the group accused Jews of having masterminded the events. By this point, the group's leader was Malik Zulu Shabazz, a Washington D.C.-based attorney. In July 2002 Shabazz announced that the NBPP intended to support accused Arab terrorist Zacarias Moussaoui, who was on trial for his role in the September 11, 2001 attacks.

WATCHDOG GROUPS: WHO MONITORS DOMESTIC TERRORISTS?

Besides the U.S. government, a number of watchdog groups, such as the Anti-Defamation League (ADL) and the Southern Poverty Law Center (SPLC), maintain a database of current domestic terrorist and militia groups. In addition to keeping abreast of the activities of these groups, the ADL and SPLC also provide education and training to reduce hate crimes, or attacks perpetrated against an individual or a group based on ethnicity, religion, or sexual preference. Groups specifically preaching hateful ideologies against homosexuals and different races include various Ku Klux Klan, neo-Nazi, black separatist, and racist skinhead organizations.

The Anti-Defamation League (ADL)

The ADL was founded in Chicago in 1913 by Sigmund Livingston with the mission "to stop, by appeals to reason and conscience, and if necessary, by appeals to law, the defamation of the Jewish people" (http://www.adl.org/). Today, the ADL is one of the nation's premier civil rights/human-relations agencies, dedicated to fighting anti-Semitism and all forms of bigotry, defending democratic ideals, and protecting civil rights for all. The ADL develops materials, programs, and services to build communication, understanding, and respect among diverse groups. It has conducted and published four national surveys and analyses of far-right extremism in the United States. The ADL Web site provides articles on a wide range of issues, including extremist groups, hate crimes, security awareness, and terrorism.

Since it was founded, the ADL has acted against groups such as the Ku Klux Klan (by circulating pamphlets and calling on Presidents William Howard Taft and Theodore Roosevelt to denounce automaker Henry Ford's anti-Semitic books) and U.S. fascist groups (by accumulating a storehouse of information on extremist groups and individuals in the United States).

In 2002 the ADL took several measures to aid the fight against terrorism. It established a partnership with the Israel-based International Policy Institute for Counterterrorism (ICT) to facilitate meetings between ICT terrorism experts and American law enforcement, government officials, media, and community groups, and to distribute ICT publications in the United States. The ADL monitored the response of extremist groups to the attacks of September 11, 2001, by posting their statements on the ADL's Web site. The ADL also issues a periodic report on international and domestic terrorism called *Terrorism Update,* which is distributed to the media, members of Congress, the presidential administration, state and local legislators, academics, and Jewish organizations.

The Southern Poverty Law Center (SPLC)

The SPLC was founded as a small civil rights law firm in 1971 by Morris Dees and Joe Levin, two local lawyers who shared a commitment to racial equality. Today, the SPLC is a nonprofit organization that combats hate, intolerance, and discrimination through education and litigation. The center is internationally known for its tolerance education programs, its legal victories against white supremacist groups, its tracking of hate groups, and its sponsorship of the Civil Rights Memorial.

In 1981, in response to the resurgence of the Ku Klux Klan, the SPLC began to monitor hate activity. In 2004 the SPLC's Intelligence Project tracked the activities of more than six hundred racist and neo-Nazi groups. In 1994, after uncovering links between white supremacist organizations and elements of the emerging antigovernment Patriot movement, the SPLC expanded its monitoring operation to include the activities of militias and other extremist antigovernment groups. Six months before the Oklahoma City bombing, the SPLC warned the U.S. attorney general that the new mixture of armed militia groups and those who hate was a recipe for disaster.

At the peak of the Patriot movement in the mid-1990s, the SPLC tracked more than eight hundred militia-like Patriot groups. As of 2004 that number had dwindled to fewer than two hundred. Using information collected by the Intelligence Project during its monitoring and investigative activities, the SPLC provides comprehensive updates to law enforcement agencies, the media, and the general public through its quarterly publication, *Intelligence Report.*

Several of the SPLC's lawsuits have reached the U.S. Supreme Court, and many have resulted in landmark rulings. The SPLC has developed novel legal strategies to shut down extremist activity and to help victims of hate crimes extract monetary damages from groups such as the Ku Klux Klan.

RESPONDING TO DOMESTIC TERRORISM

With increased attention being given to international terrorism groups such as al Qaeda, issues of domestic terrorism may seem to be on the back burner. On the contrary, many U.S. legislators recognize the problem of homegrown terror groups. On November 2, 2001, several members of Congress wrote to various environmental groups, urging them to abandon tactics of ecoterrorism. The Agroterrorism Act of 2001 fights domestic terrorism by increasing penalties against perpetrators, and the Hands Off Our Kids Act of 2001 calls for measures to stop groups like the ALF and ELF from recruiting young people for illegal activities.

Through interagency efforts, the U.S. government also developed the Concept of Operations Plan (CONPLAN) which, according to *United States Government Interagency Domestic Terrorism Concept of Operations Plan* (http://www.fema.gov/) is "designed to provide overall guidance to Federal, State, and local agencies concerning how the Federal government would respond to a potential or actual terrorist threat or incident in the United States, particularly one involving Weapons of Mass Destruction." Primary agencies involved in this plan are the Department of Justice (led by the FBI), Federal Emergency Management Agency, Department of Defense, Department of Energy, Environmental Protection Agency, and Department of Health and Human Services. These six agencies are responsible for developing coordinated tactical and strategic options to deal effectively with terrorist attacks.

CHAPTER 8
CIVILIAN NATIONAL SECURITY INFRASTRUCTURE

At the apex of the U.S. federal government is the Constitution. (See Figure 8.1.) The Constitution gives the job of providing for America's national security to the president and the executive branch of the government, as well as to the legislative branch (the U.S. House of Representatives and the U.S. Senate). It designates the president the commander in chief of the American armed forces. Executive branch entities involved in national security can be found at the White House level—for example, the White House Office, the National Security Council (NSC), and the Office of Management and Budget (OMB)—and all the way down to the subcabinet/department, independent-agency level—for example, the Central Intelligence Agency (CIA), the Defense Nuclear Facilities Safety Board, and the Nuclear Regulatory Commission. Most responsibility for day-to-day national security matters falls to the executive branch of government.

This is not to say that the legislative branch plays a passive role in national security affairs. While key powers requiring strong central direction, such as treaty making, the appointing of ambassadors, and committing armed forces to conflicts, are given to the executive branch, exercising these powers requires the approval of the Senate. Also, Congress has its own powers related to defense and national security. These include the right to declare war; to raise armies, navies, and militias; to provide money for those forces; to authorize a draft (the pressing of individuals into mandatory military service); to make rules regulating the armed forces; to make all laws "necessary and proper" for carrying out the foregoing powers; and to provide advice and consent to the executive branch in foreign affairs—for example, approval of treaties that the executive branch has negotiated and its appointments of ambassadors, ministers, and other key officers of government.

Presidential powers, especially those relating to unilaterally declaring wars, were greatly questioned during the Vietnam era. In August 1964 President Lyndon John-son had used an alleged attack by North Vietnam on the U.S. Navy in the Gulf of Tonkin to justify a massive buildup of troops in South Vietnam. As the American public grew increasingly disenchanted with its extended military adventure in the Southeast Asian nation, they decided to limit the "imperial" war-making powers of the president. Despite President Richard M. Nixon's veto, Congress passed the War Powers Act in 1973. The legislative measure called for a sixty-day waiting period before engaging in an undeclared war.

Some political scientists claim that the War Powers Act is nothing more than a mere "paper tiger"—something that looks effective on paper but that does not work in reality. Since the act's passage, U.S. presidents have violated the measure and invaded several countries (such as Lebanon, Grenada, and Panama) without congressional approval. In October 2002 Congress, under joint Resolution 114, granted President George W. Bush full authority to use any "necessary and appropriate" force against Iraq to protect America and its citizens without returning to Congress for approval.

Typically, the U.S. national security infrastructure leaves the executive branch in the position of "proposing" national security or foreign policy initiatives, such as treaties and agreements. It leaves the Congress, especially the Senate, in the position of "disposing" them, as in ratifying treaties, approving foreign-aid budgets and defense appropriations, approving the appointment of ambassadors, and providing some oversight of intelligence and covert operations.

The roles of Congress and the executive branch are occasionally reversed. However, the experience of the Constitution's framers was generally that the Congress, a "deliberative body" (a group that must debate and vote on issues before acting on them), acts slowly compared with the executive branch. Therefore, it would not be appropriate for Congress to control functions requiring strong, immediate control, such as commanding the armed forces

FIGURE 8.1

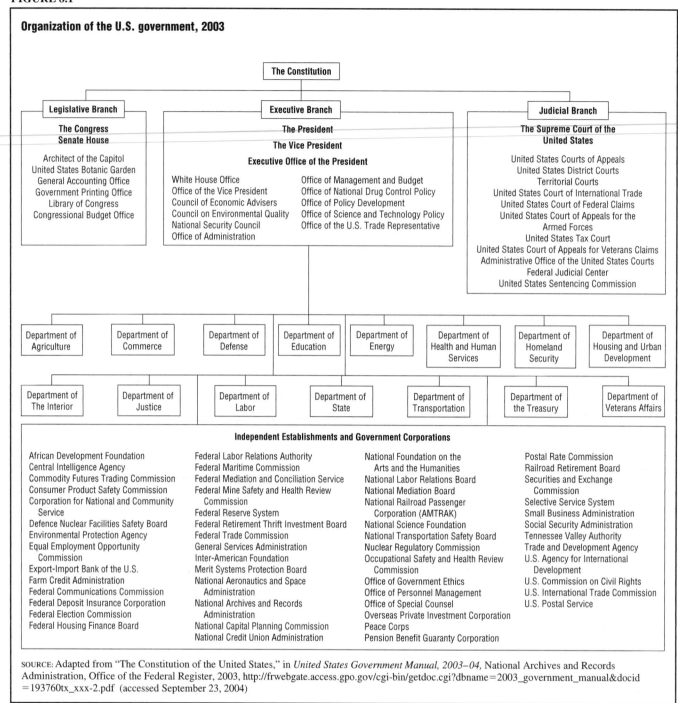

Organization of the U.S. government, 2003

The Constitution

Legislative Branch — **Executive Branch** — **Judicial Branch**

The Congress
Senate House

Architect of the Capitol
United States Botanic Garden
General Accounting Office
Government Printing Office
Library of Congress
Congressional Budget Office

The President
The Vice President
Executive Office of the President

White House Office
Office of the Vice President
Council of Economic Advisers
Council on Environmental Quality
National Security Council
Office of Administration

Office of Management and Budget
Office of National Drug Control Policy
Office of Policy Development
Office of Science and Technology Policy
Office of the U.S. Trade Representative

The Supreme Court of the
United States

United States Courts of Appeals
United States District Courts
Territorial Courts
United States Court of International Trade
United States Court of Federal Claims
United States Court of Appeals for the
    Armed Forces
United States Tax Court
United States Court of Appeals for Veterans Claims
Administrative Office of the United States Courts
Federal Judicial Center
United States Sentencing Commission

Department of Agriculture | Department of Commerce | Department of Defense | Department of Education | Department of Energy | Department of Health and Human Services | Department of Homeland Security | Department of Housing and Urban Development

Department of The Interior | Department of Justice | Department of Labor | Department of State | Department of Transportation | Department of the Treasury | Department of Veterans Affairs

Independent Establishments and Government Corporations

African Development Foundation
Central Intelligence Agency
Commodity Futures Trading Commission
Consumer Product Safety Commission
Corporation for National and Community
    Service
Defence Nuclear Facilities Safety Board
Environmental Protection Agency
Equal Employment Opportunity
    Commission
Export-Import Bank of the U.S.
Farm Credit Administration
Federal Communications Commission
Federal Deposit Insurance Corporation
Federal Election Commission
Federal Housing Finance Board

Federal Labor Relations Authority
Federal Maritime Commission
Federal Mediation and Conciliation Service
Federal Mine Safety and Health Review
    Commission
Federal Reserve System
Federal Retirement Thrift Investment Board
Federal Trade Commission
General Services Administration
Inter-American Foundation
Merit Systems Protection Board
National Aeronautics and Space
    Administration
National Archives and Records
    Administration
National Capital Planning Commission
National Credit Union Administration

National Foundation on the
    Arts and the Humanities
National Labor Relations Board
National Mediation Board
National Railroad Passenger
    Corporation (AMTRAK)
National Science Foundation
National Transportation Safety Board
Nuclear Regulatory Commission
Occupational Safety and Health Review
    Commission
Office of Government Ethics
Office of Personnel Management
Office of Special Counsel
Overseas Private Investment Corporation
Peace Corps
Pension Benefit Guaranty Corporation

Postal Rate Commission
Railroad Retirement Board
Securities and Exchange
    Commission
Selective Service System
Small Business Administration
Social Security Administration
Tennessee Valley Authority
Trade and Development Agency
U.S. Agency for International
    Development
U.S. Commission on Civil Rights
U.S. International Trade Commission
U.S. Postal Service

SOURCE: Adapted from "The Constitution of the United States," in *United States Government Manual, 2003–04,* National Archives and Records Administration, Office of the Federal Register, 2003, http://frwebgate.access.gpo.gov/cgi-bin/getdoc.cgi?dbname=2003_government_manual&docid =193760tx_xxx-2.pdf (accessed September 23, 2004)

or negotiating treaties. If Congress was responsible for these types of functions, it could potentially act incompetently, or it might not act at all.

On national security matters, the president first works with executive staff, then with certain executive departments. The executive staffs and departments involved in national security include the White House staff; the NSC and its staff; the State Department; the Department of Defense (DOD), including the secretary of defense and the Joint Chiefs of Staff (JCS); the CIA; and the OMB. This chapter will specifically deal with the civilian branches of

U.S. government that deal with national security. The military and its various aspects will be addressed in Chapter 9.

WHITE HOUSE STAFF

The office of the White House consists of personal and political assistants to the president, who serve at his request to facilitate his decisions. The White House staff has seen tremendous growth in the last several decades. The administration of President Herbert Hoover (1929–33) had three secretaries, a military and a naval aide, and twenty clerks in the office. In 1997 the White

House office of President Bill Clinton had a permanent staff of more than four hundred people.

Growth has not been the only trend evident in the White House staff's structure over the years. Another has been the evolving and expanding role of the president's national security assistant, who heads the NSC staff. Since the early days of the cold war between the United States and the Soviet Union during the Dwight D. Eisenhower administration (1953–61), the post of national security assistant (then called special assistant to the president for national security affairs) has become increasingly important. In each administration, the assistant's personal relationship with the president, and the president's wishes as to how the assistant should function, have modified the role. The national security assistant's role has also evolved as the NSC staff, which the assistant heads, has changed.

NATIONAL SECURITY COUNCIL (NSC)

The president's principal cabinet officers also serve as his closest national security advisors. The president, the vice-president, the secretary of state, and the secretary of defense make up the NSC, which was established by the National Security Act of 1947. That act mandated that the CIA director, also known as the director of central intelligence (DCI), and the chairman of the military's Joint Chiefs of Staff (JCS) also be advisors to the NSC. Present during most NSC meetings are the president, the vice-president, the secretary of state, the secretary of defense, and the president's national security assistant. Others who may attend include the president's chief of staff and counsel, the attorney general, and other senior officials depending on their responsibilities.

U.S. DEPARTMENT OF STATES

The Department of State was created in 1789. Its first secretary was Thomas Jefferson, who went on to become the third president of the United States. The State Department represents the interests of the United States and its citizens in relations with foreign countries and also serves as a principal source of advice to the president on aspects of national security and foreign affairs.

The secretary of state faces the task of managing a huge bureaucracy. (See Figure 8.2.) For the most part, the U.S. Department of State is organized by function, such as counterterrorism, intelligence and research, protocol, and public affairs. However, under the undersecretary for political affairs, it is organized regionally, by foreign "desks," a classic structure found in many departments and agencies involved in foreign affairs, including the CIA. Throughout the department, there are distinct areas of functional or regional responsibility. Functional units naturally cut across regional lines, and within the foreign bureaus are special functional "desks." Sometimes analysis of an issue or problem by a foreign desk contradicts analysis from a functional bureau.

The secretary of state is the president's principal advisor on foreign policy, but history shows that the secretary's power has been weaker or stronger depending on a particular president's own interests and activities in foreign affairs. President Nixon appointed William Rogers as secretary of state in 1969 but bypassed him systematically to take matters into his own hands. Other presidents, such as Ronald Reagan and Bill Clinton, relied more heavily on their secretaries.

In addition to its national security objectives, the State Department remains responsible for the official day-to-day presence of the U.S. government in foreign countries. The department follows the "country team" concept. The American ambassador, who is a representative of the American government, oversees all U.S. programs and personnel within a country, with the exception of American military forces in the country that may be in the field or in combat roles.

OFFICE OF MANAGEMENT AND BUDGET (OMB)

The president and his advisors control the most powerful armed forces and intelligence systems in history. These are employed to defend and secure the vital interests and security of the most powerful nation on earth. Yet, their slightest move can cost millions of dollars. The president has little choice but to structure the priorities of national defense constantly within the defense budget.

The director of the OMB assists the president in preparing the federal budget and supervising its administration in executive branch agencies, such as those dedicated to defense and national security. The OMB helps formulate the president's spending plans, which means that it assesses the cost-effectiveness of agency programs, policies, and procedures. The OMB also resolves the competing demands of defense and national security agencies to set funding priorities, and ensures that federal agency reports, rules, testimony, and proposed legislation are consistent with the president's budget policies.

In addition, the OMB oversees and coordinates the administration's procurement, financial management, information, and regulatory policies. In each of these areas, the OMB's role is to help improve administrative management, to develop better performance measures and coordinating mechanisms, and to reduce any unnecessary burdens on the taxpayer. The largest items in the national budget are defense and national security, so the president relies heavily on the OMB to set funding priorities.

THE INTELLIGENCE COMMUNITY: WHAT INFORMATION IS GATHERED, AND BY WHOM?

Organization

Below the level of cabinet members and presidential advisors is the next component of the national security

FIGURE 8.2

Organization of the U.S. State Department, 2003

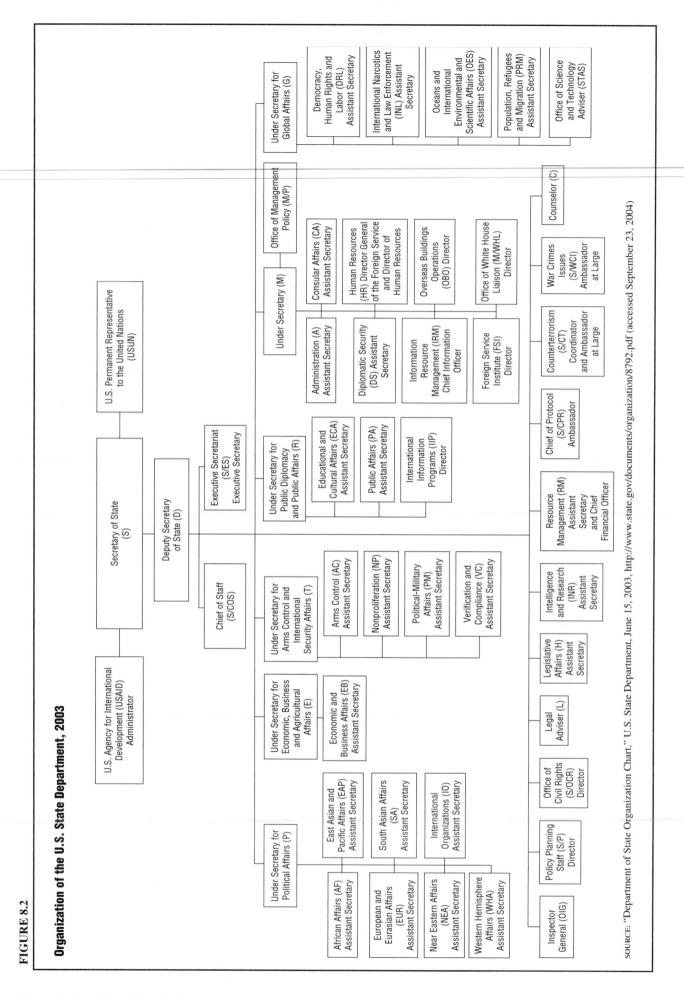

SOURCE: "Department of State Organization Chart," U.S. State Department, June 15, 2003, http://www.state.gov/documents/organization/8792.pdf (accessed September 23, 2004)

apparatus: the intelligence community. The intelligence community consists of executive branch agencies and units conducting a variety of intelligence activities in furtherance of national security.

The National Security Act of 1947 established the CIA and made the DCI an advisor to the NSC. The DCI directs not just the CIA but the intelligence community as a whole. Members of the intelligence community related to the DOD include the Defense Intelligence Agency (DIA), the National Security Agency (NSA), the National Reconnaissance Office (NRO), the National Geospatial-Intelligence Agency (NGA), and intelligence agencies of the various branches of the military. Non-DOD agencies include the CIA, State Department, Energy Department, Treasury Department, Federal Bureau of Investigation (FBI), and the Coast Guard. The CIA, DIA, NSA, NRO, and National Imagery and Mapping Agency (NIMA) are solely tasked with intelligence responsibilities, while the others are primarily concerned with other duties (such as law enforcement, border security, and the like) but deal with intelligence as a part of their mandate.

Information as Intelligence

What makes information intelligence? Intelligence is information that has a strategic value—information whose collection is instrumental in making important national security decisions by the president, the DOD, or others in government. What also makes information intelligence is that it has been gathered at a more or less central location, where it can be integrated with other data, including secret data, and carefully analyzed.

TECHNICAL INTELLIGENCE. The intelligence community refers to the collection of technical data on opposing forces' weapons systems, personnel capabilities, and other technical information as "techint." A surprising amount of relevant techint comes from open (unclassified) sources, such as foreign and domestic newspapers, magazines, government reports, technical and professional journals, news media, academic studies, and popular literature. A much smaller amount of relevant techint comes from obscure, classified, and secret sources, such as lost or stolen weapons systems and government documents, stolen classified documents, stolen weapons or classified facility blueprints, and classified maps.

HUMAN INTELLIGENCE. Intelligence starts with spying, which was certainly the most common method of intelligence gathering prior to World War II (1939–45). In general, the community refers to the cultivation of human sources, whether open or secret, as "humint," short for human intelligence. Often, humint is the best (or only) way for U.S. defense planners to find out what another country's leadership thinks of its own capabilities. It is likely that humint will become more and more useful against threats like rogue states, transnational actors, terrorists, and organized crime, whose assets are smaller and

thus less susceptible than sovereign states' forces to observation or surveillance via satellite imaging from space or by other processes.

CIA field agents are one source of humint. They collect information from open sources such as the media and recruit foreign citizens, either "defectors in place" (who volunteer their services) or "turned" informants (who are bribed or blackmailed into service). The latter can include foreign government officials or businesspeople, paid informants in terrorist cells, and members of organized crime groups, among others. Soviet Lieutenant Colonel Pyotr Semyonovich Popov, for example, contributed invaluable data on his country's missile systems during the 1950s.

Once CIA field agents have collected information, they turn it over to their superiors, the station chiefs, who send it to CIA headquarters. Unlike their counterparts in the State and Defense Departments, CIA agents in the field and analysts in the home offices do not rotate their focus on foreign areas. Rather, they are given long-term assignments to allow them to focus on specific areas.

SIGNALS INTELLIGENCE. Signals intelligence consists of the interception and processing of electronic signals—for example, missile and satellite telemetry, shortwave radio transmissions, and cell-phone exchanges intercepted via ground-, air-, or space-based eavesdropping or monitoring equipment. The government devotes large sums of money to this activity. Most estimates put the budget of the NSA, the agency mainly responsible for this function, in the billions of dollars.

The federal government also funds In-Q-Tel, a not-for-profit venture-capital project allied with the CIA, which, among other endeavors, has pursued technology to facilitate monitoring of the World Wide Web through custom information retrieval and multiple-language and anonymous search services.

IMAGERY INTELLIGENCE. Imagery intelligence is collected using photography from reconnaissance satellites and aircraft, as well as other types of photographic and image-producing processes. The satellites and aircraft used are known as "overhead platforms," one famous example of which was the U-2, a high-altitude plane with sophisticated cameras that was promoted by CIA Director Allen Dulles in the 1950s. The U-2 was used extensively to monitor military and missile sites in the Soviet Union.

Intelligence provided by U-2 photo reconnaissance proved indispensable in defusing and managing the Cuban Missile Crisis of 1962, during which the Soviet Union threatened the safety of the United States by placing ballistic missiles in Cuba which were capable of striking major American cities. The administration of President John F. Kennedy had to decide how to respond to this threat. The reconnaissance images gave Kennedy an idea of the precise nature and extent of the new missile batteries, which he used in fashioning an appropriately measured U.S. military response. After a tense standoff involving a U.S. naval blockade of Cuba, the Soviets eventually withdrew their missiles.

MEASUREMENT AND SIGNATURE INTELLIGENCE. Measurement and signature intelligence (often called "masint") is produced by collecting, storing, and analyzing atmospheric and environmental emissions, including radar, infrared, chemical, acoustic, and seismic data, usually as detected by specialized sensors. The CIA Science and Technology Directorate, for instance, employs seismic sensors to keep tabs on global military activity and has researched methods of detecting poisonous gases. Masint, according to specialists, came into being as the result of the SCUD missile hunts of the Persian Gulf War of 1991.

Covert Action

Only the president can direct the CIA to undertake a covert action. Such actions are usually based on recommendations from the NSC. U.S. foreign policy objectives may not be fully realized by normal diplomatic means, but military action would be too extreme. In these cases, the president may direct the CIA to conduct a special activity abroad in support of foreign policy in which the role of the U.S. government is neither readily apparent nor publicly acknowledged. However, once ordered to undertake the activity, the DCI must notify congressional oversight committees.

In the past few decades, covert actions have often taken the form of assistance (money, equipment, and/or advice) to operatives in foreign lands, as those forces attempt to resolve situations in ways that are favorable to the interests of the United States. Most of these operations, being low level and involving only a few people, have remained secret. Others, being larger scale and involving many more participants, have found their way into the news.

Paramilitary operations are an extreme form of covert action. In these cases, CIA operatives go beyond giving advice to opposition groups and other elements and may actually lead the charge and direct them. Such activities can be very controversial because they fall within a gray area. Even though they do not involve uniformed U.S. military personnel, and so do not come under the presidential restrictions on war making of the War Powers Act, many people think such actions amount to undeclared war.

Among the more controversial covert actions undertaken by the CIA since the 1960s have been the Bay of Pigs invasion and the Phoenix Program. In the 1961 Bay of Pigs invasion the CIA sent a force of 1,500 men to

Cuba where, denied air support by President Kennedy, they were decisively defeated. DCI Allen Dulles resigned after the disaster. The Phoenix Program, which started in 1968, was designed to lessen support for the Communist Viet Cong in South Vietnam but resulted in the deaths of at least twenty thousand noncombatants.

THE CENTRAL INTELLIGENCE AGENCY (CIA)

The CIA's Role

Congress established the CIA to serve as a central depository for the various specialized intelligence and espionage functions within the intelligence community. This leaves the CIA, per Congress's intent, better able to focus directly on the overall community's three functions: (1) collecting vital intelligence; (2) disseminating it within the executive branch; and (3) conducting and coordinating spying, covert actions, and counterintelligence as effectively as possible.

The CIA's modern-day role derives from that of its World War II predecessor, the Office of Strategic Services (OSS), which had two main wartime functions: (1) for the first time in the nation's history, centrally gathering and analyzing intelligence; and (2) conducting covert operations, such as active aid to resistance movements in Europe, as authorized by the president.

The CIA is the only agency within the intelligence community authorized (and even then, only on a case-by-case basis) to conduct spying and covert actions abroad (although the president could conceivably order other agencies to be involved). However, both the National Security Act of 1947 and Executive Order 12333—United States Intelligence Activities (1981) prohibit the CIA from spying on or acting against U.S. citizens domestically. Executive Order 12333 specifically forbids "physical surveillance of a United States person in the United States by agencies other than the FBI." Counterintelligence—monitoring and thwarting spying and intelligence activities against the United States, mostly within the United States—is thus a function assigned to the FBI domestically, with the CIA and the intelligence units of the armed services also assisting abroad.

This subject has prompted renewed interest and debate since the passage of H.R. 3162, commonly known as the Patriot Act, in October 2002. This legislative measure calls for, among other things, fewer restrictions on information sharing among intelligence agencies and law enforcement authorities on suspected terrorists, as well as greater authority for law enforcement to monitor the phone conversations and e-mail activities of such individuals.

Secret versus Public Information about the CIA

Legislation passed in 1949 provided statutory authority for the CIA's undisclosed budget and staffing levels.

The Central Intelligence Agency Act of 1949 exempted CIA funding from most of the usual appropriations procedures. Further, it allowed the agency not to divulge its "organization, functions, names, officials, titles, salaries, or numbers of personnel employed." The defense budget disguises funds intended for intelligence within the budgets of nonsecret defense agencies. Under the 1949 act, CIA funds listed in the budgets of other agencies could be moved back to the agency free of limitations placed on the original appropriations. In this way, intelligence community programs were shielded from outside evaluation, making it impossible for congressional overseers to get an idea of their cost-effectiveness and propriety.

As the cold war flourished from 1947 to 1977, the intelligence community was given unusual autonomy. Through 1977, thirty years after it was founded, the CIA was exempt from exposing and defending its budget. However, the gradual replacement of the cold war by détente (a period of new U.S.-Soviet understanding, especially about arms control, that developed in the early 1970s under President Nixon) and other developments led Congress to weaken the intelligence community's power in the mid-1970s. During the mid- to late 1970s, although the intelligence community's budget was not made public, congressional oversight committees were given more authority over the CIA's behavior, especially its espionage and covert actions.

Little official information about the CIA's size or appropriations is publicly available even now. For security reasons, the CIA keeps most of its activities and finances secret. The overall extent of the headquarters complex located in Langley, Virginia, suggests that the usual estimate of about twenty thousand employees at that location is accurate, but the number of agents, operatives, and others in various countries is unknown.

Estimates of current CIA budgets vary widely because CIA funds continue to be hidden in the annual budgets of other agencies. According to *The CIA Factbook on Intelligence* (Washington, DC: Office of Public Affairs, Central Intelligence Agency, 2004), the total intelligence budget for the U.S. government in 1998 (of which the CIA is one part) was $26.7 billion. More recent data had not been made public at the time of that publication.

Organizational Structure

A modern intelligence organization such as the CIA has a distinctive organizational chart, shown in Figure 8.3. There is a balance between functional and regional divisions as well as analytic and administrative divisions. Each of the CIA's three divisions, or "directorates," has its own deputy director. The Operations Directorate is responsible for covert actions and counterintelligence; the Science and Technology Directorate specializes in data interpretation; and the Intelligence Directorate generates reports based on analyses of raw data for the president and other members of the executive branch.

The DCI is distinct from other agency heads in that he or she serves as (1) head of the CIA; (2) head of the intelligence community; and (3) principal advisor on foreign intelligence to the NSC. Directors of other intelligence community agencies advise the DCI, in turn, by sitting on a number of specialized intelligence committees. Chief among these groups is the National Foreign Intelligence Board.

The DCI, in advising the NSC and the president, must be objective and resist political pressures that would influence his or her counsel. One way the CIA attempts to remain independent is by giving stable, lifelong careers to people who are not just competent technicians and accomplished specialists but who also pass a rigid background check, swear an oath of secrecy, and appear to possess such traits as loyalty, discretion, ingenuity, and a commitment to protecting and promoting American values. For this reason, the CIA places a high premium on trust and is often referred to by its employees as "the family." When the CIA's trust in its employees is misplaced, the consequences can be serious, as illustrated by the case of Aldrich Ames, a high-ranking CIA official who sold secrets to the Soviet Union (and later Russia) from 1985 to 1994. Not only did Ames's treachery as head of the CIA's Soviet counterintelligence unit result in flawed American policy, but it also cost several agents their lives.

The History of the CIA

THE FIRST THREE DECADES. During its first thirty years, the CIA became known as a producer and disseminator of the highest-quality intelligence. It developed economic forecasting methods that helped gauge the Soviet Union's strength; disproved the "missile gap," which assumed that American weaponry was insufficient to counter the Soviet threat; and provided useful information during the Vietnam War. However, the agency also fell short of expectations on some occasions. It failed to warn of the Suez Crisis in 1956 and the Yom Kippur War in 1973.

THE 1970s: INTELLIGENCE PROBLEMS AND CONGRESSIONAL CURBS. Congressional hearings during 1975 brought to light the CIA's role in several assassination plots against foreign leaders in Chile, the Congo, Cuba, the Dominican Republic, Haiti, and Indonesia. Charges were also made that CIA surveillance programs had been aimed at innocent foreign students, visitors to the United States, and Americans traveling abroad. Domestic practices of the CIA were attacked as illegal extensions of the CIA's foreign duties: domestic wiretaps, break-ins, and mail intercepts; infiltration of religious groups; surveillance of national political figures; training of local law enforcement in espionage techniques; and involvement in the academic world through subsidies and research contracts.

FIGURE 8.3

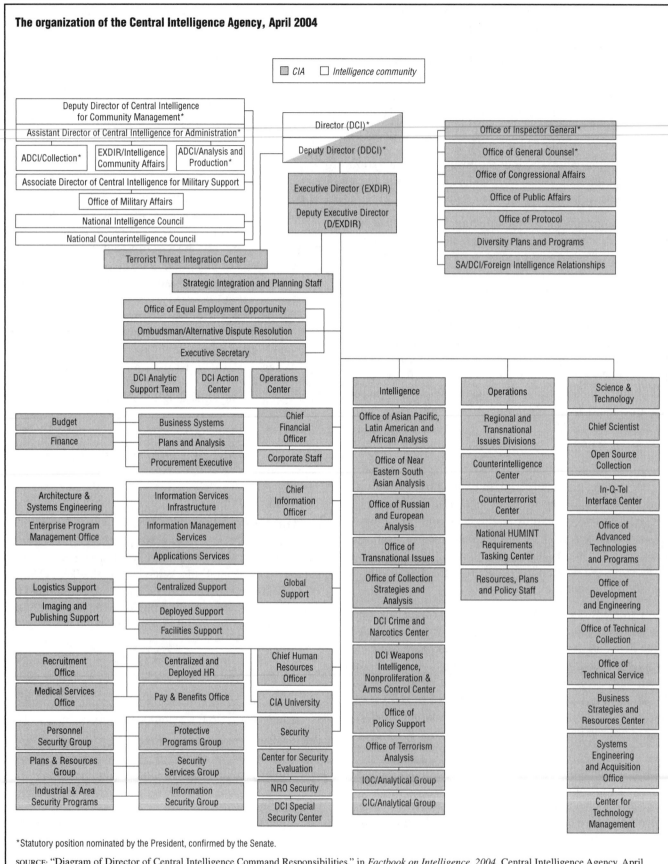

The organization of the Central Intelligence Agency, April 2004

*Statutory position nominated by the President, confirmed by the Senate.

SOURCE: "Diagram of Director of Central Intelligence Command Responsibilities," in *Factbook on Intelligence, 2004,* Central Intelligence Agency, April 2004, http://www.cia.gov/cia/publications/facttell/jpg/org_chart.pdf (accessed September 23, 2004)

Concerned about possible domestic spying, President Gerald Ford appointed the Rockefeller Commission to report on CIA activities within U.S. borders. Before long, a number of congressional and executive actions had defined and limited the CIA's activities. In 1974 the first major restriction on the CIA's activities passed: the Hughes-Ryan Amendments. They required the CIA to submit plans for covert activities to the president, who in turn had to justify them to appropriate committees of Congress as being critical to national security.

The U.S. Senate also conducted an investigation into CIA activities. The Senate Select Committee to Study Governmental Operations with Respect to Intelligence Activities, popularly called the Church Committee for its chairman Senator Frank Church, issued fourteen reports during 1975 and 1976. They documented such abuses as the assassinations of foreign leaders and the clandestine monitoring of the domestic mail of American citizens.

The House and Senate Armed Services Committees, up until this time, had loosely supervised the CIA. Stricter supervision began when President Ford established the Intelligence Oversight Board (IOB) in 1976. The IOB is authorized to investigate the legality and appropriateness of intelligence activities and directs its reports to the attorney general. The Senate Select Committee on Intelligence (SSCI) was set up in 1976, and the following year the House Permanent Select Committee on Intelligence (HPSCI) came into being.

Although the number of employees and size of the CIA's recent budgets have not been publicly disclosed, they are scrutinized by several other government agencies. Along with the SSCI and the HPSCI, the OMB and the Defense Subcommittees of the Appropriations Committees in both houses of Congress must review these details. As with all other government organizations, an examination and approval process applies to the CIA's functions.

THE 1980s: IRAN-CONTRA. The infamous Iran-contra affair of the 1980s is one of the largest scandals to have plagued the intelligence community. The Reagan administration was determined to contain what it determined was a threat by the Sandinista government of Nicaragua to export communism to nearby countries. In April 1984, when word leaked out that CIA agents had helped place mines in three Nicaraguan harbors, several congressional representatives claimed that the CIA had not informed them properly, and some were convinced that they had been deceived. As a result, Congress passed a law in 1986, known some months later as the Boland Amendment to the War Powers Act of 1973, that prohibited any military aid to the Nicaraguan government's opponents, called the contras.

Still, Reagan's NSC was eager to continue such aid. The national security advisor and his staff, taking the view that the Boland Amendment did not apply to the NSC, continued aiding the contras by other means—for example, via private funds and contributions from other nations. The administration's efforts to skirt congressional appropriations (the only legitimate funds for national security) included efforts by members of the NSC to divert funds from sales of arms to Iran (although the sales of the arms were also done in an attempt to secure the release of American hostages).

Before long, Congress became extremely disenchanted with the CIA and the NSC. It investigated the Iran-contra connection, finding that CIA personnel in Central America had rendered logistical and tactical support and assistance even after passage of the Boland Amendment. Congressional committees investigating the affair also concluded that senior officials of the CIA had misled Congress, withheld information, or failed to contradict others who they knew were giving incorrect testimony.

THE 1990s: ADDITIONAL SCRUTINY OF THE CIA. Congress chartered the Commission on Roles and Capabilities of the U.S. Intelligence Community in 1994 to play an advisory role on the use of intelligence in national security. In the years that followed, it made a number of novel recommendations. One recommendation, for example, is that the government should disclose the current fiscal year's budget for the intelligence community and the total amount requested for the next fiscal year. As the commission noted in its final report, intelligence agencies "are institutions within a democracy, responsible to the president, the Congress, and ultimately, the people. Where accountability can be strengthened without damaging national security . . . it should be."

The 1996 report of a commission appointed by the Clinton administration to investigate the intelligence community also urged the CIA to pursue its mission with less secrecy and more accountability. The report suggested that the country take a middle-ground approach to the CIA's future by neither abolishing it nor giving it more powers.

The CIA Now and in the Future

Since the end of the cold war, the intelligence community has focused on such activities as fighting global terrorism, assisting law enforcement in fighting narcotics producers and traffickers, and collecting economic intelligence. Its strong mandate has been to cooperate more closely within the intelligence community and to reduce or eliminate duplications of effort. The CIA has established special multidisciplinary centers to address such major issues as nonproliferation, counterterrorism, counterintelligence, international organized crime and narcotics trafficking, environment, and arms control intelligence.

To address the threat of terrorism against American interests abroad, for example, the CIA created the Counter-Terrorism Center (CTC) in 1986, three years after the U.S. embassy and Marine barracks bombings in Beirut, Lebanon. This interagency group includes members representing the Pentagon and FBI, as well as the CIA. Though the CTC has been criticized for failing to prevent the September 11, 2001, terrorist attacks masterminded by Osama bin Laden, it did succeed in capturing Abu Zubaydah, bin Laden's chief of operations and recruiting, on March 27, 2002.

In 2004, the National Commission on Terrorist Attacks on the United States, popularly called the 9/11 Commission, studied how the government could be reformed to prevent terrorist attacks such as those on September 11, 2001. Their report made several suggestions as to how to improve intelligence operations in particular, including the creation of a National Intelligence Director. This director would oversee all intelligence operations, regardless of the specific agency conducting them. The commission's report stirred a series of proposals and arguments. On August 2, 2004, President George W. Bush called for Congress to establish the position of National Intelligence Director, who would be "the President's principal intelligence advisor and will oversee and coordinate the foreign and domestic activities of the intelligence committee." On August 27, 2004, President Bush issued executive orders enhancing the fiscal authority of the position of CIA director, creating a new national counterterrorism center, and authorizing the creation of a presidential board on safeguarding civil liberties.

Some Americans do not trust the CIA, and the intelligence community as a whole, and are wary of the national security agencies' sweeping new powers, provided through the Patriot Act, to conduct surveillance against U.S. residents involved in terrorist activities. Other Americans state that there has to be some entity protecting the country and that, pursuant to its mission, the tasks and abilities of the intelligence community need be potentially unlimited in technical and geopolitical scope. For better or worse, an intelligence system is indispensable to protecting national security, yet a balance between security and civil rights must be maintained.

THE FEDERAL BUREAU OF INVESTIGATION (FBI)

The FBI is the part of the U.S. Department of Justice (DOJ) charged with investigating crimes and working with law enforcement agencies. According to the FBI's official mission outlined at http://www.fbi.gov, its duties are:

- To uphold the law through the investigation of violations of federal criminal law

- To protect the United States from foreign intelligence and terrorist activities

- To provide leadership and law enforcement assistance to federal, state, local, and international agencies

- To perform these responsibilities in a manner that is responsive to the needs of the public and is faithful to the Constitution of the United States

How the FBI Is Organized

The FBI is located in Washington, D.C., and is headed by a director, who holds a maximum term of ten years. The director is appointed by the president but has to be approved by the Senate. The FBI director and the Washington, D.C., office coordinate the work of fifty-six field offices, about four hundred satellite offices (called resident agencies), and forty-five foreign posts (called legal attaché offices, or legats). As of June 30, 2003, there were nearly 28,000 FBI employees. About 11,633 of these were special agents, while 15,904 held support positions.

The FBI's goals are often confused with those of several other government agencies, notably the CIA, the Drug Enforcement Agency (DEA), and the Bureau of Alcohol, Tobacco, and Firearms (ATF). The FBI is distinct from the CIA in two major ways: (1) the CIA is specifically forbidden from collecting information on U.S. citizens or corporations (it is allowed to collect information only on foreign citizens and other countries); and (2) the CIA is not a law enforcement agency but rather collects and analyzes data pertinent to national security. The FBI differs from both the DEA and the ATF in that those agencies have very specific missions (the enforcement of drug laws and the enforcement of firearms statutes, including the investigation of nonterrorist arsons and bombing incidents, respectively), while the FBI, as stated on its Web site (http://www.fbi.gov/), is the "primary law enforcement agency for the U.S. government."

The FBI's activities are monitored by a variety of government agencies. The FBI director reports directly to the U.S. attorney general. The FBI reports investigative findings to the attorney general and U.S. attorneys nationwide, and these findings are also often examined by judicial agencies. The U.S. Congress supervises FBI budget requests (in 2003, the budget was $4.3 billion), as well as its day-to-day operations and investigations.

The History of the FBI

THE EARLY YEARS: FOUNDING TO WORLD WAR II. The FBI came into being on July 26, 1908, when President Theodore Roosevelt's attorney general, Charles J. Bonaparte, ordered a group of special agents to report to Chief Examiner Stanley W. Finch. In 1909 this force was designated the Bureau of Investigation. At first, the bureau mainly investigated crimes like antitrust or naturalization violations. After the outbreak of World War I (1914–18), its mandate expanded, with the bureau gaining some responsibility in such areas as espionage, sabotage, and selective service. It monitored individuals such as anarchists, communists, trade union activists, civil rights activists, and foreign resident agitators.

However, after the war, the bureau was criticized for supposedly conducting illegal searches and denying suspects proper legal counsel.

Throughout the 1920s and 1930s the bureau expanded. In 1924 perhaps the bureau's most famous director took office—J. Edgar Hoover, who would remain bureau/FBI director for almost half a century until his death in 1972. Hoover immediately began to reform and "professionalize" the organization. The fingerprint database, which would come to be the largest repository of fingerprints in the world, was created in 1924. The FBI Laboratory was established in 1932 to analyze physical evidence. In 1934 agents gained the legal right to make arrests themselves rather than having to rely on local law enforcement officials, and in 1935 the bureau became known by its current name, the Federal Bureau of Investigation.

As World War II (1939–45) started in Europe, the FBI began focusing significant energies on such wartime issues as sabotage, and when the United States entered the war in 1941, the bureau's responsibilities increased again. This time, it was responsible for enforcing the internment of American citizens of Japanese descent, something that was done, supporters argued, for reasons of national security, despite J. Edgar Hoover's protests that it was an unnecessary measure.

CONTROVERSIAL DECADES: THE 1950s TO 1970s. In the post–World War II period, the Soviet Union gained in power and became a major rival of the United States. In the late 1940s the Soviet army occupied much of Eastern Europe. China fell to the communists in 1949, the same year the Soviet Union tested its first atomic weapon. The combined specters of communist aggression and atomic weaponry made the American public nervous. The FBI began undertaking thorough background checks of applicants for government jobs—particularly those requiring access to nuclear data or materials.

Yet during the 1950s and 1960s, the FBI overstepped its bounds, providing secret assistance to the House Committee on Un-American Activities (HCUAA) and Senator Joseph McCarthy. The HCUAA and Senator McCarthy promoted communist "witch hunts" in an attempt to expose communist sympathizers, whom they believed had infiltrated all parts of American culture and government and who threatened national security. The FBI provided the HCUAA with information from confidential files.

The 1960s were a turbulent decade, during which the FBI was involved in civil rights cases in America's South. In the summer of 1964 the FBI investigated the murders of three voter registration workers in Mississippi. In 1967 seven men connected to the Ku Klux Klan were convicted of the murders. The FBI also investigated and made arrests in the assassinations of civil rights leaders Martin Luther King, Jr., and Medgar Evers. Violent groups such as the Weather Underground, who set off bombs in the Pentagon and the U.S. Capitol, and the Black Panther Party, whose members were involved in several shootouts with police, were also investigated.

In the 1970s there were attacks on the FBI's domestic information gathering after news damaging to the FBI emerged—notably, that many of the FBI investigations of the 1950s and 1960s were illegal. At the time, public scrutiny went beyond the FBI's information activities to its program of domestic covert actions, called COINTELPRO—shorthand for "counterintelligence program." COINTELPRO was evidently intended to disrupt and discredit the leaders of certain domestic dissident groups, such as "New Left" groups (who opposed the Vietnam War), the U.S. Communist Party, the Socialist Workers Party, the Ku Klux Klan, the Nazi Party, black nationalists, the Black Panthers, and other extremist groups. It became clear that these secret disruptive activities, which dated back to the 1950s, went well beyond the law in most instances.

According to Kenneth O'Reilly (in "Federal Bureau of Investigation," *Dictionary of American History, Supplement,* New York: Scribner's, 1996), "After Hoover's death in 1972, many of the FBI's files were opened under the Freedom of Information Act. They revealed that the bureau had done much more than compile intelligence on such 'dissidents' as civil rights leader Martin Luther King, Jr. Special agents committed thousands of burglaries to gather information and ran counterintelligence programs to 'neutralize' communists and anti-Vietnam protestors." The FBI also ran illegal wiretaps and collected and distributed information for political reasons. Special committees from the Senate and House investigated these abuses, and a 1977 DOJ task force referred to these types of actions as felonious conduct. It also became known that Hoover also used the power of the FBI to more or less blackmail politicians into keeping him in office. He used FBI staff to conduct research on prominent congressional representatives and senators, then used any negative information as leverage against them.

The 1970s were also made turbulent by the Watergate scandal, which forced President Richard M. Nixon to resign. It also led to the resignation of the acting FBI director, L. Patrick Gray, because he had destroyed Watergate evidence and had leaked information on the FBI investigation to White House staff. Watergate hearings revealed that President Nixon had used the FBI to conduct illegal investigations of his political enemies.

COINTELPRO was terminated by the attorney general in 1971. In the early 1990s, the FBI pulled back on its domestic counterintelligence activities, limiting its focus to domestic terrorist and antigovernment militia groups. In a way, this newfound restraint would be rewarded: the 1996 Anti-Terrorism Law, passed by Congress in response to the Oklahoma City, Oklahoma, and Atlanta, Georgia (Olympics), bombings, gave freer rein to the FBI to conduct surveillance and counterintelligence against truly violent groups.

RECENT HISTORY: THE 1980s THROUGH TODAY. In the early 1980s the prevalence of terrorism soared, and counterterrorism became an important part of the FBI's mission. At the same time, the FBI worked with the DEA to combat drug activity. With the collapse of the Soviet Union in 1991, the United States was left as the one major superpower in the world. The FBI took this opportunity to concentrate more of its resources on domestic issues while still taking a large part in national security efforts. Some three hundred special agents were reassigned from counterintelligence work to domestic violent crime investigations. Not all of the FBI's domestic efforts were successful, and the shooting deaths in 1992 of two members of a white separatist family in Ruby Ridge, Idaho, and the eighty deaths resulting from a standoff at the Branch Davidian religious sect compound in Waco, Texas, in 1993 turned some public opinion against the FBI. Many American politicians and citizens considered the Ruby Ridge and Waco incidents evidence that the FBI could not adequately handle "crisis situations."

In addition, then–FBI director William S. Sessions was accused of numerous ethical violations, including personal use of FBI resources. A DOJ investigation later confirmed these violations, but Sessions refused to resign, so he was fired by then-President Bill Clinton. Louis J. Freeh, who became FBI director in 1993, attempted to revitalize the beleaguered bureau, streamlining and overhauling various FBI procedures. An International Law Enforcement Academy was founded in 1995.

Terrorism continued to be a major issue throughout the 1990s, and the FBI participated in investigations of the 1993 World Trade Center bombing in New York City, the 1995 Alfred P. Murrah Federal Building bombing in Oklahoma City, Oklahoma, the "Unabomber" bombings of Theodore Kaczynski, and U.S. embassy bombings in Kenya and Tanzania.

The late 1990s saw more controversy, related to supposedly sloppy work at the FBI Laboratory and the investigation of Richard Jewell, who was questioned in connection with a bombing at the 1996 Olympic Games in Atlanta, Georgia. Jewell, a security guard who was working at the Olympics, was originally the FBI's prime suspect. Jewell was never charged, and his name was ultimately cleared, but the FBI was suspected of leaking his name to the media and of interviewing him outside the guidelines of the law. The FBI Laboratories were cleared by a 1997 DOJ investigation of the most heinous charges, but the FBI did censure agents for violating Jewell's rights when they interrogated him without his lawyer present.

The FBI took a leading role in investigations relating to the September 11, 2001, terrorist attacks, including the perpetrators of the attacks, the anthrax-laced letters that followed, and the prevention of future attacks. In the immediate aftermath of the attacks, more than half of the FBI's special agents were working on issues directly related to the attacks or prevention of future attacks.

FBI Investigations

On its Web site the FBI defines its "investigative functions" as "applicant matters; civil rights; counterterrorism; foreign counterintelligence; organized crime/ drugs; violent crimes and major offenders; and financial crime." It can take on any investigation that Congress has not expressly given to another federal agency. Examples of investigations handled by other agencies include postal investigations, which are handled by the U.S. Postal Service; customs investigations, handled by the U.S. Customs Service; and counterfeiting investigations, handled by the Secret Service. The mandate of the FBI includes gathering information and evidence and making arrests (at least on U.S. soil—special agents generally do not have the power to make arrests abroad). However, the FBI has no power to prosecute or recommend prosecution for specific individuals; those decisions must come from federal prosecutors working for the DOJ.

SPECIAL AGENTS. The FBI staff who carry out investigations are called special agents. They have numerous powers to help them fulfill their duties, including the rights to carry weapons, to arrest suspects, and to subpoena witnesses of grand jury investigations. With judicial backing, they can tap telephone lines, read mail, and obtain personal documents such as tax returns and telephone bills.

SHARING AUTHORITY WITH LOCAL LAW ENFORCEMENT. In investigations with "concurrent jurisdiction" (for example, where a crime is a local, state, and federal violation at the same time), the FBI does not "outrank" the other agencies. Law enforcement agencies representing all levels of government, including the FBI, often work cooperatively on investigations. Some of the ways in which the FBI can assist local investigations include:

- Monitoring and identifying fugitives' fingerprints. The FBI maintains a Criminal Master File in its Integrated Automated Fingerprint Identification System (IAFIS) that contains the fingerprints and criminal histories of more that forty-seven million individuals.

- Entering data on local fugitives into its national database, the National Crime Information Center. In 2003 the center revealed that it processes about 3.5 million inquiries a day from law enforcement agencies across the nation.

- Providing laboratory analysis of evidence

- Pursuing and attempting to arrest fugitives who cross state lines or leave the country. According to information available on the FBI's Web site in October 2004, at any given time the FBI is searching for about twelve thousand fugitives.

Local law enforcement agencies assist the FBI by providing it with crime statistics, which are then collected in the Uniform Crime Reporting Program. These statistics are provided by about seventeen thousand agencies, and the data represent 94% of the U.S. population. The FBI works with federal law enforcement agencies as well, both on specific investigations and in ongoing task forces, and also shares information with some foreign law enforcement organizations. Training of law enforcement officers is provided by the FBI to both domestic and foreign law enforcement staff.

INTERNATIONAL AND TERRORIST THREATS. The FBI has various duties in regard to terrorism and espionage. It investigates bombings both on U.S. soil and abroad when the suspected target of the bombing is a U.S. citizen or a U.S. interest (such as an embassy). It works with other domestic and foreign agencies to share information that might be useful in combating terrorism. The FBI monitors hate groups and potential terrorist groups in accordance with guidelines set by the attorney general. Only those groups showing strong evidence of a predilection toward unlawful behavior are monitored.

Beyond terrorism, the FBI also has other duties to protect the country from international threats, including counterintelligence. According to information available on the FBI's Web site in October 2004, "the FBI is responsible for detecting and lawfully countering actions of foreign intelligence services and organizations that employ human and technical means to gather information about the United States which adversely affects U.S. national interests." This espionage can consist of "the acquisition of classified, sensitive, or proprietary information from the U.S. government or U.S. companies." The FBI estimates that espionage costs the United States $100 billion each year.

A Changing FBI in the Wake of 9/11

After the September 11, 2001, terrorist attacks against America, a new wave of criticisms were leveled at the FBI. Several different incidents provided detractors with ammunition. For one, a group of FBI counterterrorism special agents based in Minneapolis, Minnesota, learned of a new student at a Minnesota flight school—one Zacarias Moussaoui. Moussaoui piqued the agents' interest because he paid $6,200 in cash for flight training and only wanted to learn to fly, not land, Boeing 747s. When Moussaoui's visa expired in August 2001 and he continued to remain in the country, the Minneapolis agents arrested him and did a thorough background check, only to discover that his background included ties to followers of Osama bin Laden. The agents requested a special search warrant to check a computer disk owned by Moussaoui, but their request was denied. Less than a week later, the September 11, 2001, terrorist attacks rocked the nation. As of 2004, Moussaoui was being prosecuted as the alleged "twentieth" hijacker, who would supposedly have participated in the attacks had he not been in FBI custody.

A "whistle-blowing" letter from Minneapolis FBI Chief Counsel Colleen Rowley accused the FBI of deliberately obstructing the Minneapolis agents. Another FBI special agent, Kenneth Williams, wrote a memo in July 2001 warning of suspicious activity by Middle Eastern men in Arizona flight schools. The Phoenix, Arizona, agent suggested that FBI headquarters take a nationwide survey of Arab-American flight school students, but the memo was not passed along to the appropriate people and was never acted upon. The FBI director was not aware it existed until a few days after September 11, 2001. A 1998 memo out of the Oklahoma City, Oklahoma, FBI office warned of a similar phenomenon but did not receive much attention either.

As part of the wave of reforms undertaken after the 9/11 terrorist attacks, FBI Director Robert S. Mueller III instituted a major reorganization of the FBI to deal with such complaints. Director Mueller viewed his plan as an "evolving road map" that could be adjusted to meet the needs of American security. According to a June 6, 2002, statement before the Senate Committee on the Judiciary ("A New FBI Focus"), Director Mueller adapted previous strategic plans to come up with the following ten priorities for the FBI (listed here in his own words):

1. Protect the United States from terrorist attack.

2. Protect the United States against foreign intelligence operations and espionage.

3. Protect the United States against cyber-based attacks and high-technology crimes.

4. Combat public corruption at all levels.

5. Protect civil rights.

6. Combat transnational and national criminal organizations and enterprises.

7. Combat major white-collar crime.

8. Combat significant violent crime.

9. Support federal, state, municipal, and international partners.

10. Upgrade technology to successfully perform the FBI's mission.

In the same statement, Director Mueller noted that the changes to the FBI to accomplish these priorities would

FIGURE 8.4

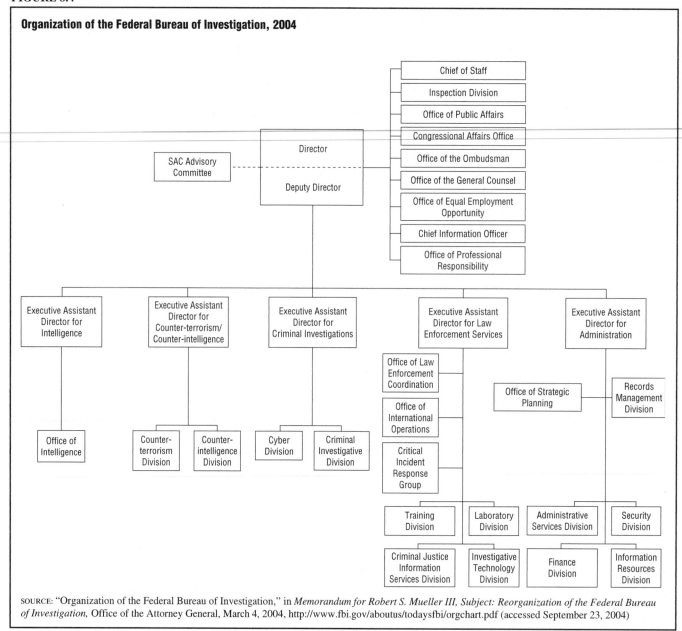

Organization of the Federal Bureau of Investigation, 2004

SOURCE: "Organization of the Federal Bureau of Investigation," in *Memorandum for Robert S. Mueller III, Subject: Reorganization of the Federal Bureau of Investigation,* Office of the Attorney General, March 4, 2004, http://www.fbi.gov/aboutus/todaysfbi/orgchart.pdf (accessed September 23, 2004)

be "built upon three key interrelated elements: (1) refocusing FBI mission and priorities; (2) realigning the FBI workforce to address these priorities; and (3) shifting FBI management and operational cultures to enhance flexibility, agility, effectiveness, and accountability."

The priority reorganization led to important changes in the FBI's Counterterrorism, Counterintelligence, and Laboratory divisions, and the establishment of a Cyber Division, as well as a Security Division, Records Management Division, and Office of Law Enforcement Coordination. Other changes included personnel reorganization, revised procedures for information sharing, and changes in the way the FBI conducts criminal investigations.

PERSONNEL CHANGES. In addition to the movement of personnel from some divisions to others, the personnel reor-

ganization led to five new executive assistant directors, who report directly to the FBI director. These assistant directors oversee the areas of intelligence, counterterrorism and counterintelligence, criminal investigations, law enforcement services, and administration. (See Figure 8.4). This relieved some of the burden formerly shouldered by the FBI's deputy director and increased accountability and oversight.

The restructuring plan also called for the nature of the FBI's workforce to change. Previously, most special agents had been generalists. After the reorganization the agency sought subject experts with extensive knowledge in such fields as information technology, foreign languages, engineering, and so forth.

INFORMATION SHARING AND THE OFFICE OF LAW ENFORCEMENT COORDINATION. In a May 8, 2002, statement before the Senate Committee on the Judiciary ("FBI Reorganization"), Director Mueller admitted that "information sharing," or FBI coordination with state and local law enforcement authorities, left something to be desired: "[Our] history of solid, personal relationships alone was not addressing the basic information needs of our counterparts. . . . Adding 650,000 state and local officers to our efforts is the only way to make this truly a national effort, not just a federal effort." Many of the changes to specific FBI divisions integrated an increase in information sharing with state and local law enforcement agencies.

The FBI's plan also created a new Office of Law Enforcement Coordination, whose purpose is to "improve relationships and information sharing with state and local police professionals and others" ("FBI Reorganization"). The emphasis on information sharing had its genesis in local law enforcement complaints that the FBI sometimes kept local agencies "out of the loop" and that FBI personnel turnover had a damaging effect on cooperation efforts.

STRONGER FBI/CIA COOPERATION. The FBI also strengthened its ties with the CIA to facilitate information sharing. In a June 27, 2002, statement before the Senate Committee on Governmental Affairs ("Homeland Security"), Director Mueller noted that the FBI/CIA "relationship has a long history, and is the subject of much contemporary comment, most of it critical. But for those commentators, I would counsel caution. The relationship has changed, and is still changing, all for the better." Under the new plan, FBI staff worked at CIA headquarters, and vice versa. Information about important security issues was exchanged between the two agencies on a daily basis.

SECURITY DIVISION CREATED. The FBI reorganization plan included the creation of a Security Division, the purpose of which is to raise the level of FBI security practices and standards. This measure was in many ways a response to the 2001 arrest of Robert P. Hanssen, a veteran FBI special agent who was charged with selling national security secrets to the Soviet Union/Russia during a fifteen-year time span. As Director Mueller's May 8, 2002, statement declared, "We need to remedy the weaknesses that the Hanssen investigation made painfully obvious."

RECORDS MANAGEMENT DIVISION ESTABLISHED. The FBI reorganization plan established a new Records Management Division. This division is charged with modernizing FBI record-keeping systems, policies, and procedures in order to prevent important records from becoming lost or misplaced.

COUNTERTERRORISM DIVISION REORGANIZED. Much of Director Mueller's plan focused on improvement of the FBI's counterterrorism investigations and programs. The FBI Counterterrorism Division work was once done by local field offices investigating groups in their own areas, with little contact between them. In December 2001 the

TABLE 8.1

FBI intelligence analysts, 2004

Year	Total	Hired
FY 2001	1023	42
FY 2002	1012	96
FY 2003	1180	250
3/4/2004	1197	72
	Grand total	**460**

SOURCE: "Intelligence Analysts," in *Report to the National Commission on Terrorist Attacks upon the United States: The FBI's Counterterrorism Program since September 2001*, U.S. Department of Justice, Federal Bureau of Investigation, April 14, 2004, http://www.fbi.gov/publications/commission/9-11commissionrep.pdf (accessed September 23, 2004)

Counterterrorism Division was reorganized into branches, sections, and units, each of which focuses on a different aspect of the terrorism threat facing the United States. (See Figure 8.5.) Each component within the division is staffed with intelligence analysts and subject matter experts who work with agents in the field, providing a real-time reaction to imminent threats.

After September 11, 2001, the number of FBI agents working on counterterrorism matters rose from 1,351 to 2,398 in February 2004. (See Figure 8.6.) The number of intelligence analysts working on counterterrorism increased from 1,023 in 2001 to 1,197, as of March 4, 2004. (See Table 8.1.) In addition, after 2001 nearly seven hundred new translators, both FBI employees and contract workers, were added. Among new employees, the increases have been particularly dramatic in the languages of the Middle East. (See Table 8.2.)

COUNTERINTELLIGENCE DIVISION RESTRUCTURED. The restructuring plan for the FBI's Counterintelligence Division instituted a new espionage section, focusing on investigations. This allows operations staff to concentrate their energies on detecting and thwarting intelligence threats. The division works more closely than previously with other government agencies and the private sector to protect U.S. secrets. An Office of Intelligence was created in December 2001 to provide a "tactical intelligence analytical capacity" ("A New FBI Focus")—in other words, to try to create a "big picture" from what may be many seemingly unrelated pieces of data.

CYBER DIVISION ESTABLISHED. The FBI established its Cyber Division in December 2001. This group, according to the FBI Web site, "coordinates, supervises and facilitates the FBI's investigation of those federal violations in which the Internet, computer systems and networks are exploited as the principal instruments of targets of criminal, foreign intelligence, or terrorism activity and for which the use of such systems is essential to that activity." The FBI works with private businesses, academia, and governmental agencies to procure the technology skills needed to conduct these high-tech investigations.

FIGURE 8.5

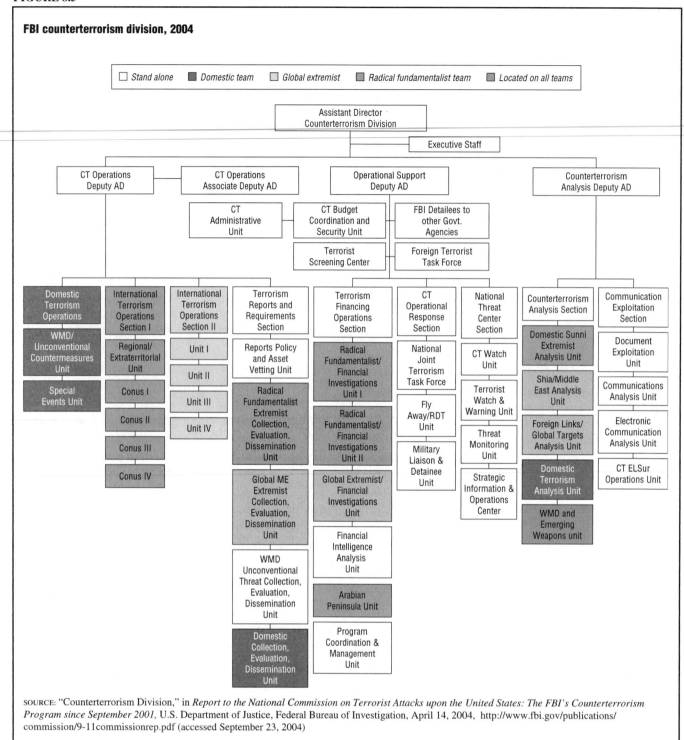

FBI counterterrorism division, 2004

SOURCE: "Counterterrorism Division," in *Report to the National Commission on Terrorist Attacks upon the United States: The FBI's Counterterrorism Program since September 2001,* U.S. Department of Justice, Federal Bureau of Investigation, April 14, 2004, http://www.fbi.gov/publications/commission/9-11commissionrep.pdf (accessed September 23, 2004)

LABORATORY FUNCTIONS DIVIDED. During the FBI reorganization, the Laboratory Division was split into two separate divisions, Laboratory and Investigative Technologies, to address questions of "mission, staffing, and funding" ("A New FBI Focus"). The new Laboratory Division collects, processes, and analyzes evidence. It also provides training and conducts forensic research and development. The Investigative Technologies Division focuses on technical support to investigators, including electronic or physical surveillance and wireless or radio communications. Like the Laboratory Division, the Investigative Technologies Division also has training and research and development functions.

CRIMINAL INVESTIGATION CHANGES. With the emphasis on improvements to the FBI's counterterrorism efforts, Director Mueller pointed out that particular care should be taken that the new operations should not eclipse the FBI's "day-to-day" criminal investigation priorities: public cor-

FIGURE 8.6

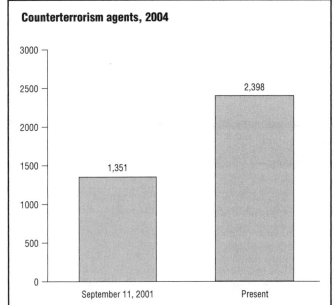

Counterterrorism agents, 2004

SOURCE: "Counterterrorism Agents," in *Report to the National Commission on Terrorist Attacks upon the United States: The FBI's Counterterrorism Program since September 2001,* U.S. Department of Justice, Federal Bureau of Investigation, April 14, 2004, http://www.fbi.gov/publications/commission/9-11commissionrep.pdf (accessed September 23, 2004)

TABLE 8.2

FBI translators

Major language	9/11/01	Present	Net change
Arabic	70	207	+137
Farsi	24	55	+31
Pashto	1	10	+9
Urdu	6	21	+15
Chinese	67	119	+52
French	16	34	+18
Hebrew	4	11	+7
Korean	18	25	+7
Kurdish	0	5	+5
Russian	78	100	+22
Turkish	2	10	+8

SOURCE: "Increased Language Translation Capabilities to Support Counterterrorism," in *Report to the National Commission on Terrorist Attacks upon the United States: The FBI's Counterterrorism Program since September 2001,* U.S. Department of Justice, Federal Bureau of Investigation, April 14, 2004, http://www.fbi.gov/publications/commission/9-11commissionrep.pdf (accessed September 23, 2004)

ruption, civil rights, transnational and national criminal organizations, major white-collar crime, and significant violent crime. While staff were transferred to counterterrorism assignments from criminal investigation areas like drug investigations, white-collar-crime investigations, and violent-crime investigations, these areas remain a priority.

However, in the short run, according to one of Director Mueller's statements before the Senate Committee on the Judiciary, the FBI

> must be prepared . . . to defer criminal cases to others, even in significant cases, if other agencies possess the expertise to handle the matter adequately. In situations where other . . . capabilities are not sufficient to handle a case or situation, SACs [Special Agents in Charge] should be prepared to step in and provide FBI resources as needed. However, once the immediate situation is under control or resolved I expect SACs to reevaluate the level of FBI commitment and make necessary adjustments.

Director Mueller pointed out that it was also crucial for FBI agents working on seemingly mundane cases to be watchful for any evidence of terrorism. In his June 2002 statement he noted, "Other terrorist investigations have revealed patterns of low-level criminal activity by terrorists. It is the duty of every FBI employee to remain vigilant for suspicious activity or informant information that could be a tip-off to a future terrorist attack." That way, even FBI agents not involved in the FBI's Counterterrorism or Counterintel-

ligence Divisions can become useful tools in the battle against terrorism.

THE DEPARTMENT OF HOMELAND SECURITY

In late November 2002 President George W. Bush signed the Department of Homeland Security (DHS) bill, thereby officially creating one of the most important domestic security agencies. The new department is the result of reorganizing twenty-two federal agencies with some 180,000 employees into the DHS, which is headed by the director of homeland security. (See Figure 8.7.) Former Pennsylvania governor Tom Ridge was named the first secretary of the DHS. Efforts to establish the DHS were spurred by the September 11 terrorist attacks of 2001. According to the National Strategy for Homeland Security, homeland security is "a concerted national effort to prevent terrorist attacks within the United States, reduce America's vulnerability to terrorism, and minimize the damage and recover from attacks that do occur." Six aspects of the department are: (1) intelligence and warning; (2) border and transportation security; (3) domestic counterterrorism; (4) protecting critical infrastructure and key assets; (5) defending against catastrophic threats; and (6) emergency preparedness and response.

The president keeps abreast of issues relating to homeland security through the director of the DHS and an Advisory Council on Homeland Security. This council is primarily divided into counterterrorism and cyberspace security divisions and features policy coordination committees that oversee plans between state and local governments. The FY2004 budget for the new department was $36 billion. For FY2005, $40 billion was proposed. According to *Securing Our Homeland: U.S. Department of Homeland Security Strategic Plan* (Washington, DC: Department of Homeland Security, 2004), the department's immediate goals include identifying and eliminat-

FIGURE 8.7

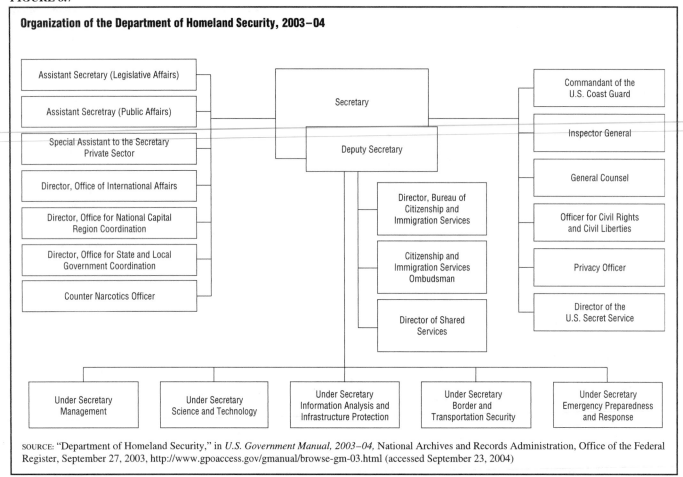

Organization of the Department of Homeland Security, 2003–04

SOURCE: "Department of Homeland Security," in *U.S. Government Manual, 2003–04,* National Archives and Records Administration, Office of the Federal Register, September 27, 2003, http://www.gpoaccess.gov/gmanual/browse-gm-03.html (accessed September 23, 2004)

ing areas of overlap and omission within the twenty-two agencies, developing results-oriented approaches, and monitoring performance so that funds are allocated to the most successful efforts.

U.S. CITIZENSHIP AND IMMIGRATION SERVICES (USCIS)

Immigrants have made the United States the strongest and most diverse country in the world, and the vast majority of legal immigrants work together to maintain the principles on which the United States was founded and make their adopted country a better place for everyone living within it. Even many illegal immigrants have a sincere loyalty to the United States and a desire to stay in the country because they believe the United States allows them to make better lives for themselves. Still, some people enter the country without the best interests of the United States in mind. They may actively seek to do harm to the nation's citizens and values. These people, many of whom are living in the country illegally, can present national security threats to the United States.

The terrorist attacks of September 11, 2001, present a good example of how illegal immigration or illegal entrance into the United States threatens national security.

Of the nineteen alien airplane hijackers who participated in the attacks, several had no immigration documents at all, and others had overstayed their visas (papers granted by the U.S. State Department, giving permission to travel within the United States). Authorities and the public are still not certain how some of the hijackers, of whom the FBI and Immigration and Naturalization Service (INS) had no record at all, actually entered the country. As a result, after the attacks, reforming the U.S. immigration system became an important issue. In 2003 the INS became a branch of the Department of Homeland Security. In the spring of 2004 it was replaced by two new agencies, U.S. Citizenship and Immigration Services (CIS) and U.S. Customs and Border Protection (CBP).

What Was the INS?

The INS began in 1933, after the immigration and naturalization functions of two different agencies of the federal government were consolidated by executive order within the Labor Department. It was headed by a commissioner who reported to the attorney general. During a period of increased international tensions prior to World War II, the INS was moved into the DOJ in 1940.

Since the INS determined who may enter the United States and enforced immigration laws with respect to those who remained, many people placed some of the blame for the September 11, 2001, terrorist events on the INS. The agency conducted immigration inspections of travelers entering (or seeking entry to) the United States; regulated permanent and temporary immigration to the United States; provided services such as granting legal permanent status, temporary status, and naturalization (the process of obtaining U.S. citizenship); controlled U.S. borders; and worked with other agencies to remove illegal aliens.

The agency also shared the responsibility for inspection of all applicants seeking admission to the United States with the U.S. Customs Service at about 250 U.S. ports of entry at land, air, and sea locations. The INS and the Customs Service prevented the entry of illegal aliens mainly by detecting fraudulent documents, including claims of U.S. citizenship or permanent resident status. Inspectors from the two agencies also seized conveyances used for illegal entry, such as cars, trucks, and boats.

It was the task of the U.S. Border Patrol, a subagency of the INS, to secure the eight thousand miles of U.S. borders—clearly a difficult and dangerous task. The Border Patrol worked to stop the influx of illegal aliens, the smuggling of aliens, and also seized illegal imports, like narcotics. In 2000 the Border Patrol consisted of some 9,200 agents. They located and deported nearly 1.7 million illegal aliens attempting to enter the United States.

Restructuring the INS

In the wake of the events of September 11, 2001, Attorney General John Ashcroft announced in November 2001 the Bush administration's plans to reorganize the INS by 2003 if approved by Congress. Under the new Homeland Security bill signed by President Bush in November 2002, INS functions were incorporated into the new DHS. On March 1, 2003, the DHS's Directorate of Border and Transportation Security (BTS) officially assumed responsibility for securing the nation's transportation systems and borders, including 317 official ports of entry, and also assumed responsibility for enforcing the nation's immigration laws. The INS's immigration enforcement functions were transferred to the U.S. Customs and Border Protection (CBP) and the immigration service functions were placed under the U.S. Citizenship and Immigration Services (CIS). Both of these agencies are parts of the DHS.

The CBP combines the duties of four separate agencies: U.S. Customs, which supervised the import and export of goods to and from the United States; the U.S. Border Patrol, which had served to stop illegal aliens and smugglers; the INS, which had dealt with those wishing to become legal citizens; and the Animal and Plant Health Inspection Service, which had worked to keep diseased livestock and infected plants from entering the country.

CBP employs some forty-one thousand people to manage, control, and protect the nation's borders. Of these, eighteen thousand CBP agents are stationed at ports of entry to carry out all of the functions previously performed by the earlier inspectional workforces. Called the "One Face at the Border" plan, this approach streamlines the enforcement of border security into a single coordinated workforce. While its primary goal is to secure the country from terrorists, the CBP continues the important traditional work of overseeing legitimate trade, stopping illegal aliens, assisting visitors and legal immigrants, and inspecting plants and animals entering the country. As CBP Commissioner Robert C. Bonner explained in a statement posted on the agency's Web site (http://www.cbp.gov/): "CBP's priority mission is preventing terrorists and terrorist weapons from entering the United States, while also facilitating the flow of legitimate trade and travel."

According to the CBP's *Performance and Annual Report, Fiscal Year 2003,* among the tasks of the CBP are:

- Determining the admissibility of people and goods

- Regulating and facilitating international trade

- Collecting duties, taxes and fees—$25 billion was collected in FY 2003

- Enforcing all laws of the U.S., including trade laws, at our borders

- Intercepting high-risk travelers while expediting the travel of low-risk travelers

- Deploying selectivity techniques, technology, and tools for the physical inspection of travelers' baggage and vehicles to enforce U.S. laws and avert high-risk situations

Among the new programs established by the CBP to meet its goals are the Custom Trade Partnership against Terrorism (C-TPAT), the Free and Secure Trade (FAST) Program, and the Container Security Initiative (CSI). The C-TPAT is a joint initiative between government and business to ensure that proper supply-chain security procedures are in place—from factory to shipping dock—to keep all shipments into the U.S. safe from tampering by terrorists. FAST allows importers, truck drivers, and commercial carriers who bring goods into the U.S. from Canada or Mexico to enjoy expedited entry into the United States if they meet specific security criteria. The CSI concerns the safety of containerized cargo shipped into the United States. Under the program, foreign ports agree to have CBP agents inspect maritime containers headed for the United States before they are loaded.

In August 2004 it was announced that the DHS was planning to expand the powers of CBP agents. They will now be allowed to deport illegal aliens caught at the border without providing the aliens the opportunity to make their case before an immigration judge. Until this change, only illegal aliens caught at airports and seaports could be deported without a hearing. The new rule expands that power to those caught along the Mexican and Canadian borders. The process of appearing in immigration court can take up to a year. The DHS stated that this long process puts a strain on detention facilities and takes too much money and manpower resources from more important duties. Only illegal aliens who are third-country nationals rather than Mexican or Canadian citizens, and who are caught within a hundred miles of the Mexican or Canadian border, will be affected by this change.

The CIS handles immigration and citizenship services for those who wish to become U.S. citizens. The CIS staff consists of approximately fifteen thousand employees and contractors. The CIS Web site (http://uscis.gov/) lists the immigrant and nonimmigrant benefits processed by the agency:

- Family-based petitions—facilitating the process for close relatives to immigrate, gain permanent residency, work, etc.

- Employment-based petitions—facilitating the process for current and prospective employees to immigrate or stay in the U.S. temporarily

- Asylum and Refugee processing—adjudicating asylum and the processing of refugees

- Naturalization—approving citizenship of eligible persons who wish to become U.S. citizens

- Special status programs—adjudicating eligibility for U.S. immigration status as a form of humanitarian aid to foreign nationals

- Document issuance and renewal—including verification of eligibility, production and issuance of immigration documents

The BTS also includes the Federal Law Enforcement Training Center (formerly a part of the Department of Treasury) and the Transportation Security Administration (formerly a part of the Department of Transportation). In time, the Federal Protective Service (formerly a part of the General Services Administration) will also become part of the BTS to perform the additional function of protecting government buildings.

CHAPTER 9

THE MILITARY, PEACEKEEPING, AND NATIONAL SECURITY

Within the executive branch of the U.S. government, the Department of Defense (DOD) works directly to deal with national security threats and keep the president's military options open. For fiscal year 2005, the DOD has an estimated budget of nearly $402 billion. Of the total 2005 budget, the army receives $97.2 billion, the navy/Marine Corps gets $119.3 billion, and the air force gets $120.5 billion, with the remainder going to various other DOD departments. The DOD employs more than 3.3 million people, including an active-duty force of approximately 1.4 million and, in addition to weapons, maintains high-tech information systems and expertise the president and others can use to make informed decisions that lead to decisive actions.

MILITARY ADMINISTRATION

Department of Defense (DOD) Organization

The secretary of defense is the president's principal defense advisor and oversees the DOD. The secretary advises the president on military strategy and policy, sets defense budgets, and administers the department. The Office of the Secretary of Defense is the secretary's staff, assisting him in directing the undersecretaries, assistant secretaries, and lower-ranking officials who populate the department, which is organized along both functional and regional lines.

The DOD below the secretary and the secretary's office is made up of the Joint Chiefs of Staff (JCS; a council consisting of the highest-ranking member of each service) and their staff, called the joint staff; the military departments (army, air force, navy/Marines); the nine unified combatant commands (the multiservice groups that directly control U.S. combat forces); and several defense agencies that provide services across the entire DOD, such as the Defense Intelligence Agency.

The Secretary of Defense and the Goldwater-Nichols Act

Originally created in 1947, the position of secretary of defense was meant to be that of a basic coordinator of

the armed services, which prior to that time were much more independent organizations. However, the Goldwater-Nichols Department of Defense Reorganization Act of 1986 (PL 99-433), sponsored by Senator Barry Goldwater (R-Arizona) and House Representative Bill Nichols (D-Alabama), attempted to reduce interservice rivalries and the services' independent organization, promoting "jointness" within the DOD. The act specified the chairman of the JCS as the "principal military advisor to the President, the National Security Council, and the Secretary of Defense." In addition to reporting JCS positions on issues and problems, the JCS chairman could now give any advice he or she thought appropriate. The act also created the new position of vice-chairman of the JCS.

To bridge differences within the separate services' personnel systems, the Goldwater-Nichols Act called for the creation of a "joint specialty," requiring the army, navy, and air force to send a share of their most outstanding officers to both the joint staff in Washington and the unified commands in the field. These officers also had to receive a specified share of available promotions. Goldwater and Nichols, the bill's congressional sponsors, had relied chiefly on analysis and policy recommendations drawn from a study on developing "jointness" and interservice cooperation by the well-known Washington think tank the Center for Strategic and International Studies.

The new measures yielded a defense secretary who, far more than a coordinator, could actively consult the president on defense policy. The military as a whole, according to specialists, became more flexible and responsive; the DOD became more centralized; and the role of defense secretary became more prominent and proactive.

The Joint Chiefs of Staff (JCS)

The members of the JCS have been called the "hinge" between the highest U.S. civilian authorities and the uniformed services. The JCS has six members: the heads of the four branches of the military—the army and air force chiefs of staff, the chief of naval operations, and the Marine Corps commandant—the chairman, and the vice-chairman. The chairman of the JCS is the highest ranking officer in the military. Collectively, they are the senior military advisors to the president, the National Security Council (NSC), and the defense secretary.

Either the chairman or the vice-chairman represents the JCS as a whole at meetings of the NSC and other inter-agency forums. The Goldwater-Nichols Act included a significant attempt to improve the relationship of the JCS to the executive branch. It enhanced the JCS role by increasing the JCS chairman's power and conferring on individual JCS members the right to go directly to the president.

JCS can be important to a president seeking the support of the U.S. populace and Congress for a controversial national security initiative. For example, the JCS supported the president's resolve to achieve ratification of the Panama Canal Treaty in 1977–78. It also supported President Ronald Reagan's decision to discard the Strategic Arms Limitation Treaty II limits on nuclear weapons in 1986 and President George H. W. Bush's reduction of combat forces in Europe by half in 1990. However, some critics believe this record shows that the JCS too often succumbs to presidential pressure instead of staking out its own position on strategic issues.

U.S. Armed Services

The DOD provides the military forces needed to avoid war and to protect national security. The military departments are separate entities, with their own secretaries and service chiefs, which all report to the secretary of defense. They are charged with organizing, training, equipping, and providing forces that will defend the nation and protect national security.

Three military departments, each with its own armed services, report directly to the DOD—the army, navy, and air force. The Marine Corps is a part of the Department of the Navy. The Coast Guard, long under the control of the Department of Transportation, is now a part of the Department of Homeland Security.

THE ARMY. The U.S. Army was created during the American Revolution by the Continental Congress on June 14, 1775, and for more than two centuries it has worked at home and abroad to protect and maintain American interests. As of 2004 the army maintained a standing force of just under 500,000 soldiers. Its main jurisdiction is land combat, and it is often the decisive force in conflicts because of its ability to attack and control large geographic areas. Although the U.S. Army ranks behind many other countries in the number of active-duty soldiers, its advantages in equipment, technology, training, and mobility make it the world's most formidable ground force.

THE NAVY. In a move to protect the freedom of the seas, the U.S. Congress created the U.S. Navy on April 30, 1798. The navy's primary objectives have been to guard American shores from foreign attack, preserve freedom of the seas for commerce, protect American interests overseas, support U.S. allies, and serve as an instrument of American foreign policy. The U.S. Navy is by far the most capable navy in the world because of its advantages in technology, training, and readiness, along with 376,000 active-duty sailors.

THE AIR FORCE. The U.S. Air Force was split off from the U.S. Army and made a separate branch of the military in 1947. It plays a crucial role in national security through its control of air and space. It deploys aircraft to fight enemy aircraft, bomb enemy targets, provide reconnaissance, and transport soldiers for the other armed services. In addition, the air force maintains the greatest portion of the country's nuclear forces and military satellites. In 2004 the air force consisted of more than 370,000 active-duty members, who crewed and supported a fleet of approximately 3,700 aircraft of all types.

THE MARINES. The first battalions of the U.S. Marine Corps were formed in November 1775 to fight in the American Revolution. Today it is a military service operating within the Department of the Navy. In 2004 the Marine Corps was made up of 177,000 active-duty soldiers. Marines are trained to fight in a combination of land, sea, and air operations and are a key element in U.S. rapid-response capability. The United States is the only country to have a Marine Corps as a truly independent fighting force.

THE COAST GUARD. Commissioned in 1790 to collect taxes from ships carrying imported goods, the U.S. Coast Guard is known today as a worldwide leader in maritime safety, search and rescue, and law enforcement operations. During peacetime, the Coast Guard operates under the Department of Homeland Security. As of July 2004, 39,000 active-duty men and women served in the Coast Guard. It maintains a fleet of approximately two hundred cutters (vessels sixty-five feet or longer), 1,400 smaller vessels, and more than two hundred aircraft.

While all of the services were heavily influenced by the terrorist events of September 11, 2001, the Coast Guard in particular has gained new responsibilities for homeland security. Prior to the attacks, its vital missions included counternarcotics/drug interdiction, migrant interdiction, fisheries enforcement, marine safety, environmental protection, and, to some degree, port security. Now, however, port security has begun to dwarf other Coast Guard roles, which have been sharply reduced.

ATTEMPTS AT "JOINTNESS." Military specialists have long granted that the army, navy, air force, Marines, and Coast Guard have distinct service identities, "personalities," and cultures; that they suffer from interservice rivalries; and that these factors have as much impact on molding the armed services as national security threats. To mitigate these conditions, the Goldwater-Nichols Department of Defense Reorganization Act of 1986 required that officers serve in joint assignments before they can rise to the rank of general or admiral. The law broke down the services' cultural barriers in other ways, as well, promoting greater "jointness" and teamwork.

Unified Combatant Commands

The president applies his constitutional authority as commander in chief of the armed forces by filtering orders and other communications down through the secretary of defense, the JCS chairman, the JCS, the heads of the military agencies, and the nine unified combatant commands. Together the president and the secretary of defense are known as the National Command Authority. The JCS chairman is not formally part of the operational chain of command but still transmits orders from the National Command Authority to the nine unified combatant commands. In this chain of command, the secretary of defense is tantamount to a deputy commander in chief, who relies on the individual chiefs' advice and assistance to implement national commands.

The unified combatant commands directly control U.S. combat forces. Each command is composed of forces from two or more services; has a broad and continuing mission; and is normally organized on a geographical basis. The number of unified combatant commands is not fixed by law or regulation and may vary from time to time. The nine commands as of August 2004, and their locations, are as follows: U.S. Northern Command, Peterson Air Force Base, Colorado; U.S. European Command, Stuttgart-Vaihingen, Germany; U.S. Pacific Command, Honolulu, Hawaii; U.S. Joint Forces Command, Norfolk, Virginia; U.S. Southern Command, Miami, Florida; U.S. Central Command, MacDill Air Force Base, Florida; U.S. Special Operations Command, MacDill Air Force Base, Florida; U.S. Transportation Command, Scott Air Force Base, Illinois; and the U.S. Strategic Command, Offutt Air Force Base, Nebraska.

Defense Agencies

Besides the various branches of the military, a number of agencies related to the DOD perform a host of tasks ranging from advanced defense modeling to logistical support. Some of the primary defense-related organizations are detailed below.

DEFENSE ADVANCED RESEARCH PROJECTS AGENCY. The primary mission of the Defense Advanced Research Projects Agency is research and development in science and technology. It takes innovative, cutting-edge research ideas and tries to develop potential military applications by creating prototypes.

DEFENSE CONTRACT MANAGEMENT AGENCY. The Defense Contract Management Agency is the main contact point for most defense contractors/suppliers working for the U.S. military. It helps to ensure that military and allied government supplies are delivered on time and meet quality standards.

DEFENSE INFORMATION SYSTEMS AGENCY. The Defense Information Systems Agency is primarily a combat support organization that helps to plan, develop, operate, and support the DOD's C4I (command, control, communications, computers, and information) elements during times of both conflict and peace. The agency makes sure that the military's C4I systems are interoperable (able to share information and communicate with each other) and secure at all times.

DEFENSE INTELLIGENCE AGENCY. Also a combat support group, the Defense Intelligence Agency is a vital component of the U.S. intelligence infrastructure. Its personnel primarily gather information on foreign military intelligence. The agency is headquartered at the Pentagon in Washington, D.C., but has a significant operational presence at the Defense Intelligence Analysis Center, the Armed Forces Medical Intelligence Center, and the Missile and Space Intelligence Center.

DEFENSE LEGAL SERVICES AGENCY. The Defense Legal Services Agency is the main organization providing legal advice and services to DOD agencies and personnel. It is headed by the general counsel of the DOD, who is appointed by the president (with the advice and consent of the Senate). The general counsel also leads the DOD in all international negotiations and treaty commitments.

DEFENSE LOGISTICS AGENCY. As its name implies, the Defense Logistics Agency is responsible for providing logistical support (supplies and services) to military personnel around the world. As of August 2004 the agency was working out of twenty-eight countries. During Operation Iraqi Freedom it provided U.S. troops with 138 million field meals and delivered 1.8 million Humanitarian Daily Rations to displaced refugees.

DEFENSE SECURITY COOPERATION AGENCY. The Defense Security Cooperation Agency helps create and maintain ties between the U.S. and foreign militaries in order to achieve common defense goals. It runs a group of programs (authorized under the 1961 Foreign Assistance Act and the Arms Export Control Act) by which the DOD and military contractors sell materials and services abroad.

DEFENSE SECURITY SERVICE. Formerly known as the Defense Investigative Service, the Defense Security Service plays an integral part in the country's security infrastructure. It conducts personnel security investigations, provides industrial security products, and holds comprehensive security training for DOD personnel.

DEFENSE THREAT REDUCTION AGENCY. The Defense Threat Reduction Agency has a crucial role in ensuring American preparedness for attacks involving weapons of mass destruction (WMD). Under the agency, all DOD resources are combined to ensure that the country is prepared for any potential WMD threat.

MISSILE DEFENSE AGENCY. Formerly known as the Ballistic Missile Defense Organization, the Missile Defense Agency has the primary mission of developing, testing, and preparing for the deployment of a system to defend the United States from nuclear missiles.

NATIONAL GEOSPATIAL-INTELLIGENCE AGENCY. The National Geospatial-Intelligence Agency (called the National Imagery and Mapping Agency until November 2003) provides geospatial intelligence, or geographic data gathered from satellite imagery (including imagery, imagery intelligence, and geospatial data and intelligence) from across the globe. Organizationally, it is divided into the Analysis and Production Directorate (intelligence analysis for policy makers), Acquisition Directorate (acquires and produces business solutions that help it advance the agency's mission), and the Innovision Directorate (forecasts future environments and trends in the science and technology industry).

NATIONAL SECURITY AGENCY. The National Security Agency is the U.S. government's foremost intelligence organization in terms of gathering and analyzing electronic intelligence, and protecting U.S. information systems and communications. Two primary missions of the agency, as outlined on its Web site (http://www.nsa.gov), are "designing cipher systems that will protect the integrity of U.S. information systems and searching for weaknesses in adversaries' systems and codes." The agency is headquartered in Fort Meade, Maryland, and employs a range of cryptographers, computer programmers, analysts, engineers, and researchers.

PENTAGON FORCE PROTECTION AGENCY. Established primarily in response to the terrorist attacks of September 11, 2001, the Pentagon Force Protection Agency is basically a police force for the Pentagon. The newly created agency incorporated the former security force for the Pentagon (Defense Protective Service) and provides law enforcement and security for the Pentagon.

PEACEKEEPING AS A DEFENSE STRATEGY

Military responses and treaties are not the only ways the United States and the world community as a whole attempt to defend themselves. In many ways, it is more desirable to prevent military problems before they start than to wait until tensions boil over. To this end, since the early 1990s the world community has placed more and more emphasis on peacekeeping efforts.

What Is Peacekeeping?

The term "peacekeeping" encompasses many different types of actions. In the landmark report *An Agenda for Peace: Preventive Diplomacy, Peacemaking, and Peacekeeping* (June 17, 1992), the secretary-general of the United Nations (UN) delineated four main "areas of action" for the UN in peace activities: preventive diplomacy, peacemaking, peacekeeping, and post-conflict peace building. Preventive diplomacy aims to keep disputes or violence from arising. Peacemaking negotiates between states or other bodies that are already in an adversarial relationship, while peacekeeping consists of UN forces (which may not necessarily be military) actually positioned and active in a given location. Post-conflict peace building acts as a follow-up to peacemaking and peacekeeping to strengthen institutions such as law enforcement and judicial systems in order to ensure a lasting peace. In addition, the U.S. executive branch uses the term "peace operations," and in DOD terminology, peacekeeping falls under "operations other than war." For the sake of simplicity, this section will generally refer to all of these activities as "peacekeeping."

According to a December 2001 U.S. General Accounting Office (GAO) report (*United Nations Peacekeeping: Issues for Congress*), a "second generation" of peacekeeping missions is coming into existence. These missions include "disarming or seizing weapons, aggressively protecting humanitarian assistance, and clearing land mines," along with "maintaining law and order (police), election monitoring, and human rights monitoring." This second generation of peacekeeping has overtaken first-generation peacekeeping missions, which involved monitoring cease-fires, reporting on situations, and, in some cases, intervention with limited means and resources.

Organizations Coordinating Peacekeeping Missions

UNITED NATIONS (UN). The UN is a multinational body most often associated with peacekeeping missions. As the GAO report *U.N. Peacekeeping: Status of Long-Standing Operations and U.S. Interests in Supporting Them* (April 1997) explains, under the auspices of the UN Security Council (led by permanent members the United States, China, France, Russia, and the United Kingdom),

> the United Nations undertakes peacekeeping operations to help maintain or restore peace and security in areas of conflict. Such operations have been employed most commonly to supervise and maintain cease-fires, assist in troop withdrawals, and provide buffer zones between opposing forces. The main objective of peacekeeping operations, according to UN and U.S. policies, is to reduce tensions and provide a limited period of time for diplomatic efforts to achieve just and lasting settlements of the underlying conflicts.

Conditions of UN peacekeeping missions are generally set by the Security Council. Missions must have the

TABLE 9.1

Ongoing United Nations peacekeeping missions, August 2004

Operation name	Acronym	Location	Number of UN personnel	Number of U.S. personnel	Start date of action
United Nations Trace Supervision Organization	UNTSO	Middle East	143	2	June 1948
United Nations Military Observer Group in India and Pakistan	UNMOGIP	India-Pakistan (Asia)	44	0	January 1949
United Nations Peacekeeping Force in Cyprus	UNFICYP	Cyprus (Europe)	1,242	0	March 1964
United Nations Disengagement Observer Force	UNDOF	Golan Heights (Middle East)	1,003	0	June 1974
United Nations Interim Force in Lebanon	UNIFIL	Lebanon (Middle East)	3,629	0	March 1978
United Nations Iraq-Kuwait Observation Mission	UNIKOM	Iraq/Kuwait (Middle East)	1,098	12	April 1991
United Nations Mission for the Referendum in Western Sahara	MINURSO	Western Sahara (Africa)	243	7	April 1991
United Nations Observer Mission in Georgia	UNOMIG	Georgia (Europe)	106	2	August 1993
United Nations Mission in Bosnia and Herzegovina	UNMIBH	Bosnia and Herzegovina (Europe)	1,530	46	December 1995
United Nations Mission of Observers in Prevlaka	UNMOP	Prevlaka Peninsula (Europe)	27	0	February 1996
United Nations Interim Administration Mission in Kosovo	UNMIK	Kosovo (Europe)	4,548	537	June 1999
United Nations Mission in Sierra Leone	UNAMSIL	Sierra Leone (Africa)	17,484	0	October 1999
United Nations Organization Mission in the Democratic Republic of Congo	MONUC	Democratic Republic of the Congo (Africa)	4,233	0	December 1999
United Nations Mission in Ethiopia and Eritrea	UNMEE	Ethiopia and Eritrea (Africa)	4,152	7	July 2000
United Nations Mission of Support in East Timor	UNMISET	East Timor (Asia)	5,847	67	May 2002
United Nations Mission in Liberia	UNMIL	Liberia (Africa)	15,174	N/A	September 2003

SOURCE: Prepared by Information Plus staff from United Nations data, 2004

consent of the parties in conflict and the host government, which also must provide complete freedom of movement to UN personnel. UN members provide the voluntary peacekeeping personnel, usually consisting of personnel from some or all of the nations in the UN. The missions do not interfere in the host government's internal affairs and try to avoid the use of force to carry out their objectives.

From 1948 to August 1, 2004, the UN was involved in fifty-nine peacekeeping efforts around the globe, sixteen of which were still ongoing (see Table 9.1 for statistics about the ongoing missions). Through June 2004 the UN had spent a total of $31.5 billion on peacekeeping operations and, as of July 2004, had suffered a total of 1,934 fatalities. Examples of the types of actions taken during UN peacekeeping missions include enforcing cease-fires, improving living conditions for minority groups, observing and verifying national elections, disarming warring factions, assisting in the formation of unified national governments, assisting legitimate governments in reestablishing their authority, working for the release of political prisoners and detainees, and assisting with refugee repatriation.

The total number of UN military observers, civilian police, and troops to the sixteen ongoing missions as of July 2004 was 58,741. Only 427 members of the UN force, mostly civilian police, were Americans (see Table 9.1); the United States ranked twenty-sixth in the total number of participating UN peacekeeping forces among countries worldwide. Other Western powers contributed similar numbers of personnel: the United Kingdom (567), Canada (564), France (561), and Germany (297). In contrast, the five largest suppliers of contributors were Pakistan (with a force of 8,544), Bangladesh (7,163), Nigeria (3,579), Ghana (3,341), and India (2,934). Between them, these five countries comprised over 40% of the total UN peacekeeping force worldwide in mid-2004. (Poorer

nations often find that contributing troops for UN duty, and allowing the UN to maintain those troops overseas, takes financial strain off their military budgets.)

THE NORTH ATLANTIC TREATY ORGANIZATION (NATO) IN THE BALKANS. The North Atlantic Treaty Organization (NATO) is another organization that has been taking on peacekeeping operations around the world. This is an expansion of NATO's original mandate, which was much more defense oriented. A September 2000 NATO fact sheet ("What Is NATO?") explained the change in focus: "following the momentous changes which occurred in Europe in the 1990s, [NATO] has become a catalyst for extending security and stability throughout Europe." Two of its most high-profile peacekeeping missions have been the Stabilization Force in Bosnia and Herzegovina (SFOR) and the Kosovo Force (KFOR). According to *European Security: U.S. and European Contributions to Foster Stability and Security in Europe* (Washington, DC: General Accounting Office, November 2001), the United States provided the largest single national contingency to both of these NATO operations.

SFOR, which began in December 1995, marked the first time NATO had really played a leading role in peacekeeping. The SFOR mission, according to the NATO fact sheet "Bosnia and Herzegovina—Facts and Figures" (March 8, 2001), "is related to the maintenance of a secure environment conducive to civil and political reconstruction." Some of the programs SFOR is implementing or assisting with include collecting and destroying unregistered weapons, investigating and apprehending war criminals, assisting in the processing of property claims of returning refugees and displaced persons, maintaining and repairing roads to ensure freedom of movement, and both participating in the removal of mines and training others to do so.

According to NATO's *SFOR Fact Sheet* (May, 2004), there were a total of seven thousand NATO-led forces in

SFOR. The number of personnel in the SFOR operation has been steadily dropping: a May 10, 2002, NATO press release listed the total number of participating troops at about nineteen thousand. In line with President George W. Bush's desire to reduce the U.S. presence in the Balkans, the number of U.S. troops has also dropped.

The KFOR operation began after NATO air strikes designed to end the conflict between Serbian forces and Kosovar ethnic Albanians. In January 2000, according to *Balkans Security,* KFOR had a total of 45,700 military personnel, 7,000 of whom were from the United States. As with SFOR, the number of participants in KFOR has been steadily dropping: As of October 2004 the KFOR Web site (http://www.nato.int/kfor/kfor/kfor_hq.htm) disclosed that more than seventeen thousand troops from thirty-six nations made up the force. As with SFOR, U.S. personnel numbers had fallen to an estimated two thousand by 2004. More than half of KFOR's manpower is dedicated to the protection of Serbs and other ethnic minorities, many of whom are refugees now returning to their homes. The KFOR force is also collecting and destroying weapons and helping establish a local civil emergency force, among other activities.

The Debate about U.S. Involvement in Peacekeeping Missions

Peacekeeping is a topic of frequent and vehement debate among American politicians, military leaders, and citizens. Some people believe that the United States should maintain a policy of noninvolvement and refuse to get drawn into conflicts that do not directly threaten U.S. interests. Others argue that the stability of the entire world does, in fact, directly relate to U.S. national security, so peacekeeping in distant nations is in our own best interests.

Other arguments center around the role of the United States in the community of nations. Some people suggest that since the United States is the leading superpower in the world, it should set an example for other nations by taking a leading role in peacekeeping operations. In addition, participating in peacekeeping operations with other nations can give the United States an idea, before a conflict breaks out, as to how its allies will perform in battlefield situations. According to former Secretary of Defense William S. Cohen (in "Creating an Environment for Peace, Stability, Confidence," *U.S. Foreign Policy Agenda,* December 1999), peacetime military cooperation can also "yield increased levels of trust, confidence-building, and rapport that far outlive any operation. . . . In the Department of Defense, we refer to them as 'force multipliers,' and they can make substantial contributions to success during times of war."

Others argue, as former army officer John Hillen does in an article in *NATO Review* (Summer 2001), that "superpowers don't do windows." In other words, America's

strength is "large-scale war-fighting," and it should take a leadership role in that capacity, allowing other nations to take larger roles in peacekeeping operations.

A February 2002 report from the Peace Through Law Education Fund (*A Force for Peace and Security: U.S. and Allied Commanders' Views of the Military's Role in Peace Operations and the Impact on Terrorism of States in Conflict*) reports the results of interviews with more than thirty American and allied generals about peacekeeping. They found that, generally, the commanders agreed that participation in peacekeeping is "in our national interests and will be a key ingredient in the war against terrorism." (Terrorists often thrive in countries without strong police and judicial systems, feeding off of organized crime networks to distribute materials and cash. As peacekeeping missions strengthen law enforcement institutions, those areas become less and less appealing to terrorists.) The United States needs to be heavily involved in peacekeeping, as does the UN. The military leaders generally agreed that peacekeeping missions teach leadership and other valuable skills—including skills useful in the war against terrorism—to participating troops.

U.S. INTERESTS IN UN PEACEKEEPING MISSIONS. The 1997 GAO report *U.N. Peacekeeping: Status of Long-Standing Operations and U.S. Interests in Supporting Them* closely analyzed U.S. participation in the eight UN peacekeeping missions that were then ongoing and more than five years old. Two of these dated back to the 1940s. Summarizing the results of its analysis, the report states:

> Despite the long-standing operations' cost and mixed performance in carrying out their mandates, U.S. policymakers support continuing these operations because, in their view, they help to stabilize conflicts that could threaten U.S. foreign policy objectives. In their judgment, ending these operations—or even modifying them substantially—would risk renewed conflict and damage future peacemaking efforts.

The report goes on to explain that the costs of potential conflicts in strategically important areas, including the Middle East, the Persian Gulf, southern Europe, southern Africa, and southwest Asia, would be greater than the costs of maintaining the peacekeeping missions. Operations in the Middle East reduce tensions and keep Israel secure, both of which help keep the possibility of a peace settlement between Israel and the Palestinians alive. The Persian Gulf operations safeguard oil reserves and impede aggression from Iraq. Southern European operations maintain peace and stability in all of Europe. Therefore, the report concludes, while UN peacekeeping operations are by no means ideal, there are no better substitutes.

Yet debates surrounding peacekeeping missions are always present. Whether a current presidential administration is more focused on engaging the international community or maintaining an isolationist stance, U.S. relations with the outside world are directly pertinent to any peace-keeping involvement. Many Americans question the necessity of sending their troops abroad and getting soldiers killed in battles that have no importance (or consequence) to national security interests. In recent years, such sentiments were especially noticeable after eighteen U.S. soldiers were killed and eighty-four more wounded while enforcing Operation Restore Hope in Mogadishu, Somalia, in 1993. The losses suffered in Somalia led the American public to question the importance of fighting wars for others and continued support for UN operations. It was also a significant factor in American lack of intervention while ethnic massacres were being carried out in Rwanda in the mid-1990s.

A CHANGING NATIONAL SECURITY STRATEGY. The Clinton administration's *A National Security Strategy for a New Century* (White House, Washington, DC, December 1999) outlines some of the reasons the United States participated in peacekeeping missions in the 1990s. The GAO's *Balkans Security* applies these reasons specifically to SFOR and KFOR. According to the National Security Strategy, national interests can either be classified as vital, important, or humanitarian. Vital interests are those that are "of broad, overriding importance to the survival, safety, and vitality of the United States," while important interests are those that "do not affect the survival of the United States but do affect national well-being and the character of the world in which Americans live." Humanitarian interests that might merit U.S. military involvement include "(1) natural and manmade disasters; (2) promoting human rights and seeking to halt gross violations of those rights; and (3) supporting democratization, adherence to the rule of law, and civilian control of the military."

Peacekeeping missions can fall into any of these three categories. Vital interests include such things as the security of Europe, which might be threatened by instability in Bosnia/Herzegovina and Kosovo. Thus, the United States has an interest in participating in NATO's KFOR and SFOR operations. Vital interests, as listed in the National Security Strategy, specifically include NATO's operations in the Balkans. Humanitarian interests may be less specific and harder to pin down. They are not specifically listed in the National Security Strategy as reasons for participating in peacekeeping in the Balkans, but many U.S. government officials have informally mentioned humanitarian interests as a reason for involvement in SFOR and KFOR.

Released in September 2002, the Bush administration's *The National Security Strategy of the United States of America* changed the way the United States saw its role in keeping the world's peace. The new National Security Strategy saw a fundamental change in the world situation. With the cold war over, the United States stood alone as the world's only superpower. There was now a chance to promote political and economic freedom in areas of the

world previously closed off because of the cold war stand-off. "We will work to translate this moment of influence into decades of peace, prosperity, and liberty," according to the report. To achieve these goals, the United States will:

- champion aspirations for human dignity;
- strengthen alliances to defeat global terrorism and work to prevent attacks against us and our friends;
- work with others to defuse regional conflicts;
- prevent our enemies from threatening us, our allies, and our friends, with weapons of mass destruction;
- ignite a new era of global economic growth through free markets and free trade;
- expand the circle of development by opening societies and building the infrastructure of democracy;
- develop agendas for cooperative action with other main centers of global power; and
- transform America's national security institutions to meet the challenges and opportunities of the twenty-first century.

The Costs of Peacekeeping

MONETARY COST OF MISSIONS. One factor that plays a large role in the debate about peacekeeping missions is their cost. According to *U.N Peacekeeping: Estimated U.S. Contributions, Fiscal Years 1996–2001* (U.S. General Accounting Office, February 2002), the United States spent $3.45 billion to support UN peacekeeping from fiscal years 1996 through 2001. Indirect U.S. contributions include unilateral peacekeeping activities in such countries as Sierra Leone and Haiti. (See Figure 9.1.) According to DOD estimates, the United States provided 30% of all funding for UN peace operations in 1999 (this does not count NATO operations such as SFOR and KFOR).

Another issue with UN peacekeeping missions and finances is that the United States refused for several years to pay its share. The United States demanded that the United Nations adopt certain reform measures, including a reduction of the U.S. peacekeeping assessment rate and the creation of results-oriented standards to measure the success of its peacekeeping missions. U.S. officials claimed that UN peacekeeping efforts were often disorganized and ineffectual. According to *U.N Peacekeeping: Transition Strategies for Post-Conflict Countries Lack Results-Oriented Measures of Progress* (GAO-03-1071, Washington DC: U.S. General Accounting Office, September 2003), "the U.N. Department of Peacekeeping Operations acknowledges that it needs better indicators by which to measure the progress peacekeeping operations are making in attaining sustainable peace."

In regards to the NATO missions in Kosovo and Bosnia/Herzegovina, estimates are that the U.S. spent some $18 billion for stabilization measures (including air strikes) in

FIGURE 9.1

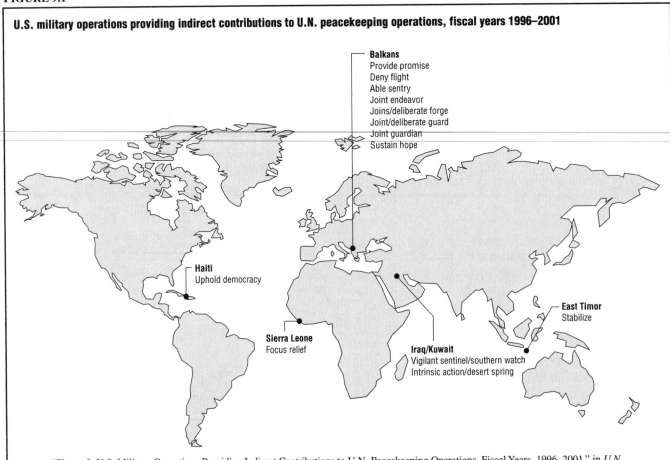

U.S. military operations providing indirect contributions to U.N. peacekeeping operations, fiscal years 1996–2001

Balkans
Provide promise
Deny flight
Able sentry
Joint endeavor
Joins/deliberate forge
Joint/deliberate guard
Joint guardian
Sustain hope

Haiti
Uphold democracy

East Timor
Stabilize

Sierra Leone
Focus relief

Iraq/Kuwait
Vigilant sentinel/southern watch
Intrinsic action/desert spring

SOURCE: "Figure 3. U.S. Military Operations Providing Indirect Contributions to U.N. Peacekeeping Operations, Fiscal Years, 1996–2001," in *U.N. Peacekeeping: Estimated U.S. Contributions, Fiscal Years 1996–2001*, GAO-02-294, U.S. General Accounting Office, February 2002, http://www.gao.gov/new.items/d02294.pdf (accessed September 23, 2004)

Bosnia and Kosovo from 1992 through 2000. Table 9.2 shows estimated 2001 costs. While not all this funding may be directly related to SFOR and KFOR, the numbers do demonstrate the importance the United States has placed on stability in this volatile region of Europe since the mid-1990s.

CONTRIBUTING PERSONNEL. Manpower is another of the concerns surrounding peacekeeping missions. Peacekeeping missions, despite their name, often take place in volatile locations, with the personal safety of the peacekeeping force often as at risk as that of actual combat troops. In UN peacekeeping operations, for example, 1,965 people have died since 1948, according to January 2005 figures posted at the UN Web site (http://www.un.org/Depts/dpko/fatalities/totals_annual.htm). The total number of deaths of UN peacekeeping forces since the turn of the century are: fifty-two fatalities in 2000; sixty-four each in 2001, 2002, and 2003; and eighty as of October 2004.

According to *Allied Contributions to the Common Defense,* the United States contributed 11,138 people to UN and major non-UN peace operations in 2000. This was 6.8% fewer than were contributed in 1999, but it was still by far the largest number contributed by any country and made up about 18% of the total worldwide peace

TABLE 9.2

Estimated costs for Department of Defense's contingency operations, fiscal year 2001

Dollars in millions

	Military personnel	Operation and maintenance	Miscellaneous procurement	Total
Bosnia	153.4	1,234.4	20.8	1,408.6
Kosovo	194.5	1,455.9	62.6	1,713.0
Total Balkans	347.9	2,690.3	83.4	3,121.6
Southwest Asia	144.8	913.7		1,058.5
East Timor	0	3.9		3.9
Total	**492.7**	**3,607.9**	**83.4**	**4,184.0**

SOURCE: "Table 1. Estimated Costs for DoD's Contingency Operations, Fiscal Year 2001," in *Defense Budget: Need for Continued Visibility over Use of Contingency Funds,* U.S. General Accounting Office, July 2001, http://www.gao.gov/new.items/d01829.pdf (accessed September 23, 2004)

operation force. U.S. peacekeepers totaled 0.0001% of the U.S. labor force. When this peacekeepers-to-labor-force ratio is compared with that of other countries around the world, the United States ranks nineteenth. The nations providing the highest percentages of their labor force to peace operations in 2000 were United Arab Emirates

(0.0008%), Norway (0.0005%), Greece (0.0005%), Denmark (0.0004%), and Italy (0.0004%).

POSSIBLE DETERIORATION OF COMBAT READINESS. According to *Peacekeeping: Issues of U.S. Military Involvement* (Washington, DC: Congressional Research Service, 2003), "The [George W.] Bush Administration's desire to reduce the commitment of U.S. troops to international peacekeeping parallels the major concerns of recent Congresses: that peacekeeping duties are detrimental to military 'readiness'; i.e., the ability of U.S. troops to defend the nation." Proponents of this argument note that military training teaches troops to be fighters, while peacekeeping skills are better learned through a law enforcement background. In addition, some argue, military action requires quick, decisive force, while peacekeeping generally calls for restraint. Can a soldier be both a good fighter and a good peacemaker?

Increasingly, many answer yes. As noted earlier, peacekeeping missions inherently require military skills to allow a quick and appropriate response to unforeseen risks. In addition, peacekeeping proponents argue, if the purpose of peacekeeping is to prevent conflict, who better to deter conflicts from starting than well-trained soldiers?

The demands of peacekeeping, though, differ from those of actual combat participation. While troops engaged in peacekeeping activities may be able to expand their skills in such areas as intelligence, leadership, logistics, transportation, and engineering, according to some critics their skills in more combat-oriented arenas, such as shooting, combined arms skills, and unit maneuverability, may degrade. *Peacekeeping: Issues of U.S. Military Involvement* points out that many efforts are now being made to provide peacekeepers with ongoing combat training to reduce this deterioration. According to Christopher Bellamy, professor of military science and doctrine at Cranfield University in the United Kingdom, local populations respond best to peacekeeping troops who are "also unmistakably professional soldiers," but not too "heavy-handed" (*NATO Review,* Summer 2001).

CHAPTER 10

GLOBAL DYNAMICS OF NATIONAL SECURITY: ALLIANCES AND RESOURCES

Interdependence is one of the key words of foreign policy in the post-cold-war era. States are increasingly relying on each other, as well as nongovernmental and multinational entities, to accomplish their stated political and economic goals. Interdependence is complicated because it does not rely on ideological loyalties, as the communist and democratic blocs each did during the cold war. Ideologies are now being replaced by such motivations as money and regional dominance in interstate alliances.

ALLIANCES

The United States faces a unique set of threats from different parts of the globe and continues to make alliances to suit its tactical and strategic national security goals. Such alliances are usually formal agreements that two or more parties enter into in order to defend their collective security goals. Some of the key allies for the United States in the new world order include the European nations (primarily the United Kingdom), Israel, and certain Persian Gulf states in the Middle East.

Since World War II (1939–45), the United States has signed a number of treaties assuring protection to states that needed military assistance, often to fight the communist threat. However, Congress has always been hesitant about committing U.S. resources for long periods of time. Besides formal alliances that have to be ratified by the U.S. Senate, the president can also enter into executive agreements that commit U.S. resources internationally but do not have to be ratified by the Senate, which gives the president greater flexibility in foreign affairs. Such agreements are usually initiated at the executive level of government, and terms are negotiated by a representative. The secretary of state authorizes the negotiator to sign the agreement. The Senate needs to be notified by the executive branch within sixty days of signing an executive agreement, and to be implemented it requires a simple majority vote of the House and Senate.

Many agreements require implementing bills to be passed by both chambers before they can take force. Congress can express its opposition to any particular executive agreement by withholding the necessary implementing legislation. The president's authority to negotiate executive agreements flows from two sources: the power granted to him or her in the Constitution as chief executive, and/or specific powers delegated by earlier acts of Congress. Instances of presidential initiatives involve the 1991 Persian Gulf War coalition, support for anti-Vietnam forces in Kampuchea, and aiding the mujahideen against Soviet forces in Afghanistan during the 1980s.

U.S. policy makers often have to deal with states that are not necessarily considered close allies in any ideological sense. This would include authoritarian regimes and dictatorships that are not democratic and that may even (intentionally or unintentionally) support anti-U.S. entities. However, support from countries such as these (for example, Saudi Arabia and Kuwait) might be useful for America in terms of strategic regional goals or commercial interests (for example, oil). The United States also supported various substate groups opposing their respective governments in Central America and the Caribbean (such as El Salvador, Chile, Nicaragua, and Haiti) and Africa during the 1980s. Mixed public reaction to actions such as these clearly demonstrates that national security is pursued through a variety of channels and that security agendas are not necessarily always clear-cut.

The United States pursues its national security agenda through international organizations, as well as state-level ties with allies and other countries. It must be kept in mind that any alliance entered into by a state or states requires certain commitments on behalf of all parties involved. These obligations can potentially constrain America's ability to shift policies and make some decisions. Thus, it is important for the United States to consider the flexibility of any commitment it makes, as the national security environment is constantly changing. Also, most security commitments are designed to be honored by each succeeding U.S. administration, unless major changes in the security environment have occurred.

UNITED NATIONS (UN)

The United Nations (UN) is one of the leading players in the international arena and deals with a host of subjects, ranging from human rights to weapons of mass destruction (WMD) nonproliferation. The multinational organization can trace its roots back to the days immediately following World War II. At a 1945 conference in Yalta in the Crimea (then part of the Soviet Union), the leaders of the United Kingdom, the United States, and the Soviet Union decided that the UN was to be an international entity, with five permanent powers with veto authority in its Security Council: China, France, the Soviet Union, the United States, and the United Kingdom.

Creation of the UN was finalized at the San Francisco conference that same year, when the charter of the organization was signed and ratified by several countries. The UN charter sets forth the organization's rights and obligations and establishes its procedures. According to the UN Web site (http://www.un.org), the primary functions of the UN "are to maintain international peace and security; to develop friendly relations among nations; to cooperate in solving international economic, social, cultural and humanitarian problems and in promoting respect for human rights and fundamental freedoms; and to be a centre for harmonizing the actions of nations in attaining these ends."

The two-year 2004–05 proposed budget for the UN is $2.9 billion, or about $1.45 billion per year, which is raised primarily by contributions of member countries. Each individual contribution is determined by the capability of a country, measured through its gross national product. In addition to membership fees, countries are also assessed for the costs of peacekeeping operations. All told, the UN system spends some $12 billion a year, including operating expenses, the costs of UN peacekeeping operations, and all of the organization's programs, funds, and specialized agencies. In addition, through the World Bank, International Monetary Fund, and International Fund for Agricultural Development, the UN loans out billions of dollars each year to help developing countries.

According to its various functions, the UN is divided into six principal organs:

1. The General Assembly is the legislative arm of the UN and is broken down into six committees: Disarmament and International Security; Economic and Financial; Social, Humanitarian and Cultural Issues; Special Political and Decolonization; Administrative and Budgetary; and Legal Matters.

2. The Security Council has fifteen members: ten elected by the General Assembly for two-year terms and five permanent members. Each member of the Security Council has one vote. Decisions on procedural matters require at least nine of the fifteen members voting in favor; substantive issues require nine positive votes, including one from each of the permanent members. There are two standing committees in the Security Council: one dealing with rules and procedures and another with admission of new members. Ad hoc committees are established as needed, as well as working groups on various issues.

3. The Economic and Social Council is responsible for promoting higher standards of living, employment, and economic and social progress around the world. It facilitates cultural and educational cooperation, deals with social and health problems, and encourages respect for global human rights and fundamental freedoms. The council coordinates the work of fourteen specialized UN agencies, ten functional commissions, and five regional commissions.

4. The Trusteeship Council was established to supervise and administer trust territories. These were territories that were formerly part of Western colonial empires, that the UN wished to aid in their development toward full and effective self-governance. The council suspended its operations as of November 1994, with the independence of Palau, the last remaining UN trust territory.

5. The International Court of Justice, headquartered in The Hague, Netherlands, is the principal judicial organ of the UN. The court is charged with settling legal disputes submitted to it by state parties, as well as giving advisory opinions on questions referred to it by international entities. It is composed of fifteen judges, elected to nine-year terms by the General Assembly and the Security Council.

6. The Secretariat is composed of an international staff carrying out the day-to-day maintenance work of the organization. It is headed by the secretary-general, who is appointed by the General Assembly for a five-year term.

THE NORTH ATLANTIC TREATY ORGANIZATION (NATO)

On April 4, 1949, the United States and Canada signed the North Atlantic Treaty. This entered them into a political and military alliance with ten European nations: Denmark, France, Iceland, Italy, Portugal, Norway, Great Britain, Belgium, the Netherlands, and Luxembourg. The North Atlantic Treaty Organization (NATO) was essentially created to protect Europe from potential Soviet aggression and create a balance of power between the communist and democratic states. In 1952 Greece and Turkey joined the treaty, followed by the Republic of Germany, which joined in 1955. In 1982 Spain became a member of NATO. After communism fell in Eastern Europe in the late 1980s and 1990s, NATO expanded its membership rapidly, starting with the reunified Germany in 1990. By 1999 member states also included the Czech Republic, Hungary, and Poland. On March 29, 2004, Bulgaria, Estonia, Latvia, Lithuania, Romania, Slovakia, and Slovenia also joined the organization, the largest round of enlargement in NATO's history. As of 2004 three countries—Albania, Croatia, and Macedonia—were in the process of meeting NATO standards for possible future membership.

NATO is primarily a multinational alliance, promoting collective defense while allowing states to maintain their individual sovereignty. According to the NATO handbook, NATO has the following fundamental tasks:

- It provides an indispensable foundation for a stable security environment in Europe, based on the growth of democratic institutions and commitment to the peaceful resolution of disputes. It seeks to create an environment in which no country would be able to intimidate or coerce any European nation or to impose hegemony (leadership or dominance of one state over another) through the threat or use of force.

- It serves as a transatlantic forum for allied consultations on any issues affecting the vital interests of its members, including developments that might pose risks to their security.

- It provides deterrence and defense against any form of aggression against the territory of any NATO member state.

- It preserves a strategic balance in Europe.

These security undertakings have gone through a transformation since the 1990s. With the collapse of the Soviet Union, NATO had to redefine its security goals to fit the changing security environment. After the Prague Summit of 2002, it developed a list of points that helped the organization shift its strategies to better fit the new millennium. The highlights of the list, as posted on the NATO Web site, include:

- The NATO Response Force will be a technologically advanced, flexible, deployable, interoperable (able to operate between different branches and locations), and sustainable force including land, sea, and air elements ready to move quickly to wherever needed.

- NATO's command structure will be made leaner, more efficient, more effective, and more deployable, in order to meet the operational requirements for the full range of NATO missions. There will be two strategic commands: one operational (the strategic command for Operations, based in Europe) and one functional (the strategic command for Transformation, based in the United States).

- In the Prague Capabilities Commitment, individual allies have made firm and specific political commitments to improve their capabilities in areas key to modern military operations, such as strategic air-and-sea lift and air-to-ground surveillance.

- To defend against new threats like terrorism, five specific initiatives in the area of nuclear, biological, and chemical weapons defense were endorsed to enhance NATO's defense capabilities against such weapons. NATO's defense against cyber attacks will be strengthened, and a missile defense feasibility study will be initiated.

THE MIDDLE EAST

Israel

Preserving the security of the state of Israel while supporting the Arab-Israeli peace negotiations has been, and continues to be, an important policy for the United States. The United States has been a strong ally of Israel since the country was established in 1948 because of the two countries' shared political values, a historical relationship, and shared cultural and personal ties. Over time, the two states have also shared similar security threats, including Soviet aggression and, more recently, threats from radical Islamic fundamentalists and WMD.

The Persian Gulf and the Gulf Cooperation Council (GCC)

Perhaps one of the most significant regions for U.S. foreign policy is the Persian Gulf (also known as the Arabian Gulf). Both Iran and Iraq have always been major powers in the Persian Gulf region in terms of size, population, resources, and military capabilities. But the Iranian Revolution in 1979, the subsequent oil crisis, and the Iran-Iraq war (1980–88) left many of the other Gulf States feeling vulnerable. On May 25, 1981, Bahrain, Kuwait, Oman, Qatar, Saudi Arabia, and the United Arab Emirates met in Abu Dhabi in the United Arab Emirates to form an alliance known as the Cooperation Council for the Arab States of the Gulf, or the Gulf Cooperation Council (GCC). The six GCC countries are tied together by their religious, cultural, and social mores. The GCC is headquartered in Riyadh, Saudi Arabia, and holds meetings annually. The main bodies of the organization are the Supreme Council, the Ministerial General, and the Secretariat General.

The Peninsula Shield Force, created in 1982, was designed to increase the interoperability of GCC states' militaries, but its strength and validity were strongly questioned during the 1991 Gulf War. As of 2001 the cumulative strength of personnel in the GCC militaries (273,730) fell far short of Iranian totals, which stood at 424,600. Interestingly, a defense pact was never mentioned in either the charter or the framework of the GCC. It is generally believed that the states specifically chose to omit the terms "defense alliance" or "military cooperation" in order not to upset Iran or Iraq (when the dictator Saddam Hussein was still in power).

Instead, the purpose of the GCC, as stated on its Web site (http://www.gcc-sg.org/), is:

> [To bring about] inter-connection between Member States in all fields, strengthening ties between their peoples, formulating similar regulations in various fields such as economy, finance, trade, customs, tourism, legislation, administration, as well as fostering scientific and technical progress in industry, mining, agriculture, water and animal resources, establishing scientific research centers, setting up joint ventures, and encouraging cooperation of the private sector.

In the aftermath of the Iraqi invasion of Kuwait in the early 1990s, however, the GCC countries adopted a pact underlining the interconnectivity of their security. On December 31, 2000, at their annual meeting in Bahrain, the six countries resolved to come to on another's defense if necessary, stating that aggression against one meant aggression against all. Even though GCC states agreed to come to each other's aid in the face of aggression, this pact had not been ratified as of 2004.

Overall, the GCC aims to strengthen its political, economic, and strategic position in the region. Its member states seek to alleviate economic and population problems and increase trade flow to the area. Commercial, social, and even political alliances cannot be achieved if there is strategic regional instability. As a consequence, increasing military cooperation and securing defensive capabilities are priorities for the GCC.

U.S. ALLIES IN THE GULF. Since the decline of British authority in the Persian Gulf in the early 1970s and the end of the cold war in the early 1990s, the United States has played a strong role in the Persian Gulf theater. Its primary regional interests include protecting its national interests, protecting allies' security, and guarding the international oil supply.

In addition to the Peninsula Shield Force, each of the GCC states relies heavily on the United States for military protection, and the United States has dozens of military bases throughout the region. Among the most important American bases in the Gulf is the headquarters for the U.S. Navy's Fifth Fleet in Bahrain. The Fifth Fleet is primarily responsible for all naval activities in this theater. Oman, which retains strong military ties with the United Kingdom, hosts U.S. airbases in Seeb, Thumrait, and Masirah. Qatar hosts the forward headquarters for the U.S. Army's Central Command. Many of the U.S. bases

in the GCC states played important roles in the 2003 invasion of Iraq. Many additional bases have been established within Iraq to support the U.S. presence there.

The United States has individual formal defense agreements with each GCC state except Saudi Arabia. Because of internal opposition, Saudi Arabia has not signed a formal defense pact with the United States but continues to have strong defense ties (including weapons procurement and training exercises) to its Western ally. Maintaining strategic stability in the six GCC states is of great importance to the United States because these countries' support is vital to U.S. presence in the region.

ENERGY SECURITY: THE IMPORTANCE OF OIL

An important element to consider while studying U.S. alliances and the global dynamics of national security is the heavy Western dependence on energy resources from around the world. The United States and most other developed countries do not produce enough petroleum to meet domestic demand and must therefore import oil from other nations. As a steady supply of oil is essential to the functioning of a modern economy, this dependence on foreign oil exposes the United States to danger and plays a significant role in U.S. defense and foreign policies.

The Persian Gulf Peril

SHARE OF WORLD OUTPUT. The main issue for national security planners in the early twenty-first century regarding oil is less that the world is running out of it than that there is an increasing concentration of supply from one region: the Persian Gulf. The Persian Gulf producers work through the Organization of Petroleum Exporting Countries (OPEC) to control oil prices. OPEC membership includes Algeria, Indonesia, Iran, Iraq, Kuwait, Libya, Nigeria, Qatar, Saudi Arabia, United Arab Emirates, and Venezuela. The primary mission of OPEC is to coordinate and unify the petroleum policies of its member countries and to determine the best strategy for protecting their individual and shared interests.

World oil reserves (yet-to-be-tapped sources of supply) in 2003 were 1,147.7 billion barrels. The Persian Gulf producers and OPEC, while sitting on top of mammoth untapped supplies, have occasionally held back production, as they did noticeably in 2000. Indeed, they have usually produced at a rate lower than the maximum possible to limit supply and bolster prices. By contrast, the American oil industry's goal is to produce a full 7% of an oil field's underground capacity each year. Industry analysts have said that if this practice were applied worldwide, it would, in theory, yield a capacity of 190 million barrels per day, more than twice the expected worldwide demand in 2010.

As calculated by BP Amoco in its *Statistical Review of World Energy, 2004,* the eight main Middle East/Persian Gulf oil-producing states—Saudi Arabia, Iraq, United Arab Emirates, Kuwait, Iran, Qatar, Yemen, and Oman—were responsible for almost 30% (22.6 million barrels per day) of the world's daily production in 2003. Proven Middle Eastern reserves (726 billion barrels) accounted for 63.3% of the world's unproduced sources of supply at the end of 2003.

Nearly two-thirds of the world's global petroleum supplies lie in the Persian Gulf. Should the price of oil remain relatively low, U.S. dependence on Persian Gulf oil may increase—historically, Gulf oil has been the cheapest oil to produce. In the future, non–Persian Gulf producers, such as Venezuela, Russia, and Mexico, may supply as much as forty-seven to fifty-seven million barrels per day, or 62%–65% of demand. Still, if world oil demand comes in at its DOE estimate of about ninety-five million barrels per day by 2010, and if non–Persian Gulf production remains at forty-seven million barrels per day, then the Persian Gulf states might be supplying 50% or more of world oil demand by the end of the twenty-first century's first decade. Such a high level of dependence on one region, and especially Saudi Arabia, would leave the world and U.S. economies vulnerable.

As resource-conflict specialist Michael T. Klare noted in *Resource Wars* (New York: Metropolitan Books, 2001), "A significant share of the additional petroleum will have to come from the Gulf—there is simply no other pool of oil large enough to sustain an increase of this magnitude. All projections of future supply and demand assume that the Persian Gulf will account for an ever-expanding share of the world's oil requirements: from 27% in 1990 to 33% in 2010 to 39% in 2020."

ARMS, WAR, AND SECURITY CONCERNS. Such large reserves of oil in the Persian Gulf actually increase the likelihood of interstate conflict there. They give the nations in the region the means to procure huge quantities of sophisticated modern weapons, and so when warfare breaks out, the scale and intensity of the fighting are elevated. For example, the war between Iran and Iraq of 1980–88 yielded an estimated one million casualties and over $100 billion in property damage.

The arms that Gulf States have acquired from the United States alone have been substantial. According to the Congressional Research Service (*Conventional Arms Transfers to Developing Nations, 1995–2002*), from 1999 to 2002 the value of U.S. arms-transfer agreements with the Persian Gulf states of Bahrain, Kuwait, Oman, Saudi Arabia, and the United Arab Emirates came to $16.8 billion.

WORLD OIL TRANSIT CHOKEPOINTS. U.S. national security concerns regarding Persian Gulf oil extend to those areas that do not themselves hold large petroleum supplies. These are sea passages and straits used to ship oil by tanker or pipeline. Because several of these areas adjoin areas of recurring conflict, the DOE has dubbed them "world oil transit chokepoints." Figure 10.1 and Table 10.1 provide details on each of these chokepoints, including the major concerns should closures occur. These six passages carried over thirty-five million barrels of oil per day in 2004—more than 45% of global consumption. This list illustrates the importance of the volatile Middle East and Persian Gulf to petroleum supplies—four of the six chokepoints (the Strait of Hormuz, Bab el-Mandeb, the Suez Canal/Sumed Pipeline, and the Bosporus/Turkish Straits) lie in these regions.

The Former Soviet Union: A Challenge to the Persian Gulf?

Some industry experts suggest that Russia may be in a position to pose a challenge to Saudi Arabia's role as top worldwide oil producer. As Edward L. Morse, executive advisor at Hess Energy Trading Company and former assistant secretary of state for international energy policy, and James Richard, portfolio manager at Firebird Management, point out in "The Battle for Energy Dominance" (*Foreign Affairs,* March/April 2002), before the breakup of the former Soviet Union, state-owned oil production had reached 12.5 million barrels per day, well beyond the largest amount reached by Saudi Arabia at its production height. Currently, Russia keeps a much larger amount of its oil for internal use than does Saudi Arabia, so Saudi exports are still substantially higher than Russia's.

Significant, and so far unresolved, difficulties associated with Russian oil production include issues of sufficient investment, management, construction and maintenance of pipelines, and ownership/development disputes among the countries bordering the oil-rich Caspian Sea. Morse and Richard predict that the Caspian area could become the source of enough oil to supplant Saudi Arabia as the West's primary source of oil within four years, but the DOE is more cautious. It predicted in *International Energy Outlook 2004* that the former Soviet Union's net oil production would increase to 17.2 million barrels per day by 2025. OPEC production is expected to grow by an annual rate of 2.6% through 2025.

Although Russia had agreed in a deal with OPEC to cut its output, in mid-2002 it announced that it would abandon that agreement. Russian president Vladimir Putin also promised to keep Siberian oil flowing during any Middle East crisis.

China, Oil, and U.S. Interests

China has gone from a net petroleum exporter to a net petroleum importer. It has taken vigorous steps to grow its economy and, as part of that goal, has tried to promote private automobile ownership by individuals since 1993 (which could lead to twenty-five million more cars in the country by 2015). Even if it develops the oil fields located

FIGURE 10.1

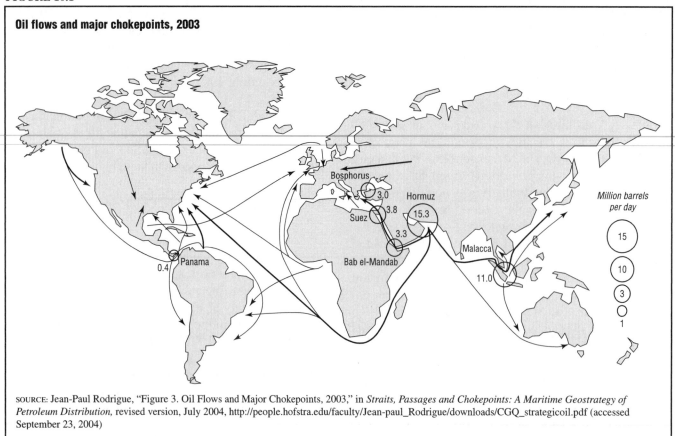

Oil flows and major chokepoints, 2003

SOURCE: Jean-Paul Rodrigue, "Figure 3. Oil Flows and Major Chokepoints, 2003," in *Straits, Passages and Chokepoints: A Maritime Geostrategy of Petroleum Distribution,* revised version, July 2004, http://people.hofstra.edu/faculty/Jean-paul_Rodrigue/downloads/CGQ_strategicoil.pdf (accessed September 23, 2004)

in its isolated interior regions, it is likely to import at least two million barrels per day by 2015.

China's discomfort with reliance on world markets for this vital resource could translate into a political alliance with one or more oil-exporting states in the hope that this would mean a more secure source of oil. The problem for the United States with such an alliance is that China's partners in the Middle East will most likely not be American allies but those considered rogue states, such as Iran. China's friendly relationship with Iran has given Iran plenty of access to Chinese advanced-weapons technology. One worst-case scenario for the United States would be that Iran would take an aggressive stance, armed with nuclear weapons obtained from the Chinese, or enter into some sort of defense pact with the Chinese.

Continued U.S. Oil Supply and the Strategic Petroleum Reserve (SPR)

Petroleum reserves are vital to modern economies, but controlling them does not guarantee prosperity or security. As with other mineral resources and raw materials, petroleum is distributed within a well-developed market, one that allows almost any country access to the commodity, even during times of conflict, as long as adequate worldwide supplies exist within a reasonable and customary price range. Once the oil-supply chain is upset by events such as civil wars in developing countries, how-

ever, prices may become volatile. Additionally, increased production can destroy land and the environment, and eventually overdevelopment may cause large migrations of displaced people, for example in Africa. A widely accepted summary of the state of oil in the future is that this resource is finite, production will peak well before the middle of the century, and an alternative must be found to avoid widespread dislocations in modern life.

One strategy for dealing with the risk or threat of disruption in petroleum supplies has been to encourage public and private stocking. Planners generally give the government the role of creating its own strategic reserves and establishing incentives for such stocking. Because the stocks would help ensure the flow of oil, they could reduce the U.S. need to intervene during a crisis—or at least the need to intervene quickly. These stocks could provide time needed to bring alternative energy sources online—to shift, say, from oil to coal for electricity generation in dual-fuel-capable boilers, or to start pumping oil wells that were temporarily idle. This has been the strategy adopted by the U.S. government and the International Energy Agency (IEA).

The Arab oil embargo of 1973–74 and later price spikes motivated the United States to create the Strategic Petroleum Reserve (SPR), a series of underground salt caverns along the Gulf of Mexico coast with a capacity of 560 million barrels. The SPR was authorized in 1975 and

TABLE 10.1

Chokepoint capacity, limitations, and threats, 2004

Chokepoint	Usage (ships/day, 2003)	Additional capacity	Limitation	Threat
Hormuz	50	Limited	Narrow corridors	Iran/terrorism
Suez	38	Some	200,000 dwt* and convoy size	Terrorism
Bosporus	135	Very limited	Ship size and length; 200,000 dwt*	Restrictions by Turkey; navigation accident
Malacca	600	Substantial	300,000 dwt	Terrorism/piracy
Panama	35	Limited	65,000 dwt	No significant

*dwt=deadweight ton

SOURCE: Jean-Paul Rodrigue, "Table 1. Chokepoints: Capacity, Limitations and Threats," in *Straits, Passages and Chokepoints: A Maritime Geostrategy of Petroleum Distribution,* revised version, July 2004, http://people.hofstra.edu/faculty/Jean-paul_Rodrigue/downloads/CGQ_strategicoil.pdf (accessed September 23, 2004)

began operation in 1977. It is the first line of defense against an interruption in petroleum imports. If necessary, the reserve can be drawn down at a rate of 4.3 million barrels per day, equal to about 40% of daily U.S. oil imports.

At the same time as the reserve was established, the industrial nations agreed to hold reserves equal to ninety days' imports. They also agreed to coordinate their responses, in the event of an interruption in oil supplies, through the twenty-three member countries of the IEA.

In November 2001 President George W. Bush issued a directive to fill the Strategic Petroleum Reserve to its capacity. Since then, the SPR has been adding to its reserve. As of May 2003 the oil stockpile passed the 600-million-barrel mark, a new high. The SPR's goal is an eventual stockpile of 700 million barrels. According to the Department of Energy, Office of Fossil Energy Web site (http://www.fe.doe.gov/programs/reserves/), storage capacity as of 2004 was 727 million barrels.

Emergency use of the SPR occurred in January 1991, at the start of Operation Desert Storm, the U.S. military campaign to oust Iraqi forces from Kuwait. According to the Institute for National Strategic Studies (*Strategic Assessment 1999: Priorities for a Turbulent World,* Washington, DC, 1999), "The mere announcement of SPR sales had a considerable stabilizing effect on world markets. Only 17 million barrels were actually sold before market conditions returned to normal."

Reserve oil was also loaned to southern refineries after the disruption of normal offshore production and import deliveries during the hurricane seasons of 2002 and 2004. Energy Secretary Spencer Abraham noted in an October 2004 press release that "the SPR was designed to protect American consumers against supply disruptions, including natural disasters."

U.S. Internal Oil Production versus Conservation

The ultimate exhaustion of the world's oil reserves—which would be preceded by hefty price increases that could occur well before the year 2050—constitutes a long-term national security problem. Because oil is a non-

renewable resource (i.e., there is only a limited amount of it), many believe that Americans should begin seriously to reduce their use of energy, particularly energy from oil-based sources. Although recent public-opinion polls have shown that most Americans support the idea of decreasing our dependency on foreign oil, especially in the wake of the terrorist attacks of September 11, 2001, Americans continue to use great amounts of energy—one illustration of this is the continued popularity of heavily gas-consuming sport utility vehicles (SUVs).

Conservation efforts, while one option, are not the only route to decreasing U.S. dependence on foreign oil. Increasing production by exploiting known reserves in various regions of the United States has also been proposed. Such regions include the Great Lakes, the Gulf Coast, and, particularly, the 19.6-million acre Arctic National Wildlife Refuge (ANWR), the largest national wildlife refuge in the United States. Oil companies have long been interested in ANWR and, along with their political supporters, have challenged its protected status. According to Secretary of the Interior Gale Norton in testimony before the House Committee on Resources in March 2003, there are an estimated 10.4 billion barrels of oil in the ANWR coastal region. Conservationists and others who feel the ecological and environmental damage caused by drilling would be significant, even disastrous, have been fighting off these challenges.

Until September 11, 2001, public support tended to be on the side of the conservationists, but immediately after the terrorist attacks, U.S. opinion changed radically. When *The Wirthlin Report* asked in July 2001 whether the positives of producing oil and natural gas by drilling in ANWR outweighed the negatives, only 39% of Americans surveyed agreed. When asked the same question soon after September 11, however, 61% felt the positives outweighed the negatives. In August 2001 the House of Representatives passed a bill that would have allowed drilling within ANWR, but the Senate rejected this proposal in April 2002, thereby continuing the area's protected status. It is likely that drilling in ANWR will continue to be a matter of debate for some time.

THE FUTURE OF RESOURCE CONFLICTS: THE AFRICA PATTERN

Resource stresses abroad may well keep the United States on its toes in the next few decades, especially because they may make conflicts among regional powers more likely and more intense. Although involved parties may seek economic sanctions before resorting to military force, transborder resource conflicts are likely to occur.

The Middle East is of special concern to U.S. security planners. The great oil riches of the Persian Gulf states have inflamed border disputes between them. Water scarcity is also a problem in the Middle East, especially in the westernmost part of the region where most water comes from the Jordan and Nile rivers. Here, national populations are expected to soar in size, making water supplies increasingly scarce. The existing political ill will between nations and groups may easily increase and lead to conflict. Now that the United States is maintaining a presence in these regions, the low likelihood of success of an outright resource or territorial grab might discourage sovereign states from contemplating aggression.

The Middle East is not the only region where resource conflicts are likely. Civil wars in some of the African states have been caused by wars over natural resources, including water, land, diamonds, oil, minerals, and timber. These wars have taken their toll in the form of millions of deaths. For example, coltan is a mineral used in cell phones, DVDs, and other electronic products. Mined in the Democratic Republic of Congo and other African nations, it has created problems as conflicts arise over ownership of rights. Governments, rebel factions, and other entities have confronted Americans and others involved in importing coltan, diamonds, and other resources with increasing resentment.

In Nigeria, where several U.S. oil companies have operations, there have been serious conflicts involving protesters concerned with environmental and health damages they say have been caused by oil-related activities. Protesters also feel that, despite what they were told about the economic benefits of oil production for their villages, corrupt local governments and the oil companies have been the only beneficiaries. In the meantime, they have received little or no compensation for environmental damage and ill health that has resulted from oil-production activities.

Additionally, human-rights abuse charges have been made by Nigerian citizens and by international human-rights organizations, which believe that the oil companies have been complicit in the violent repression of protesters. If not directly involved in such abuses, the oil companies have failed in their responsibility to prevent or publicly oppose such abuses, opponents say. In one particular incident on January 4, 1999, for example, Nigerian soldiers using a Chevron helicopter attacked villagers in two communities, killing at least four people and destroying most of the village structures and homes. (In March 2004 a U.S. District Court judge ruled that ChevronTexaco can be held liable for its subsidiary's involvement in the Nigerian raids.) Other protests, including the takeover of a ChevronTexaco oil plant by protesters in 2002 and the kidnappings of foreign workers in oil-producing areas, demonstrate the potential threats to Americans in the region and to U.S. interests abroad.

CHAPTER 11
NEW ARENAS: ORGANIZED CRIME AND EMERGING TECHNOLOGIES

The concept of U.S. national security is constantly evolving and adapting to the changing global security environment. There are a variety of emerging trends and threats with which America has not had to deal before. At the same time, the domestic national security infrastructure itself is changing.

ORGANIZED CRIME AND NATIONAL SECURITY

Transnational threats to American national security include not only hostile states and terrorism but also organized crime, which is associated with a host of illicit activities. Drug trafficking, financial fraud, environmental crimes, and contraband smuggling, to name just a few, are not only threats to Americans and their communities but to U.S. business and financial interests as well as global stability and security.

Such crimes are of special concern to U.S. policy makers because in most cases there is no clear identifiable enemy to target. This is different from interstate conflicts, in which parties have well-defined targets, and wars, in which armies and rules of engagement are obvious to all sides. Any battles waged against the generic front of "drug trafficking" or "money laundering" are extremely hard to fight and require significant international cooperation.

The Nature of Organized Crime

As noted on its Web site (www.fbi.gov), the Federal Bureau of Investigation (FBI) defines organized crime as "any group having some manner of a formalized structure and whose primary objective is to obtain money through illegal activities." Organized-crime groups have several characteristics in common. Much as it does with business, financial gain drives and sustains organized crime. Most groups carry out more than one type of crime. Although not an absolute requirement, many groups require their members to be of the same family or ethnic background in order to ensure loyalty and to pursue a common goal or scheme. Most organized-crime groups have become successful in one way or another by corrupting government officials. Another common characteristic is a hierarchical structure, with defined leadership-subordinate roles. In many cases, organized-crime groups are permanent and do not depend on the participation of one or a few individuals to exist, and the groups usually have influence over large areas of a region, country, or countries.

Organized-crime groups undertake a wide range of illicit global activities. They traffic in explosives, arms, narcotics, humans, metals, minerals, endangered flora and fauna, and Freon gas. They conduct extensive money laundering, fraud, graft, extortion, bribery, economic espionage, smuggling of embargoed goods, multinational auto theft, international prostitution, industrial and technological espionage, bank fraud, financial market manipulation, counterfeiting, corruption, and contract murder.

Of these activities, corruption is perhaps the biggest threat to states. Crime groups greatly compromise and jeopardize governments when they use corruption to achieve their aims. Organized criminals co-opt officials and leaders with a combination of bribery, graft, collusion, and/or extortion. Organized crime has successfully targeted such countries as Colombia, Italy, Thailand, Mexico, Russia, and Japan with payoffs or threats to justice officials to alter charges, change court rulings, lose evidence, or simply lose interest. By undercutting justice systems, these groups undercut society. Sometimes when organized crime targets members of police forces and armed forces, and those members do not cooperate, they become targets of hired assassins.

There is increasing interdependency among crime groups. The now largely defunct Medellin drug cartel in Colombia at one time joined with Russian and Italian mobsters to smuggle cocaine into Europe. In addition to conspiring with one another, crime groups also often fight with one another, which can be equally disruptive to the state. The Colombian and Mexican drug-smuggling rings have clashed more with each other than they have collaborated. Rival drug dealers and suppliers battle in New York, south Florida, and many European cities.

RUSSIA. Organized-crime groups originating in Russia and areas nearby are a growing concern. There are an estimated eight thousand Russian/Eastern European/Eurasian criminal groups, 150 of which are ethnic oriented. These include Chechens, Georgians, Armenians, and Russian-ethnic Koreans. At the time of the *International Crime Threat Assessment* (December 2000), Russian organized-crime groups were a major force in that nation's industrial and financial sectors. Automobile manufacturing, coal mining, and oil were among the industries they had penetrated. Russian organized crime groups are believed to maintain close ties to established American criminal groups and drug-trafficking organizations. They participate in complex criminal activity such as gasoline tax fraud, cyber-security breaches, bankruptcy fraud, insurance fraud, and health care industry fraud.

Illegal Drugs

The international networks that underpin the drug trade are a complex network of drug producers, processors, traffickers, and street vendors, orchestrated by organized-crime groups—often more than one. For example, one network arranged for hashish originating in Pakistan to be transported to Mombasa, Kenya. There, it was added to a cargo of tea and reshipped to Haifa, Israel, by way of Durban, South Africa. Then the drugs went to a ship that took cargo to Constanza, Romania, every two weeks. From there, via Bratislava, Slovakia, it went to Italy, where it was sold. The head of the network was a Ugandan native who became a German citizen and worked for a Romanian company. When some of the perpetrators were apprehended in Constanza, they revealed the network.

Illegal drug trafficking is big business. No one knows precisely how much money is involved in the drug trade but it is probably in the hundreds of billions of dollars annually. This makes drug criminals very powerful, especially in poorer parts of the world. Governments of countries like Colombia, Peru, and Bolivia have largely been unable to significantly reduce their countries' production and export of drugs. The U.S. federal government's drug control budget for 2004 is over $12 billion, and state and local governments also spend billions of dollars fighting drugs. Drug use has mostly plateaued in the United States, but narcotics trafficking worldwide continues to grow because of increasing demand elsewhere.

Leaders of countries often view international crime, such as drug production, as domestic legal concerns. Because criminal groups are primarily concerned with making money, their political objectives, if any, may not seem significant. Leaders also can view transnational criminals, because they operate across international borders, as other countries' problems. The U.S. government has historically considered organized crime as a law enforcement issue, not a national security threat. However, the United States has become increasingly aware that international organized crime is much more than an extension of domestic crime. Highly organized illegal enterprises operate internationally, with scant regard for state boundaries. They become larger, more complex, and grow in number. They penetrate borders and operate with relative impunity in several states. Within national borders, they pollute the integrity of domestic governments. Their willingness to use violence is often more destabilizing than the activities of revolutionary or terror groups alone. In fact, there is a fine line between the two, and occasionally organized-crime groups may operate as both, or have ties to terrorist groups.

DRUGS AND TERRORISM. Steven W. Casteel, assistant administrator for intelligence for the Drug Enforcement Administration (DEA), testified before Congress in May of 2003 about the many links between terrorist groups and drug smugglers—a phenomenon labeled narco-terrorism. In Afghanistan under the Taliban, drug money raised from the opium trade helped the fundamentalist Islamic government to support and protect Osama bin Laden and the al Qaeda terrorist group. In Colombia, three revolutionary groups routinely sell or trade cocaine with international crime organizations for guns and ammunition. Casteel testified that fourteen (or 39%) of the State Department's current list of thirty-six designated foreign terrorist organizations are connected in some way with the drug trade.

Money Laundering

Countering money-laundering efforts has taken on increased importance in the wake of the September 11, 2001, terrorist attacks on America. Money laundered through legitimate companies and nonprofit organizations has been tracked to various terrorist activities. In January 2001 the U.S. Treasury Department issued a new money-laundering guidance system. The system primarily calls for private businesses and citizens to be more aware of their banking practices and to "apply enhanced scrutiny to their private banking and similar high dollar value accounts and transactions where such accounts or transactions may involve the proceeds of corruption by senior foreign political figures, their immediate family or close associates."

This issue gained increasing importance in late 2002, when money was believed to have been laundered for terrorist organizations through the bank of a Saudi Arabian princess. The *2003 National Money Laundering Strategy* (Washington, DC: U.S. Department of the Treasury, 2003) reported that since 2001 "over 315 terrorist related entities have been designated and over $136 million in assets frozen." In November of 2003 the prominent Muslim leader Abdul Rahman al-Amoudi was accused of laundering money through front groups he operated, including the American Muslim Council and the American Muslim Foundation, and sending the money to the Hamas and al Qaeda terrorist groups. In July 2004 seven officers of a Texas-based Muslim charity called the Holy Land Foundation for Relief and Development were charged with providing $12.4 million to Hamas. Efforts to further tighten regulation of financial and charitable institutions around the world are a top priority for the administration of President George W. Bush. The United States is not alone in this effort. Many other nations have also taken strong measures to stop international criminal organizations and terrorists from misusing their own banking institutions. (See Table 11.1.)

The United Nations and Organized Crime

In November 2000 the United Nations Convention against Transnational Organized Crime and its Protocols was enacted. The convention is aimed at creating greater international cooperation against criminal organizations operating across national boundaries. Member states that ratify the convention agree to mutual legal assistance, extradition, law-enforcement cooperation, and technical assistance and training. Among the specific crimes addressed in the convention are money laundering and the smuggling of migrants.

The U.S. Response to Transnational Crime

The United States has adopted widely publicized policies for countering transnational threats such as terrorism and organized crime. These include Presidential Directive 62, signed in May 1998, which establishes a systematic approach to counterterrorism. An International Crime Control Strategy has also been created. Each year, a U.S. National Drug Control Strategy is adopted. Other legislative steps undertaken by the U.S. Congress to counter various transnational threats include: the 2001 USA Patriot Act ("Uniting and Strengthening America by Providing Appropriate Tools Required to Intercept and Obstruct Terrorism"), the Civil Asset Forfeiture Reform Act of 2000, the Money Laundering and Financial Crimes Strategy Act of 1998, and the Controlled Substances Trafficking Prohibition Act of 1998. The challenge now is to implement these strategies effectively.

In response to international crime, the FBI has three distinct strategies: first, provide an active overseas presence to establish relationships with foreign law enforcement agencies; second, train foreign law enforcement officers in both basic and advanced investigative techniques and principles to promote cooperation; and third, build an institution to help promote the rule of law in newly democratic republics, which will protect U.S. interests and citizens in these countries and bring stability to their regions.

INFORMATION TECHNOLOGY (IT) AND NATIONAL SECURITY

Many functions of national security, including the use of computers and communications to thwart or attack an enemy, are evolving rapidly. With each new improvement in information technology (IT), information warfare (IW) and computer security become more important issues for security planners. Federal, state, and local agencies involved in national security rely extensively on computer systems and electronic data. All computer systems, however, contain weaknesses and vulnerabilities that put critical operations and security assets at risk of compromise or disruption.

In 1998 an infrastructure-protection strategy was outlined in Presidential Decision Directive 63 to safeguard government and privately controlled systems from computer-based attacks. In December 2003, President George W. Bush issued Homeland Security Presidential Directive 7 (HSPD-7), which superseded Presidential Decision Directive 63 and established a national policy for the federal government to identify and prioritize critical infrastructure and key resources and to protect them from terrorist attack. *The National Strategy for Homeland Security* (Washington, DC: Department of Homeland Security, July 2002) identified the critical infrastructure sectors in need of protection from terrorist attack. (See Table 11.2.)

As part of the ongoing effort to implement HSPD-7, the Department of Homeland Security in February 2004 announced the creation of the Protected Critical Infrastructure Information (PCII) Program. This program enables private businesses voluntarily to submit confidential details of their critical infrastructure to the federal government. The government will identify potential security risks and thereby reduce their vulnerability to terrorist attack. Any widespread vulnerability issues discovered can be handled quickly at a national level through the centralized program. Critical infrastructure is defined as systems that, if disrupted, would threaten our national security, public health, and safety. They include utilities and hospitals. Since an estimated 85% of the nation's critical infrastructure is privately owned, the PCII Program is designed to create a system to improve cooperation between the public and private sectors on infrastructure security issues.

In another step meant to assist businesses in protecting their critical infrastructure, the Department of Homeland Security has also set up Information Sharing and Analysis Centers. Each center shares preventative measures, security information, and potential threats within a particular field, such as public health, information and telecommunications, banking and finance, and the food industry. The center for banking and finance, for example, has available that industry's first database of electronic security threats, vulnerabilities, incidents, and solutions.

TABLE 11.1

Efforts to combat money laundering worldwide, 2003

Actions by governments	Criminalized drug money laundering	Criminalized beyond drugs	Record large transactions	Maintain records over time	Report suspicious transactions (NMP)	Financial intelligence unit	System for identifying/forfeiting assets	Arrangements for asset sharing	Cooperates with international law enforcement	International transportation of currency	Mutual legal assistance	Non-bank financial institutions	Disclosure protection "safe harbor"	States party to 1988 UN convention	Criminalized financing of terrorism	International terrorism financing connection
Government/Jurisdiction																
Afghanistan	N	N	N	N	N	N	N	N	N	N	Y	N	N	Y	N	Y
Albania	Y	Y	N	Y	M	Y	Y	N	Y	N	Y	Y	Y	Y	Y	Y
Algeria	N	N	N	N	M	N	N	N	N	Y	N	N	Y	Y	Y	Y
Andorra	Y	Y	Y	Y	M	Y	Y	N	Y	N	Y	Y	Y	N	N	N
Angola	Y	N	N	N	M	N	N	N	N	N	N	N	N	N	N	Y
Anguilla	Y	Y	N	Y	M	Y	Y	N	Y	Y	Y	Y	Y	Y	Y	Y
Antigua & Barbuda	Y	Y	Y	Y	M	Y	Y	N	Y	Y	Y	Y	Y	Y	Y	Y
Argentina	Y	Y	Y	Y	M	Y	Y	N	Y	Y	Y	Y	Y	Y	N	N
Armenia	Y	Y	Y	Y	M	Y	Y	Y	Y	Y	Y	Y	Y	Y	N	N
Aruba	Y	Y	Y	Y	M	Y	Y	N	Y	Y	Y	Y	Y	Y	N	Y
Australia	Y	Y	Y	Y	M	Y	Y	Y	Y	Y	Y	Y	Y	Y	Y	Y
Austria	Y	Y	Y	Y	M	Y	Y	N	Y	Y	Y	Y	Y	Y	Y	Y
Azerbaijan	Y	N	N	N	N	Y	N	N	N	N	N	N	N	Y	N	N
Bahamas	Y	Y	Y	Y	M	Y	Y	N	Y	Y	Y	Y	Y	Y	Y	Y
Bahrain	Y	Y	N	Y	M	Y	Y	N	Y	N	N	N	N	Y	Y	Y
Bangladesh	Y	N	Y	Y	M	Y	Y	N	Y	N	Y	Y	N	Y	N	N
Barbados	Y	Y	Y	Y	M	Y	Y	N	Y	N	Y	Y	Y	Y	N	N
Belarus	Y	Y	Y	Y	M	N	Y	N	Y	Y	Y	Y	N	Y	Y	Y
Belgium	Y	Y	Y	Y	M	Y	Y	N	Y	Y	Y	Y	Y	Y	Y	Y
Belize	Y	Y	Y	Y	M	N	Y	N	Y	Y	N	Y	Y	Y	N	N
Benin	Y	N	Y	N	N	N	N	N	N	N	N	N	N	Y	N	N
Bermuda	Y	Y	Y	Y	M	Y	Y	N	N	N	Y	N	Y	Y	Y	Y
Bolivia	Y	Y	N	Y	M	Y	Y	N	Y	N	Y	Y	Y	Y	N	N
Bosnia & Herzegovina	Y	Y	Y	Y	M	Y	Y	N	Y	Y	N	N	N	Y	N	N
Botswana	Y	Y	Y	Y	M	N	Y	N	Y	Y	N	N	N	Y	N	Y
Brazil	Y	Y	Y	Y	M	Y	Y	N	Y	Y	Y	Y	Y	Y	Y	Y
British Virgin Islands	Y	Y	N	Y	M	Y	Y	N	Y	N	Y	Y	Y	Y	Y	Y
Brunei Darussalam	Y	N	Y	Y	M	Y	Y	N	Y	N	Y	Y	Y	Y	N	N
Bulgaria	N	Y	Y	Y	M	Y	N	N	Y	N	Y	Y	Y	Y	N	N
Burkina Faso	Y	Y	Y	Y	N	Y	Y	N	Y	N	N	Y	N	Y	Y	Y
Burma	N	N	N	N	N	N	Y	N	N	N	N	N	N	Y	N	Y
Burundi	N	N	N	N	N	N	N	N	N	N	N	N	N	N	N	N
Cambodia	N	Y	N	Y	Y	N	N	N	Y	N	Y	N	Y	Y	N	Y
Cameroon	N	N	N	Y	N	N	N	N	Y	N	Y	Y	N	Y	N	Y
Canada	Y	Y	Y	Y	M	Y	Y	Y	Y	Y	Y	Y	Y	Y	Y	Y
Cayman Islands	Y	Y	Y	Y	M	Y	Y	Y	Y	Y	Y	Y	N	Y	Y	Y
Chad	Y	N	Y	N	M	Y	Y	N	N	N	N	N	N	N	N	N
Chile	Y	Y	Y	Y	M	Y	Y	N	Y	Y	Y	Y	Y	Y	N	Y
China (PRC)	Y	Y	Y	Y	M	N	Y	N	Y	Y	Y	Y	Y	Y	Y	Y
Colombia	Y	Y	Y	Y	M	Y	Y	Y	Y	Y	Y	Y	Y	Y	N	Y
Comoros	N	N	N	N	N	N	N	N	N	N	N	N	N	N	N	N
Congo (Dem. Republic)	N	N	Y	N	N	N	N	N	N	N	+	N	N	Y	N	N
Cook Islands	Y	Y	N	Y	N	N	N	N	Y	Y	N	N	N	Y	N	Y
Congo (Republic)	Y	Y	N	N	M	N	N	N	Y	N	Y	N	N	N	N	Y
Costa Rica	Y	Y	Y	Y	M	Y	Y	Y	Y	Y	Y	Y	Y	Y	N	Y
Cote D'Ivoire	Y	Y	Y	Y	M	N	Y	Y	Y	Y	Y	Y	Y	Y	N	Y
Croatia	Y	Y	N	Y	P	N	Y	N	N	Y	N	N	N	Y	N	Y
Cuba	Y	Y	Y	Y	N	Y	Y	N	Y	Y	Y	Y	N	Y	N	Y
Cyprus	Y	Y	Y	Y	M	Y	Y	N	Y	Y	Y	Y	Y	Y	Y	Y
Czech Republic	Y	Y	Y	Y	M	Y	Y	N	Y	Y	Y	Y	Y	Y	N	N

TABLE 11.1

Efforts to combat money laundering worldwide, 2003 [CONTINUED]

Actions by governments **Government/Jurisdiction**	Criminalized drug money laundering	Criminalized beyond drugs	Record large transactions	Maintain records over time	Report suspicious transactions (NMP)	Financial intelligence unit	System for identifying/forfeiting assets	Arrangements for asset sharing	Cooperates with international law enforcement	International transportation of currency	Mutual legal assistance	Non-bank financial institutions	Disclosure protection "safe harbor"	States party to 1988 UN convention	Criminalized financing of terrorism	International terrorism financing connection
Denmark	Y	Y	Y	Y	M	Y	Y	N	Y	Y	Y	Y	Y	Y	Y	Y
Djibouti	Y	Y		Y	M	N	N	N	Y	N	Y	Y	Y	Y	N	N
Dominica	Y	Y	N	Y	M	Y	Y	Y	Y	Y	Y	Y	Y	Y	Y	N
Dominican Republic	Y	Y	Y	Y	M	Y	Y	N	Y	Y	Y	Y	Y	Y	N	N
East Timor	N	N	Y	N	N	N	Y	N	N	N	N	N	N	N	N	Y
Ecuador	Y	N	Y	Y	M	Y	Y	N	Y	N	Y	Y	Y	Y	N	N
Egypt	Y	Y	N	Y	M	N	Y	N	Y	Y	Y	N	N	Y	Y	Y
El Salvador	Y	Y	Y	Y	M	Y	Y	Y	Y	Y	Y	Y	Y	Y	N	N
Eritrea	N	N	N	N	N	N	N	N	N	N	N	N	N	N	N	Y
Estonia	Y	Y	N	Y	M	Y	Y	N	Y	Y	Y	Y	Y	Y	N	N
Ethiopia	N	N	Y	N	N	N	N	N	N	N	N	N	N	N	N	Y
Fiji	Y	Y	Y	Y	M	Y	Y	N	Y	Y	Y	Y	Y	Y	Y	N
Finland	Y	Y	Y	Y	M	Y	Y	Y	Y	Y	Y	Y	Y	Y	Y	Y
France	Y	Y	Y	Y	M	Y	Y	Y	Y	Y	Y	Y	Y	Y	Y	Y
Gabon	N	N	N	N	N	N	N	N	N	N	N	N	N	N	N	N
Gambia	Y	Y	N	Y	M	Y	Y	N	Y	Y	Y	Y	Y	Y	Y	N
Georgia	Y	Y	N	N	N	Y	N	N	N	Y	N	N	N	Y	N	Y
Germany	Y	Y	Y	Y	M	Y	Y	Y	Y	Y	Y	Y	Y	Y	Y	N
Ghana	N	N	Y	N	N	N	N	N	N	N	N	N	N	N	N	Y
Gibraltar	Y	Y	Y	Y	M	Y	Y	Y	Y	Y	Y	Y	Y	Y	Y	Y
Greece	Y	Y	Y	Y	M	Y	Y	N	Y	Y	Y	Y	Y	Y	Y	Z
Grenada	Y	Y	Y	Y	M	Y	Y	N	Y	Y	Y	Y	Y	Y	Y	Y
Guatemala	Y	Y	N	Y	M	Y	Y	N	Y	Y	Y	N	N	Y	N	N
Guernsey	Y	Y	Y	Y	M	Y	Y	Y	Y	Y	Y	Y	Y	Y	Y	Z
Guinea	Y	Y	N	N	N	N	N	N	N	N	N	N	N	Y	Z	Z
Guinea-Bissau	N	N	N	N	N	N	N	N	N	N	N	N	N	N	Z	Z
Guyana	Y	Y	Y	Y	M	Y	Y	N	Y	Y	Y	Y	Y	Y	Y	Z
Haiti	Y	Y	Y	Y	M	Y	Y	N	Y	Y	Y	Y	Y	Y	Z	Y
Honduras	Y	Y	N	Y	M	N	Y	N	Y	N	Y	Y	Y	Y	Z	N
Hong Kong	Y	Y	Y	Y	M	Y	Y	N	Y	Y	Y	Y	Y	Y	Y	Y
Hungary	N	Y	Y	N	M	Y	Y	Y	Y	Y	Y	N	N	Y	Z	N
Iceland	Y	Y	Y	Y	M	Y	Y	N	Y	Y	Y	Y	Y	Y	Y	Y
India	Y	Y	Y	Y	M	Y	Y	N	Y	Y	Y	Y	Y	Y	Y	N
Indonesia	Y	Y	Y	Y	M	Y	Y	N	Y	Y	Y	Y	N	Y	Y	Y
Iran	N	N	N	N	N	N	N	N	N	N	N	N	N	Y	Z	Y
Ireland	Y	Y	Y	Y	M	Y	Y	Y	Y	Y	Y	Y	Y	Y	Y	Y
Isle of Man	Y	Y	Y	Y	M	Y	Y	Y	Y	Y	Y	Y	Y	Y	Y	N
Israel	Y	Y	Y	Y	M	Y	Y	N	Y	Y	Y	Y	Y	Y	Y	Y
Italy	Y	Y	Y	Y	M	Y	Y	Y	Y	Y	Y	Y	Y	Y	Y	N
Jamaica	Y	Y	Y	Y	M	Y	Y	N	Y	Y	Y	Y	Y	Y	Z	N
Japan	Y	Y	Y	Y	M	Y	Y	N	Y	Y	Y	Y	Y	Y	Y	Y
Jersey	Y	Y	Y	Y	P	Y	Y	Y	Y	Y	Y	Y	Y	Y	Y	Y
Jordan	Y	N	N	Y	P	Y	Y	Y	Y	Y	Y	Y	Y	Y	Z	N
Kazakhstan	Y	N	N	N	N	N	N	N	N	N	N	N	N	Y	Z	Z
Kenya	N	N	N	N	N	N	N	N	N	N	Y	N	N	Y	Z	Y
Korea (DPRK)	N	N	N	N	N	N	N	N	N	N	N	N	N	N	Z	Y
Korea (Republic of)	Y	Y	Y	Y	M	Y	Y	Y	Y	Y	Y	Y	Y	Y	Y	Y
Kosovo	N	N	N	Y	P	Y	N	N	N	N	N	Y	Y	N	Z	Z
Kuwait	Y	Y	Y	Y	M	N	Y	N	Y	Y	Y	Y	Y	Y	Z	Z
Kyrgyzstan	N	N	N	N	P	N	Y	N	N	N	Y	N	Y	Y	Z	Y
Laos	N	N	N	N	N	N	N	N	N	N	N	N	N	N	Z	N

TABLE 11.1

Efforts to combat money laundering worldwide, 2003 [CONTINUED]

Actions by governments — Government/Jurisdiction	Criminalized drug money laundering	Criminalized beyond drugs	Record large transactions	Maintain records over time	Report suspicious transactions (NMP)	Financial intelligence unit	System for identifying/forfeiting assets	Arrangements for asset sharing	Cooperates with international law enforcement	International transportation of currency	Mutual legal assistance	Non-bank financial institutions	Disclosure protection "safe harbor"	States party to 1988 UN convention	Criminalized financing of terrorism	International terrorism financing connection
Latvia	Y	Y	Y	Y	M	Y	N	N	Y	N	Y	Y	Y	Y	N	Y
Lebanon	Y	Y	Y	Y	M	Y	Y	N	N	N	Y	Y	N	Y	Y	N
Lesotho	N	N	Y	N	M	N	N	N	N	Y	N	N	N	N	N	Y
Liberia	Y	N	N	Y	N	N	N	N	N	N	N	N	N	N	N	Y
Liechtenstein	Y	Y	Y	Y	M	Y	Y	Y	Y	N	Y	Y	Y	Y	Y	Y
Lithuania	Y	Y	Y	Y	M	Y	Y	N	Y	N	Y	Y	Y	Y	Y	Y
Luxembourg	Y	Y	Y	Y	M	Y	Y	N	Y	N	Y	Y	Y	Y	Y	N
Macau	Y	Y	N	Y	M	Y	Y	N	Y	N	Y	Y	N	Y	Y	Y
Macedonia	Y	Y	N	Y	M	N	N	N	N	Y	Y	Y	N	N	Y	N
Madagascar	N	N	N	N	N	N	N	N	N	N	N	N	N	N	N	Y
Malawi	Y	N	N	Y	P	N	N	N	Y	N	N	N	N	N	N	N
Malaysia	Y	Y	N	Y	M	Y	N	N	Y	N	N	Y	N	Y	Y	Y
Maldives	N	N	N	N	N	N	N	N	N	N	N	N	N	N	N	Y
Mali	N	N	N	N	N	N	N	N	N	N	N	N	N	N	N	Y
Malta	Y	Y	Y	Y	M	Y	Y	Y	N	Y	Y	Y	Y	Y	Y	Y
Marshall Islands	Y	Y	N	Y	M	N	Y	N	N	N	Y	Y	N	Y	Y	N
Mauritius	Y	Y	N	Y	M	Y	Y	N	N	N	Y	Y	Y	Y	Y	Y
Mexico	Y	Y	Y	Y	M	Y	Y	Y	Y	N	Y	Y	Y	Y	Y	Y
Micronesia	Y	Y	N	Y	N	N	Y	N	N	N	Y	Y	N	Y	Y	N
Moldova	Y	Y	N	Y	M	N	Y	N	N	N	Y	Y	N	Y	Y	Y
Monaco	Y	N	N	Y	M	N	N	N	N	Y	Y	Y	N	Y	Y	N
Mongolia	N	N	N	N	N	N	N	N	N	N	N	N	N	N	N	N
Montenegro	Y	Y	Y	Y	M	Y	Y	N	N	N	Y	Y	N	Y	Y	Y
Montserrat	Y	Y	N	Y	M	N	N	N	N	N	Y	Y	N	Y	Y	Y
Morocco	N	N	Y	Y	M	N	Y	N	N	N	Y	Y	N	N	Y	N
Mozambique	N	N	N	Y	M	N	N	N	N	N	Y	N	N	Y	Y	Y
Namibia	N	N	N	Y	M	N	Y	N	N	N	Y	Y	N	N	N	Y
Nauru	N	N	N	N	N	N	N	N	N	N	N	N	N	N	N	N
Nepal	N	N	N	N	N	N	N	N	N	N	N	N	N	Y	N	Y
Netherlands	Y	Y	Y	Y	M	Y	Y	N	Y	N	Y	Y	Y	Y	Y	Y
Netherlands Antilles	Y	Y	N	Y	M	Y	Y	Y	N	N	Y	Y	Y	Y	Y	N
New Zealand	Y	Y	Y	Y	M	Y	Y	Y	Y	Y	Y	Y	Y	Y	Y	Y
Nicaragua	Y	Y	N	Y	N	N	Y	N	N	N	Y	Y	Y	Y	Y	Y
Niger	Y	N	Y	N	P	N	N	N	N	N	Y	N	N	N	N	N
Nigeria	Y	Y	N	Y	M	N	Y	N	N	N	Y	Y	N	Y	Y	Y
Niue	Y	Y	Y	Y	M	N	Y	Y	N	Y	N	Y	N	NA	Y	Y
Norway	Y	Y	Y	Y	M	Y	Y	Y	N	Y	Y	Y	Y	Y	Y	Y
Oman	Y	Y	N	Y	M	N	Y	N	N	N	Y	Y	N	Y	Y	N
Pakistan	Y	Y	N	Y	M	N	Y	N	N	N	Y	Y	Y	Y	Y	Y
Palau	Y	Y	Y	Y	M	Y	Y	Y	N	N	Y	Y	N	Y	Y	Y
Panama	Y	Y	Y	Y	M	Y	Y	Y	N	Y	Y	Y	Y	Y	Y	Y
Papua New Guinea	N	Y	N	N	M	N	Y	N	N	N	Y	Y	N	Y	Y	Y
Paraguay	Y	Y	Y	Y	M	Y	Y	Y	Y	N	Y	Y	Y	Y	Y	Y
Peru	Y	Y	Y	Y	M	Y	Y	N	N	Y	Y	Y	Y	Y	Y	Y
Philippines	Y	Y	Y	Y	M	N	Y	N	N	N	Y	Y	Y	Y	Y	N
Poland	Y	Y	Y	Y	M	Y	Y	N	Y	N	Y	Y	Y	Y	Y	Y
Portugal	Y	Y	Y	Y	M	Y	Y	Y	N	N	Y	Y	Y	Y	Y	Y
Qatar	Y	Y	Y	Y	M	Y	N	Y	N	N	Y	N	Y	Y	Y	Y
Romania	Y	Y	Y	Y	M	Y	Y	N	N	N	Y	Y	Y	Y	Y	Y
Russia	Y	Y	N	Y	M	Y	Y	N	Y	N	Y	N	Y	Y	N	N
Rwanda	N	N	N	N	N	N	N	N	N	N	N	N	N	Y	N	Y

TABLE 11.1

Efforts to combat money laundering worldwide, 2003 [CONTINUED]

Actions by governments / Government/Jurisdiction	Criminalized drug money laundering	Criminalized beyond drugs	Record large transactions	Maintain records over time	Report suspicious transactions (NMP)	Financial intelligence unit	System for identifying/forfeiting assets	Arrangements for asset sharing	Cooperates with international law enforcement	International transportation of currency	Mutual legal assistance	Non-bank financial institutions	Disclosure protection "safe harbor"	States party to 1988 UN convention	Criminalized financing of terrorism	International terrorism financing connection
Samoa	Y	Y	N	Y	M	N	Y	Y	Y	Y	Y	Y	Y	N	Y	Y
San Marino	Y	Y	N	Y	M	N	Y	Y	Y	Y	Y	Y	Y	Y	N	Y
Sao Tome & Principe	N	N	N	N	N	N	N	N	N	N	N	N	N	Y	N	N
Saudi Arabia	Y	N	N	Y	M	N	Y	N	Y	N	Y	N	N	Y	N	N
Senegal	Y	N	Y	Y	M	N	N	N	Y	N	Y	N	N	Y	N	Y
Serbia	Y	Y	N	Y	M	N	N	N	N	N	Y	N	N	Y	N	Y
Seychelles	Y	Y	N	Y	M	N	N	N	N	N	N	N	N	Y	N	N
Sierra Leone	N	N	N	Y	P	N	N	N	N	N	N	N	N	Y	N	N
Singapore	Y	Y	N	Y	M	Y	Y	Y	Y	Y	Y	Y	N	Y	N	Y
Slovakia	Y	Y	Y	Y	M	Y	Y	Y	Y	Y	Y	Y	N	Y	N	Y
Slovenia	Y	Y	N	N	M	Y	Y	Y	Y	Y	Y	Y	N	N	Y	N
Solomon Islands	N	N	N	N	M	N	N	N	N	N	N	N	N	Y	N	Y
South Africa	Y	Y	N	Y	M	Y	Y	Y	Y	N	Y	Y	N	Y	N	Y
Spain	Y	Y	Y	Y	M	Y	Y	Y	Y	Y	Y	Y	N	Y	Y	Y
Sri Lanka	Y	N	N	N	M	N	N	Y	Y	N	Y	N	N	Y	N	Y
St Kitts & Nevis	Y	Y	N	Y	M	N	N	N	N	N	Y	N	N	Y	N	Y
St. Lucia	Y	Y	N	Y	M	N	Y	N	Y	N	Y	N	N	Y	N	Y
St. Vincent/Grenadines	Y	Y	Y	Y	M	Y	Y	Y	Y	Y	Y	Y	N	Y	N	N
Suriname	Y	Y	N	Y	M	N	Y	N	N	N	Y	N	N	Y	N	Y
Swaziland	Y	Y	N	Y	M	N	Y	Y	Y	N	Y	N	N	Y	N	Y
Sweden	Y	Y	N	Y	M	Y	Y	Y	Y	Y	Y	Y	N	Y	Y	Y
Switzerland	N	N	Y	Y	M	Y	Y	Y	Y	Y	Y	Y	Y	N	N	Y
Syria	N	N	N	N	M	N	Y	Y	N	N	Y	N	Y	Y	N	Y
Taiwan	Y	Y	Y	Y	M	Y	Y	Y	Y	Y	Y	Y	Y	NA	N	Y
Tajikistan	Y	Y	N	N	P	N	Y	Y	Y	Y	N	N	Y	Y	N	Y
Tanzania	Y	N	Y	Y	M	N	Y	Y	Y	Y	N	Y	Y	Y	N	Y
Thailand	Y	Y	Y	Y	M	Y	Y	N	Y	Y	Y	N	N	Y	Y	Y
Togo	N	N	N	Y	N	N	Y	N	N	N	N	N	N	Y	N	Y
Tonga	Y	Y	Y	Y	M	N	Y	Y	Y	Y	N	Y	N	Y	N	N
Trinidad & Tobago	Y	Y	Y	Y	M	Y	Y	Y	Y	Y	Y	Y	N	Y	Y	Y
Tunisia	N	N	Y	N	N	N	Y	N	N	N	N	N	N	Y	N	Y
Turkey	Y	Y	Y	Y	M	N	Y	Y	Y	N	Y	Y	N	Y	N	Y
Turkmenistan	Y	N	N	N	N	N	Y	N	N	N	N	N	N	Y	N	Y
Turks & Caicos	Y	Y	Y	N	N	N	Y	Y	Y	N	Y	N	N	Y	Y	N
Uganda	Y	N	N	Y	N	N	N	N	Y	N	N	N	N	Y	N	Y
Ukraine	Y	Y	Y	Y	M	Y	Y	N	Y	N	Y	N	N	Y	N	Y
United Arab Emirates	Y	Y	Y	Y	M	Y	Y	N	Y	Y	Y	Y	N	Y	Y	Y
United Kingdom	Y	Y	N	Y	M	Y	Y	Y	Y	N	Y	Y	N	Y	Y	Y
United States	Y	Y	Y	Y	M	Y	Y	Y	Y	Y	Y	Y	N	Y	Y	Y
Uruguay	Y	Y	Y	Y	M	Y	Y	Y	Y	N	Y	N	N	Y	Y	Y
Uzbekistan	Y	Y	Y	Y	M	N	Y	N	Y	N	Y	Y	N	Y	Y	N
Vanuatu	Y	N	Y	N	M	N	Y	Y	N	N	Y	N	N	N	Y	Y
Venezuela	Y	N	Y	Y	M	Y	Y	N	Y	N	Y	Y	N	Y	Y	Y
Vietnam	Y	Y	N	Y	N	N	N	N	N	N	N	N	N	Y	N	N
Yemen	Y	Y	N	Y	M	N	N	N	N	N	N	N	N	Y	N	Y
Zambia	Y	Y	N	Y	M	N	Y	N	N	N	Y	Y	N	Y	N	N
Zimbabwe	Y	N	N	N	N	Y	Y	N	N	N	N	N	N	Y	Y	N

SOURCE: "Comparative Table," in *International Narcotics Control Strategy Report 2003*, Bureau for International Narcotics and Law Enforcement Affairs, March 2004, http://www.state.gov/g/inl/rls/nrcrpt/2003/vol2/html/29928.htm (accessed September 23, 2004)

TABLE 11.2

Critical infrastructure sector identified by the National Strategy for Homeland Security and Homeland Security Presidential Directive 7 (HSPD-7), 2004

Sector	Description	Sector-specific agencies
Agriculture	Provides for the fundamental need for food. The infrastructure includes supply chains for feed and crop production.	Department of Agriculture
Banking and finance	Provides the financial infrastructure of the nation. This sector consists of commercial banks, insurance companies, mutual funds, government-sponsored enterprises, pension funds, and other financial institutions that carry out transactions including clearing and settlement.	Department of the Treasury
Chemicals and hazardous materials	Transforms natural raw materials into commonly used products benefiting society's health, safety, and productivity. The chemical industry represents a $450 billion enterprise and produces more than 70,000 products that are essential to automobiles, pharmaceuticals, food supply , electronics, water treatment, health, construction and other necessities.	Department of Homeland Security
Defense industrial base	Supplies the military with the means to protect the nation by producing weapons, aircraft, and ships and providing essential services, including information technology and supply and maintenance.	Department of Defense
Emergency services	Saves lives and property from accidents and disaster. This sector includes fire, rescue, emergency medical services, and law enforcement organizations.	Department of Homeland Security
Energy	Provides the electric power used by all sectors, including critical infrastructures, and the refining, storage, and distribution of oil and gas. The sector is divided into electricity and oil and natural gas.	Department of Energy
Food	Carries out the post-harvesting of the food supply, including processing and retail sales.	Department of Agriculture and Department of Health and Human Services
Government	Ensures national security and freedom and administers key public functions.	Department of Homeland Security
Information technology and telecommunications	Provides communications and process to meet the needs of businesses and government.	Department of Homeland Security
Postal and shipping	Delivers private and commercial letters, packages, and bulk assets. The U.S. Postal Service and other carriers provide the services of this sector.	Department of Homeland Security
Public health and healthcare	Mitigates the risk of disasters and attacks and also provides recovery assistance if an attack occurs. The sector consists of health departments, clinics, and hospitals.	Department of Health and Human Services
Transportation	Enables movement of people and assets that are vital to our economy, mobility, and security with the use of aviation, ships, rail, pipelines, highways, trucks, buses, and mass transit.	Department of Homeland Security
Drinking water and water treatment systems	Sanitizes the water supply with the use of about 170,000 public water systems. These systems depend on reservoirs, dams, wells, treatment facilities, pumping stations, and transmission lines.	Environmental Protection Agency

SOURCE: "Table 18. Critical Infrastructure Sectors Identified by the National Strategy for Homeland Security and HSPD-7," in *Technology Assessment: Cybersecurity for Critical Infrastructure Protection,* General Accounting Office, May 2004, http://www.gao.gov/new.items/d04321.pdf (accessed September 23, 2004)

The Growth of IT: Processor/Chip Development

The fundamental driving force of the information revolution continues to be the rapid and consistent rate at which silicon-based devices, such as computer chips and microprocessors, are developing. Since 1981 processor speeds for personal computers have risen several hundredfold. Personal computers experienced similar increases in memory, hard-drive storage systems, and modem speeds. Bandwidths—communications line capacities—have also increased. Increased bandwidths have enabled leaps in the speed and convenience of common software functions such as scrolling text and transferring graphics. In addition, this type of technology has become more and more accessible to the general public, making life convenient but also creating dangerous tools.

These increased capabilities have transformed the way the U.S. government and military use technology and the way in which enemies of the United States are able to access information and potentially cause the country harm. In addition, the spread of technology has created an increasing interconnectivity between computer systems, which, though useful for many purposes, also creates substantial risks. According to the U.S. General Accounting Office (GAO) report *Critical Infrastructure Protection: Federal Efforts Require a More Coordinated and Comprehensive Approach for Protecting Information Systems* (July 22, 2002), the interconnectivity of government computer systems allows individuals or groups to launch attacks across a span of these systems or computers, making it easy to disguise identity, location, and intent. In turn, this can make it difficult to find the attackers. Potential risks include the compromise of confidential material, disruption of communications and computer-assisted operations, and corruption of the integrity of data. Table 11.3 lists some of the computer-related threats to the U.S. government that the GAO has observed.

The terrorists that struck the United States on September 11, 2001, made use of easily obtained technology, such as e-mail and cell phones, to orchestrate the attacks. By using public computers, such as those in Internet cafés and public libraries, and cell phones, potential attackers can decrease law enforcement and intelligence agencies' ability to find or stop them.

TABLE 11.3

Threats to critical infrastructure, 2004

Threat	Description
Criminal groups	International corporate spies and organized crime organizations pose a threat to the United States through their ability to conduct industrial espionage and large-scale monetary theft and to hire or develop hacker talent.
Hackers	Hackers sometimes crack into networks for the thrill of the challenge or for bragging rights in the hacker community. While remote cracking once required a fair amount of skill or computer knowledge, hackers can now download attack scripts from the Internet and launch them against victim sites. Thus, while attack tools have become more sophisticated, they have also become easier to use. According to the Central Intelligence Agency (CIA), the large majority of hackers do not have the requisite tradecraft to threaten difficult targets such as critical U.S. networks. Nevertheless, the worldwide population of hackers poses a relatively high threat of an isolated or brief disruption causing serious damage.
Hacktivists	Hacktivism refers to politically motivated attacks on publicly accessible Web pages or e-mail servers. These groups and individuals overload e-mail servers and hack into Web sites to send a political message. Most international hacktivist groups appear bent on propaganda rather than damage to critical infrastructures.
Insider threat	The disgruntled organization insider is a principal source of computer crimes. Insiders may not need a great deal of knowledge about computer intrusions because their knowledge of a victim system often allows them to gain unrestricted access to cause damage to the system or to steal system data. The insider threat also includes outsourcing vendors.
National governments and foreign intelligence services	Several nations are aggressively working to develop information warfare doctrine, programs, and capabilities. Such capabilities enable a single entity to have a significant and serious impact by disrupting the supply, communications, and economic infrastructures that support military power—impacts that could affect the daily lives of U.S. citizens across the country. The threat from national cyber warfare programs is unique because they pose a threat along the entire spectrum of objectives that might harm U.S. interests. According to the CIA, only government-sponsored programs are developing capabilities with the prospect of causing widespread, long-duration damage to U.S. critical infrastructures.
Terrorists	Terrorists seek to destroy, incapacitate, or exploit critical infrastructures to threaten national security, cause mass casualties, weaken the U.S. economy, and damage public morale and confidence. However, traditional terrorist adversaries of the United States are less developed in their computer network capabilities than other adversaries. Terrorists likely pose a limited cyber threat. The CIA believes terrorists will stay focused on traditional attack methods, but it anticipates growing cyber threats as a more technically competent generation enters the ranks.
Virus writers	Virus writers are posing an increasingly serious threat. Several destructive computer viruses and worms have harmed files and hard drives, including the Melissa Macro Virus, the Explore.Zip worm, the CIH (Chernobyl) Virus, Nimda, Code Red, Slammer, and Blaster.

SOURCE: "Table 6. Threats to Critical Infrastructure," in *Technology Assessment: Cybersecurity for Critical Infrastructure Protection,* General Accounting Office, May 2004, http://www.gao.gov/new.items/d04321.pdf (accessed September 23, 2004)

The Expanding Scope of Information Warfare (IW) against the U.S. Government

The scope of IW can be defined by the "players" and three dimensions of their interactions: their nature, level, and arena (means of interaction). Nation-states or combinations of nation-states are not the only players. Nonstate actors (including political, ethnic, and religious groups; organized crime; international and transnational organizations; and even individuals, empowered by laptops and fast Internet connections) are able to launch information attacks and develop information strategies to achieve their desired ends. They can pose information threats, engage in information attacks, and develop digital warfare strategies, such as the introduction of digital "viruses" and "worms," to achieve their ends. (A digital virus is usually passed from computer to computer via e-mail in attachments sent to unsuspecting people. Digital worms are a type of computer attack that propagates through networks without user intervention.) Some examples of particularly problematic attacks are shown in Table 11.4.

Attacks on information systems are a fact of life in the information age. Only a small portion of these attacks result in significant loss or damage—the vast majority do not. These are the computer equivalents of crimes such as trespassing, public nuisance, minor vandalism, and petty theft. Yet large companies and the government are at risk from attacks against their computer systems and networks, as well as espionage committed with IT.

CLASSIFYING THREATS. As with any national security concern, the first task of those who would undertake information warfare defense (IWD) is to identify and classify threats. Some planners refer to such threats within a spectrum known as "the threat space." The consequences of failing to counter attacks in the range of threats on one end of the spectrum are isolated and limited, but on the other end are potentially catastrophic consequences. Planners divide the threat space into three main areas, or regions: (1) everyday—troublesome challenges that exact a price in vigilance but do not pose a threat to national security; (2) potentially strategic—may or may not have national security implications; and (3) strategic—have definite national security implications.

In the category of everyday threats are attacks on commercial targets, which include information-age versions of fraud, theft, and white-collar crimes, combined with some transformations of violent crime into virtual form. Some of these attacks can amount to bank robbery, when money is transferred out of accounts. However, attacks by competing commercial organizations typically do not target money but rather vital information, also known as trade secrets. Still, theft of trade secrets has the potential for more serious consequences than isolated thefts or embezzlement. Such attacks may constitute "economic spying" or commercial espionage and can become a potential strategic threat (part of the middle area of the threat space) when foreign companies target key industries.

TABLE 11.4

Types of cyber attacks, 2004

Type of attack	Description
Denial of service	A method of attack that denies system access to legitimate users without actually having to compromise the targeted system. From a single source, the attack overwhelms the target computer with messages and blocks legitimate traffic. It can prevent one system from being able to exchange data with other systems or prevent the system from using the Internet.
Distributed denial of service	A variant of the denial-of-service attack that uses a coordinated attack from a distributed system of computers rather than a single source. It often makes use of worms to spread to multiple computers that can then attack the target.
Exploit tools	Publicly available and sophisticated tools that intruders of various skill levels can use to determine vulnerabilities and gain entry into targeted systems.
Logic bombs	A form of sabotage in which a programmer inserts code that causes the program to perform a destructive action when some triggering event occurs, such as terminating the programmer's employment.
Sniffer	Synonymous with packet sniffer. A program that intercepts routed data and examines each packet in search of specified information, such as passwords transmitted in clear text.
Trojan horse	A computer program that conceals harmful code. A Trojan horse usually masquerades as a useful program that a user would wish to execute.
Virus	A program that "infects" computer files, usually executable programs, by inserting a copy of itself into the file. These copies are usually executed when the infected file is loaded into memory, allowing the virus to infect other files. Unlike the computer worm, a virus requires human involvement (usually unwitting) to propagate.
War-dialing	Simple programs that dial consecutive phone numbers looking for modems.
War-driving	A method of gaining entry into wireless computer networks using a laptop, antennas, and a wireless network adaptor that involves patrolling locations to gain unauthorized access.
Worms	An independent computer program that reproduces by copying itself from one system to another across a network. Unlike computer viruses, worms do not require human involvement to propagate.

SOURCE: "Table 9. Types of Cyber Attacks," in *Technology Assessment: Cybersecurity for Critical Infrastructure Protection,* General Accounting Office, May 2004, http://www.gao.gov/new.items/d04321.pdf (accessed September 23, 2004)

The second area of the threat space is potentially strategic threats, attacks on the country's national or international physical/monetary infrastructures. These include attacks on systems and services related to public safety, energy, finance, and communications. Hackers mount the vast majority of these attacks. Their motives run the gamut from financial to entertainment to sociopathy to terroristic.

Only a small number of such lone-perpetrator attacks are likely to have strategic consequences, although they can clearly result in significant data loss, interrupted services, and stolen assets. It is conceivable that a hacker attack could somehow mushroom into a national security concern, though unlikely. A well-planned and coordinated infrastructure attack would be another matter, however, and would probably qualify as digital warfare, with strategic consequences.

The strategic category of the threat space contains a relatively small number of threats that must be defended against with great vigor. These would include attacks against U.S. systems that control and safeguard weapons of mass destruction and the country's minimum essential emergency communication network. Other systems and/or networks in this category would be associated with the National Command Authority; command, control, communications, and intelligence; and intelligence, especially information regarding sources and methods.

Information attackers have their choice of the time, place, medium, and method of attack. The technology edge also goes to the attacker—it is difficult to perfect defenses at an affordable cost. Defense planners know that IW is a learning environment, with attackers learning from undetected attacks, whether successful or not, and both sides learning from detected attacks, whether successful or not.

In 1988 the Defense Advanced Research Projects Agency, a Department of Defense (DOD) agency, established a Computer Emergency Response Team (CERT) to address the computer security concerns of research users. Based on incidents reported to CERT, an estimated 90% of IW attacks are perpetrated using readily available tools and techniques. Only one attack in twenty is noticed by the victim. Corresponding results of a Defense Information Systems Agency study show that only one in twenty IW attacks may even be reported, and similar findings have been reported by others.

Despite the lack of reporting, in a small number of cases IT technology used against major corporations and the U.S. government can cause major financial and national security costs. According to CERT (in the GAO report *Critical Infrastructure Protection*), the number of cyber attacks against critical infrastructure has risen remarkably since the terrorist attacks of September 11, 2001. In the first six months of 2002 alone, information-security incidents rose to almost 45,000, with all incidents for the entire year of 2001 numbering about 55,000. In 2003 over 137,000 incidents were reported. (See Figure 11.1.)

COMPUTER CRIMES AND THE GOVERNMENT. By the 1990s computer-assisted crime became a major part of white-collar crime, and it has had an impact on the way the government works. Computer crime is faceless and

FIGURE 11.1

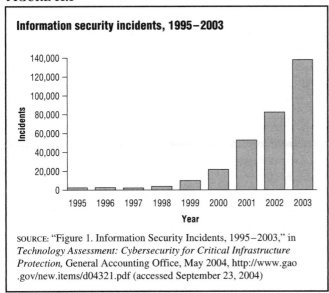

Information security incidents, 1995–2003

SOURCE: "Figure 1. Information Security Incidents, 1995–2003," in *Technology Assessment: Cybersecurity for Critical Infrastructure Protection,* General Accounting Office, May 2004, http://www.gao.gov/new.items/d04321.pdf (accessed September 23, 2004)

bloodless, and the financial gain can be huge. The National Institute of Justice defines different types of computer criminal activity as:

- Computer abuse—A broad range of intentional acts that may or may not be specifically prohibited by criminal statutes. Any intentional act involving knowledge of computer use or technology . . . if one or more perpetrators made or could have made gain and/or one or more victims suffered or could have suffered loss.

- Computer fraud—Any crime in which a person uses the computer either directly or as a vehicle for deliberate misrepresentation or deception, usually to cover up embezzlement or theft of money, goods, services, or information.

- Computer crime—Any violation of a computer-crime law.

A common computer crime involves tampering with accounting and banking records, especially through electronic funds transfers. These electronic funds transfers, or wire transfers, are cash-management systems that allow the customer electronic access to an account, automatic teller machines, and internal banking procedures, including online teller terminals and computerized check products. Money could potentially be taken from the U.S. government through these methods.

Computers and associated technology (printers, modems, computer bulletin boards, e-mail) are used for credit card fraud, counterfeiting, bank embezzlement, theft of secret documents, vandalism, and other illegal activities. The *2004 E-Crime Watch Survey* (CSO magazine/U.S. Secret Service/CERT Coordination Center, May 25, 2004) estimated a loss from computer crimes of $666 million for 2003. This estimate includes the cost of

time lost because of computer operations being shut down. That figure may be an underestimate because many victims try to hide the crime. The government and businesses may not want to admit that their computer security has been breached and their confidential files and accounts are vulnerable. No centralized databank exists for computer-crime statistics, and computer crimes are often written up under other categories, such as fraud and embezzlement.

In 1986 Congress passed the Computer Fraud and Abuse Act (PL 99-474), which makes it illegal to carry out fraud on a computer. The Computer Abuse Amendments of 1994 (PL 103-322) make it a federal crime "through means of a computer used in interstate commerce or communications . . . [to] damage, or cause damage to, a computer, computer system, network, information, data, or program . . . with reckless disregard" for the consequences of those actions to the computer owner. This law refers to someone who maliciously destroys or changes computer records or knowingly distributes a virus that shuts down a computer system.

The 2004 Computer Crime and Security Survey was conducted by the Computer Security Institute in San Francisco, California, with the participation of the FBI's San Francisco Computer Intrusion Squad. According to the survey results, of 481 computer security practitioners, government agencies, major U.S. corporations, financial and medical institutions, and universities, 53% had detected computer-security breaches in the previous twelve months. Over $141 million was reported lost because of computer-security breaches, down from $201 million in 2003.

According to the survey, the most serious financial losses ($26 million) resulted from denial of service. Still, despite the significant amount of financial losses, only 20% of respondents stated that they reported the computer intrusions to law enforcement. This low level of reporting may have to do with an unwillingness to reveal how vulnerable the company is to computer-security breaches or that valuable proprietary information stolen.

Survey respondents reported various types of attacks on, or unauthorized uses of, their computer systems. Some 59% of respondents stated they had detected employee abuse of Internet access privileges, such as downloading pornography or pirating software. Almost 80% reported the detection of computer viruses. Ten percent of respondents reported the theft of proprietary information, and less than 5% reported financial fraud.

For the government, one type of computer crime involves the sabotage or threatened sabotage of its computer systems and networks. It is almost impossible to determine how often this happens because few incidents are reported. In the computer age, several new scenarios of sabotage involving employee threats have come into being. A disgruntled employee might want to take revenge on the government. A systems administrator responsible for running computer systems might feel unappreciated. A discontented employee might create a "logic bomb" that explodes a month after he or she has left and destroys most or all of the computer records, bringing operations to a halt.

Although infrequent, charges have sometimes been brought against those who destroy a company's computer system. In February 1998 the U.S. Department of Justice (DOJ) brought charges against a former chief computer network program designer of a high-tech company that did considerable work for the National Aeronautics and Space Administration (NASA) and the U.S. Navy. The designer had worked for the company for eleven years. After he was terminated, it was alleged that in retaliation he "intentionally caused irreparable damage to Omega's computer system by activating a 'bomb' that permanently deleted all of the company's sophisticated software programs." The loss cost the company at least $10 million in sales and contracts. Such crimes committed directly against government agencies could have the potential for even greater damage.

COMPUTER HACKING: EASY ENOUGH FOR KIDS? Illegal accessing of a computer, known as hacking, is a crime juveniles frequently commit. When it is followed by manipulation of the information of private, corporate, or government databases and networks, it can be costly. Another means of computer hacking involves creation of a "virus" program. The virus resides inside another program and is activated by a predetermined code to create havoc in the host computer. Virus programs can be transmitted either through the sharing of disks and programs or through e-mail.

Cases of juvenile hacking have been around for at least two decades and have included teens getting into more than sixty computer networks, including the Memorial Sloan-Kettering Cancer Center and the Los Alamos National Laboratory in 1983; several juvenile hackers accessing AT&T's computer network in 1987; and teens hacking into computer networks and Web sites for NASA, the Korean Atomic Research Institute, America Online, the U.S. Senate, the White House, the U.S. Army, and the DOJ in the 1990s.

In 1998 the U.S. Secret Service filed the first criminal case against a juvenile for a computer crime. In 1997 computer hacking by the unnamed perpetrator shut down the airport in Worcester, Massachusetts, for six hours. The airport is integrated into the Federal Aviation Administration's traffic system by telephone lines. The suspect got into the communication system and disabled it by sending a series of computer commands that changed the data carried on the system. As a result, the airport could not function. (No accidents occurred during that time.) According to the DOJ, the juvenile pled guilty in return for two years' probation, a fine, and community service. U.S.

Attorney Donald K. Stern, lead attorney for the prosecution, observed: "Computer and telephone networks are at the heart of vital services provided by the government and private industry, and our critical infrastructure. They are not toys for the entertainment of teenagers. Hacking a computer or telephone network can create a tremendous risk to the public and we will prosecute juvenile hackers in appropriate cases."

On September 21, 2000, a sixteen-year-old from Miami pled guilty and was sentenced to six months' detention for illegally intercepting electronic communications on military computer networks. The juvenile admitted that he was responsible for August and October 1999 computer intrusions into a military computer network used by the Defense Threat Reduction Agency (DTRA), an arm of the DOD. The DTRA is responsible for reducing threats against the United States from nuclear, biological, chemical, conventional, and special weapons.

On December 6, 2000, eighteen-year-old Canadian Robert Russell Sanford pled guilty to six felony charges of breach of computer security and one felony charge of aggravated theft in connection with cyber attacks on U.S. Postal Service computers. Sanford was placed on five years' probation but could have been sentenced to up to twenty years in prison. Sanford was also ordered to pay over $45,000 in restitution fines for the cyber attacks.

THE VULNERABILITY OF THE U.S. DEPARTMENT OF DEFENSE (DOD). In 1998 hackers broke into unclassified Pentagon networks and altered personnel and payroll data. Two teenaged Americans and a twenty-one-year-old Israeli later pled guilty to the incidents. According to *Military and Cyber-Defense: Reactions to the Threat* (Center for Defense Information, Washington, DC, November 8, 2002), Department of Defense computer systems were being attacked 250,000 times a year in the mid-1990s. In response to this threat, measures were taken to increase the cyber security of military computer systems, including the ability to identify intruders before they reach a Department of Defense computer system at all. The numbers of cyber attacks had dropped by 2001, when only 14,500 probes were made by hackers; only 70 of these attempts were able to breach security measures and get in. In 2003, the *Guidelines for FBI National Security Investigations and Foreign Intelligence Collection* (http://www.usdoj.gov/olp/nsiguidelines.pdf) was revised to specifically authorize the FBI to investigate "foreign computer intrusion" from hackers employed by foreign governments to disrupt or destroy American military computer systems.

IT ESPIONAGE. In a computerized global economy, where any advantage given to the competition can mean success or failure for a company, trade secrets, copyrighted information, patents, and trademarks become important. The collapse of a large corporation because of the loss of such information could have widespread effects on the U.S. and world economies. In addition, companies that create military supplies, weapons, and the like for the government may have information that could be deadly in the wrong hands. Although most major companies have developed sophisticated security systems to protect their secrets, the American Society for Industrial Security and PriceWaterhouseCoopers *Trends in Proprietary Information Loss Survey* (http://www.asisonline.org/newsroom/surveys/spi2.pdf) estimated that potential losses to American businesses from thefts of proprietary information were $59 billion in 2002.

The theft of classified corporate information has become a major issue for national governments worldwide. Many governments have begun to use their national intelligence organizations to protect local companies from espionage by foreign companies or governments. In the United States the Central Intelligence Agency has tried to convince Congress that the agency could be useful in protecting American companies from foreign industrial spies. The Economic Espionage Act of 1996 (PL 104-294) made it a federal crime to steal trade secrets for another country.

OTHER THREATS. Most of the crimes listed above were committed with readily available tools. Of most concern to the U.S. government are attacks that might move beyond these easily available tools and techniques to cause significant damage and disruption to the U.S. information infrastructure, compromising the integrity of vital information. Analysts have been able to identify groups, domestic and global, with the motivation and opportunity to launch such attacks. Given the present vulnerabilities of many U.S. computer systems, a well-planned, coordinated strategic IW attack could have major consequences. Such an attack, or the threat of such an attack, could thwart U.S. foreign policy objectives, degrade military performance, result in significant economic loss, and undermine citizens' confidence in government. In light of such threats, the U.S. government is taking a proactive approach to defense.

Finding Solutions: Protecting the United States

Because both attackers and defenders make adjustments after every IW attempt they make or perceive, defense against such threats is not a one-time effort but a continuous activity. Collection and analysis of information about attacks is vital if defenders are to stay on par with the attackers. Defenders must be proactive and anticipate future methods of attack so that timely defenses can be developed.

Over the years, several executive orders, presidential directives, and acts have focused on, or mentioned activities related to, protecting cyber-critical infrastructure. Since the attacks of September 11, 2001, the USA Patriot Act and the Aviation and Transportation Security Act have been enacted. To deal specifically with cyber attacks against the nation's computer systems, *The National Strategy to Secure Cyberspace* (Washington, DC: White House, February 2003) called for government and the private sector to work together to secure cyberspace: "The purpose of this document is to engage and empower Americans to secure the portions of cyberspace that they own, operate, or control, or with which they interact." As part of that effort, in January 2004 the Department of Homeland Security announced the creation of the National Cyber Alert System, designed to alert Americans with timely and actionable information to better secure their computer systems from potential attack. Under this system, interested computer users can subscribe to several e-mail alert options:

- Cyber Security Tips: Targeted at nontechnical home and corporate computer users, the biweekly tips provide information on best computer security practices and "how-to" information.

- Cyber Security Bulletins: Targeted at technical audiences, bulletins provide biweekly summaries of security issues, new vulnerabilities, potential impact, patches and work-arounds, as well as actions required to mitigate risk.

- Cyber Security Alerts: Available in two forms—regular for nontechnical users and advanced for technical users—Cyber Security Alerts provide real-time information about security issues, vulnerabilities, and exploits currently occurring. Alerts encourage all users to take rapid action.

Other methods of protection using new and emerging technology include facial recognition software and biometric fingerprinting systems, both of which might be used to identify suspect criminal aliens at ports of entry or in other places. In 2001 alone three government agencies (the State Department, DOJ, and DOD) had received over $10.6 million by June to research such facial recognition technology. In the spring of 2004 the National Institute of Standards and Technology tested eighteen commercially available biometric fingerprinting systems. After a two-month testing period, *Government Security News* (August 2004) reported that the best systems available are 99.9% accurate in identifying individuals based on a scanned copy of their fingerprints.

Experts at the National Defense University have developed a defense-in-depth strategy against IW attacks to deal with the previously mentioned threat levels. The strategy suggests three successively stronger defensive levels corresponding to everyday threats, potentially strategic threats, and strategic threats. Basic to this thinking is that more sophisticated threats will come from fewer sources. The first two lines of defense are designed to identify and separate the most skilled, resourceful, and persistent adversaries. The last line of defense is meant to fully repel them. Intelligence and monitoring efforts are concentrated on a smaller population, increasing the chances of a successful defense. Layered on top of these lines of defense are "information first" and "security first" approaches: For everyday threats, the goal is to protect against access to information; for more strategic threats, the goal is to keep hackers out by restricting access and/or connectivity.

USING IW TO PROTECT THE UNITED STATES. U.S. national security strategies now recognize and use IW as an instrument of national power either independent of or complementary to U.S. military operations. American IW against an adversary usually bears a resemblance to classic methods of competition, conflict, and warfare but also uses more recent methods. It can run the gamut from propaganda campaigns (including media war) to attacks (both physical and nonphysical) against commanders, their information sources, and their means of communicating with their forces.

The Persian Gulf War of 1991 has become known as America's first information war. In that war, the power of IT was used to leverage information, significantly improving all aspects of warfare, including logistics, command, control, communications, computers, intelligence, surveillance, and reconnaissance. The victory of the United States and its allies in the Gulf War deterred potential adversaries from taking on the United States in the same manner as Iraq and fostered much thinking about new strategies for countering conventional forces. Thus, IW has become a strategy for our time: Potential adversaries want to engage in it, as the United States does, to achieve some of the objectives of conventional warfare.

IWD embodies actions taken to defend against information attacks, especially those against decision makers, their information, and their communications. These attacks can be launched during peacetime at nonmilitary targets by nonmilitary groups, both foreign and domestic. National security planners attempt to defend against many different kinds of information attacks, with a focus on attacks against the U.S. information infrastructure.

IW has some potential characteristics that traditional military planners strive for, including low-cost precision and stealth. IW can threaten the ability of a state's military to interpose itself between its population and "enemies of the state," causing what defense planners term a "loss of sanctuary"—just what the United States strives to achieve with its air, sea, and missile defense systems. Sanctuary can be defined as a working space, or buffer, between the population and territorial intrusions by alien enemies.

Information attacks can be very effective in destroying the image of sanctuary. Repeated attacks create a perception of vulnerability and loss of control and can cause a loss of public confidence in the state. These impacts can far exceed any actual damage. This makes the problem of IW challenging.

How will the United States respond to information attacks? As of 2004 there was no consensus. Yet, given that IW can be an instrument of power for niche competitors and nonstate actors, it needs to be taken seriously.

Some software engineers and others believe the country is not as vulnerable to information attacks as has been claimed. They point to overlaps and duplications that would make it very hard to completely disrupt a given set of services or functions.

Battlefield Systems Technology

NEW TECHNOLOGY AND ITS LIMITS. The roots of information-based battlefield and computer warfare go back several decades, along with technical and signals intelligence collection by satellites and sensors. Digital information of many kinds is of increasing importance in battlefield warfare, including command and control, mission planning, simulation, intelligence, and psychological operations. Indeed, every aspect of physical war and of gauging the threat of war is being transformed by the ever-quickening speed and ever-lessening cost of collecting, processing, and transmitting information.

The most important new battle technologies of modern warfare are precision-guided munitions, long-range airborne and space-based sensors, tandem global positioning systems (GPS), and inertial navigation systems (INS). With these new technologies, almost any target or source of information that can be located and identified can be engaged and disabled.

Because of this, a military offense must spend more time seeking targets than it previously did, and a military defense must spend more time and resources hiding them. Hiding can be done by getting them to mimic background or civilian objects and masking their "signatures"—the distinctive visual, radar, or spectroscopic profiles that, when recognized by a weapons system, enable it to identify the object as the target and destroy it. Traditional principles of battle warfare, such as firepower and maneuvering, lessen a bit in importance with these new technologies.

GLOBAL POSITIONING SYSTEMS (GPS). National security specialists know that it will be difficult to maintain the current American advantage in IW in coming decades. Relevant technologies increasingly spring from the commercial marketplace, not the military, often becoming available without restriction to prospective enemies. Sophisticated, well-funded opponents may be able to buy or lease an array of advanced communications and control technologies from around the world: for example, GPS, surveillance, communications, direct broadcast, internetworking, cryptography, and air-based imaging systems.

The costs of such purchases will likely decrease, as will the costs of IW in general.

GPS, in particular, is rapidly becoming commercially and universally available, with devices costing just a few hundred dollars able to receive signals. For example, accurate GPS data can enable rocket attacks against U.S. forces deployed in smaller contingents. In addition, GPS, coupled with surveillance data and other equipment, can place nearly any fixed facility at risk, including most U.S. logistics dumps, barracks, and command headquarters. These cannot be well hidden and thus can be identified and located if someone knows their general vicinity. If the facilities are public, a terrorist could target it with a portable GPS device on-site.

Overhead surveillance can locate fixed facilities with an accuracy of within a meter or two. With the fall of the Soviet Union, a vigorous market developed in such Russian imagery. In the next few decades, the sale of satellites with similar capabilities will permit many countries to acquire and transmit such imagery nearly in real time.

It is conceivable that in a time of crisis, the American military could degrade GPS signals worldwide so that U.S. forces could determine locations far more closely than their adversaries. However, as a practical matter, three factors make this option difficult to implement. First, the U.S. government has promoted the use of GPS for civilian purposes, most notably commercial aviation. Only a major and prolonged crisis could justify the global degradation of information upon which others rely for their safety. Second, GPS may be complemented by other navigation systems, or what specialists call "communications constellations." Third, the development of differential GPS means that if a set of fixed points near a target can be located with precision, the target itself can be located with similar precision. So-called differential correction systems have also come online throughout North America, Europe, and East Asia. Their accuracy often exceeds that of military systems without differential correction.

UNMANNED AERIAL VEHICLES (UAVS). Additional information-gathering capability comes from the use of digital video cameras on unmanned aerial vehicles (UAVs). UAVs, also known as drones, do the battlefield and reconnaissance tasks formerly assigned to manned aircraft, but without a pilot aboard. Instead, they are piloted remotely from the ground by radio links. The advantages are that human life is not placed in harm's way and the vehicle can be designed without having to safeguard and support an onboard human pilot.

Since the mid-1990s commercially available digital-imaging systems mounted on UAVs have been able to collect high-resolution imagery fifty miles to each side of themselves with real-time data links to ground locations. Although the initial resolution of such systems was imprecise, higher-resolution digital cameras are becoming more widely available. In addition, digital cellular telephony is already available through several technologies. Within several years, such technology may be widely available globally and have bandwidth high enough to transmit imagery directly.

BLOCKING COMMUNICATIONS AND HIDING TROOP MOVEMENTS. In UAV and satellite imaging American forces might attempt to deny an enemy communications capability by blocking access to third-party satellites. Such an attempt, however, could present several political obstacles. Commercial satellites have a variety of corporate owners in different countries, and not every satellite owner would necessarily cooperate with U.S. forces. If cooperation were incomplete, an opponent's access to satellite links would not be entirely blocked. Jamming signals to and from geosynchronous satellites also usually requires being in their line of sight; thus, it is probably not feasible for the United States to jam all signals to and from them.

Global low-earth-orbit cellular systems would make it even more unlikely that the United States could block an adversary's communications that were handled by a third party. System managers could refuse to transmit signals into or out of a region, but doing so would limit or eliminate local service and service to nonbelligerent neighboring states. It would be hard to shut down a system used by a terrorist group operating inside a friendly country or to interrupt a more primitive system they were using, based on, say, a citizen's band (CB) radio. Similarly, in any attempt to disrupt another nation's air traffic control network, it would be difficult not to interfere with international air traffic control operations in the general vicinity.

Increasing global satellite connectivity also decreases the chance for military activity to go undetected. Daylight infantry movements can now be kept secret only to the extent an area is not electronically connected to the outside world. But as even the most remote sites become ever more tied to the global communications network, such movements are more likely to be noticed—and counteracted. The predicted marriage between digital video cameras and digital cellular means that many more military movements will potentially be liable to detection.

The communications and information revolutions have tended to knit the world together. More than ever, to disrupt an enemy's communications is to disrupt those of one's friends. Increasing global communications connectivity—thanks to new technologies such as advanced semiconductors, advanced computers, fiber optics, cellular technology, satellite technology, and advanced networking (including the Internet)—has empowered individuals, governments, and armies, making U.S. national security tasks in this area much more difficult and complex.

IMPORTANT NAMES AND ADDRESSES

Anti-Defamation League
(202) 452-8320
FAX: (202) 296-2371
E-mail: Washington-dc@adl.org
URL: http://www.adl.org

Brookings Institution
1775 Massachusetts Ave. NW
Washington, DC 20036
(202) 797-6000
FAX: (202) 797-6004
E-mail: webmaster@brookings.edu
URL: http://www.brookings.edu

**Carnegie Endowment for
International Peace**
1779 Massachusetts Ave. NW
Washington, DC 20036-2103
(202) 483-7600
FAX: (202) 483-1840
E-mail: info@ceip.org
URL: http://www.carnegieendowment.org

Center for Defense Information
1779 Massachusetts Ave.
Washington, DC 20036-2109
(202) 332-0600
FAX: (202) 462-4559
E-mail: info@cdi.org
URL: http://www.cdi.org

Center for Nonproliferation Studies
460 Pierce St.
Monterey, CA 93940
(831) 647-4154
FAX: (831) 647-3519
E-mail: cns@miis.edu
URL: http://www.cns.miis.edu

**Center for Strategic and
International Studies**
1800 K St. NW, Suite 400
Washington, DC 20006
(202) 887-0200
FAX: (202) 775-3199

E-mail: webmaster@csis.org
URL: http://www.csis.org

**Centers for Disease Control
and Prevention**
1600 Clifton Rd.
Atlanta, GA 30333
(404) 639-3311
Toll-free: 1-800-311-3435
URL: http://www.cdc.gov

Central Intelligence Agency (CIA)
Office of Public Affairs
Washington, DC 20505
(703) 482-0623
FAX: (703) 482-1739
URL: http://www.cia.gov

**Chemical and Biological Arms
Control Institute**
1747 Pennsylvania Ave. NW, 7th Floor
Washington, DC 20006
(202) 296-3550
FAX: (202) 296-3574
E-mail: cbaci@cbaci.org
URL: http://www.cbaci.org

**Computer Security Resource Center
Computer Security Division
National Institute of Standards
and Technology**
100 Bureau Dr., Mail Stop 8930
Gaithersburg, MD 20899-8930
(301) 975-6478
URL: http://csrc.nist.gov

Congressional Research Service
101 Independence Ave. SE
Washington, DC 20540-7500
(202) 707-5000
URL: http://www.loc.gov/crsinfo

Council for a Livable World
322 4th St. NE
Washington, DC 20002

(202) 543-4100
E-mail: clw@clw.org
URL: http://www.clw.org

Federal Bureau of Investigation (FBI)
935 Pennsylvania Ave. NW, Room 7972
Washington, DC 20535
(202) 324-3000
URL: http://www.fbi.gov

Federation of American Scientists
1717 K St. NW, Suite 209
Washington, DC 20036
(202) 546-3300
FAX: (202) 675-1010
E-mail: fas@fas.org
URL: http://www.fas.org

Hoover Institution
Stanford University
Stanford, CA 94305-6010
(650) 723-1754
FAX: (650) 723-1687
E-mail: horaney@hoover.stanford.edu
URL: http://www.hoover.stanford.edu

Institute for Defense & Disarmament
675 Massachusetts Ave.
Cambridge, MA 02139
(617) 354-4337
FAX: (617) 354-1450
E-mail: info@idds.org
URL: http://www.idds.org

**Institute for Science and
International Security**
236 Massachusetts Ave. NE, Suite 500
Washington, DC 20002
(202) 547-3633
FAX: (202) 547-3634
E-mail: isis@isis-online.org
URL: http://www.isis-online.org

**International Institute for Strategic
Studies—U.S.**
1747 Pennsylvania Ave. NW, 7th Floor

Washington, DC 20006
(202) 659-1490
FAX: (202) 296-1134
E-mail: taylor@iiss.org
URL: http://www.iiss.org

The International Policy Institute for Counter-Terrorism
P.O. Box 167
Herzlia, Israel 46150
FAX: 972-9-9513073
E-mail: info@ict.org.il
URL: http://www.ict.org.il

Joint Chiefs of Staff
Public Affairs Office
9999 Joint Staff Pentagon, Room 2E857
Washington, DC 20318-9999
(703) 697-4272
URL: http://www.dtic.mil/jcs

National Center for Policy Analysis
601 Pennsylvania Ave. NW
Suite 900, South Building
Washington, DC 20004
(202) 628-6671
FAX: (202) 628-6474
URL: http://www.ncpa.org/newdpd/index.php

National Defense University Press
300 5th Ave.
Ft. McNair, DC 20319-6000
(202) 685-4210
FAX: (202) 685-4608
E-mail: ndupress@ndu.edu
URL: http://www.ndu.edu

National Guard Bureau
1411 Jefferson Davis Highway
Arlington, VA 22202-3231
(703) 607-3162
URL: http://www.ngb.army.mil

National Institute of Allergy and Infectious Diseases (NIAID)
6610 Rockledge Dr., MSC 6612
Bethesda, MD 20892-6612
(301) 496-5717
URL: http://www.niaid.nih.gov/default.htm

National Rifle Association
11250 Waples Mill Rd.
Fairfax, VA 22030
(877) 672-2000
URL: http://www.nra.org

National Security Agency
(301) 688-6524
FAX: (301) 688-6198
E-mail: nsapao@nsa.gov
URL: http://www.nsa.gov/home_html.cfm

North Atlantic Treaty Organization (NATO)
Blvd. Leopold III
1110 Brussels, Belgium

E-mail: natodoc@hq.nato.int
URL: http://www.nato.int

Nuclear Control Institute
1000 Connecticut Ave. NW, Suite 410
Washington, DC 20036
(202) 822-8444
FAX: (202) 452-0892
E-mail: nci@mailback.com
URL: http://www.nci.org

Nuclear Threat Initiative
1747 Pennsylvania Ave. NW, 7th Floor
Washington, DC 20006
(202) 296-4810
FAX: (202) 296-4811
E-mail: contact@nti.org
URL: http://www.nti.org

RAND Corporation
P.O. Box 2138
1700 Main St.
Santa Monica, CA 90407-2138
(310) 393-0411
FAX: (310) 393-4818
E-mail: correspondence@rand.org
URL: http://www.rand.org

Southern Poverty Law Center (SPLC)
400 Washington Ave.
Montgomery, AL 36104
(334) 956-8200
URL: http://www.splcenter.org/

Stockholm International Peace Research Institute (SIPRI)
Signalistgatan 9
SE-169 70 Solna, Sweden
46-8-655 97 00
FAX: 46-8-655 97 33
E-mail: sipri@sipri.org
URL: http://www.sipri.org

United Nations (UN)
Public Inquiries Unit GA-57
New York, NY 10017
(212) 963-4475
FAX: 212) 963-0071
E-mail: inquiries@un.org
URL: http://www.un.org

U.S. Air Force
Office of the Secretary of the Air Force
Public Affairs Resource Library
1690 Air Force Pentagon, Room 4A120
Washington, DC 20330-1690
(703) 697-4100
URL: http://www.af.mil

U.S. Army
Chief of Public Affairs
1500 Army Pentagon
Washington, DC 20310-1500
(703) 602-5201
URL: http://www.army.mil

U.S. Coast Guard
2100 2nd St. SW
Washington, DC 20593
(202) 267-2229
URL: http://www.uscg.mil

U.S. Department of Defense
Director for Public Inquiry and Analysis
1400 Defense Pentagon, Room 3A750
Washington, DC 20301-1400
URL: http://www.dod.gov

U.S. Department of Energy, Office of Fossil Energy
Forrestal Building
1000 Independence Ave. SW
Washington, DC 20585
(202) 586-6503
FAX: (202) 586-5146
E-mail: fewebmaster@hq.doe.gov
URL: http://www.fe.doe.gov

U.S. Department of Homeland Security
Nebraska Avenue Complex
Washington, DC 20528
URL: http://www.dhs.gov

U.S. Department of State
2201 C St. NW
Washington, DC 20520
(202) 647-4000
E-mail: askpublicaffairs@state.gov
URL: http://www.state.gov

U.S. House Permanent Select Committee on Intelligence
H-405 U.S. Capitol Building
Washington, DC 20515
(202) 225-4121
Toll-free: 1-877-858-9040
E-mail: intelligence.hpsci@mail.house.gov
URL: http://intelligence.house.gov

U.S. Humanitarian Demining Program
U.S. Army RDECOM CERDEC NVESD
(AMSRD-CER-NV-CM-HD)
10221 Burbeck Rd., Suite 430
Fort Belvoir, VA 22060-5806
URL: http://www.humanitarian-demining.org/demining/default.asp

U.S. Marine Corps
Director of Public Affairs
2 Navy Annex
Washington, DC 20380-1775
(703) 614-6251
URL: http://www.usmc.mil

U.S. Navy
Office of Information
1200 Navy Pentagon, Room 3E335
Washington, DC 20310-1200
(703) 697-9020
URL: http://www.navy.mil

U.S. Senate Select Committee on Intelligence
211 Hart Senate Office Building
Washington, DC 20510-6475
(202) 224-1700
URL: http://intelligence.senate.gov

The White House
1600 Pennsylvania Ave. NW
Washington, DC 20500
(202) 456-1111
FAX: (202) 456-2461
URL: http://www.whitehouse.gov

RESOURCES

The U.S. government provides useful nonclassified information on national security. Government sources include the U.S. Departments of State and Defense, along with such government agencies as the Central Intelligence Agency (CIA), the Federal Bureau of Investigation (FBI), and the Bureau of Citizenship and Immigration Services (CIS) in the U.S. Department of Homeland Security.

The U.S. Department of State's annual *Patterns of Global Terrorism* report, mandated by Congress and issued every spring (the latest version being *Patterns of Global Terrorism 2003*, April 2004), offers detailed assessments of significant terrorist acts. The report also highlights a watch list of countries that have repeatedly provided state support for international terrorism.

In addition to the *Patterns* report, the State Department's *Foreign Terrorist Organizations Designations* are compiled every two years when the secretary of state designates, by mandate of Congress, approximately thirty groups as global terrorist organizations. *Significant Incidents of Political Violence against Americans* is another State Department report, published annually by the Bureau of Diplomatic Security, Office of Intelligence and Threat Analysis. It examines terrorism-related acts and other instances of violence affecting Americans.

The U.S. Department of Defense (DOD) has an excellent educational institution that serves as a resource: the National Defense University (NDU), and in particular the NDU's Institute for National Strategic Studies (INSS). The INSS's *The Global Century: Globalization and National Security* (2001) is a two-volume anthology on global security and defense issues. The institute's 2015 Project, which forecasted global defense and national security conditions through the year 2015, yielded the anthology *2015: Power and Progress*, a good source on demographic, pollution, and resource stresses; coalitions; and information technology and warfare.

An extremely useful report from the DOD is *Proliferation: Threat and Response* (January 2001). The report updates information about the worldwide proliferation of nuclear, biological, and chemical weapons with good figures and tables, especially concerning ballistic missile ranges, and focuses on DOD policies and programs countering such threats. The DOD's congressionally mandated Quadrennial Defense Review (QDR), also known as the *Report of the Quadrennial Review,* broadly describes future defense policies.

The U.S. General Accounting Office (GAO), the investigative arm of Congress, produces many documents relating to national security, especially those addressing domestic threats. One such document used in this book is *Bioterrorism: Public Health and Medical Preparedness* (2001), the testimony on October 9, 2001, of Janet Heinrich, director of Health Care-Public Health Issues, before the Senate's Subcommittee on Public Health, Committee on Health, Education, Labor and Pensions.

On preparations for potential domestic bioterrorism, the premier government source is the U.S. Department of Health and Human Services' Centers for Disease Control and Prevention (CDC) in Atlanta, Georgia. CDC sources used in this book include "Biological and Chemical Terrorism: Strategic Plan for Preparedness and Response: Recommendations of the CDC Strategic Planning Workgroup," in *Morbidity and Mortality Weekly Report* (April 2000) and *A National Public Health Strategy for Terrorism Preparedness and Response, 2003–2008* (March 2004).

U.S. sources provide abundant information on illegal immigration that could affect national security. The former U.S. Immigration and Naturalization Service (INS), until early 2003 an agency of the U.S. Department of Justice but now rolled into the new U.S. Department of Homeland Security, publishes a number of useful titles.

Its annual *Statistical Yearbook of the Immigration and Naturalization Service* (2002) is a complete statistical resource on immigrants, illegal aliens, and refugees who come to the United States. Demographic data on the foreign-born population in the United States is available from the U.S. Census Bureau in *The Foreign-Born Population in the United States, 2003* (August 2004).

The Central Intelligence Agency (CIA) is an invaluable resource for information on the WMD potential of foreign countries. Among their publications on this subject are *Unclassified Report to Congress on the Acquisition of Technology Relating to Weapons of Mass Destruction and Advanced Conventional Munitions, 1 January through 30 June 2003* (November 2003) and *Comprehensive Report of the Special Advisor to the DCI on Iraq's WMD* (September 2004). The CIA also publishes *The CIA Factbook on Intelligence* (2004).

The Congressional Research Service (CRS) is the research arm of the Library of Congress in Washington, which serves members and committees of Congress but makes many of its findings available to the public as well. The CRS reports consulted in preparing this book included *Intelligence Issues for Congress* (Richard A. Best, updated July 2002), *China and Proliferation of Weapons of Mass Destruction and Missiles: Policy Issues* (August 8, 2003), and *Disarming Libya: Weapons of Mass Destruction* (April 22, 2004).

Information Plus sincerely thanks all of the organizations listed above for the invaluable information they provide.

INDEX

Page references in italics refer to photographs. References with the letter t following them indicate the presence of a table. The letter f indicates a figure. If more than one table or figure appears on a particular page, the exact item number for the table or figure being referenced is provided.

A

ADL (Anti-Defamation League), 90–91
Afghanistan
  Operation Enduring Freedom, 75
  al Qaeda, 69
Africa
  natural resource conflicts, 130
  nongovernmental organizations and conventional weapons trade, 16
African Americans, 79
Agencies and organizations
  Border and Transportation Security, 111–112
  Central Intelligence Agency (CIA), 97–102, 100f, 107
  Citizenship and Immigration Services, 110–112
  combating biological and chemical terrorism, 47
  Commission on Roles and Capabilities of the U.S. Intelligence Community, 101
  Computer Emergency Response Team (CERT), 140
  Congress, U.S., 2
  Counter-Terrorism Center (CTC), 101–102
  Customs and Border Protection, 111–112
  Executive Office of the President, 1–2
  FBI offices and divisions, 107–108
  Federal Bureau of Investigation (FBI), 102–109
  Homeland Security, Department of, 109–110
  House Permanent Select Committee on Intelligence (HPSCI), 101
  Immigration and Naturalization Service, 110–112

Intelligence Oversight Board (IOB), 101
  Joint Chiefs of Staff, 113–114
  key U.S. government agencies, 2
  National Commission on Terrorist Attacks on the United States, 102
  National Security Council (NSC), 95
  North Atlantic Treaty Organization (NATO), 124–125
  Office of Management and Budget (OMB), 95
  peacekeeping, 116–118
  Senate Select Committee on Intelligence (SSCI), 101
  State Department, 95, 96f
  United Nations, 123–124
  White House, Office of the, 94–95
An Agenda for Peace: Preventive Diplomacy, Peacemaking, and Peacekeeping (UN Secretary General), 116
Agriterrorism, 53–55, 53t, 54t, 55t
Air Force, U.S., 114
Airport security, 74
ALF (Animal Liberation Front), 84–86
Alfred P. Murrah Federal Building bombing, 80
Alliances, 123–126
Allied Contributions to the Common Defense, 120–121
ABM (Antiballistic Missile) treaty, 29
American Airlines flight routes, September 11, 2001, 67f, 68(f6.6)
Amherst, Lord Jeffrey, 21
Animal and Plant Health Inspection Service (APHIS), 53, 55
Animal diseases, transmissible, 53–55, 54(t5.5)
Animal Liberation Front (ALF), 84–86
Anthrax attacks, 21, 23, 80–81
Anti-Defamation League (ADL), 90–91
Anti-land mine campaign, 17
Antiabortionist terrorism, 86
Arab/Israeli conflict, 62–63
Arctic National Wildlife Refuge (ANWR), 129
Arms Export Control Act, 12–13
Army, U.S., 114

Atlanta, Georgia, 80, 104
Attitudinal research
  internal oil production *vs.* conservation, 129
  terrorism and homeland security, 75–78
Aum cult, 25–26
Australia Group, 29

B

Balance of power, 6–7
Ballistic missiles
  China, 33–35
  Iran, 36, 38f
  Iraq, 42
  limits, 30–31
  Pakistan, 42f
  possession and range by country, 33t
  See also Nuclear weapons; Weapons of mass destruction
Bashar al-Assad, 10
Battlefield systems technology, 144–145
Bay of Pigs invasion, 98
Bin Laden, Osama, 69, 75
Biological and chemical weapons
  agriterrorism, 53–55, 53t, 54t, 55t
  characteristics, 23–24
  China, 34–35
  classes, 24–26, 25t
  Dalles, Oregon incident, 49
  definition, 47
  Egypt, 35
  examples of chemicals used, 23t
  India, 35
  Iran, 36
  Iraq, 44–45
  Israel, 36, 38
  status of country programs, 24f, 27f
  Syria, 41–42
  terrorism preparedness, 48–53
  threat delineation, 49–51, 50t, 52t
  Tokyo attack, 25–26
  World War I, 21, 23
  See also Weapons of mass destruction
Biological and Toxin Weapons Convention (BTWC), 28–29
Birmingham, Alabama, 79

Black separatist groups, 86, 88–90, 89*t*–90*t*
Blister agents, 25
Blood agents, 25
Blue Lantern program, 14, 14*f*
Bombings, domestic terrorism, 79–80
Border and Transportation Security, 111–112
Bosnia and Herzegovina, 117–118, 119–120
BTWC (Biological and Toxin Weapons Convention), 28–29
Bush, George W.
    ABM treaty withdrawal, 29
    homeland security efforts, 73–75
    information technology infrastructure security, 133
    9/11 Commission Report, response to the, 102
    Operation Iraqi Freedom, 45–46
    peacekeeping, 119
    Proliferation Security Initiative (PSI), 31
    on state-sponsored terrorism, 62
    Strategic Petroleum Reserve, 129
    strategy for national security, 1

C

Castro, Fidel, 10, 65
CDC (Centers for Disease Control and Prevention), 48–53
Centennial Park bombing, 80
Centers for Disease Control and Prevention (CDC), 48–53
Central Intelligence Agency (CIA)
    covert action, 98
    FBI/CIA cooperation, 107
    funding, 98–99
    future of, 101–102
    history, 99, 101
    intelligence types, 97–98
    organizational structure, 99, 100*f*
    oversight, 99, 101
    role of, 98
Chechnya, 18, 66
Chemical weapons. *See* Biological and chemical weapons
Chemical Weapons Convention (CWC), 29, 34–36, 40
Cherry, Bobby Frank, 79
China
    conventional weapons trade, 14
    defense expenditures, 8, 9*f*, 9*t*
    oil, 127–128
    as transition state, 8–9
    weapons of mass destruction, 33–35
China and Proliferation of Weapons of Mass Destruction and Missiles: Policy Issues (Congressional Research Service), 35
Choking agents, 25
Chronology
    domestic terrorist incidents, 85*t*–86*t*
    weapons of mass destruction use and control, 22*t*
Church Committee, 101
CIA. *See* Central Intelligence Agency (CIA)
Cities Readiness Initiative, 52–53
Citizenship and Immigration Services, 110–112
Civil liberties, 75

Clinton, William J., 119
Coast Guard, U.S., 114
COINTELPRO, 103–104
Cold war
    conventional weapons trade, 11
    end of, 6
    funding, 99
    policy milestones, 4–6
Collective security, 7
Commission on Roles and Capabilities of the U.S. Intelligence Community, 101
Computer chip development, 138
Computer Crime and Security Survey, 2004 (Computer Security Institute), 141
Computer crimes, 140–142
Computer Emergency Response Team (CERT), 140
Computer Fraud and Abuse Act, 141
Computer hacking, 142
CON-PLAN (Concept of Operations Plan), 91
Concept of Operations Plan (CON-PLAN), 91
Concluding Act of the Negotiation on Personnel Strength of Conventional Armed Forces in Europe, 18–19
Concurrent jurisdiction investigations, 104–105
Conference on the Illicit Trade in Small Arms and Light Weapons, UN, 15–16
Congressional hearings, 99
Congressional powers, 2, 93–94
Conservation *vs.* oil production, 129
Constitution, U.S., 93, 94*f*
Convention against Transnational Organized Crime, 133
Convention on the Prohibition of the Use, Stockpiling, Production and Transfer of Anti-Personnel Mines and on Their Destruction, 17
Conventional Arms Transfers to Developing Nations, 1995–2002 (Congressional Research Service), 12, 127
Conventional weapons trade, 11–15, 12*f*, 13*t*, 14*f*, 17–19, 127
Cooperative Defense Initiative, 15
Corporate information theft, 142–143
Costs
    computer crime and security, 141
    Defense Department Contingency operations, 120*t*
Counter-Terrorism Center (CTC), 101–102
"Countering State-Sponsored Terrorism" (Ganor), 61–62
Counterintelligence, FBI, 103–104, 107
Counterterrorism
    Customs and Border Patrol efforts, 111–112
    Federal Bureau of Investigation, 107, 108*f*, 109*f*
Covert action, 98
Criminal investigations, 108–109
Critical information technology infrastructure security, 133, 138*t*, 139*t*
Crop diseases, 53–55, 55*t*
Cruise missiles
    Iran, 38*f*

possession and range by country, 33*t*
CTC (Counter-Terrorism Center), 101–102
Cuba
    as state of concern, 10
    state-sponsored terrorism, 65
Cuban Missile Crisis, 5–6, 97–98
Customs and Border Protection, 111–112
CWC. *See* Chemical Weapons Convention (CWC)
Cyber attacks, 139–144, 140*t*
Cyber Division, FBI, 107

D

The Dalles, Oregon, 49
Defense, U.S. Department of
    Computer Emergency Response Team (CERT), 140
    computer hacking incidents, 142
    organizational structure, 113–116
    peacekeeping, 120–121, 120*f*, 120*t*
Defense Advanced Research Projects Agency, 115
Defense agencies, 115–116
Defense Contract Management Agency, 115
Defense expenditures, Chinese, 8, 9*f*, 9*t*
Definitions
    national security, 1
    organized crime, 131
    peacekeeping, 116
    terrorism, 57–58
    weapons of mass destruction, 21
Desert Fox air strikes, 42
"Détente," 6
Developing countries, 11–12, 13*t*
Director of central intelligence, 99
Disarming Libya: Weapons of Mass Destruction (Congressional Research Service), 39
Domestic terrorism
    activity by region, 84*f*
    anthrax attacks, 80–81
    antiabortionists, 86
    black separatist groups, 86, 88–90, 89*t*–90*t*
    bombings, 79–80
    chronology of incidents, 85*t*–86*t*
    ecoterrorism, 84–86
    government prevention and preparedness efforts, 91
    by group class, 83*f*
    incidents and preventions, 82*f*
    left-wing organizations, 81, 83
    right-wing organizations, 83, 87*t*–88*t*, 91
    watchdog groups, 90–91
Drug trafficking, 132
Dunlea, Mark, 46

E

Earth Liberation Front (ELF), 84–86
Economic and Social Council, United Nations, 124
Ecoterrorism, 84–86
Egypt, 35
Eisenhower, Dwight D., 5
ELF (Earth Liberation Front), 84–86
Environmental issues
    ecoterrorism, 84–86

internal U.S. oil production, 129
Espionage, 105, 142–143
Europe, conventional weapons trade in, 14, 18–19
Executive branch, role of the, 93–95
Executive Office of the President, 1–2
Expenditures and funding
  biological and chemical terrorism preparedness and research, 47–48, 48
  Central Intelligence Agency (CIA), 98–99
  Chinese defense expenditures, 8, 9f, 9t
  Homeland Security, Department of, 109
  peacekeeping, 119–121, 120t
  terrorism, 69, 72, 73f, 132–133
  United Nations, 124
Exports of conventional weapons, 11–15, 13t, 14f, 127

F

Farrakhan, Louis, 88–89
FBI. See Federal Bureau of Investigation (FBI)
Federal agencies and organizations. See Agencies and organizations
Federal Bureau of Investigation (FBI)
  counterterrorism agents, 109f
  history, 102–104
  intelligence analysts, 107t
  organizational structure, 102, 106–109, 106f
  reforms, 105–109
  translators, 109t
  transnational crime response, 133
Federal government organization, 93–94, 94f
Federal Law Enforcement Training Center, 112
Fissile material, 26–28
Flexible response policy, 5–6
Flight routes of September 11th hijacked planes, 67f, 68f
Foot-and-mouth disease (FMD), 55
A Force for Peace and Security: U.S. and Allied Commanders' Views of the Military's Role in Peace Operations and the Impact on Terrorism of States in Conflict (Peace Through Law Education Fund), 118
Foreign Assistance Act of 1961, 13
Foreign policy
  alliances, 123–126
  conventional weapons exports, 12–13, 14f
  factors influencing, 7–9
  Israel, 125
  milestones, 4–6
  Persian Gulf and Gulf Cooperation Council, 125–126
Former Soviet Union, 27–28
Freedom of Access to Clinical Entrances Act (FACE), 86
French and Indian War, 21
Funding. See Expenditures and funding

G

Gallup Poll, 78
Ganor, Boaz, 60, 61–62

General Accounting Office, 47
General Assembly, 124
Global positioning systems (GPS), 144–145
Goldwater-Nichols Department of Defense Reorganization Act, 113–115
Government and computer crime, 140–142
Gulf Cooperation Council (GCC), 15

H

Hacking, computer, 142
Hamza, Khidir, 43, 44
Harris Polls, 77–78
Health and Human Services, Department of, 47–48
Helder, Luke J., 80
Hillen, John, 118
Hirose, Kenichi, 26
Hiroshima bombing, 26
History
  Central Intelligence Agency (CIA), 99, 101
  Federal Bureau of Investigation (FBI), 102–104
  North Atlantic Treaty Organization (NATO), 124–125
  policy milestones, 4–6
  United Nations, 124
  weapons of mass destruction, 21–23, 22t
Hoaxes, anthrax, 21, 23, 81
Homeland Security, Department of
  creation and organization, 73–74, 109–110, 110f
  information technology infrastructure security, 133
  National Cyber Alert System, 143
Hoover, J. Edgar, 103
House Permanent Select Committee on Intelligence (HPSCI), 101
Human intelligence, 97

I

ICBL (International Campaign to Ban Landmines), 17
Illegal drug trade, 132
Imagery intelligence, 97–98
Immigration and Naturalization Service, 110–112
Imports of conventional weapons, 15
In-Q-Tel, 97
India
  as transition state, 9
  weapons of mass destruction, 35
Industrial spying, 142–143
Information sharing, 106–107
Information technology
  battlefield systems technology, 144–145
  computer hacking, 142
  critical infrastructure security, 133, 138t, 139t
  espionage, 142–143
  information security incidents, 141f
  information warfare, 139–144, 140t
  processor/chip development, 138
  warfare and, 144–145
Information warfare, 139–144, 140t
Inspections, weapons
  Chemical Weapons Convention, 29

Iran, 36
Iraq, 42–45
Libya, 39
North Korea, 39
Treaty on Conventional Armed Forces in Europe, 18
Institute for Physics and Power Engineering (IPPE), 28
Intelligence community
  intelligence types, 97–98
  organization, 95–96
  overview, 2–3, 3f
  See also Central Intelligence Agency (CIA); Federal Bureau of Investigation (FBI)
Intelligence gathering
  FBI intelligence analysts, 107, 107t
  Federal Bureau of Investigation (FBI), 103–104
  history, 5–6
Intelligence Oversight Board (IOB), 101
Intelligence types, 97–98
Intercontinental ballistic missiles. See Ballistic missiles; Weapons of mass destruction
Interim Agreement between the United States and the Union of Soviet Socialist Republics on Certain Measures with Respect to the Limitation of Strategic Offensive Arms (SALT I), 30
International Action Network on Small Arms (IANSA), 16
International Atomic Energy Agency (IAEA), 36
International Campaign to Ban Landmines (ICBL), 17
International Code of Conduct against Ballistic Missile Proliferation, 31
International Court of Justice, 124
International Criminal Police Organization (Interpol), 16
International government organizations (IGOs), 15–16
International terrorism
  agriterrorism, 53–55, 53t, 54t, 55t
  American public opinion, 75–78
  biological and chemical weapons, 48–53, 50t, 52t
  critical infrastructure security, 133, 138t, 139t
  definition, 57–58
  drug trade and, 132
  FBI role in investigating, 105
  financing, 69, 72, 73f
  homeland security efforts, U.S., 73–75
  incidents, 59–60, 59f, 60f, 76f
  information technology, use of, 138
  money laundering, 132–133
  motivation, 58
  September 11 attacks, 66–69, 67f
  state-sponsored, 60–65
  against the United States, 58–60, 61f
Interpol, 16
Investigations
  Central Intelligence Agency (CIA) activities, 99, 101

Federal Bureau of Investigation (FBI), 104–105
Investigative Technologies Division, FBI, 108
IOB (Intelligence Oversight Board), 101
Iran
    Iranian hostage crisis, 61
    nuclear sites, 37*f*
    as state of concern, 10
    state-sponsored terrorism, 63
    weapons of mass destruction, 35–36
Iran-contra affair, 13–14, 101
Iraq
    Operation Iraqi Freedom, 45–46, 75
    state-sponsored terrorism, 63
    weapons of mass destruction, 42–46, 43*t*
Iraq Liberation Act, 45
Islam, 58–59, 62–63
Israel
    alliance with United States, 125
    Arab/Israeli conflict, 62–63
    nuclear stockpile estimates, 39*f*
    weapons of mass destruction, 36, 38, 39*f*

J

Japan, 21, 26
Jewell, Richard, 104
Joint Chiefs of Staff, 113–114

K

Kaczynski, Theodore, 80
Kamel, Hussein, 43
Kennedy, John F.
    Cuban Missile Crisis, 97–98
    flexible response policy, 5–6
KFOR (Kosovo Force), 117–120
Khamenei, Ali, 10
Khan, Abdul Qadeer, 39, 40
Kim Jong-il, 10
Klare, Michael T., 127
Kosovo Force (KFOR), 117–120

L

Laboratory Division, FBI, 108
Laboratory response networks, 51–52
Land mines, 16–17
Latin America
    nongovernmental organizations and conventional weapons trade, 16
    state-sponsored terrorism, 65
Left-wing organizations, 81, 83
Legislation and international agreements
    Arms Export Control Act, 12–13
    Biological and Toxin Weapons Convention (BTWC), 28–29
    Chemical Weapons Convention (CWC), 29, 34–36, 40
    Computer Fraud and Abuse Act, 141
    Concluding Act of the Negotiation on Personnel Strength of Conventional Armed Forces in Europe, 18–19
    Convention against Transnational Organized Crime, 133
    Convention on the Prohibition of the Use, Stockpiling, Production and Transfer of

Anti-Personnel Mines and on Their Destruction, 17
    Foreign Assistance Act of 1961, 13
    Freedom of Access to Clinical Entrances Act (FACE), 86
    Goldwater-Nichols Department of Defense Reorganization Act, 113–115
    Iraq Liberation Act, 45
    National Security Act, 97
    Nuclear Nonproliferation Treaty (NPT), 29, 33–34
    Patriot Act, 74–75, 98
    Proliferation Security Initiative (PSI), 31
    SALT I agreement, 30
    SALT II agreement, 30
    Strategic Offensive Reductions Treaty (SORT), 31
    Treaty between the United States and the Russian Federation on Further Reduction and Limitation of Strategic Offensive Arms (START II), 31
    Treaty between the United States and the Soviet Union on the Limitation of Antiballistic Missile Systems (ABM Treaty), 29
    Treaty between the United States of America and the Union of Soviet Socialist Republics on the Reduction and Limitation of Strategic Offensive Arms (START I), 30
    Treaty on Conventional Armed Forces in Europe, 18
    Treaty on Open Skies, 30–31
    Treaty on the Principles Governing the Activities of States in the Exploration and Use of Outer Space, Including the Moon and Other Celestial Bodies, 29
    War Powers Act, 93
    Wassenaar Arrangement on Export Controls for Conventional Arms and Dual-Use Goods and Technologies, 19
Legislative branch, 93–94
Libya
    as state of concern, 9–10
    state-sponsored terrorism, 64
    weapons of mass destruction, 38–39
Local law enforcement, 104–105
Lockerbie, Scotland, 64
Long-range missiles, 45

M

Mailbox bombings, 80
Marine Corps, U.S., 114
Marshall Plan, 5
"Massive retaliation," 5
Materials Protection, Control, and Accounting Program (MPCA), 28
McGee, Michael, 89–90
McVeigh, Timothy, 80
Measurement and signature intelligence, 98
Middle East
    natural resource conflicts, 130
    oil, 126–128, 130
    state-sponsored terrorism, 62–63
    terrorism, 59
Military
    organizational structure, 113–116

peacekeeping operations, 120–121, 120*f*, 120*t*
    personnel strength, by region and country, 70*t*–72*t*
    technology and warfare, 144–145
Military observers, 117
Militia groups, 83, 87*t*–88*t*, 91
Missile Defense Agency, 116
Missile Technology Control Regime (MTCR), 29
Missiles. *See* Ballistic missiles; Cruise missiles; Nuclear weapons; Weapons of mass destruction
Mogadishu, Somalia, 119
Money laundering, 132–133, 134*t*–137*t*
Motivation
    domestic terrorism, 81, 83–86, 88–90
    international terrorism, 58
    Operation Iraqi Freedom, 46
Moussaoui, Zacarias, 105
MTCR (Missile Technology Control Regime), 29
Mueller, Robert S. III, 105–109
"Multilateral interdependence" theory of world power, 7
Multiple independently targetable reentry vehicles (MIRVed) missile limits, 30–31
"Multipolar" theory of world power, 6–7
Muslims. *See* Islam

N

Nagasaki bombing, 26
Nation of Islam, 88–89
National Commission on Terrorist Attacks on the United States, 102
National Cyber Alert System, 143
National Defense University, 143–144
National Geospatial-Intelligence Agency, 116
National laboratories, 51–52
National Security Act, 5, 97
National Security Agency, 116
National Security Council (NSC), 95
A National Security Strategy for a New Century (White House), 119
The National Security Strategy of the United States of America (White House), 119
The National Strategy to Secure Cyberspace (White House), 143
NATO. *See* North Atlantic Treaty Organization (NATO)
Navy, U.S., 114
Nerve agents, 25
New Black Panther Party for Self Defense, 89–90
"New Look" policy, 5
Nichols, Terry L., 80
Nigeria, 130
Nixon, Richard M., 6
Nongovernmental organizations (NGOs), 16
Nonlethal chemical agents, 25
North Atlantic Treaty Organization (NATO)
    creation and functions, 124–125
    peacekeeping, 117–118
    Treaty on Conventional Armed Forces in Europe, 18
North Korea
    demilitarized zone, 17

as state of concern, 10
state-sponsored terrorism, 65
weapons of mass destruction, 39–40, 39*t*, 40*f*
NPT. *See* Nuclear Nonproliferation Treaty (NPT)
NSC (National Security Council), 95
Nuclear Nonproliferation Treaty (NPT), 29, 33–36, 39–40, 43–44
Nuclear Suppliers Group (NSG), 29
Nuclear Threat Initiative (NTI), 35
Nuclear weapons
China, 33–35
Egypt, 35
Hiroshima and Nagasaki bombing, 26
India, 35
Iran, 35–36, 37*f*
Iraq, 42–44, 43–44
Israel, 36, 39*f*
Libya, 38–39
North Korea, 39–40
Pakistan, 40
physics of, 26–27
Syria, 40–42
trafficking, 27–28
Nye, Joseph, 7

O

Office of Defense Trade Controls (ODTC), 12, 14
Office of Homeland Security, 73
Office of Law Enforcement Coordination, 107
Office of Management and Budget (OMB), 95
Office of the White House, 94–95
Oil
Africa, resource conflicts in, 130
China, 127–128
internal oil production, U.S., 129
Operation Iraqi Freedom, 46
Persian Gulf, 126–128, 130
Russia, 127
Strategic Petroleum Reserve (SPR), 128–129
world oil transit chokepoints, 127, 128*f*
Oklahoma City bombing, 80
Olympic Park bombing, 80, 104
OMB (Office of Management and Budget), 95
OPEC (Organization of Petroleum Exporting Countries), 126–127
Operation Enduring Freedom, 75
Operation Iraqi Freedom, 45–46, 75
Operation Restore Hope, 119
Organization of Petroleum Exporting Countries (OPEC), 126–127
Organization for Security and Cooperation in Europe (OSCE), 16
Organizational structure
Defense Department, 113–116
Federal Bureau of Investigation, 106–109, 106*f*
federal government, 93–94, 94*f*
Homeland Security, Department of, 110*f*
North Atlantic Treaty Organization (NATO), 125

United Nations, 124
Organizations. *See* Agencies and organizations
Organized crime, 131–133, 134*t*–137*t*

P

Pakistan, 40, 42*f*
Palestinians, 62–63
Pan Am Flight 103 bombing, 64
Paramilitary operations, 98
Patriot Act, 74–75, 98
Patriot groups, 83, 87*t*–88*t*, 91
Patriot missiles, 15
Patterns of Global Terrorism (State Department), 62
PCII (Protected Critical Infrastructure Information) Program, 133
Peacekeeping, 116–121, 117*t*, 120*f*, 120*t*
Peacekeeping: Issues of U.S. Military Involvement (Congressional Research Service), 121
Peninsula Shield Force, 125–126
Pentagon Force Protection Agency, 116
Performance and Annual Report, Fiscal Year 2003 (Customs and Border Patrol), 111
Persian Gulf and Gulf Cooperation Council, 125–126
Persian Gulf countries, conventional weapons trade with, 14–15
Phoenix Program, 98
Pilots. *See* Airport security
Politics and terrorism, 58
Port security, 74
Post-September 11th attacks
Federal Bureau of Investigation, 105–109
Immigration and Naturalization Service, 110–112
Postal facilities and anthrax attacks, 80–81
Power, measuring, 4
Preparedness
biological and chemical terrorism preparedness funding, 48
Centers for Disease Control and Prevention, role of the, 49–53
domestic terrorism, 91
Presidential powers, 93–95, 123
Prevention
domestic terrorism, 90–91
information warfare, 143–144
Preventive Defense: A New Security Strategy for America (Carter and Perry), 73
Programme of Action, UN, 16
Project BioShield Act, 48
Proliferation Security Initiative (PSI), 31
Protected Critical Infrastructure Information (PCII) Program, 133
Public opinion
internal oil production *vs.* conservation, 129
terrorism and homeland security, 75–78

Q

Qadhafi, Muammar, 10, 64
al Qaeda
financing, 132–133
September 11 attacks, 66–69, 67*f*
Quick Reaction Demining Force (QRDF), 17

R

Rajneeshees, 49
Reagan, Ronald, 13–14, 101
Reconnaissance images, 97–98
Records Management Division, CIA, 107
Regional Centers of Excellence for Biodefense and Emerging Infectious Diseases Research, 47–48
Regional conflicts and terrorism, 65–66
Religion and terrorism, 58
Research, biological and chemical terrorism, 47–48
Reserve oil, 128–129
Restructuring
Defense Department, 113
Federal Bureau of Investigation, 106–109
Immigration and Naturalization Service, 110–112
Rickettsiae, 26
Ridge, Tom, 73, 109
Right-wing organizations, 83, 87*t*–88*t*, 91
Rogue states. *See* States of concern
Rowley, Colleen, 105
Ruby Ridge incident, 104
Rudolph, Erick Robert, 80
Russia
conventional weapons trade, 12, 14
Materials Protection, Control, and Accounting Program (MPCA), 28
oil production, 127
organized crime, 132
as transition state, 9
treaties concerning weapons of mass destruction, 30–31, 31
Treaty on Conventional Armed Forces in Europe violation, 18

S

Saddam Hussein
chemical weapons, 44
fall of, 45
nuclear weapons program, 43–44
state-sponsored terrorism, 64–65
SALT I agreement, 30
SALT II agreement, 30
Sanctions
against China, 35
against Iraq, 42–43
states of concern, 7, 10
Sanford, Robert Russell, 142
Sarin attack, Tokyo, 25–26
Satellite imaging, 145
Scud missiles, 41–42
Secretariat, United Nations, 124
Secretary of Defense, 113
Security Council, United Nations, 124
Security Division, FBI, 107
Senate, U.S.
Central Intelligence Agency (CIA) investigation, 101
international agreements, 123
Senate Select Committee on Intelligence (SSCI), 101
September 11th attacks
flight routes of hijacked planes, 67*f*, 68*f*
9/11 Commission, 102
overview, 66–69, 67*f*

public opinion, 77–78
SFOR (Stabilization Force in Bosnia and Herzegovina), 117–120
Signals intelligence, 97
Single-issue terrorism, 84–86, 88–90
Sky marshal program, 74
Somalia, 119
SORT treaty, 31
South Korea, 17
Southeast Asia, 66
Southern Poverty Law Center (SPLC), 91
Soviet Union
    breakup of the, 6–7
    flexible response policy towards, 5–6
    "massive retaliation" policy towards, 5
    policy milestones, 4–6
    Strategic Arms Limitations Talks, 6
    treaties concerning weapons of mass destruction, 28–31
Special agents, FBI, 104
Special-interest terrorism, 84–86, 88–90
SPLC (Southern Poverty Law Center), 91
SSCI (Senate Select Committee on Intelligence), 101
Stabilization Force in Bosnia and Herzegovina (SFOR), 117–120
START I treaty, 30
START II treaty, 31
State, U.S. Department of, 14, 14f, 95, 96f
State-sponsored terrorism, 60–65, 69
States of concern, 9–10
Statistical information
    Chinese defense expenditures, 9f, 9t
    conventional weapons trade, 12f, 13t
    Defense Department contingency operations costs, 120t
    domestic terrorism, 82f, 83f
    FBI counterterrorism agents, 109f
    FBI intelligence analysts, 107t
    FBI translators, 109t
    information security incidents, 141f
    international terrorism, 59f, 60f, 61f, 76f
    military personnel strength, by region and country, 70t–72t
    peacekeeping missions, United Nations, 116–117, 117t
    unfavorable Blue Lantern checks, 14f
Strategic Arms Limitations Talks, 6
Strategic National Stockpile, 48
Strategic Offensive Reductions Treaty (SORT), 31
Strategic Petroleum Reserve (SPR), 128–129
Studies, reports and surveys
    An Agenda for Peace: Preventive Diplomacy, Peacemaking, and Peacekeeping (UN Secretary General), 116
    Allied Contributions to the Common Defense, 120–121
    China and Proliferation of Weapons of Mass Destruction and Missiles: Policy Issues (Congressional Research Service), 35
    Conventional Arms Transfers to Developing Nations, 1995–2002 (Congressional Research Service), 12, 127

Disarming Libya: Weapons of Mass Destruction (Congressional Research Service), 39
    A Force for Peace and Security: U.S. and Allied Commanders' Views of the Military's Role in Peace Operations and the Impact on Terrorism of States in Conflict (Peace through Law Education Fund), 118
    Gallup Poll, 78
    Harris Polls, 77–78
    The National Strategy to Secure Cyberspace (White House), 143
    9/11 Commission report, 102
    Peacekeeping: Issues of U.S. Military Involvement (Congressional Research Service), 121
    The 2004 Computer Crime and Security Survey (Computer Security Institute), 141
    U.N. Peacekeeping: Estimated U.S. Contributions, Fiscal Years 1996-2001 (GAO), 119
    U.N. Peacekeeping: Status of Long-Standing Operations and U.S. Interests in Supporting Them (GAO), 116, 118
    "U.S. Policy on Small/Light Arms Export" (Lumpe), 12
    "Vulnerability of U.S. Strategic Air Power" (RAND), 5
    The Wirthlin Report: Current Trends in Public Opinion from Wirthlin Worldwide (Wirthlin Worldwide), 77
Substate terror groups, 65
Sudan, 64–65
Sugimoto, Shigeo, 26
Suspicious activity reports (SARs), 72, 73f
Syria
    as state of concern, 10
    state-sponsored terrorism, 64
    weapons of mass destruction, 40–42, 42t

T

Taiwan, 8–9
Technical intelligence, 97
Technology and warfare, 144–145
Terrorism. See Domestic terrorism; International terrorism
16th Street Baptist Church bombing, 79
Threats
    biological agents, 49–51
    chemical agents, 51
    defining, 3–4
    information warfare threats classification, 139–140, 140t
Tokyo sarin attack, 25–26
Toxins, 26
Toyoda, Toru, 26
Trade issues
    conventional weapons, 11–15, 12f, 13t, 14f, 127
    oil, 126–128
Transition states, 8–9
Translators, FBI, 109t
Transmissible animal diseases, 53–55, 54(t5.5)
Transnational terror groups, 65

Treaty between the United States and the Russian Federation on Further Reduction and Limitation of Strategic Offensive Arms (START II), 31
Treaty between the United States and the Soviet Union on the Limitation of Antiballistic Missile Systems (ABM Treaty), 29
Treaty between the United States and the Union of Soviet Socialist Republics on the Limitation of Strategic Offensive Arms (SALT II), 30
Treaty between the United States of America and the Union of Soviet Socialist Republics on the Reduction and Limitation of Strategic Offensive Arms (START I), 30
Treaty on Conventional Armed Forces in Europe, 18
Treaty on Open Skies, 30–31
Treaty on the Principles Governing the Activities of States in the Exploration and Use of Outer Space, Including the Moon and Other Celestial Bodies, 29
Truman Doctrine, 4–5
Trusteeship Council, United Nations, 124

U

UAVs (unmanned aerial vehicles), 145
U.N. Peacekeeping: Estimated U.S. Contributions, Fiscal Years 1996-2001 (GAO), 119
U.N. Peacekeeping: Status of Long-Standing Operations and U.S. Interests in Supporting Them (GAO), 116, 118
UN Conference on the Illicit Trade in Small Arms and Light Weapons, 15–16
UN Monitoring, Verification, and Inspections Commission (UNMOVIC), 42–43
Unabomber, 80
Unified combatant commands, 115
"Unipolar hegemony" theory, 6
United Airlines Flight 93 flight route, September 11, 2001, 68(f6.7)
United Nations
    Conference on the Illicit Trade in Small Arms and Light Weapons, 15–16
    Convention against Transnational Organized Crime, 133
    conventional weapons sales regulation, 15–16
    creation, function and structure, 124
    peacekeeping, 116–117, 117t
United Nations Special Commission (UNSCOM), 42–45
United States
    accusations of terrorism by, 66
    alliances, 123–125
    arms exports, 12–14, 14f, 127
    Cooperative Defense Initiative, 15
    homeland security efforts after September 11th attacks, 73–75
    internal oil production, 129
    land mines policy, 17
    Materials Protection, Control, and Accounting Program (MPCA), 28

Operation Iraqi Freedom, 45–46
public opinion on terrorism and
    homeland security, 75–78
sanctions against Iraq, 42–43
September 11 attacks, 66–69, 67*f*
Strategic Petroleum Reserve (SPR),
    128–129
terrorism and the, 58–60, 61*f*
transnational crime response, 133
treaties concerning weapons of mass
    destruction, 27–31
See also Domestic terrorism
Unmanned aerial vehicles (UAVs), 145
UNMOVIC (UN Monitoring, Verification,
    and Inspections Commission), 42–43
UNSCOM (United Nations Special
    Commission), 42–45
U.S. Humanitarian Demining Program, 17
"U.S. Policy on Small/Light Arms Export"
    (Lumpe), 12
USA Patriot Act, 74–75, 98

V

Viruses, 26
Vital interests, 119
"Vulnerability of U.S. Strategic Air Power"
    (RAND), 5

W

Waco Incident, 104
War on Terrorism, 75
War Powers Act, 93
Wassenaar Arrangement on Export Controls
    for Conventional Arms and Dual-Use
    Goods and Technologies, 19
Watchdog groups, 90–91
Weapons. *See* Conventional weapons;
    Weapons of mass destruction
Weapons of mass destruction
    China, 33–35
    Egypt, 35
    history, 21–23, 22*t*
    Iran, 35–36

Iraq, 42–46, 43*t*
Israel, 36, 38
Libya, 38–39
North Korea, 39–40, 39*t*, 40*f*
Pakistan, 40, 42*f*
states of concern, 7, 10
Syria, 40–42, 42*t*
treaties and international agreements,
    28–31
Weather Underground, 79
Whistle-blowers, 105
White House, Office of the, 94–95
*The Wirthlin Report: Current Trends in
    Public Opinion from Wirthlin Worldwide*
    (Wirthlin Worldwide), 77
World Health Organization (WHO), 16
World oil transit chokepoints, 127, 128*f*
World Trade Center bombing, 1993, 66
World War I, 21
World War II, 4

ELEVENTH EDITION	JANUARY 2014
Editor	LINA GUILLEN
Production	SUSAN PUTNEY
Proofreading	SUSAN CARLSON GREENE
Index	JANET MAZEFSKY
Printing	BANG PRINTING

Woodhouse, Violet, 1948- author.
  Divorce & money : how to make the best financial decisions during divorce / Violet Woodhouse, Matthew J. Perry. -- 11th edition.
     pages cm
  Summary: "Make savvy, informed financial decisions during divorce. Learn about dividing debts, setting alimony, negotiating a fair settlement, and more. With Divorce & Money, you'll find out how to: - avoid tax problems - handle alimony and child support - reduce risks to your investments "-- Provided by publisher.
  ISBN 978-1-4133-1995-8 (pbk.) -- ISBN 978-1-4133-1996-5 (epub ebook)
  1. Marital property--Valuation--United States--Popular works. 2. Divorce settlements--United States--Popular works. 3. Divorce--Law and legislation--United States--Popular works. I. Perry, Matthew J., author. II. Title. III. Title: Divorce and money.
  KF524.W66 2013
  346.7301'664--dc23

                                        2013031504

This book covers only United States law, unless it specifically states otherwise.

Please note

We believe accurate, plain-English legal information should help you solve many of your own legal problems. But this text is not a substitute for personalized advice from a knowledgeable lawyer. If you want the help of a trained professional—and we'll always point out situations in which we think that's a good idea—consult an attorney licensed to practice in your state.

Dedication

We dedicate this book to…

Those grappling with the enormity of divorce and, thus, most in need of guidance.

Our families and friends, who kept us sane through the writing process.

Larry Woodhouse, for his continued friendship, and our children, Brooke and Tyler, all of whom have taught me the value of communication, negotiation, and compromise.

All our children, including Brooke and Tyler, and our clients, who teach us the value of communication, negotiation, and compromise.

Thank You

Lina Guillen, our editor at Nolo, who was most patient as well as wise in the way she dispensed legal and editorial acumen.

Victoria F. Collins and M.C. Blakeman, who contributed so much to the earlier editions on which this work is based and who deserve recognition for their hard work, creativity, and cooperation.

Special Acknowledgments

To the late Dale Fetherling, who we always will remember for his gifts of balance, tolerance, organizational skills, and for his boundless respect for our readers.

A special thank you to the staff at Violet F. Woodhouse, APC, for its competence, integrity, and humanity, and especially to my legal administrator, Tamyko Furman, who routinely makes possible the impossible.

About the Authors

Violet P. Woodhouse is a trial attorney and negotiator, as well as an educator, media consultant, and author. She is certified as a legal specialist in family law by the State Bar of California, Board of Legal Specialization. In addition to California, Ms. Woodhouse has been admitted to practice before the United States District Court, Central District of California, and the United States Supreme Court. She is a Certified Financial Planner® and has been ranked one of the nation's top financial advisors by *Worth* magazine for five consecutive years since the list's inception. She is also a frequent guest on national radio and television programs, and has written for or been interviewed by numerous national publications, including the *Wall Street Journal, New York Times, Working Woman, Money Magazine, Kiplinger's Personal Finance Magazine, USA Today*, and the *Los Angeles Times*, as well as legal and trade journals. Ms. Woodhouse is a member of the State Bar of California, the Taxation and Family Law Sections of the American and Orange County Bar Associations, the Family Law Section of the Los Angeles Bar Association, and is a member of the Association of Certified Family Law Specialists. *California Super Lawyers Magazine* named Ms. Woodhouse a Southern California Super Lawyer for eight consecutive years, and one of the top 50 Orange County California lawyers in 2004. Ms. Woodhouse provides divorcing clients with specialized legal representation through Violet Woodhouse, APC, in Newport Beach, California.

Matthew J Perry is a writer who has edited, ghostwritten and collaborated on many books, primarily in the genre of money and finance. He lives with his wife and son in New York City.

Table of Contents

Introduction:
Your Financial Companion During Divorce............1

If You Are Feeling Overwhelmed….............4

Managing Your Divorce in Troubled Financial Times.............5

How to Use This Book.............8

A Note on Same-Sex Marriage and This Book.............9

1 Legal vs. Financial Realities of Divorce.............11

Lessons in Legal Reality.............13

Your Best Strategy: Think Financially—Act Legally.............20

Legal vs. Financial Stages of Divorce.............21

2 Financial Realities No One Talks About.............25

In Divorce, Everything Takes Longer and Costs More.............28

You're at Risk When You're Financially Connected
to Another Person.............29

The IRS Is Watching Your Divorce.............30

Cash Is King.............31

You're Playing for Keeps.............33

3 Emotional Divorce: Managing the "Money Crazies".............35

Reduce Stress Whenever You Can.............39

Safeguard Your Sanity.............40

Watch Out for Sore Spots.............42

Be Prepared for the Worst..42

Develop a Financially Focused Mental Attitude 44

Avoid the "All at Once" Syndrome...44

Manage the Ebb and Flow of Emotions 45

Don't Let Financial Tasks Overwhelm You...................................47

4 The Hardest Part: Is My Marriage Really Over?..............51

5 The Separation: What Happens When One Spouse Moves Out?..55

The Separation Date...57

Building a Better Credit Report and Credit Score 65

Social Security Benefits After Divorce 68

Questions to Ask an Attorney...69

6 Closing the Books: What Do We Do With Joint Property?..71

Joint Account Checklist...72

Questions to Ask an Attorney...83

7 Getting Help: Who Can I Turn To? 85

Questions to Consider When Seeking Outside Help......................86

Selecting Professionals to Assist You..88

8 Financial Fact-Finding: What Must I Know and When Must I Know It? 111

Advice to the Terminally Disorganized 113

If You Think Your Spouse May Be Hiding Assets 116

Don't Forget the "Easy-to-Forget" Assets 118

The W-2 and the Tax Return 122

What About the Gifts You Gave Me? 127

Financial Facts Checklist 128

Questions to Ask an Attorney 128

9 Facing the Future: What Must I Plan For? 135

Major Upcoming Life Events.. 136

Anticipated Financial Commitments.. 137

Major Goals That Will Cost Money... 138

Where Does the Money Come From? 141

10 Protecting Against Risks to Life, Health, and Property 145

Insurance 146

Property and Estate Protection.. 152

Questions to Ask an Attorney 155

11 Taxes: How Do I File and Pay? 157

Get a Rough Estimate of Your Tax Bill 158

What Status Is Better to Use When Filing Tax Returns? 160

When in Doubt, File Separately.. 161

What to Know If You File Jointly... 165

Dividing the Joint Tax Liability—Or the Refund 168

Tax Issues Involving Temporary Alimony or Child Support................. 171

Get Your Tax Agreement in Writing... 172

Questions to Ask a Tax Adviser.. 172

12 Property and Expenses: Who Owns and Who Owes What? 173

Who Owns What—Marital Property and
the Laws of Your State 174

Who Knows What—Using Legal Discovery 177

Net Worth—What Do You Own and
What Do You Owe? 179

The Difference Between Assets and Income 191

Cash Flow—Where Does the Money Go? 193

Questions to Ask an Attorney 195

13 What Will Happen to the House? 207

Financial vs. Legal Realities 209

The House—Keep It, Transfer It,
or Sell It? Now or Later? 210

Steps Toward Settling the House 210

Questions to Ask an Attorney or Financial Adviser 247

14 Retirement Benefits: Who Gets What? 249

Understanding Retirement Plans 253

Qualified Domestic Relations Orders 261

The Legal Value of Your Retirement Plans 265

The Financial Value of Your Retirement Plans 268

Calculating the Financial Value of Plans 273

Additional Financial Factors Affecting Retirement Plan Divisions 280

The Division Decision: Now or in the Future 284

Questions to Ask an Attorney 288

15 Financial Investments: How Do We Divide the Portfolio Pie? 291

Concepts to Consider 293

Steps to a Settlement 302

16 Evaluating Employee Benefits and Stock Options 341

Employee Benefits 342

Stock Options and Nonqualified Deferred Compensation Plans 348

Questions to Ask an Attorney or a Tax Specialist 357

17 How Will We Divide Debts? 359

General Rules on Who's Responsible for Debt 360

If You Live in a Community Property State 362

Listing Your Debts 363

Marital Debts and Bankruptcy 363

Dividing Debts at Divorce 375

Dividing Debts When There's Nothing to Fight Over 376

Questions to Ask a Divorce Attorney 377

Questions to Ask a Bankruptcy Attorney 378

18 Child Support and Alimony: What Might I Pay or Receive? 379

Child Support—Legal, Financial, and Emotional Realities 381

Steps to a Settlement 384

Alimony—Legal, Financial, and Emotional Realities 408

Steps to a Settlement 410

Questions to Ask an Attorney 428

19 Negotiating and Finalizing the Best Possible Settlement 431

Have You Done Your Financial Homework? 433

Tallying Your Marital Balance Sheet 433

How Are the Offers and Counteroffers Made? 438

How Do You Finalize the Settlement? 441

Divorce Ceremonies 443

20 After the Divorce: How Do I Get From "We" to "Me"? 445

How Do I Finish the Business of Divorce? 446

Can I—Or My Ex-Spouse—Change the Settlement? 451

What Do I Want to Do With My Life? 453

If You Find a New Love, Protect Your Old Assets...and Your Alimony 461

How Can I Move Beyond the Divorce? 465

A Appendix 469

Law Libraries 470

Lawyers and Typing Services 472

Online Legal Resources 475

Additional Resources 475

Present Value Factors 480

Future Value Factors 486

List of Professional Advisers 492

Index 497

Introduction:
Your Financial Companion During Divorce

If You Are Feeling Overwhelmed... 4

Managing Your Divorce in Troubled Financial Times 5

    Alternatives to Traditional Litigation 6

    Flexibility on Key Financial Issues 7

How to Use This Book 8

A Note on Same-Sex Marriage and This Book 9

The week when you take the first steps to end your marriage has been the worst week of your life. Suddenly, the person you're divorcing calls and demands copies of your tax returns for the past five years.

You brown-bag your lunch and take the bus to work so you can make ends meet while the divorce settlement is pending. But then, you receive an unexpected bill for a joint credit card you never canceled. Your soon-to-be ex-spouse charged an expensive vacation on your joint credit card, and now you have to pay for it.

Money often is the last thing you want to deal with when your marriage is ending. But there may never be a more crucial time to take control of your assets.

In fact, making financial decisions is possibly the most important job you have when you're dealing with divorce.

How can you make the right decisions? How can you deal with tedious financial details when you're going through such a stressful event? Unfortunately, it's hard to find answers to those questions.

Legal advice is plentiful, and therapists or support groups can help you through emotional upheavals. But who can show you how to make sense of your financial life?

That's where this book comes in. It explains what you must know to avoid the potential financial disasters of divorce. Specifically, this book draws on our years of experience as attorneys and financial planners to help you understand:

- what you own and owe
- how divorce affects you taxwise
- how best to divide your property, investments, and other assets
- what can happen to your retirement nest egg or your business when you divorce
- what to do about alimony and child support
- how to prepare to negotiate a final settlement, and
- how to gain financial stability in your new life as a single person or single parent.

Each divorce is different, of course, so we cannot promise to provide complex financial advice for every possible situation you may encounter. But we *can* let you know where—and to whom—you can turn to find answers for yourself.

As we see it, divorce is a crash course in managing personal finances—a course you need to complete whether you signed up for it or not. However daunting this course may seem, the first lesson can be stated plainly: There is a big difference between legal reality and financial reality.

The financial truth about divorce is this: Just because you "get it in writing" doesn't mean you will get it for real. In other words, even though your settlement may be perfectly legal and fair, it can still be costly in financial terms. For example:

How will you pay for debts that mounted during the marriage if your spouse refuses to come up with any part of the money—even if you were not supposed to pay those bills under your settlement agreement?

If you leave the division of your property up to the courts, the tax consequences of your transactions will most likely be ignored. But the IRS will still collect those taxes.

What will you do if your spouse agreed to pay child support but is consistently late or fails to pay?

As you divorce, you must be on the alert for the limitations of the law. Throughout this book, we will show you how to craft a strong financial settlement without ignoring the blind spots of the legal system.

Keep in mind that divorce does not have to be messy and expensive. Even if you have a great deal of property, you need not spend a fortune. One couple we worked with had $5 million in assets, yet they spent only $5,000 in legal fees. This couple evaluated their property, analyzed tax consequences, and negotiated fairly.

By contrast, another couple, whose property totaled only $100,000, ended up spending $20,000 on attorneys' fees alone—all because they could not be realistic about their personal finances and instead let their emotions interfere with reason.

Whether you and your spouse have a little money or a lot, it is important that you look at your divorce from the perspective of financial reality. This means protecting your financial well-being, whether or not

your spouse cooperates. Without this perspective, you may one day find yourself painted into a financial corner, saddled with debts, or burdened with assets you cannot afford or hope to sell.

The choice is yours. Divorce is painful enough. Don't let money problems make it worse.

If You Are Feeling Overwhelmed...

Congratulations. You're normal. Feeling overwhelmed is a very common experience during divorce. Even extremely capable people may suddenly find it difficult to balance a checkbook or perform other simple tasks.

If that happens to you, slow down. You can't expect to consider every financial aspect of divorce at once.

Although every divorce is different, the process seems to have one universal effect: temporary paralysis. Whether you're a hard-driving executive or a stay-at-home spouse, a traumatic separation can cause otherwise competent people to become immobilized by the emotion of the situation and the gravity of the financial questions they face. Should the house be sold? How much alimony will be awarded? What will happen at tax time? Who will support the children?

Feeling overwhelmed, confused, angry, or depressed—these are normal responses to the stress of divorce. Our advice, as with any major undertaking, is to break it down into small, manageable pieces. And because the hectic pace of life today puts so many demands on your time, we present this plan in step-by-step segments. When you complete several small goals each day, you are less likely to become overwhelmed and immobilized. That feeling of accomplishment helps you stay out of the crisis mode and works to your benefit both emotionally and financially.

Managing Your Divorce in Troubled Financial Times

Money problems cause terrible stress in the best of times. When divorce is added to those problems, the mix is often personally and emotionally devastating.

The years following the massive recession of 2008 have been far from the best for millions of people. Layoffs, foreclosures and lingering economic uncertainty have brought enormous pressure on couples contemplating a split.

The family home used to be the most valuable asset most divorcing couples had. But with many home values still worth a fraction of their pre-2008 values, the home often is an illiquid, toxic asset, carrying only debt and liabilities for both parties.

Given the downturns in many industries and persistent, long-term unemployment, many unhappy couples are thinking twice about divorce, feeling it would only further deplete their already-shrunken assets. As strange as it seems, these difficult circumstances can provide some financial benefit to couples that might otherwise make decisions about their marriage based exclusively on their emotions.

As you read this book, you often will be encouraged to get past the emotional import of your situation before you make a financial or legal decision. It always makes sense to consider the financial implications of splitting up and, if possible, measure them against of the emotional costs of staying together. Before you even embark on that difficult calculus, you should look hard at your marriage, especially if you have children. Have you done all you would reasonably expect a peer to do if his or her marriage was in crisis? Are there emotional wounds that might be healed through counseling? Extended work with a marriage therapist is not cheap, but we can assure you that the cost of divorce is almost always greater.

Of course, in some cases, particularly those where domestic violence, addiction, alcoholism, or child abuse is a factor, divorce may be the only way out. We do not advocate staying in a marriage if you feel unsafe.

Only you can know when there's no turning back. But be aware that divorce does not repair feelings of hurt, sadness, disappointment, or failed expectations. We strongly advise that you attempt to address your problems through other means before you consider a divorce. At a time when millions are living on the edge of financial hardship, with underwater mortgages, diminished pensions, high medical costs, and a tough job market, you should make a concentrated effort to separate your emotional concerns from financial realities.

Alternatives to Traditional Litigation

If you do decide that divorce is your best option, be aware that the traditional, bare-knuckles "see you in court" approach is the most expensive route you can take, especially if your case ends up in a courtroom. Moreover, with budgetary cutbacks in many states making judicial calendars overcrowded, it can be a painfully long process as well.

If the decision to divorce is mutual, and you have good reason to believe that your spouse will be truthful and forthcoming with important information concerning joint and separate assets, debts, income, and expenses, the following are options you might explore. (There's more about them in Chapter 7.)

Mediation. For parties who are able to be realistic and relatively cooperative, working with a mediator is an option that should be on the table even in good economic times. In this scenario, a couple reaches an agreement with the help of a mediator, who serves as a neutral facilitator—decisions are made by the divorcing couple, not the mediator. Now, more than ever, it could pay to compromise. Mediation saves money and allows you to avoid spending wasteful hours, days, or weeks in our clogged court system.

Private judging/arbitration. Arbitrators, unlike mediators, hear arguments from each side and make decisions regarding issues in dispute. Those decisions then are formalized into a court judgment. It's usually quicker and less expensive to pay an arbitrator or private judge than to go through the normal court process.

Limited-scope representation. You may be able to minimize legal fees by hiring a lawyer to do just a limited amount of work, such as one court appearance to help you argue about the amount of support you should get or about the custody schedule. The lawyer would not be responsible for your case, just for working with you on that specific issue(s). Limited scope can be appealing to couples that, because of income restraints or prior training, are able to undertake a do-it-yourself approach to divorce.

Collaborative divorce. Each party has a lawyer who is committed to resolving the issues without litigating. All meetings include both spouses and both attorneys, and all parties commit to full disclosure of financial documents. The spouses seek to reach a written agreement that's then taken to a judge for a final decree.

Flexibility on Key Financial Issues

Being creative and nimble on certain divisive items can help minimize financial loss for both spouses.

Rethinking living arrangements. The most common ways divorcing couples deal with the family home is a sale to a third party or a buyout in which one spouse pays the other and keeps the house. But with many homes now subject to mortgages higher than property values, keeping a home until its value increases is becoming more attractive to some. For more on house options, see Chapter 13.

Hold off cashing out assets. Divorcing spouses are often eager to get a fresh start, putting bad memories and poor decisions quickly behind them—and many often think this means assets must be cashed out. But when the values of homes, businesses, and retirement accounts have fallen, spouses should consider dividing holdings in kind and waiting for them to regain some value. For more on these choices, look at Chapters 7, 13, and 14.

Be flexible with spousal and child support. If one or both spouses are unemployed, it may be wise to defer a decision on alimony and child support for a few months in hopes the situation will improve. You will save money if you include a provision in the divorce agreement that allows you to revisit child or spousal support later without having to file additional paperwork. See Chapter 18 for more about support.

Stay nimble as to who claims the kids as a tax deduction. In some cases it might make sense for the noncustodial parent to claim the children if, for example, the custodial spouse is unemployed and has no income. Chapter 18 details the tax consequences of child support.

How to Use This Book

The advice in these pages is gender neutral. With more and more women taking on breadwinner roles and more men running households or earning less than their wives, the power dynamics between the sexes continue to shift. Meanwhile, the traditional divorce between a working husband and homemaker wife, while still common, is outdated as a template. Women who earn more than their husbands and men who have spent years caring for children must be careful not to identify with the wrong parent when shaping a divorce strategy. This also applies to same-sex marriages and domestic partnerships, where the duties of earning income and child rearing are not evenly divided.

Before you jump ahead and try to figure out what your house is worth or what you will need to do about alimony or child support, we recommend that you read Chapters 1 and 2, which give the framework for the legal and financial reality that you *must* understand as your divorce progresses.

For tips on dealing with emotional upheavals, read Chapter 3.

Chapters 4 through 7 cover "The First 30 Days" and are especially important if you are in the initial stages of divorce. Chapters 8 and 9 help you start gathering financial data and thinking about the future.

In Chapters 10 through 12, you begin to seriously evaluate your assets, debts, income, and expenses. Whether you are a "do-it-yourselfer" or working with an attorney, you must figure out what you own and owe, and what it costs you and your family to live. We've provided several worksheets in Chapter 12 to help you.

Chapters 13 through 16 cover specific assets—family home, retirement benefits, investments, and other employee benefits. Chapter 17 deals with

your debts. And in Chapter 18, you will evaluate the likelihood of paying or receiving alimony and child support—and in what amounts.

Your analyses and calculations from Chapters 12 through 18 are brought together in Chapter 19 to help you structure your final settlement. Finally, Chapter 20 gives you guidelines for establishing a healthy financial life once the marriage has officially ended.

A Note on Same-Sex Marriage and This Book

On June 26, 2013, the United States Supreme Court issued two historic decisions:

- Section 3 of the Federal Defense of Marriage Act (known as "DOMA") barring the U.S. government from recognizing same-sex marriages is unconstitutional. Now, the federal government must recognize the marriages of same-sex couples that are married and live in any of the 14 U.S. jurisdictions that recognize gay marriage. Such same-sex married couples are entitled to the same federal benefits opposite-sex couples receive, including Social Security benefits, immigration status, and federal tax benefits.

- The Supreme Court dismissed the California Proposition 8 case, *Hollingsworth v. Perry*, on a legal technicality. The court declined to determine whether the proposition (a same-sex marriage ban) is constitutional and instead ruled that Prop. 8 supporters (private citizens) lacked standing to appeal the lower court's decision that Prop. 8 violates constitutional rights. The result is that the lower court's decision stands, and California may resume issuing same-sex marriage licenses.

As of these rulings, same-sex couples can marry in 13 states, plus D.C.. In California, D.C., Hawaii, Illinois, Nevada, New Jersey, Oregon, and Washington state, domestic partnerships or civil unions allow same-sex couples to enter into relationships that are the legal equivalent of marriage. Wisconsin provides some relationship recognition for same-sex couples, though not a marriage equivalent.

While national marriage equality may be some way off, the trend towards more rights for same-sex couples is undeniable. Many questions remain, however, since the rules for receiving benefits—including Social Security, health insurance, taxes, bankruptcy protection, and other topics covered in this book—vary from federal agency to agency. What's more, couples residing in states that do not recognize same-sex marriage may or may not be eligible for benefits, even if they were married in states where their marriages were legally recognized. It will take time for these complications to be sorted out.

If you are in a same-sex marriage and you're splitting with your partner, this book will not help you figure out the odd patchwork of benefits between the states. If you have since moved to a state that does not recognize your marriage, you will need to consult a lawyer who is current with developments in this still developing legal territory.

But you can use this book to get the lay of the land, begin making important decisions about your financial future, and start moving forward in your divorce process. If your marriage is legally recognized and you have assets to protect, remember that federal regulations may now provide your spouse with a legal claim to property and income.

RESOURCE

Learn more about same-sex relationships. For details on same-sex marriage, civil unions, and domestic partnership rights, see *Making It Legal: A Guide to Same-Sex Marriage, Domestic Partnerships, and Civil Unions* and *A Legal Guide for Lesbian & Gay Couples*, both by Frederick Hertz and Emily Doskow (Nolo).

Get Updates and More Online

When there are important changes to the information in this book, we'll post updates online, on a page dedicated to this book:

www.nolo.com/back-of-book/DIMO.html

You'll find other useful information there, too, including author blogs, podcasts, and videos.

Legal vs. Financial Realities of Divorce

Lessons in Legal Reality..13

    Most Divorces Are Settled Out of Court...14

    Divorce Law Is Local 15

    Don't Expect the Legal System to Take Care of You...........................17

    It's Easier to Write Laws Than to Enforce Them...............................19

Your Best Strategy: Think Financially—Act Legally 20

Legal vs. Financial Stages of Divorce..21

Sooner or later during your divorce, you will discover one insight that is central to this book *and* to the successful outcome of your settlement:

Legal reality and financial reality are fundamentally different.

A seemingly simple idea—but you'd be surprised how long it takes to sink in. To help you understand why this concept is so important, take a few moments to consider the following real-life divorce stories. In each, read the Legal Reality first. Then, see the true outcome in the Financial Reality side.

Legal Reality	Financial Reality
Jonathan and Penny were married for five years before they divorced. During that time, Penny frequently ran their credit cards to the limit buying clothing and had trouble balancing their checkbook.	After the divorce, Penny didn't pay off the credit cards, and creditors began hounding Jonathan for the money. Jonathan ended up footing the bills, because a divorce settlement assigning debts—even one included in a divorce judgment—cannot change a couple's original joint obligation to their creditors.
When they reached the final settlement hearing, Jonathan was greatly relieved when the court made Penny solely responsible for paying the $10,000 in credit card debts she had accumulated during their marriage. The settlement was included in the final divorce judgment, which made Jonathan feel safe.	Had Jonathan raised the issue before their settlement was finalized, he could have demanded more property in exchange for paying Penny's debts or insisted that they sell some jointly held property to pay off their creditors.

Moral of the story:
Getting something "in writing" from the court
doesn't always mean you'll get it for real.

Legal Reality	Financial Reality
During their 15-year marriage, Sharon and Bill were committed to building up a good portfolio of stocks and mutual funds for their retirement.	Sharon paid attention to basic financial facts that Bill ignored: costs and taxes that decrease the value of an asset.
Because Sharon avidly followed the market, she wanted to keep a batch of stocks she had recently purchased and asked Bill to take stocks of equal value, which they had purchased early in their marriage. After negotiating over a few other assets, Bill and Sharon reached an agreement in which each of them would receive the exact same dollar amount in cash or assets at the end of the divorce. The court accepted the terms of their settlement, and the books on their marriage were quickly closed.	Sharon wisely picked the stocks most recently purchased. Because these stocks had not increased substantially in value, the taxable capital gains were low. Bill, however, blithely accepted the older stocks, which had gone up a lot in value since the time of purchase. Even at a capital gains rate of 20%, he owed substantial taxes when he sold the stocks. Had he taken the time to calculate his potential tax burden before agreeing to the settlement, he could have suggested splitting the stocks so that each spouse took half of the older stocks and half of the newer stocks.

Moral of the story: A 50-50 settlement isn't always equal.

Lessons in Legal Reality

The legal and financial perspectives on divorce are two distinct views of the same event. Ending your marriage with no assets or huge debts is the hard way to learn about the difference between these two realities. But if you are proactive from the get-go, your lessons don't have to be so costly.

If you intended to master Chinese cooking, you wouldn't begin by picking up Julia Child's *Mastering the Art of French Cooking*, even though it covers the same *general* topic of cooking. Similarly, you must

take the time to consider the crucial differences between the legal and financial perspectives on divorce.

These four basic guidelines help define the legal perspective of divorce:

- Most divorces are settled out of court.
- Generally, divorce law is local.
- Don't expect the legal system (or a lawyer) to take care of you.
- It's easier to write laws than to enforce them.

Most Divorces Are Settled Out of Court

You may imagine that your divorce will be capped by a courtroom trial in which your "good guy" lawyers win the day and settle everything in an hour. Perhaps you're eagerly anticipating that your day in court will be an opportunity to explain to a wise, kindly judge exactly how your spouse wronged you.

Don't count on it. *L.A. Law* and other legal TV scripts do not reflect reality.

An estimated 90% of divorce cases are settled without a court trial. Most of your settling will be done through meetings between you and your spouse or between your lawyers, often on the courthouse steps. As the trial date nears, you will quite likely be rushed into conferences in the courtroom hall or coffee shop. In these frantic meetings, your spouse and/or the attorneys may confront you, demanding instant decisions on issues that will affect the rest of your life.

Most divorce courts today are primarily concerned with money, not morals. The main job of the legal system is to resolve property disputes and to ensure the welfare of children. Spousal misconduct, of course, could affect custody, and economic mischief (such as hiding assets) can change the outcome of the final settlement. But judges don't have time to allow you to vent in open court about your mate, when numerous other cases, with more urgent problems, also need to be heard that day. Emotional pain is not part of the legal or financial reality of divorce.

The impersonal atmosphere of the legal world may baffle or intimidate you, but this isn't necessarily a bad thing. A divorce that ends up in court is a protracted, expensive affair that many couples are wise to avoid. Often, it is to your advantage to stay out of the courtroom.

As long as you and your spouse work toward a settlement without involving the court, you can trade property, negotiate terms, and still maintain some measure of control over your destiny. *If you cannot reach a settlement and must have a trial, however, you put your fate into the hands of a judge—a stranger who knows nothing about your children or property.* Yet, you'll have to live with whatever that judge decides.

Even in cases that are resolved before trial, there often are numerous court hearings, such as those for temporary orders, child custody, child or spousal support, and attorney or expert fees and costs, to mention a few. These hearings alone can be incredibly expensive, yet many could be settled beforehand, with a little more effort at compromise.

Divorce Law Is Local

Divorce laws not only differ from state to state, but interpretations of divorce law can vary from judge to judge. Whether you ultimately hire an attorney and whether or not you have a court trial, you should ask a local lawyer to assess the most likely outcome of a divorce like yours. Ask about the types of settlements local judges tend to approve. Find out how the local courts and individual judges view mediation. Also ask about judges' attitudes and the prevailing mood regarding, for example, joint custody, moving away with the children, or alimony.

Divorce courts are unlike other courts. They are called "courts of equity," which means that the judge has wide discretion in making decisions. While a lawyer can educate you on the law, a lawyer cannot ethically or realistically promise what the judge will do in your case.

You may not like what you hear when you see a lawyer, who may tell you that the things you want in your divorce are impossible to get. Another attorney may promise you everything, but ultimately deliver nothing. Interviewing several people for a cross section of opinion can give you a more accurate picture of your situation. (See Chapter 7 for information on hiring an attorney.)

If you know how individual judges normally rule in your locale, your expectations will be more realistic. Even if you don't have a trial (remember, 90% of cases are resolved before a trial), divorce lawyers tend

to give advice that is consistent with local court rulings. Granted, it's hard to ignore sensational newspaper stories about big-dollar divorces in other parts of the country. But those cases are irrelevant. You must concentrate on what happens in your backyard, because that is where your divorce and your financial future will be decided.

Affording Attorneys

Throughout this book, we may tell you to "check with an attorney" on various questions. We say this because the individual circumstances of your divorce may require legal information beyond the scope of this book.

But who is going to pay for this costly legal advice?

Attending a brief consultation with a lawyer should not bankrupt you. Organize your thoughts and questions before you seek legal advice so you can save time when the attorney's meter is running. At an initial visit, spend an hour—not a day. Then go home and think about how much more you may have to spend on an attorney.

In some cases, your spouse may be required to pay all or some of your lawyer's fee. The lawyer should be able to advise you about this at the initial consultation.

If you have very little money and feel you truly need advice from an attorney, consider borrowing money, holding a garage sale, or talking a short-term second job to help pay for legal fees. Check with local law schools or bar associations, which may run divorce clinics where students assist low-income clients. Some law firms will provide pro bono services to clients who fall below certain income levels. Check with your local court, and visit its website, which may provide help in filling out paperwork. Many will also provide valuable explanations of different aspects of the divorce process.

No matter what other decisions you make, here's an important rule: *Do not* let your spouse's attorney or anyone else chosen by your spouse be the one to represent you. If your spouse is represented by an attorney, you should be also. Except in very limited circumstances, one attorney cannot represent both parties in a divorce action. For more information on finding—and working with—lawyers, see Chapter 7.

Don't Expect the Legal System to Take Care of You

Always remember that the professionals you hire to represent you in a divorce are simply working for you: They are not living your life, and they expect you to be an active advocate for your own interests. Even if you have the money to hire expert negotiators, don't fall into a passive attitude. You should be thinking like an entrepreneur during a divorce, not like a victim.

If you were starting a business, you would first sit down with accountants and financial professionals to crunch the numbers and determine whether your new venture has a good chance of turning a profit. Only then, after the financial aspects have been examined, would you consult an attorney about potential legal problems. Likewise, when you know you're heading towards divorce, you should determine the value of all separate and marital assets and try to reconstruct a paper trail of financial information *before* you consult a divorce attorney.

Don't Be Afraid to Ask Questions

Don't be intimidated or afraid to ask questions if something is unclear. One divorcing woman admitted that some of her troubles resulted from her own unwillingness to appear ignorant. In an interview with sociologist Terry Arendell in the book *Mothers and Divorce*, the woman recalled her divorce and commented, "Part of the problem was my own fault. I gave the appearance of being knowledgeable. I knew more about buying property and bank accounts than my lawyer did, but I didn't understand all the tax things. And so I was reluctant to ask some of the things I should have asked."

However, becoming your own financial expert may be a very tall order—especially for spouses who aren't accustomed to managing finances and are starting out from a position of no control. If your spouse has held the financial reins tightly, don't despair. You can get up to speed with some expert assistance. If you're a financial neophyte, you

should immediately seek legal help to understand your rights and begin preparing for the negotiations to come.

Use the checklists and forms in this book to learn the basics of documenting your separate and joint assets. As you assemble documents, find a safe place (*not* in your home) to store them. Never underestimate the value of financial statements in a divorce.

You must also accept from the beginning that your future is in your hands, not the court's. The legal system is not designed to help you with your finances once the divorce is granted. For example, you may legally and fairly split the benefits of a pension plan in a divorce settlement, but when the time comes to retire, you may have less income than you need to live on. The court cannot anticipate or resolve that problem for you.

Further, the implications of future taxes on property are not taken into account in settlement agreements in most states. Generally, only existing—or impending—taxes can be factored into a division of assets. Anything beyond these taxes is considered speculation—and speculation is not normally welcome in the courtroom.

For example, if, as part of the divorce settlement, you and your spouse will sell $40,000 of stock at a profit, you could agree that the taxes owed on the profit will be factored into the settlement and split between the two of you. But, if you decide instead to keep all of the stocks and your spouse gets another asset of the same value in exchange for his or her portion of the stocks, the court will not award you more at the time of trial to cover whatever amount of taxes you may owe in the future.

You will need to call the shots today to keep a grip on your future. You can't leave complex, speculative questions about your financial future to the one-dimensional perspective of divorce law.

Don't expect the legal system to make the best financial decisions for you. Only you can do this.

Definition: Settlement Agreement

Throughout this book, we refer to marital settlement agreements, property settlement agreements, settlement agreements, or simply agreements. They all mean the same thing. A settlement agreement is a written contract between you and your spouse outlining how you will divide your property and debts, the amount of alimony and child support and who will pay it and who will receive it, who will have custody of or visitation with the children, and other major issues. If you and your spouse are unable to reach an agreement on these issues (on your own or with the help of a mediator), you will have to go to court to have a judge resolve them. Once these issues are finalized, they are incorporated into the final divorce decree.

It's Easier to Write Laws Than to Enforce Them

Laws concerning child support payments are among the most stringent on the books. Yet every year, millions of parents don't receive the money to which they are legally entitled. Nonenforcement of court orders is one aspect of legal reality for which you must prepare yourself. As you go through each step of your negotiations, ask, "How will I handle this if my ex refuses to abide by the agreement or the judge's orders? What options do I have to enforce this agreement?" And most important, "How much will enforcement cost me?"

When you recognize these risks ahead of time, you can take steps to minimize them.

To enforce your divorce agreement, you will probably have to go to court. The process is expensive, time-consuming, and emotionally draining. If at all possible, keep animosity to a minimum after the divorce so that both parents' custodial or visitation time with the children goes smoothly. In turn, that may make it more likely that your ex will make alimony and child support payments on time.

Covenant Marriages

Three states (Arizona, Arkansas, and Louisiana) provide marrying couples with the option of entering into "covenant" marriages. A covenant marriage makes it more difficult to get a divorce and is an alternative to a traditional marriage permitting divorce without restrictions. In addition to undergoing premarital counseling prior to their wedding, a couple wanting to end a covenant marriage must wait a full two years before proceeding, and may divorce only for reasons such as adultery or alcoholism. Parties in states offering covenant marriage are free to enter into a traditional marriage if they prefer. Opponents claim covenant marriages will force some women to stay in an abusive relationship.

Arkansas was the last state to adopt covenant marriage in 2001. Since then, many other states have considered covenant marriage laws; legislation was pending in Oklahoma as recently as February 2013. Most estimates indicate that a small number of couples—between 1 and 3 percent—have opted for covenant marriages in the states where they're legal.

Your Best Strategy: Think Financially—Act Legally

Make your financial concerns the centerpiece of your divorce, and work within the framework of the law. That is the most powerful position you can take. If you think financially and act legally, you will be able to anticipate risks and assess your needs, before a financial disaster hits.

No one wants to negotiate for an asset in a divorce and then be unable to sell it because they'd owe too much in taxes. Why should you go through the nightmare of settlement negotiations only to end up losing everything you fought for six months after the divorce is over?

Remember: The legal process of divorce is something you will live *through*—but the financial reality is what you will have to live *with* for the rest of your life.

In a divorce, it's not what you get that counts—it's what you keep.

Legal vs. Financial Stages of Divorce

Use the following table to help understand the relationship between the legal and financial stages of divorce. These stages will be explored in more detail throughout the book. Keep in mind that few divorces will follow the steps in this exact order.

 WEB RESOURCE

Nolo's website, www.nolo.com and www.divorcenet.com, contain legal information and resources that will help you through your divorce. In addition, several sites specialize in divorce information. Typically, these sites include forums, message centers, and chat rooms on topics such as custody, mediation, parenting, child support, debt, grandparent issues, and divorce in the military. Some can also help you find professionals—lawyers, therapists, mediators, or financial advisers—to assist you in your divorce. Divorce information online has exploded in recent years. You will need to be discriminating when you search. Information available on the sites of private firms, for example, may be technically accurate, but this content is often more of a marketing tool than a reliable legal source. Government or nonprofit sites (which typically end in .gov or .org) such as those belonging to states, local courts, state bar associations, and law schools tend to be more reliable. These sites often offer online self-help centers, links to law libraries, clinics, and general information about the law in your county or district:

- www.aaml.org
- www.divorceasfriends.com
- www.divorcecare.com
- www.divorcecentral.com
- www.divorcedirectory.com
- www.divorceinfo.com
- www.divorcesource.com
- www.divorcesupport.com
- www.divorcewithoutwar.com, and
- www.lifemanagement.com/flyingsolo.

Legal Stages	Financial Stages
Consult an attorney or do some research at a law library to learn about your legal rights and responsibilities. In particular, investigate how your state's laws regarding separation affect custody, alimony, child support, debts incurred after separation, and changes in the value of marital assets after separation.	Gather together your financial papers. Investigate the financial impact of separation and complete a cost–benefit analysis of your options. Make copies of all documents relating to assets, liabilities, income, and expenses whether or not you believe they're important. Use the Financial Facts Checklist (Chapter 8) as a tool (but not a limitation) to what you copy. Close or freeze access to joint accounts. (See Chapter 6.) Open accounts in your own name before filing for divorce. Visit the Federal Trade Commission's website to begin the process of obtaining a free copy of your credit report (www.consumer.ftc. gov/articles/0155-free-credit-reports).
Physically separate. For some couples, this means moving apart. For others, it's living in different parts of the house and no longer sleeping together. Additionally, your state law may use its own criteria to define the date of marital separation. Consult with an attorney to determine the rules in your state.	Keep track of debts incurred before and after separation, joint bills paid, and improvements made to property during separation. Keep receipts for moving and other expenses. Update insurance as necessary. Think about whether you will file taxes jointly or separately.
One spouse files a complaint or petition requesting a divorce. This begins the formal divorce proceedings. The other spouse must file an answer or response.	

Legal Stages	Financial Stages
One spouse files a request for temporary orders regarding custody, visitation, alimony, or child support. The request may also ask that the other spouse pay both partners' attorneys' fees.	Document all temporary alimony payments made, and write down your agreements about alimony. These payments may be tax deductible as long as there is an agreement in writing or a court order concerning the payments.
Conduct legal discovery (the procedures used to obtain information during a lawsuit) or win spouse's cooperation to share documents. Determine the amount of alimony, child support, and attorneys' fees you will pay or receive, if applicable.	Conduct financial fact finding. Complete the net worth and cash flow statements in Chapter 12. Hire a forensic accountant if necessary to search for hidden assets. Analyze your assets and debts—use appraisers, accountants, tax advisers, actuaries, and others to help you assess values, tax consequences, and other risks of keeping or giving up property.
Begin settlement negotiations, using one of these possible scenarios: • Use mediation to negotiate the settlement. • Negotiate between yourselves. • Negotiate through your attorneys. • If you are unable to settle certain issues, bring those issues to a judge.	Before settlement negotiations, make a list of all items you want the agreement to cover. Be sure to carefully analyze the tax ramifications and other financial pitfalls of each offer and counteroffer. Reduce attorneys' fees by doing much of the legwork on your own, settling without an attorney, keeping anger out of your negotiations, and avoiding a trial. Remember that a trial can be very expensive. You'll have to pay lawyers' fees as well as the fees of the experts (accountants, actuaries, and the like) whom you bring in to testify.
Draft your marital settlement agreement to incorporate terms of the settlement or the court order. The agreement is incorporated into the final judgment of divorce.	If you settle by agreement, carefully check it against your wish list.

Financial Realities No One Talks About

In Divorce, Everything Takes Longer and Costs More 28

You're at Risk When You're Financially Connected to Another Person 29

The IRS Is Watching Your Divorce 30

Cash Is King 31

You're Playing for Keeps 33

The true cost of divorce is rarely discussed openly.

You're probably better prepared to negotiate a car purchase than to negotiate a divorce. Most of us accept that the advertised "sticker" price of the car is not the final price. With a new car, there's often a "dealer markup," taxes and other fees that will be tacked on before you drive the car off the lot. If the car is preowned, you may have the opportunity to negotiate a lower price. You'll also need to factor in the cost of insurance, registration fees, repairs, and maintenance.

Many people enter divorce looking for a "sticker price," which will be final and all-inclusive. But there's just no way to predict the cost of divorce. There are too many unforeseeable variables, even for spouses that know their net worth. Costs and fees, such as those related to refinancing and transferring property, conducting research, and hiring experts, can turn what you thought was a great deal into one with devastating financial consequences.

You will need to keep a sharp eye out for additional amounts you may be charged, but were unaware of when you agreed to a settlement. Divorce clients that focus primarily on the legal or emotional implications of their situation, often fail to consider important financial angles of their settlements.

Take the family home as one example. You have to consider much more than the price you paid for it and the eventual sale price. In addition you must:

- Make sure property taxes are paid and current.
- Search for contractors' liens on the property.
- Make sure no state or federal taxing authority has a lien on the property.
- Keep the mortgage payments current and free of late charges.
- Confirm there are no judgments on record, which must be paid before title can be cleared for transfer.

If you were going to sell your house under conditions other than a divorce, you would have to consider and negotiate the cost of necessary

repairs or inspections and a real estate agent's commission (unless you sold the house yourself). You would also consider the implications of a capital gains tax. Under current tax law, you must pay taxes on any profit beyond the first $250,000, or $500,000 if you are married and file jointly with your spouse (a 3.8 percent Medicare tax may apply to the transaction if your adjusted gross income (AGI) is $200,000 filing separately or $250,000 filing jointly).

In our experience, these important considerations are rarely discussed during a divorce. Many courts must, by law, consider only the equity value of a home. The costs of sale, federal and state taxes and other subsidiary expenses are beyond their scope. Spouses often pay their exes for a full share of a home's equity, while totally ignoring the other normal costs and benefits of the house sale.

Only after the sale is final does one spouse recognize the benefits or risks of carrying those costs alone. And most divorcing couples *do* sell their family homes—either at the time of judgment or within five years after the divorce.

At divorce, you and your spouse are essentially "selling" everything you own—either by way of transfer to the other or to an outsider.

It is critical that you approach divorce the way you would any other business transaction. Look at your situation from a realistic perspective that takes all predictable financial events into account, especially those with tax-related consequences.

Financial reality begins where legal reality ends. To help you close the gap between the two, we provide you with worksheets and instructions on specific issues. We also suggest that you keep in mind five general observations on financial reality. These are the basic "truths" about money and divorce you should refer to whenever you face your toughest financial decisions:

- In divorce, everything takes longer and costs more.
- When you're connected to another person financially, you're at risk.
- The IRS is watching your divorce—even when you're not.
- Cash is king.
- You're playing for keeps; don't sell off tomorrow for today.

In Divorce, Everything Takes Longer and Costs More

One of the universal misconceptions about divorce is that it will end quickly so that all parties move on with their lives, free of the animosity that brought their marriage to an end. This is rarely the case, especially when a court gets involved.

In fact, your divorce can cost more money and take a longer time to settle than you ever imagined. For many couples, the process takes one to two years—even simple divorces that both parties thought would take only six months. The cost can range from several hundred dollars to several thousand. (One couple we knew took three years, spent $1 million, and still hadn't settled when last seen....)

To understand the costly nature of divorce, you must recognize the high price of splitting one economic unit in half. On the surface, an equitable property division would seem to mean each person walks away with half of what was shared by two and is, therefore, left with enough to support one.

But in the mathematics of divorce, that's not how the equation works out. Most spouses have unequal salaries and earning potential. Even during times of economic hardship, many people live beyond their means. When it comes time to divide one household into two, there is rarely enough money to go around. That holds true as much for young married couples with little property as it does for wealthy couples with assets accumulated over many years. In a divorce, dismantling these arrangements is not like performing a surgical procedure; it's more like experiencing a natural disaster. When couples separate, their fragile economic bases are torn apart, much as an earthquake loosens a house from its foundation, leaving everything in disarray. Spouses, children and extended families are left fearful of the unknown.

Recognizing that everything takes longer and costs more can help you through those moments when you are suddenly faced with an unexpected debt or an unwanted delay in your divorce.

Accept the Fact That Money Is Tight

Maybe you have gotten used to the idea that your divorce will cost you dearly. What do you do with that knowledge?

Be willing to accept a change in your lifestyle. Face it, you may have to borrow money, move, get a second job, or buy a used car instead of a new one. Be prepared for an economic pinch—at least for a while.

You will have to resist the very understandable temptation to splurge when your marriage is ending. The fact that you have to prepare for lean times also means that you must pay special attention to each financial decision you will make in divorce. In the end, the expense of the divorce itself means you must begin conserving your financial resources now.

You're at Risk When You're Financially Connected to Another Person

This truth seems obvious, yet it's often ignored. Divorcing partners tend to resist or "forget" the fact that they must cut or minimize the financial ties that bind them, or pay for those ties long after the marriage ends.

You're at risk any time you hold a joint interest with your spouse, are jointly responsible for obligations related to that interest, or are financially dependent on your spouse or ex-spouse. You have no control over their health, disability, death, employment, ability to earn income, or any third parties they may bring into their lives (and yours). If your former spouse defaults on payments, commits fraud, goes bankrupt, dies or becomes disabled, your future could be jeopardized unless you have taken deliberate steps to completely sever your financial relationship.

These risks often are overlooked—even through a divorce—because of the common human aversion to change. If you've maintained a joint checking account for 20 years, it may feel strange or wrong to close it. Reorganizing your financial position often forces people to accept that their marriages are really over. It's a difficult adjustment, but it's critical that you see it through.

A spouse may feel overly responsible for the other partner's welfare —and so continue paying for something like a car long after it's no longer appropriate. Resentments over an affair or some other perceived wrongdoing, such as lying, losing money, gambling, or abusing alcohol, can also cause one partner to remain financially entangled with the other.

People often leave themselves at risk with their former partners because they simply do not think about it. As you work through this book and your divorce, we will point out the most common financial risks divorcing couples face and give you tools for protecting yourself. Because your divorce is unique, however, you must stay alert to the risks in your particular case. Identify the areas where you and your spouse could remain connected financially, then work to break those connections whenever possible.

You will need courage and support from people you trust. But you should also seek the professional help of qualified attorneys, CPAs, or financial planners. You should base your decisions on their advice rather than horror stories you might hear via word of mouth.

Start thinking like a problem solver. Avoidance is not the problem solver's method. Avoidance only creates more risk, uncertainty, and fear.

The IRS Is Watching Your Divorce

You must consider the tax implications of each major financial move you make during your divorce. To you, selling your property may seem like a simple business transaction—but to the Internal Revenue Service, your actions may create something called a "taxable event." Be sure you're managing those events to your best advantage. Factor in tax costs anytime you make a financial decision in your divorce, because someday, the IRS will want to be paid.

As a general rule, a transfer of property between spouses related to a divorce is a nontaxable event. However, there may be major tax consequences to face in the future when that asset is sold. With anything you keep as part of your settlement, you become solely responsible for the taxes due on all gain (profit) on that asset from the time the two of you originally purchased it.

It is in your interest to consult with tax accountants or other financial professionals who can help you calculate the impact of taxes on your property division.

You will also need to decide how you will file tax returns once a divorce is under way. If you expect refunds on joint returns, decide how you will split the money. Be sure you know how the tax bill will be paid if money is due. You'll also need to understand the tax consequences of alimony or support payments.

The point is not to become paralyzed with worry about the IRS, but to recognize that your changing marital status definitely affects your tax status. Throughout the rest of the book, we give you guidance and formulas regarding tax matters.

WEB RESOURCE

The IRS continues to improve its resources so they are more user-friendly. The IRS website (www.irs.gov) offers plain-English answers to many difficult tax questions, and keyword searches will produce many articles and bullet point lists explaining how a divorce can affect both spouses' taxes.

Cash Is King

Quick—which would you rather have:

- a retirement plan that promises to pay $125,000 in ten years, or $100,000 today?
- a Mercedes currently valued at $35,000, or a mutual fund with a current market value of $35,000?
- a secured note from your spouse promising to pay you $500 a month for the next five years, or a lump sum payment of $30,000 today?

The simple answer is: Take the money and run.

Like the proverbial "bird in the hand worth two in the bush," a dollar you get today is worth more than a promise of one tomorrow. In the examples above, the retirement plan may be riskier than you think, the Mercedes will depreciate and could be expensive to repair, and your spouse might never make good on the promised payments. (While the

$35,000 mutual fund is not exactly in cash, a mutual fund is a liquid asset and can be converted to cash much more quickly than a car that is depreciating in value.) "Cash is king" holds true whether you are speaking of actual cash or cash-equivalent liquid assets.

Three basic financial concepts explain why cash is king: inflation, the time value of money, and the risk-free nature of cash.

Inflation. This is the rise in the prices of goods and services. In brief, your dollar may not be capable of buying as much tomorrow as it will today. Suppose you currently have $1,000. With an inflation rate as small as 3.5% per year, your $1,000 would only have the power to buy $503 in goods in 20 years.

What if you and your spouse agree that you will be paid $10,000 as part of your settlement? If you're paid at the time of the divorce, you'd get $10,000 in today's dollars. If you have to wait three years for your money, that $10,000 will be worth less. Consequently, if you are offered a divorce settlement payment to be collected in the future, you must account for inflation.

Time value of money. Quite simply, if your money is working for you today—by growing in a quality mutual fund, producing profits in your business, or even earning interest in a bank account—you are getting more value than if you have to wait until tomorrow to put that money to work. Why should you have to wait, and give your spouse the use of your money for months or years to come? Better to get your payments in cash at the time of the divorce than to miss the opportunity of using your money to best advantage.

Risk-free nature of cash. The value of a lump sum payment doesn't fluctuate and is not taxed upon receipt when it's paid as part of a divorce settlement. That may make it attractive enough for you to accept cash in exchange for getting less than the full amount to which you might otherwise be entitled.

You're Playing for Keeps

In the rush to reach a settlement, some divorcing couples seek the quickest way out—regardless of the future cost. When the dust settles, mistakes may surface that end up costing you greatly in money and time.

To manage money effectively, you need alternatives to help solve unexpected problems and to meet your goals. When you trade away long-term options for short-term needs, you're forcing yourself into a corner. You literally take future choices away from yourself. Accepting property without considering maintenance costs and your future lifestyle can strap you with debt for years to come. Agreeing to pay your spouse's bills just to hasten the divorce process may take care of some emotional needs, but it probably does not make sense financially. Nor should you calculate your finances based on a situation or arrangement that hasn't yet come to fruition or that may prove only temporary.

> **EXAMPLE:** Marla was divorcing and calculating her cost of living. During the divorce, Marla's mother provided free child care, and Marla assumed that would continue indefinitely. She did not stop to think of what would happen should her mother become unable to babysit. After Marla's divorce was final, her mother found a new part-time job and could no longer care for her grandson. Marla spent a great deal of time and money renegotiating her support payments to reflect her true cost of living.

When you are in the middle of a divorce, you may overlook obvious problems unless you apply long-term thinking to your divorce decisions. Anytime you feel rushed or pressured, either by your spouse or the legal calendar of your divorce proceedings, take a moment to step back and look at your divorce from a future-oriented perspective. That view should help you withstand the pressures and make more informed decisions.

Money Is Not Math

Anxieties and phobias about math might hamper your ability to make financial decisions when a marriage ends. Do not feel embarrassed if you are intimidated by columns of numbers and fine-print paperwork. We have purposely simplified the formulas in this book so that everyone can understand the financial information needed at divorce.

Don't worry about mastering math. You need only gather current and accurate information about your financial situation. Getting the right numbers to plug into the formulas is much more important than learning how to make complicated tax or other calculations.

Years ago, computer scientists coined the term "GIGO," which stood for "garbage in/garbage out." If they fed the computer the wrong data (garbage in), they would get the wrong answers (garbage out). Likewise, if you try to calculate your cost of living or the value of your property using outdated information or guesswork, you will end up with "garbage" in your final settlement. So gather the correct data. If you don't want to plug the numbers into the formulas yourself, you can hire a bookkeeper, an accountant, or a financial planner to work with you.

Emotional Divorce:
Managing the "Money Crazies"

Reduce Stress Whenever You Can 39

Safeguard Your Sanity 40

Watch Out for Sore Spots 42

Be Prepared for the Worst 42

Develop a Financially Focused Mental Attitude 44

Avoid the "All at Once" Syndrome 44

Manage the Ebb and Flow of Emotions 45

Don't Let Financial Tasks Overwhelm You 47

To make the best financial decisions during divorce, you must be alert to the emotional states that can sabotage your settlement negotiations.

Be on guard if you find yourself attempting to:
- get even
- get it over with, or
- get back together.

These three "gets" show the influence of strong emotions, which will hamper your ability to see both the upside and downside of your decisions. Anger, resentment, and running away are *reactions*. But in a divorce, you must not simply react; you need to be thinking carefully—and coolly—whenever you can.

Start by checking your attitude. Do any of the following sound like you?

"I'm going to get even no matter what it takes. You're going to pay for what you did to me. Just you wait. I'll see you in court."

You may feel that this attitude conveys strength, but the "get even" approach rarely leaves people satisfied, even when their divorce settlements or the court's orders are totally equal according to the numbers on paper. Bitter feelings linger long after the settlement, especially if you use poor financial judgment while motivated by revenge. The hostility of warring parents can have terrible consequences for their children, which may linger for years.

What's more, the attorneys' fees of vengeful spouses often soar. If you insist on going to court to get even, a trial will likely cost you at least three times as much as an out-of-court settlement. Besides that, you are unlikely to get what you want from the judge; even if you do, there is no guarantee that an ex-spouse will comply with court orders.

While you should certainly pursue your legal rights, the judicial system is no place to satisfy your emotional demands.

"Holding on to anger is like grasping a hot coal with the intent of throwing it at someone else. You are the one who gets burned." (Variously attributed)

"I don't care what happens. I just want to get it over with."

Stop right there. In a divorce, running away or becoming defenseless is the equivalent of playing the victim, and being victimized is perhaps the worst strategy you can employ when it comes to your finances. The decisions you make during your divorce will affect you (and your children) for the rest of your life. It's no time to choose the path of least resistance. Your financial survival depends on participating in each step of your divorce settlement. And if you take too many shortcuts, you could create unexpected money problems that won't take effect until later. In a year or two, your emotional life will be different. The financial agreements you make during divorce, however, affect you permanently.

"Maybe if I don't cause problems financially, we can work it out. I won't make waves. I'll just give in so we can get back together."

Certainly this attitude is understandable when you're hoping to reconcile or if both of you are committed to a fair and equitable settlement. But it's futile—and dangerous—to think you can save the relationship by surrendering your financial leverage. Avoiding conflict is often another expression of victimhood or passive aggressiveness, neither of which will set things right. Why should someone come back into a faltering relationship if you are agreeing to everything anyway? Meanwhile, if you and your partner do get back together, your relationship will be much healthier if you are on equal footing—financially and emotionally.

All this advice may make sense, but in reality, how do you deal with money when you are emotionally involved with your spouse?

That is perhaps the most important question to ask when you embark on a divorce, and it is central to the purpose of this book.

The primary concept can be stated simply: The better you manage your emotions, the better equipped you'll be to manage financial decisions during divorce. But applying this rather simple concept can be a very difficult task. The emotional toll of divorce is so high that some therapists and counselors use the label "crazy time" to describe it.

Ironically, you'll need to make calm, rational financial decisions that affect your future while you're under intense emotional pressure. Your two tasks—the logical and linear management of money and the release of emotional tension—pull you in opposite directions. Balancing between these poles is an insane proposition. No wonder you feel as if you have the "money crazies."

You will need to be cool and focused while you manage your finances during divorce, which means you need to limit the hours you spend wrestling with your emotions. When you take steps to understand the divorce process in emotional terms, you may avoid some of the common pitfalls that lead to victimization and unchecked anger. Forewarned is forearmed.

Depression During Divorce

Depression often begins quietly, but it can linger long after your divorce. The CDC has estimated that one in every 10 adults experiences depression in a given year. We are not aware of any figures for people undergoing divorce, but our experience suggests the percentages climb much higher for this group.

If left untreated, depression can be devastating to your physical as well as mental health. There are many possible symptoms, common ones being:

- anxiety
- no longer enjoying things that made you happy
- difficulty making decisions (an especially severe symptom in divorce)
- feelings of hopelessness or unworthiness, and
- suicidal thoughts.

Depression is a treatable condition. You should not hesitate to seek medical help if the emotional and financial considerations leave you feeling overwhelmed. Being responsible and efficient when you're at your worst may not seem fair, but remember that you *will* get beyond the pain of splitting up, and you'll do it faster if you can summon the strength now.

Reduce Stress Whenever You Can

No matter how eager you may be to get on with a divorce, it's probably going to be a struggle for you to achieve the financial-emotional balance necessary to see it through. That struggle means stress, perhaps more than you've ever experienced. Research indicates that only the death of a spouse creates more stress than divorce. Some counselors contend that when other factors are considered—separation anxiety, moving from a long-term residence, and loss of status, to name just a few—divorce is *the* most stressful life event you can ever face.

Money is a difficult topic—taboo—for most people; even happily married couples can experience stress when discussing it. If you are now handling money on your own, especially if your spouse took exclusive charge of the family finances during the marriage, the unknowns can seem overwhelming. Worries about financial survival, the divorce settlement, finding a job after staying home for years, and paying attorneys can be crushing, especially when you can no longer "talk things through" with your spouse.

Stress can manifest in dozens of ways. Sleeplessness, fatigue, loss of appetite, disruption of work patterns, lack of focus, and emotional explosions over frustrations, large or small, are all common symptoms.

It's critical that you do whatever is necessary to take care of yourself. If you have children, good self-care is even more important. Your frustrations are less likely to spill over onto the children if you're attending to your own needs.

If you have disposable income, schedule a massage, go out to dinner or a movie, or even arrange a weekend away. These quick fixes can take the edge off during a rough patch. But even if money is tight, there is no excuse for ignoring your stress. Practicing meditation or simply sitting under a tree or watching a sunset costs nothing, and the interval of quiet can help ease the strain on your bruised emotions.

Even if you don't feel like it, try to keep your eating patterns close to normal. Good nutrition is essential to good health, and it also is good medicine for tense nerves.

Regular exercise is one of the smartest and most efficient antidotes to divorce stress. It will combat fatigue and provide an essential outlet for physical and emotional tension.

These stress reduction techniques sound simple and can be applied to many life events, but make no mistake: They are crucial in your fight against the money crazies. The stress of divorce can last months and even years—you have to learn to manage it.

Safeguard Your Sanity

During a divorce, it is important to safeguard your emotional well-being. Spend more time with people who can give you positive reinforcement and less with those who may be critical of you or your divorce. Join a support group or see a therapist if necessary. In the long run, any money you spend on your mental health will be less than you'd lose if you got fired from your job or went on a shopping spree to deal with the money crazies.

Rely more on your voicemail and email. Not only can you screen nuisance calls or messages if your spouse is hostile, but you can let others know that you are unreachable when you need to be alone.

There will be days when you feel like talking and other days when it is the last thing you want to do. You may have just calmed down, and then, when you run into a friend and retell your story, emotions rise again—the catch in the throat, the anger, the frustration. That may be a healthy experience, but if you're feeling that it's too much, limit your personal conversations just as you would your phone calls.

Take control of your home environment. Clean a closet, paint a room, or move the furniture. There's an almost universal tendency during divorce to think about the past—obsessively and excessively. Most people find it is simply easier to put away the family pictures, the mementos, and the visual cues that can trigger feelings of loss and depression.

You are not only affected by what you see, but also by what you hear. Instead of sitting back and passively allowing a favorite old song to make you sad, reach over and snap off the radio. The same goes for those heart-tugging commercials on television featuring happy couples—change the channel or turn off the TV. You'll be amazed at how good it can feel to exert control over something when so much of your life seems to be out of control.

"No, We Can't Afford It Now"

Few parents enjoy denying their children toys, gifts, or other pleasures. During divorce, you may find yourself repeating the phrase, "No! We can't afford it now!" more times than you care to count. Disappointing your children can only add to your frustration. But while you may want to protect them from harsh realities, you must also be honest and help your family adjust to the new circumstances.

Make sure discussions about money are age appropriate, and make sure your children are mature enough to understand the difference between making hard choices (such as choosing to fund one after-school activity instead of two) and a real crisis (losing the family home to foreclosure). If your children are too young to understand the concept of money, you'll need an extra supply of patience to handle their confusion and mood swings.

Take heart in the fact that millions of children have survived divorce, and that many benefit from learning about how to live well under a tight budget and watching their parents manage a divorce courageously. And don't forget: the better you handle your divorce today, the more likely your children's future will be grounded in a positive and healthy lifestyle, instead of negative thinking.

We Need to Talk: Money & Kids After Divorce by Linda Leitz, CFPA (Bright Leitz Publishing, LLC, 2012), contains useful advice for broaching these tough subjects with your children.

Watch Out for Sore Spots

You probably already know which financial issues are likely to get you upset. All couples have sore spots—resentments about an unpaid bill, spending sprees, bounced checks, interfering in-laws, loans to relatives, or gifts to a lover.

If you know that you will be dealing with an issue that's been a particular long-term irritation, prepare yourself for it instead of merely reacting to the inevitable pressure. Get a good night's sleep before tackling such a problem, or postpone a confrontation until you feel strong enough to be at your best. Taking these basic steps can go a long way toward reducing the tension in your life. Remember, reducing stress means increasing your ability to make sound financial decisions.

Be Prepared for the Worst

Some emotional problems of divorce offer no easy solutions—when people are under stress, they tend to revert to their base, survival behaviors. This can lead to excessive drinking, shouting matches, and other bad scenes. If someone tends to be absent-minded, selfish, or controlling, those tendencies will be exaggerated when a marriage is ending.

During divorce, couples are facing the psychological task of separating, and consciously or unconsciously look for ways to break their connections to each other. During courtship, partners do every-thing possible to build the relationship, while in divorce, energy goes into destroying it. Amplified by stress, these destructive impulses create many of the horror stories of divorce. The sweet irrational gestures of romance are reversed by the equally ridiculous acts of "vengeance" during separation.

Even if you believe that you and your spouse are capable of proceeding with a "civilized" divorce, it helps to prepare for the worst-case scenarios. Remember, survival instincts can also stimulate awareness, which is something you'll need when it comes time to think about money.

Don't Let Threats Throw You

In the process of divorce, your soon-to-be ex-spouse may try to threaten you in some of the following ways.

"Unless you play this my way, you'll never get a dime."

Keep calm. Of course the person making such a threat wants to scare and intimidate you and to continue the power plays that have worked in the past. When you hear such statements, tell yourself that coercion won't work. The property will be divided fairly, and support will be awarded in accordance with schedules.

"I'll go to jail before I'll pay you a dime of support."

Don't panic. Know that support can be enforced through a wage assignment, which means that your check will come directly from your ex-spouse's employer. If your ex-spouse doesn't work for someone else and falls behind in support payments, there are a number of enforcement methods. Ultimately, failing to pay support can mean a jail term, but most people pay voluntarily before going to jail.

"I'll quit my job before I'll pay you that kind of money."

Try to get witnesses to this kind of comment. If you can show a court that your ex-spouse quit a job to avoid support obligations, the court will probably continue support at the same level, and your ex will have to find a way to comply.

"I will reconcile with you only if you put everything in my name."

Nice try. If you are going to reconcile, why do you need an agreement that is lopsidedly in favor of the person making such a demand? Be extremely suspicious of such statements, and don't sign anything without consulting your attorney.

If you are on the receiving end of a threat, do not be intimidated to the point of giving in. Ignore threats as best you can and continue to work for a reasonable resolution.

If you anticipate your spouse may cut off any connection to you—by wiping out the checking account, running off to another state, or ruining the family's credit—you may be able to short-circuit potential damage. Even if the worst-case scenarios never come to pass, you will have done yourself a favor by simply drawing up a plan of action.

Whatever happens, you must do your best not to take bizarre or vengeful behavior personally. This will help you to remain detached from the short-term drama of divorce, and therefore better able to concentrate on the long-term money questions that demand your attention.

Develop a Financially Focused Mental Attitude

To combat a devious spouse during divorce, you must develop a strong mental attitude and a solid legal strategy. Certainly you are entitled to feelings of outrage and betrayal. You can utter the cry, "How can you do this to me?!?" as much as you want to. In fact, that phrase is a common chorus in divorce cases. Yes, many things that happen in divorce *are* unfair.

But the fact remains that you will have to keep going and keep fighting for your financial life regardless of the injustices that may be perpetrated by your spouse or your spouse's attorney. You can stand up for yourself. If you stay focused on the future and the financial realities of your life, you will be in a much stronger position as your divorce progresses.

Avoid the "All at Once" Syndrome

Most likely, you will experience a wide range of feelings and moods: anger, hatred, elation, excitement, sadness, loss, depression, bitterness, rejection, loneliness, guilt, and hostility. Sometimes it will feel as though you're experiencing those feelings all at once; this phenomenon seems to be a hallmark of the divorce money crazies.

You may also feel that everything is coming apart all at once. Just as soon as you get the car fixed, the washer or the stove breaks down. One explanation is that you simply notice problems more because of your stressed state. Moreover, by the time most people separate, a great deal of their energy has been focused on the relationship—not on the normal chores that can keep a household running. You may not be in the mood to defrost the refrigerator or check the oil in the car as usual. That lack of maintenance can catch up with you as the machinery in your life begins breaking down.

The all-at-once syndrome in divorce can manifest itself in other ways as well. Some people decide that while they are changing a mate, they should change everything else in their lives as well. They try to lose weight, quit smoking, get a new job, and redecorate the house, all at the same time. This is never helpful. Give yourself time. Go slowly. You will have your hands full just getting through the divorce.

Manage the Ebb and Flow of Emotions

Keep in mind that emotions tend to be experienced in waves. One day you may feel fine; the next day, life is awful. Such fluctuations are common in divorce, and eventually the wave action subsides. Do not attempt to handle important money tasks on the bad days—wait until the storm passes.

Because the level of tension in a divorce is often compared to the stress that accompanies the death of a loved one, some psychologists claim the grief processes are also similar. Elisabeth Kubler-Ross's groundbreaking work on death and dying pinpointed five stages of grief: denial, anger, bargaining, letting go, and acceptance. These stages do not necessarily occur in this order. How you go through these stages is unique to you, but it is important that you experience them.

Marriage counselors note that people who avoid grieving by jumping into a new relationship too quickly are only prolonging the process. Those who do allow themselves to grieve for the marriages they had (or the marriages they wished they had) actually get through their divorces

more quickly—and carry less psychological baggage when the marriage is over. As a consequence, they are often able to manage their financial lives better, too.

While the grief processes are similar in death and divorce, some important differences do exist. Social attitudes have changed greatly, but widows and widowers still get more sympathy than those who have divorced. With divorce, too, the door to your relationship does not close with the same finality. You may still have contact with your former spouse for years and go through the grieving process many times as you make contact and separate.

Ambivalence—the love-hate relationship with your former spouse—is common when marriages end. Actually, the push and pull in and out of the relationship characterizes the separation process that began before the actual divorce. Commonly, one person in the relationship becomes discontented before the other, and this "initiator" is usually the one to ask for the divorce.

If your partner was the initiator, don't be surprised if he or she seems to have already moved on. In fact, the initiator has a head start on the psychological work needed to separate. Just don't let yourself lag behind financially, or get pushed into doing something you're not ready for.

The initiator may already be out the door and pressuring you to "get on with it"—to sign over the condo, to sell the house, to separate the silverware. Meanwhile, you may still be at the "starting gate," emotionally shocked and dazed, unable to finalize a process you have only begun.

If you initiated the divorce, realize that it is probably in your best financial and family interests to be sensitive to your partner's pace of accepting the separation. Your negotiations are likely to be more successful if you present your financial proposals to someone who has had a chance to assimilate what is happening in the relationship on a personal level.

Psychologists have found that the hardest part of the divorce process tends to be the time just before the actual physical separation. Tensions often mount dramatically until the real break occurs. Some couples separate and reconnect several times in an attempt to save the marriage.

Whatever is happening to you emotionally, continue taking steps to put your financial life in order. Should you and your partner ultimately reunite, your relationship will be stronger if you have not used money as a weapon against each other.

Mixed Emotions—Working With Attorneys

Controlling your emotions is important—particularly when it comes to working with an attorney on your divorce settlement. Remember, your attorney is not your therapist or confidante. As you proceed through this book and through your divorce settlement, be honest with yourself about whether you are being a model client (and one who will save attorneys' fees) or a client whom your attorney would like to avoid. The following questions will help you make that determination:

- Have I given my attorney accurate and complete information on time?
- Have I thought through the objectives of my divorce agreement?
- Do I waffle in making decisions or compromises?
- If I'm not clear on a legal issue, do I ask for clarification?
- If I have a disagreement with my attorney, am I willing to talk it through to resolution?
- Do I neglect to cover issues and later blame my attorney for forgetting certain items?
- On a continuum between being a pest and being unavailable, where am I?
- Am I paying my bills on time?

Don't Let Financial Tasks Overwhelm You

As you move through your divorce, it will be easy to become overwhelmed with the financial details. You may suffer from math anxiety or money phobias. Or perhaps you and your partner shared a complicated financial life, one that will take effort to untangle.

Whatever your feelings, you can get some relief by breaking down your tasks into small steps. Reward yourself for completing an item. Hire a math whiz to help you when necessary. If you are upset, do not hesitate to see a counselor or join a therapy group. That strategy makes much more sense, and is less costly, than using your attorney as a therapist or asking your friends to help you understand a situation that calls for expert analysis.

Anything you can do to build your self-esteem is important now. Divorce tends to make you feel worse about yourself, and it is easy to confuse money issues with your sense of self-worth. Besides the stress-management techniques already described, you may also find it helpful to keep a journal. Having a notebook handy to jot down feelings throughout the day helps dissipate tense moments. This record can serve as an "invisible calculator" to tally the emotional costs of financial decisions. Tracking feelings and reactions in this way can make it easier to reach those bottom-line decisions.

While it may seem as though the inner turmoil from your divorce will never end, it is a finite process that only takes time. For some people, the process lasts one or two years, while for others, recovery from a divorce may take longer. You are very likely to become a different person in that time, with different needs and attitudes. Keep that in mind as you manage the money crazies of your divorce—and make sure the financial choices you make today will work for you tomorrow.

Divorce—On a Spiritual Level

For some people, divorce stress is more than emotional. It becomes a religious or spiritual crisis. We cannot attempt to address such a crisis in this book. You can, however, contact local churches, synagogues, or clergy members for referrals to divorce ministries.

One note of caution: No matter how you feel about your divorce from the spiritual point of view, you cannot ignore its legal and financial consequences.

For example, one client insisted on working exclusively with attorneys and other professionals who shared her faith. While there is nothing inherently wrong with this, it didn't work in her case. She didn't judge the professionals she hired on their competence or abilities. In the end, she had to go through the costly process of finding a new attorney and accountant to help her complete the divorce.

For more information on selecting professionals, see Chapter 7.

The Hardest Part:
Is My Marriage Really Over?

Legally, a final decree from the court marks the termination of your married life.

motionally, it may be years before you feel complete peace of mind about the end of your marriage. But from a strictly financial standpoint, the marriage ends when you or your spouse begin to take unilateral actions without considering the effect on the other person.

It's that moment when one spouse stops acting like a trusted friend or partner and starts putting the other spouse in economic jeopardy. One spouse might empty a bank account, leaving the other with nothing to pay the bills or hire an attorney. Or a soon-to-be ex might run up joint debts unbeknownst to the other.

A great deal of financial damage can take place in the early stages of divorce. You must protect your interests whether or not you feel sure the marriage is over.

Couples often go through several phases of splitting up and getting back together before the final break. While you are going through this push and pull of separation, you may find yourself walking a tightrope between conflicting demands and emotions—watching out for your personal property without antagonizing your spouse, or taking care of your individual needs while not letting your spouse off the hook for joint responsibilities.

Eventually, you will recognize when the marriage has reached a point of no return. Even then, you may hold out hope for a reconciliation. There's nothing wrong with hope—as long as you continue taking care of business. That means knowing precisely where your financial interests and risks lie, and what to do about them. The following stories illuminate the dilemma you are in at this stage of divorce—and the danger of ignoring its consequences.

A Happy Ending	An Unhappy Ending
When you're in the midst of a crumbling relationship, it may be hard to believe that resolving money problems can sometimes help clear up other troubling issues in a marriage. Alice came in for a financial consultation and said she was considering a divorce. Although she had always worked outside the home, her husband managed the finances and kept her in the dark. Unwilling to tolerate the situation any longer, Alice began educating herself and learning about the couple's cash flow, taxes, and investments. She also consulted an attorney to assess her legal rights. Secure with her knowledge, she confronted her husband and demanded a greater financial role in the marriage. They argued bitterly and almost divorced at several points along the way. Through patience and counseling, however, they resolved their differences and kept the marriage intact.	The infamous Palm Beach divorce of Roxanne and Peter Pulitzer in the 1980s demonstrates what can happen if both parties are not equally committed to a reconciliation. Roxanne claimed that in trying to save her marriage, she lost her financial power and leverage as a mother. Upon receiving divorce papers concerning hearings and proceedings, she said she followed her husband's instructions to throw them in the garbage because he did not intend to end the marriage. Doing just that placed Roxanne at a great disadvantage when her husband did, in fact, continue the divorce proceedings. "I lost the case before it even started," she said once it was over. Instead of focusing on the reality of her situation, she put her energy into reconciliation—and lost.

No one can really predict which marriages will come apart or stay together, because every couple, and the dynamic between them, is unique. If you are still questioning whether you have a viable marriage or not, the following options are available to you.

Counseling. Now is the time to find out whether your marriage is worth saving. Some people cannot—and perhaps should not—let go until they have explored every avenue of reconciliation. But if your spouse refuses to cooperate, you may have little choice but to accept the inevitable and move on with the divorce.

Temporary separation. For some couples, time apart becomes a time for healing. At the same time, it's important to keep your financial house in order. Make a list of your joint obligations and expenses. Through mediation or counseling, you and your spouse may be able to reach agreements about dealing with the practical demands of life while you work on your emotional relationship.

After you give yourself some time to consider this information and your options, you must confront the main decision presented in this chapter: Is your marriage really over? This question is not meant to rush you toward divorce, but rather to help you clarify your position. You can afford to leave some questions open in life—but this isn't one of them. Once you've made this decision, it will be significantly easier for you to tackle the hard tasks ahead. But if you're unclear on this, all decisions that follow become much harder to make.

Okay—take a deep breath. Write out your answer below, or speak it out loud to someone you trust. Either way, the time has come to make the decision that only you can.

Is my marriage really over?

Whatever Stage You're in, Don't Ignore Legal or Financial Realities

Whether you attempt counseling, a trial separation, or some other method of resolving conflicts in your marriage, do not ignore your financial responsibilities and potential liabilities. Just as you would consider legal notices or documents important at any other time in your life, so, too, must you recognize the serious implications of any legal actions in the early stages of divorce—even if your spouse tries to convince you otherwise. Do not ignore court orders or papers. And, if you have any reason to suspect your spouse is moving assets or taking funds from your joint accounts, you should immediately read Chapter 6 for information on how to protect your interests.

The Separation: What Happens When One Spouse Moves Out?

The Separation Date..57

   Debts and Credit..61

Building a Better Credit Report and Credit Score.................................65

   Retirement Plans and Pension Benefits..66

   Income and Income Taxes..67

   Investments and Business Assets..67

   Alimony...67

Social Security Benefits After Divorce...68

Questions to Ask an Attorney..69

"I told him I wanted him out by the end of the month."

"I can't stand it another minute—I'm getting out of here."

"It's over and it would be easier on all of us if you would just go."

Don't be surprised if you find yourself saying—or hearing—any of these statements before you separate. Feelings run high, and emotions seem to dictate the schedule of events in your life. Old arguments and long-held resentments commonly surface in this most difficult period, just before or during the actual physical separation. Power plays may become more troublesome as one of you may try to force the other out of your home.

No matter how you feel about your spouse, neither of you necessarily has to be the one to move—unless you or your children would be in physical danger if your spouse stayed. In such a case, you can get a restraining order from the court, prohibiting your spouse from remaining in the family home. In many states, you may be able to find a kit you can use to do this yourself, without a lawyer's assistance. In other states, however, court clerks must assist you in completing papers to apply for a restraining order. Many women's legal clinics also provide help—including assistance in applying for restraining orders—for victims of domestic violence. (See the appendix.)

Even when violence is not an issue in your marriage, practical reasons might dictate that one spouse move out rather than the other. For example, many divorcing couples who live under the same roof find making agreements difficult. If one of the spouses moves out, the divorce often proceeds more smoothly and with less pain.

In the midst of ending a marriage, you may not want to look at your situation in the cold, harsh light of financial reality. But that denial could ultimately cost you thousands of dollars. Before you insist on leaving or having your spouse move out, take a little time to consider the financial consequences of the separation date.

Whether you view this time period as a trial separation (while you continue to work on your marriage) or a separation for real (as in the beginning of divorce), the following information is important to you.

Protect Your Rights to Custody of Your Children If You Plan to Move Out

If you have children, the primary caretaker might want to stay in the family home while the other spouse rents a place, housesits for someone else, or moves in with a friend. However, by moving out of the family home without your children, you could jeopardize your chances of getting physical custody of them. This is because few judges like to change the status quo when it comes to kids. So before you move out, be sure to make arrangements with your spouse as to the conditions under which you will each see the children. Define each parent's schedule with the children, including who will drive them to and from school, softball practice, ballet lessons, and worship services.

If you have any question or doubt about your rights as a parent or feel that you are being pushed away from your children by moving out of the house, or if you're considering moving out of state, be sure to consult an attorney who can help you establish custody arrangements before you move.

If either parent moves out of state with the children for six months or longer before filing for divorce, your divorce case could end up being split between two states, with the monetary aspects determined in one state and the custody aspects determined in another. Be sure to consult an attorney and think carefully before making such a move or agreeing to a move by your spouse.

The Separation Date

Determining who will move out of the family home may be the first, really difficult question you'll have to tackle. The next question—when—is also extremely important. Depending on your state's laws, the

date you and your spouse formally separate can affect your credit, health insurance, pension benefits, and other assets.

Before the date of separation, you and your spouse are still married and subject to the same laws you've been living under since you first wed. After that date, however, you enter a gray area of financial and legal reality that won't be fully clarified until the divorce is final. You'll have no control over the actions of your spouse during this time, yet you are still connected in some ways—and any connection might result in your being held responsible for your spouse's bills, even bills that were racked up after your legal separation.

In addition, the value of a retirement plan, dividends and other assets can be thousands of dollars more or less at settlement time, depending on whether your state defines marital property from the date of separation or the date of divorce.

The definition of the separation date varies between states, but generally speaking, it may be defined in the following ways:

- the date on which one spouse formally announces that he/she intends to file for divorce
- the date on which the couple begins to conduct life differently (such as when one spouse starts sleeping in a guest bedroom or occupies another section of the house)
- the date on which divorce papers are filed, or
- the date on which one spouse physically relocates from a shared dwelling.

Note that in all, but the last case, couples may continue living together after the separation date. While this can be to your advantage if neither can afford to move right away, the advantage can be tempered by the loss of privacy, the increasing hostilities as the divorce battle heats up, and the effect any parental conflicts may have on the children.

The date that's determined to be the date of separation may have financial advantages for one spouse or the other. If you are going to be responsible for the credit card charges your spouse incurred during the marriage, for instance, you'd want the date of separation to precede your spouse's shopping spree.

A competent divorce attorney can help to explain the laws of your state and their implications for your finances. This expense can be far outweighed by the benefits of being prepared for possible credit scares as you disentangle your finances from those of your spouse.

In six months to a year, you will have the advantage of hindsight and a different perspective, but with some planning and research now, you can anticipate the most critical effects separation will have on your financial outlook.

Separation Agreements in Writing

Even if you are clear about your state's definition of the separation date, you may need a formal, written separation agreement to ensure that you and your spouse are legally in accord concerning your divorce procedure, how the bills will be paid, who moves out, and how temporary child custody or spousal support will be structured. Relying only an oral agreement is never a good idea. Unwilling spouses sometimes claim that verbal separation agreements never happened or were not legally binding. If your spouse relocates, he or she could still list your shared home as a primary address or attempt to claim that the separation was an experiment, or a "time-out," to think things through.

If you do not trust your spouse to act in good faith, make certain that you understand the details of your written separation agreement and that it supports your version of events. Both parties must enter the agreement voluntarily (never under duress) and understand the consequences of signing it. This is another reason why it often makes sense to seek out legal advice before either of you files for divorce. Remember: It's you, not the court, who should be the steward for your financial well-being.

If You Must Move

Although you may not like the idea, it's possible that you will have to move from your home. You can make life easier by preparing for the pressures you will face. Moving is stressful at any time; divorce only compounds the stress. Not only are you losing a mate, but you're letting go of a home and all that it has meant to you.

Feelings of separation and abandonment can be exacerbated by anxiety about where you will live next. To cope with your emotional reactions, look at your situation in practical terms and separate real concerns from unfounded fears. Here are a few things to remember:

- Movers often charge a premium for their hectic summer season, while offering lower prices between late September and April. The first and last days of the month are also busy times for movers, so you may be able to save money by moving midmonth or midweek. The American Movers Conference suggests that you question a mover carefully about rates, liability, pickup and delivery, and claims protection. If you have time, get estimates from at least three movers so you can determine if their prices and terms are fair. To move locally, you'll probably be charged an hourly rate or a flat fee. Long-distance movers usually charge by weight and mileage.

- In addition to the cost of the move itself, you'll need money to hook up utilities, phones, and other services. You'll have other "new house" costs like buying kitchen staples, cleaning products, and other basics. Count on eating out more than usual, too, before you restock your kitchen or unpack your dishes and pots and pans.

- If you move your children out of state without your spouse's consent, a judge may order you to return the children to the state immediately and could ultimately deny you custody. Judges do not look favorably on parents who move their children away from the other parent, school, friends, and community ties. Before making such a move, consult an attorney to make sure you do not jeopardize your rights.

- If you move out of the house during the divorce, it can still be considered your principal residence for tax purposes, which affects capital gains taxes. This happens when ownership of the house is transferred to one spouse in a divorce and that spouse moves out. In other words, if you have to move out of the house during the divorce, you can still claim it as your principal residence for tax purposes, as long as you and your spouse have a written agreement about your intention that the house will remain your primary residence.

D-Days
During your divorce, you may see or hear these abbreviations used to refer to certain milestones: DOM........Date of marriage DOS..........Date of separation DOD.........Date of divorce

Debts and Credit

Bill collectors do not suddenly stop sending their statements just because you are going through a divorce. When you enter the limbo of separation, pay particular attention to its impact on your debts and credit standing.

Debts

In general, both spouses are responsible for paying debts generated during the marriage. Once the divorce is final, neither is responsible for the debts created by the other. Debts incurred during separation, however, can be tricky. The general rule is that debts incurred after the separation date but before the divorce is final are the responsibility of the spouse who incurred them and must be paid by that person.

One exception is for "family necessities." For example, if, during the period of separation, one spouse incurs a debt for food, clothing, shelter, or medical care, the other spouse may be obligated to pay a portion of that bill. Children's expenses, too, usually fall into this category. The law views a spousal support or child support obligation as more important than one to a creditor.

The general rule—that debts generated during separation must be paid by the person who incurred them—does not always protect you, however. If your partner defaults or simply refuses to pay, the creditors no doubt will come after you for payment. Because you were still married when the debts were created, the creditor will assert a right to

collect from you. Your only remedy may be to try to get reimbursed by your ex-spouse.

> **EXAMPLE:** Janet's husband, Jim, had dental work done after he moved out of their apartment, but he never paid the bill. Janet did not know anything about his dental work, and Jim left town as soon as the divorce was final. The dentist turned the bill over to a collection agency, which then went after Janet for the money. Eventually she settled the bill, spending lots of her time, energy, and money to resolve the problem.

Credit

Remember the warning given in Chapter 2: When you're financially connected to someone else, you're at risk. You are responsible for any use of a credit or charge card by either cardholder on a joint account.

Most of the time it is best to close joint credit accounts, because you cannot control your spouse's use of them. But before angrily closing all joint credit accounts, check first to see how your individual credit will be affected by such an action. If you do not have credit in your own name, be sure to establish it before you separate.

There may be certain circumstances in which you'll need joint credit. For instance, if you'll both be buying maintenance items for your house, rental property, or children, it may be convenient to have a joint checking account to which both you and your spouse contribute. You can make a formal written agreement as to how the account will be used and even require that all checks written on the account must have both spouses' signatures. Keep in mind, however, that even though you may need to retain joint accounts, you should not consider starting any new joint financial ventures.

RESOURCE

Extensive information on debts, creditors, establishing credit in your own name, and other similar issues can be found in *Solve Your Money Troubles: Debt, Credit, and Bankruptcy* and *Credit Repair*, both by Margaret Reiter and Robin Leonard, (Nolo). You can also find a variety of informative articles on this subject on Nolo's website at www.nolo.com.

Check Your Credit File

During a separation, it's a good idea to get a copy of your credit report. You're entitled to one free credit report per year. You can begin by visiting the Federal Trade Commission's article on credit reports (www.consumer.ftc.gov/articles/0155-free-credit-reports). From there, you can link to the Annual Credit Report Service, which is a site maintained by the three nationwide credit agencies, and submit an order for your report. You can order a copy of your report online or by phone or mail.

Annual Credit Report Service

P.O. Box 105283

Atlanta, GA 30348-5283

877-322-8228

www.annualcreditreport.com

You must provide your name, address, Social Security number, and date of birth when you order. You also may be required to provide information that only you would know, such as the amount of your monthly mortgage payment.

If you want additional copies of your credit report within a year, or you need to correct information in your credit report, you'll probably need to contact a credit bureau directly. Contact any of the following:

Equifax

P.O. Box 740241

Atlanta, GA 30374

800-685-1111

www.equifax.com

Experian National Consumer Assistance Center

P.O. Box 2002

Allen, TX 75013

888-397-3742

www.experian.com

TransUnion

2 Baldwin Place

P.O. Box 1000

Chester, PA 19022

800-888-4213

www.transunion.com

Check Your Credit File (cont'd)

If you want to get a combined report that shows what information each of the reporting companies has on you, you can go to any of their websites or to www.myfico.com. You can get combined reports and credit reports for a fee.

Your credit report will provide you with valuable information about your outstanding debts. In addition, you'll have the chance to make corrections should there be any errors in your report. The credit bureau will enclose information with your report on how to dispute incorrect information in your credit file, or you can find that information on the website.

Debt Collection After Divorce

It's important to remember that while divorce means the end of your legal relationship with your spouse, creditors may continue to see you and your ex as joint debtors. In community property states, both spouses may be liable for a credit debt, even when only one of them signed the credit agreement. The rules vary from state to state, but it's best to be cautious, and assume that any expense incurred for a joint purpose may result in a collection attempt against you.

If you carry debt into your separation, you should be aggressive in monitoring your credit status. Freeze all joint credit card accounts and home equity lines of credit (HELOCs) with carried balances to make sure the bleeding stops. You should both call your creditors to deactivate accounts and write letters confirming the dates and times you called, who you spoke with, and the details of what each of you said. For more on handling credit card debt, see Chapter 6.

Building a Better Credit Report and Credit Score

A good credit report and a high credit score can go a long way toward making your financial life easier after divorce. A credit *report* is a file that shows potential creditors what credit accounts you have open, your bill-paying habits, and other information affecting your creditworthiness, such as whether you've declared bankruptcy or have been sued or arrested. Credit-reporting companies collect this information and sell it to creditors, insurers, employers, landlords, and others who may wish to do business with you.

A credit *score* is a number that encapsulates information about you and your credit experiences. Using statistical formulas, creditors compare your information with the credit performance of customers with similar profiles. One commonly used scoring system, known as FICO, scores using a range from 300 to 850. The median FICO score is 723, and a score below 600 suggests a particularly risky borrower.

Having a good credit report and high credit score means it will be easier for you to get credit as well as to benefit from, for instance, loans with lower interest rates and smaller monthly payments.

You're entitled to free copy of your credit report once every 12 months. (See "Check Your Credit File," above.) You can get your credit score from the three nationwide credit-reporting agencies, but you will need to pay a fee for it. Other companies offer credit scores for sale alone or as part of a package of products.

Although consumers often believe—and the media suggests—there's just one score per consumer, actually many different credit scores may exist for any individual. For instance, a car dealer may use a score designed for auto lending while others may exist for mortgage lending and insurance purposes as well as for specific types of lenders, such as credit unions or traditional banks.

If you need to improve your credit score, paying your bills on time is the most obvious way. But also important are the amount of your debt and your credit utilization rate—that is, how much of your credit limit you're using up by making charges on your credit cards.

If you get a copy of your credit report—and you *should*—you need to check it carefully. If you find *inaccurate negative* information, be sure to tell the credit reporting company in writing what you believe is false, and submit documents that support your position. If you prevail, the credit agency must correct your report and send copies of the revised report to you and to anyone who's requested the report during the past six months.

If you find *accurate negative* information, such as a bankruptcy or a foreclosure, understand that only the passage of time can ensure its removal. A credit agency can report accurate negative information for several years and bankruptcy information for ten years.

Retirement Plans and Pension Benefits

In most situations, the marital portion of your retirement plan is defined by the date of your divorce. This means you'll probably be entitled to any growth in the account that occurred during the entire marriage, even if you separated months earlier.

Many companies provide an accounting of retirement benefits only on an annual basis. The amount in a pension fund, a profit-sharing plan, or another type of benefit program could differ significantly from one part of a year to the next. By the time your divorce becomes final, a year or two could have passed since your separation, and the retirement fund may have grown by thousands of dollars.

To safeguard your interest, get a copy of the benefit brochure issued by your own or your spouse's employer if possible. It's also helpful to have copies of the actual retirement plan itself or a summary of it. (Chapter 14 covers retirement plans in detail. It includes a letter you can send to request information and give notice of your interest in your spouse's retirement benefit plans.) Because so many factors affect your retirement benefits, it is best to consult an attorney, a financial adviser, a pension plan expert, or an actuary who can analyze your particular situation. (See Chapter 7 for more information on seeking advice from a professional.)

Income and Income Taxes

Under the laws of many states, the income that you earn after the date of separation is yours and yours alone. Therefore, you are solely responsible for the taxes due on this income. Also, the date of your separation may influence your decision to file your income taxes jointly or separately. (See Chapter 11.)

Investments and Business Assets

Many states value assets at the date the divorce is final, not the date of separation. This means that if one spouse moves out of the house and the assets appreciate between that date and the date the divorce becomes final, the spouses will share in the appreciation. The valuation of investments and business assets should be double-checked, however, as the rules do vary from state to state.

> **EXAMPLE:** Having assets valued from the date of divorce instead of the date of separation cost Harry a bundle. On the date of separation, his extensive portfolio was valued at $5 million. By the date of the divorce, his holdings had almost doubled because of the stock market boom during the period between separation and divorce. So instead of having to split $5 million with his soon-to-be ex-wife, Celeste, he had to share the full $10 million.

Alimony

Be aware that your separation date can have a significant effect on alimony you pay or receive. In many states, a nonworking, dependent, or lower-wage-earning spouse is presumed to be in need of alimony for a longer period of time after the end of a long-term marriage (usually one lasting at least ten years from the date of marriage to the date of separation). If you're a few months shy of the ten-year presumptive period for alimony (or the time length mandated by your state), you may want to stay around to avoid jeopardizing those future payments. Read more about alimony in Chapter 18.

Social Security Benefits After Divorce

If you were married for at least ten years and are at least 62 years old, you are entitled to Social Security benefits based on your ex-spouse's contributions. These benefits become available after you have been divorced for at least two years. You are entitled to these benefits regardless of whether your ex-spouse claims benefits. Your receipt of benefits will not reduce the benefits payable to your former spouse (or any other former spouse of your former spouse).

A few additional facts:

- Benefits are based on your ex-spouse's total contributions, not just those made during your marriage.
- If you were married (and divorced) twice and both marriages lasted more than ten years, you can claim benefits through the ex-spouse with the larger Social Security account.
- If you remarry before you reach the age of 60 and stay married to your second spouse, you cannot claim benefits based on your first spouse's account—but you can get benefits on the record of your new spouse.
- If you receive disabled, divorced, widow's, or widower's benefits, your benefits will continue if you remarry when you are age 50 or older.

CAUTION

Divorce can alter your Social Security benefits and options. For instance, a surviving divorced spouse who plans to remarry before age 60 might consider delaying the wedding until turning 60, to avoid the remarriage penalty. Also, delaying a divorce until the marriage has lasted ten years can mean the difference between getting Social Security benefits and missing out. Check out the Social Security website or contact the Social Security office to learn more.

Because each worker's Social Security file is confidential, you may not be able to find out the status of your spouse's benefits. You can try to get this information by asking your spouse or by contacting the local or national office of the Social Security Administration (SSA). Call

800-772-1213 to get a Request for Earnings and Benefits Estimate Statement (Form SSA-7004). You can also visit the SSA website at www. ssa.gov. The site includes downloadable copies of many publications and forms, including Form SSA-7004. Once you have the form, complete it and send it back to the SSA.

If these methods fail, you will probably need a lawyer's help to get information about your spouse's benefits. (See Chapter 12 for information on working with lawyers. For more on different types of retirement benefits, see Chapter 14.)

Before You Move

Before you move out of the family home, you should make copies of all documents that may have any relevance to your financial settlement. When in doubt, copy the document first, and evaluate it later. Refer to the Financial Facts Checklist in Chapter 8 for the most important documentation. Even if you don't have your spouse's cooperation in this project, it is important that you get the copies. If you must resort to the legal process of discovery (see Chapter 12), your legal bill may run sky high—and there is no guarantee that you will get all the information you may ultimately need.

You should also do an inventory of all shared or individual property. Use photos or video to document the condition of property at stake in your divorce. (For more information on dealing with property, see Chapter 6.)

Questions to Ask an Attorney

As you can see, you'll need to assess your legal and financial positions carefully before separating. Because the legal definition of the separation date varies from state to state, we cannot explain the nuances of your state's law. But below, we provide you with a list of important questions for you to ask an attorney or research at your local law library.

1. Who should move out of the house? Can I be forced to move?
2. In our state, what defines the date of separation? Is it when my spouse:
   - moves out?
   - files for divorce?
   - declares the marriage is over?
3. What other legal or financial ramifications of the date of separation should I know about?
4. Which debts—incurred after the date of separation but before the divorce becomes final—is each spouse responsible for?
5. What date is used—separation date, date of final negotiations, or date closest to the final divorce—to calculate the value of assets?
6. Can the court award alimony to a spouse for an indefinite period? If not, what end date is the court likely to apply to spousal support? Does the court consider the length of the marriage in awarding spousal support?
7. What is the average length of an alimony award in our state?
8. Who should make payments on joint debts during the divorce?
9. Do I have the right to reimbursement for payments I make on joint obligations during the divorce?
10. Can I get divorced before our property is fully divided?

Closing the Books:
What Do We Do With Joint Property?

Joint Account Checklist..72

    Credit Cards..72

    Equity Credit Lines...75

    Joint Checking, Savings, and Other Deposit Accounts 76

    Joint Title to Property 77

    Joint Investments and Other Holdings 79

    Safe-Deposit Boxes... 80

    Shared Property and Special Collections.................................81

Questions to Ask an Attorney...83

Your spouse may not want to ruin you—but his or her actions can still leave you at risk.

When you and your mate are in the midst of splitting up, you come face to face with a hard fact we've stated before: When you are connected to another person financially, you are at risk. Does that mean you need to rush out in a panic and close all your joint accounts? Not necessarily—it is never wise to act in panic, and very few spouses deliberately destroy each other financially. But even if your spouse doesn't purposely ruin your financial life, you can end up vulnerable as a result of your spouse's actions.

One way to gauge how much financial damage your spouse could inflict is to measure the level of hostility between you. For example, if you're no longer on speaking terms—or you speak only about the children—you will probably want to act quickly to separate your financial lives.

Even if your divorce is "friendly," don't count on that goodwill to determine what you should do about joint accounts. Look at your situation objectively. Apply your own common sense and instincts. You should be moving toward eliminating the financial obligations you share with the person you are divorcing. Because income is often reduced—and expenses increased—by a divorce, it may not be possible to cut all connections. Nevertheless, you can still reduce your risks.

Joint Account Checklist

The material that follows should alert you to common trouble spots.

Credit Cards

Before closing joint credit card accounts, be sure you have established credit in your own name, based on your own income. A department

store, for instance, may open an individual account for you when you close a joint one. You may be tempted to include your ex-spouse's income when applying for separate credit. Don't do it. Because you won't be able to use your ex's income to repay your bills, it would border on fraud.

If you don't qualify for credit based on your income, you can apply for a secured credit card. Many banks will give you a credit card if you secure it by opening a savings account. Your credit limit equals a percentage of the amount you deposit into your account. Depending on the bank, you'll be required to deposit as little as a few hundred dollars or as much as a few thousand. Many secured credit cards have a conversion option. This option lets you convert the secured card into a regular credit card (one not tied to a savings account) after a certain time period, if you use the card responsibly.

To find the names of banks issuing secured credit cards, call local banks or check the Internet. Websites that list and evaluate available credit cards are www.bankrate.com and, for more sophisticated financial folks, www.cardweb.com. Whatever you do, don't call a "900" or even an "800" number service that claims "instant credit—no questions asked."

Once you have established your own credit, write simple letters to close any jointly held credit cards or revolving loan accounts. Even if a credit card account hasn't been used for years, close it to prevent your ex-spouse from using it in the future without your authorization. Such letters may not fully protect you from obligation for your spouse's later-incurred debts, but they put the burden on the creditors to take action.

Your credit reports should give you all the information you need to complete the letter suggested below. (See Chapter 5 for information on how to get a copy of your credit report.) Call the lender (a toll-free number is usually on your statement) to confirm the address. Then fill in the creditor's name and address, your account number, your Social Security number, and your signature. Send all letters to creditors by certified mail, return receipt requested, and keep copies for your files.

Sample Letter for Closing Joint Accounts

[Date]

To:

Re: Account number _____

Card member _____ [your name]

Social Security number _____

Dear_____:

This letter is to inform your company that the above-referenced account is to be closed, effective immediately. At this time, my (husband/wife), [name of spouse], and I are seeking a divorce.

I am requesting a "hard close" of the account so that neither party may incur new charges. If [spouse's name] is the primary signer on the account, s/he does not have my permission to reestablish the card at any time using my name and/or credit.

If you are unable to close this account with only my authorization, you are instructed to terminate my relationship with this credit card account immediately, and any charges made to the account after the date of this letter will be the responsibility of [spouse].

If the account has an outstanding balance, you may keep the account open for billing purposes only, but I request that you keep it inactive so that neither party can incur new charges. Please advise me immediately of any outstanding charges on the account.

Please notify the three major credit bureaus that this account has been "closed by consumer" and kindly send me written confirmation that you have done so.

Thank you for your assistance.

Sincerely,

[Your name]

Most of the time, you will have little trouble terminating joint accounts. On occasion, however, a creditor may demand that you pay off the balance first. If you can afford to pay the bill, do so. That's better than leaving the account open and allowing your spouse to run up more debts.

If you can't pay off the entire balance, you can still ask that the account be made inactive while you make payments. This strategy prohibits further charges from being made on the account. Once the balance is paid, the creditor should close the account completely. Most customer service representatives will take care of your request. If you can't make satisfactory arrangements through a customer service representative, however, ask to speak to a supervisor, and continue up the chain of command. Explain that you are going through a divorce and that you are closing out your joint accounts. Remember to put all communications with credit card companies in writing, keep copies, and document your payments.

During your separation, if your spouse uses a joint credit card account (before you get a chance to close it) to charge basic items such as food or clothes, you might be responsible for the payments. (See Chapter 5.)

Equity Credit Lines

A frequently overlooked area of joint financial liability involves equity lines of credit. An equity credit line is an open-ended loan made by a lending institution—usually a bank, savings and loan, or thrift company—against the equity in your home. The lender gets a secured interest in your home, and if you don't repay the loan, the lending institution can force the sale of your house.

Some equity credit lines supply a checkbook that can be used just like a joint checking account. Whichever spouse has the checkbook has access to the money.

If you and your spouse have an equity credit line, pay a visit to your lender. Take a letter with you requesting that the account be closed or frozen because of your divorce. If you have a checkbook, be sure to turn it in or destroy it, and request that no additional checks be issued. By leaving an equity line of credit open during the divorce, you risk losing your home.

Similar to an equity line of credit is a margin account, offered by stock brokerage firms. You may not want to close a margin account right away, because opportunities to profit from trading may arise during the course of the divorce. Instead, you and your spouse should make agreements about what kind of activity is allowed, including an agreement that neither spouse may take any funds out of the account without either the written consent of the other spouse, or a court order. If one spouse normally does the trading, you can authorize that spouse to continue trading under specified circumstances. Whatever you agree, convey it in a letter to the brokerage firm. When you're closer to finalizing your divorce, you can divide the asset and close the account completely.

> CAUTION
>
> **Beware of fraudulent loans.** If your spouse claims to have borrowed money during your marriage from friends or relatives, proceed with caution. It may be a scheme to snatch joint assets. Here's how it works: One spouse falsely claims to have borrowed money during the marriage, unbeknownst to the other spouse, from a close friend or relative. The false loan is then paid back from joint assets. The friend or relative who receives the money then returns it surreptitiously once the divorce is final, and voilà! What was once a joint asset now belongs only to one spouse. The way to prevent this is to review records showing the flow of money in and out, and check income tax returns to determine whether interest was paid on purported loans to relatives.

Joint Checking, Savings, and Other Deposit Accounts

As with joint credit cards, you may not want to close joint checking or saving accounts unless you have an account in your own name, or will open one immediately. If you don't close a joint account, however, you run the risk that your spouse could come in and empty it without your knowledge.

You have a few options:

- Ask the bank to freeze the account and not allow either you or your spouse to move money in or out without both of your signatures.

- Take out half the money in the account and place it into your individual account.
- Close the account and place all the money into a dual-signature account. You'll probably need an agreement with your spouse to use this option.

CAUTION

If your spouse takes all the money from your joint accounts, your settlement—or a court—will probably provide for your reimbursement. That repayment, however, could be months or years away. Until then, you have to deal with losing your share of a joint account. If you don't have a personal savings or checking account, you may have to borrow money to cover the shortfall.

If You Are Broke

If you're out of money, you're probably safe in withdrawing half of the cash in joint accounts to pay for necessities. But, generally, you'll be required to disclose the liquidation and document what you used it for, and you may have to pay it back. (If you have any doubts, ask an attorney.) Of course, there are also the old standby methods of holding a garage sale or calling Mom and Dad. You could also ask for temporary alimony. (See Chapter 9.)

Joint Title to Property

How you hold title to property can be important. There are two ways that are the most common for married people to hold title.

Joint Tenancy

Many couples hold property, especially real property, in joint tenancy. Joint tenancy's major advantage is that when one joint owner dies, the remaining owner or owners automatically inherit the deceased's portion of the property without having to go through the long, expensive probate

process. If you don't want your spouse to automatically inherit your share, then change your joint tenancy ownership to tenancy in common.

Tenancy in Common

With tenancy in common, married couples may own equal shares, but upon the death of one spouse, that spouse's share goes to whomever is named in a will or another estate planning document. If you die without a will or another estate plan, your property passes under the laws of your state. Usually the property is divided between the spouse and/ or children, if any; if there is no spouse and no children, it goes to the parents or siblings.

The best way to convert property from joint tenancy to tenancy in common is to secure your spouse's agreement and then change the title together. For example, a couple could transfer their house "from Herbert and Donna Walker as joint tenants to Herbert and Donna Walker as tenants in common." You can obtain a blank deed from a title insurance or real estate office. In California, homeowners can use *Deeds for California Real Estate*, by Mary Randolph (Nolo).

If you cannot find a blank deed, you could simply draft a statement saying, "Herbert and Donna Walker hereby transfer their real property located at 4555 Ellison Boulevard, Fargo, North Dakota, from Herbert and Donna Walker as joint tenants to Herbert and Donna Walker as tenants in common." Be sure you both sign the statement before a notary public. Then record the deed or statement in the office where your original house deed is recorded.

If your spouse will not consent to the change, you can complete the deed or statement yourself. Write, "Donna Walker hereby transfers her interest in the real property located at 4555 Ellison Boulevard, Fargo, North Dakota, from Donna Walker joint tenant to Donna Walker tenant in common." Any one person can unilaterally terminate his or her interest in a joint tenancy. Again, sign it before a notary and record it in the office where your original house deed is recorded.

CAUTION

Whether the change in property ownership is unilateral or jointly agreed upon, make sure you update your estate plan after any changes. In addition, be sure your title change follows all of your state's laws regarding notice to your spouse. And if you're in the midst of a divorce, make certain you're not violating any order restraining you from transferring title.

CAUTION

By closing accounts or serving your spouse with a restraining order, recognize that you may be upping the ante by increasing the degree of conflict in your divorce. You (and any professional advisers you use) must assess the risks and the consequences, because sooner or later, you will meet your spouse at the negotiating table.

RESOURCE

For more details on holding title, see Chapter 12.

Joint Investments and Other Holdings

Make a list of any investments or other joint assets you can think of. Immediately call your broker, your discount brokerage house, and any other financial institutions where you have holdings. Tell them that no assets or money should be moved or transferred without the knowledge and approval of both you and your spouse. Be sure to tell the broker to make a note to your file on the computer. That way, when your spouse speaks to anyone else at the firm, the computer will show that the account has been frozen. Also, if the account has check writing or credit card privileges, specify whether the freeze applies to those transactions. Follow up the call with a letter identifying the person with whom you spoke, noting the date and time of the call, and giving a summary of the conversation.

When it comes to joint investments, you will have to move quickly. Many stock transactions are handled over the phone and, increasingly, electronically through email or the Internet. Your spouse may be the

one who normally deals with the broker or discount brokerage house. As a result, that one phone call and a matter of minutes, could mean that all of your joint holdings are sold and the money transferred out of the account. If you call first to freeze the accounts, you stand a good chance of protecting your interests.

Sample Letter to Broker

November 27, 20xx

Rodney Washington
Bull and Bear Investment Company
3400 Financial Square, Suite 1740
Boston, MA 02000

Dear Mr. Washington:

This letter is to inform you that my spouse and I are in the process of divorce. Please freeze our account, #12345-67890, so that no transactions occur until further notice. Please make a note to this effect on the firm's "online" file on the computer. This freeze is to be effective immediately, and this letter confirms my instructions to you by telephone on November 24, 20xx.

These instructions apply to the assets in our account as well as the check writing and/or the credit card privileges we currently have with this account.

Yours truly,

Lenore Kwong

Safe-Deposit Boxes

Just as one spouse can remove all money from a joint checking account, so too can a spouse take all of the items out of a joint safe-deposit box. Unfortunately, safe-deposit boxes are one of those items in which "get there first" is the operative rule. Try to take possession of both keys to the safe-deposit box if at all possible. If your spouse empties the box before you do, you may have little chance of reclaiming the contents.

The best way to legally protect yourself is to get a restraining order from the court and give a copy to the bank. The order would prevent either spouse from having access to the box until you reach a settlement or the court removes the restraining order.

You might possibly protect yourself if you can get to the box, write an inventory of the contents, and take photos of the items. Then have the bank officer sign the list documenting your inventory. If your spouse removes anything later, you'll have some proof of what was taken.

Shared Property and Special Collections

While your furniture, appliances, and knickknacks around the house may not be as valuable as other assets and investments, this property can cause more irritation and antagonism than almost anything else. Emotions flare up when you reach for a favorite tool only to find it missing or look for your mother's antique brooch and discover it is gone.

To avoid future problems, do an inventory of your home and possessions similar to the kind you would conduct for insurance purposes. Obviously, it will be simpler to do such an inventory while you are still living in the family home and have access to your property—so do it before you move out.

List the contents of each room and photograph or videotape the area for documentation. Your ultimate goal is to divide these items as fairly as possible. Few couples split the small items exactly 50-50, but you still want to have any valuable property appraised so your division is not horribly one-sided. You can get simple estimates of household property values by asking furniture or appliance dealers and auctioneers to visit your home and give you written estimates. Any coin, gold, or other collection should be appraised by a reputable dealer. Realize that these items should be valued at the amount the dealer is willing to pay—not the retail value.

While retail values are useful as a starting point, in reality, collections and property are only worth what you can sell them for. If you are valuing property on your own, use "garage sale" values—that is, if you saw your furniture, microwave, or other property at a garage sale, how

much would you really pay for it? The person who will not be getting the household items may only remember the original price tag, not the current value, so you have to insist on using the garage sale price.

In the meantime, jointly owned household property should not be disposed of unless both spouses agree, as neither one owns an item outright until the settlement. You only own your personal, separate property. Your children's property, such as furniture or toys, should be given to the children and not divided in the divorce settlement.

Financial First Aid for Divorce Emergencies

If a stranger stole your wallet with all the credit cards inside, you'd probably have more protection than if your spouse wipes out your joint accounts. Almost nothing can stop one spouse from closing a joint account without the knowledge of the other. During a divorce, it can come down to who gets there first—to the bank account, the computer files, the brokerage office, or the safe-deposit box.

Many states include automatic restraining orders in the initial court paperwork, so that both spouses are prohibited from transferring property without the other's permission. If your state does not do this, and if you are concerned that your spouse might empty out a joint account or otherwise take assets to which you are entitled, you may need to obtain a temporary restraining order prohibiting the removal, sale, transfer, or other use of the property. Not only can restraining orders be placed against your spouse, but they can also be served on third parties, such as bank managers and others with control over your assets.

Restraining orders are also used to bar a violent or abusive spouse or parent from having contact with the other spouse or children. Domestic violence organizations can help you obtain that kind of restraining order. See the appendix for more help.

Questions to Ask an Attorney

Closing accounts isn't always a straightforward procedure and can vary from state to state. Below is a list of important questions for you to ask an attorney or research at your local law library.

1. Do any state laws prohibit me from withdrawing half of the cash in our joint accounts or borrowing money on joint accounts?

2. Can I unilaterally change property from joint tenancy to tenancy in common, or must I have my spouse's consent?

3. Would it be advisable to close my joint accounts and move the funds to an account solely in my name or into a joint account requiring both signatures on the checks?

4. Do any state laws prohibit me from canceling or changing beneficiaries on my life or medical insurance policies during divorce?

5. What would I need to do to prevent my spouse from taking money out of our home equity line of credit? ●

Getting Help: Who Can I Turn To?

Questions to Consider When Seeking Outside Help.. 86

Is the Person Competent and Suited to Handle My Specific Case?.............. 86

How Is This Professional Paid?..87

What's the Best Use of This Professional's Services? 87

Am I Evaluating This Professional Objectively?................................... 88

Selecting Professionals to Assist You 88

Lawyers...89

Limited-Scope Representation or "Unbundling" 89

Tips on Working With Your Lawyer 91

Mediators.. 94

Collaborative Divorce..97

Arbitrators and Private Judges.. 99

Financial Planners 100

Divorce Financial Analysts 102

Stockbrokers and Money Managers.. 103

Accountants 103

Bankers 104

Insurance Agents... 104

Real Estate Agents 105

Therapists.. 105

Credit Counselors.. 106

Typing or Paralegal Services 107

Actuaries 108

Business Appraisers 108

Private Investigators.. 109

Online Do-It-Yourself Divorce Services 109

During divorce, you're going through a time of transition unlike any other in your life. Everything is changing. No matter how capable you are, you'll need help to get through this transition.

Even if you have little money and few resources, it's important to get support. Check with local libraries, churches, community organizations, law schools, colleges, or government agencies for low-cost services. In some areas, divorce specialists and centers can advise you for a small fee.

The kind of help you seek will depend on what you need. For some, a personal counselor or therapist is crucial; others want only a temporary adviser to handle financial details. As your divorce progresses, you may also need to call on a variety of specialists ranging from attorneys to real estate agents, credit counselors to appraisers. Because your future is at stake, don't cheat yourself; get the best you can afford. Comparison shop and check references before signing contracts for outside services.

Questions to Consider When Seeking Outside Help

While you need to ask different, specific questions of each service provider you consult, the following four questions are useful ones to ask yourself when choosing professionals in general.

Is the Person Competent and Suited to Handle My Specific Case?

Obviously, you want to work with someone who does a good job and can be trusted. Beyond those basic qualities, it's essential to determine that the person's training and services match your needs. In other words, you don't hire a bookkeeper if your situation calls for an accountant to make complex calculations or compute taxes. By the same token, an experienced divorce attorney is better than a lawyer who specializes in business transactions. Interview potential consultants on the phone, or use introductory sessions to outline your case and find out exactly what to expect, before you hire anyone.

> ⚠ CAUTION
>
> **Accepting a free session with a lawyer or another professional may mean that you end up being pressured into hiring that person.** You may be sold a service that you do not want or need. To avoid such problems, clearly state that you are only gathering information in the initial session and do not intend to make a hiring decision until after you have had a chance to sort through the information you receive. Remember, too, the adage that if something sounds too good to be true, it probably is. Be wary of any lawyer who seems to promise you the world, making guarantees that you can get anything you want in your divorce. There are no guarantees.

How Is This Professional Paid?

How a professional is paid—by the hour, on a retainer, or through commissions—inevitably affects you. This is a statement of fact, not a matter of philosophy or morals. If you will be paying by the hour, you must understand how that time will be billed. Does the professional bill in ten-minute, 15-minute, or longer intervals? Are you going to be charged for time on the phone? Does the clock begin to run the minute you walk in the door? Find out.

When people work on commission, their livelihood depends on selling you something. Again, there's nothing wrong with that—you just need to recognize that reality and act accordingly. For instance, when you call for information about insurance coverage, the agent may try to sell you a new policy or different kind of coverage. Only you can decide whether it's appropriate to make such a purchase. Because of the emotional strain of divorce, take extra time to make decisions, or insist on getting a second opinion before you buy anything.

What's the Best Use of This Professional's Services?

Like many people going through a divorce, you may sometimes feel that you don't know where to turn. As personal and financial pressures mount, you could go to the wrong people for the wrong services. Trying to use your lawyer as a therapist or your banker as a tax consultant is an

exercise in frustration. Think about your needs first and be sure you are using each professional's expertise appropriately.

Recognize the limits of each adviser you contact. Your local real estate agent, for example, may not be able to appraise the value of commercial property or of a vacation house located in a different city. If necessary, ask for referrals to specialists, and keep asking questions until you find the right person to answer them. You can save time and money by gathering much of the data the professional will need, especially financial information. Chapter 8 details the kind of information you'll need.

Am I Evaluating This Professional Objectively?

It's fine to ask friends and associates for referrals, but do not give up your own objectivity and accept a referral blindly, even if it comes from someone you trust. When you first meet with the professional, ask questions to determine if the person is knowledgeable, capable and a good match for you and the unique facts of your case. If you feel uncertain, take your concerns seriously.

Similarly, do not feel compelled to use your sister just because she is an attorney or your uncle just because he is an accountant. Objectivity—on their parts and yours—is very important. Your relatives may have your best interests in mind and be very trustworthy, but they may not have the expertise and/or experience you need. You may not want to disclose personal but important facts to relatives, or they may miss creative solutions and the big picture because they are too close to your situation. Empathy may feel soothing, but a professional's clear eye for detail and probing for the facts is what matters most.

Selecting Professionals to Assist You

Besides providing quality service, a good professional should explain potential risks in your situation and offer realistic appraisals of results and consequences. Keep that in mind as you review these descriptions of professionals who can help you during divorce. (For handy reference, you can enter names, addresses, and phone numbers of your professional advisers at the end of the appendix.)

Lawyers

For assisting during a divorce, lawyers charge hourly rates. These fees vary widely across the country and locally, too, mostly based on the lawyer's experience level. Be sure to find one that fits your wallet. Be aware, too, that most lawyers require that you pay an up-front retainer before they do any work, sometimes amounting to thousands of dollars. As the lawyer spends time working for you, the bill is paid out of your retainer.

When the retainer is used up, your lawyer will either bill you directly or ask for another lump sum from which to draw. If you hire a lawyer who requests a retainer, be sure that you get monthly statements showing the precise work done, the amount billed, the amount deducted from your retainer, and the amount left. If there's a balance remaining at the end of the case (or if you change lawyers), you're entitled to get back any unused money.

Financial negotiations and lines of authority remain clearest when each party is represented by an independent attorney. In some states the court can order your wealthier spouse to pay your lawyer fees, so if you really can't afford an attorney you can check into whether that's a possibility. You can also look into limited-scope representation, described below.

Limited-Scope Representation or "Unbundling"

Hiring a lawyer to represent you in a divorce can be enormously expensive, way beyond the reach of many people. In California, for instance, about 70 percent of divorce petitions involve self-represented parties. But many of the folks representing themselves in family court could benefit greatly from even a little bit of legal advice.

The House of Delegates of the American Bar Association (ABA) reports that 41 states have adopted a rule allowing lawyers to limit the scope of representation if it's reasonable to do so under the circumstances. Many states now authorize attorneys to provide limited legal services to their clients—a practice that was frowned upon in the past. This is called limited-scope representation or "unbundled" legal services. While aspects of limited-scope representation are still controversial, the American Bar Association now encourages its members

to offer unbundled services, and many advocates believe that in divorce cases, it can be particularly useful.

With limited-scope representation, instead of an attorney representing you in all aspects of your case, you and the attorney can agree that you will perform some of the work in your case yourself and the lawyer will help with tasks you can't or don't want to do. Examples of how you can use unbundled legal services are:

- consulting a lawyer for legal information and advice about your case when you come upon something you don't understand, or for help with the more complicated parts of your case (such as discovery and legal research) while you do the simpler tasks yourself
- hiring a lawyer to represent you on certain issues in your case (such as child custody, child or spousal support, or dividing property) while you do the rest yourself
- hiring a lawyer to prepare forms and court documents, which you then file yourself and follow up by appearing in court
- hiring a lawyer to make a court appearance for you, or
- hiring a lawyer to review documents you've prepared, coach you on how to represent yourself at court hearings, or instruct you on how best to present evidence in court.

The cost of limited-scope representation varies. Attorneys generally charge their regular hourly rate for the hours they work. Whether you pay a retainer depends on the scope of services requested.

CAUTION

States have not developed uniform rules about the nature and extent of limited-scope services attorneys are allowed to offer—and some states don't allow it at all. The American Bar Association website (americanbar.org) includes a pro se/unbundling resource center that can help you determine how your state treats limited-scope representation.

If you hire a lawyer to provide unbundled services, it is critical that you discuss in detail which tasks each of you will undertake, and sign a written agreement spelling that out. Be sure to ask questions about issues that are not clear to you. Your success depends on working together as a team.

Not all lawyers are comfortable offering limited-scope representation, so you may have to look around to find someone who will work cooperatively with you.

Tips on Working With Your Lawyer

Whether or not your lawyer is a specialist in divorce, you should expect any lawyer to:

- provide you with a clear understanding of how you will be billed and what costs to expect, including retainer, court costs, expert witness fees, and other expenses
- provide you with copies of all documents related to your case
- maintain detailed time records and send you accurate and understandable bills
- return your phone calls promptly
- explain your rights and how the lawyer will go about securing them for you
- keep you up to date on settlement negotiations, and
- put significant effort into trying to settle your case out of court and, if that is not possible, prepare your case—and you—for trial.

Remember, communication is a two-way street. Just as your attorney should provide you information, you must provide complete and honest information to your attorney. This includes filling out requested forms and gathering requested documents. (This is covered in detail in Chapter 8.) If your attorney has incomplete or inaccurate information, the advice you receive may be inappropriate for your particular circumstances.

Do not be concerned if your attorney is friendly with the opposing counsel. In fact, this often works to your advantage. Even if the attorneys are friends, they will each work for their client's interest. Attorneys who cooperate with each other are better able to negotiate and settle out of court, which will result in lower legal fees. In addition, if the attorneys are enemies and drawn into an emotional fight, the end result will be higher fees and a lower quality of representation for you. An attorney in such a situation will not be able to focus clearly and objectively on the facts of your case.

Expect your lawyer to be pleasant but not to act as your de facto therapist. The lawyer's job is to guide you through the legal maze, not to listen to how your marriage might have turned out better if only you or your spouse had done this or that. If, as is often the case, you're suffering emotionally because of the breakup, you *should* see a therapist (see below). Therapists usually don't charge as much as lawyers, and they're much better equipped to deal with emotions.

Does Gender Matter in Picking a Lawyer?

With the growth of feminism and the fathers' rights movement, there's also been a growth in law firms that specialize in representing either primarily men or primarily women in divorce cases (especially those involving domestic violence or child custody). No statistics exist on the number of such firms, but at least one mother- or father-focused firm can be found in most large U.S. cities.

Some legal observers say firms focusing on one gender can foster needless confrontation. Others believe parties may find comfort in believing their attorney understands and empathizes with their gender's particular needs and concerns.

Probably the personal and professional qualities mentioned in "What to Look for in a Lawyer," below, are the best indication of who would represent you well. But if you wish to explore the possibility of hiring a lawyer who specializes in representing one sex, among the websites men may want to visit are www.dadsrights.com (Chicago), www.socaldadslaw. com (Southern California), and www.fathersrightsinny.com (New York City). Women may be interested in www.dawnforwomen.com, a nationwide network of divorce attorneys for women.

What to Look for in a Lawyer

Choosing the right attorney is one of the most important decisions you will make. The right lawyer can help you achieve financial security as well as ease the strain of the divorce process. But the wrong lawyer can add to your woes.

When selecting an attorney, make sure your schedules are compatible. If they're not, you will have difficulties together. Because your lawyer is working on more than one case at a time, don't expect to always be able to connect on the first phone calls—but the lawyer should return your calls promptly. You must also promptly return phone calls from your attorney or his or her office staff and respond to any correspondence to keep your case moving along.

Take the time and make the effort to choose a lawyer who is strong and experienced, yet sensitive to your needs. You need to resist the urge to quickly choose a lawyer just because a friend or relative says "so-and-so is good." Instead, interview at least three lawyers and compare them using the following criteria:

- **Knowledge and experience in family law.** An attorney who's practiced five years or more and whose practice is completely or primarily devoted to matrimonial law would be ideal. In addition, ask how much courtroom experience the attorney has, how many contested custody cases the attorney has handled, and whether the attorney has given lectures or written articles on divorce (often a sign of in-depth knowledge). Does the attorney belong to the Academy of Matrimonial Lawyers, a professional group that certifies experts? In some states, such as California, consider hiring a certified family law specialist. These attorneys must pass an examination that tests their specific knowledge of divorce law. Most keep current on developments in divorce law by attending continuing legal education programs. They may know more about divorce law—including support, custody, and property division questions—than nonspecialists, and they may charge more as well.

- **Interest in your case.** Does the lawyer seem interested and excited, or bored and preoccupied? A divorce case tests any lawyer's zest for facts and details. You want someone who seems energized and up to the challenge.

What to Look for in a Lawyer (cont'd)

- **Availability of a team.** Does the attorney have the assistance of other lawyers and paralegals? Particularly in a complex case, the lawyer may need backup. You might ask how many others in the firm handle divorce cases and ask to meet with them as well. Similarly, because divorce cases often involve other fields of law—such as real estate, taxes, trusts, and the like—you should ask whether the firm has other attorney resources available for consultation if needed.
- **Coherent strategy.** All legal cases involve a plan of action that's based on goals on which the lawyer and client concur. After you explain your situation, ask the prospective attorney: What results can I reasonably expect? How will we achieve those results? How will we decide whether to try the case or settle? If we do try the case, how much will that cost?
- **Rapport.** Ask yourself: What's my gut feeling about this person? Charm is no substitute for competence. But you are going to be spending lots of time with your attorney, so you'll need a certain comfort level. Look for a lawyer who is pleasant but also confident and assertive.

In short, you want a lawyer who is going to solve problems, not create them. To get such an able advocate means you need to choose well, not necessarily quickly or cheaply. Take your time and be thorough.

For further help finding a lawyer, see the appendix.

Mediators

A private mediator is a neutral third person, trained in conflict resolution and family law, who works with both parties together to help them reach agreement on some or all of the issues involved in the divorce, including property division, child custody, and child and spousal support.

In some counties, mediators associated with the family courts work with divorcing couples on child-related issues only—they don't deal with support or property issues. This type of court-connected mediation is often mandatory, even if you are also seeing a private mediator.

In mediation, you and your spouse are in control of the decisions about how to divide property and deal with alimony and child custody and support. The mediator won't give legal advice or advocate for either party's position as a lawyer does. Instead, the mediator—along with your lawyers, if you have them—can help you explore different options and find a resolution that works for both of you. (For more about when mediation is appropriate, see "When to Use Private Mediation," below.)

Mediators are often lawyers themselves; sometimes they are therapists, social workers, or clergypeople. You pay for mediation by the hour, and the rates will be what a lawyer, therapist, or social worker normally charges. If the mediator is an attorney with lots of experience in family law, you can expect to pay the same rate that you are paying your own lawyer. In fact, you could pay more—but keep in mind that mediation tends to be an efficient, cost-effective way to settle cases. Generally, the parties share the cost of mediation equally, unless they agree otherwise.

 RESOURCE

To find a private mediator, ask friends and professional colleagues for referrals. The court clerk at your local family or domestic relations court may be able to give you some referrals. You can also contact the national Association for Conflict Resolution (ACR) for a list of mediators in your area. ACR's website, www.acrnet.org, lists mediators all over the country.

Used properly, mediation can be much less costly than a prolonged legal battle—even when you have a lawyer as well. Often, couples who are mediating their divorces use consulting attorneys to help them consider their options and make proposals for settlement, and to draft or review the agreement that comes out of mediation. If you don't use a lawyer during your mediation, you may still need one to help you at the end. In some cases, the mediator will draw up a "memorandum of understanding" that outlines your agreement—then you can ask a lawyer to spend a relatively short time looking over the agreement and putting it into the proper form to be submitted to the court and made into a final judgment.

When to Use Private Mediation

As mediation becomes more and more common, most people going through a divorce will consider using the process. If you're not sure whether mediation is for you, consider some of the following questions:

- Do you and your spouse both want to resolve your divorce with the least conflict possible? If so, then mediation should be your first stop.

- Are both of you able to stand up for yourselves? If so, mediation can work. But if you are easily intimidated by your spouse and end up agreeing to things just to avoid an angry confrontation, then you might be better off with an attorney advocate.

- Was the decision to divorce mutual, and are you both clear that reconciliation isn't an option? The less conflict there is about who left whom and whether the marriage is over, the more likely you'll be able to cooperate in mediation. If either one of you is still dealing with very intense feelings of anger or having difficulty just being in the room with the other person, it will be much more difficult. But even if the answer to this question isn't an unqualified yes, mediation can still work.

- Do you respect and support your spouse's relationship with your kids? If you have children, it's enormously important that you and your spouse be able to communicate effectively and make decisions together. Mediation is a great way to practice this with the help of a skilled professional.

- Is there a history of violence or emotional abuse by one spouse, or does either spouse abuse drugs or alcohol? In these circumstances, it's probably better for both parties to have attorneys advocating for them.

- Are you willing to share all financial information, and do you trust your spouse to do the same? To be successful, mediation requires full disclosure of financial information by both parties.

There may be other factors that seem important to you. The following is by Gary Friedman, author of *A Guide to Divorce Mediation: How to Reach a Fair, Legal Settlement at a Fraction of the Cost* (Workman Publishing):

When to Use Private Mediation (cont'd)

"Central to determining mediation's appropriateness is whether or not the parties can deal fairly with one another.

"Differences in the parties' openness to the process, the tendency of one party to dominate the other and inequalities in ability (or willingness) to deal with the subject matter must all be examined. Before jumping into mediation, ask yourself several questions:

- Are we both open to reaching a result that is fair to the other?
- Are we able to communicate clearly with each other? Can we each express ourselves and hear the other?
- Can we each identify what is important to us as a realistic and solid base for making choices? Can we each express that?
- Is either of us unwilling to seek outside support or unable to use it effectively when it is needed?

"The decision whether or not to mediate should be an informed one, and one that can be reconsidered. Either party's hesitation should be taken seriously; both must be willing for the process to be meaningful. A decision not to mediate or to stop the mediation does not mean that adversary litigation is the sole choice. You can bring in a co-mediator, use a type of mediation more protective of the parties or work with lawyers committed to collaborating with their clients."

Collaborative Divorce

If it doesn't seem like mediation is for you because you want the support of having a lawyer of your own, but you are committed to the peaceful resolution of your divorce, consider using a process called collaborative law.

In a collaborative process, each spouse is represented by an attorney who is committed to a fair resolution of all the issues in the divorce. To begin the process, both spouses and their attorneys sign an agreement that calls for the following:

- Each spouse agrees to fully disclose and exchange all financial documents that will allow for fully informed financial decisions.
- Both spouses and their attorneys agree to maintain absolute confidentiality during the process, creating a safe environment for spouses to freely express needs and concerns.
- The spouses agree to reach a written agreement on all issues and not to seek a judge's intervention, and agree that the attorneys may use the written agreement to obtain a final court decree.
- Both spouses and both attorneys agree that if either party ends the collaborative process before they've reached an agreement, both attorneys must withdraw from the case and can't represent their respective clients in a court fight. To go to court, the spouses must hire new lawyers.

Central to the collaborative law process is the commitment by everyone involved to resolving instead of litigating. While the attorneys advocate for their respective clients, they also support the collaborative process, participate in full disclosure of all assets and liabilities, and encourage agreements for both spouses and their children. The "no court" agreement is critical, as it provides a major incentive to settle and avoid incurring the cost of starting over again with a new attorney.

During the process, the parties and their attorneys meet in a neutral setting to begin negotiations. All meetings include both spouses and both attorneys. All participants must contribute to positive discussions. At some point the spouses may agree to jointly retain experts, such as forensic accountants, to prepare accounting analyses, value assets, and assist with other technical issues. Other professionals may provide financial and tax advice, and in some cases, the spouses may hire a child specialist and even a mental health professional to provide ways to cope with the emotional stresses of divorce. What assistance is needed will vary from case to case.

Pros and Cons of Collaborative Divorce

There are some important benefits to the collaborative process:

- You receive support from your attorney's advocacy, expertise, and problem-solving and negotiating skills, while still retaining control over creating a negotiated agreement.
- You stay focused on finding positive solutions instead of getting derailed by threats of going to court.
- You negotiate a comprehensive settlement, instead of receiving an order made by a judge with too little time to understand the issues important to you or the nuances of your case.
- You increase the likelihood of compliance with the terms of a negotiated agreement when each spouse had a stake in settlement. Working together cooperatively creates a positive environment for coparenting and can improve your relationship.

On the other hand, collaborative divorce has its downsides. It can be expensive and also cumbersome if too many professionals become involved with too many issues. And it can be unfair to the less powerful spouse who may not have the resources to start over with another lawyer in a court battle if the collaboration is aborted.

RESOURCE

For more detail about divorce mediation and collaborative practice, take a look at *Divorce Without Court: A Guide to Mediation & Collaborative Divorce,* **by Katherine E. Stoner (Nolo).**

Arbitrators and Private Judges

Arbitrators differ from attorneys, who represent the interests of one spouse, and from mediators, who represent the interests of neither but act as neutral facilitators with no decision-making power. Instead, arbitrators and private judges (often retired from the bench) hear a case and make decisions regarding disputed issues. These decisions are then incorporated into a court judgment.

Arbitrators charge hourly or daily rates. Typically, these fees are some-what more than what a top-paid attorney receives. However, arbitration

is usually far less expensive than going to trial with an attorney, because fewer attorney hours are required.

Arbitration can be a better option than going to court. It can be most useful in resolving specific issues on which you cannot agree, such as what a business is worth. Each side presents its case, and the arbitrator decides. Arbitration is not about negotiation and, in fact, may come up as an option after all negotiating efforts have stalled. You may want to consider this relatively quick and effective way to reach a decision.

RESOURCE

The best way to find an arbitrator is to ask your professional adviser—attorney, CPA, financial adviser, or insurance agent—for a referral. You can also contact the American Arbitration Association (AAA) for lists of arbitrators, at www.adr.org.

Financial Planners

As attorneys begin to recognize how critical financial planning is to the decisions their clients make during divorce, they are turning more and more to financial planners to provide advice and recommendations to clients on dividing assets.

A financial planner can help you prepare for your financial future as a single person. A planner may be able to help you identify potential risks in your settlement agreement and evaluate the potential tax consequences of decisions you are contemplating. Planners can review your net worth and cash flow statements as well as your retirement benefits, investment portfolio, and business holdings.

There are three types of financial planners. Fee-only planners charge as much as several thousand dollars to develop a full financial plan addressing every aspect of your finances or offer consultations to answer your specific questions for a fee based on an hourly rate.

Commission-only planners derive all of their income from 3%–5% sales commissions on the investments they sell you. While they may give

good advice, if they earn the majority of their income selling products, they may have a conflict of interest that can bias their recommendations.

Lastly, there are financial planners who are compensated by a combination of fee and commission, and who may also have a conflict of interest when they advise you to buy certain products.

To find a financial planner, ask for a referral from an accountant, a banker, or an attorney you trust. You can also contact the Financial Planning Association at 800-322-4237, or visit its website at www.fpanet.org.

How Financial Planners Can Help With Your Divorce

As you go through your divorce, a financial planner can help you:
- set long-term and short-term goals
- develop an action plan to reach those goals
- identify cash flow needs, sources of income, and expenses
- put yourself on a budget
- evaluate current investments in terms of risks and objectives
- determine which assets are most appropriate for you to keep
- develop an investment portfolio for your needs and goals
- learn the costs and tax consequences of keeping or not keeping your family home
- determine whether the proceeds from a retirement plan should be rolled over to an IRA or a Roth IRA
- determine whether to take an immediate or a deferred distribution from a retirement plan
- evaluate whether your current insurance coverage is appropriate and cost-effective
- decide how to invest for education funding, and
- make estate planning decisions.

To be certain you are dealing with a reputable professional, check references and credentials. Planners who have undergone a rigorous

program offered through the College for Financial Planning in Denver earn the title of Certified Financial Planner (CFP). Chartered Financial Consultant (ChFC) is another credential to look for. It's given to financial planners who pass a series of courses at The American College, located in Bryn Mawr, Pennsylvania.

In hiring a financial planner, look for someone who can analyze the financial impact of divorce on you—not someone whose primary focus is on selling investments. You also want someone who can communicate with your attorney, because your planner should help you identify and assess risks so you can protect yourself. Few of us manage our finances as well as we could, but you will find it to your benefit if you choose a planner who is willing to tell you what you would rather not hear.

Divorce Financial Analysts

The growing popularity of assets that can be tricky to value, such as hedge funds, stock options, private-equity ventures, and other sophisticated financial investments, has led to a new brand of adviser known as the "certified divorce financial analyst." Previously known as "certified divorce planners," these specialists help couples wade through the fiscal intricacies of splitting up. In fact, many have a financial title, such as certified public accountant (CPA) or certified financial planner (CFP), in addition to their divorce financial analyst designation.

In recent years, some 2,500 of these divorce specialists have been trained, according to the Institute for Divorce Financial Analysts, and some big financial houses, like Merrill Lynch and Morgan Stanley, have divorce financial analysts on staff. For fees of $150 to $250 an hour, such advisers help couples and individuals navigate the economic aspects of divorce while leaving the legal issues, like custody, to divorce lawyers.

Divorce financial analysts are most effective working with clients on specific, complex financial issues, like avoiding tax snafus, dividing a stock portfolio, obtaining health insurance after a split, and deciding which assets to fight for and which to concede. While they're open to all clients, it's usually higher-asset couples (with estates of at least $250,000)

who have been married more than ten years—so that alimony is a major issue—who can make the best use of their services.

To find a divorce financial analyst, contact the Institute for Divorce Financial Analysts, the body that certifies these professionals, at 800-875-1760 or on the Web at www.institutedfa.com.

Stockbrokers and Money Managers

Stockbrokers tend to focus on individual stock, bond, mutual fund, and insurance transactions. Money managers look at the overall performance of investment portfolios and evaluate how they meet short-term and long-term needs for growth, income, and safety.

In gathering information on your financial condition, you may have to contact your stockbroker for copies of brokerage accounts or other items. Again, as with other professionals who work on commission, recognize that your stockbroker may try to sell you products when you call for financial information.

Stockbrokers receive commissions on sales charges, which are paid by the client—usually 1% to 5% of the amount of money invested in stocks, bonds, mutual funds, or annuities. Money managers, on the other hand, receive a fee for the service they provide, which is designing and managing investment portfolios. They charge a percentage of the value of the money they manage—usually 1% to 1.5% on accounts of $200,000 to $1,000,000 and 1% to 3% on smaller accounts.

Accountants

Accountants can provide a variety of services ranging from auditing a business to assessing employee benefits packages. Accountants charge by the hour or by the project. A certified public accountant (CPA) can work as a tax consultant—but don't assume that all CPAs can provide personal tax services. If you have income tax concerns, you may want to start by consulting an enrolled agent (EA), a tax specialist qualified to practice before the IRS. If you think your spouse is hiding assets or covering up financial facts, you may want to hire a forensic accountant.

These specialists can trace marital property, evaluate financial reports, and value businesses.

To find out about enrolled agents, and to find one in your area, you can contact the National Association of Enrolled Agents at 202-822-6232 or www.naea.org.

Bankers

Your bank and banker can prove quite useful in helping you with various financial issues of divorce, such as canceling or opening lines of credit, establishing new accounts, or getting loans. Do not expect a banker, however, to advise you on investments or other complex issues. Also, while bankers work on salaries, they often function as salespeople for their own institutions, promote their own products, or sell mutual funds with a commission. Your banker may advise you to park your money in a certificate of deposit even though your long-term goals might be better served by investing in stocks or mutual funds. Get information from several different sources or consult with your financial planner before making any long-term decisions.

Insurance Agents

Insurance is one way to protect yourself from risks, and during divorce it's important to review your coverage. Contact your insurance agent to double-check your policies for life, health, disability, property, and business. Also check the cash and surrender values of the policy—they may be different from the amount stated in your contract because the actual rate of interest paid may differ from the rate of interest assumed when the policy was originally bought. You might also be charged a fee if the policy is cashed in.

Divorce is a good time to check the cash value of any life insurance policies because this value represents marital property to be divided. In addition, the policy can provide emergency cash if necessary; but be aware that the surrender value may be lower than the cash value. When you divide the family cars, be sure to check your auto insurance to make

sure you are paying only to insure yourself (and perhaps your driving-age children) on the car(s) you keep.

Insurance agents are paid commissions. The commissions are paid by the insurance company, but the company prices its policies to include the commission—so ultimately, the customer pays the commission.

If you do not have an insurance agent, shop around and talk to several to find one with whom you feel comfortable. Be aware that insurance agents hold sales positions and may ask you to buy a new policy or expand an old one when you call for information. Resist their pressures and be sure you absolutely need any new insurance you purchase.

Real Estate Agents

When a real estate agent is involved in the sale of a home, that agent usually receives a commission from the seller, even if the agent has been working with the buyer. An agent who appraises a house usually will do so for free, in hopes of getting your business later when you sell.

Because your house may be the most valuable asset in your divorce settlement, it's important to get an accurate appraisal of its current market value. Try to get estimates from several real estate agents or from a certified appraiser so you can negotiate with your spouse or a third-party buyer from a knowledgeable position. The best agents to ask are those familiar with the neighborhood. If you used an agent when you bought the house, that person is often well suited to appraise the property now.

Certified appraisers are covered in Chapter 13. You can get referrals from real estate agents, escrow companies, and lending institutions.

Therapists

In the stress of divorce, a therapist or counselor can be a valuable resource for helping resolve personal issues. To find someone, ask friends for referrals, or contact a local mental health association, church, or family service agency.

Therapists charge by the hour, with 50 minutes counting as an hour. Especially if money is tight, find out whether your medical insurance might reimburse you for all or part of your therapy sessions.

As with any professional, it's important to check credentials, but also make sure the therapist's approach feels appropriate to you. Don't hire a Freudian who does long-term psychoanalysis if you want short-term counseling to help you get through the divorce. If you can't afford one-on-one therapy, you may be able to find help and encouragement through a support group, a community service agency, or a program sponsored by a religious group.

Credit Counselors

Credit advisers can help you negotiate with your creditors and set up systems to manage your finances. These agencies generally receive most of their funding from major creditors, such as department stores, credit card companies, and banks.

Use the services of credit counselors who are affiliated with nonprofit organizations, such as the United Way, a local "Y," or a church- or synagogue-run association. A few other nonprofit credit counselors operate nationwide. The best are Myvesta, also known as www.getoutofdebt.org, and the National Foundation for Credit Counseling. These nonprofit organizations generally charge nothing or a nominal fee for setting up a payment plan to get your commercial credit accounts paid off.

To use a nonprofit credit adviser to help you pay your debts, you must either have some disposable income or be willing to sell some of your property. A credit adviser contacts your creditors to let them know that you've sought assistance and need more time to pay. Based on your income and debts, the adviser, with your creditors, decides on how much you pay. You then make one or two direct payments each month to the agency, which in turn pays your creditors. A credit adviser can often get wage garnishments revoked and interest and late charges dropped. These agencies also help people make monthly budgets.

Typing or Paralegal Services

Typing and paralegal services cannot give legal advice or represent you on legal matters, but they can offer document preparation services at low cost. If you want help in preparing your divorce papers, these services may be able to help you. Typing and paralegal services charge per project (for typing forms) or, sometimes, per page.

In selecting a typing service, you must separate the good (honest and competent) from the bad (dishonest and/or incompetent). Good services will provide you with a written contract that describes the services they plan to provide, states the total price you will be charged, and explains their complaint procedure and refund policy.

Unlike attorneys, paralegals are neither state licensed nor rated by their peers or professional associations.

Here are some things to look for when choosing a typing or paralegal service:

- **An established or recommended service.** Few services stay in business unless they provide honest and competent services. A recommendation from a social service agency, friend, or lawyer is probably a good bet.
- **Reasonable fees.** The fee should be based on the amount of work a task requires, the specialized nature of the task, and reasonable overhead. For example, if the task is straightforward and takes just 30 minutes of typing, the fee should reflect the rate charged by basic typing services with similar overhead—about $10 a page or $20 an hour.
- **Access to quality self-help publications.** Good typing services provide ready access to reliable self-help materials, either for free or at a reasonable price.
- **Trained staff.** One indication of a business's commitment to providing good service is whether or not they have undertaken skills training. Appropriate training is available through independent paralegal associations and continuing education seminars for financial planners.

Actuaries

If a pension or retirement plan is at stake in your divorce, you may need the services of an actuary, a specialized financial professional. For details, including fees charged, see Chapter 14.

Business Appraisers

To determine the financial value of any business—including a sole proprietorship or professional practice—a business appraiser conducts a detailed analysis.

Call the IRS and request a copy of Internal Revenue Rulings 59-60 and 83-120 if you or your spouse operates a business or is a professional— doctor, lawyer, accountant, psychotherapist, and the like—with a private practice in the form of a closely held corporation. You will want to know the value of the business or practice. IRS 59-60 deals with methods of valuing closely held corporations; 83-120 concerns valuation of stock.

Are Fees Paid to Professionals Tax Deductible?

You can deduct from your income taxes certain fees paid to professionals. Not many fees are deductible, but especially during divorce, every bit can help. Here are the rules:

- If you receive alimony, you can deduct all attorneys' fees you pay to secure or collect that alimony. If your spouse is paying your legal fees, the fees are not deductible. Nor are fees paid to lawyers by the person paying alimony.
- All fees paid to any professional—including attorneys, accountants, financial planners, and stockbrokers—for advice on tax consequences arising from a divorce are deductible from your income taxes.

If you plan to deduct fees, your adviser must itemize your billing showing exactly how much you were charged for the advice regarding alimony collection or tax issues.

Be sure to ask your lawyer, financial planner, and/or accountant about tax-deductible fees to be certain you've covered them all.

Private Investigators

Private investigators, or private detectives, can sometimes be helpful in a divorce by exposing fraud or locating missing funds or other assets. Contrary to the "gumshoe" image spawned by countless movies and TV shows, a private detective is much more likely to wield a database than a pistol. Computers allow access to massive amounts of information from probate records, motor vehicle registrations, credit reports, association membership lists, and other sources.

There are no formal education requirements for most private detectives, but many have experience in the military, law enforcement, insurance, collections, or the security industry. Most states require that private investigators be licensed, and a growing number of states specify training requirements. California, for example, requires 6,000 hours of investigative experience, a background check, and a qualifying score on a written exam; 4,000 hours and a college or law degree; or 5,000 hours and an AA degree in police science, criminal law, or justice.

A good investigator is curious, aggressive, persistent, assertive, and a good communicator. A skilled private detective is able to think on his or her feet and has strong interviewing or interrogation skills. Most bill on an hourly basis and may ask for reimbursement of expenses.

Online Do-It-Yourself Divorce Services

State and county court systems and bar associations have done a great deal to provide the public access to the legal system, especially in family law matters. Their websites are generally user-friendly and more reliable than most Internet sources that cover the legal process of divorce. Many have self-help centers where you can prepare for an interview with a lawyer, or begin the divorce proceedings yourself (if money is tight).

A growing number of websites provide legal forms for divorce and offer varying degrees of help in completing them. For fees ranging from $50 to $300—a fraction of what most lawyers charge even for an uncontested divorce—these sites offer to help you prepare the paperwork and represent yourself in your divorce.

Much as over-the-counter remedies may help clear up minor health problems but can't replace a doctor when it comes to serious health concerns, these sites may be useful for the simplest of divorce procedures. But if the spouses disagree on any important issue and/or if they have children or complex finances, there's a substantial risk to relying on these sites alone, without consulting a lawyer.

CAUTION

When researching on the Internet, stick to governmental sites in your state. Since divorce law is local, you need to stay focused on your regional courts, one of which ultimately will process and perhaps render a judgment in your case.

RESOURCE

More about online resources. There's an entire chapter on how to find divorce information yourself in *Nolo's Essential Guide to Divorce,* by Emily Doskow (Nolo).

Financial Fact-Finding: What Must I Know and When Must I Know It?

Advice to the Terminally Disorganized 113

If You Think Your Spouse May Be Hiding Assets 116

Don't Forget the "Easy-to-Forget" Assets 118

The W-2 and the Tax Return 122

What About the Gifts You Gave Me? 127

Financial Facts Checklist 128

Questions to Ask an Attorney 128

Frequently, one partner takes care of the paperwork and financial decision making. This is practical when you're still together, but a potential disaster during divorce.

The answers to the questions posed in this chapter's title are straightforward: You must know everything that can be known about your finances, and you need to know it as soon as you possibly can, preferably before filing for divorce.

No one can make proper financial decisions without adequate information. How can you tell whether to keep the family home if you do not know what it costs to maintain it? On what will you base requests for alimony or child support if you are in the dark about living expenses? Both you and your spouse will need to document your financial life as you go through the divorce. Now, in the early stages, is the best time to start.

If you left the bookkeeping to your partner, you may feel as though you are starting from scratch. It's important not to panic. Take control by finding out as much as you can about the finances of your marriage.

Linda, a 35-year-old mother of two, recalled that when she first separated from her husband, she was petrified at the thought of confronting her financial life. As she kept at it, however, the job became easier. "Now I can look at all of these complicated papers and forms and not feel scared," she reported.

SKIP AHEAD

If you were the family bookkeeper, you will be in much better shape than your spouse who did not participate in managing the finances. You can move on to "Don't Forget the 'Easy-to-Forget' Assets," below, and begin verifying the financial information you will need during your divorce. Otherwise, continue reading.

Whatever role you played previously, by the time the divorce is over, you will probably know more about the details of your financial life than

you ever cared to. You gain this knowledge by going through these three steps:

1. **Information gathering.** Find information and sort files, receipts, bill statements, and other papers.

2. **Information analysis.** After sorting your papers, review your files, appraisals, and valuations. It's also important to calculate taxes and other costs—either by yourself or with an accountant or financial advisor.

3. **Decision making.** Once you analyze what you own and owe and what it costs to live, you can begin to decide what to do with your property in the divorce.

Start investigating your financial position now. Do not wait until you are at the bargaining table to determine the worth of your house or the value of your spouse's retirement benefits. When you enter the negotiations, you want to have as much up-to-date information as you can.

In this beginning stage of the divorce process, focus on information gathering only. Even if you do not understand all of it, do not get sidetracked from your fact-finding mission. Later on, you can concentrate on decision making. For now, your time is best spent on documenting your economic situation. Staying focused may seem hard, but it gets easier as you go along. Besides, it will help you:

- save time and money, especially once you hire an attorney
- distinguish your separate property from the marital assets
- negotiate a settlement from a position of strength, and
- begin building a foundation for a secure financial future.

Advice to the Terminally Disorganized

Ideally, your books and accounts are up to date and in order. More likely, however, you will have to hunt for records and receipts. You are not alone. Few people are as on top of their personal finances as they would like to be—or pretend to be.

Fortunately, you have a number of options in finding what you need. Tax returns, for instance, are gold mines of information in a divorce. The "Financial Facts Checklist" at the end of this chapter will give you other clues for finding financial data. But before you get there, here are some ideas to get you started.

Tax Returns. You will need the information found in your tax returns from as far back as you can go, with the past three years as the bare minimum. This is one time when you will be grateful that the IRS keeps records (often electronically) of everything you send them.

You can get complete copies of your returns by using IRS Form 4506. However, retrieving these copies requires considerable time and money.

The IRS will provide you with the "transcript" of any tax return you filed within the last three years, free of charge. A transcript will display most of the line items in the return and is often sufficient for divorce purposes. It can be helpful to start with these free transcripts and then decide whether you need to pay for copies of your actual returns. Transcripts can be useful when time is a factor, as they are generally delivered within five to ten days. You can order transcripts of your returns online, using Form 4506-T or 4506-T-EZ.

You may also order a transcript of your tax account, which will show any adjustments that have been made (by you or the IRS) to your returns. Both these transcripts can be ordered free of charge on the IRS website or by calling 800-908-9946.

If you require an exact duplicate of a complete tax return for any given year (or if no transcript is available), it can be ordered for a fee of $57. Copies of tax returns may take up to 60 days to be processed.

The IRS generally sends your tax information to the address shown on your previous return. If you've moved or are concerned about arousing your spouse's anger or suspicion, ask that the forms be sent to your office, your attorney, a friend, or a post office box. Once you have copies of the tax returns, you can review them with a trusted accountant or financial planner who can show you how to find basic financial information on the forms.

Loan applications and account statements. Check recent copies of credit or loan applications, canceled checks, and bank, brokerage, and credit union statements for information on assets or debts.

Insurance policies. Look for the names of the beneficiaries on any insurance policies. Determine whether the insurance is term or whole life. If it is whole life, find out the cash value.

Estate plans. You may find a wealth of information about property you and your spouse own by reviewing wills, trusts, and other estate planning documents.

Retirement plans. Keep track of work-related pension or savings plans, both your own and your spouse's.

Financial professionals. Make a list of accountants, lawyers, and other financial advisers you or your spouse have been using.

Credit reports. You can get a copy of your credit report periodically from one of the major credit bureaus. (See Chapter 5.)

Title reports. Visit your county land records office and conduct a title search, or have an escrow company do one for you, if you suspect there might be liens against your home or other real estate. (See Chapter 13 for details on liens.)

If you know you'll have trouble staying organized once you begin collecting your paperwork, you can hire a bookkeeping service, a virtual assistant, or a professional organizer. Before hiring one of these professionals, make sure the person has references and listens to you and asks what you want, instead of telling you what you need.

No matter what approach you take in organizing your documents, keep it simple. You can buy an inexpensive three-ring binder to hold your various papers. As your stack of information grows, create categories of documents by using the divider tabs that come with the binder. The "Net Worth" and "Cash Flow" statements in Chapter 12 will give you an idea of the kinds of divisions you can make.

When you begin to feel overwhelmed, remember that you are not just shuffling papers to get through your divorce. You are doing the groundwork for your financial decision making and security for many years to come—and that is worth the effort it takes to become informed and get organized.

Taxes and Power of Attorney

You may have a tax return transcript sent to a third party by completing the instructions on Form 4506-T. This can be useful when you're working with an accountant, an actuary, a financial manager or an attorney in your divorce.

However, there are some circumstances in which you may need greater involvement from a third party, such as your attorney. IRS Form 2848 provides authorization for a third party to communicate with the agency about your tax filings. In some cases, when a client is impaired or simply overwhelmed, an attorney in good standing with his or her state bar may be authorized to practice with the IRS and receive documents on the client's behalf.

This type of authorization would be helpful, for example, where a client's husband prepared all tax returns for his spouse over several decades, but never provided her with copies of the forms. If the client becomes subject to collection, she may want to sign a limited power of attorney that allows her lawyer to receive documents from the IRS and resolve the collection attempt.

Make sure your attorney is a professional in good standing with the state bar before you grant them even limited power to uncover previously unknown documents and negotiate on your behalf.

If You Think Your Spouse May Be Hiding Assets

Unfortunately, divorce can be the occasion for a game of financial cat and mouse in which one spouse hides assets or makes other unethical moves. In some cases, the gaps in the financial information you get from your spouse occur because of honest mistakes. Many people are not detail oriented when it comes to personal finances and, in the stress of divorce, may forget some specifics.

On the other hand, some spouses purposely conceal the value of a business or switch funds into secret accounts. If the judge finds out, the spouse who hides assets may be deemed guilty of "economic

misconduct." This misconduct is likely to affect the final settlement—the misbehaving spouse will probably get less than the innocent spouse.

Winning Big?

One California woman learned the hard way how seriously judges take the duty to disclose assets to the other party. In 1996, Denise Rossi won a $1.3-million lottery jackpot and 11 days later filed for divorce from her husband of 25 years. She kept her lottery winnings secret from her husband during the divorce proceedings. Two years later, he inadvertently learned of her windfall when he received a misdirected piece of mail. In 1999, a Superior Court judge ordered that the entire jackpot be given to the husband after determining that the wife violated state disclosure laws and acted out of malice or fraud. News reports indicate that the wife filed for bankruptcy in early 2000.

If you have any reason to believe your spouse is being less than candid with you about income or other assets, you will have to do a little detective work on your own.

If your own review of records like old income tax returns doesn't yield results, consider hiring a special type of certified public accountant called a forensic accountant. These specialists will comb through your records and can usually create an accurate picture of your spouse's financial position. Their services, however, can be expensive. You must weigh the cost of hiring a forensic accountant against the dollar amount at stake in your divorce.

The legal process of "discovery" can be used to force a reluctant spouse to turn over records and statements. Discovery is a formal information-gathering process used in lawsuits. Spouses (usually through lawyers) send papers to each other asking that questions be answered or documents turned over. Another procedure used in discovery is taking depositions—that is, orally asking questions.

Discovery is usually time-consuming and costly and may ultimately turn up nothing of importance. Therefore, gather information informally if possible.

No matter how you feel about your soon-to-be ex-spouse, you have an obligation to be financially responsible as long as you are legally married. In other words, each spouse must handle the family finances in a prudent manner. Financial misconduct is a breach of a spouse's fiduciary duty, and a court may find that a spouse engaged in "waste-dissipation" if the person squanders or destroys marital resources—including misusing credit. Each spouse is also responsible for making all financial information available to the other.

In some states (for example, California, Maine, Colorado, and Arizona), as soon as divorce papers are served, neither party may transfer, encumber, conceal, sell, or dispose of property (whether marital or separate), unless there's a written agreement or court order. However, most other states require a formal request before a court will restrict either spouse's right to control property and are reluctant to prohibit the use or transfer of marital assets unless there's a history of past dissipation or recent threats to dissipate assets or incur debt.

Every divorce court will consider a party's financial misconduct when dividing property. A spouse who engages in financial conduct that is detrimental to the other spouse may find the judge dividing property in favor of the other spouse.

Don't Forget the "Easy-to-Forget" Assets

Many assets divided during the divorce are not difficult to account for: the house, cars, retirement plans, and stocks. For other assets, you may have to search a little if your spouse is trying to hide them, as discussed above. Still other assets are in plain view, but you may overlook them nevertheless. While these "easy-to-forget" assets may not be worth a great deal of money, they can turn out to be valuable when you and your spouse are negotiating over who gets what in your divorce settlement. They could include such things as country club memberships, points for hotel, airline, or credit card programs, or equity in an auto lease.

Searching for Hidden Assets

This checklist was adapted from one created by Ginita Wall, a CPA and certified financial planner in San Diego, California. It includes common ways in which a spouse may undervalue or disguise marital assets. You may have difficulty finding some items or getting the proof you need to show they exist. As mentioned, a forensic accountant, formal discovery procedures, or both may help. Hidden assets may involve:

- collusion with an employer to delay bonuses, stock options, or raises until after the divorce, or salary paid to a nonexistent employee with checks that will be voided after the divorce. You might find this information by taking the deposition of your spouse's boss or payroll supervisor, but more likely you'll need a forensic accountant

- money paid from the business to someone close—such as a father, mother, girlfriend, or boyfriend—for services that were in fact never rendered. The money will no doubt be given back to your spouse after the divorce is final

- a custodial account set up in the name of a child, using the child's Social Security number

- delay in signing long-term business contracts until after the divorce (although this may seem like smart planning, if the intent is to lower the value of the business, it is considered hiding assets)

- skimming cash from a business

- antiques, artwork, hobby equipment, gun collections, and tools that are overlooked or undervalued (look for lush furnishings, paintings, or collector-level carpets at the office, reflecting income that is unreported on tax returns and financial statements; document any of the cash expenses you know your spouse has incurred)

- debt repayment to a friend for a phony debt

- expenses paid for a girlfriend or boyfriend, such as gifts, travel, rent, or tuition for college or special classes, and

- cash kept in the form of traveler's checks and money orders. You may be able to find these by tracing bank account deposits and withdrawals.

Who Gets the Miles?

Many millions of Americans belong to frequent flyer programs. With the average active member logging 12,000 miles yearly, the number of banked miles is growing by 400 billion per year. In addition to the usual credits from airline flights, car rentals, and hotel stays, miles are offered for use with certain credit cards and other purchases.

Often one spouse accumulates most of the couple's frequent flyer miles through business travel, though they may be used for the couple's joint vacations. Thus, the miles may become an issue in divorce, with one spouse asserting that the other's accumulated miles should be split upon dissolution of the marriage.

Some divorce and estate lawyers place a value on air miles—two cents apiece—that approximates what the airlines charge when they sell miles to their corporate partners. Thus, 500,000 miles could be worth $10,000.

But valuation can be more complicated than that. For example, one spouse may be a present or former airline employee who as a retiree will be entitled to fly free. How can that privilege be valued and divided? That would depend on the number of years until retirement, the average present cost of total flights, and the presumed inflation rate between now and retirement.

Further complicating the issue of frequent flyer miles is that some airlines do not permit division of this asset. If a significant number of frequent flyer points (more than 25,000) are at issue in your divorce, you or your attorney should review the airline's latest frequent flyer program statement. Be sure to specify in any agreement or court order the points being transferred and a deadline for the transfer before the expiration date of the points. Similar steps should be taken for other bonus plans, such as those sponsored by hotels, credit card issuers, banks, rental car companies, and department stores.

Here are some other commonly overlooked assets and some suggestions for how to handle them.

Stock options. The right to purchase the employer's stock at a bargain price some time in the future may have significant value. Usually, the spouse who owns the stock options keeps them in exchange for an asset the other spouse wants to keep. Alternatively, you can agree to exercise the option jointly in the future if the company's buy-sell agreement permits this. (See Chapter 16 for more on stock options.)

Tax refunds. If you file a joint tax return, the tax refund check will be in both names. Your divorce agreement should clearly state how the refund will be split. (See Chapter 11.)

Property taxes. If property taxes are paid partly for the past and partly for the future, the spouse paying those taxes for a period in which both spouses used the house may be entitled to a reimbursement of half of the taxes paid during that period of joint use.

Prepaid insurance. Because all insurance is paid for in advance, including life, disability, and casualty insurance, consider these prepayments when you are valuing and dividing property.

Vacation pay. The value of accumulated vacation hours should be based on the spouse's pay rate. If the employed spouse will be compensated now or later for unused sick pay, it may also be valued.

Frequent flyer points. To split these points, you can divide the miles themselves, have free tickets issued in the other spouse's name, or determine the monetary value of the travel benefits and compensate the other spouse accordingly (see "Who Gets the Miles?" above).

Season tickets. For some couples, season tickets for a hot sports team or important cultural events can be practically priceless. Whether you have tickets for the 50-yard line or a box at the opera, be sure you know their financial value and whether they can be replaced or replicated. Otherwise, you could find yourself losing an asset you cherish—or fighting for something that's not worth the effort.

Timeshares. A resort timeshare is often worth less than the amount still owed on it. You must decide whether one of you will accept it at a lesser value or whether you will continue to own it jointly, sell it, or let it go into foreclosure.

Professional dues and magazine subscriptions. Discounts are frequently offered for two- and three-year memberships in professional associations or for subscriptions to academic journals or other magazines. A spouse preparing for divorce may make substantial prepayments using money from marital funds in order to avoid using separate funds in the future. The amounts spent may not be much, but every bit matters when you're dividing it all up.

The W-2 and the Tax Return

Your financial fact-finding process should include an examination of your own and your spouse's W-2 forms, as well as your marital tax return or the returns you and your spouse file separately. There are several good reasons for doing this. First, it is never too soon to educate yourself on your finances, and the W-2 and the tax return are two primary tools. In addition, these documents provide information about how much income is available for alimony and child support payments (see Chapter 18). Finally, if you think it's possible your spouse might be concealing assets, the information contained there may point you toward sources you were not aware of.

Early every year, employers are required by law to issue a W-2 to each employee for the tax (calendar) year that has just ended. The employee attaches copies of the W-2 when filing federal and state tax returns.

> **WEB RESOURCE**
>
> **The W-2 form and the information included may change from year to year.** For comprehensive information and instructions on the W-2 form, go to the IRS website www.irs.gov and, under "Forms and Publications," type in "W-2."

Of the numbered boxes on the W-2, the most important for this discussion are:

- **Box 1: Wages, tips, other compensation.** This is where the total of straight wages earned during the year appears. Also included are some employee business expense reimbursements, moving expense

reimbursements, and employer contributions to certain fringe benefits, such as stock option compensation (for more on stock options, see Chapter 16). However, Box 1 does not include wages that the employee has deferred under certain employer-sponsored retirement plans, such as contributions to a 401(k) plan or a Simple IRA (see Chapter 14).

- **Box 5: Medicare wages and tips.** This is usually a larger number than Social Security wages. Medicare wages are gross wages and are not reduced by contributions to salary deferral and other tax-deferred employee benefit plans. Generally, all income reported in Box 5 is considered gross wages, and is the amount used in calculating child and spousal support.
- **Box 11: Nonqualified plans.** Money that the taxpayer has received from certain retirement plans and certain deferred-compensation plans appears here.

As for income tax returns, the format of the forms varies. Sources of funds may be marital income or separate income, and a joint return may reflect marital property, separate property, or both.

There are several items on a tax return and its supplementary schedules that are especially important:

- **Schedule A: Itemized Deductions.** This schedule may include several types of deductions that can figure into the earning-power equation for support. When you read the schedules filed by you and your spouse, look for entries that reveal items such as:

  - **Medical and dental expenses.** Most taxpayers may claim a deduction only for expenses that exceed 10% of adjusted gross income. If there is a large number here, it may indicate that medical expenses will be an important part of a supported spouse's needs or that a supporting spouse may be unable to pay.

  - **Real estate taxes.** An entry here indicates home ownership or investment property. Holdings you didn't know about may be reflected here.

  - **Home mortgage interest.** An entry here suggests a family residence or possibly a vacation home. If home mortgage interest is being paid, request a copy of the note. Whether the note is for fixed or variable interest could affect support calculations. Also check whether the property has been refinanced.

- **Schedule B: Interest and Ordinary Dividends.** The entries on this schedule reflect interest from banks, money market accounts, brokerage accounts, and the like, as well as dividends paid by corporations to stockholders. These amounts are generally considered income, and thus they are available for child and spousal support. On this schedule, you'll see entries for:

  - **Interest earned.** This indicates the size of the current balance of invested funds. Thus, this may be helpful in assessing disclosure statements in other aspects of the divorce process. Whether the interest is taxable or tax free will affect the support calculation.

  - **Dividends.** These are funds paid by securities. An entry here indicates that one or both spouses hold corporate stock or other securities. But many securities do not pay dividends, and those that don't may not be reflected on the tax return.

- **Schedule C: Profit or Loss From Business.** Unless the two of you can come to an agreement, the job of placing a dollar value on a business owned by spouses at divorce is best done by a forensic accountant or a business appraiser. But the annual tax return can reveal clues about the earning power of a sole proprietorship or a sole practice.

  You won't necessarily get easy, clear information from a business tax return. As recent scandals involving Fortune 500 companies have shown, business accounting can be an inexact science, with much room for fudging. A businessperson can manipulate figures by delaying the deposit of checks; overstating travel, meals, and entertainment ostensibly related to business; putting nonworking friends or relatives on the payroll to inflate "expenses"; buying personal items with business money; or writing off automobile expenses. If you suspect such practices, you may wish to hire a forensic accountant to do a detailed cash flow analysis and to value the business.

  If you decide to wade into the world of business accounting, there are a few things you need to know. Different businesses use different accounting methods. Some operate on a *cash basis*: They report to the IRS only cash income received, not money that their customers owe them, and they deduct expenses actually paid during the year. Others work on an *accrual basis*: They report income when it is earned, not necessarily received, and they deduct expenses when incurred, not necessarily when paid. Understanding the type of accounting method used is essential to determining a business's profit or loss and, thus, its cash flow for purposes of support.

- **Schedule D: Capital Gains and Losses.** On this schedule, the taxpayer notes the sale or exchange of a capital asset, which is an asset owned and used personally or kept as an investment. Almost everything you own, including your house, household furnishings, car, jewelry, and personal collections, is a capital asset. If an item is a capital asset, its sale generates a capital gain. If you exercise a stock option, that's also capital gains income. (For more on capital gains and losses, see Chapter 11 and Chapter 13. For stock options, see Chapter 16.)

Many capital gains don't count as income for support purposes. The determining factor may be whether the gain is considered a one-time gain or a recurring gain. For example, if the periodic sale of chunks of land from a large parcel has been funding the lifestyle of the parties and there's a reasonable expectation that this pattern will continue, the sale proceeds may be considered income. In such a case, the tax effect of the capital gains portion of income may have a significant impact on the after-tax amounts available to both the supporting and the supported spouse.

Most state courts have said that income realized from the exercise of stock options is income available for child support. One Ohio court specifically held that the value of unexercised stock options (their increase in value from the date they were granted to the date they could have been exercised) is gross income for child support purposes.

- **Schedule E: Real Estate, Royalties, Partnerships, Trusts.** In Schedule E, the taxpayer lists income derived from the sources mentioned. Depending on what's listed, the items here may be the source of significant cash flow that may be used in calculating support. In particular, this schedule may show:

    - **Rental revenue.** Rental property could be a significant source of cash flow. Think about more than just the monthly rent when evaluating this figure for cash flow purposes. For example, if rental property is leased to a business owned by one or both of the spouses, does it rent at market rates? When was it acquired, and how much is owed on it? Was it ever refinanced? If a taxpayer is paying debt service, there will be less cash available for support.

    - **S corporation and partnership income.** S corporations and partnerships do not pay federal income tax. All taxable income, whether distributed or not, must be reported individually by the shareholders and partners based on their ownership percentages and is reflected on what's called a K-1 form. If an S corporation or a partnership is involved, get a copy of the corporate or partnership income tax return, together with a

K-1, to understand significant details about what constitutes income and assets.

- **Form 1099-MISC.** This lists miscellaneous income, such as payments made to nonemployees (for example, consultants or independent contractors) and interest and dividend income reinvested but not actually received. If the income is for services rendered, you may find it on the Schedule C form.

How Long Should You Keep Tax Returns?

The short answer: Forever.

As the only official record of your earnings history, the tax return may be your most important financial document, period. It is crucial not only in a divorce but also for other financial purposes, such as setting your level of Social Security benefits.

Some advisers suggest you keep all tax returns for three years. That's just plain bad advice. Our strong recommendation: *Never* throw them away.

What About the Gifts You Gave Me?

It's sad but true that some of the most heated arguments during a divorce center on gifts—whether the gifts are an art collection and silverware, or the bread machine and Tupperware. Determining how to attribute financial gifts from family members, such as money used toward the down payment on the house or to buy a car, can cause a great deal of friction during the divorce.

As a general rule, property acquired by gift or inheritance is the separate property of the spouse who receives it. But it may not always be the case. For instance, in a situation involving one of our clients, a husband transferred real estate holdings to his wife to protect himself against creditors. At divorce, she claimed the property was a gift and her husband had no interest in it. She was dismayed when the transfer was deemed a transfer to defraud creditors. The transfer was voided, and the property was subject to the claims of her husband's creditors.

Financial Facts Checklist

At the end of this chapter is a list of the items you should gather, whether you are doing your divorce yourself or working with an attorney or other legal or financial professionals. We recommend getting this information together yourself—you'll save time and money and gain a sense of control over your divorce.

Most of this information should be in your files at home or your office, or possibly in a safe-deposit box. If you don't have a clue about where to find a particular item, watch the mail. You may be able to glean the names of insurance companies, banks, brokerage houses, and other institutions that you can contact to get information. Your insurance agent should be able to answer questions on policies, and the personnel or benefits office of your (or your spouse's) employer can shed some light on retirement benefits. For information on real estate, contact the agent who sold you the house or the company that holds the mortgage.

Instructions: Financial Facts Checklist

Put a check (✓) in the blanks next to items you have. Enter a zero for items that neither you nor your spouse has—and when you find them, enter a check mark within the zero. For documents you don't have, make a note about which of you will be responsible for getting them. Some of the terms in the chart may be unfamiliar to you. Don't worry. You will understand them better as you continue reading. For now, use the list to jog your memory—or your accountant's—about what documents to look for.

Questions to Ask an Attorney

1. Is there any hint that my spouse is hiding income? If so, how can we expose that?
2. What documents do I need to prove our real income?
3. What liability do I have if I find our real income is greater than the income we reported on our tax returns?
4. Are there any obvious sources of other income on my spouse's tax return, such as income from passive activities like rentals, business interests, dividends, and capital gains or losses?

Financial Facts Checklist

Tax returns (past five years)

☐ U.S.

☐ State

Pre- or postnuptial agreements

☐ husband

☐ wife

Previous divorce documents, including property settlement and alimony and child support orders

☐ husband

☐ wife

Wills

☐ husband

☐ wife

Trust documents

☐ husband

☐ wife

Pay stubs (from January 1 of current year)

☐ husband

☐ wife

Bank records

☐ **savings accounts**

    in whose name:

    account number:

    balance:

    latest statement

    deposit slips

☐ **checking accounts**

    in whose name:

    account number:

    balance:

    latest statement

    cancelled checks

    deposit slips

☐ **children's custodial accounts**

☐ **certificates of deposit**

Other accounts (including all accounts closed within past two years)

☐ **annuities**

    in whose name:

    account number:

    interest/dividends:

    balance:

☐ **cash management accounts**

    in whose name:

    account number:

    interest/dividends:

    balance:

Financial Facts Checklist (cont'd)

❏ **brokerage accounts**

  in whose name:
  account number:
  interest/dividends:
  balance:

❏ **credit union accounts**

  in whose name:
  account number:
  interest/dividends:
  balance:

❏ **commodities accounts**

  in whose name:
  account number:
  interest/dividends:
  balance:

❏ **mutual funds**

  in whose name:
  account number:
  interest/dividends:
  balance:

Tax-free investments

❏ **municipal bonds**

  in whose name:
  account number:
  interest:
  balance:

❏ **tax-exempt money market accounts**

  in whose name:
  account number:
  interest/dividends:
  balance:

Insurance policies (either spouse)

❏ life

❏ homeowners' or renters'

❏ liability

❏ health

❏ children

❏ motor vehicles

❏ personal umbrella

❏ disability

❏ parents

Employee/group insurance benefits (either spouse)

❏ medical

❏ disability

❏ life

❏ long-term care

Other employee benefits

❏ auto allowances

❏ sick pay

❏ **travel benefits**

  ❏ frequent flyer miles

  ❏ rental car bonus points

  ❏ hotel bonus points

❏ **severance pay**

❏ expense account

❏ stock options

❏ **salary bonuses**

❏ **deferred compensation**

❏ **military/VA benefits**

❏ **workers' compensation claim/award**

❏ **vacation pay**

Financial Facts Checklist (cont'd)

Retirement plans

- ☐ money purchase plans
- ☐ ESOPs (Employee Stock Ownership Plans)
- ☐ Keoghs
- ☐ thrift plans
- ☐ IRAs or SEP-IRAs
- ☐ SIMPLE IRAs

- ☐ 401(k) plans
- ☐ TSAs (tax-sheltered annuities)
- ☐ profit-sharing plans
- ☐ defined-benefit plans
- ☐ Roth IRAs

Financial statements

- ☐ net worth—balance sheet or assets/liabilities statements
- ☐ cash flow or income/expense statement
- ☐ records documenting collections—gold, coins, stamps, etc.
- ☐ inheritances (include copy of will or other verification of amount received)
- ☐ gifts received
- ☐ personal injury awards
- ☐ prior court judgments
- ☐ personal loans payable
- ☐ **debts**

- ☐ anticipated inheritances/gifts
- ☐ litigation claims
- ☐ loan applications
- ☐ personal loans receivable

    - ☐ lists of outstanding debts on date of separation
        - ☐ husband
        - ☐ wife
        - ☐ joint
- ☐ **credit card accounts**

    in whose name:

    account number:

    balance:
- ☐ ticket/subscription rights to sports/cultural events
- ☐ memberships in country clubs, fraternal orders, social/charitable groups

Financial Facts Checklist (cont'd)

Real estate records

- ❏ current value of home(s)
- ❏ original purchase price(s)
- ❏ property tax statements
- ❏ deeds
- ❏ bills for home improvements
- ❏ original mortgage(s)
- ❏ last monthly statement
- ❏ escrow closing statements
- ❏ appraisals

Safe-deposit box

- ❏ location/number/key
- ❏ inventory and/or copies of contents

Documents identifying other property, such as:

- ❏ patents
- ❏ royalties
- ❏ copyrights
- ❏ license agreements

Miscellaneous personal property, such as:

- ❏ boat
- ❏ camper
- ❏ other motor vehicles
- ❏ furs
- ❏ airplane
- ❏ motorcycle
- ❏ jewelry
- ❏ art work

Information on businesses

- ❏ filing papers
- ❏ percentage of ownership
- ❏ tax returns
- ❏ bank statements
- ❏ financial statements (last five years)
- ❏ credit card records
- ❏ current list of owners
- ❏ officer/director positions held by spouse/self
- ❏ retirement plan documents
- ❏ loan applications
- ❏ business agreements (partnership, buy-sell, etc.)
- ❏ appointment diaries
- ❏ documents on business-owned real estate (deeds, appraisals, leases, etc.)
- ❏ accounts payable/receivable

- ❏ **Household furnishings (attach itemized list)**

Miscellaneous personal documents

- ❏ passport
- ❏ power of attorney
- ❏ resume (for self and spouse)
- ❏ appointment diaries

Financial Facts Checklist (cont'd)

❑ bills for living expenses (child care, phone, electric, Internet access, cleaning, yard work, laundry, etc.)

❑ check stubs for other payments, such as disability income, unemployment compensation, or Social Security benefits

Any other document that will help establish

❑ net worth	❑ income	❑ liabilities
❑ husband	❑ husband	❑ husband
❑ wife	❑ wife	❑ wife
❑ joint	❑ joint	❑ joint

Facing the Future: What Must I Plan For?

Major Upcoming Life Events.. 136

Anticipated Financial Commitments.. 137

Major Goals That Will Cost Money 138

Where Does the Money Come From? 141

If you're thinking about a divorce or beginning one, you might believe you see only problems ahead. Nevertheless, you need to begin changing your outlook from "we" to "me"—from being part of a couple to being on your own.

It's helpful at an early stage to set basic goals for yourself. In most goal-setting exercises, you're asked to think about where you'd like to be in a few years—an effective approach for some people. In counseling divorcing clients, however, we've found it helpful to turn the question around. If you have trouble looking at the future, answer this:

*Where **don't** you want to be five years from now?*

Speaking financially, few of us want to be in debt, burdened with property we can't sell, or working two jobs just to survive. Yet that could happen to you if you drift through a divorce without planning ahead. By thinking about the kinds of financial problems you'd like to avoid, you can narrow your focus and clear the way for what you *do* want to accomplish.

As an initial planning exercise, use the columns in the chart that follows to jot down the immediate problems you foresee, as well as your goals. (Use the examples to give you ideas about your own needs.)

Once you have a basic idea of where you'd like to go in the future, take a few moments to reexamine where you are now. Use these questions to help you think about your own situation.

Major Upcoming Life Events

You can't predict the future, but there are major upcoming events that will affect what you will and will *not* be able to do with your money. Some you've anticipated for years. For instance, if your teenagers will begin college in a year or two, you may need a smaller house and a larger paycheck. Or perhaps you're reaching retirement and must prepare for living on a reduced income.

Other events, although not certainties, also warrant planning. If you're in an industry that's downsizing or a company that's being restructured, consider what it might cost to undertake a career shift or

pursue more education. Even if you're young and healthy, you should carry insurance in case of an accident. If you have children, you should map out a support system that will kick in should you die suddenly or become incapacitated—consider the finances of these arrangement as well as maintenance of the children's overall well-being.

Whatever your situation, remember that a proactive approach is always better than a reactive one. When you anticipate and prepare for major financial events, you can increase your odds of staying afloat by having a backup plan in place to cover related expenses.

Anticipated Financial Commitments

Take a moment to outline what's happening in your life and your anticipated financial commitments. Your list should include a timeline from start to finish. To whom are you committed and for how long? Have you agreed to support aging parents or another family member? How will the commitment be funded—monthly payments or annual installments? What happens if you can't afford the commitment you made? What other resources are available to you? Will you need money for a business investment or your educational goals? Do you have an adjustable rate or balloon payment mortgage? These types of game-changer events will challenge your ability to shape your financial future.

By getting this information down on paper, you will not only clarify your financial needs, but you can also reduce the fear factor in your vision of the future. You can determine whether a previous commitment, or one you want to make, is realistic. This is an important step to take when you're making key financial decisions in your divorce.

Once your list is complete, mark your top three priorities, in order. These are the issues that should immediately command your time and attention. We recommend tackling only three items at a time, because while you're working on them, your financial condition is likely to change, and your priorities may shift accordingly. This is not a one-shot exercise; you can repeat it any time you feel stuck in deciding what to do next in your divorce.

Major Goals That Will Cost Money

We all have personal goals, such as to lose weight, be great parents, and learn new skills. Here, we're listing the important life goals that are likely to affect your finances—and vice versa. These are big-ticket items like owning your own home, gaining the education necessary to improve your employability, or fulfilling an aspiration that's essential to your happiness and sense of accomplishment.

List those goals and estimate their corresponding costs.

Where You Do/Don't Want to Be in the Future	
In the Next Six Months: **Financial Problems to Avoid**	**Financial Goals to Meet**
Examples:	**Examples:**
Letting bills become overdue Paying only interest due on credit cards Using credit cards for all purchases Not tracking expenses	Pay off credit cards, or at least reduce balances by 50% Create a monthly budget calendar to avoid overdue bills Save 5% of income each month in emergency reserve Start monthly investing into XYZ mutual fund
Problems to Avoid:	
	Your Goals:

Where You Do/Don't Want to Be in the Future

In the Next Year:

Financial Problems to Avoid

Examples:

Not enough money to pay mortgage

Being late on mortgage

No medical insurance

Borrowing from friends or Mom

Your Problems to Avoid:

Financial Goals to Meet

Examples:

Sell house and prepare to buy my own—
set aside $20,000 toward down payment

Obtain disability insurance policy

Your Goals:

What's Happening in Your Life

Major Upcoming Life Events

Examples:

Daughter beginning college in two years

Move to new home in three years

Job change in spring because of recent layoffs

Son's wedding next June

Events in Your Life:

Anticipated Financial Commitments

Examples:

Pay for Dad's health care

New car—$300 per month

Remodel kitchen—$15,000

Your Commitments:

Major Goals That Will Cost Money	
Major Goal	**Expected Price Tag**
Educational:	
Example: Earn M.B.A. degree	$50,000 over two years
Example: Earn certificate in computer programming	$3,500 over six months
Your Educational Goals:	
Job:	
Example: Acquire contractor's license	$2,000
Example: Expand catering business by getting second truck	$18,000
Your Job Goals:	
Other:	
Example: Get certified as scuba diver	$1,000
Example: Go on African safari	$5,000
Your Other Goals:	

Where Does the Money Come From?

You've taken a look at what you absolutely *must do* and what you'd *like to do* next in your life. But how are you going to pay for it? Where will the money come from? And what are the consequences of how you bring money into your household?

During Your Separation, Be Aware of Where Your Money Comes From

While separated, you can get money in the same ways you did while married. Funds are available to you through:

- a job (salary) or business (income)
- borrowing/credit
- selling property or assets
- your spouse—in the form of temporary alimony, or
- your spouse's payments of marital debts.

But in this time period—between today and the final settlement—you are in a unique situation. You are setting the stage for the settlement, and you do not want to make financial decisions that could work against you. You must be sure to keep detailed, accurate records.

For instance, a woman who had never worked during her marriage got a low-paying job in a retail store and left her children in the care of her parents when her husband filed for divorce. In the final settlement, she received no alimony because she was already working to support herself—even though she wanted to go back to school to get training for a better job. Had she borrowed money to live on instead of getting the retail job, she might have been able to make a strong case for alimony.

In another case, the husband dutifully paid the household bills and temporary alimony for the two years that elapsed between the separation and the final settlement. He made the payments based on an informal arrangement with his spouse and did not document the arrangement in a court order or in a written agreement. In his final settlement, he received no tax deduction for the alimony he paid out.

As these cases illustrate, you must take care to anticipate the consequences of how you handle your money.

If you borrow or use credit cards. Under normal circumstances, financial planners would not recommend that you go into debt. During divorce, however, you may need to borrow money simply to survive. If you must use credit cards, try to find those with the lowest interest rates possible. If you borrow from family or friends, ask whether they will let you start repaying them when you're back on your feet.

It's crucial for you to document any money spent to support yourself and your children. If you paid for "common necessities of life"— food, shelter, clothing, or medical care—you may be able to obtain reimbursement for half of the costs. (See Chapter 5.)

If you sell property or assets. Before selling any jointly owned property without your spouse's knowledge or consent, ask your attorney whether you are legally permitted to do so. You will not be able to sell real estate, motor vehicles, or other assets with title documents that need both spouses' signatures. If you do sell any joint assets, be sure to set aside half of the proceeds for your spouse, or keep clear records of the money you received so that your spouse gets similarly valued property in the settlement.

If you need temporary alimony. If you need help paying the bills while the divorce is pending, you can request that your spouse pay temporary alimony. If your spouse is unwilling to do so voluntarily, you will need to request a hearing to have the issue resolved by a judge.

The judge looks at the needs of the party making the request— particularly that spouse's ability to be self-supporting—and the other spouse's ability to pay the temporary alimony. In some places, court officials have issued guidelines for judges to use in setting alimony. (For more information on alimony, see Chapter 18.)

If you will be *paying* alimony, be sure to get your agreement either ordered by a court or in writing. If you don't, you cannot deduct the payments from your income tax. If you will be *receiving* alimony pursuant to a court order or written agreement, you must report it as income on your tax returns. (See Chapter 11.)

Enforcing—And Making Permanent—Temporary Support Agreements

If your temporary alimony agreement is in writing, but has not been ordered by the court, you may want to take that extra step to obtain a court order. Without a court order, you probably won't be able to enforce the agreement if your spouse doesn't pay.

As explained in Chapter 11, getting a court order is not difficult. If you and your spouse agree to the amount of temporary alimony, you can put your agreement into writing as a "stipulation." You—or your attorney—can then take a few simple steps to get that stipulation turned into a court order. It involves asking the judge to sign the stipulation. You should not have to appear in court, though you may have to pay a filing fee for the stipulation.

Protecting Against Risks to Life, Health, and Property

Insurance 146

    Health Insurance 146

    Life Insurance 149

    Disability Insurance 152

    Business Continuation Coverage 152

Property and Estate Protection 152

    Title Documents 152

    Wills, Trusts, and Other Estate Planning Documents 154

Questions to Ask an Attorney 155

*If you had no choice but to drive through a snowstorm on a curvy road,
you would no doubt put chains on the tires and check the antifreeze. In
navigating through your divorce, you must also take precautions if you are to
survive financially.*

U se this chapter to "check the antifreeze"—that is, to make sure
you are doing what you can to reduce risks and protect yourself
and your property.

Insurance

Insurance is designed to protect you from financial disaster. But when
you divorce, your coverage itself may be at risk. Because policies can
make for boring reading, and almost no one thinks about this subject
until it's too late, insurance needs are commonly forgotten during
divorce.

To demonstrate the importance of insurance in your life, ask yourself:
*If I were in an accident or became ill, how would I pay my medical
expenses?*

Of course, being aware of what can happen without insurance and
doing something about your situation are two different things. Review
the steps below, then follow up.

Health Insurance

Find copies of the policy under which you are currently covered. Check
with your insurance agent to be certain that you and your children are still
covered. Find out how long your current policy will remain in effect and
what the costs will be for the next six months to one year.

If you are covered under a group policy, you may need to speak to the personnel office or a special agent who handles the company's health care insurance. If your family is covered by your spouse's insurance, you will be able to keep your coverage for 36 months after the divorce. (See "Keeping Health Care Coverage Through Your Spouse's Employer Under COBRA," below.)

Parents with custody can get health insurance for their children through the *noncustodial* parent's employment-related insurance program (or other group insurance plan), according to federal law. Neither the noncustodial parent, that parent's employer, or that parent's insurance company may deny coverage for any of the following reasons:

- The child does not live with the noncustodial parent.
- The child is not claimed as a dependent on the noncustodial parent's federal income tax return.
- The child lives outside of the plan's service area.

One challenge that the parent with custody may face is the noncustodial parent's reluctance to turn over money received for reimbursed medical claims. To avoid this problem, the custodial parent can obtain a Qualified Medical Child Support Order, known as a QMCSO.

The QMCSO is a court order that provides for health coverage for the child of the noncustodial parent under that parent's group health plan. When this court order is implemented, it requires the noncustodial parent's employer to enroll the child in the group plan. Further, the QMCSO requires that the insurance company reimburse the custodial parent directly for health care expenses incurred by the custodial parent for the child's benefit. This guarantees that custodial parents actually receive reimbursement for out-of-pocket contributions for a child's health care expenses. (For more on QMCSOs, see Chapter 18.)

Keeping Health Care Coverage Through Your Spouse's Employer Under COBRA

If your health insurance is provided through a plan offered by your spouse's employer, and 20 or more people work for that employer, you can continue your health care coverage for a period of time after the divorce. Under the federal COBRA (Consolidated Omnibus Budget Reform Act) law, divorced spouses of employed medical plan participants can pay for their own coverage for up to 36 months after the divorce is final. Under the law, the premiums you're charged for the COBRA coverage may not exceed 102% of what the employer pays to cover you. But you should check the costs of other health plans before deciding to get COBRA coverage through your spouse's employer. (For more information, contact your local office of the federal Department of Labor and ask for a copy of the booklet entitled *Continuing Health Care Coverage* or check the department's website, www.dol.gov.)

Your spouse's employer must inform you of your right to continue the coverage when your divorce becomes final. Don't leave it to your ex-spouse to notify the employer of that date. You should contact your ex-spouse's employer (probably the human resources department) and request the continued health care coverage as soon as your divorce is final.

Remember, the maximum length of time you can continue the coverage is 36 months. If you are healthy, start looking into individual coverage immediately. It may cost less than the continued group coverage. (Ask whether the provider of your continued coverage offers conversion to individual coverage. Be aware that such plans are usually expensive and limited in benefits.) Also, if you get a new job that provides group health insurance or if you remarry and your new spouse's employer offers group coverage, your continued coverage will usually be terminated.

Keep in mind that if you don't obtain new insurance before the end of the 36-month COBRA period, you risk becoming uninsurable if you fall ill.

Even if your ex-spouse's employer has fewer than 20 employees, state law may require the employer to provide COBRA-type benefits. For more information, contact your state Department of Labor, the personnel office at your spouse's job, or talk to your attorney or financial adviser.

Life Insurance

When you divorce, you need to reconsider who should be named the beneficiary of any life insurance policies. Often, divorcing people forget to change beneficiaries. Then, years later, when such a person dies, the *ex*-spouse could get all the benefits—even if there is a second spouse or if the deceased spouse wanted someone else, such as a child, to be the beneficiary.

In some divorce settlement agreements, the spouses may identify the beneficiaries of their life insurance policies. In certain cases, the judge might make a similar order. If your settlement agreement or divorce decree says who the beneficiary will be, but you forget to change the policy, the beneficiaries named in the agreement or decree may be protected when you die. But they may also end up in a messy fight with the insurance company or your former spouse's heirs, with no guarantee they will win. Don't rely on the settlement agreement or divorce decree to name the beneficiary.

At the same time, don't automatically remove your spouse from your policy when you separate. You may be prohibited from doing so until after your divorce is final. Also, to provide for your children's support and college education, you may want to either establish a trust or keep your soon-to-be ex-spouse as the beneficiary of your policy. This way, your spouse will receive a lump sum of cash to rear your minor children if you die before they reach adulthood.

 RESOURCE

For more on trusts, see *Make Your Own Living Trust*, **by Denis Clifford (Nolo).**

If you are paying alimony, your ex-spouse may insist on being the beneficiary of a life insurance policy. That will keep your ex-spouse from losing the stream of income you provide if you die prematurely.

Sometimes, one spouse may want to name the children as beneficiaries rather than have any money go to the children's other parent. Think carefully before you make such a move. If the children are under

18 when you die, the probate court that oversees the distribution of your assets will require that a guardian be appointed to manage the insurance proceeds—usually in a bank account—until the children reach 18. If the guardian wants to remove money from the account, called a blocked or suspended account, the guardian will need court approval—an expensive, time-consuming, and cumbersome procedure. And even when the money is distributed to the children at age 18, keep in mind that young adults are rarely mature enough to handle large amounts of money.

If you are receiving alimony or child support, don't overlook the importance of maintaining life insurance on your spouse so that your income stream is protected. Keep existing life insurance in force or add a new policy on the life of the person paying support. The paying spouse can deduct insurance premiums if the person receiving the alimony is the owner of the policy (meaning they have the right to control or change the policy or the beneficiaries). (See Chapter 18 for more on alimony.)

Look After Your Life Insurance

Would you want your soon-to-be ex-spouse to benefit from your life insurance? When you divorce, you both must make sure that the disposition of any life insurance policies you own is clearly specified in your agreements. These questions can help you clarify what needs to be done with these policies:

- Who will own the policy if it is transferred?
- To whom may it be transferred?
- Are there any restrictions or conditions of transfer?
- Will the original owner get the policy back at some future time (for example, upon remarriage, termination of spousal/child support, or death of the person to whom the policy was transferred)?
- Are there any outstanding policy loans? If yes, who will pay this loan?

Look After Your Life Insurance (cont'd)

- Will the policy owner have the right to designate beneficiaries? If yes, are there any restrictions on who is named a beneficiary? Must the policy owner designate a beneficiary (that is, an irrevocable beneficiary)?
- If so, does the restriction end upon a remarriage, termination of spousal support, or a death?
- If the beneficiary is a child, does this designation end when the child support obligation ends?
- If a policy matures during the lifetime of the insured person (as can happen with an annuity or an endowment policy), who will receive the maturity proceeds?
- What other policy rights will be affected by ownership or beneficiary restrictions? (These rights might include cash surrender, policy loans, or dividends.)
- Who will pay the premiums, and for how long?
- What happens if premiums are not paid? (While negotiating your settlement, you may want to require the payer spouse to provide evidence of payment to the other spouse at least ten days before the end of the policy's grace period.)
- Does a policy provision state what will happen if a premium is not paid—for example, that it will be paid via an automatic premium loan, paid-up insurance premiums, or extended term insurance?

If your settlement agreement or judgment makes changes to the life insurance policy, you must notify the insurance company and complete the necessary forms. Ideally, your agreement should give the spouse who will keep the policy until a certain date to complete the forms; if that spouse does not comply, the agreement should permit the court to complete the forms on that spouse's behalf.

(© Northwestern Mutual Life Insurance Company. Reprinted with permission.)

Disability Insurance

Another risk to your income is that you—or the ex-spouse who is paying support—may become disabled. If you're dependent on support, either spousal or child, you may want to consider obtaining disability insurance on your paying spouse. Talk to the benefits department at your spouse's company to see whether disability insurance is available. Or contact your insurance agent to determine the cost of getting disability coverage either for you or the paying spouse.

Especially if you're a single parent, you may hesitate to buy disability insurance due to the expense and your perceived lack of need. Consider the harsh reality your children would face, however, if you became disabled and unable to work.

Business Continuation Coverage

Do you or your spouse own a business? Check to see if you have insurance referred to as "key man" coverage. This coverage is designed so that the business can cover its operational costs in the event of the death of the business owner or another key person.

Property and Estate Protection

While the ultimate fate of your property may be unknown at the moment, you should nevertheless protect your interests by double-checking title documents and reviewing your will, trusts, and other documents that dispose of your property at death.

Title Documents

In order to divide assets and negotiate your divorce settlement, you must know how title (history of ownership) to those assets is legally recorded, regardless of whether that property is ultimately considered your separate property or part of the marital assets. Real estate title documents (deeds) are located at the county records office, usually called

the county recorder. If you can't find the deeds at the recorder's office, a title company can help you. The recorder may have a website where you can order a copy of a recorded document (expect to pay a fee). You can also ask a real estate agent for a "property profile" (usually free) that shows how title is held, or ask your mortgage company for loan documents, including any recorded deeds.

You should take a look at all recorded deeds to make sure your spouse has not unilaterally removed your name from property the two of you held jointly. In most states, either spouse can change title held in joint tenancy into tenancy in common, and you want to be sure your spouse hasn't done this without your consent.

Joint Tenancy and Tenancy in Common

Joint tenancy property is owned with each joint tenant having a "right of survivorship." This means that if either joint tenant dies, the other owner automatically inherits the deceased's portion, even if a will or another document states otherwise. To sell joint tenancy property, all owners must agree, or the owner who wants to sell must change title from joint tenancy to a different form of ownership, called tenancy in common, and then sell that share. Tenancy in common does not have the right of survivorship. Either owner can leave—or sell—that owner's share. (For more on joint tenancy property, see Chapter 6.)

Title documents for other types of property—for example, car registration forms and bank account and investment statements—are located in different places. To protect yourself from having your spouse unilaterally change title, you should write to the Department of Motor Vehicles or your bank or stockbroker to ask that no title changes take place until the divorce is settled.

Wills, Trusts, and Other Estate Planning Documents

If you've made a will or living trust or taken other measures to determine what will happen to your property after your death, your divorce most likely alters those plans dramatically.

In many states, a final divorce judgment automatically revokes part or all of your will. It's common for divorce to cancel all parts of your will that leave property to your ex-spouse; in some states, the entire will is wiped out. In any case, you should make a new will.

If you die before your divorce is final, your spouse may be entitled to a share or all of your property—even if your will leaves everything to other beneficiaries. To be certain about the distribution of your property, you can consult a Nolo estate planning publication (see the resource list at the end of this chapter) or an attorney.

CAUTION

Take steps to ensure that your property is distributed according to your wishes. If you don't have a will and haven't arranged to transfer your property by other means (for example, joint tenancy or a living trust), your property will be distributed at your death according to state law. Usually this means your spouse and children will get your property. If you're divorced and have no children, your property will probably go to your parents, siblings, and siblings' children. If that's not what you want, make sure you take the time to make a new will.

Planning for Your Children

If you have children, it's especially important to have an estate plan. At the least, consider:

- who will physically care for your children if you can't
- how your children will be supported if you're not around, and
- who will manage the property you leave them if they are under 18.

In your will, you can name someone to be the personal guardian of your children and have physical custody. And you can name the same person, or someone else, to manage the property your children inherit from you. As for figuring out how to support the children after your death, you may want to consider term life insurance, which provides a preset amount of money if you die while the policy is in force. It's a relatively cheap way to provide quick cash for beneficiaries.

Questions to Ask an Attorney

1. Can I change the beneficiary of my life insurance from my spouse to my children during the divorce? How and when can I do that?
2. Can the divorce settlement include a provision for life insurance on the person who provides support, to protect the support order or alimony?
3. As additional alimony or a separate provision, can the settlement include a provision for health insurance? Disability insurance?

Nolo's Estate Planning Tools

Nolo publishes several books and software products on wills, living trusts, and avoiding probate:

- *Quicken WillMaker Plus* is interactive software that helps you make a sophisticated will. For example, with *WillMaker* you can choose among three ways to provide property management for children should you die before they are competent to handle property themselves. In addition, *WillMaker* allows you to express your last wishes for your funeral and burial and contains a health care directive (living will), and durable power of attorney for finances, all valid in your state.

- *Plan Your Estate,* by Denis Clifford, is a comprehensive estate planning book, covering everything from basic estate planning (wills and living trusts) to sophisticated tax-saving strategies (AB trusts and much more).

- *Quick & Legal Will Book,* by Denis Clifford, is a primer for the do-it-yourself-minded who have the confidence and thoroughness to create detailed wills without an attorney's oversight. The book also provides the necessary forms for married and single individuals. *Nolo's Online Will* has a similar scope, and also includes tools to ensure that the will meets your state's legal criteria.

- *8 Ways to Avoid Probate,* by Mary Randolph, explains important and frequently overlooked ways to avoid probate, an often lengthy process of certifying a will. It is now possible to avoid probate for many kinds of property without creating a living trust. If you suspect you should be paying attention to probate avoidance, but dread thinking about it, start with this small and thorough book.

- *Make Your Own Living Trust,* by Denis Clifford, explains in plain English how to create a trust yourself, without hiring a lawyer. It contains all the forms you need to create your own living trust with step-by-step instructions for filling them out.

Taxes: How Do I File and Pay?

Get a Rough Estimate of Your Tax Bill 158

What Status Is Better to Use When Filing Tax Returns? 160

When in Doubt, File Separately 161

What to Know If You File Jointly 165

    Your Responsibility When Signing a Joint Return 166

    What If Your Spouse Refuses to Sign a Joint Return? 167

    Other Concerns When Filing a Joint or Separate Return 167

Dividing the Joint Tax Liability—Or the Refund 168

Tax Issues Involving Temporary Alimony or Child Support 171

Get Your Tax Agreement in Writing 172

Questions to Ask a Tax Adviser 172

When you're divorcing, you can't wait until just before April 15 to think about your taxes.

While you do not have to confront every small detail in your tax papers when you're in the midst of a separation, you can and should take basic precautions that will save you money and prevent hassles later.

For instance, suppose you assume that you and your divorcing spouse will file income taxes jointly. At tax time, however, you discover your spouse has already filed separately. How will you pay the IRS for taxes you did not expect to owe? Better to grapple with the filing question now than to make costly assumptions.

The following guidelines will help you address the income tax questions that are most important at this stage of your divorce. As you move through the rest of the divorce process (and this book), you will face more complicated tax issues regarding your property. For now, stick to the basics.

Get a Rough Estimate of Your Tax Bill

You don't want any nasty surprises next April 15. To avoid problems, make a rough calculation of your federal and state taxes. Ask your accountant or tax preparer to help you if need be. Figure out the amount you would pay if you filed separately and the amount you would pay if you filed with your spouse. Even if you haven't done all the paperwork or don't have all the information necessary to find the exact amount, it's useful to get a tax estimate. The estimate will at least give you a guide for deciding how to file. You'll also be able to use the information if you eventually need to convince your spouse to file jointly.

If possible, discuss the decision on tax filing with your spouse, lawyer, accountant, or tax adviser. Your goal is to reach an agreement about filing and then document that agreement in writing.

If you don't have access to tax information. Often, one spouse does most of the bookkeeping for a family. If your spouse played that role and does not want to give you the information, you may have to get a court

order or use legal discovery (evidence-gathering) techniques. For either, you'll probably need the help of an attorney. Consult the appendix for resources. Before obtaining legal help, however, be sure to point out to your spouse that withholding information will only increase both of your legal bills, and you'll get the information eventually.

You can contact your accountant for copies of previous returns, or ask the Internal Revenue Service for Form 4506. By completing the form and returning it to the IRS, you should be able to get the tax returns you need—as long as you signed the return when it was filed. The easiest way to get forms from the IRS is to visit the website at www.irs.gov. From this address, you can link to state tax forms, tax tables, and publications, and search for divorce-related tax topics by a word search. Upon request, the IRS will send you an account transcript, created from information on your tax returns, from as far back as ten years.

The IRS has several publications that can help during a divorce. The most important are:

- Publication 17, *Your Federal Income Tax for Individuals*
- Publication 501, *Exemptions, Standard Deduction, and Filing Information*
- Publication 504, *Divorced or Separated Individuals*
- Publication 505, *Tax Withholding and Estimated Tax*
- Publication 552, *Recordkeeping for Individuals,* and
- Publication 555, *Community Property.* If you live in the community property states of Arizona, California, Idaho, Louisiana, Nevada, New Mexico, Texas, Washington, or Wisconsin, many federal tax rules are applied differently than in the other states, which are called equitable distribution states.

If you have to wait for copies of previous years' returns and financial information, you can work with a tax accountant to estimate the amount you will owe the IRS. You can find out about deductions in several ways. Banks and lenders can tell you how much interest has been paid on a mortgage, and charities can give you the dollar amount you or your spouse donated. Call your spouse's employer to verify salary information and check copies of any loan applications you have. Chapter 12 will walk you through preparing *Net Worth* and *Cash Flow* statements, which can also help clarify your tax situation.

What Status Is Better to Use When Filing Tax Returns?

One of the most important decisions you will need to make is whether to file a joint return with your spouse or to file separately. The table below summarizes the types of filing status available to you and the requirements for each.

Filing Status	Requirements
Married filing jointly	You must be married up to and including the last day of the tax year (December 31) even if you are living apart.
Married filing separately	You must be married up to and including the last day of the tax year
Head of household	If you are single: • you must provide more than half of the costs of maintaining your household, and • your household must be the principal home of at least one dependent. If you are married and have lived physically apart since before July 1 of the tax year: • you must file a separate return • you must maintain your home and have your child, stepchild, or other dependent living there for more than half the tax year (temporary absences such as vacations, time in school, and time when a child is absent under a custody agreement do not count as time spent away) • you must be able to claim the child as a dependent or release that claim (provide your spouse with a waiver—see Form 8332), and • you must furnish more than half the cost of maintaining the household during the tax year. (In figuring that cost, include clothing, education, medical treatment, mortgage principal payments, vacations, life insurance, transportation, and the value of services performed in the household by the taxpayer. Exclude property taxes, mortgage interest, rent, repairs, utilities, home insurance, and food eaten in the home.)
Single	You must be unmarried, legally separated, or have your marriage annulled as of the last day of the tax year and be ineligible for head of household status.

When in Doubt, File Separately

If you and your spouse are unable to agree on how to handle your taxes, file separately. Generally, you can amend two separate returns (yours and your spouse's) to file jointly if you change your minds. You cannot, however, amend a joint return after the due date for that return in order to file separately. If you initially file separately, you have only three years to amend your returns and file jointly. Although filing separately may actually cost more in dollars, it could save you a great deal of grief.

You will, of course, want to file separately if it will save on taxes—but you should also file this way if your spouse misrepresents income or expenses. If you file jointly, you could be liable for back taxes, interest, and penalties on a joint return. By filing separately, you will lessen your risk and keep your options open. Bear in mind that if you file separately, you forfeit the right to certain deductions. These include education credits, child or dependent care expenses in most instances, the earned income credit, and interest paid on student loans.

Except in community property states, if you file a separate return, you generally report only your own income, exemptions, credits, and deductions. If your spouse had no income, you can claim an exemption for your spouse. If your spouse refuses to provide the information necessary to file a joint return, you should file a separate return.

Special Rules for Taxpayers Living in a Community Property State

As explained in detail in Chapter 12, married couples in Arizona, California, Idaho, Louisiana, Nevada, New Mexico, Texas, Washington, and Wisconsin live under community property laws. (Alaska gives married couples the option of identifying property held in trust as community property.)

Although these states' community property laws generally all operate in the same way, there are differences among the states. One significant variation is how they treat income from separate property. Arizona, California, Nevada, New Mexico, and Washington treat it as separate income. Idaho, Louisiana, Texas, and Wisconsin consider it community income even though it comes from separate property. (See "Questions to Ask a Tax Adviser" at the end of this chapter.)

Broadly, these rules provide that income earned and assets acquired during marriage are owned 50-50, regardless of which spouse actually brought home the paycheck. Before filing a separate tax return in a community property state, check with a tax adviser. You will probably have to report half of the combined community income and be permitted to take half of the joint deductions, in addition to reporting your separate income and claiming your deductions. As straightforward as this sounds, it's not always clear whether a deduction (expense) is community or separate. It basically depends on whose money was used to pay it and the character of the debt. If you paid an expense out of community funds, you must divide it equally and report half of it on each spouse's return.

Community property laws also affect exemptions for dependents. When community income supports a dependent child, the exemption may be claimed by either spouse, but it cannot be divided. If you have two or more dependent children, you can divide the exemptions (each claiming a child), but you cannot divide any single exemption in half. If you both try to take the exemption, the IRS will assume that each of you provided exactly half of the support and won't permit either of you to claim the exemption.

Avoiding Liability for Your Spouse's Dishonesty With the IRS

The IRS Restructuring and Reform Act of 1998 provides relief from joint liability when a spouse incorrectly reports items on a jointly filed tax return. The three kinds of relief and the factors influencing them are shown below. Innocent Spouse Relief is for all joint filers. It relieves you of some or all of the responsibility for paying tax, interest, and penalties if your spouse did something wrong on your joint tax return. The Separate Liability Election permits you to limit your liability for any deficiency to the amount that would be allocated to you if you had filed a separate return. Rules for Equitable Relief give the IRS discretion to relieve you of liability when you cannot qualify for relief under the rules for Innocent Spouse Relief or the Separate Liability Election.

There are two additional forms of relief for spouses in community property states. If one spouse fails to inform the other of community property income, the IRS can force that spouse to pay 100% of the tax due on the income that was kept secret. Also, if a couple has lived separately for an entire calendar year and each spouse has filed a separate return, the community income would be taxed to the person who earns it.

If you believe you may qualify for relief, go to the IRS website, www.irs.gov. The material there includes a section called "Spousal Tax Relief Eligibility Explorer." From the IRS home page, under the "Individuals" heading, click on "Tax Information for Innocent Spouses." The site will guide you through the qualifying factors one question at a time and, if you appear to qualify for the relief, will download the appropriate application forms. You can also download Publication 971, *Innocent Spouse Relief*. Before filling out the forms, however, consult your tax adviser.

Avoiding Liability for Your Spouse's Dishonesty With the IRS (cont'd)

Factors	Rules for Innocent Spouse Relief	Rules for Separate Liability Election	Rules for Equitable Relief
Type of liability	You must have filed a joint return that has an understatement of tax due to an erroneous item of your spouse.	You must have filed a joint return that has an understatement of tax due, in part, to an erroneous item of your spouse.	You must have filed a joint return that has either an understatement or underpayment of tax.
Marital status	No requirements	You must be no longer married, be legally separated, or have not lived with your spouse in the same house for an entire year before you file for relief.	IRS considers whether you are legally or physically separated or divorced.
Knowledge	You must establish that at the time you signed the joint return you did not know, and had no reason to know, that there was an understatement of tax.	If the IRS establishes that you actually knew of the item giving rise to the understatement, then you are not entitled to make the election to the extent of actual knowledge.	Intent to commit fraud. You must not have transferred property for the purpose of avoiding tax.
Other qualifications	You must file Form 8857 or similar statement within two years of the first date of collection activity.	You must file Form 8857 or similar statement within two years of the first date of collection activity.	You must file Form 8857 or similar statement within two years of the first date of collection activity.
Unfairness	You must be able to show that it would be unfair, based on the facts and circumstances, to hold you liable for the understatement.	Not applicable	The IRS must determine that it would be unfair to hold you liable for the underpayment or understatement of tax taking into account all the facts and circumstances.
Refunds	Yes, your request can generate a refund.	No, your request cannot generate a refund.	Yes, for amounts paid from your own funds between July 22, 1998, and April 15, 1999, and for amounts paid pursuant to an installment agreement after the date the request for relief is made. No refunds where relief is available under separate liability election.

A Note on Tax Rates

The American Taxpayer Relief Act, signed into law in early 2013, preserved most of the tax provisions that were introduced in the first decade of the century. That said, we can be no more certain than anyone else that the tax rate figures used in this book will be current by the time you read them.

In 2013, a new 39.6% tax rate applied to top income earners. In addition, new Medicare taxes are now applied to regular and investment income above certain thresholds.

In the interest of sanity and clarity, we've used all the 2013 rates in our examples and stuck with the current brackets—generally 10%, 15%, 20%, 25%, 33%, 35%, and 39.6%, with taxes on capital gains at 0%, 15% and 20% (plus a 3.8% Medicare tax). The planning principles outlined in this book apply regardless of any future rate adjustments, but in the event of new tax laws, you may need to plug new numbers into the worksheets you find here.

The changes to the tax code make advance planning for your tax filings—always a sound strategy—even more important if you are facing the prospect of divorce in the near future.

What to Know If You File Jointly

Filing a joint return is often to your economic advantage. A married taxpayer can claim the child and dependent care credit and the earned income credit (for low-income taxpayers) only on a joint return. In addition, certain deductions—such as the dependency exemption for a spouse and the deduction for a spousal IRA contribution—can be taken only on a joint return.

In deciding whether to file a joint return, you will want to calculate —or talk to your tax adviser about—how each of these factors will affect your tax bill:

- your tax rate
- any deductions or credits you may lose
- tax losses from partnerships or business losses
- tax benefits available by filing a joint return as compared to filing separately, and

• your liability for your spouse's dishonesty in reporting income and taking deductions when you file a joint return. (See below.)

Liability on Joint Tax Returns

If you file a joint return, each spouse is liable for 100% of the taxes due on the return as well as any penalties and interest assessed on that return. Unless you qualify as an innocent spouse (see "Avoiding Liability for Your Spouse's Dishonesty With the IRS," above), you are generally liable for any unpaid taxes, even after your divorce is final. (In some cases, if you sign a joint return under duress or your spouse forges your name, the IRS will not consider you jointly liable.)

If you believe your spouse is not being honest in reporting income or claiming deductions and credits, file a separate return—even if it is to your economic disadvantage. If it turns out that your spouse was being honest, you can always amend your returns and file jointly within three years of the original due dates. You cannot, however, amend a joint return and file separate returns.

The federal government increasingly gets involved in collecting past-due child support and alimony payments. If the IRS withheld a tax refund on your joint return because your spouse was behind in child support or alimony payments, you may qualify as an "injured" spouse and be able to recover your share of that refund. There's an easy process for doing this. Simply go to the IRS website, www.irs.gov, download and complete Form 8379, and return it to the IRS.

Additional factors to consider include your responsibility when signing a joint return and liabilities you could incur if your spouse signs your name to a return or refuses to sign the return.

Your Responsibility When Signing a Joint Return

According to the IRS, you may not ignore the information on any income tax return you sign. Legally, you must examine the return and

make certain that any transaction of which you are aware is properly reported.

What If Your Spouse Refuses to Sign a Joint Return?

If a spouse refuses to file a joint return and instead files separately, the other spouse must also file a separate return.

One spouse can file a joint return using only that spouse's name if the other spouse meets all the following requirements:

- refuses to file a joint return
- doesn't file a separate one
- had no income or earnings, and
- always filed a joint return in the past.

This works only if it is clear that the couple intended to file a joint return. If the refusing spouse feels the prepared return is fraudulent and won't sign as a result, the other spouse cannot file a joint return without the refusing spouse's signature.

Other Concerns When Filing a Joint or Separate Return

You may have several other concerns regarding your spouse and the IRS, such as:

What If Your Spouse Settles a Tax Bill With the IRS?

Suppose one spouse makes an agreement (called an offer in compromise) to pay the IRS less than the actual taxes due, plus interest and penalties, on a jointly filed return. That does not prevent the IRS from collecting money from the other spouse for the balance due. If the ex-spouse who is not covered in the settlement remarries, that person's new spouse also could be subject to unforeseen tax liabilities.

For instance, in one case, a widow had reached a settlement with the IRS to take care of significant tax liabilities that were attributable to her deceased husband. When she remarried, she had not finished paying the amount mandated by the settlement. She also transferred property she received from her first husband's estate to her new husband. The IRS sought to collect the balance of the settlement from the second

husband, who tried to argue that he should not be responsible for the payments because he did not participate in the original agreement. The court found, however, that because the property had been transferred to him, he was effectively a party to the agreement and therefore had to pay up. (Although this example involved a widow, the same principle would apply had she been divorced and transferred her divorce settlement property to her new husband.)

Will an Indemnity Clause in a Divorce Agreement Protect You From the IRS?

An indemnity clause is a section in a marital settlement or separation agreement in which one spouse agrees to pay the tax liability and to hold the other spouse harmless—in other words, ensure that the other spouse doesn't have to pay. This clause, however, isn't binding on the IRS. Except in unusual circumstances, the IRS can grab a joint taxpayer's property even after it has been divided in divorce, to satisfy a joint tax debt. In that case, the indemnity clause means that the spouse who ended up paying can collect reimbursement from the other spouse.

What If Your Marriage Is Annulled?

If your marriage is legally annulled, the IRS takes the position that it never existed. If you filed joint tax returns during your marriage, you will have to file corrected returns (as a single person) for the preceding three years. You must do this even if you had no separate income to declare.

Dividing the Joint Tax Liability—Or the Refund

Once you have prepared a joint income tax return, you will discover that you owe additional taxes or are due a refund. To avoid animosity and misunderstandings, you and your spouse must find a fair way to share the tax liability or divide the refund.

Of course, you need not do all of this work by yourself; a tax preparer can make these calculations for you. No matter who does the math, plug the numbers into the formula below. To do so, you must first figure out

what you each would have owed (or would have received as a refund) if you had filed separately.

Example: How to Calculate Your Share of a Refund on a Joint Tax Return

	Example	You
1. Your individual separate tax liability (married filing separately)	$3,000	
2. Your spouse's individual separate tax liability (married filing separately)	$7,000	
3. Total separate tax liability [Step 1 + Step 2]	$10,000	
4. Joint tax liability on return (married filing jointly)	$6,000	
5. Your share of the total tax liability [Step 1 ÷ Step 3]	0.30	
6. Your spouse's share of the total tax liability [Step 2 ÷ Step 3]	0.70	
7. Your share of joint tax liability [Step 4 x Step 5]	$1,800	
8. Your spouse's share of joint tax liability [Step 4 x Step 6]	$4,200	
9. Total tax payments made by couple	$12,000	
10. Total tax payment made by you [in Step 9]	$3,000	
11. Your share of the refund [Step 10 − Step 7]	$1,200	

Step 1.   Prepare a separate tax return for yourself using the "married filing separately" status to find the amount you would owe individually if you and your spouse filed alone.

Step 2.   Prepare a separate tax return for your spouse using the "married filing separately" status to find the amount your spouse would owe individually if you and your spouse filed alone.

Step 3.   Add together each spouse's individual separate tax liability to determine the total amount of your separate tax liabilities.

Step 4.   Prepare a joint tax return using the "married filing jointly" status to determine your total joint tax liability.

Step 5. Determine the percentage of the total taxes you owe by dividing your individual separate tax liability by the total separate tax liability.

Step 6. Determine the percentage of the total taxes your spouse owes by dividing your spouse's individual separate tax liability by the total separate tax liability.

Step 7. Determine the amount of the total taxes you owe by multiplying your joint tax liability by the percentage of the total taxes you owe.

Step 8. Determine the amount of the total taxes your spouse owes by multiplying your joint tax liability by the percentage of the total taxes your spouse owes.

Step 9. Calculate the total tax payments made by each spouse by adding all federal tax withheld, as shown on the W-2 forms, and all estimated tax payments made. If this amount is less than the amount in Step 4, more money must be paid to the IRS; if this amount is more than the amount in Step 4, the IRS will send you a refund.

If you owe more to the IRS:

Step 10. Subtract the total amount of tax payments made by each spouse (Step 9) from each spouse's separate share of the joint tax liability (Steps 5 and 6).

Step 11. Send a check to the IRS for the amount due. The spouse who pays the IRS is entitled to be reimbursed by the other spouse, who must make up the difference between what he or she has already paid individually (through withholding or estimated tax payments) and that spouse's share of the amount the couple still owes the IRS.

If you are entitled to a refund:

Step 10. Determine the amount of the total tax payment that should be refunded to each spouse by subtracting each spouse's individual share of the joint tax liability (Steps 7 and 8) from the amount each spouse has already paid to the IRS (through withholding or estimated tax payments).

Step 11. The spouse who receives the refund from the IRS must reimburse the other spouse as determined in Step 10.

> ⓘ **CAUTION**
>
> **If you and your spouse anticipate receiving a tax refund on your joint return, don't forget that this money is likely to be considered marital property.** That means it will be assigned or divided between the spouses as part of the property division upon divorce. For more on dividing up property, see Chapter 12.

Tax Issues Involving Temporary Alimony or Child Support

Often in divorce negotiations or mediation, spouses reach informal agreements about temporary alimony or child support payments. But sometimes they neglect to get the agreement in writing or to get a court order outlining the agreement. Do not make this mistake. (Read Chapter 18 before entering into any agreement to pay or receive alimony or child support.)

Getting a court order is not difficult. If you and your spouse agree on the amount of temporary alimony and/or child support, you can put your agreement in writing as a "stipulation." You (or your attorney) can then take the simple steps needed to get that stipulation turned into a court order. It involves asking a judge to sign the stipulation. You and your spouse should not have to appear in court.

You may resent paying alimony—but at least there is one bright spot. Alimony, including temporary alimony, is tax deductible to the payer. For alimony to be deductible, however, the court must order it or the parties must agree to it in writing. If you pay alimony or temporary alimony based on an informal, oral agreement, you cannot deduct those payments on your tax return.

For alimony to be deductible, the payer and the recipient of alimony cannot file a joint return.

If you're receiving alimony, you must report it as income on your tax return. And beware: Alimony payments made after the legal obligation to pay alimony ends—that is, voluntary payments—are not tax deductible to the payer, but they could be taxable income to the former spouse who received the payments.

Child support, unlike alimony, is neither deductible for the parent paying it nor taxable as income to the parent receiving it. However, regardless of whether you and your spouse are living apart, you will want to make arrangements for child support so that your children are adequately cared for. For more on child support agreements, see Chapter 18.

Get Your Tax Agreement in Writing

Whatever you and your spouse decide to do about taxes, be sure to get the agreement in writing. Have it notarized or possibly even reviewed by an attorney. Be certain, too, to check with an accountant or a financial planner for advice on the tax implications of your divorce.

RESOURCE

For more information on dealing with the IRS, see *Stand Up to the IRS*, by Frederick W. Daily (Nolo).

Questions to Ask a Tax Adviser

1. In my state, is income from separate property considered separate or marital income?
2. What income from separate property is considered to be separate property?
3. If my spouse and I ran a business while we were married, under what circumstances must I report the income and deductions on my own tax return now?
4. Is the income I earned after we separated but before the divorce becomes final considered separate or marital income?

Property and Expenses: Who Owns and Who Owes What?

Who Owns What—Marital Property and the Laws of Your State 174

    Jointly Owned Property 175

    Separately Owned Property 176

    Debts 177

Who Knows What—Using Legal Discovery 177

Net Worth—What Do You Own and What Do You Owe? 179

The Difference Between Assets and Income 191

Cash Flow—Where Does the Money Go? 193

Questions to Ask an Attorney 195

What—exactly—is at stake in your divorce? To answer that question, you must get a handle on your property and living expenses.

In this chapter, you define which property is legally yours and which assets you may have to struggle for in the tug-of-war of divorce. You will also develop a much-needed picture of your cash flow. And you will learn about how to use legal discovery procedures if you have any trouble getting the information you need. The financial fact-finding you did in Chapter 8 (you *did* do it, didn't you?) will be immensely helpful here.

By completing the tasks that follow, you will lay the foundation for the analysis and decision making that must be done to reach your final settlement. Think of this stage as developing your database or stocking your pantry with the items you will need throughout the divorce process.

The four basic issues you must address at this point are covered in the following sections:

- Who Owns What—Marital Property and the Laws of Your State
- Who Knows What—Using Legal Discovery
- Net Worth—What Do You Own and What Do You Owe?
- Cash Flow—Where Does the Money Go?

Who Owns What—Marital Property and the Laws of Your State

Before you begin to negotiate your divorce settlement, you must know what property you own alone—as opposed to what property is owned solely by your spouse or by the two of you together. Keep in mind that the rules that follow are general. To get the specifics on the property laws of your state, you will have to consult a lawyer or do some legal research.

Read this section for an overview of ownership, and then do whatever additional research is needed. Once you have determined which of your property is separate and which is marital, you will list this information in the "Net Worth Statement" in "Net Worth—What Do You Own and What Do You Owe?" below.

Jointly Owned Property

The property that couples accumulate during marriage is called, straight-forwardly enough, marital property. Depending on which state you live in, however, marital property will take different forms.

Community Property: Arizona, California, Idaho, Louisiana, Nevada, New Mexico, Texas, Washington, and Wisconsin. (Alaska also gives married couples the option of identifying property held in trust as community property.) In community property states, all earnings during marriage and all property acquired with those earnings are considered community property owned jointly by the couple. Even if one spouse earned a salary during the marriage while the other one kept house, the salaried spouse's earnings are shared equally. Separate property in these states—that is, property that is not "community"—consists for the most part of money and property that the individual spouses owned before marriage, and gifts and inheritances that either spouse receives during marriage.

At divorce, community property is divided equally (in half) between the spouses. A spouse who contributed separate money to an item bought with community funds—for instance, a wife who contributed her $5,000 inheritance toward a $20,000 down payment for the family home—may be entitled to reimbursement for that contribution. Conversely, a spouse who added community property money to a separate property item belonging to the other may be reimbursed.

In most community property states, a court has the discretion to divide the property equitably (fairly), if dividing the property in half would result in unfairness to one party. Additionally, in some community property states, a spouse whom the court considers at fault in ending the marriage may be awarded less than 50% of the community property. (These exceptions do not apply in California.)

Equitable Distribution of Property: All Other States. In the District of Columbia and the 41 states that follow equitable distribution principles, assets and earnings accumulated during marriage are divided equitably (fairly) at divorce. In theory, equitable means equal, or nearly so. In practice, however, equitable often means that as much as two-thirds of the property goes to the higher wage earner and as little as one-third goes

to the lower (or non) wage earner—unless the court believes it is fairer to award one or the other spouse more. In some equitable distribution states, if a spouse obtains a fault divorce, the "guilty" spouse may receive less than a full share of the marital property upon divorce.

Separately Owned Property

In all states, a married person can treat certain types of earnings and assets as separate property. At divorce, this separate property is not divided under the state's property distribution laws, but rather is kept by the spouse who owns it.

In community property states, the following is considered separate property:

- property accumulated before marriage
- property accumulated during marriage with premarital earnings (such as income from a pension that vested before marriage) or with the proceeds of the sale of separately owned premarital property
- gifts given to only one spouse
- inheritances, and
- property acquired after permanent separation.

In the equitable distribution states, separate property includes:

- property accumulated by a spouse before marriage
- gifts given to only one spouse, and
- inheritances.

In addition, some equitable distribution states consider wages kept separate from other marital property to be separate property.

Separate Is Separate Unless It's Mixed

In both community property and equitable distribution states, the separate property of a spouse generally remains separate property. However, if it's mixed with marital property or the other spouse's separate property, you may not be able to keep your separate property separate, unless you can trace the origin of your separate property and show how it was commingled.

Debts

Debts incurred during a marriage are usually considered joint debts—that is, during the marriage, both spouses are legally responsible for them. When a couple divorces, however, responsibility for marital debts is allocated in accordance with the property division laws of the state.

This usually means that the debts are divided equally or equitably (fairly, though not in half), especially when they were incurred for food, shelter, clothing, and medical care (called necessities). The court also considers who is better able to pay the debts (the spouse with the higher income or lower living expenses). If a couple has many debts but also has a great deal of property, the spouse better able to pay the debts may be ordered to assume the payments and also receive a larger share of the property.

Who Knows What—Using Legal Discovery

As described in Chapter 8, there are several ways to gather financial information during a divorce. Doing it yourself not only keeps down your lawyer's bill but also educates you about your family's finances and prepares you to negotiate from a position of strength.

If you can't find the information yourself, however, and all efforts to collect it informally fail, you can have your lawyer conduct "discovery." Discovery is the term for formal procedures used to obtain information during a lawsuit. As with all matters in your divorce, remember that the attorney works for you. You should be the one making the decisions about how extensive discovery procedures should be.

If you've never seen the checkbook and have signed tax returns for years without reviewing them, then your lawyer may have to do a lot of digging to create a picture of your financial life. Likewise, if you have participated in budgeting and bookkeeping, but your spouse owns a cash-and-carry business that you suspect is being used to hide assets, discovery may be worth every penny you spend.

On the other hand, if you and your spouse have few assets and you have a pretty good idea of what you're both worth, then using discovery to ferret out more information may simply be a waste of money.

Apply the maxim: Think financially and act legally. If you do, you should be fine.

In a divorce, the common discovery devices include the following:

Deposition (or examination before trial). A deposition is a proceeding in which a witness or party must answer questions orally under oath before a court reporter. In divorces, many lawyers want to take the deposition of the other spouse in order to ask about potentially hidden assets.

Interrogatories. Interrogatories are written questions sent by one party to the other to be answered in writing under oath. Interrogatories are often used to ask a spouse to list all bank accounts, investment accounts, and other assets ever held by that spouse.

Request (or notice) for production of documents. This is a request to a party to hand over certain defined documents. In divorce cases, spouses often request from each other bank statements, profit and loss statements, tax returns, pay stubs, and other documents showing earnings, assets, and debts.

Request (or notice) for inspection. This is a request by a spouse to look at items and documents in the possession or control of the other spouse. Items commonly inspected are original financial documents, houses, and cars. In divorces, the request often comes up regarding house appraisals. For example, Bill and Bernice are divorcing. The court orders Bill out of the family home to allow Bernice to stay with the children. They cannot agree on the value of the house, and Bernice won't give Bill's appraiser access in order to evaluate it. Bill must request an inspection.

Subpoena and subpoena duces tecum. A subpoena is an order telling a witness to appear at a deposition (or at trial). A subpoena duces tecum is an order to provide certain documents to a specific party. These devices are commonly used to get documents or testimony directly from banks, insurance companies, stockbrokers, and the like.

Net Worth—What Do You Own and What Do You Owe?

Generally, couples divide marital net worth at divorce. Subtracting what you owe from what you own reveals that net worth. It is important to know that net worth number because if you go to trial the judge will attempt to see that you each end up with your fair share, depending on the laws of your state. But it's also important for you to know specifically—item by item—what there is to divide. If, as is usually the case, you are able to negotiate a property settlement and avoid a court battle (perhaps with the help of attorneys or a mediator), you'll want to know exactly who owns what and how much it's worth. Only by knowing that can you negotiate the fairest deal.

The "Assets and Liabilities Worksheet" in this chapter will help you figure this out. It will also help you complete the "Marital Balance Sheet" in Chapter 19 that will set out your joint property, the equity in it, and its before-tax value. Your home, for example, might have a fair market value of $400,000. But if the loan is $250,000, you have just $150,000 in equity. Further, you or your spouse may have contributed unequal amounts of separate cash for the down payment. That kind of information must be factored in if you are to have a realistic view of what true assets there are to divide.

While it may seem laborious to compile all these facts and figures, your effort will pay off in succeeding chapters. We will refer back to these lists often. Patience and attention to detail now will be rewarded later.

To figure out your net worth, use the worksheets on the following pages. If you can't fill in everything in each category, don't worry. Gather as much information as you can. If you eventually hire someone to help you, you'll still be far ahead and will have saved money by gathering some of the information yourself.

Do You Hate Filling Out Forms?

Almost no one enjoys the task of writing information into small blanks on financial forms. In your divorce, however, you must get your financial facts straight. Any attorney or other professional you consult will want the information you are documenting in your Net Worth and Cash Flow statements. Most important, you will be organizing the information you must have to make the best decisions in your divorce.

Get help or meet with others who are dealing with the same issues so you can assist each other when the work gets tedious. You are not completing these forms as a mere exercise—you are using them to lay the groundwork for your new financial life.

Make copies of these forms so you can use them throughout your divorce and make changes as events change. Also, keep copies of the documents you consulted while you were filling out the forms.

Instructions: Net Worth Statement Assessing Assets

To determine your net worth, you need only estimate the current value (last column) of each asset you own. Complete the other columns if the information is available. List all marital property—not just the property in which you have an interest. These columns will help you understand as much as possible about your assets so you can make informed decisions when it comes time to divide your marital property.

As you fill in the information, subtotal the amounts in the last (far right) column for various asset categories (such as real estate or personal property). You will then transfer these subtotals to the "Net Worth: Balance Sheet Summary" that follows.

Column 1: Title [or Owner]. List the owner of each item of property, and note how ownership is listed on any title document. You or your spouse may have an ownership interest in an item of property even if title is in the other spouse's name only. This could happen if marital money or labor went toward the purchase of, payments on, or improvements to the property. Several types of property—such as houses, cars, and stocks— come with title documents (the house deed, car registration, and stock

certificate). When property is owned jointly, these title documents specify exactly how title is held—meaning how the property is owned. The possible ways to hold title to property are:

- *Joint tenancy.* Co-owners of joint tenancy property own the property in equal shares. When two or more persons own property as joint tenants, and one owner dies, the remaining owner(s) automatically inherits the share owned by the deceased person. This is termed the "right of survivorship." For example, if a husband and wife own their house as joint tenants and the wife dies, the husband ends up owning the entire house, even if the wife attempted to give away her half of the house in her will.

  In most states, a joint tenant can terminate a joint tenancy and change title to tenancy in common at any time, even without a spouse's knowledge. You only need to do this if you don't want your spouse to inherit your share if you die. The easiest way to do this is to write up a deed stating something like the following:

  Leslie Matthews transfers her one-half share of the home at 1312 Lincoln Drive from Leslie Matthews joint tenant to Leslie Matthews tenant in common.

  When such a change is made, the right to survivorship no longer holds—but if you want the property to go to someone else before your divorce is final, make sure you have a will.

- *Tenancy by the entirety.* Tenancy by the entirety is a way married couples can hold title to property in some noncommunity property states. Tenancy by the entirety is very similar to joint tenancy in that upon the death of one of the spouses, the property automatically passes to the surviving spouse, regardless of contrary provisions in the will. Unlike joint tenancy, however, one person cannot unilaterally sever the tenancy by the entirety. After divorce, former spouses cannot hold title as tenants by the entirety. Instead, title will be tenancy in common unless you specify joint tenancy.

- *Tenancy in common.* Tenancy in common is a way for any two or more people to hold title to property together. Each co-owner has an "undivided interest" in the property, which means that no owner holds a particular part of the property and all co-owners

have the right to use all the property. Each owner is free to sell or give away his or her interest. On death, an owner's interest passes through a will or by the automatic inheritance laws if the owner had no will. Divorcing spouses who plan to own property together after they split usually hold the property as tenants in common. Occasionally, ex-spouses will hold property in joint tenancy.

- *Community property.* Married couples in community property states can take title as community property. Community property is divided equally at the end of their marriages. Property owned jointly after divorce cannot be held as community property.

- *Separate title.* Sometimes title will be in only one person's name, even if both spouses own the property. The property will be divided between the husband and wife if the spouse whose name is not on the title document can prove an ownership interest. For example, if the house is in the wife's name only, but the canceled checks for the mortgage payments are written on the husband's separate account, the property will be divided.

Column 2: Source [of Asset]. Specify where the money to purchase the asset came from. Be sure to note whether the source is separate or marital property or income, and keep copies of all records that show the disposition of separate property.

Column 3: Date of Purchase/Acquisition. Put the date you bought items or received a gift or an inheritance. For items bought over time—such as shares of the same stock—list all dates and the number of shares bought each time. Use a separate sheet of paper if necessary.

Column 4: Current Balance or Market Value. For deposit accounts and similar investments, list the balance from your last statement. Keep copies of each statement.

Market value means what you could sell the item for now, not your purchase price or replacement cost. For valuable assets, such as art, jewelry, collections, antiques, furs, china, and silver, list the wholesale value. To determine these values, you might go to pawnshops, antique stores, jewelers, art galleries, auctioneers, or sterling silver replacement firms; you might also visit www.eBay.com to find sale prices on the

Web. For automobiles, use the *Kelley Blue Book* value. (Check out their website, www.kbb.com.) For household furniture, furnishings, and personal effects, give your best estimate of garage sale value.

Important: Use these amounts to calculate your subtotals.

Column 5: Debt. How much do you still owe on this item?

Column 6: Equity or Legal Value. Equity is the actual value of the item after the debt is paid off. Subtract Debt from Current Balance/Market Value to get this figure.

Instructions: Net Worth Statement Calculating Liabilities

The other half of determining your net worth is listing your liabilities (that is, what you owe). The Balance Due in the last column on the worksheet shows what you currently owe in each liability category. But you should also fill in the information in the other columns because it will be useful in working on your Cash Flow Statement and in evaluating assets in your settlement negotiations.

Column 1: Original Amount. Enter the amount you originally borrowed. Don't include interest. For your credit cards, put the amount of your purchases or cash advances.

Column 2: Date Incurred. Put the date the debt was incurred. For credit cards, note the various dates you charged items.

Column 3: Interest Rate. List all interest rates. If the rate is variable (as many mortgages and credit cards are), list the present rate and note that it is variable.

Column 4: Loan Term. How many years do you have to repay the loan or debt?

Column 5: Monthly Payment. List the amount you actually pay each month.

Column 6: Balance Due. Enter how much it would cost to pay off the loan in full. Check with the lender or creditor to determine the loan payoff amount. This balance is your current liability. The subtotal for each liability will be the amount you transfer to the "Net Worth: Balance Sheet Summary" chart, which follows.

Net Worth Statement: Assets and Liabilities Worksheet

Assets	Title H/W/ J/C [1]	Source S/M [2]	Date of Purchase/ Acquisition	Current Balance or Market Value	Debt	Equity or Legal Value
Cash & Cash Equivalents						
Cash						
CDs						
CDs						
Checking						
Checking						
Savings						
Credit Union						
Money Market Funds						
Money Market Funds						
Money Market Funds						
Other Liquid Assets						
Cash & Cash Equivalents Total						

[1] H=husband, W= wife, J=jointly, C=community property

[2] S=separate property, M=marital property

Net Worth Statement: Assets and Liabilities Worksheet (cont'd)

Assets	Title H/W/ J/C [1]	Source S/M [2]	Date of Purchase/ Acquisition	Current Balance or Market Value	Debt	Equity or Legal Value
Marketable Assets & Investments						
Life Insurance (cash values)						
Life Insurance (whole life policies)						
Stocks						
Bonds						
Mutual Funds						
Loans/ Accounts Receivable						
General Partnerships						
Limited Partnerships						
Liens & Judgments Due You						
Mortgages & Notes Receivable						
Contract Rights						
Marketable Assets & Investments Total						

[1] H=husband, W= wife, J=jointly, C=community property

[2] S=separate property, M=marital property

Net Worth Statement: Assets and Liabilities Worksheet (cont'd)

Assets	Title H/W/ J/C [1]	Source S/M [2]	Date of Purchase/ Acquisition	Current Balance or Market Value	Debt	Equity or Legal Value
Retirement Plan Assets						
Keogh or Self-Employment Plan						
IRA						
IRA						
Roth IRA						
Employee Stock Owner-ship Plan						
Pension/Profit Sharing/401(k)						
Annuity Plan						
Retirement Plan Assets Total						
Real Estate						
Residence						
Vacation Home						
Income Property						
Unimproved Real Property/ Lots						
Real Estate Total						

[1] H=husband, W= wife, J=jointly, C=community property

[2] S=separate property, M=marital property

Net Worth Statement: Assets and Liabilities Worksheet (cont'd)

Assets	Title H/W/ J/C [1]	Source S/M [2]	Date of Purchase/ Acquisition	Current Balance or Market Value	Debt	Equity or Legal Value
Business Interests						
Closely Held Private Stock						
Sole Proprietor						
Other Business Assets						
Business Interests Total						
Equity/Employee Benefit-Based Compensation						
Restricted Stock Awards						
Restricted Stock Units						
Stock Options/ Cash Value						
Life Insurance, purchased through company						
Equity/Employee Benefit-Based Compensation Total						
Personal Property						
Vehicle						
Vehicle/Motor Home						
Motorcycle/Off-Road Vehicle						
Home Furnishings (inc. art, antiques, silver, heirlooms)						
Jewelry/Furs						

Net Worth Statement: Assets and Liabilities Worksheet (cont'd)

Assets	Title H/W/ J/C [1]	Source S/M [2]	Date of Purchase/ Acquisition	Current Balance or Market Value	Debt	Equity or Legal Value
Collections/ Coins/Stamps						
Boat/Plane						
Horses/Other Livestock						
Other (specify):						
Personal Property Total						

[1] H=husband, W= wife, J=joint tenancy, C=community property

[2] S=separate property, M=property

Liabilities	Original Amount	Date Incurred	Interest Rate	Loan Term (mos)	Monthly Payment	Balance Due
Mortgages/Real Estate Loans						
Mortgage						
Mortgage						
Equity Loans						
Mortgages/Real Estate Loans Total						
Vehicle Loans						
Vehicle:						
Vehicle:						
Vehicle Loans Total						
Other Loans (Be sure to note if loans are secured by an asset)						
Bank Loan						
Bank Loan						

Net Worth Statement: Assets and Liabilities Worksheet (cont'd)

Liabilities	Original Amount	Date Incurred	Interest Rate	Loan Term (mos)	Monthly Payment	Balance Due
Loan Against Retirement Plan						
Loan Against Brokerage Account						
Private Loan						
Student Loans						
Student Loans						
Other:						
					Other Loans Total	
Credit Accounts						
Credit/Charge Account						
Credit/Charge Account						
Credit/Charge Account						
Credit/Charge Account						
					Credit Accounts Total	
Miscellaneous Debts						
Claims						
Unsettled Damages						
Liens & Judgments Owed						
Leases						
Taxes Owed						
					Miscellaneous Debts Total	

Net Worth: Balance Sheet Summary

Instructions: Enter the subtotal for each asset and liability category from the last (far right) column of the asset and liability worksheets. Add those figures to get total assets and liabilities. To find your net worth, subtract Total Liabilities from Total Assets.

Summary of Assets (from Assets Worksheets)

Cash and Cash Equivalents Total	
Marketable Assets/Investments Total	
Retirement Plans Total	
Real Estate Total	
Business Interests Total	
Equity/Employee Benefits-Based Compensation Total	
Personal Property Total	
Total Assets:	

Summary of Liabilities (from Liabilities Worksheets)

Mortgages/Real Estate Loans Total	
Vehicle Loans Total	
Other Loans Total	
Credit Accounts Total	
Miscellaneous Debts Total	
Total Liabilities:	
Net Worth (Assets minus Liabilities)	

The Difference Between Assets and Income

You've just inventoried your assets and how much they're worth. In the next section, you'll be examining your income—the stream of money that enables you to pay your expenses.

Does the difference between assets and income matter much? After all, in the end they're both your money, right? Well, when you're divorcing, the distinction is crucial. *Assets* are what get divided in the course of your divorce settlement. *Income* is what a court primarily looks at in determining how much alimony and child support you're entitled to or how much you have to pay.

Some items are considered both income and assets. And both terms cover more things than you might think. Now is the time to be as accurate as possible about what your income and assets both are.

You can begin to define income by using the IRS's rules—if you have to pay taxes on it when filing your returns, it's income. The government generally taxes you on your salary and wages, including tips, commissions, bonuses, vacation pay, sick leave, and compensatory time, as well as distributions from retirement or deferred compensation plans, and income from investments. You also owe taxes on income paid for lost wages and salary, such as severance pay, workers' compensation, unemployment insurance, disability insurance, and some disability pensions.

But for purposes of determining alimony and child support, income isn't just what the IRS says it is. A judge will consider virtually every possible source of funds in calculating the amount. (For more on the legal, financial, and emotional realities of alimony and child support, see Chapter 18.)

As for assets, certain ones are obvious: cash, house, car, bank accounts, and so on. But others are more obscure and easy to miss. (See Chapter 8 for a rundown of easy-to-forget assets.)

"Sources of Compensation," below, lists various types of compensation and assets—the items of value you're likely to be dealing with when ending a marriage. The table indicates whether each one is treated as income for purposes of determining child and spousal support or whether it is considered marital property that may be divided when the marriage ends. As you know, sometimes an item can be both.

Sources of Compensation

Source	Income for Support Purposes?	Asset Subject to Division?
Capital gains from sale of assets	Yes	Yes
Social Security benefits	Yes	No
Veterans' benefits	Yes	No
Military personnel fringe benefits, such as overseas housing allowance	Yes	No
Gifts	Generally no	No
Prizes	Possibly	Usually yes, if won during marriage
Educational grants and loans	No, at least in some states	No
Income of new spouse or partner	No, in most states	No
Reimbursement for employment expenses	Yes, to the extent it reduces cost of living	Yes, if expenses were paid out during marriage
Personal loans from corporations	Yes	Yes
Retained earnings of private corporation	Yes	Yes
Depreciation on equipment	Yes	Equipment is treated as property.
Depletion allowance from oil and gas interests	Yes	Interests are treated as property.
Personal expenses paid by a business	Yes	Payment increases value of marital business
Voluntary contributions to pension, retirement, and savings plans	Yes	Yes
Voluntary debt reduction	Yes	Yes
Stock options	Yes, whether exercised or not, as long as they are exercisable and have increased in value	Yes, generally
Accrued vacation time or sick time payable in cash	Yes	Yes

Cash Flow—Where Does the Money Go?

Cash flow is how money comes in and goes out. Regardless of how your settlement winds up, a divorce forces *two* households to exist—at least for a while—on the same income that had supported only *one*. So you need to know as precisely as possible what it costs you to live. And, of course, you must take the needs and expenses of your children into account.

The cash flow worksheets on the next few pages outline your sources of income and day-to-day living expenses. While filling out the worksheets, be realistic. And do not worry if you do not have enough information for every line. Do the best you can.

How Much Do Children Cost?

Once you have completed the worksheets, you will have a good idea of your estimated annual expenses. But what proportion of that goes for raising your kids? Knowing that figure could be important as you negotiate child support.

So you need to estimate how much you spend on your kids. To find that sum, do the following simple calculation:

Total monthly expenses	$_____
Subtract total child care expenses (monthly × 12)	$_____
Subtotal	$_____

Now multiply that subtotal by the percentage of your household that's composed of children. (For example, if you have two children living with you, that's 2 out of 3 or 67%; if three out of four residents are children, that's 75%, and so on.)

(Subtotal) _____ × (percentage representing children) _____ = _____

That's the children's pro rata share of non-child-care expenses. Now add back in the amount of child care expenses.

Child care expenses	$_____

And your total is the child care expenses plus the children's pro rata share of other expenses.

Total estimated child-related expenses	$_____

Instructions: Cash Flow Statement Adding Up Income

Income generally comes from three sources:
- earned salary or wages from your job as an employee
- earnings from your own business as a self-employed individual, and
- miscellaneous other sources, such as job bonuses, interest earned from investments, and income from a disability insurance policy.

Record all sources of income for you and your spouse, completing only those categories that apply. For investments from marital property such as interest income, rental property, and notes and trust deeds, divide the monthly income by two and enter half in your column and half in your spouse's column.

If an investment or asset is not owned 50-50, enter the appropriate percentages that apply. For instance, suppose a second house that produces rental income was bought partially with one spouse's separate property and partially with marital property (the income earned during marriage). In that case, the split might be 75-25 instead of 50-50.

If either you or your spouse holds more than one job or owns a self-run business, total your income from all sources on a separate sheet of paper and enter the total on this worksheet.

Instructions: Cash Flow Statement Estimating Expenses

Designed to create a realistic picture of your cash flow, this worksheet is divided into 15 categories, such as residence, child care, installment debt, transportation, and insurance.

To determine your current expenses, record the actual amount you spent for the past month for each item that applies. Multiply this monthly average by 12 and enter the amount at the bottom of the page. Don't forget to include once-a-year expenses, such as insurance premiums or real estate taxes. Finally, estimate how much of each expense is allocated for your children, and enter it in the box at the beginning of this section.

Listing Credit Card Expenses

To get a clear picture of your spending, list your various credit card expenditures under appropriate categories. For example, if your Visa bill is $198, with $120 spent for clothes and $78 for meals out, and you pay the bill off in full each month, enter the $120 under clothes and $78 under food expenses. If, however, you carry a balance on your credit card bill—for example, the bill totals $4,000 now and you make monthly payments of $300—enter the amount of the monthly payment under installment debt, and do not list the payment under the specific expense category.

Certain payments could be listed under more than one category. For example, a car payment could be considered a loan payment under the Installment Debt category, but may also be listed under the Transportation category. To help develop a picture of where your money actually goes, expense items are specified whenever possible—as with the car payment. When you complete the worksheet, double-check it to be sure you have not listed the same expenses under more than one category.

Income and expenses must be considered together. If you're spending too much, you've got to reduce expenses or increase your income. Otherwise, you'll find yourself facing serious debt problems.

Questions to Ask an Attorney

1. In my state, does the way title is held determine the disposition of property at divorce? Does it determine whether property is characterized as marital or separate?
2. Does interest income or appreciation on separate property belong to "me" or to "us"?
3. What happens if an asset loses money?
4. Are debts in my spouse's name separate property or marital property?
5. Am I liable for my spouse's credit card debt even though I did not sign the credit application?

Cash Flow: Income and Expenses

Income	Self		Spouse	
	Per Pay Period	Per Month	Per Pay Period	Per Month
Salary & Wages Pay period is: ☐ weekly ☐ biweekly ☐ monthly				
Gross income (including commissions, allowances, bonuses, and overtime)				
Subtract: Federal & state taxes				
FICA-Social Security				
Other: medical or other insurance, pension, stock plans, union dues, and the like				
Net Income from Salary & Wages				
Self-Employment				
Gross income				
Subtract: Federal & state taxes				
Self-employment taxes				
Net Income from Self-Employment				

Cash Flow: Income and Expenses (cont'd)

	Self		Spouse	
Income	**Per Pay Period**	**Per Month**	**Per Pay Period**	**Per Month**
Other Sources				
Dividends				
Interest income				
Rental property				
Royalties				
Notes and trust deeds				
Annuities, pensions/401(k)				
Alimony				
Child support				
Social Security				
Income tax refund				
Other (miscellaneous)*				
Subtract: Any deductions				
Net Income From Other Sources				

* Include income from trusts, disability benefits, unemployment insurance, welfare, and other public assistance programs.

Your Own Net Income from Salary & Wages	
Spouse's Net Income from Salary & Wages	
Your Own Net Income from Other Sources	
Spouse's Net Income from Other Sources	
Total Net Monthly Income From All Sources	
Estimated Net Annual Income (Total Net Monthly Income From All Sources x 12)	

Cash Flow: Income and Expenses (cont'd)

Monthly Expenses	Amount
Residence (family home and vacation home, if applicable)	
Rent/mortgage	
Property taxes	
Homeowners' insurance	
Savings for repairs	
Maintenance, general repairs, painting	
Lawn/garden service	
Supplies	
Pool care	
Housecleaning	
Pest control	
Homeowners' association fees	
Other (specify):	
Other (specify):	
Other (specify):	
Other (specify):	
Total Residence Expenses	
Clothing	
For yourself	
For children	
Total Clothing Expenses	

Cash Flow: Income and Expenses (cont'd)

Monthly Expenses	Amount
Medical and Dental Care (expenses not covered by insurance; include children)	
Doctors	
Hospital	
Counseling/psychiatrist	
Prescriptions	
Optometrist/eye care	
Dentist/orthodontist	
Lab costs	
Vitamins	
Other (specify):	
Other (specify):	
Other (specify):	
Total Medical/Dental Expenses	
Insurance (do not include property insurance from Residence Expenses)	
Life insurance	
Disability insurance	
Long-term care insurance	
Personal property insurance	
Umbrella policy	
Other (specify):	
Other (specify):	
Other (specify):	
Total Insurance Expenses	

Cash Flow: Income and Expenses (cont'd)

Monthly Expenses	Amount
Child Care Expenses	
Babysitter/day care	
Nursery school	
Lessons (art, sports, music, etc.)	
Allowances	
School tuition	
School lunches	
School transportation	
School books and supplies	
School uniforms/clothing	
Extracurriculars (art, sports, music lessons, etc.)	
Other (specify):	
Other (specify):	
Total Child Care Expenses	
Food at Home/Household Supplies	
Groceries/household supplies	
Liquor/wine	
Other (specify):	
Other (specify):	
Total Food at Home/Household Supplies Expenses	
Food Outside Home	
Restaurant meals	
Socializing outside home	
Total Food Outside Home Expenses	

Cash Flow: Income and Expenses (cont'd)

Monthly Expenses	Amount
Entertainment Expenses	
Movies	
Videos	
Plays	
Concerts	
Sports events	
Gym/club dues	
Entertaining/catering	
Records/tapes/CDs	
Books/magazines/newspapers	
Family trips/vacations	
Hobby supplies	
Other (specify):	
Total Entertainment Expenses	
Transportation/Auto Expenses	
Gas	
Oil/antifreeze	
Repairs/tires	
Car insurance	
Registration/fees	
Auto club	
Parking	
Taxi/train/bus/rentals	
Savings for new car	
Total Transportation/Auto Expenses	

Cash Flow: Income and Expenses (cont'd)

Monthly Expenses	Amount
Utilities	
Gas and electric	
Water	
Garbage	
Cable TV	
Internet service	
Cell phone	
Home phone	
Other (specify):	
Other (specify):	
Other (specify):	
Total Utilities Expenses	
School/Adult Education Expenses	
Tuition/fees/books	
Transportation	
Lunches	
Total School/Adult Education Expenses	
Laundry and Cleaning	
Dry cleaning and laundry (clothes)	
Dry cleaning and laundry (rugs, drapes, and furniture)	
Other (specify):	
Total Laundry and Cleaning Expenses	

Cash Flow: Income and Expenses (cont'd)	
Monthly Expenses	**Amount**
Incidentals	
Pet care (food/veterinarian)	
Gifts (birthdays, holidays)	
Donations and charities	
Personal care (hair care/cosmetics/manicures/massages)	
Stationery (paper/postage/computer supplies)	
Tobacco products	
Bank service charges	
Plants (florist/houseplants)	
Storage	
Church/synagogue contributions (tithes)	
Miscellaneous	
Other (specify):	
Other (specify):	
Other (specify):	
Total Incidentals Expenses	
Installment payments (Important: Include here only charged items not listed in other expense categories. Also note which party acquired the debt.)	
Creditor:	
Creditor:	
Creditor:	
Creditor:	
Total Installment Payments	

Cash Flow: Income and Expenses (cont'd)

Monthly Expenses	Amount
Other Expenses	
Estimated tax/income tax	
Tax preparation	
Nonreimbursed business expense	
Alimony paid	
Child support paid	
Legal/accounting (nonbusiness)	
Savings and investments	
Support of relatives not living in your home	
Deductible paid for auto accident or other claim	
Other (specify):	
Total Other Expenses	

Cash Flow: Income and Expenses (cont'd)

	Category Amount	Monthly Amount	Annual Amount
Estimated Net Annual Income			
Monthly Expenses			
Total Residence Expenses			
Total Clothing Expenses			
Total Medical/Dental Expenses			
Total Insurance Expenses			
Total Child Care Expenses			
Total Food at Home/Household Supplies Expenses			
Total Food Outside Home Expenses			
Total Entertainment Expenses			
Total Transportation/Auto Expenses			
Total Utilities Expenses			
Total School/Adult Education Expenses			
Total Laundry and Cleaning Expenses			
Total Incidentals Expenses			
Total Installment Payments			
Total Other Expenses			
Total Monthly Expenses From All Sources			
Annual Expenses (Total Net Annual Expenses From All Sources x 12)			
Cash Flow (subtract total annual expenses from total net annual income)			

What Will Happen to the House?

Financial vs. Legal Realities...209

The House—Keep It, Transfer It, or Sell It? Now or Later?.................210

Steps Toward Settling the House...210

Step 1: What Is Your Total Housing Cost Per Month?..........................215

Step 2: What Is the Current Fair Market Value of Your House?....................221

Step 3: What Is the Equity Value of Your House?.................................224

Step 4: How Much Will It Cost to Sell the House?...............................226

Step 5: What Is the Tax Basis for Your House?....................................229

Step 6: What Will the Financial Value of Your House Be If You Sell
   It With Your Spouse?...232

Step 7: What Will the Financial Value of Your House Be If You Sell
   It in the Future?...234

Step 8: What Is Your Share of the House?...239

Step 9: What's Your Best Option Regarding the House? 240

Step 10: How to Purchase the House From a Spouse or
   Former Spouse 243

Questions to Ask an Attorney or Financial Adviser...........................247

If you're a homeowner, a divorce raises some difficult questions.
Before making a decision, think through all the available options.

SKIP AHEAD

If you and your spouse don't own a house, skip ahead to Chapter 14.

A home can represent the partnership between husband and wife, a place where the children grew up, and a refuge from the demands of life. Perhaps your house stands for a connection to the community—or to the future generations of your family to whom you had hoped to bequeath it.

Whatever your house has meant to you, you're threatened with its loss during divorce. But rather than thinking you will "lose" your house, consider at least four basic options available to you:

Option A: Buy out your spouse's share and either keep it or sell it in the future.

Option B: Sell your share of the house to your spouse.

Option C: Sell the house together and split the proceeds with your spouse.

Option D: Continue to own the house jointly with your spouse and either buy your spouse's interest or sell yours in the future to your spouse or a third person.

To ease your anxieties, it helps to look at the numbers to find the true value of your home so you can make an informed choice. In the following sections, we explain the financial factors you need to consider before choosing the best option. Since the housing market is still sluggish in some parts of the country, divorcing couples are facing harder decisions and challenges than homeowners of 15 or 20 years ago. Make sure that any appraisal of the value of your home is a current one, not an estimate from months or years earlier.

CAUTION

The laws in your state and the circumstances of your particular case may affect your options regarding the house. For example, if you cannot reach an

agreement concerning the house, judges may presume that a custodial parent will stay in the family home until the children reach maturity. In that case, you would not be able to sell the house and split the proceeds with your spouse (Option C). Rather, the custodial parent would have to buy out the other parent's interest, or you might sell the house some time in the future (Option A or D). If you are unsure of your state's laws regarding custody and the right to sell the house, consult with an attorney before completing the calculations in this chapter.

Financial vs. Legal Realities

To understand your options better, consider home ownership from both legal and financial perspectives.

Suppose that during your marriage, you and your spouse bought a house using income earned from the marriage. You might assume that if you sell the house, the proceeds will be split 50-50. But the following table compares the legal and financial realities.

Legal Reality	Financial Reality
In the 41 states that follow the principles of equitable distribution, it's quite possible that the proceeds of the house won't be split down the middle. The 41 equitable distribution states are all the states except Arizona, California, Idaho, Louisiana, Nevada, New Mexico, Texas, Washington, and Wisconsin. (In Alaska, a married couple can elect to designate property as community property when it is within a trust. See Chapter 12 for a full discussion.) In equitable distribution states, judges are supposed to divide property "fairly." "Fair" doesn't necessarily mean "equal." In reality, "fair" could mean that one spouse receives a greater share of the proceeds.	If you and your spouse can reach your own settlement without going to court, you can each negotiate the best deal for yourself, rather than have a judge impose one on you. Rarely is anyone happy with the settlement a judge carves out. The judge won't factor in most financial consequences of keeping or giving up certain assets. A negotiated settlement, though not always perfect, is usually more palatable.

The House—Keep It, Transfer It, or Sell It? Now or Later?

If you've been in a house any length of time, you're probably accustomed to paying a set amount each month on your mortgage. Many people going through divorce simply focus on that monthly payment, which often looks lower than anything they would have to pay if they moved into an apartment or bought another home.

But looks are deceiving. A low or moderate monthly payment can keep you from seeing many of the other costs of owning a home. Suppose you do keep the house. First of all, a refinancing process may mean that your monthly payment increases, especially if you take cash out to buy out your spouse's share. In addition, will you be able to cover the cost of insurance, property taxes, repairs, cleaning, painting, and the like? What if you decide to sell the house in a few years? Are you prepared to cover the costs of the sale? Can you afford to buy out your partner's share of the house and still have money to live on? Unless you know the true financial costs of keeping a home, you can't answer these questions.

Steps Toward Settling the House

Here's a bit more about the most common scenarios for dealing with the family home at divorce.

Option A: Buy out your spouse's share and either keep it or sell it in the future.

This option can hold the most unforeseen financial risks. Therefore, you'll want to take into consideration as many known factors as possible. You must determine whether you can afford your monthly housing costs. You also want to estimate how much you can expect to make if you sell the house, and recognize the tax implications of a future sale.

A Tax Break on Capital Gains

Before May 6, 1997, the rules for determining the financial value of the family home often made it an asset of questionable value in a divorce settlement, because any profit realized from its sale could ultimately be eaten up in taxes. With changes to the tax code since that time, however, many homes are now sold without incurring capital gains taxes. The key change, for purposes of this discussion, is that capital gains—such as profit from selling the house—are not taxable up to $250,000.

Most capital gains that exceed $250,000 are taxed at a maximum capital gain tax rate of 15%. But for some homeowners, gains between $250,000 and $500,000 are also tax free. To qualify for the $250,000 exclusion, you must have owned and occupied the residence for two of the five years prior to the sale or exchange of the house. (Even if you sell sooner than in two years, you may be able to reduce or eliminate capital gains taxes on your profits under certain circumstances. Those exceptions would include being forced to sell the house for health reasons, a change in your place of employment, or because of an "unforeseen circumstance." The IRS defines the latter as war, man-made or natural disasters, losing your job, death of a spouse, and divorce or legal separation.)

If you are still married (or are remarried) when you sell the house, you and your spouse will owe no taxes on the first $500,000 of gain if you file a joint tax return and all of the following are true:
- either spouse owned the house two of the last five years.
- both spouses used the house two of the last five years.
- neither spouse used the exclusion during the prior two years.

If you hold on to your house and sell it after remarrying, you may have to pay taxes on gains over $250,000 if, for example, your new spouse doesn't live in the house at least two years, or if your new spouse used the exclusion on the sale of a different house during the previous two years.

And, you want to compare what you would receive if you and your spouse sold the house now. Of course, if you plan to keep the house for life, you don't need to worry about future sales costs or income taxes.

Option B: Sell your share of the house to your spouse.

To determine whether or not your spouse is making a good offer for your share of the house, you must understand its true financial value. The steps below will help you figure that out.

Option C: Sell the house together and split the proceeds with your spouse.

Before choosing this option, you need to know not only your current housing costs and equity value—you also need to know the costs of selling the house and the amount of profit you will realize after the taxes are paid.

Option D: Continue to own the house jointly with your spouse and either buy it in the future from your spouse or sell it in the future to your spouse or a third person.

When a divorcing couple holds on to their house, it's usually either because a custodial parent is staying with the children or because the housing market is so weak that the couple doesn't want to sell the house yet. However, owning the house together can be emotionally and legally tricky. If you do it, make sure you have a written agreement ensuring that the out-of-house spouse's $250,000 capital gains exclusion is preserved. This agreement must be pursuant to a "divorce or separation instrument"—meaning a divorce judgment or related document that discusses your arrangement about the home; a written separation agreement; or a decree requiring one spouse to make payments for the support or maintenance of the other.

If the agreement doesn't meet the strict requirements of the Internal Revenue Code, the out-of-house spouse could lose the exclusion. (See "A Tax Break on Capital Gains," above, for a general discussion of the capital gains exclusion.) It is also possible to buy the house from your spouse in the future. But before making such an agreement, ask your tax adviser about the consequences.

STEP BY STEP

Finding the true value of your house will take several calculations. Follow the steps below to answer these questions:

1. What is your total housing cost per month?
2. What is the current fair market value of your house?
3. What is the legal equity value of your house?
4. How much will it cost to sell the house?
5. What is the tax basis for your house?
6. What would be the financial value (after-sale/after-tax value) of your house if you sell it with your spouse?
7. What would be the financial value (after-sale/after-tax value) of your house if you keep it now and sell in the future?
8. What is your ownership share of the house?
9. Can you afford to purchase the house from your spouse?
10. What is your best option regarding the house?

CAUTION

You may be tempted to skip certain steps, knowing that you favor one option over the others. You are strongly advised to work through each step before making a decision. Completing all of the calculations will give you the most comprehensive picture of your financial choices—and can alert you to factors you may have overlooked.

Is Defaulting on the Morgage an Option?

With joblessness rising and home prices falling, the pressure on divorcing couples to do *something* about what's probably their biggest debt—the home mortgage—is greater than ever. You've probably read about millions of homeowners defaulting on their mortgages—that is, failing to pay their monthly mortgage payments or paying less than the amount due. In some cases, this is a strategy they've chosen; many other times, it's seen as the only way to survive financially. Short-term, of course, defaulting will lower your monthly expenses. But long-term, it can have significant costs.

For one thing, defaulting on your mortgage can add to the amount you already owe. For example, you'll be charged late fees for nonpayment, and that can add hundreds of dollars to your mortgage bills. Similarly, once you're in default, your loan servicer may charge you for "default-related services" such as property inspections, property preservation services like maintenance, and foreclosure costs, such as attorneys' fees and title searches. In addition, even one late payment will damage your credit score, affecting your ability to get a loan in the future. (See Chapter 5 for more on credit scores.)

If you are in default, your lender may start the foreclosure process. This will add to the costs you will need to pay to bring your account current. It's also a matter of public record, meaning it'll be tougher for you to get credit and buy another home in the future.

If you aren't able to bring your account current or work out another solution, your home could be sold at a foreclosure auction. And in many states, you may be responsible for paying what's called a "deficiency judgment"—the difference between what you owe on the loan and the amount the mortgage holder gets at the foreclosure auction.

Given all of these facts, you want to do everything you can to avoid default and foreclosure. If you're struggling to make your mortgage payments or you're in default, contact your mortgage lender right away. You may be embarrassed to talk about your payment problems, and you may hope that your financial situation will soon improve. But keeping the lines of communication open is a key to resolving your problem.

Is Defaulting on the Morgage an Option? (cont'd)

Talk to your lenders about the options, which may include loan modifications, repayment plans, or a temporary reduction or suspension of payments. Even if you're not eligible for any of these options, your lender may be able to help you find a solution short of foreclosure, including a possible short sale—selling the house at a price below what's owed and getting the bank to accept that as payment in full—or a voluntary transfer of the property through a "deed in lieu of foreclosure."

Staying in touch with your lender also could save you money. For instance, some may not order property inspections or property preservation work if you let them know each month that you're still living in the home and maintaining it. They're also more likely to delay a foreclosure sale if you are working with them to find a better solution.

If you have a hard time reaching your loan servicer, call 1-888-895-HOPE or go to www. hopenow.com for free personalized guidance from housing counseling agencies certified by the U.S. Department of Housing and Urban Development.

Step 1: What Is Your Total Housing Cost Per Month?

Your mortgage does not represent the entire amount you spend on housing. (And remember, if you refinance the house in order to raise cash to buy out your spouse, you're likely to end up with a larger loan and probably bigger monthly payments.) You must also consider costs like utilities and maintenance to get a true picture of expenses.

Use these steps to find your net monthly housing costs:

a. List your monthly housing costs.

b. Total your gross monthly housing costs.

c. Find your net monthly housing costs by subtracting tax savings from your gross monthly housing costs.

List Your Monthly Housing Costs

List the amount you pay each month for the items on the "Monthly Housing Costs" chart that follows. You already have this information on the "Cash Flow Statement" you completed in Chapter 12. If you don't know the monthly cost of an item, or if you don't incur an expense every month, estimate the amount you spend per year and divide by 12. To estimate what you'll spend on significant repairs and maintenance needs, look around the house and note any major work that needs to be done. Check previous invoices to get an idea of what you'd normally spend on these expenses—or call a plumber, an electrician, or another person who can give you an estimate.

Monthly Housing Costs		
Item		**Your Monthly Costs**
Mortgage		$
Second mortgage (or equity loan payment)		$
Property taxes		$
Maintenance		
General repairs & painting	$	
Lawn/garden service & supplies	$	
Pool care	$	
Housecleaning	$	
Pest control	$	
Homeowner's association fees	$	
Other	$	
Total Maintenance		$
Insurance		$
Utilities		$
Major repairs		$
Total Gross Housing Costs Per Month		$

As this chart reveals, the price of owning and maintaining your house is higher than your monthly mortgage alone. Even if certain expenses aren't being incurred now—or haven't been for a while—do not be lulled into thinking they no longer exist.

This chart includes the cost of repairs because those repairs will be your sole responsibility if you opt to keep the house. You may be able to get your spouse to split immediate repair costs with you when you negotiate the settlement. But because you don't know for sure that your spouse will agree, figure that you will have to come up with the money yourself.

Total Your Gross Monthly Housing Costs

Once you have filled in the information on the "Monthly Housing Costs" chart, add up all of the items in the "Your Monthly Costs" column. That figure is your gross monthly housing costs. Write that amount here.

$ _____ *Gross monthly housing costs*

Find Your Net Monthly Housing Costs

In considering your monthly housing costs, you should not overlook the tax benefits of home ownership. Under current tax laws, you can deduct your mortgage interest payment and the property taxes you've paid, so your actual housing costs may be *lower* than the sum of your gross monthly housing costs. At the end of each year, you'll receive a Form 1099 from the bank or lender who holds your mortgage showing how much you paid in mortgage interest during the year.

If you take out a loan to purchase your spouse's share of the house, interest on that loan won't be deductible unless you use the house as collateral, you are legally liable for repayment, and the mortgage or deed of trust is recorded. Thus, interest on an informal loan from parents or friends may not be deductible, unless you give the lender a secured note and record it.

In addition, you may be able to deduct points (prepaid interest) in the year you purchase a home if you take out a mortgage to buy it.

However, points are not deductible if you refinance your house to buy out your spouse's interest. Points on loans for home improvements are deductible in the year in which the loans are taken out. If you refinance the mortgage on your house, your deductions for the points are spread out over the life of the new loan. Additionally, any prepayment penalty you incur because you refinance the mortgage is deductible as interest in the year paid. Any points not previously deducted can be deducted as interest in the year of the refinancing.

To calculate your tax savings, follow these instructions:

a. Add together the total amount of interest paid on your mortgage during the year and the amount you paid in property taxes.

b. Multiply that number by your tax bracket—we assume a 28% federal tax bracket. (Don't forget your state tax bracket if your state allows a deduction for mortgage interest.)

c. Divide this calculation by 12 to get a monthly average.

d. Subtract the monthly average from your gross monthly housing costs to get your net housing costs.

Although the result may not be the exact amount you can deduct, you'll get a pretty close estimate for your housing costs. For precise figures, consult your accountant or tax adviser. The example and step-by-step formula that follow will make it easier for you to find your net monthly housing costs.

> **EXAMPLE:** Mitch and Candy's house is worth $150,000. Their mortgage is $1,200 per month ($14,400 per year), and they pay $272 a month ($3,264 per year) on a second mortgage. Their annual mortgage payments are $17,664 ($14,400 + $3,264); $14,000 of those payments is interest. They pay property taxes of $2,250 a year. Their federal tax bracket is 28%, giving them an annual tax savings of $4,550 or $379 a month. Assume their gross monthly housing costs equal $2,502.

Annual mortgage interest		$14,000
Annual property taxes	+	2,250
Deduction	=	16,250
Tax bracket	×	0.28
Tax savings	=	4,550
Monthly tax savings	÷ 12 =	379
Gross monthly housing costs		$2,502
Monthly tax savings	−	379
Net monthly housing costs	=	$2,123

To apply this formula to your situation, fill in the blanks below. You will need to know your annual mortgage interest, annual property taxes, and tax bracket.

$ _____ *Annual mortgage interest*

To find your mortgage interest, contact your lender and request a calculation of the annual mortgage interest on your loan. Or, ask the lender for the amortization schedule for your loan. On either document, you will see a monthly listing of how much of the mortgage payment repays principal (your debt or balance due) and how much pays interest on the loan.

$ _____ *Annual property taxes*

To find your property taxes, refer to your federal tax returns, call the county tax assessor, or call your lender if your property taxes are paid through an escrow account with your lender. You can subtract taxes assessed for street, sidewalk, utility, curb, sewer, and other property improvements.

Some high-income taxpayers must reduce their annual mortgage interest and property taxes by the lesser of 3% of their adjusted gross income over a certain amount or 80% of the amount of the itemized

deductions otherwise allowed for the year. (Internal Revenue Code § 68.) Your tax preparer can help you make the calculation if you're in that bracket.

_____ % *Tax bracket*

To find your tax bracket, check the income tax schedules contained in the IRS tax packet or ask a tax adviser to verify your current tax bracket. Or simply use 25%, the federal tax bracket into which most people fall. Remember to use only your anticipated income as a single person, including any alimony you might receive or pay, as the basis for your tax bracket. And your filing status will be either "single" or "head of household." (For help in figuring out which category to use, see Chapter 11.)

Now enter those numbers in the "Net Monthly Housing Costs" chart. Using the formula in the chart, calculate your net monthly housing costs. You will refer to this number later when making your decision on what to do with the house.

Net Monthly Housing Costs		
Annual mortgage interest*	=	$
Annual property taxes*	+	$
Deduction**	=	$
Tax bracket	×	
Tax savings	=	$
Monthly tax savings	÷ 12 =	$
Gross monthly housing costs		$
Monthly tax savings	–	$
Net monthly housing costs	=	$
$		Net monthly housing costs

* Some high-income taxpayers may not be able to fully deduct mortgage interest and property tax payments. See section on annual property taxes, above.

** Remember: Your interest and therefore your interest deduction may be higher if you refinance to get the money to buy your spouse's share of the house. Ask yourself: "Can I afford to keep the house if I include the additional costs of buying out my spouse?"

Appraiser Qualifications

In every state, a real property appraiser must meet the minimum national qualifications set by the Appraiser Qualifications Board (AQB) and may be limited in the services he or she can provide. Some are only qualified to appraise residential real estate; others are certified general appraisers, whose services may extend to all types of real property.

If you have both residential and commercial property to divide, make certain that the appraisers you plan on using have the necessary credentials to assess all the real property in your possession. While the AQB sets minimum national standards for real estate appraisers across the country, each state may set its own standards, which may be even more restrictive.

Step 2: What Is the Current Fair Market Value of Your House?

The current fair market value of your house is the amount you can realistically expect to get when you list your house for sale on the open market. You can obtain fair market value estimates from real estate agents free of charge. You might want to ask the agent who helped you buy the house originally or an agent who consistently markets in your neighborhood.

You can also get appraisals or opinions of fair market value from a certified appraiser, but be prepared to pay. For a referral to a certified appraiser, ask your banker or lender or contact the Appraisal Institute (www.appraisalinstitute.org, 1-888-756-4624). If you and your spouse cannot agree on the value of the house, ask your attorney to select an appraiser. This person will testify to the house's value if the issue is contested in court.

Real estate appraisers generally look at a house from several financial angles and take an average of these values to estimate what it is worth. The two primary views they take are:

- what the market will bear—that is, the sales prices of comparable houses in the area, and
- your house's replacement value (the cost of building materials and the like).

 WEB RESOURCE

Real estate sites that may be helpful in determining your home's value:

- **www.dataquick.com.** Offers detailed information to help you compare sales prices for many areas of the country based on information from county recorders' offices and property assessors
- **www.realtor.com.** Can help you find a home or neighborhood, learn about home values, find lenders and movers, make calculations, compile checklists, and locate real estate reference materials, and
- **www.zillow.com.** One of the best real estate sites, an online real estate service that allows you to quickly access the "value" on your home or others you may be interested in, as well as providing mortgage information.

Once you have a good estimate of the fair market value of your house, write it here:

$ _____ *Fair market value*

Remember that the estimates you receive are just that: estimates. The precise value of your house will only be known once the property is sold to a neutral third party—one that has no interest in your divorce. This is why valuation experts often disagree over property values.

In the event you and your spouse both hire appraisers who don't agree on the value of your house, be prepared to ask the professional you've hired a few questions:

- Where are the areas of disagreement?
- What is the estimated value of those areas of disagreement?
- How might the differences be resolved?

Since you will probably have to pick and choose your battles in divorce, you should undertake a cost-benefit analysis of any appraisal disagreements. Try to determine how much you stand to lose in the disagreement and weigh it against the (considerable) fees of litigation, including attorneys' fees, expert witness fees, court costs, recording fees, and any wages you might lose during the process.

Loan Modification Programs

If you are intent on keeping your home after a divorce, but fear that monthly payments will be too high, you should investigate various government-sponsored programs designed to help people stay in their homes during economic hardship.

The Treasury Department, in partnership with the U.S. Department of Housing and Urban Development, has developed the Making Home Affordable Program (MHA) as a response to the wave of foreclosures that crested after the economic downturn of 2007-2008. Homeowners who cannot meet their current payments may be eligible for a loan modification through this program.

The Home Affordable Refinance Program (HARP) is available to some homeowners who are current with their mortgage payments, but who have not been able to take advantage of lower interest rates because their home values have dropped or because they lack sufficient equity. This program is available only to homeowners whose mortgages are owned or guaranteed by federally regulated Fannie Mae or Freddie Mac.

You can find out more about these programs and their benefits by visiting the MHA website (www.makinghomeaffordable.gov/) or calling 1-888-4673.

Think Like a Buyer

Before paying for an appraiser, you might ask a real estate agent who regularly markets your neighborhood to prepare a comparative market analysis (CMA). This is typically provided at no charge.

After physically inspecting your house, the agent researches the market for recent sales, how long property takes to sell, currently listed homes, pending sales, and expired listings (unsold homes taken off the market).

Don't hesitate to ask more than one agent to provide you with a CMA. In addition to getting a consensus about what your home is worth, you will also learn about the quality of agents with whom you may do business later.

In valuing your house, think like a buyer, not a seller. Specifically, do not rely on the following as indicators of a sales price for your home:

- the amount you spent on improvements (you may have spent more than the home prices in your neighborhood will support)
- the price you paid for your home; markets are dynamic and the market value of your home may have gone up or down since you purchased it, or
- the assessed value shown on your property tax statement—local tax assessments are often out of date and rarely reflect the actual market value.

Step 3: What Is the Equity Value of Your House?

The equity value of your house is its fair market value (the figure you obtained in Step 2) minus debt connected to the house. Remember—if a court were to calculate each spouse's share of a house, the court would consider only the amount of equity. That number, however, does not reflect the sale costs or taxes—that is, the financial reality of owning or selling a house.

The debt connected to your house equals what you owe to any lenders (first mortgage, second mortgage, or home equity loan or line of credit)

plus any liabilities on the house. All liabilities must be paid before any cash disbursements can be made from the sale of a house. Liabilities include income or property tax liens, child support liens, judgment liens (if someone sues you and obtains a judgment, that person can put a lien on your house), or mechanic's or materialman's liens placed by a contractor who did work on your home but wasn't paid.

Ask your lender for the balance on your mortgages and equity loans. The lender should offer you two figures—a principal balance and a payoff balance. The principal balance is simply the amount of principal remaining to be paid on your loan. The payoff balance is the principal balance plus an additional month's interest and any prepayment charges you must pay to close out the loan. Use the payoff balance in calculating the debt on your house.

You may not be aware of all liens on your property. Tax liens and mechanic's liens, for example, often appear without the owner's knowledge. To find all liens on your house, you'll have to do a title search. You can hire a title insurance company to conduct the search, or you can visit the records office in the county where your deed is recorded. A clerk in that office can show you how to search for any liens on your house.

Possible Debts Against Your House	
Mortgage balance	$
Second mortgage balance	$
Equity loan or line of credit balance	$
Property tax lien	$
Income tax lien	$
Child support lien	$
Judgment lien	$
Mechanic's or materialman's lien	$
Other liens	$
Total Current Debt on House	$

Now subtract the debt on the house from its fair market value to obtain the equity value.

Fair market value (Step 2)	$ _____
Total current debt on house (Step 3, above)	– $ _____
Equity value of your house	= $ _____

Step 4: How Much Will It Cost to Sell the House?

To determine how much money it will take to actually sell your house, you can do a quick ballpark estimate, or you can use the calculations that follow to find a more precise figure. To find a precise figure, you'll have to include not only the costs of sale, but also the amount it will take to fix up the house and prepare it for sale.

Quick Ballpark Estimate of Cost of Sale

Take the fair market value figure you determined in Step 2. To get a ballpark amount for the sales cost, multiply that amount by .07. Obviously, some sales will yield higher or lower costs of sale, but 7% will do for a rough estimate. Enter that number in the *Total cost of sale* blank, below.

Precise Valuation of Cost of Sale

To obtain a more accurate cost of sale figure, you need to total the following amounts:

$ _____ *Agent's commission*

This represents the amount a real estate agent charges you for selling a house. Generally, it's set at 6% of the final selling price. The fee can sometimes be negotiated down. Also, if you list your house as a FSBO— for sale by owner—you avoid the agent's commission altogether. But be sure to educate yourself about the process before you list your house as

a FSBO. You can find information on the topic at www.nolo.com in the "Real Estate" section of the website. If you live in California, check out *For Sale By Owner in California*, by George Devine (Nolo).

$ _____ *Closing costs*

These costs can include escrow fees, abstract and recording costs, appraisal fees, surveys, title insurance, transfer and stamp taxes, loan charges, and miscellaneous expenses. All this can add up to several thousands of dollars. A lender or real estate agent can help you figure the exact amount.

$ _____ *Attorney's fees*

In a few states, you will need to hire an attorney to help you "close" the sale of your house. If you paid an attorney when you bought the house, the cost should be similar, though you may need to adjust it upward to account for inflation and the passage of time. Otherwise, ask a real estate agent for an estimate.

$ _____ *Fix-up costs*

Fix-up expenses are not considered when calculating capital gains, but they are a very real expense when selling a home. Estimate here the amount of money you will have to spend to prepare your house for sale. Does it need major repairs or just a paint job? Does the foundation, roof, or plumbing need work? Will you need to have the house inspected? Must you get a termite inspection? (If you need help in getting a handle on what repairs are necessary, the American Society of Home Inspectors' website at www.ashi.com provides a list of inspectors near you, or you can ask your realtor for referrals.)

$ _____ *Total cost of sale*

Add these amounts together to get a reasonably close estimate of what it will cost to sell your house.

What Happens When the House Secures a Spouse's Separate Property Debt?

Generally, a spouse should not keep a house that secures the debts of a soon-to-be-former spouse. Before agreeing to keep a house securing your spouse's separate property debt, be certain your spouse has paid all such debts, and request proof of payment.

A California case underscores the point. The husband forged his wife's signature on a quitclaim deed (giving him sole ownership) and then used the house as security while running up additional debts. His creditors got a judgment against him and attached a lien to the house. At the divorce trial, the court ordered the husband to pay the judgment. The wife chose to keep the house and buy out his share. In calculating the value of the house, however, the court did not reduce it by the amount of the lien. Thus, the wife had to pay the husband one-half of the value of the house—and also pay off the lien plus interest.

Moral of this story: Once a lien, always a lien—until the creditor is paid. A creditor's right to payment is not affected by who is ordered to pay marital debt. If you want to keep the house, deduct all liens, plus interest, when calculating the value. If your spouse refuses, don't keep the house.

If you decide to sell the house together, make sure that all separate property debts secured by the house are identified. Instruct the escrow company to pay each spouse's separate debts out of that spouse's share of the proceeds. For instance, if the net sale proceeds total $100,000 after costs of sale, each spouse is entitled to $50,000. However, if there is a separate property lien of $25,000 against one spouse, the escrow company should issue checks as follows: $50,000 to the spouse without the separate property debt, $25,000 to the spouse with the separate property debt, and $25,000 to the creditor.

Step 5: What Is the Tax Basis for Your House?

The tax basis of an asset is a little-mentioned financial concept that no one thinks about until it's time to sell the asset—or until a divorce.

The tax basis is the dollar amount the IRS uses to determine whether you've made or lost money on an asset. Essentially, your tax basis is the original purchase price plus the cost of any improvements you have made, minus any tax benefits you have realized. Knowing the tax basis lets you calculate your profit to determine whether you'll owe taxes when your house is sold during or after divorce. If the house isn't sold until after the divorce, the person who gets the house gets the tax liability.

To figure your tax basis, enter the appropriate amounts in the blanks below.

$ _____ *Purchase price*

The purchase price should be on the original purchase contract or closing statement—check your files or safe deposit box. If you still can't find it, call your lender or the escrow company that processed your purchase.

$ _____ *Acquisition costs*

These are costs associated with the purchase, such as recording fees, title insurance, and document fees listed on the escrow statement. Do not include loan principal, interest, or points.

(If you acquired the house before May 6, 1997, check to see whether you filed IRS Form 2119 with your tax return. The "Adjusted Basis of the New House" shown on the form is equal to the purchase price plus the acquisition costs. If you used Form 2119, you don't need to add in the acquisition costs.)

$ _____ *Capital improvements*

Capital improvements are items that add value to your house, such as a new bathroom or den, a security system, or extensive landscaping. The costs of these changes are not fix-up or repair costs. You will need to go through your files to find receipts that document the improvements made to your house.

The following are examples of capital improvements:

- Additions
  - bedroom
  - bathroom
  - deck
  - garage
  - porch
  - patio
  - storage shed
  - fireplace
- Lawns and grounds
  - landscaping
  - driveway
  - walkway
  - fence
  - retaining wall
  - sprinkler system
  - swimming pool
  - exterior lighting
- Communications
  - satellite dish
  - intercom
  - security system
- Heating and air conditioning
  - heating system
  - central air conditioning
  - furnace
  - duct work
  - central humidifier
  - filtration system
- Electrical
  - light fixtures
  - wiring upgrades

- Plumbing
  - water heater
  - soft water system
  - filtration system
- Insulation
  - attic
  - walls
  - floors
  - pipes and duct work
- Interior improvements
  - built-in appliances
  - kitchen modernization
  - bathroom modernization
  - flooring
  - wall-to-wall carpet
- Miscellaneous
  - storm windows and doors
  - roof
  - central vacuum system.

$ _____ *Tax benefits*

Tax benefits refer to the rollover of gain (profit) from homes sold up to August 5, 1999 as well as deductions taken for a home office. To find your tax benefits, refer to your tax returns or call your accountant. *For simplicity's sake, you can calculate the tax basis without subtracting any tax benefits.*

Use the figures you entered earlier in this section to find the tax basis of your house.

Purchase price	$ _____
Acquisition costs (do not include fix-up costs)	+ $ _____
Capital improvements	+ $ _____
Tax benefits	+ $ _____
*Tax basis of house**	= $ _____

* If you have Form 2119, you can use the Adjusted Basis for this figure if you add capital improvements.

For Those With Home Office
Depreciation After May 6, 1997

Claiming a tax deduction for use of a home office is a growing trend. However, the law allowing you to exclude $250,000 of gain (profit) on the sale of a house does not include the amount attributed to depreciation for home office use. So if you depreciate your house for home office use, the amount of depreciation will be taxed as a capital gain (25%).

The following formula allows you to figure the amount not qualifying for the exclusion:

Total home office depreciation taken since May 6, 1997	= $	_____
× 25% capital gains rate	×	0.25
Taxes not qualifying for exclusion	= $	_____

Step 6: What Will the Financial Value of Your House Be If You Sell It With Your Spouse?

If you profit from the sale of a house, you may exclude up to $250,000 ($500,000 if you file a joint return) of that profit once in a 24-month period. You must consider this exclusion in order to figure out the financial value (after-sale/after-tax value) of your house and analyze your future options concerning house purchases and tax deferrals. Remember, too, that you will need to know your capital gains rate to determine the financial value of the house. As a general rule, if you are in the 25% tax bracket your capital gains rate is 15%, and if you're in the 15% bracket, your capital gains rate is 5%. (Don't forget that you may owe state taxes as well.)

To find the financial value (after-sale/after-tax value) of your house, enter the following information from the previous steps:

$ _____ *Fair market value (from Step 2)*

$ _____ *Current debt (from Step 3)*

$ _____ *Cost of sale (from Step 4)*

$ _____ *Fix-up expenses (from Step 4)*

$ _____ *Tax basis of the house (from Step 5)*

$ _____ *Taxes to pay from home office use (from Step 5)*

_____ *% Capital gains rate (including state)*

Now take the following steps:

1. Find the amount realized by subtracting the cost of sale from the fair market value.

2. Find the gain or loss on the potential sale by subtracting the tax basis from the amount realized. If you had a loss, use zero as the amount of gain and go directly to Step 5.

3. Find the taxable gain by subtracting your exclusion from the gain. (Taxable gain is that portion of gain exceeding the maximum gain you can exclude.) The maximum exclusion is $250,000 for individuals and $500,000 for many married couples filing jointly. The net gain is the portion of the gain that will be taxed. If your net gain does not exceed the maximum exclusion allowed, go directly to Step 5.

4. Find the taxes on the taxable gain by multiplying the taxable gain by the capital gains rate.

5. Find the financial value by subtracting the current debt on the house, taxes on taxable gain, cost of sale, fix-up expenses, and taxes on home office use from the amount realized.

EXAMPLE:

1. Find the amount realized:

	Your House	Example
Fair market value	_____	$400,000
Cost of sale	_____	$ 28,000
Amount realized	_____	$372,000

2. Find the gain or loss:

	Your House	Example
Amount realized	_____	*$372,000*
Less tax basis	_____	*$100,000*
Gain (or loss)	_____	*$272,000*

3. Find the taxable gain:

Gain	_____	*$272,000*
Exclusion	_____	*$500,000*
Taxable gain	_____	*$0*

4. Find the taxes on taxable gain:

Taxable gain	_____	*$0*
Capital gains rate	_____	*× 15%*
Taxes on taxable gains	_____	*$0*

5. Find the Financial Value:

Amount realized	_____	*$372,000*
Current debt on house	_____	*$90,000*
Taxes on Taxable Gain	_____	*$0*
Taxes on home office use	_____	*$0*
Fix-up expenses	_____	*$6,000*
Financial value	_____	*$276,000*

Step 7: What Will the Financial Value of Your House Be If You Sell It in the Future?

Of all the options you may pursue regarding the house, the choice of selling it in the future can be the most difficult to analyze. Take your time and get help in determining real estate values, if necessary.

No one can predict what will happen in the future to the real estate market or the tax laws. But real estate brokers can make calculated guesses. Assuming you would not keep a house you expected to be

worth less in the future, it's important that you know what its future value is likely to be.

A technical aid, the *Future Value Factor* table, lets you calculate changes in the value of an asset like a house. We've included a *Future Value Factor* table in the appendix. Use it, along with the formula below, to get an idea of what your house will be worth in the future. But remember that this will be only an estimate.

To use the *Future Value Factor* chart, you must decide how long you plan to stay in your house (called the holding period), and you must select a rate of inflation. Financial planners often figure an annual inflation rate of 3% or 4% when estimating housing values. Depending on the state of the economy and your local housing market, you may be tempted to use an inflation rate higher than 3% or 4%. Certainly many houses increase in value at a rate substantially higher than the inflation rate. Nevertheless, we recommend that you be conservative when speculating on the future value of your house. It's better to expect less and get more than vice versa.

You will see that the *Future Value Factor* chart lists numbers of years vertically and rates of inflation horizontally. Estimate the number of years you expect to hold on to the house, select an anticipated inflation rate, and find where the selected row and column intersect. That will give you the future value factor. For example, if you expect to hold the house another seven years and you assume inflation won't go above 4% during that period, your future value factor is 1.3159.

To find the future financial value (after-sale/after-tax value), enter the following information for your house:

_____	*% Inflation (rate you anticipate)*
_____	*Holding period (how many years you plan to keep the house)*
_____	*Future value factor (from appendix)*
$ _____	*Fair market value (Step 2)*
$ _____	*Current debt on house (Step 3)*

$_____   *Cost of sale (Step 4)*

$_____   *Fix-up expenses (Step 4)*

$_____   *Tax basis (Step 5)*

_____   *% Capital gains rate (Use the state rate. Virtually all states follow the federal law for exclusion regarding the house, but many have their own capital gains rules.)*

$_____   *Taxes to pay from home office use (see "For Those With Home Office Depreciation After May 6, 1997," above)*

Now take the following steps:

1. Find the future value factor (see the appendix) by estimating the holding period and the rate of inflation.
2. Find the future sale value by multiplying the fair market value by the future value factor.
3. Find the potential amount realized by subtracting the cost of sale from the future sale value.
4. Find the potential gain or loss on the potential sale of your house by subtracting the tax basis from the amount realized. If you had a loss, use zero as the amount of gain and go directly to Step 7.
5. Find the potential taxable gain by subtracting your exclusion from the gain. Taxable gain is that portion of gain exceeding the maximum gain you can exclude. If you are single, the maximum gain you can exclude is $250,000. (Remember: To qualify for the exclusion, you must own and reside in the home for two of the last five years prior to sale, and you must not have sold or exchanged a principal residence in the two years prior to the sale.) If your taxable gain does not exceed the maximum exclusion allowed you, go directly to Step 7.
6. Find the taxes on the taxable gain by multiplying the taxable gain by your capital gains rate.

7. Find the financial value (after-sale/after-tax value) by subtracting what you expect your debt on the house will be, as well as taxes on taxable gain, cost of sale, fix-up expenses, and taxes on home office use, from the amount realized calculated in Step 3.

EXAMPLE:

1. Find the future value factor—See appendix.
For the purpose of this example, assume a holding period of five years and an inflation rate of 5%—the future value factor is 1.2763.

2. Find the future sale value

Fair market value	*$400,000*
Future value factor	*× 1.2763*
Future sale value	*= $510,520*

3. Find potential amount realized

Future sale value	*$510,520*
Cost of sale	*– 35,736*
Potential amount realized	*= $474,784*

4. Find potential gain or loss

Amount realized	*$474,784*
Less tax basis	*– 100,000*
Gain (or loss)	*= $374,784*

5. Find the potential taxable gain

Gain	*$374,784*
Exclusion	*– 250,000*
Potential taxable gain	*= $124,784*

6. Find the potential taxes on taxable gain

Taxable gain	*$124,784*
Capital gains rate	*× 15%*
Potential taxes on taxable gain	*= $18,717*

7. Find the future financial value

Amount realized	*$474,784*
Anticipated debt on house	*– 90,000*
Taxes on taxable gain	*– 18,717*
Taxes on home office use	*– 0*
Fix-up expenses	*– 6,000*
Potential financial value	*= $360,367*

Now fill in the numbers for your house.

1. Find the future value factor

   Enter number from appendix: _____

2. Find the future sale value

   Fair market value _____

   Future value factor × _____

   Future sale value = _____

3. Find potential amount realized

   Fair market value _____

   Cost of sale – _____

   Potential amount realized = _____

4. Find potential gain or loss

   Amount realized _____

   Less tax basis – _____

   Gain (or loss) = _____

5. Find the potential taxable gain

   Gain _____

   Exclusion – _____

   Potential taxable gain = $ _____

6. Find the potential taxes on taxable gain

Taxable gain _____

Capital gains rate × *15%*

Potential taxes on taxable gain = _____

7. Find the future financial value

Amount realized _____

Anticipated debt on house − _____

Taxes on taxable gain − _____

Taxes on home office use − _____

Fix-up expenses − _____

Potential financial value = _____

Step 8: What Is Your Share of the House?

You may have expected your marriage to be a 50-50 proposition, and you may believe that your divorce will be the same—a simple division in which each person gets half. When it comes to the house, however, those percentages may not hold up.

In this step, you must determine what share of the house belongs to you. That number will help you choose your best option regarding the house and inform your negotiation of the final settlement. When you determine the total value of the house under different scenarios, you need to know your share so you can figure out how much of that total value you can expect to receive. For instance, if the total value of your house is $100,000 and you are each entitled to 50% of it, you'd get $50,000. But if you are only a one-fourth owner of that house, you'd be entitled to $25,000, not $50,000.

The laws of your state and the origin of the money used to buy your house determine the amount of your share. Refer to Chapter 12 for an overview of marital and separate property. If one spouse was the sole owner of the house when you entered the marriage, the other spouse may nevertheless be entitled to a share at divorce. In some states, when the nonowning spouse or the marital earnings pay for mortgage

payments and improvements (such as additions and renovations), that action usually gives the nonowner some ownership interest.

Once you have determined your estimated ownership share, enter it here.

_____ *% Separate property*

_____ *% Marital property*

Your house can be worth different amounts under different circumstances, depending on your ownership share and the option you ultimately choose. Multiply the percentage of marital property by the equity or legal value and place that figure on the "Marital Balance Sheet" in Chapter 19. Then, using the same process, figure out the amount of the home that is separate property and place that in the "Marital Balance Sheet," noting that that separate-property share must be bought out as part of the ultimate negotiations.

Step 9: What's Your Best Option Regarding the House?

Now consider your options in light of your calculations in the previous steps. The advantages and disadvantages of each option are recapped below. If you choose Option A or B and do not think you can afford to buy out your spouse's interest, see the suggestions in Step 10, below.

Option A: Buy out your spouse's share and either keep it or sell it in the future.

Advantages: For the sake of children, or stability in your own life, it may serve you to keep the family home. You can take advantage of the IRS exclusion that lets you sell a house and not pay taxes on the first $250,000 of profit.

Disadvantages: When you eventually sell the house, you will be solely responsible for costs of sale and for taxes on the profit over $250,000. There is also a chance that the real estate market in your area could crash and that the house could sell for less money in the future.

Is this a viable option? _____ Yes _____ No

Option B: Transfer your share of the house to your spouse.

Advantages: You do not have to pay taxes on the money you receive because the IRS does not tax money received as a property buyout during a divorce.

   Disadvantages: You may not earn enough on the sale of the house to afford to buy a new one. If your income is not high enough, you may not qualify for financing.

Is this a viable option for me? _____ *Yes* _____ *No*

Option C: Sell the house together and split the proceeds with your spouse.

Advantages: You share the cost of sale and the tax liability with your spouse. Also, you can each take advantage of the IRS provision that lets you exclude up to $250,000 of gain on the sale of the house ($500,000 if you've remarried and meet certain conditions; see "A Tax Break on Capital Gains," above).

   Disadvantages: You'd be selling the family home, to which you may have strong emotional ties. Further, the amount you receive from the sale of the house may not be enough for you to afford to buy another house, or you may not qualify for financing to buy another home.

Is this a viable option for me? _____ *Yes* _____ *No*

CAUTION

If you sell the house to a third person during the calendar year before the year in which your divorce is final, and taxes are due on the sale, consider filing taxes using a "married filing separately" taxpayer status. Otherwise, you could be responsible for your spouse's liability if your spouse is dishonest with the IRS (see Chapter 11). Alternatively, you could set up an escrow account to cover potential taxes due.

Checking Out Sex Offenders

When evaluating a new place to live, you may be particularly concerned with safety, especially if you have children. The Dru Sjodin National Sex Offender Public Website (www.nsopw.gov) is the first national search site that allows parents and other concerned residents to identify sex offenders not only in their neighborhoods but in adjacent communities as well—even if the neighboring community is across a state line. Users may search all, or selected, sex offender registries by name, state, county, city/town, or ZIP code. Coordinated by U.S. Justice Department, the site connects sex offender lists in all 50 states, the District of Columbia, and Guam. It's named for a University of North Dakota student who was abducted and murdered in 2003 by a convicted sex offender who'd been released by a neighboring state.

Option D: Own the house jointly with your spouse and buy it in the future from your spouse or sell it in the future to your spouse or a third person.

Advantages: You (and your children) will continue having a familiar place in which to live. The house may bring a higher price if real estate values rise. You and your spouse will jointly share the eventual costs of sale and taxes. While you're living in the house, your spouse may agree to pay part of the mortgage (in the form of alimony or child support), repair costs, or maintenance costs.

Disadvantages: Major misunderstandings between divorced spouses can arise as to responsibility for repairs, maintenance, improvements, taxes, and the like. Nonresident spouses often are unable to purchase homes for themselves while they're continuing to share responsibility for the family home. As a safeguard, spouses who jointly own a home while one party uses it should make written agreements stating the rules for repairs and other financial contributions. In order for the out-of-house spouse to keep the capital gains tax exclusion of $250,000, the parties must agree in writing, or the court must order, that the in-house spouse have use of the residence.

Is this a viable option? _____ *Yes* _____ *No*

Continued Home Ownership and Estate Planning

If you decide to hold on to your house with your spouse and sell in the future, you will need to plan your estate and possibly acquire life insurance to cover each spouse's obligations in the event of the other spouse's death.

You should also evaluate how title to the house is held. (See Chapter 12.) For example, couples who currently hold their property in tenancy by the entirety or as community property will no longer have that option after divorce. In addition, if you held your house in joint tenancy while you were married—an arrangement whereby the other automatically inherits if one owner dies—that joint tenancy may no longer be appropriate. In some states, divorce may automatically sever joint tenancy. Converting ownership to tenancy in common would let you leave your share of the house to whomever you wish when you die.

Carrying life insurance on your former spouse provides you with ready cash to purchase the home from the estate if your ex dies before the house is sold. An agreement outlining your rights and responsibilities for co-ownership also can protect you in the event your ex-spouse dies prematurely.

Regarding the house, the following option is the best one for me:

☐ *Option A: Buy out my spouse's share and either keep it or sell it in the future.*

☐ *Option B: Sell my share of the house to my spouse.*

☐ *Option C: Sell the house together and split the proceeds with my spouse.*

☐ *Option D: Own the house jointly with my spouse and sell it in the future to my spouse or a third person.*

Step 10: How to Purchase the House From a Spouse or Former Spouse

Now that you have chosen an option, you need to face one more financial reality: paying for your choice. Typically, individuals going

through divorce do not have sufficient cash to purchase a spouse's interest in the family home. The following suggestions may help you to find a way to finance your best option on the house.

Refinance the house and pay proceeds to selling spouse.

Refinancing the home provides a source of cash to buy out the other spouse. As a practical matter, you may refinance in order to make a buyout happen, so that you become the only holder of the loan. If refinancing does not generate enough cash to buy out your spouse, you'll need another loan (called a second mortgage) secured by the house.

Refinancing a home mortgage can reduce your monthly mortgage payment and lower the amount of interest you pay over the life of the loan. It also may provide an opportunity to lock in a fixed mortgage rate. You generally need three things to qualify for refinancing:

- an adequate income
- a credit score that shows your ability and willingness to handle debt responsibly, and
- some equity in your home.

If you face challenges in meeting those requirements, don't give up. Some refinancing plans offer more flexibility, so it pays to shop around. Talk to lenders, mortgage brokers, and credit unions about what programs they offer. If you don't qualify for refinancing now, make an effort to find out what you'll need to do to qualify in the future. If you do qualify, explore whether a 15-year mortgage would make sense. If you can afford the higher monthly payment, you'll build up equity significantly faster than with a 30-year loan.

When looking for the best refinancing deal, be sure to shop, compare, and negotiate. Contact several lenders—commercial banks, mortgage companies, and credit unions—to be sure you're getting the best price. Specifically, ask about:

- **Rates.** Is the quoted interest rate fixed or adjustable? Is it the lowest available? What's the full annual percentage rate (APR)? This takes into account not only the interest rate but also other charges expressed as a yearly rate.

- **Points.** These are fees paid to the lender or broker and are often linked to the interest rate—usually the more points you pay, the lower the rate. Ask for the points to be quoted to you as a dollar figure.
- **Fees.** The broker or lender should be able to give you an estimate of fees, which can include loan origination or underwriting fees, broker fees, and settlement or closing costs. Some are paid when you apply; others are paid at closing, and many are negotiable. Often you can include the money to pay these fees in the loan, but that will increase your loan amount and total costs. Ask for an explanation of any fee you don't understand.

Sell the nonresident's interest to a third person.

If you're going through a divorce, a relative or friend may be able to help you acquire your home while also getting a tax benefit. Your relative or friend would purchase your spouse's interest in the house under what is called "an equity share financing arrangement." You would then co-own the property with that person and probably would pay rent. If you were to sell the property, the two of you would share in any profit.

You and your relative or friend will need an agreement spelling out your respective rights and responsibilities. To be sure the agreement complies with IRS rules and that you will both obtain available tax benefits, see a tax adviser. For more about sharing property as well as many other ways to save money and meet daily needs, see *The Sharing Solution: How to Save Money, Simplify Your Life, and Build Community*, by Janelle Orsi and Emily Doskow (Nolo).

Sell to a spouse using an installment loan.

Spouses often agree, and some courts order, that the custodial parent will continue to live in the family home with the minor children. The custodial parent could purchase the home from the other, in exchange for a secured installment note with payments over a fixed period of time. An installment note secured by the home allows the buyer to deduct interest and protects the seller in case the buyer defaults.

Keep the house in exchange for a release of alimony.

If an ex-spouse would be entitled to a sizable amount of alimony, consider exchanging it for the house. You will want to discuss this fully with a tax specialist to determine the present and after-tax value of total support payments. Courts may not allow all future support payments to be waived; check with an attorney on this point.

Keep the house in exchange for other assets.

Most often, couples trade assets and debts depending on needs and affordability. If you choose this option, see "Tallying Your Marital Balance Sheet" in Chapter 19.

Borrow from your retirement plan to finance the buyout.

Consider your retirement plan account as a possible source of financing for a home purchase. The costs of borrowing from a retirement account are usually lower than borrowing from a traditional lender. And because you are borrowing from yourself, loan payments are credited to your retirement account. Plus, loan processing is often quicker than with a traditional lender.

However, if you do not repay the loan within a reasonable time period, it is considered a distribution from your retirement plan. In addition to having to report the withdrawal as income on your tax return (and pay ordinary income taxes on it), you'll be subject to a 10% IRS early withdrawal penalty unless certain conditions are met. (See Chapter 14 for details on retirement plans.)

You can only borrow the money if both of the following apply:
- the loan provides for both principal and interest payments, and
- you agree to repay the money within a "reasonable" time, generally assumed to be five years.

To be able to deduct the mortgage interest paid, the loan must be secured by your residence.

You will not be able to deduct the mortgage interest if either of the following is true:

- the loan is secured by your employer's contributions to your retirement plan, or
- you are considered a key employee of the company, which means you:
  - are an officer or the employer
  - are one of the ten largest owners
  - own more than 5% of the company during the current year or preceding year, or
  - own more than 1% of the company and earn more than $150,000 per year.

Questions to Ask an Attorney or Financial Adviser

CAUTION

The issue of the house raises several important legal and financial questions. Below are some important questions for you to ask an attorney or research at your local law library, as well as questions for a financial adviser.

Attorney Questions

1. Is a portion of the house considered separate property? How does the court treat a separate property interest in the family home in my state?
2. Am I entitled to a credit for mortgage payments I make while my spouse lives in the family home after we separate? Is this credit reduced by the value of tax benefits received from the mortgage interest deduction?
3. What happens if the bank forecloses on this loan?
4. What is my income tax liability, if any, if the title to the house transfers to me and there is a foreclosure?
5. If there is a foreclosure, can the lender get a judgment against me for any difference between the foreclosed value of the house and the mortgage balance that I owe?
6. Should I get divorced before or after the house sells?

7. If we continue to hold the house as joint tenants, does the right of survivorship continue after the divorce?

8. If we want to defer sale of the house, what are the consequences for me? What steps would we need to take?

Financial Adviser Questions

1. What penalties apply for prepayment of our mortgage?

2. Do we have any unamortized points from our existing mortgage?

3. Am I adequately insured?

4. If our house sells and we are still married, should we file a joint tax return?

5. If we own it jointly and then I buy out my spouse's share, how do I report this for tax purposes?

6. Who gets the mortgage interest deduction and the property tax deduction if the title is held as tenants in common? If it's held as joint tenants?

7. If I refinance the mortgage, will the interest and points, if any, be deductible?

Retirement Benefits: Who Gets What?

Understanding Retirement Plans 253

    Retirement Plans Are Not Created Equal 259

    Some Plans Promise Better Tax Breaks Than Others 259

    Divorce Leads to Division Decisions 261

Qualified Domestic Relations Orders 261

    QDRO 261

    What Constitutes a QDRO 261

    Interim QDROs 262

The Legal Value of Your Retirement Plans 265

    Defined Contribution Plans, Personal Annuities, and IRAs 265

    Defined Benefit Plans 265

    Determining the Marital Portion 266

The Financial Value of Your Retirement Plans 268

    Income Tax Considerations 269

    Charges and Penalties 272

Calculating the Financial Value of Plans 273

    Defined Contribution Plans 273

    Defined Benefit Plans 275

Additional Financial Factors Affecting Retirement Plan Divisions 280

    Vesting 280

    Benefits in Payout Status 281

    Double Payments 282

    Cost of Living Adjustments 282

    Access to Cash 282

    Tax Consequences in Dividing Benefits 283

The Division Decision: Now or in the Future...284

   IRAs...285

   Defined Contribution Plans and TSAs...286

   Defined Benefit Plans..287

Questions to Ask an Attorney...288

Whether your retirement benefits come from your own job, your spouse's job, or other retirement plans, those benefits must be divided when you divorce.

SKIP AHEAD

If neither you nor your spouse has a retirement plan, skip ahead to Chapter 15.

Under the best of circumstances, planning for your retirement is difficult. You may not have given much thought to retirement planning. Perhaps you hoped it would somehow take care of itself. But it won't. You've got to give some serious consideration to when you will retire, how much income you'll need during retirement, the sources for that income, and how you will divide the retirement benefits with your ex-spouse.

Retirement planning is a relatively new concept. Generations ago, few people had to worry about late-life security because most just didn't live that long. Today, an entire industry has emerged to provide retirement planning services to aging baby boomers. The pressure to create sound retirement plans can be intense, and divorce adds another layer of complexity.

In most states, retirement plans are property to divide at divorce. Naturally, you may be emotionally attached to the retirement benefits that you worked so hard to earn or feel that you are entitled to as part of your commitment to the marriage. Take heart. Retirement benefits can serve as significant bargaining chips when spouses trade assets in the final property settlement.

Planning for Your Retirement

No matter how you and your soon-to-be ex-spouse divide the retirement benefits, it is important now to plan for your own retirement. Here's why, and what you will need to do:

- Americans are living longer. You should plan for an income stream through at least age 90.
- The responsibility for providing retirement income is being shifted slowly but surely by employers onto the shoulders of employees. Take advantage of any and all retirement plans available to you now through your employer. Once your credit cards are paid off, contribute more—the maximum allowed if possible.
- You have more investment choices now than ever before in 401(k) plans and self-directed plans. Learn what you need to know about available mutual funds, as well as their objectives and performance, so you can make intelligent decisions.
- If you are not part of the "triple-squeeze" generation—supporting an aging parent, putting a child through college, and trying to save for your own retirement simultaneously—you may be soon. Be sure to consider these factors as you make your decisions about splitting retirement plans in divorce.

 RESOURCE

We only introduce the concept of retirement planning in this chapter. For more information, here are some resources: *Work Less, Live More,* by Bob Clyatt (Nolo); *Investing For Retirement: Make Good Choices Without a Ph.D. in Finance,* by Ralph Warner (Nolo); *IRAs, 401(k)s & Other Retirement Plans: Taking Your Money Out,* by Twila Slesnick, Ph.D., EA, and John Suttle, CPA (Nolo).

Understanding Retirement Plans

Almost every book and expert on the subject of retirement planning uses a different system to categorize plans. In the pages that follow, we separate plans into several major categories.

The table on the next several pages gives detailed information on many major categories of retirement plans. Once you've determined which plans you and your spouse have, you can focus on the sections in this chapter that apply to you.

The table also specifies how, under the law, particular plans may be divided at divorce. To divide some types of plans, you must get a special kind of court order called a QDRO (pronounced "quadro")—a Qualified Domestic Relations Order (see below). Other retirement assets can be divided by a court order or judgment other than a QDRO. And some kinds of retirement assets can't be divided at divorce.

One distinction we make in this chapter is between defined benefit plans and defined contribution plans. In defined benefit plans, also called pension plans, an employer provides a fixed amount to an employee at retirement, usually in monthly payments that continue until the employee's death. In defined contribution plans, which include 401(k) plans, the employee and often the employer contribute money into the employee's retirement account, which grows over time. IRAs, or Individual Retirement Accounts, have similarities to defined contribution plans, although dividing the assets at divorce doesn't require a QDRO.

If you or your spouse has a defined benefit plan, it's difficult to determine just how much the plan is worth today. That's because the plan is designed to pay out in the future, and inflation causes today's dollars to have a different value than they'll have 20 or 30 years from now. It takes an expert, such as an actuary, to find the current value of this kind of plan. That makes dividing the asset a complicated task, as we'll see. It's a lot easier to tell how much a defined contribution plan or an IRA is worth today: As the account grows, the current value is reported to the account holder in regular statements.

Employer-Sponsored Defined Contribution Plans

Type of Plan	Description	Type of Order Necessary to Divide Plan
Salary Savings/401(k), 403(b), and 457 plans	Employee contributes a portion of salary on a pretax basis which may be matched in full or in part by the employer. The employee may defer up to $17,500 annually (as of 2013) or $23,000 if age 50 or older.	QDRO
Thrift Plan	Employee contributes a portion of salary on an after-tax basis, which may be matched in full or in part by the employer. Because the employee's contribution is made on an after-tax basis, distributions (payouts) may be partially tax free.	QDRO
Profit-Sharing Plan	Employer contributes to the employee's account only if company is profitable. Amount of contribution is based on either a fixed or a discretionary formula.	QDRO
Money Purchase Plan	Employer contributes a fixed percentage of salary to employee's account every year. Employer contribution is mandatory, regardless of whether company makes a profit.	QDRO
Employee Stock Ownership Plan (ESOP)	Contributions are a percentage of salary and are used to purchase company stock. The stock is held in trust for the employee, who receives the accumulated interest in the plan at termination of employment in the form of stock or cash. The plan provisions define the rights of the employee to exercise stock options.	QDRO
Tax-Sheltered Annuities (TSAs)	TSAs are a retirement vehicle allowing teachers, public school system employees, and employees of nonprofit organizations to contribute a portion of salary into the annuity. If your TSA is through the Teachers Insurance and Annuity Association–College Retirement Equities Fund (TIAA-CREF), you can get detailed information, including account values, on the Web at www.tiaa-cref.com.	QDRO

Employer-Sponsored IRAs

Type of Plan	Description	Type of Order Necessary to Divide Plan
Simplified Employee Pension IRA (SEP-IRA)	SEP-IRAs are retirement plans in the form of individual retirement accounts. An early withdrawal penalty of 10% applies to SEP-IRAS as well as traditional IRAs and Roth IRAs. The maximum contribution to a SEP-IRA stood at 25% or $51,000 for 2013.	A regular court order or judgment divides these plans.
Salary Reduction-Simplified Employee Pension IRA (SAR-IRA)	These retirement plans were available prior to 1997 and included a salary reduction agreement in which the employee could elect to have a portion of salary contributed to a SEP-IRA on a pretax basis.	A regular court order or judgment divides these plans.
SIMPLE IRA (Simplified Incentive Match Plans for Employees)	SIMPLE IRA involves a written salary reduction agreement between the employee and employer that allows the employee-participant to reduce compensation and have the employer contribute the difference to a SIMPLE IRA on the employee's behalf. (Employees may elect not to contribute.) The employer makes matching contributions or nonelective contributions. Nonelective contributions are made for each eligible employee, even if the employee does not elect to contribute. The total employee contribution (from salary reduction) allowed is $11,500 (workers over age 50 are allowed to contribute another $2,500). Nonelective contributions are limited to 2% of compensation. The matching contribution is the lesser of the employee contribution or 3% of the employee's annual salary. The distribution rules applicable to a traditional IRA also apply to a SIMPLE IRA, except that if the employee withdraws money prematurely within two years after beginning participation, the 10% penalty is increased to 25%. In addition, during the first two years of plan participation, a rollover (transfer of funds) can be made only to another SIMPLE IRA plan. A SIMPLE IRA cannot be designated as a Roth IRA.	A regular court order or judgment divides these plans.

Defined Benefit Plans

Type of Plan	Description	Type of Order Necessary to Divide Plan
Business/ Corporate Pension	Businesses or corporations offer these defined benefit plans. The maximum annual benefit is $205,000 as of 2013.	QDRO

Government-Sponsored Retirement Plans

Type of Plan	Description	Type of Order Necessary to Divide Plan
Government/ Military Pension	These defined benefit plans are offered to civil service workers, government employees, and military personnel. (Veterans can get additional information on their benefits at www.va.gov.)	Specific governmental regulations and procedures apply for implementation of a court order to divide these plan benefits between spouses.

Personal Retirement Plans/Personal Annuities

Type of Plan	Description	Type of Order Necessary to Divide Plan
Deferred Annuity	A deferred annuity is an annuity contract in which an insurance company promises to make payments at some future date. Taxes on appreciation are deferred until the money is withdrawn.	A regular court order or judgment divides these plans. Most insurance companies have forms that permit the transfer of ownership.
Traditional IRA	Individuals may accumulate up to $5,000 per year (workers over age 50 are allowed to contribute another $1,000 per year). Withdrawals before age 59½ are subject to a 10% penalty. The early withdrawal penalty does not apply to a withdrawal because of disability or a withdrawal for certain medical insurance premiums, educational expenses, or first-time homebuyer expenses up to a maximum of $10,000. The account value is untaxed until funds are withdrawn.	A regular court order or judgment divides IRAs.
Roth IRA	Individuals may contribute up to $5,000 per year (workers over age 50 are allowed to contribute another $1,000 per year). Contributions are not deductible when you make them, but qualified distributions of the contributions and earnings are tax free when withdrawn. In general, you can take tax-free withdrawals if, after holding the account for five years, you meet certain conditions. You must be at least 59½, disabled, or using the money (up to $10,000) toward the purchase of your first home, or you must be the beneficiary of the account. If you do not meet any of these conditions, an early withdrawal penalty will apply.	A regular court order or judgment divides IRAs.

Miscellaneous Retirement Benefits

Type of Plan	Description	Type of Order Necessary to Divide Plan
Nonqualified Deferred Compensation Plan	A small percentage of executives and key upper-level employees have deferred compensation plans. Generally, such a plan provides that a portion of the employee's compensation is deferred until the future. Agreements to participate in the plans are made directly between the employer and the employee, which makes requesting documentation from the employer important. (For more on these plans, see Chapter 16.)	Often the plans do not permit the asset or income to be divided, making them legally complicated. In such a case, the court might order the employee spouse to pay the nonemployee spouse when the employee spouse receives the benefit. Because the employee spouse pays taxes upon receipt of benefits, the order should include a provision that the employee spouse will deduct the nonemployee spouse's share of taxes on the benefit.
Social Security	If you are divorced, you can receive benefits at age 62 based on your former spouse's earnings if the marriage lasted at least ten years, you are unmarried, and the benefit based on your own earnings is not greater than 50% of your former spouse's benefit. If your former spouse is at least age 62 and not receiving benefits but can qualify for them, you can still receive benefits if you have been divorced for at least two years and meet the other requirements. To apply for benefits on your former spouse's record, you will need either the spouse's Social Security number or date and place of birth and parents' names. Your benefits as an ex-spouse do not reduce or otherwise affect benefits available to your former spouse. You can find out about Social Security benefits by calling 800-772-1213 or visiting www.ssa.gov. For more on Social Security, see Chapter 5.	Social Security benefits are generally not considered marital property and are not divisible at divorce.

WEB RESOURCE

The Internet is a terrific resource for gathering additional information.
Sites such as www.morningstar.com and www.yahoo.com offer valuable
insights into and analysis of many different types of retirement funds. Be sure
to check whether your retirement plan has information available online. Many
company and government plan sponsors provide up-to-date account values and
information through their websites or through automated phone systems.

Retirement Plans Are Not Created Equal

Retirement plans vary tremendously. Even if two plans look equal on
paper, fine-print plan provisions can make one plan more valuable than
another. For instance, one plan may allow for a lump sum payout upon
retirement, while another may not. Getting your retirement dollars in a
lump sum may provide more flexibility than getting monthly payments
in dollars whose value erodes over time or disappears entirely if the plan
becomes insolvent in later years.

Retirement plans must comply with federal regulations, but the way
they are apportioned at divorce may differ from state to state. One state
may define retirement benefits as marital property (an asset to be divided
at divorce), while another may rule that benefits are income (and not to
be divided). Be sure to ask your lawyer how your state treats retirement
benefits upon divorce. (In a worst-case scenario for the employee spouse,
a state may regard retirement benefits as both marital property and
income, allowing "double dipping" for the nonemployee spouse.)

Some Plans Promise Better Tax Breaks Than Others

One major attraction of most retirement plans is that you defer paying
income taxes on the money that accumulates in the plan. In some plans,
though, you must pay income taxes on the money before it's deposited.
One task during your divorce will be to determine the tax position
of your benefits. Normally, you want to defer paying the taxes until
you withdraw the money from the plan and presumably are in a lower
income bracket or have a lower annual income.

Retirement Plan Lingo

Before you start making decisions, familiarize yourself with the terms you are likely to hear when you investigate your retirement plans. You have a better chance of getting answers to your questions if you know what to ask for.

- **Account balance.** For plans such as IRAs and defined contribution plans, you want to know how much is in the account. This amount may simply be called the balance.

- **Accrued benefit.** The amount that has been earned (or has accrued) in a retirement plan as of a particular date.

- **Annuity contracts.** A contract between a life insurance company and the investor or owner of the annuity. Typical provisions include the amount of interest (both the guaranteed minimum and the current), surrender charges (fees for ending the contract prematurely), and penalties and options at retirement.

- **Defined benefit plan.** A plan in which the employee's retirement benefit is fixed, based on a specific formula.

- **Defined contribution plan.** A plan in which the employee and sometimes the employer contribute to a retirement account, often at regular intervals.

- **Employee benefit statement.** A summary telling you the amount of benefits (in dollars) that have accumulated in your name or your spouse's.

- **Full plan.** You may or may not need a copy of the full plan, depending on how well the summary plan description (see below) is written. Do not hesitate to ask for the full plan if you have questions or need to check details.

- **Procedures and model orders for division.** Many companies have written procedures for dividing up company retirement plan benefits. Request a copy of these materials.

- **Qualified plan.** A retirement plan that qualifies for certain benefits described in Section 401 of the U.S. Tax Code; also known as an ERISA plan because it is governed by the Employee Retirement Income Security Act.

- **Summary plan description.** Employers are required to issue a description (annually in most cases) outlining the status and terms of retirement plans.

- **Trust/custodial agreement.** A document spelling out specific provisions of the retirement plan and the agreements between the legal owner of the plan, usually the plan administrator (called the "trustee"), and the employee (the beneficial owner). Understanding these agreements is particularly important with IRAs because they describe the costs of establishing, maintaining, and terminating IRA accounts.

Divorce Leads to Division Decisions

Will you split the retirement money now or later? A present division means you divide the value of the plan now, at divorce. A future division means the plan is divided at retirement or at the point in the employee spouse's career when the benefits would normally be paid out. When and how you divide the plan depends on the type of plan you have, payout provisions, valuation methods, and other factors. Your division decision at divorce has a direct impact on your future financial security, so do not make this choice lightly.

Qualified Domestic Relations Orders

QDROs, interim QDROs, and other kinds of court orders are used to divide various kinds of retirement plans.

QDRO

QDRO stands for Qualified Domestic Relations Order. It is an order from the court to the retirement plan administrator spelling out how the plan's benefits are to be assigned to each party in a divorce. It's critical that your QDRO be complete and accurate, because any assets or items that the order omits, even accidentally, cannot be reinstated later.

What Constitutes a QDRO

For a document dividing a retirement plan to be considered a QDRO, it must specify:

- the name and last known mailing address of the employee spouse (the participant)
- the name and mailing address of the nonemployee spouse (the alternate payee)
- the amount or percentage of the participant's benefits to be paid to the alternate payee, or the manner in which the amount or percentage is to be determined

- the number of payments or the period to which the order applies, and
- the name of the plan, such as the Walton Company Pension Plan, to which the order applies.

A QDRO cannot:

- require a plan to pay any benefit or option not otherwise provided by the plan
- require a plan to provide benefits that exceed the value of the participant's interest as determined by an actuary
- require the payment of benefits to an alternate payee if those benefits are already payable to another alternate payee (that is, another ex-spouse) under a previous qualified order, or
- specify who will be responsible for payment of taxes.

Plans subject to QDROs include:

- qualified plans
- defined contribution plans
- defined benefit plans, and
- tax-sheltered annuities (TSAs).

Plans that QDROs do not cover include:

- deferred annuities
- traditional IRAs
- Roth IRAs
- SEP-IRAs
- nonqualified deferred compensation plans (government and church plans)
- plans administered by a plan participant (you or your ex-spouse), and
- plans under which you do not receive benefits from your ex-spouse's retirement plan.

Sometimes an order concerning the retirement plans is in a document called a Domestic Relations Order, Marital Settlement Agreement, or Divorce Settlement Agreement. These kinds of orders are not QDROs.

Interim QDROs

If you are not ready to divide the retirement plan assets and prepare a QDRO, you can still protect your rights to survivor benefits under your

spouse's retirement plan by filing what is called an interim QDRO. Later, when retirement assets are divided, you can get a final QDRO. If you divorce before dividing your property, it's very important to have an interim QDRO prepared and signed by a judge, simultaneous with the judgment that changes your marital status to legally separated or divorced. Here's what an interim QDRO accomplishes:

- It treats the nonemployee spouse as a surviving spouse. Even if the employee remarries, the former spouse's interest in survivor benefits is paid before the new surviving spouse receives anything.
- It prohibits the employee from electing to receive retirement plan benefits without permission of the former spouse or a court order, except for any minimum required distribution.
- It sets the former spouse's rights at 50% of the community portion of the employee's benefit.

An attorney should be able to advise you on filing an interim QDRO. But don't wait very long to obtain the actual QDRO. A delay can mean a loss of benefits. Be sure the QDRO is accepted by the employee spouse's plan administrator. A divorced ex-spouse without a QDRO does not have the same protection available under the plan during marriage.

CAUTION

You should *always* consult a lawyer when you need a QDRO, and make sure the lawyer is comfortable preparing a QDRO or uses a reputable specialist to outsource the work. This is not something you can do yourself; even some very experienced attorneys find QDROs painfully complicated.

If you're representing yourself, you'll need help protecting your right to share in retirement plans. This may require filing papers to "join" the retirement plan as a party in your divorce, and other important paperwork. Make sure you get the help you need to do it right.

Benchmarks to Retirement

Most people think of age 65 as the standard retirement age. But other birthdays are also important in preparing for the after-work years. By noting these key ages ahead of time, you may be able to minimize taxes and maximize benefits.

Be sure to review your company's retirement plans. Each plan has its own rules about the age at which retirement benefits can begin. In some circumstances you may choose to receive your retirement benefits early, but your monthly benefit will be lower. Your benefits administrator can give you more information on what retirement benefits are available to you and the ages at which you can begin receiving these benefits. Most employers also provide employees with a statement showing an estimate of retirement benefits at early and regular retirement ages. Ages to note include:

Age 55   If you retire, quit, or are fired and are age 55 or older, you may receive benefits from an employer-sponsored qualified plan without having to pay a 10% penalty. This exception does not apply to either personal or employer-sponsored IRA distributions.

Age 59½   You are allowed to withdraw funds from personal and employer-sponsored IRAs and retirement plans without paying the 10% penalty.

Age 60   Widows and widowers become eligible for Social Security benefits.

Age 62   You may be eligible to receive Social Security benefits, but benefits will be less than if you retire at age 65.

Age 65   Retirees qualify for Medicare benefits.

Age 67   Full Social Security benefits are available for anyone born in 1960 or later. The actual age at which full Social Security benefits are available depends on the year in which you were born. For those born before 1938, full Social Security benefits are available at age 65. For those born later, the retirement age is between 65 and two months and 67, depending on your birth year.

Age 70   Full Social Security benefits are available, even if you continue to work full-time.

Age 70½   You're required to begin taking distributions from IRAs (except Roth IRAs) by April 1 of the year following the calendar year in which you reach age 70½. For other employer-sponsored qualified retirement plans, distributions must begin by April 1 of the year following age 70½ or the year of retirement, whichever is later. If the participant is at least a 5% owner of the business, distributions must begin at age 70½.

The Legal Value of Your Retirement Plans

As we've often stressed in this book, the value of an asset as recognized by a court may be vastly different from what the asset is worth outside the courtroom. So it is with retirement plans. Normally, the value of a plan as determined by the court—its "legal value"—is the following:

- For defined contribution plans, personal annuities, and IRAs, the legal value is the amount that appears on your statement as the current value of the account, less any loans outstanding.
- For defined benefit plans, the legal value is what the plan administrator or an actuary tells you it's worth.

You need to know the legal value because it is often the only dollar value lawyers and judges will take into consideration. This value represents the amount you and your spouse will argue over if you disagree about the value of the benefits. However, the legal value does not account for important financial realities: taxes, early withdrawal penalties, early termination fees, and other assessments.

Defined Contribution Plans, Personal Annuities, and IRAs

If you have a defined contribution plan, a personal annuity, or an IRA, determining the legal value is simple. The plan statement gives the value. If you can't find the information you need, consult the plan administrator or get a copy of the annuity contract.

Defined Benefit Plans

If you have a defined benefit plan, you'll need to do some work to find its legal value. Because these plans promise a payment in the future, they are harder to value in the present.

The legal value of a defined benefit plan is the plan's present or actuarial value. The plan administrator or the company may be able to give you this value. Another option is to hire an independent third-party actuary to calculate the plan's present value. This should be relatively inexpensive—perhaps $300 to $500. The bill could run higher if your spouse hires a separate actuary to argue that your actuary's assumptions about the future are wrong, and you find yourself having to call the actuary as an expert witness to testify in a court trial. Disputes can arise because the value of these plans lies in the future and is therefore speculative.

After you get the actuarial value for each plan, you will need to determine what percentage of the plan qualifies as marital property to be divided at divorce. (Some actuaries will do this at the same time they establish the value of the plan.)

Determining the Marital Portion

To simplify the process of calculating the marital portion of your retirement plan, use the marital interest percentage method (total months of participation while married divided by total months of plan participation). This method is generally used in dividing defined benefit plans.

The problem with using this method is that it does not accurately value the higher contributions made during the later years of employment. In the early years of plan participation, the contribution is smaller because contribution is based on salary, which is likely to have increased over time. For a more precise calculation, you may need to talk to your actuary, CPA, financial planner, or pension administrator.

How Can a Pension Worth Hundreds of Thousands Later Be Worth So Little Now?

Actuaries commonly look at several factors when calculating the present value of retirement plans. For example, plans often assign benefits based on the highest three years of earnings. But during divorce, current salary is used in the calculation.

Similarly, calculations of the future worth of a plan will differ if a person anticipates retiring at age 62 rather than age 65. They will also vary with the rate of inflation that's predicted and how the payouts are structured. Some plans include a COLA—cost of living adjustment—that affects the plan's value. And, finally, some actuaries assume a consistent rate of growth—say, 5% of retirement plan investments—and others use a different figure.

For example, take the case of John and Andrea. An actuary tells Andrea that the marital portion of John's plan will pay out about $100,000 during John's retirement but is worth only about $38,000 in today's dollars. So if she and John divide the plan equally at divorce, she will get one-half, or $19,000. If she waits until John retires, she could get $50,000. Andrea, understandably, wants to know why something supposed to be so valuable tomorrow can be worth so little today.

Here's why. John's pension plan shows that he will get $10,000 per year during retirement. He's supposed to retire at age 65 and, according to actuarial tables, live ten years after that. In theory, then, he will receive $100,000.

John, however, is only 45 and won't be retiring for another 20 years. Assuming an inflation rate of 4%, the $10,000 he will be paid in the first year of his retirement is equivalent to $4,564 today. The second year of his retirement (21 years from now) is worth even less.

The "Present Value Factors" chart in the appendix compares time value with inflation and can help you make a rough estimate of the value of benefits. The left-hand column on that chart represents years, while the percentages across the top of the page are rates of inflation. Follow the 20-year line (the number of years after which we assumed John would retire) to the 4% predicted rate of inflation and you find the present value factor of .4564. Multiply $10,000 by .4564 and you get the "value today" of $4,564. You can follow the rest of the 21-to-30-year lines to see how the rest of the value of John's plan was calculated.

How the Separation Date Can Affect a Retirement Plan's Value

A doctor in California asked the pension plan administrator of his medical group to tell him the value of his plan. The doctor had separated from his wife four years earlier, and they were ready to divide the retirement plan. The administrator valued the plan currently at $225,000. Wanting a second opinion on this obviously valuable asset, the doctor went to an actuary for another valuation of the plan. The actuary valued the plan at only $175,000. Why the difference? Under California law, the marital period of the doctor's retirement plan stopped on the day of separation four years before. The doctor's wife was entitled only to a portion of the amount that had accrued from their wedding day to the date they separated. By contrast, the plan administrator had calculated the entire value of the plan, including the four-year period that the couple had been separated. Different states define the applicable period differently.

For more on the significance of the separation date, see Chapter 5.

The Financial Value of Your Retirement Plans

The financial value of a retirement plan is almost always lower than its legal value. Financial value encompasses everything from unexpected bonuses (a cash settlement upon divorce) to nasty surprises (a penalty from the IRS).

Although retirement benefits are often years away from being paid out and you may not see the financial return for some time, you nevertheless want to compare all assets on similar terms during your divorce. Just as you may be contemplating a sale of the family home, so, too, must you assume the "sale" (or liquidation) value of a retirement plan. When negotiating the overall settlement, you'll want to be able to compare each asset's legal value and financial value.

To find the financial value of your retirement plans, you'll have to look at income tax consideration as well as charges and penalties. Below is an overview.

Tax Brackets and State Income Taxes

Throughout this book, we assume that you fall into the 25% federal income tax bracket. Because some states tax a percentage of IRA deposits and other retirement plans (thrift plans and employee stock ownership plans), you'll need to ask a tax adviser whether you might owe state taxes on your benefits. You won't owe state income taxes in Alaska, Florida, Nevada, South Dakota, Texas, Washington, or Wyoming, states that impose no income taxes. (New Hampshire and Tennessee tax only interest income and dividends.)

Income Tax Considerations

Uncle Sam does not let you simply accrue money for retirement and then use it, without receiving his share sooner or later. Your job is to find out when taxes on benefits are paid—either before the money goes into the retirement plan or when it comes out during retirement. The plan will contain this information.

Suppose your ex-spouse offers to keep the house but to give you all the retirement plan benefits—your own and your spouse's. Should you take the deal? You cannot answer that question unless you know what income tax you will owe when the benefits begin paying out.

At first glance, your share of the retirement benefits and your share of the house may equal the same dollar amounts. But those amounts probably reflect legal reality only. You may have to pay substantial income taxes when you begin receiving the retirement benefits, while your spouse could take the house and possibly avoid any tax liability.

When dividing plans at divorce, look to keep the plan in which the contributions—or at least some of the contributions—were made with after-tax dollars. When plans are funded with after-tax dollars, part of the money you withdraw at retirement will not be taxed.

Qualified plans—such as defined contribution plans and defined benefit plans—must meet certain IRS regulations for employers to get tax benefits. Tax-sheltered annuities (TSAs), although not considered qualified plans, tend to be affected by the same regulations.

In determining the taxability of money contributed to a retirement plan, the IRS asks two questions:

- Did the employer contribute money? If the answer is yes, when you begin collecting your retirement benefits, you'll have to pay taxes on the amount your employer contributed and the interest earned on that contribution.
- Was the contribution made with pretax or after-tax dollars? Retirement plans are sometimes referred to as contributory or noncontributory. Contributory plans are those to which the employee makes a contribution, and employers often make a matching contribution. For instance, some employers chip in 50 cents for every dollar the employee contributes. Some even match the employee contribution dollar for dollar, often to a limited amount, such as $1,000.

The IRS may seem harsh, but it doesn't intend for you to pay taxes twice on your income. If you are currently reporting as income the money you are contributing to your plan, you will not have to pay taxes on those contributions when the money comes out of the plan at retirement.

If you do not know the percentages of your retirement plan contributions made with pretax dollars and after-tax dollars, you can:

- contact the plan administrator, personnel manager, or benefits coordinator and ask for an accounting of pre- and after-tax contributions, or
- check the annual benefits statement, which should break down pretax and after-tax contributions.

The income taxes you will have to pay on retirement benefits are based on a number of factors: your tax bracket, the tax basis of the plan, your status as either the plan participant (employee) or the alternate payee (spouse of plan participant), and the type of plan you or your spouse have.

The tax basis of a retirement plan consists of money contributed that's already been taxed as income—that is, contributions for which you weren't permitted to take a deduction when you filed your taxes. Most employer contributions, plus interest earned on all contributions—from both employer and employee—are not included in basis because those

dollars are not taxed until they are paid out. The plan administrator should be able to help you find the tax basis.

For IRAs and deferred annuities, special tax considerations apply.

IRAs. When you deposit money into a traditional IRA, you do not report that money as income—that is, you pay no taxes on it. This assumes that neither you nor your spouse is an active participant in a qualified plan. If both of you are active participants in qualified plans, you can deduct your $3,000 IRA contribution if your income falls within specified limits.

Roth IRA. Unlike with regular IRAs, money you contribute to a Roth IRA is not tax deductible. But you receive the benefits tax free when you collect at retirement. This makes Roth IRAs particularly valuable to retain in a divorce.

In general, you can also take tax-free withdrawals if you have held the account for five years or more and are under age 59½, become disabled, use up to $10,000 toward the purchase of your first home, or are the beneficiary of the account.

Until recently, you could convert your traditional IRA to a Roth only if you had an adjusted gross income of less than $100,000. But now anybody—even spouses filing separately—can convert retirement assets to a Roth IRA, though they'll owe taxes on the amount converted.

Deferred annuities. The money put into a deferred annuity usually comes from your savings or from income on which you have already been taxed. The tax basis of your annuity is the total purchase price less unpaid loans and any other tax-free amounts you received. You will pay taxes on the difference between the current account value and the tax basis. From the standpoint of income tax liability, the Roth IRA and deferred annuities are similar. Both will have a tax basis, because contributions were made with after-tax dollars.

Charges and Penalties

Participants in retirement plans often face charges and penalties when there is an early withdrawal—that is, when money is taken out before a certain date. Check your plan documents carefully or ask your plan administrator to see if these charges or penalties apply, to be sure you don't unknowingly incur them.

If a participant in a qualified retirement plan takes money out before retiring, the IRS may assess an early withdrawal penalty (also called an excise tax) as high as 10% of the amount withdrawn. There may be a state tax penalty in addition to the federal penalty.

The holder of an IRA or an annuity who withdraws money before reaching age 59½ will pay a 10% penalty to the IRS unless certain exceptions are met.

Insurance companies and some investment companies impose surrender charges to discourage participants from withdrawing funds before the date specified in the contract. These can take the form of deferred sales charges on mutual funds, surrender charges on annuities, early withdrawal charges on CDs, and brokerage or transaction costs.

Another tax wrinkle concerning IRAs: If the participant spouse transfers all or part of an IRA to the other spouse, neither will owe taxes on the transfer. If the participant spouse cashes out the IRA and transfers the money to the other spouse, however, the participant spouse will owe taxes on the money withdrawn. It's therefore crucial to be sure that the money goes directly from one account to another in what's called a "trustee-to-trustee" transfer or "assignment of ownership."

The 10% early withdrawal penalty tax does not apply to any distribution to an alternate payee pursuant to a QDRO, regardless of alternate payee's age when he or she receives it and whether the alternate payee is the spouse, former spouse, child, or other dependent. However, once the distribution is rolled over into an individual retirement account, it becomes subject to the penalty if a later distribution occurs. (Note: An IRA is not a qualified plan and is not subject to a QDRO.)

Pros and Cons of Early Distributions

It is becoming increasingly common for employees to break into retirement accounts early or take out loans on them. When there is a real economic necessity—such as a medical emergency or buying a first home—this may be a prudent decision. As we pointed out in Chapter 13, it also can make sense to take out a loan on your 401(k) to purchase a share of your house. The 10% tax on early distributions is sometimes waived for this purpose (visit www.irs.gov to research exceptions to the early distribution rule).

That said, early withdrawal penalties and taxes can severely diminish the value of a 401(k). If you and your spouse are considering an early distribution as you move forward in a divorce settlement, be certain that you do the math to determine what your account will be worth. If you're in your 40s or 50s, the consequences of losing value of your retirement plan in a taxable—and avoidable—event can be devastating.

Calculating the Financial Value of Plans

As we've said, any retirement plan has a financial value different from the legal value that a court assigns it. The financial value takes into account real-world factors like tax consequences and early withdrawal penalties. Doing the calculation that follows and entering it into the chart may seem complicated. But knowing the financial value of a plan will help you make a better-informed decision about where the assets should go in your divorce settlement.

The calculation is different depending on whether the plan is a defined benefit plan or a defined contribution plan, a TSA, or an IRA.

Defined Contribution Plans

For each defined contribution plan, TSA, or IRA plan, follow the steps below to complete the chart that follows this calculation:

1. In Column A, enter the total number of months you or your spouse participated in the plan.

2. In Column B, enter the number of months in which you or your spouse participated in the plan while married.

3. In Column C, enter the marital interest percentage by dividing the number in Column B by the number in Column A. For example, if you participated in a plan for 120 months and were married 72 months while participating in the plan, you would have a marital interest percentage of 60% ((72 ÷ 120) x 100.)

4. In Column D, enter the current dollar value of the account. (The amount should be on your statement.)

5. In Column E, enter the marital portion of the account by multiplying the amount in Column D by the percentage in Column C.

6. In Column F, enter the amount of any loans you and your spouse have taken against this retirement account. (The amount should be on your statement.) Do not include any separate loans.

7. In Column G, enter the legal value of your retirement plan by subtracting the amount in Column F, if any, from the amount in Column E.

8. In Column H, enter the tax basis of the retirement plan—the after-tax contributions, including nondeductible contributions made to an IRA.

9. In Column I, enter the marital portion of the tax basis by multiplying the figure in Column H by the percentage in Column C.

10. In Column J, enter the financial value of this account. You do this as follows:

Marital account value (Column E less Column I) $_____

*Early withdrawal penalty**   – $_____

Taxes due on distribution (marital account value × your tax bracket)   – $_____

Other surrender charges or penalties   – $_____

Financial value $_____

* The early withdrawal penalty is the amount in Column D x 10%, or x 25% for employer-sponsored Simple IRAs you have participated in for two years or less. Be sure to add any state early withdrawal penalties.

Defined Benefit Plans

For a defined benefit plan, follow the steps below to complete the chart that follows:

1. In Column A, enter the total number of months you or your spouse participated in the plan.
2. In Column B, enter the number of months in which you or your spouse participated in the plan while married.
3. In Column C, enter the marital interest percentage by dividing the number in Column B by the number in Column A. For example, if you participated in a plan for 120 months, and were married 72 months while participating in the plan, you would have a marital interest percentage of 60% ((72 ÷ 120) × 100).
4. In Column D, enter the actuarial value of the plan. (You'll need to consult with the plan administrator or an actuary for this information.)
5. In Column E, enter the marital portion of the account by multiplying the value in Column D by the percentage in Column C.
6. In Column F, enter the amount of any loan you and your spouse have taken against this retirement account. (The amount should be on your statement.) Do not include any separate loan.
7. In Column G, enter the legal value of your retirement plan by subtracting the amount in Column F, if any, from the amount in Column E.
8. In Column H, enter the tax basis of the retirement plan—the after-tax contributions.
9. In Column I, enter the marital portion of the tax basis by multiplying the figure in Column H by the percentage in Column C.
10. In Column J, enter the financial value of this account. You do this as follows:

Marital account value (Column E less Column I)	$_____
Taxes due on distribution (marital account value × your tax bracket)	– $_____
Other surrender charges or penalties	– $_____
Financial value	$_____

Once you have completed the steps above for each retirement plan, add all the numbers in Columns D, E, F, and J and enter these results in their respective columns on the "Total" line. Then, be sure to transfer the totals to the "Net Worth: Assets & Liabilities Worksheet" in Chapter 12.

Also transfer the legal value to the "Marital Balance Sheet" in Chapter 19. Any separate property (including a separate property loan) should be entered in the appropriate section on the "Marital Balance Sheet".

Calculating the Financial Value of Plans										
Type of Plan	A: Months of Plan Participation*	B: Months of Marriage During Plan	C: Marital Interest	D: Dollar Amount of Account	E: Marital Portion of Account	F: Marital Loans Against Account**	G: Legal Value	H: Tax Basis	I: Marital Portion of Tax Basis	J: Financial Value***
Employer-Sponsored Retirement Plans										
Qualified Plans—Defined Contribution Plans										
Business or Corporate Defined Benefit Plan										
Salary Savings or 401(k) Plans										
Thrift Plans										
Profit-Sharing Plan										
Money Purchase Plan										
ESOPs										
TSAs										
Employer-Sponsored IRA										
SEP IRA										
SAR-SEP										
Simple IRA										

*Check state laws to determine the date retirement plan benefits become separate property.

**If separate, do not include balance here, but be sure to include separate property loan balance on the marital balance sheet in Chapter 19.

*** If your state awards unvested benefits, you may wish to subtract the unvested portion.

Calculating the Financial Value of Plans (cont'd)

Type of Plan	A: Months of Plan Participation*	B: Months of Marriage During Plan	C: Marital Interest	D: Dollar Amount of Account	E: Marital Portion of Account	F: Marital Loans Against Account**	G: Legal Value	H: Tax Basis	I: Marital Portion of Tax Basis	J: Financial Value***
Qualified Plans—Defined Benefit Plans										
Business or Corporate Defined Benefit Plan										
Government-Sponsored Retirement Plans										
Government Plans or Military Pensions										
Personal Retirement Plans										
Deferred Annuities										
Personal IRA										
Traditional IRA										
Roth IRA										
Total										

*Check state laws to determine the date retirement plan benefits become separate property.

**If separate, do not include balance here, but be sure to include separate property loan balance on the marital balance sheet in Chapter 19.

*** If your state awards unvested benefits, you may wish to subtract the unvested portion.

Rolling Over Your Distribution From Your Ex-Spouse's IRA

Under IRA rules, any money you withdraw from an IRA prematurely (before age 59½) must be reinvested (referred to as a "rollover") within 60 days or it will be subject to an IRS early withdrawal penalty.

You can defer the income tax liability on the distribution by rolling over the distribution to another IRA within 60 days of the divorce. If you decide to roll over the distribution, keep in mind the following:

- You cannot roll over any portion of the distribution that is attributable to after-tax contributions of the plan participant. In other words, any nondeductible contributions that your ex-spouse made to an IRA can't be rolled over into your account. But if part of the distribution is attributable to after-tax contributions and part is attributable to before-tax contributions, you will be able to roll over a portion of the distribution.
- If you take possession of funds from your ex-spouse's IRA and *then* roll them over into your own IRA, 20% will be withheld for taxes.
- However, if the plan distribution goes directly to the new plan, then you avoid the 20% withholding on the distribution.

Even if you have someone else managing your account, it is still your responsibility to see that it is done correctly. In a recent case, a taxpayer's account was being handled by a trustee. The trustee mistakenly transferred a rollover distribution to a brokerage account instead of an IRA. When the trustee discovered its mistake, the amount was transferred to a qualified plan, but the 60-day limit had expired. The IRS ruled that the distribution was not rolled over within the 60 days, and therefore the distribution was taxable even though the taxpayer was not at fault.

The lesson to be learned is that it is your responsibility to follow up on how your accounts are handled. Mistakes by trustees can and do occur. Failure to properly time your rollover can lead the IRS to assess penalties.

Additional Financial Factors Affecting Retirement Plan Divisions

Often, spouses keep their own retirement benefits in exchange for other assets. As a simplified example, if each spouse's share of the house is $75,000 and the nonemployee spouse's share of the other's retirement plan is $75,000, then the employee would keep the retirement plan while the other spouse would take the house.

Fairly dividing assets during a divorce, however, is not as simple as the above example. To truly exchange assets of equal value, you must take a variety of financial factors into account. Consider the following factors before you make a proposal about dividing retirement benefits with your spouse.

Vesting

One retirement plan concept that is especially important at divorce is vesting. You've probably heard people say they'll be "fully vested" after a certain number of years or that they'll "be vested in another year." Vesting means that you are entitled to the retirement benefits your employer has contributed to the plan for you. Even if you quit your job or are fired, you are entitled to those benefits.

Being fully vested means that you are entitled to all the benefits your employer has contributed. Being partially vested means that if you were to begin receiving the benefits from your employer's contribution to the plan, you'd be entitled to only a particular percentage. If the plan paid $1,000 a month upon retirement and you were 30% vested, you'd receive $300.

Vesting affects only the portion of your benefits that your *employer* contributed toward your retirement plan and its earnings. The contributions you make to your plan, and the earnings attributable to those contributions, always belong to you.

If you are not fully vested, consider consulting an attorney to determine whether your state characterizes nonvested retirement benefits as marital property. If it does not, then division of your plan in that state

is not affected by vesting. If the state does characterize the benefits as marital property, however, and you divorce before you fully vest, you could face the awful prospect of paying your spouse a portion of the retirement benefits even though you might *never* receive them yourself. If you are not fully vested and might encounter this costly dilemma, be sure you know your rights and obligations before negotiating over your retirement plan as part of your divorce settlement. (See "Questions to Ask an Attorney," below.)

For example, in a state where nonvested benefits are considered marital property, the total value of your retirement plan is calculated and that amount is divided. If you want to keep the retirement plan, you might have to buy out your spouse's share or exchange it for another asset. If you're fired or change jobs, you might not receive your employer's contributions to your retirement plan, even though you paid your spouse for them.

To reduce these risks, carefully consider whether to divide your plan in the future or in the present (see "The Division Decision," below).

If you are fully vested (and therefore entitled to your employer's contribution and required to divide all retirement benefits at divorce), double-check the details of your benefits package to be sure you know about any provisions that could be affected by divorce or death. In some cases, when a single (unmarried or divorced) employee dies before the age of 55, that person's estate may not be entitled to the benefits even though the person is fully vested.

Benefits in Payout Status

Even after pension benefits vest, the employee may still have to wait before actually receiving the benefits. Usually, vested pensions do not pay out ("mature") until the worker reaches a certain age, such as 65. Some employees choose to continue to work even after their pensions mature because pension payments increase the longer an employee works past the maturity date. But if you continue to work to prevent your ex from getting matured benefits, a court may order you to retire (or to pay your former spouse the amount the ex would receive if you did retire).

Double Payments

Be aware of the risk of double payments. You should not have to pay your spouse twice for the retirement benefits you receive as income. If your spouse is given a portion of your retirement benefits as part of the property settlement, you do not want your retirement benefits to be considered part of your income when alimony is calculated. To protect yourself against double-dipping, remember the following: You can divide the plan or divide the stream of income, but do not divide both.

This double-dipping of payments is one of the most common ways in which retirement benefits are lost at divorce. Don't let it happen to you. One way to protect yourself is to include a provision in your settlement exempting your retirement payments from being considered income for the purposes of alimony. Check with an attorney to discuss your options.

Cost of Living Adjustments

Cost of living adjustments (COLAs) are provisions in retirement plans that provide for an increase in benefits based on your life expectancy and the formula by which benefits are paid out. A plan with a cost of living provision is usually more valuable than a plan without one. Check with the plan administrator to see whether your retirement plan or your spouse's includes a cost of living provision.

Access to Cash

In getting cash out of a retirement plan at divorce, the alternate payee (nonemployee spouse) holds the advantage. If you are the alternate payee, you can request that the QDRO provide for a payout when a marital property settlement becomes final. The plan administrator must withhold 20% of the amount distributed to you to offset some of the income taxes you will owe, even if you deposit the money into an IRA or any other qualified pension plan within 60 days of receiving it. If the plan administrator directly rolls over the plan assets into your IRA (or another qualified plan), however, the 20% mandatory withholding rule does not apply.

If money is tight (as it often is during divorce), you may want to open a savings account with all or part of the money you receive from a retirement plan distribution. Although you'll have to pay taxes on this money, you will have easy access to the funds should you need them. Most important, you won't be hit with the 10% early withdrawal penalty that would apply if you put it into an IRA and then needed to withdraw money to live on.

If you are the plan participant (employee spouse), you probably won't get an early payout from your plan even though your spouse may. Your divorce does not change your status with respect to your benefit payments. It's rare that you will be able to get to your benefits before the date on which the plan specifies you are to receive them. The exceptions are if you become disabled, if you quit your job or get fired, or if the plan itself ends.

Tax Consequences in Dividing Benefits

When dividing matured benefits in a divorce, the employee spouse must be aware of the tax risks. Many divorce settlements specify that the nonemployee spouse gets a share of the other spouse's retirement benefits when the employee spouse retires. Usually, the pension plan administrator sends a check to the ex-spouse. But this means that only the retired employee pays income taxes on the benefits.

To reduce your tax burden, make sure the QDRO orders the pension plan administrator to send two checks—one to you and one to your ex-spouse. Each person pays income taxes on the benefits that person receives. If your plan is not subject to a QDRO, it may be possible to reduce your tax liability by using the domestic relations order, marital settlement agreement, or divorce agreement to specify how the benefits are to be paid.

The Division Decision: Now or in the Future

Taking into account all the financial factors that can affect your retirement benefits, you are ready to move on to the division decision itself. Generally, you can divide a retirement plan or take a payout in one of the two following ways.

Present Division. You divide the value of the plan now, at divorce, instead of later, at retirement.

Future (Deferred) Division. You divide the value of the plan at retirement or at that point in the employee's career when benefits would normally be paid out. Future divisions are most common with defined benefit plans, which promise payment in the future and are difficult to place a current value on.

Remember—no matter which approach you take, everything must be spelled out in the QDRO.

Survivor Benefits

A defined benefit (pension) plan must offer survivor benefits to the spouse of a vested participant employee who dies before retirement age. The surviving spouse is entitled to payments known as a qualified preretirement survivor annuity. An ex-spouse can be treated as the participant's current spouse for the purpose of receiving these survivor benefits. But a provision to that effect must be included in the QDRO, or the benefits will be lost.

If the participant dies prior to retirement, the preretirement survivor annuity begins paying out to the survivor at the point when the participant would have reached the earliest retirement age provided for in the plan.

Debts and Retirement Benefits After Divorce

After a divorce, some former spouses find it necessary to file for bankruptcy to discharge their debts. With a QDRO, retirement benefit payments are made directly from the plan to the nonemployee ex-spouse. This eliminates the risk that the employee might be able to avoid the obligation to pay retirement benefits to the nonemployee ex-spouse by filing for bankruptcy. If, however, there is no QDRO, then the obligation to pay retirement benefits is a personal obligation, and the employee may be able to avoid it by filing for bankruptcy.

Even if a former spouse doesn't need to file bankruptcy, it's easy to fall behind on support obligations. While money cannot be forcibly taken from most retirement plans to pay support, some courts have ruled that IRA funds are not insulated if an ex-spouse or the IRS seeks to collect unpaid alimony or child support.

Furthermore, once divided, retirement plan assets are not insulated from the IRS if taxes are due on a joint return. That's true even if the parties agreed that one spouse was liable for payment of those taxes.

Here are some pros and cons of dividing each of the three main types of retirement plans either at the time of the divorce or later at retirement age. You'll see that the considerations can vary depending on whether you're the alternate payee (nonemployee spouse) or the plan participant (employee spouse).

IRAs

IRAs are not subject to QDROs and can be divided any way the court orders. They can also be transferred from one spouse to another.

Distributing the proceeds from an IRA, however, triggers a 10% early withdrawal penalty for those under age 59½. The proceeds are also fully taxable to the participant spouse.

Transferring an IRA to a spouse without a divorce or separation agreement or a court order creates tax liability for the participant spouse.

The only way to avoid the tax is for the court to order a direct transfer of the IRA from one plan to a new one for the receiving spouse.

I would like to divide the IRAs as follows:

_____

_____

_____

_____

Defined Contribution Plans and TSAs

There are advantages in splitting this kind of plan now, and the advantages are the same whether you are the alternate payee or the plan participant. If you divide the benefits now, you'll cut an economic tie to your spouse and won't need to worry about it later. In addition, consider that:

- Deferred divisions are difficult for the court to supervise.
- Each spouse gains control over a share of funds.
- It's obvious how much the plan is now worth—you don't need an actuary to figure out its future value.

But there are also advantages in waiting before you divide the plan. For both the plan participant and the alternate payee, waiting allows time for any unvested benefits to become vested. Additionally, if you have insufficient property or cash to buy out your spouse's share, deferred division may be the only way to even up a property split. Also, because the plan participant won't receive money from the plan until later anyway, why divide it early? If you have unvested benefits and you divide now, you may end up paying more to your spouse than you receive from the plan.

I would like to divide the defined contribution plans or TSAs as follows:

_____

_____

_____

_____

Defined Benefit Plans

The advantages of splitting this kind of plan at divorce rather than later depend on whether you are the plan participant or the alternate payee. For the alternate payee, the advantages are:

- You can get the money now.
- The risk of not vesting is eliminated.
- There is no 10% early withdrawal penalty for distribution.
- You avoid the risk that the plan will suffer large investment losses or that the company or plan will go bankrupt.

The plan participant should consider that:

- Because the value is based on present earnings, your ex-spouse won't receive the benefit of your future work and pay increases.
- If you outlive the mortality tables, you may end up receiving more income than the actuary predicted you would.

There are also advantages in waiting to divide this type of plan. Because benefits are based on the participant's future earnings, waiting allows the alternate payee to enjoy the fruits of those increased earnings. For the plan participant, the vesting risk is shared with the ex-spouse.

I would like to divide the defined benefit plans as follows:

_____

_____

_____

_____

Questions to Ask an Attorney

Retirement plans raise many legal and financial questions. Below is a list of important questions for you to ask an attorney or research at your local law library.

1. Does my state treat retirement benefits upon divorce as marital property to be divided or as income that may belong only to the earner? (Benefits treated as income could affect the amount you pay or receive in alimony or child support.)

2. Is a nonqualified deferred compensation plan considered income or property subject to division?

3. What is the marital portion of our retirement benefits?

4. Must I (or my spouse) be fully vested to receive my (or my spouse's) employer's contribution to a retirement plan?

5. Does my state consider nonvested retirement plans to be marital property?

6. Does my state consider disability pension benefits to be income or property subject to division?

7. How much experience do you have in writing Qualified Domestic Relations Orders? Do you use another firm to write them? (Some firms specialize in this area.)

8. Does my spouse's pension plan direct the distribution of assets upon my death in the event I die before receiving all of the benefits?

9. Will the divorce deprive me of any medical benefits currently covered under my spouse's retirement package? (This is particularly pertinent if the spouse is/was employed by a federal, state, or municipal government.)

To get details on retirement plans held by your spouse, speak with the employer and, if it's a large company, the head of the human resources department. That person may refer you to the plan administrator or the trustee/custodian of the plan. The plan administrator keeps the books and records on the retirement plan, and the trustee/custodian is authorized to administer the plan assets. Begin by writing a letter to your spouse's employer. A sample is below.

Re: Marriage of _____

       Case No. _____

       Employee's Name: _____

       Employee's Social Security #: _____

       Employee's Date of Birth: _____

To Whom It May Concern:

This letter is to serve as notice that I am pursuing an interest in any retirement or other deferred compensation plans available through your organization or its successors for the benefit of employees belonging to [name of your spouse] .

At your earliest possible convenience, I would appreciate receiving the following:

1. Summary of all deferred compensation plans or stock plans that are available to employees, directors, officers, or other personnel of your company and in which my spouse is eligible to participate, regardless of whether my spouse participated in the plan.

2. A copy of the full text of all plans for which my spouse was eligible, regardless of my spouse's participation therein.

3. A copy of all benefit statements for my spouse for the last five years.

4. Documentation of all stock or stock options owned by or held for the benefit of my spouse.

5. Acknowledgment that your organization has received this Notice. An acknowledgment for the Plan Administrator is below.

6. A copy of your written procedures used in determining your requirements for a Qualified Domestic Relations Order.

7. The formal names of the plan or plans.

8. The names and addresses of all plan administrators.

9. Any other documentation or information that you believe may be of assistance in order to establish a Qualified Domestic Relations Order.

Sample Letter to Spouse's Employer (cont'd)

To save everyone the cost of formal discovery, I ask that you respond as soon as possible.

If you have any questions, please feel free to contact me at your convenience.

Sincerely,

[_your signature_]

I consent to the release of this information.

Date: _____

Employee/Plan Participant Spouse: [_your spouse's signature_]

ACKNOWLEDGMENT OF RECEIPT OF NOTICE OF ADVERSE INTEREST

I, _____, hereby acknowledge receipt of the letter dated _____, from _____, which claims an interest in any retirement, pension, or other deferred compensation plan or accounts of _____.

Dated: _____

Title: _____

Organization: _____

_____

If you don't get the answers you need from this letter, write to the plan administrator, company president, or anyone else who could help you get the information. You might engage the services of an attorney if you're having particular problems. However you do it, persist until you get what you need. Retirement plans are too complex not to leave a "paper trail," so do not let your spouse or spouse's attorney intimidate you into thinking that the information you need does not exist. ●

Financial Investments:
How Do We Divide the Portfolio Pie?

Concepts to Consider 293

An Investment From Marriage May Make Little Sense
When You're Single 293

Don't Take an Investment You Can't Live With 293

Use Your Divorce to Cut Your Losses 294

No Investment Is Risk Free 294

Asset Options: Sell Now, Keep It Forever, or Take It for a Loss 295

Balance Security, Income, and Growth 297

If You Don't Know the Tax Basis, the Tax Bill May Shock You 298

Beware of Inflation and Inflated Claims 301

Look to Total Return, Not Yield 301

Steps to a Settlement 302

Step 1: What Investments Do You and Your Spouse Hold? 310

Step 2: Who Owns Each Investment? 310

Step 3: What Is the Legal (Equity) Value of Each Investment? 310

Step 4: What Is the Financial (After-Tax/After-Sale)
Value of Each Investment? 313

Step 5: Which Assets Should You Keep? 337

Step 6: How Will You and Your Spouse Divide the Investments? 338

When you divide the investments you and your spouse made during marriage, it's doubtful you'll get exactly what you want. But you should get what you need.

 SKIP AHEAD

If neither you nor your spouse has investments, skip ahead to Chapter 16.

The first step in dealing with financial investments is to analyze your portfolio objectively—that is, the stocks, bonds, mutual funds, real estate investments, gold, collectibles, and the like that make up your investment holdings. Then work to get the assets that can put you in the best position to build toward your future after divorce.

How do you know which investments will serve you best?

Ask a dozen financial planners and you'll get the same answer: "That depends."

Choosing the investments that will serve you best depends on several factors—your future plans, current age, risk tolerance, income needs, and investment experience. You should aim for investments that meet your individual needs, allow you to sleep at night, and keep your taxes to a minimum.

As you read this chapter and evaluate your investments, keep in mind the following two questions:

- Is the investment performing well?
- Is the investment appropriate for me?

Whether you are a novice or veteran investor, use the formulas and tips in this chapter to select the investments that will best support you after the divorce. Even if you are completely unfamiliar with the investments you and your spouse share, you need not be intimidated—if you do your homework.

In some instances, you may need to consult with financial experts before making decisions, because certain investments can carry

complications due to tax laws or illiquidity (not being able to sell an investment at a reasonable price). Tax advisers, stockbrokers, and financial planners may be particularly useful. Whether you get help from experts or not, you can save time and money by doing some groundwork and information gathering.

Concepts to Consider

To get started, here are a few basic investment concepts to help you divide investments when a marriage ends.

An Investment From Marriage May Make Little Sense When You're Single

While married, the two of you may have chosen certain investments because they met your shared goals—such as a mutual fund to help save for the down payment on a house or an annuity to increase the retirement kitty. Once you divorce, however, your goals, tax bracket, and ability to withstand losses may change. Perhaps as a couple you could afford to hold on to a stock with great potential, even when its value fell. As a single person, however, such a stock may be totally unsuitable if you're looking for stable income. Similarly, a spouse in a high-income tax bracket may not want to keep an investment that generates substantial taxable interest.

Don't Take an Investment You Can't Live With

Does the mere mention of Wall Street give you the jitters? Can you bear to sit by and watch what happens to the bond market? Do you have the time, energy, and interest to keep up with the performance of your investments? Think about these questions carefully as you look at each asset. Like a spouse, you should only keep an investment if you can stand living with it. Don't take an asset if you don't understand its risks.

Use Your Divorce to Cut Your Losses

When it comes to investments, no one likes to admit to making a bad call or a poor decision—especially if your spouse warned you not to buy the loser in the first place. But you wouldn't fight to keep the awful picture Aunt Lucille gave you for a wedding present, would you? Why hold on to investments that are not doing well? Because of the emotional distress of divorce—or simple "investment inertia"—some people hang on to investments they ought to sell. Divorce is the perfect time to review *all* of your investments and unload those that drain your portfolio value or don't meet your long-term financial goals.

No Investment Is Risk Free

Even in a booming economy, no capital investment comes with a 100% guarantee of success. Whether you're working with banks, hedge funds, brokerage firms, or investment companies or buying U.S. government-issued bonds, everything carries risk.

The booms, busts, and investment scandals and swindles of the last two decades should be enough to bury the notion of a "risk free" investment. Yet, this great illusion of the American marketplace lives on. You should not be taken in, no matter how persuasive an argument seems. The fact is, regardless of what you might see, hear, or want to believe, every investment carries risk. (Even what were once the most trusted places for money—banks, savings and loans, and insurance companies—have shown this to be true.) Ignoring risk is common among today's investors who have been hit by a virtual avalanche of new financial products, each promising rewards that may or may not materialize. Take time to ask questions about the downside of any investment you plan to keep as part of your settlement. Keep in mind one general rule: *The greater the potential for gain, the greater the risk.*

Never keep an investment that carries a greater risk than you can afford. Doing so increases the risk of loss, because you will probably sell the investment when its value drops rather than ride out its ups and downs.

EXAMPLE: In the property settlement, Cheryl received 300 shares of WOW Mutual Fund, a fund investing in small-capitalization computer companies. The fund had a high return of 22% in the prior year. Her husband, Jim, convinced her that with the $14,000 worth of WOW shares, she would make more money than with the lower-return Treasury bonds he was taking as part of the settlement. Cheryl quickly learned, however, that the greater the potential for gain, the greater the risk; her $14,000 investment dropped to $8,700 the next year.

Who Do You Trust To Handle Your Investments?

It's common for people to seek help and financial advice in their community and social networks. But as the Madoff case and many other investment scandals made clear, financial fraud often hinges on the victims' trust and vulnerability. When you're in crisis—such as during a divorce—it's easy to put your faith in someone who seems wise and caring, but may not be.

For your protection and peace of mind, you should engage a third-party financial adviser to help handle your investments. If you house your investments with a third-party custodian—and you should—it's easier to monitor the health of the investment and access advice from a reliable source. Check with your local bank or an established investment management firm to see how you might work with a financial adviser.

Trust, but verify, is a sound approach when you're allowing someone else to handle your investments.

Asset Options: Sell Now, Keep It Forever, or Take It for a Loss

"Assume sale." Those two words sum up the best advice you can follow when deciding what to do with assets at divorce. So often, cherished investments you thought you would keep forever end up on the auction block after divorce. Ideally, you should keep only those investments that

financially fit your goals and needs—and you should sell or transfer to your spouse those that do not. If you sell the investment at divorce, you share the selling costs and potential tax bite with your soon-to-be ex-spouse. Waiting until after the divorce to unload assets means you incur all those expenses yourself.

Too often, divorcing couples overlook the simple fact that an asset they are eventually planning to sell can be sold *prior* to the divorce. They get caught up in making trade-offs to reach a quick or supposedly fair property settlement and accept assets they don't want or intend to keep. If, however, you assume sale, you're forced to consider taxation and other costs. Then your negotiations can be based on the real cost of keeping the asset and not an illusory value. And if you do keep the asset, you know what it's truly worth.

An investment asset may provide you with a tax loss when you sell it. If it would, and if you have another asset on which you expect a gain, it may make financial sense to take the loss-producing asset in the settlement. Then you can sell the "loser" after the divorce to offset the tax gain from the sale of the other asset, and reduce your overall tax liability. It pays to know which stocks, bonds, or mutual funds carry a tax loss with them. For example, limited partnerships and rental real estate, which produce a loss because of depreciation, actually increase your tax liability. Be sure to consult a tax advisor before taking one of these items that show a loss.

> **EXAMPLE:** Suppose you bought a stock for $2,000 and sold it for $1,500. Your loss is $500. If you have another stock you bought for $1,500, held for 12 months and then sold for $2,500, your gain is $1,000. If you didn't sell the first stock, you'd pay a capital gains tax of $150 on the $1,000 gain. If you sold both assets, however, the loss partially offsets the gain and your tax liability would be 15% of $500, or $75.

Losers Can Be Winners

When you incur a capital loss at sale, remember that this loss can offset capital gains. If your capital loss exceeds $3,000, you can carry over the excess losses until they are used up, either by offsetting capital gains or taking the maximum $3,000 deduction from your gross income.

Gains and losses on jointly owned or community property are generally divided equally between spouses upon the divorce judgment or settlement. This is important, since a loss greater than $3,000 will have tax implications in years after you and your spouse divorce. If one spouse has a greater need to declare such losses, then they become an item of value in negotiation, and an agreement for their division should always be included in the final divorce settlement or judgment.

Sell Now

Elliot accepted his wife's offer of their coin collection even though he did not particularly care about the coin market and didn't really want the collection. Nevertheless, he accepted it with the illusion that by making this concession to his wife, he would speed up the settlement. Not only did the settlement proceedings drag on, but after the divorce, when he sold the collection, Elliot paid a hefty sales commission and substantial taxes on the gain—which meant Elliot received an asset of lower value than he thought he was getting. Coin collections and other collectibles did not get the benefit of a reduced capital gains rate under the Taxpayer Relief Act of 1997 and are still taxed at 28%.

Balance Security, Income, and Growth

Your investments should give you more than headaches. In fact, a portfolio should balance three financial needs over your lifetime: security, income, and growth.

During your divorce, analyze whether or not your investments meet these needs—and in what proportions. The degree to which you need security, income, and growth will change depending on what phase of life you're in. Generally, you will want more growth than income while you are young and more income than growth when you get older.

Security. Conventional wisdom among financial advisors holds that you should have an emergency fund in place before you even consider investing. This fund should equal three to six months of your cost of living expenses such as mortgage, rent, utilities, and food. It should be readily available in a money market fund or short-term bond fund. Disability or life insurance can also provide important sources of income or cash in the event of an emergency. Once your emergency needs are met, you can consider investments for income or growth.

Income. Income investments pay monthly, quarterly, semiannual, or annual dividends or interest to supplement your salary or provide an income during retirement.

Growth. These types of investments should "make your money work for you," offering steady growth beyond what you would earn from conservative vehicles like certificates of deposit or money markets.

Growth investments include individual stocks, equity mutual funds, and real estate.

If You Don't Know the Tax Basis, the Tax Bill May Shock You

"Buy low and sell high." That cliché of the marketplace leaves out an important piece of information: When you sell high, you owe Uncle Sam taxes on the profit. Just how much you owe is determined by the amount of profit you realized on the investment. That profit can be determined once you know the tax basis of the investment.

For your mutual funds and stock investments, your tax basis is the original purchase price plus the value of any dividends you reinvested in the funds or stocks, minus any commissions you paid to purchase. The tax basis for real estate is the original purchase price plus the cost

of improvements you made to the property, minus any tax benefits received. Specific formulas for finding the tax basis are given throughout this chapter.

Income vs. Growth Needs

Susan, a young executive, had just received a raise when her husband decided he wanted a divorce. Being a conservative investor, her husband had invested their money in a large number of utility stocks, which provided a steady but small flow of dividends. Because Susan's salary increase gave her a more than adequate income, she wanted cash for her share of the stocks rather than the stocks themselves. With cash in hand, she invested in several high-tech stocks that paid no dividends (and therefore did not increase her tax liability for ordinary income), but gave her what she wanted and needed: potential for appreciation or growth.

EXAMPLE: Leslie bought 100 shares of stock at $12 per share in May and another 100 shares of the same stock in December for $30 per share. Three years later, when the stock soared in value to $80 per share, Leslie decided to sell it, after holding all of the stock more than 12 months. Assuming a 15% capital gains tax rate, here's what Leslie's tax bill would look like:

May Stocks		December Stocks	
Sell 100 shares @ $80	$ 8,000	Sell 100 shares @ $80	$ 8,000
Bought 100 shares @ $12	− 1,200	Bought 100 shares @ $30	− 3,000
Gain	$ 6,800	**Gain**	$ 5,000
(times capital gains tax rate)	× 0.15	(times capital gains tax rate)	× 0.15
Taxes due	$ 1,020	**Taxes due**	$ 750

The tax basis for the stocks purchased in May is the $12 per share price. Upon sale of these stocks, there is a greater gain, which in turn creates a bigger tax bill. For the stocks purchased in December, with the $30 per share purchase price, the gain is less, and so are the taxes. If Leslie had to sell half of the shares, she would be better off selling the December stocks than selling those she purchased in May. By doing so, she saves $270 in taxes ($1,020 – $750 = $270).

Suppose Leslie's spouse offers to split the 200 shares of stock. Should Leslie take the May stocks or the December stocks? Leslie would be better off taking the December stocks, on which the taxes owed are lower. The fairest settlement would be for each spouse to take 50 of the May stocks and 50 of the December stocks. Another possibility would be to sell the stocks and split the tax bill equally. In that case, the calculations would look like this:

200 shares sold @ $80 per share, held more than 12 months	*$16,000*
Tax Basis	
100 shares @ $12 = $1,200	
100 shares @ $30 = $3,000	
–	*4,200*
Total Gain =	*$11,800*
Capital Gains Tax Rate ×	*15%*
Total Taxes on Gain =	*$1,770*
Each Spouse's Share of Taxes Due =	*$885*
Each Spouse's Share of Proceeds from Sale of Stock =	*$5,015*

Keep in mind that losses and gains are netted to determine the final tax liability, and commissions paid to brokers reduce the final tax liability. This example assumes that this was the only asset Leslie and her spouse sold this year.

Beware of Inflation and Inflated Claims

Nothing takes the fun out of investing faster than losing money. Yet investors often overlook the simple factor of inflation, which constantly erodes the value of their holdings. To determine the true amount of return you earn from an investment, you must consider not only taxes, but inflation as well.

Take the simple example of a $10,000 certificate of deposit paying 3% interest at a time when inflation is running 3.5% annually. At the end of a year, the CD would seem to be worth $10,300 (the $300 earned via the 3% interest payment).

Assuming a federal tax bracket of 25%, the amount due the government would equal $75, reducing the $10,300 to only $10,225. Now factor in inflation. The $10,000 loses $350 (inflation of 3.5%) a year. At the end of a year, instead of $10,300, the CD really has the purchasing power of only $9,875 ($10,225 – $350), for a stunning loss of almost 1% on the original $10,000.

Be on guard, too, for inflated or misunderstood claims. Quite commonly, investors consider only one dimension, such as the interest rate an investment pays. They often forget about the numerous other factors that affect value, such as inflation, sales charges, management fees, dividend payments, and other unforeseen risks.

Look to Total Return, Not Yield

The commonly misunderstood differences between yield and total return illustrate the problem of one-dimensional analysis. *Yield* represents income in the form of dividends and interest. *Total return* is the yield plus the percentage of appreciation (growth) or loss in the per share value of an investment.

> **EXAMPLE:** Ron saw an advertisement for a mutual fund offering a yield of 9%. Later that day, his broker called to tell him about another fund with a 5% yield. Which should he buy?
>
> Ron must look beyond yield to total return. The yield represents only the dividends or interest he'll receive on his shares. The total

return takes into account both the yield and the percentage increase (appreciation) or decrease (loss) in the value of the shares.

Ron's broker explained that the shares in the fund yielding 9% decreased in value by 20% the previous year, meaning the total return was a minus 11% (9% – 20%). The shares in the fund yielding 5%, on the other hand, increased 8.3%, for a total return of 13.3%. Note that the higher-yield fund actually produced a lower total return.

Tax Basis and the IRS

When spouses divorce, information concerning the tax basis of an asset must be given to the spouse who receives the asset when title (ownership) changes. Because tax consequences are so fundamental to the selection of an asset, however, you should know the tax basis before you divide your assets, not after.

Steps to a Settlement

Now that you have an overview of the concepts to consider when making investment choices, it is time for a plan of action for reaching decisions. Answer these questions to formulate your plan:

1. What investments do you and your spouse hold?
2. Who owns each investment?
3. What is the legal value or fair market value of each investment?
4. What is the financial or after-tax/after-sale value of each investment?
5. Which assets should you keep?
6. How will you and your spouse divide the investments?

To answer these questions, complete the "Investment Chart" that follows. Begin by listing your investments in the far left column, using suggestions offered in Step 1, below. As you move on through Steps 2, 3, and 4, you will find detailed instructions on how to fill in the remaining columns. Finally, when you reach Step 6, you will return to the chart to show how you want to divide assets with your spouse.

Investment Chart							
(1)	**(2)**	**(3)**	**(4)**				
Investments	Title *	**Legal or Equity Value** (Fair Market Value – Debt)	**Financial** (After-Tax/After-Sale) **Value**	I Keep	Spouse Keeps	Sell and Split Proceeds	
Cash and cash equivalents							
Bank checking accounts							
Bank savings accounts							
Certificates of deposit							
Money market funds							

* H=husband, W= wife, J=jointly, C=community property

Investment Chart (cont'd)

			(2)	(3)	(4)			
Investments	(1)		Title *	**Legal or Equity Value** (Fair Market Value – Debt)	**Financial** (After-Tax/After-Sale) **Value**	I Keep	Spouse Keeps	Sell and Split Proceeds
Mutual funds								
Personal notes payable to you								
Other								
Stocks and bonds								
Common stocks								

* H=husband, W= wife, J=jointly, C=community property

Investment Chart (cont'd)

(1) Investments	(2) Title *	(3) Legal or Equity Value (Fair Market Value – Debt)	(4) Financial (After-Tax/After-Sale) Value	I Keep	Spouse Keeps	Sell and Split Proceeds
Preferred stock						
Mutual funds						
Treasury bills						
Government bonds (such as Ginnie Maes)						
Municipal bonds (include mutual funds, unit trusts, and individual bonds)						

* H=husband, W= wife, J=jointly, C=community property

Investment Chart (cont'd)

(1) Investments	(2) Title *	(3) Legal or Equity Value (Fair Market Value – Debt)	(4) Financial (After-Tax/After-Sale) Value	I Keep	Spouse Keeps	Sell and Split Proceeds
Corporate bonds						
EE U.S. Savings Bonds						
Zero coupon bonds						
Other						

* H=husband, W= wife, J=jointly, C=community property

Investment Chart (cont'd)

Investments	(1)	(2) Title *	(3) Legal or Equity Value (Fair Market Value – Debt)	(4) Financial (After-Tax/After-Sale) Value	I Keep	Spouse Keeps	Sell and Split Proceeds
Real estate							
Income or rental properties							
Real estate investment trusts (REITs)							
Raw land							
Other							
Insurance investments							
Cash value of life insurance policies							

* H=husband, W= wife, J=jointly, C=community property

Investment Chart (cont'd)

(1) Investments	(2) Title *	(3) Legal or Equity Value (Fair Market Value – Debt)	(4) Financial (After-Tax/After-Sale) Value	I Keep	Spouse Keeps	Sell and Split Proceeds
Annuities						
Other						
Limited partnerships						
Real estate						
Oil and gas						
Equipment leasing, etc.						
Other						

* H=husband, W= wife, J=jointly, C=community property

Investment Chart (cont'd)

(1) Investments	(2) Title *	(3) Legal or Equity Value (Fair Market Value – Debt)	(4) Financial (After-Tax/After-Sale) Value	I Keep	Spouse Keeps	Sell and Split Proceeds
Collectibles						
Gold and silver						
Coins, jewelry, art, and other valuables						
Other						

* H=husband, W= wife, J=jointly, C=community property

Step 1: What Investments Do You and Your Spouse Hold?

Using Column 1 of the "Investment Chart", list all investments owned by you, your spouse, and the two of you together. If you're unsure of certain investments, check old tax returns, or call your financial planner, stockbroker, brokerage firm, accountant, or other tax adviser. You should also check online, if you or your spouse has used electronic brokerage services.

Step 2: Who Owns Each Investment?

In Column 2, show whether the investment is owned by husband (H), wife (W), or jointly (J). Investments owned jointly—except those that are separate property—are divided at divorce. Remember: The name of the person on the investment does not necessarily determine who owns it. Most states divide marital property, even when assets are only in one spouse's name.

List the investments owned separately by you and your spouse even though they won't be divided in the settlement. As you negotiate the settlement, it is possible that your or your spouse's separate investments will be considered in terms of income needs or for alimony.

If you need more information on what constitutes separate and marital property, see Chapter 12.

Step 3: What Is the Legal (Equity) Value of Each Investment?

With investments, the equity or legal value basically equals the fair market value or face value—that is, the amount the investment is worth, less any debts. (For cash or cash-like assets—such as CDs or money market accounts—the legal value is the same as the account balance.) The legal value is generally the dollar amount attorneys use when discussing investments in trying to settle your divorce. In Steps 4 and 5, below, you will take a closer look at the financial value of your investments.

To find the legal or equity value of your investments, you often need to do no more than place a phone call.

You may, however, run into problems determining which date to use in valuing the account. As mentioned in Chapter 5, many states value portfolio assets, like houses, as close to the *date of divorce or settlement* as possible, not the *date of separation*. You'll need to find out your state's practice.

Even if your state legally values assets at the date of separation, you and your spouse can agree to another date—and you should. Stocks, bonds, mutual funds, and other liquid assets, particularly, should be valued as close as possible to the date you take control of them. Otherwise, if you value assets as of the date of separation, but then don't transfer them until months or years later, you may incur losses because of market forces.

Ultimately, to get through your divorce, you will probably need to find the account balance at different times. But you can begin by filling in the current amounts for the legal value of investments in Column 3 of the Investment Chart. Use the following information to help you calculate or locate the legal value for each investment. Some of this information may be listed on your Net Worth Statement from Chapter 12.

Cash and cash equivalents. For your bank checking and savings accounts, certificates of deposit, and money market funds, call the bank or look at your statement. For money market mutual funds, call the fund manager or administrator. The value of personal notes will probably be found among your personal papers. Find out whether the account is pledged as security for any debt. Be sure to investigate your spouse's business, if any, and any personal loans.

Stocks and bonds. For most stocks and bonds, you can check with your broker, your discount brokerage firm, or an online service. If you call a brokerage firm, ask for the trading department to check the current price of stocks or mutual funds. You can also look at the business section of the newspaper. For stocks, look in the stock market tables for the price in the column labeled "Close." That's the price the stock was selling at when the market closed the day before. Multiply this number by the number of shares you own to get a ballpark value. Get an estimate of bonds by

calling a broker or discount brokerage firm. Banks can give you the value of savings bonds. For mutual funds, look in the business section of the newspaper in the column labeled "NAV," the Net Asset Value.

Real estate. For income or rental property, you will probably need an appraiser. (See Chapter 13.) For the value of real estate investment trusts (REITs), ask a broker or brokerage firm or check the published values in a newspaper. To find the value of raw land, check the purchase price and ask a few real estate agents to find out whether land values have gone up or down in your area. It's also helpful to know whether there is a balance due on the mortgage or whether the property is free and clear of debt.

> (!) **CAUTION**
>
> **Two identical pieces of property can have different values if the rents charged in one building are not the same as the rents charged in the other.** Rent control laws can also affect (bring down) the value of property. A spouse trying to artificially lower the value of rental property may rent units at below-market rates on a month-to-month basis, and then raise rates (and the property's value) after the divorce. If you suspect this is happening, check previous tax returns to see what rents were charged and ask realtors how much rent is charged for comparable units in the area.

Insurance. Call or write the insurance company or broker and ask for the policy's current value and surrender value. If you're not the owner of the policy, the insurance company may refuse to give you the information. Call your spouse, or have your attorney call your spouse's attorney to get what you need.

Limited partnerships. Investments in businesses that operate as limited partnerships are more difficult to value than, say, investments in stocks or bonds. That's because, unlike stocks, there is no conventional trading market for interests in limited partnerships.

Certain "secondary markets," however, do buy and resell limited partnership interests. But these markets may assign widely divergent values to limited partnership interests. Ask a stockbroker or financial planner for the names of several secondary market firms. In your survey,

ask for the price at which the partnership shares could be sold and whether additional costs would be incurred.

There will probably be a difference between what secondary markets say the partnership interest is worth and the amount you will actually realize from its sale. And there will be an even larger difference between your original purchase price and what you could sell it for now. Generally, these markets buy partnership shares at a deeply discounted price from the amount you originally paid or might receive if held until maturity. Use the amount the firms give you to fill in the "Investment Chart."

Limited partnerships can be complicated investments, especially when you're attempting to value and divide your interests. For additional discussion, see below.

> **CAUTION**
>
> **Do not consider the value of a limited partnership as being equal to what you paid for it.** Most limited partnerships, very popular investments in the 1980s, are worth significantly less now. (In fact, some are worthless.) If you take that asset at an inflated value, you'll be getting less in your settlement than you are entitled to.

Collectibles. Gold and silver values are listed, per ounce, in the daily newspaper. Coins, jewelry, art, and collectibles must be appraised. Be sure to find the wholesale or resale price, not the retail price.

Step 4: What Is the Financial (After-Tax/After-Sale) Value of Each Investment?

Remember: You will need to know the tax basis of your investments to calculate gains, losses, and your potential tax liability when you sell. In addition, the IRS requires that when you transfer property at divorce, you must provide tax basis information to the party who will keep the asset. If you are in the dark about investments, you may need to press matters with your spouse to get the information to figure the tax basis. You cannot make informed decisions about which investments to accept

or sell in the settlement unless you know your potential tax liabilities *before* any property transfer takes place.

Knowing the tax basis will help you find the financial value (or after-sale/after-tax value) of your investments. For most investments, you will probably enlist the services of a professional tax adviser, but the following formulas and tips should help you get a ballpark estimate of your taxes and other costs.

After calculating the financial value of each investment, enter that value in Column 4 of the "Investment Chart." But first, talk to a broker or financial planner or another investment professional to find out whether your investments carry any additional or hidden costs. If they do, subtract those amounts from the financial value before entering the values in Column 4. Be on the lookout for:

- **Account charges.** An amount charged for maintaining an account with a broker, brokerage house, or firm
- **Redemption fees.** What you pay to redeem or sell the investment
- **Trading fees.** A commission for buying or selling stocks or bonds, and
- **Early withdrawal penalties.** Penalties incurred for early withdrawals on CDs, life insurance annuities, and other investments.

Cash and Cash Equivalents

Because the tax basis of cash and cash equivalents is essentially what you will liquidate them for, there are no taxes due when you cash them in. To the extent that you earn interest on checking accounts, savings accounts, money market accounts, or certificates of deposit, you must report that interest as income and pay taxes on the interest the year you earned it. To estimate the interest you will earn this year, call your banker. Then simply use the legal value as the financial value, and write that amount in Column 4 of the "Investment Chart."

Stocks, Bonds, and Mutual Funds

Calculating taxes and other costs associated with stocks and bonds will require you to do some tedious paperwork. You will have to gather old statements to reconstruct past trading activity (perhaps, even, over

several years). Only you can decide whether it's worth it to do that work yourself rather than hiring someone else for the job. No matter who does it, if you sell stocks and bonds you will need this information to complete your tax return. There are good software packages and websites that can help you make this calculation. Several investment calculators may be found at www.bankrate.com. Some brokerage or financial planning services companies may be able to determine this figure for you using their own specialized software programs.

Bonds

See an accountant or a tax adviser to analyze the tax aspects of bonds. The Internal Revenue Code has numerous complicated rules depending on when the bonds were purchased and the type of bonds you own. For instance, some bonds, such as municipal bonds, are tax exempt. This means their interest is free from taxes. But if you bought them at a discount, when you redeem them or they mature you will have to pay capital gains tax on the difference between the purchase price and maturity value.

Also keep in mind these general rules:

- **EE U.S. Savings Bonds.** You can pay taxes on the interest generated on EE bonds each year, as with most investments, or when you redeem the bonds. You can defer paying taxes on the interest if you roll the bonds forward into HH bonds.
- **HH Bonds.** These are government bonds that pay you interest semiannually.
- **Zero Coupon (Corporate or Municipal) Bonds.** On corporate bonds, you must pay taxes on interest earned each year even though you don't receive the interest payment until the bond matures.

Once you talk to an accountant or other tax adviser about your bonds, enter their financial value in Column 4. This value can be determined using the formula for stocks and mutual funds, below. But keep in mind that the value of bonds fluctuates with interest rates. If interest rates in the United States have risen since you purchased your bond, the value will be slightly less than when you bought it. If interest rates have declined, the opposite will be true: Your bond, if you sell

it, will be worth more than when you bought it. Check your monthly statement from your brokerage account or ask your broker for the amount you will get if you liquidate your bond or bonds.

Stocks and Mutual Funds

Use the following formula and instructions to find the financial value for your stocks and mutual funds. To complete the calculations, you will need to know the following:

$ _____   *Purchase price*

Check your confirmation statement or call your broker. Be sure to include the initial investment plus any subsequent purchases you made.

$ _____   *Dividends reinvested*

These should be shown on your monthly statements from the brokerage firm or mutual fund company. You can also look on tax Form 1099, sent annually to you from the company, showing dividends or interest earned each tax year.

$ _____   *Commissions and transactions fees*

These are fees paid when you purchase stocks and mutual funds unless the mutual fund is a "no load" fund, meaning it has no commission. The amount is on your confirmation statement; if it's not, call your broker or discount brokerage firm. For "load" mutual funds, the commissions, which are automatically charged, are reflected in the purchase price. Ask your broker whether the fund has a redemption charge or surrender charge, a fee you pay when the fund is sold.

$ _____   *Equity value*

Enter from Column 3 of the "Investment Chart."

$ _____   *Tax rate*

You will need to know your income tax bracket and how long you have held the investment to know what to put here. If you held the stock or mutual fund for more than 12 months, enter 15% as your tax rate. If you held it less than 12 months, check with your tax preparer for your tax bracket.

To find the financial (after-sale/after-tax) value of stocks and mutual funds, use the following formula:

1. Find the tax basis by adding the purchase price, reinvested dividends and commissions, and transaction fees (paid at purchase).
2. Find the adjusted sales price by subtracting commissions and transaction fees (to be paid when you sell) from the fair market value.
3. Find the gain (or loss) by subtracting the tax basis from the adjusted sales price.
4. Find the taxes due (or tax refund) by multiplying the gain (or loss) by the capital gains rate based on your tax bracket.
5. Find the financial value by subtracting the taxes due (or adding the tax refund) from (or to) the adjusted sales price.

> **EXAMPLE:** Let's assume you bought 100 shares of XYZ at $10 per share and it has grown to $22 per share.
>
> 1. Find the tax basis by adding the purchase price, reinvested dividends, and commissions and transaction fees (paid at purchase).
>
> | | |
> |---|---|
> | *Purchase price* | *$1,000* |
> | *Dividends reinvested* | + 60 |
> | *Commissions & transaction fees* | + 65 |
> | *Tax basis* | = *$1,125* |
>
> 2. Find the adjusted sales price by subtracting commissions and transaction fees (to be paid when you sell) from the equity value.

Equity value	$2,200
Commissions & transaction fees	– _____ 68
Adjusted sales price	= $2,132

3. Find the gain (or loss) by subtracting the tax basis from the adjusted sales price.

Adjusted sales price	$2,132
Tax basis	– 1,125
Gain (or loss)	= $1,007

(If you incur a loss on this asset, you can net it against a gain on another asset to reduce your capital gains tax liability.)

4. Find the taxes due by multiplying the gain by the capital gains rate based on your tax bracket.

Gain	$1,007
Capital gains tax rate	× 15%
Taxes due	= $151

5. Find the financial value by subtracting the taxes due from the adjusted sales price.

Adjusted sales price	$2,132
Taxes due	– 151
Financial value	= $1,981

To apply this formula to your situation, fill in the blanks below. Copy this page to use as a worksheet or to calculate the value of additional investments.

1. Find the tax basis by adding the purchase price, reinvested dividends, and commissions and transaction fees (paid at purchase).

Purchase price	$ _____
Dividends reinvested	+ $ _____
Commissions & transaction fees	+ $ _____
Tax basis	= $ _____

2. Find the adjusted sales price by subtracting commissions and transaction fees (to be paid when you sell) from the equity value.

Equity value		$ _____
Commissions & transaction fees	−	$ _____
Adjusted sales price	=	$ _____

3. Find the gain (or loss) by subtracting the tax basis from the adjusted sales price.

Adjusted sales price		$ _____
Tax basis	−	$ _____
Gain (or loss)	=	$ _____

4. Find the taxes due by multiplying the gain by the capital gains rate based on your tax bracket.

Gain		$ _____
Capital gains tax rate	×	$ _____
Taxes due	=	$ _____

5. Find the financial value by subtracting the taxes due from the adjusted sales price.

Adjusted sales price		$ _____
Taxes due	−	$ _____
Financial value	=	$ _____

Keep in mind that the above formula calculates only federal income taxes. In most states, you will have to pay state taxes as well.

Dividing Stocks, Bonds, and Mutual Funds: Consider the Tax Basis

As noted at the beginning of this section, the IRS requires that you give information about the tax basis of an asset to the person receiving it at the time of transfer. It's very important for you to understand the tax basis of an asset. If you don't, you could decide to keep investments that give you a higher tax burden than those your spouse will carry. At the least, you ought to share the tax burden equally.

Once you know the tax basis, you can calculate the taxes that would be due if you were to sell the assets. Knowing the potential tax liability enables you to compare the true value of investment assets and other assets. This knowledge will also influence your tax strategy after the divorce, as you consider whether to sell or keep the investments.

The IRS allows three different methods to calculate the taxes. In these examples, a gain is assumed. If you have experienced a loss on your investments, the investment may be of value if you have a gain elsewhere—but only if you sell the asset immediately.

First In–First Out. If you've purchased shares of stock over several years at different prices, this method lets you sell the first batch you bought. For example, Mindy and Ryan bought 300 shares of stock at $13 per share in 1994 and another 300 shares of the same stock at $20 per share in 1996. The stock is now worth $25 per share. They decide to sell 300 shares before they divorce. But which 300?

Using the first in–first out option, they would sell the 300 shares bought in 1994. They would have to pay taxes on the gain of $12 per share ($25 fair market value – $13 purchase price = $12 gain). Had they sold the 1996 shares, they'd owe taxes on a gain of only $5 per share ($25 fair market value – $20 purchase price = $5 gain). But by selling the shares that will give them a larger profit, they can split the taxes owed.

As an alternative to selling shares before the divorce, Mindy and Ryan consider splitting the 600 shares in half. It's fair on the surface. If one spouse keeps the 1994 shares, and the other keeps the 1996 shares, however, the division becomes unfair because the tax liabilities differ. They could even things up again if they divide other property to cover the tax difference, or if they each take half the 1994 shares and half the 1996 shares.

Specific Identification. This time, assume that Mindy and Ryan bought their 600 shares at various times paying prices ranging from $13 to $20 per share. If they decide to sell 300 shares before their divorce and use the specific identification option, they would pinpoint precisely which 300 shares to sell—perhaps 100 bought at $13, another 100 bought at $18, and the final 100 purchased at $20 per share.

If they split the shares as part of their divorce settlement, they should consider the tax liability for the shares bought at each price. If Mindy wants 300 of the $20 stocks—leaving Ryan with all of the stocks purchased at $13 and $18 per share—Ryan will pay more in taxes than Mindy.

Average Basis. Mindy and Ryan's third option is the average basis. As the name implies, this option lets them take an across-the-board average tax basis on all the shares of stock, no matter when purchased and at what price per share. They would simply take the total amount paid (and dividends reinvested) and divide it by the total number of shares owned.

Because of the difficult calculations and the fact that supporting documentation regarding reinvestments and commissions is often missing, many divorcing couples use the average basis to divide stocks and mutual funds. Before settling on it, however, compare it to the first in–first out and specific identification options. Incorporate the calculations from this section on tax basis options with financial values when you read Steps 5 and 6, below.

Document Dividends and Reinvestments

When Evelyn divorced in 2004, she got $5,000 worth of shares in a mutual fund as part of her settlement. A savvy investor, she sold the shares in the last week of September 2007 for $10,000. Initially, it appeared Evelyn would pay taxes on the $5,000 gain: $10,000 (sales price) – $5,000 (original price) = $5,000 (gain).

But Evelyn had kept records showing that she reinvested all the dividends she received. In adding up each reinvestment, she found that she had reinvested total dividends of $1,600 into the fund. This finding lowered her tax liability. Remember: Her tax basis is the original purchase price plus reinvested dividends. Her original $5,000 investment, plus the dividend reinvestments of $1,600, totaled $6,600. The amount of gain she had to pay taxes on, therefore, was only $3,400, not $5,000.

Real Estate

In valuing real estate investments, you can consider many of the factors raised in Chapter 13, assuming you own a family home. If you haven't read that material, refer to it for basic formulas and an explanation of real estate terms.

Unlike the family home, investment properties can be hard to value. The value of the property is affected by the local economy, rental vacancy rates, rent control or stabilization laws, property management, and other market forces. Nevertheless, you can estimate the financial (after-sale/after-tax) value of your real estate investments by using these guidelines.

$ _____   *REITs (real estate investment trusts)*

A real estate investment trust company invests in a variety of holdings, which can range from apartments and hotels to office buildings and shopping centers. Because most REITs are publicly traded, their value is easily identified. Refer to the stock tables in the newspaper, or ask your broker for current values. To find the financial value, subtract any commissions or fees from the fair market value.

$ _____   *Raw land*

Real estate agents and appraisers normally base the value of raw (undeveloped) land on current comparables, that is, the amount at which comparable parcels of land are currently selling. Other factors to consider in determining the real financial value of raw land are:
- how land may be affected by a city's general plan
- the kind of building permits that have been issued in the area, and
- how land surrounding your land may be used.

Real estate agents can help you answer these questions. Also, talk with the county building permit office and the city office regarding permits and future land use. Then, make your best estimate of the financial value.

$ _____   *Income or rental properties*

For most income-producing property, use the formula in Chapter 13, Step 6, to determine the financial value. With rental property, you must also consider total depreciation: the amount you've deducted in taxes to account for the declining value of the property.

If you own real property, such as a residential rental, know that there is an exception to the lower capital gains rates that went into effect under the Taxpayer Relief Act of 1997. Not all the net capital gain that you have on this kind of property will qualify for the lower rate of 15% (or 5% for those in the 15% bracket). Instead, to the extent that you have taken depreciation on the property, your gain will be taxed at a maximum rate of 25%.

> **EXAMPLE:** Jim and Carol own a rental property, which they bought in 2000 for $100,000. Over the years, they have taken $15,000 in depreciation deductions, so the property's adjusted tax basis is now $85,000. They sell the property in 2005 for $150,000, realizing a $65,000 long-term capital gain ($150,000 – $85,000 = $65,000). Assuming Jim and Carol are in the 35% tax bracket, $15,000 of their gain (the amount of the depreciation they have taken) will be taxed at 25%. The rest of their gain ($50,000) will be taxed at the lower rate of 15%.

To find the financial value of rental real estate, you will need to know the following:

$ _____   *Fair Market Value*

Single-family homes are somewhat easier to value than commercial, industrial, or multifamily dwellings. For single-family residential real estate, you can ask several real estate agents or appraisers for current comparable values. With other types of rental units, however, you will need to consult an appraiser or a broker who specializes in evaluating the type of property you have. See Chapter 13 for information on finding an appraiser.

An appraiser will use one—or a combination—of these three recognized calculations:

Cost Approach. The fair market value equals the amount it would cost to replace a building, plus the land's value. This approach usually applies to single-family dwellings.

Market Data Approach. Also used with single-family dwellings, this approach compares properties that all provide similar cash flows to determine the fair market value.

Income Approach. The fair market value is based on the cash flow a building currently generates plus an estimate of the amount it will generate in the future. Commercial, industrial, and multifamily dwellings are often appraised with this approach.

Ask the appraiser which approach is being used to value your property—and why. Just as a retirement plan can be worth different amounts when appraisers use different assumptions, so too can real estate values vary when different approaches are used. Avoid disputes by making sure everyone agrees on what basis is being used to value the property.

When you know the fair market value of your properties, use the chart below to keep track of those values. Write the address or the description by which you and your spouse will refer to the property in the first column, and the fair market value in the second column. The remaining columns will be addressed as you continue reading this section.

$ _____ *Debt on property*

The debt connected to real estate equals what you owe to any lenders (mortgages) plus any liabilities (liens) on the house. All liabilities must be paid before any cash disbursements can be made from the sale of real estate. The mortgage due is sometimes called the "payoff balance."

To find the current debt on the house, talk to your lender and check the property's title for any liens you're unaware of. (See Chapter 13.) List the debt in Column 3 of the "Real Estate Values" chart.

$ _____ *Total capital improvements or depreciation*

To find total capital improvements made or depreciation taken on rental properties, check your records or previous tax returns, or talk to your accountant or tax adviser. List the totals in the fourth column of the "Real Estate Values" chart.

Real Estate Values				
Property Address or Description	**Fair Market Value**	**Current Debt on Property**	**Total Capital Improvements**	**Total Depreciation Taken**
Rental house, 323 Third Street, Rye, NY	$100,000	$60,000	$20,000	$40,000
Duplex, Sun Court, Surf City, CA	$280,000	$100,000	$0	$32,000

$ _____ *% Your tax bracket*

By now, you have probably established your tax bracket, either by checking income tax schedules or talking with your accountant. You will need this information in the upcoming formula.

$ _____ *Cost of sale*

You can do a quick ballpark estimate of what it will cost you to sell your rental property or you can follow these steps to find as precise a figure as possible. To find this number, you'll have to include not only the costs of sale, but also the amount it will take to fix up the property and prepare it for a sale.

To get a ballpark estimate of the sales cost, multiply the fair market value by .07. This is an accepted estimate of what it costs to sell a house. Obviously, some sales will yield higher or lower costs of sale, but 7% will do.

To obtain a more accurate cost-of-sale figure, total the following amounts:

$ _____ *Agent's commission*

The amount a real estate agent charges you for selling property, this generally averages 6% of the final selling price. This fee can sometimes be negotiated down.

$ _____   Closing costs

Included in these costs are escrow fees, recording costs, appraisal fees, and miscellaneous expenses that can add up to thousands of dollars. A lender or real estate agent can help you figure the exact amount.

$ _____   Attorney's fees

In a few states, you will need to hire an attorney to help you "close" the sale of real property. If you used an attorney when you bought the property, base your figure on that amount, taking into account the fact that most lawyers raise their fees over time. Otherwise, ask a real estate agent for an estimate.

$ _____   Fix-up costs

Estimate here the amount of money you will have to spend to prepare the property for sale.

Now add all of these figures and fill in the cost of sale.

To find the financial or after-tax/after-sale value of rental real estate, use this formula:

1. Find the equity by subtracting the debt from the fair market value.
2. Find the tax basis by adding total capital improvements to the purchase price and then subtracting total depreciation taken.
3. Find the adjusted sales price by subtracting the cost of sale from the fair market value.
4. Find the gain (or loss) on the potential sale of the property by subtracting the tax basis from the adjusted sales price. If you experienced a loss, you may have no tax liability and you can probably offset other gains with your loss to reduce your taxes due.
5. Determine the portion of gain related to depreciation taken. That portion will be taxed at 25%, while the remainder of the gain will

be taxed at either 15% (if you are in a tax bracket above 15%) or 5% (if you are in the 15% tax bracket).

6. Find the financial or after-tax/after-sale value by subtracting costs of sale from the equity and then subtracting the taxes due.

EXAMPLE:

1. Find the equity by subtracting the debt from the fair market value.

Fair market value		*$100,000*
Debt	−	*60,000*
Equity	=	*$40,000*

2. Find the tax basis by adding total capital improvements to the purchase price and then subtracting total depreciation taken.

Purchase Price		*$50,000*
Total improvements	+	*5,000*
	=	*55,000*
Total depreciation	−	*12,500*
Tax basis	=	*$42,500*

3. Find the adjusted sales price by subtracting the cost of sale from the fair market value.

Fair market value		*$100,000*
Cost of sale	−	*7,000*
Adjusted sales price	=	*$93,000*

4. Find the gain (or loss) on the potential sale of the property by subtracting the tax basis from the adjusted sales price.

Adjusted sales price	=	*$93,000*
Tax basis	−	*42,500*
Gain (or loss)	=	*$50,500*

5. Find the taxes due by multiplying the depreciation portion of the gain by 25% and the remaining gain by 15% (assuming you are in the 25%+ tax bracket). Total these numbers for the taxes due.

		$12,500
	×	25%
Gain from depreciation	=	$ 3,125
Gain on sale		$50,500
	−	12,500
Remaining gain	=	$38,000
	×	15%
	=	$5,700
		$3,125
	+	5,700
Taxes due	=	$8,825

6. Find the financial or after-tax/after-sale value by subtracting the cost of sale from the equity and then subtracting the taxes due.

Equity	$ 40,000
Cost of sale	− 7,000
Taxes due	− 13,860
Financial or After-tax/after-sale value	= $ 19,140

To apply this formula to your situation, fill in the blanks below:

1. Find the equity by subtracting the debt from the fair market value.

Fair market value	$	_____
Debt	− $	_____
Equity	= $	_____

2. Find the tax basis by adding total capital improvements to the purchase price and then subtracting total depreciation taken.

Purchase price	$	_____
Total improvements	+ $	_____
	= $	_____
Total depreciation	− $	_____
Tax basis	= $	_____

3. Find the adjusted sales price by subtracting the cost of sale from the fair market value.

Fair market value		$ _____
Cost of sale	−	$ _____
Adjusted sales price	=	$ _____

4. Find the gain (or loss) on the potential sale of the property by subtracting the tax basis from the adjusted sales price.

Adjusted sales price		$ _____
Tax basis	−	$ _____
Gain (or loss)	=	$ _____

5. Find the taxes due by multiplying the depreciation portion of the gain by 25% and the remaining gain by 15% (assuming you are in the 25%+ tax bracket). Total these numbers for the taxes due.

		$_____
	×	_25%_
Gain from depreciation		$_____
		$_____
	−	$_____
		$_____
	×	_15%_
Remaining gain		$_____
		$_____
	+	$_____
Taxes due		$_____

6. Find the financial or after-tax/after-sale value by subtracting cost of sale from the equity and then subtracting the taxes due.

Equity		$ _____
Cost of sale	−	$ _____
Taxes due (or tax refund)	−	$ _____
After-tax/after-sale value	=	$ _____

Who Pays the Rent

If you or your spouse moves from the family home, you can make temporary arrangements for paying the mortgage and other bills. But what if you own rental property and the renters move? Who will pay the mortgage? Or suppose you and your spouse decide to sell the property. Who will pay for repairs? Who will be reimbursed for capital improvements to the property? You must settle these questions. If you pay these costs alone, keep good records of every expense so that you will have a chance to get reimbursed as part of the final settlement.

Insurance

Buying more insurance than you need and paying too much in insurance premiums are common mistakes. At divorce, you have the opportunity to take a good look at your policies and correct these mistakes if necessary.

Whole life or universal life insurance not only provides income in the event of someone's death, but also has some investment benefits, although these benefits may be minimal compared with other investments. As you pay your premiums, a cash reserve builds up. This reserve increases over and above the premiums you pay because you earn interest on your premiums. The reserve also grows because the expenses, such as commissions paid to brokers, decrease over time. Be aware, though, that the cash value decreases if you borrow money against the policy.

Term policies, on the other hand, provide death benefits only. They have no cash buildup and therefore have no growth potential.

Keep only those policies you truly need for insurance. Never surrender an essential policy or let it lapse without having replacement insurance, but do get rid of unnecessary policies. Life insurance companies select policyholders with care. If you have to apply for insurance in the future, you may be denied coverage because of a yet-unknown health problem. If you surrender policies with cash value—that is, you turn them in and collect the cash—you will have to pay taxes on the interest earned.

You figure the interest by subtracting the total premiums paid and the surrender charges from the cash value. The difference is the interest, and that amount is what you'd have to report to the IRS as income.

Annuities

Annuities are investments, offered by insurance companies, that are meant to provide income at retirement. Annuities are discussed in Chapter 14.

Cash Value Policies

Any policy with a cash value is often called, logically enough, a "cash value policy." If you decide that your policy is not the most cost-effective for you, you can use the cash value to purchase another insurance policy that is more suitable. The benefit of buying another policy rather than taking the cash out of your current policy is that you can defer the taxes you would owe on the accrued interest. (Internal Revenue Code § 1035.) The only drawback is that you will have to pay a commission when you purchase the new policy, and these costs will decrease the initial cash value of the new policy.

To quickly estimate the financial value of your insurance (a detailed analysis is not necessary), you need to obtain the following:

$ _____   *Cash value*

Take a look at the most current policy statement and see if it shows the cash value. If it doesn't, ask your insurance agent.

$ _____   *Surrender charges*

Again, this information should be on the current policy, or you can get it by checking with your agent.

$ _____   *Premiums paid*

Check the most current policy statement or ask your agent.

Following the example in the "Value of Insurance Policy" chart, below, use the formula shown to find the financial or after-tax/after-sale value on cash value insurance policies.

Value of Insurance Policy

	Example	Your Policy
Cash value	$ 10,000	
Surrender charge	– 500	
Cash surrender value	$ 9,500	
Premiums paid	– 6,000	
Gain (or loss)	$ 3,500	
Tax bracket	× 0.25	
Taxes due	$ 875	

Cash surrender value	$ 9,500	
Taxes due	– 875	
Financial or after-tax/after-sale value	$ 8,625	

Limited Partnerships

You may have difficulty finding an accurate value for your interests in a limited partnership. It's often hard to tell how much was invested and how much has been received in tax benefits (write-offs). We do not provide detailed information on figuring the financial value of limited partnerships. Instead, we urge you to consult your accountant to get an understanding of the value of these assets and to ask your broker for information on companies that specialize in secondary market sales of limited partnership units.

Also, call the general partner (and any other limited partners) to find out everything you can about the partnership, especially its economic health. Be alert to these potential problems with limited partnerships:

- You may not be able to divide your interest. General partners may not be willing to divide the investment in the limited partnership between you and your spouse. If you can't divide your interest,

and neither of you wants to give up your interest to the other in exchange for other property, be sure you and your ex-spouse have a carefully written agreement about how taxes, bookkeeping, and distribution checks will be handled after the divorce.

- A limited partnership can generate taxable income that you never receive, called phantom income. Ask the general partner whether phantom income is expected and how much of this income will be taxable.

- Some limited partnerships make requests for additional investments (called capital calls) from the limited partners. Before keeping a limited partnership, check the original investment document, contact your accountant, or ask the general partner whether you may be liable for additional investments.

- All limited partnerships are risky investments, because the general partner has almost complete control over the assets of the partnership. A general partner's decision to sell an asset or allow a foreclosure can create a tax liability for you. You might call your broker or the general partner to determine when the sale of assets in the partnership is anticipated and what the tax consequences would be.

- Bookkeeping can be difficult. When the partnership sends out its IRS reporting form (K-1) each tax year, it sends only one form to each investor. The partnership usually sends only one distribution check as well. If you and your spouse are unable to divide the partnership interest, then only one of you will receive the form and the check. You need to agree on who that will be and how you'll handle it—for example, what the deadline will be for the spouse who receives the check and information to pass along the other spouse's share.

Unless you've been lucky, limited partnerships are more trouble to keep after divorce than they are usually worth.

Collectibles

Your biggest problem in determining the financial value of collectibles (like art, special plates, or even baseball cards) will be finding a valuation on which you and your spouse can agree. You may hear any number of conflicting values depending on who is making the appraisal.

To find the financial value of collectibles, first list the following:

$ _____   *Purchase price*

Check original receipts, look at old tax returns, or consult with a broker.

$ _____   *Acquisition costs*

Did you drive to a distant city to buy the piece? Stay overnight in a hotel? Pay admission to an antique fair? All of these could add to the taxable cost basis of the antique, meaning you'll owe less tax when you sell it.

$ _____   *Improvements*

Did you restore the piece or do anything to it—such as reupholstering, refinishing, or refurbishing—that increased its value?

$ _____   *Selling price or wholesale value*

Ask a dealer, an auctioneer, or a consignment shop to give you an estimate of the wholesale value of your holdings. Many people mistakenly assume that some items, such as diamonds, are worth a great deal because they consider only the retail value. But jewelers buy loose diamonds at wholesale prices and will not pay you more than that price when they buy from you.

$ _____   *Commissions*

Talk with a broker, a wholesaler, or anyone familiar with the market for your item. With collectibles, the commission can be as high as 30% of the sale price.

$ _____   *% Tax bracket*

Consult an accountant or previous tax schedules.

To find the financial or after-sale/after-tax value of collectibles, use this formula:

1. Find the tax basis by adding commissions (paid at purchase), acquisition costs, improvements, and purchase price.
2. Find the adjusted sale price by subtracting commissions (due at sale) from the selling price or wholesale value.
3. Find the gain (or loss) by subtracting the tax basis from the adjusted sales price.

    If you incur a loss on this asset, you can net it against a gain on another asset to reduce your capital gains tax liability.
4. Find the taxes due by multiplying the gain times the capital gains tax rate of 15%.
5. Find the financial value by subtracting (or adding) taxes due (or tax benefit) from (or to) the adjusted sale price.

> **EXAMPLE:** Janet and Tom purchased a painting during their marriage that they now wanted to sell. It had gone down in value, so they had a loss they could net against other gains.
>
> 1. Find the tax basis by adding commissions (paid at purchase), acquisition costs, improvements, and purchase price.
>
> | Purchase price | | $10,000 |
> |---|---|---|
> | Acquisition costs | + | 250 |
> | Improvements | + | 250 |
> | Commissions | + | 500 |
> | Tax basis | = | $11,000 |
>
> 2. Find the adjusted sale price by subtracting commissions (due at sale) from the selling price or wholesale value.
>
> | Selling price or wholesale value | | $ 7,000 |
> |---|---|---|
> | Commissions | − | 500 |
> | Adjusted sales price | = | $ 6,500 |
>
> 3. Find the gain (or loss) by subtracting the tax basis from the adjusted sales price.

Adjusted sales price		$ 6,500
Tax basis	–	11,000
Gain (or loss)	=	$(4,500)

4. Find the taxes due by multiplying the gain times the capital gains tax rate of 15%.

Gain (or loss)		$(4,500)
Capital gains tax rate	×	15%
Taxes due (or benefit)	=	$(675)

5. Find the financial value by subtracting the tax benefit from the adjusted sale price.

Adjusted sales price		$ 6,500
Taxes due (or benefit)	–	($675)
Financial value	=	$ 7,175

If loss is not netted against gains, you can only use up to $3,000 in losses in one year. So you would carry forward the $1,500 ($4,500 loss – $3,000 = $1,500) to the next year.

To apply this formula to your situation, fill in the blanks below:

1. Find the tax basis by adding commissions (paid at purchase) and purchase price.

Purchase price		$ _____
Commissions	+	$ _____
Tax basis	=	$ _____

2. Find the adjusted sale price by subtracting commissions (due at sale) from the selling price or wholesale value.

Selling price or wholesale value		$ _____
Commissions	–	$ _____
Adjusted sales price	=	$ _____

3. Find the gain (or loss) by subtracting the tax basis from the adjusted sales price.

Adjusted sales price		$ _____
Tax basis	−	$ _____
Gain (or benefit)	=	$ _____

4. Find the taxes due by multiplying the gain times the capital gains rate of 15%.

Gain		$ _____
Capital gains tax rate	×	*15%*
Taxes due (or benefit)	=	$ _____

5. Find the financial value by subtracting the taxes due or benefit from the adjusted sale price.

Adjusted sales price		$ _____
Taxes due	−	$ _____
Financial value	=	$ _____

Step 5: Which Assets Should You Keep?

Once you know the financial (or after-tax/after-sale) value of investments, you have a starting point for negotiating with your spouse. The values are useful whether you want to sell your assets to the person you're divorcing, buy them for yourself, or divide them between you.

The ultimate decision to keep or sell an investment depends on many factors including your age, risk tolerance, income needs, and investment experience. Asking yourself the following simple questions will help you make sound decisions about your investments.

1. *Has this investment performed well?*

   If Yes, go to Question 2.

   If No, consider these options:

   - Sell the asset before the divorce so you and your soon-to-be ex-spouse share the costs of sale, taxes, and other expenses.
   - Take the asset for yourself, but only if it provides you with the benefit of a tax loss. That is, it can offset a gain you may have

on another asset. You should consult a tax expert to determine whether you need to take any losses.

- Offer to sell the asset to your spouse in the divorce settlement.

2. *Is this investment appropriate for me—is it providing the income, growth, liquidity, or security I need?*

If Yes, how long do you plan to keep it?

- Long-term—buy out your spouse's interest and keep this investment.
- Short-term—consider selling before your divorce becomes final, sharing the proceeds, taxes, and other costs with your spouse, and then using the proceeds to purchase a similar investment.

If No, consider these options:

- Sell before your divorce is final, sharing the proceeds, taxes, and other costs with your spouse.
- Buy out your spouse and sell immediately to take advantage of tax losses.
- Let your spouse buy out your share or interest in the investment.

Step 6: How Will You and Your Spouse Divide the Investments?

You are now ready to prepare your position for negotiating over investment assets. Look at the "Investment Chart" you completed earlier in this chapter and use it to show how investments will be handled in your divorce. For each investment listed in Column 1, put a check mark in one of the last three columns to show whether you intend to keep the asset, have your spouse keep it, or sell it and split the proceeds. If you have cash value life insurance or rental property, be sure to read the sections that follow before making a final decision about these assets.

Special Problems With Cash Value Life Insurance

The financial value of the policy isn't the only consideration with cash value life insurance. Be sure to consider the following before choosing to keep this asset in your divorce:

- What is the quality of the insurance company? You may be able to evaluate the quality of an insurance company by checking its ratings by one of the three main rating services, Standard and Poor's, Moody's, and A.M. Best. But these ratings provide no guarantees. Before it folded in 1990, for example, Executive Life received an A+ rating from Best, considered the top rating company.
- What interest rate are you receiving on the premiums accruing as your cash value? Would alternative—yet safe—investments, such as corporate bonds, municipal bonds, or federal government bonds, yield you more?
- Are there any loans outstanding on the policy? If so, the amount paid at death will be reduced by the outstanding loan. Before taking this asset at divorce, be sure the loans are paid back so the full amount is paid, or be prepared for the death benefits to be reduced.

Special Problems With Rental Property

The financial value of the property isn't the only consideration with rental property. Be sure to ask yourself the following questions before choosing to keep this asset in your divorce:

- Where is the property located? Is it on a busy street? On the beach? Given its location, what is its potential for appreciation?
- What is the quality of the neighborhood? Is it improving or deteriorating?
- How will the property be affected by changes in the local economy?
- Are newer and nicer buildings directly competing for your renters?
- Are your rents above or below market?
- What does it cost to maintain and manage the property? Is your cash flow positive or negative? Can you afford the property when renters vacate and the income is down?

- What are the tax incentives or disincentives to keeping the property?
- What would it cost to improve the property?
- How much income can you expect the property to generate?
- When might you sell? Can you determine your rate of return over that time?
- What is the vacancy rate?
- What kind of loan do you have—fixed or adjustable rate? If it's adjustable and interest rates rise, can you still afford it?

Your answers will help you reach one of three conclusions:

1. The property has strong potential (the location is great, there's little rental competition, and operating costs and vacancy rates are low) and it should appreciate faster than other investments I could make using the same amount of money.

    It would be a good asset to keep.

2. The property has some potential, but it's uncertain. Other investments using the same amount of money may outperform this one.

    We should probably sell now and split the proceeds,
    or my spouse can take it.

3. This property has little potential (operating costs are high, the location is bad, and/or vacancy rates are high) and other investments have a higher probability of meeting my needs.

    We should probably sell now and split the proceeds,
    or my spouse can take it.

Now review all your investments—stocks, bonds, rental real estate, limited partnerships, and others—and write your decisions on the "Investment Chart" at the beginning of this chapter. ●

Evaluating Employee Benefits and Stock Options

Employee Benefits..342

   Health Insurance 343

   Group Term Life Insurance..344

   Accidental Death or Dismemberment Insurance..344

   Health Savings Accounts 344

   Dependent Care Assistance Programs...345

   Cafeteria Plans..346

   Severance Plans...346

   Educational Assistance Programs 347

   Group Legal Services...347

   Employee Assistance Programs...347

Stock Options and Nonqualified Deferred Compensation Plans.................348

   Types of Stock Options..349

   What Are Stock Options Worth at Divorce?..354

   Nonqualified Deferred Compensation Plans..356

Questions to Ask an Attorney or a Tax Specialist.......................................357

As companies get more creative in compensating their employees, the stakes are raised in the divorce process.

Today's employees receive more than what shows up on the pay stub or tax form. Benefits such as stock options, insurance, retirement plans, and other fringes are forms of compensation. And to the extent they can be considered income to you or your spouse, they can help determine how much spousal and child support you are eligible to receive, or must pay.

For divorce purposes, the key question is whether to classify various types of employee compensation as:

- **income**—a continuing source of money that can affect decisions about alimony and child support
- an **asset**—property of some sort that can be apportioned between the spouses in the settlement
- **neither income nor an asset,** or
- **both income and an asset.** (For more on the distinction between income and assets, see Chapter 12.)

This chapter looks at different forms of employee compensation in light of their value to you and your spouse in the divorce process.

Employee Benefits

Employee benefits (also known as nonwage compensation) have mushroomed in recent years as companies and nonprofit organizations have sought to attract and retain employees. Start-up businesses, especially, use nonwage compensation because they may not yet generate the cash to pay competitive salaries. Employee benefits include retirement plans—both pension plans and deferred compensation plans; insurance coverage; disability and death benefits; employee assistance plans; educational assistance; and perks such as company cars, health club memberships, and discounts at stores or hotels.

This aspect of divorce law is far from settled. Courts can and do disagree about how to classify particular benefits at divorce. One court

may say that use of a company car lowers an employee's living expenses, freeing up more money for spousal and child support; another may not. So it's important to talk to your attorney about the role that employee benefits may play in your case.

Various types of retirement plans, employer sponsored and otherwise, are explained in Chapter 14. In the following sections, we provide an overview of other common employee benefits and how they could affect your divorce settlement. The question that's usually at issue is whether a particular benefit can be considered taxable (or gross) income. If so, a court may take it into account in setting child or spousal support. In other instances, a benefit may be regarded as a marital asset—something that gets allocated between spouses at the time of divorce.

Health Insurance

The employee benefits that most affect day-to-day living are in the insurance area. Medical insurance, especially, is one of the most important benefits that are subject to a divorce court's powers. Health insurance can also cover prescriptions, therapy, dental and vision care, and the like.

Many health insurance plans provide benefits through an insurance company or a health maintenance organization (HMO). Other plans are paid directly by the employer (known as "self-funded" or "self-insured" plans).

An employer's contributions for health insurance are *not* considered taxable income to the employee. Nor does an employee report as taxable income amounts paid by the insurance company for covered medical expenses.

A court may order a parent to provide health coverage for children as part of a child support order (see Chapter 18). But this benefit is not considered an asset to be divided at divorce.

Disability Plans

These plans protect employees against the risk that illness or injury will interrupt their ability to work. Typically, the employer offers coverage for disability under an accident or health plan.

Usually, the employer's contribution is excluded from the employee's gross income on the tax return. In the event of disability, however, amounts the employee receives are included as gross income.

Employer-sponsored disability insurance is not an asset to be divided between spouses at divorce. But a contribution paid by the employer may be factored into the court's calculation of child and spousal support.

Group Term Life Insurance

This is an insurance policy carried by an employer to provide death benefits for employees and, occasionally, for retirees. Unless an employee is upper management, this insurance is generally available only during the term of employment.

The employee generally excludes from taxable income the employer's contribution to the premium up to $50,000. In the event of death, the beneficiary is not taxed on the death benefits received.

Group term life insurance is not an asset that a court divides. However, the court may order a party to maintain all employer-provided life insurance, naming a former spouse or children as the beneficiary (see Chapter 10).

Accidental Death or Dismemberment Insurance

Again, employer contributions to premium payments and insurance proceeds received by the employee are excluded from the employee's income. If the employee pays the premium, the amount paid may be a factor the court considers in a child support order; however, the policy itself is not an asset to divide.

Health Savings Accounts

A health savings account (HSA) is a tax-advantaged medical savings account to which contributions may be made by employers, employees, or both. Funds in an HSA account may be invested in a manner similar to a 401(k) account or an IRA. Those funds later can be withdrawn tax free for qualified medical expenses, such as deductibles, coinsurance, dental,

vision, and chiropractic care, and other items not necessarily covered under a traditional health insurance plan.

Each HSA account must be combined with a separate catastrophic health insurance policy with a high deductible. Medical expenses must hit $1,200 for an individual and $2,400 for a family before coverage kicks in. Once the deductible is met, the insurance usually covers 100% of health care costs. The idea is that Americans will consume less care and be more vigilant about health care costs if they see a relationship between medical costs and their bank accounts.

Unused funds can be rolled over into the next year's account. (HSAs, which were launched in 2004, effectively replace medical savings accounts, or MSAs, because employers of all sizes can offer HSAs while MSAs were limited to employers of 50 or fewer workers.)

There's some controversy about whether HSAs favor high-income individuals and some dispute over how quickly employers are adopting HSA plans. But there's no doubt that payments the employer makes to the plan are not included in the employee's gross income, and if an employee contributes, that contribution is deductible from gross income on tax returns. Thus, these accounts have the potential to accumulate significant balances over years of contributions and investment gains.

The employee spouse gets to keep all amounts in the HSA account on the date of separation or divorce. The employee spouse can be ordered to pay existing marital medical expenses with money on deposit. (See Chapter 10 for a discussion of retaining health care coverage through a spouse's employer.)

Dependent Care Assistance Programs

Dependent care programs provide child care or related services for employees who need this help in order to work. Frequently, employers offer child care benefits as a choice under a cafeteria plan (see below). The program may offer direct payments to the employee, or it may involve an employer-sponsored child care center.

The employee's gross income does not include amounts contributed by the employer. This benefit is not subject to division as an asset. But the

court may consider the program's value as compensation and may order an employee to provide this benefit as part of a child support order.

Cafeteria Plans

Cafeteria plans offer participants a choice of certain benefits—such as medical insurance, group term life, disability insurance, and child care—or cash. Employees may also be given the option under a cafeteria plan to use, sell, or buy additional vacation days.

Typically, in a cafeteria plan the employee elects before the beginning of the year to take a salary reduction. The amount of the reduction is then applied to the employee's share of the insurance premium or to cover other expenses such as child care.

The value of the benefits the employee receives is excluded from the employee's gross income. But any cash the employee receives under the plan is included in gross income.

Severance Plans

Severance involves payments made to an employee who is terminated by the employer. Whether written or informal, severance plans have two goals: to reward an employee's past service and to help the employee make ends meet during retraining or a job search. Sometimes an employer offers severance pay in exchange for the employee's promise not to sue for wrongful termination.

Severance plans, where they exist, vary from employer to employer. Some offer one week's pay for each year of service. Others have a more generous "golden parachute" policy, typically for executives. Eligibility for severance and the amount received often depend on the employee's position in the firm as well as length of service.

Severance may be paid in a lump sum, periodic payments, or a combination of the two. Generally, severance payments are included in gross income.

Educational Assistance Programs

These programs reimburse employees for tuition, fees, and attendance at outside educational programs or institutions or provide tuition and fees for in-house seminars. Courses need not be related to a person's work. Such expenses paid for by an employer are not included in an employee's gross income.

Group Legal Services

These plans offer employees low-cost legal services such as referrals, advice, and representation. They allow a group to use its collective purchasing power to negotiate lower rates from lawyers.

The employer and/or the employee pays for the cost through union dues or payroll deductions. Participation may be automatic or voluntary, and a plan may be offered independently or through a cafeteria plan.

The benefits received under group legal services plans used to be excluded from employees' income. That's no longer the case; employees must now pay taxes on any such benefits received.

Employee Assistance Programs

These programs cover a variety of services for employees and their families, including help for drug and alcohol abuse, stress, anxiety, depression, family problems, money and credit problems, and legal problems. Usually, employers contract with a third party to provide counseling and referral services with a social worker or family counselor.

Benefits or reimbursements an employee receives under an employee assistance program may or may not be included in gross income, depending on the nature of the benefit and the employee's income level. Employers should be able to provide information on the tax consequences of benefits offered under their program.

Stock Options and Nonqualified Deferred Compensation Plans

SKIP AHEAD

If neither you nor your spouse has stock options, stock incentives, or nonqualified deferred compensation plans (a kind of benefit usually reserved for upper management), skip ahead to the next chapter.

Stock options and other kinds of corporate incentive and compensation programs present a challenge for divorcing spouses who are trying to sort out and take control of their finances—not just because the plans themselves can be complicated, but also because there's still a lot of uncertainty over how to deal with them in the divorce process.

This section provides an overview of different kinds of plans and explains some key terms that come up in this area. If you or your spouse has such benefits, questions to keep in mind are:

- How much is the benefit worth now?
- Is the benefit something that can be divided between spouses?
- Can the benefit be transferred from one spouse to the other?
- Is the benefit considered income?
- If the benefit is income, will a court take it into account in calculating alimony and/or child support?

In many cases, there are no sure answers. But you need to be aware of these issues so that you can discuss them with an attorney and factor them into your negotiations.

SEE AN EXPERT

If you or your spouse has stock options or other types of compensation covered in this section, you should talk to your divorce lawyer as well as a tax attorney or certified public accountant. Many tax considerations can be involved.

Types of Stock Options

Stock options are an important part of the compensation program of many corporations. The aim, in addition to compensating employees, is to encourage employees to work hard to achieve success for the firm and thus increase the value of the company's stock.

Stock options have been common perks for senior management for many years. But in recent years, stock options have become part of the compensation package of many rank-and-file workers. It's been estimated that in publicly traded companies, between a quarter and a third of all employees own stock options; in technology companies, the percentage likely tops 50%. Yet, other studies have shown that large numbers of these workers do not understand the stock options they hold—or their tax implications.

A stock option is a right granted to an employee to purchase a certain number of shares in the company's stock at today's price sometime in the future. The stock purchased can be sold later, when the market value may be higher.

Usually, the employee cannot exercise the right to purchase and sell stock as soon as the option is received. Instead, the employee must remain with the firm for a minimum period before the options vest (that is, become exercisable). When they do vest, the stock's market value presumably will have risen, thus giving the employee an automatic profit in return for having stayed with the firm through the vesting period.

As a result, employees can find themselves with potential wealth tied up in an employer's option plan. In fact, options nationally are said to represent about $1 trillion in unexercised wealth. But there is also great risk in unexercised options. That's because while the expectation is that the value of the stock will go up, sometimes it goes down quickly and significantly.

Thus, while stock options are an increasingly common marital asset and often an item in contention in a divorce, they can entail substantial losses as well as gains. A divorcing spouse may have to pay additional support when options go up in value, but receive no relief if they go down.

Option plans are complex, and they differ from company to company. What's more, divorce courts differ greatly on how to value options, what portion—if any—of them should be characterized as marital property, and how to treat them for purposes of support.

Stock Option Documents You Need to Get

To determine what kind of stock options are involved in your case, you should obtain as much information on the options as possible. Much of this information can be obtained from the employer that granted the options. Important documents include:

- the original stock option plan statement and all amendments
- grant agreements between employer and employee (one for each grant)
- employer's letter of intent to employee offering a stock option contract
- an option summary statement reflecting the status of each grant, including vesting, exercise, and sales as well as changes in the number of shares resulting from stock splits (generally, the employer provides such statements annually or quarterly)
- a list of the company's monthly stock prices and dividend payments (if it's a publicly traded firm), and
- all SEC Form 10K filings, annual proxy statements, or other filings with any governmental or regulatory agency during the term of employment. (These often are available at the EDGAR website at www.sec.gov; look under the EDGAR heading. EDGAR stands for Electronic Data Gathering Analysis and Retrieval, an information system of the Securities and Exchange Commission.)

There are various kinds of stock options, as we discuss below. However, they all have these similarities:

- Options may be exercised (the shares may be purchased) during a specified period of time and at a fixed price.
- After the vesting period, an employee can wait to exercise the option until—and if—the stock's market value exceeds the price at which the employee can buy it, thus guaranteeing a profit.

Stock Option Lingo

Even to the most sophisticated investors, stock options can be confusing. Here are simple definitions of some of the key concepts:

- **Stock option.** A contractual right, not an obligation, to buy a certain number of shares (exercise the option) at a point in the future (the vesting date) at a specific price (the grant price).
- **Grant.** The issuance by an employer of the right to acquire stock at a specific time in the future at a predetermined price.
- **Grant price.** The price at which the shares are purchased when the option is exercised, regardless of their market price when exercised. Also known as strike price, option price, or exercise price.
- **Vesting.** The time that must elapse before the employee acquires the right to exercise an option or a percentage of the option granted (for example, 25% of all total options granted over four years).
- **Exercise.** The exchange of an option for actual shares of the company. The employee must pay the grant price as well as the payroll taxes associated with the exercise. And when the employee sells the stock, he or she may be taxed on the profit made (capital gains) on the difference between the grant price and the sale price.
- **Exercise date.** The date after vesting when the employee purchases shares at the predetermined price under the terms of the option agreement.
- **Holding period.** The period of time that an employee is required to keep the stock after purchasing it and before selling it. The holding period begins on the exercise date and ends on the day before the stock is sold.
- **Expiration date.** The last day the employee can exercise the option under the terms of the contract.
- **Cashless exercise.** This occurs when the employee simultaneously exercises an option and sells the stock just acquired, covering the cost of the options and the taxes due. The employee takes the remainder in cash instead of stock shares.
- **Spread.** The difference between the grant price and the market value on the exercise date.
- **Gain.** The difference between the grant price and the market value on the day the stock is sold.

- The employee has no risk during the vesting period because there's no obligation to buy the stock and no money invested yet. If the value of the stock declines, the employee simply elects not to exercise the option.

Now let's look at the two broad categories of stock options: statutory and nonstatutory (also known as nonqualified stock options, or NQSOs).

Statutory stock options are so called because they are subject to a strict set of regulations under federal statutes. For example, to qualify for a favorable capital gain tax rate, an option holder must either exercise the option within two years after the employer grants it, or hold on to the stock for at least a year after exercising the option.

There are two types of statutory stock options: *incentive stock options* (ISOs) and *employee stock purchase plans*. ISOs are governed by a written plan approved by the company's board of directors and stockholders. The plan lists the employees who are eligible and the number of shares available for these employees.

When an employee is granted an ISO, that's not considered income on which the employee can be taxed. Nor is it taxable income when it's exercised. After the employee sells the stock, the profit is considered a capital gain and is taxed at that rate.

Can the employee transfer the ISO to an ex-spouse upon divorce? The law says that the option itself—the right to buy the stock—can't be transferred. If it is, the option becomes a "nonqualified" option (see below). However, if the employee exercises the option first, the ISO stock the employee has bought can be transferred to the employee's spouse.

Besides an ISO, the other type of qualified option is an employee stock purchase plan. Such a plan allows an employee to set aside money from regular paychecks to purchase company stock, either at a discount or at market value on the date of the grant. The stock is held in trust for the employee, who receives the holdings in stock or cash when employment ends. As with an ISO, the option to buy stock under an employee stock purchase plan cannot be transferred from one divorcing spouse to another.

Nonstatutory (or nonqualified) stock options have fewer restrictions, and the tax consequences for the employee are not as favorable. When an employee exercises an option and makes money by buying stock at lower-than-market rates, that money is subject to federal income tax as well as FICA, Medicare, and state and local taxes.

An NQSO can be transferred from an employee to an ex-spouse at divorce. The ex-spouse exercising the option is liable for income tax just as the employee spouse would be.

Restrictive stock and restrictive stock units (RSDs) are stock options that fall outside the statutory and nonstatutory categories because of how they're taxed. Rather than the date when the stock is exercised or sold, it's the *vesting schedule* that determines the tax consequences for these types of employee stock. Both have become more popular in recent years as companies try harder to tie compensation to work performance by creating a vested interest in the health of the company's stock.

Restrictive stock is typically granted to executives who have detailed, or "insider" knowledge of their companies. As such, these options are subject to Security Exchange Commission regulations. Employees generally receive these stocks up front, but are subject to forfeiture rules until fully vested.

Restrictive stock units represent an agreement, which promises that amounts of stock will pass into the employee's possession at some date(s) in the future. The number of shares in each unit varies from agreement to agreement.

In divorce, the value of restricted stock and RSUs may depend on whether they are defined as compensation or as an asset, which may have vested during the marriage or will become fully vested after the divorce is finalized. It is important to remember that in both cases, you cannot guarantee that these stock options will be received by the employee until they are fully vested.

What Are Stock Options Worth at Divorce?

How can stock options be valued and apportioned among divorcing spouses? That question presents complex problems for family lawyers and courts, as well as for divorcing spouses.

Among the issues: Are stock options property or income, or both? Are options that vest after the couple's separation date considered marital property? If the employee spouse doesn't exercise the option until years after the divorce, does the ex-spouse deserve a portion? Is there a formula courts can use for dividing options? Should a court be able to raise levels of alimony and child support if the value of the stock increases, and reduce the support if the value falls? If an employee's option hasn't yet vested because the employee hasn't worked at the company long enough, how does that affect the value of the option?

There's not a lot of legal precedent to go on, and there are no hard and fast rules that apply to all stock option plans and in all states.

There are, however, some broad guidelines. Stock options granted to an employee spouse before the separation date generally are seen as marital property. Those granted after separation tend to be viewed as separate property of the employee spouse. In a series of California cases, the judges found that options that were granted and exercised before the separation date were community property—whereas options that were granted during marriage but not exercised until after separation entitled the ex-spouse to a partial interest, based on the amount of time elapsed. About half the states have followed these general principles.

Courts also generally have held that recurring capital gains—including profits from exercised stock options—are income and can be used for calculating support.

What about potential income from *unexercised* options? Some courts have said that unexercised options are part of gross income and therefore can be used to help determine support payments. Courts have also said that an item can be considered both income for support purposes and property for purposes of equitable distribution. If an option is both unexercised and *unvested*, that raises more valuation questions that have no clear answer.

Courts have developed various methods for apportioning options. Sometimes, as we've seen, options can be transferred. Some courts have required the parties to hold stock options as tenants in common (for more on tenancy in common, see Chapter 12).

Still other courts have applied other methods to determine the options' present value, then ordered one spouse to give the other assets from the estate to offset the value of the options. The risk here is that later the stock's value can either fall or skyrocket, leaving each spouse feeling blessed or chagrined.

How stock options are handled in divorce depends on many factors. There is no guarantee your situation will be treated the same as someone else's or that the court will come to the same conclusion in your case as it has in other cases or as courts have in other jurisdictions. You may need to talk at length with a lawyer about stock option issues in your case. See "Points to Include in the Options Agreement," below, for some key points that should be covered in any settlement reached by you and your spouse.

WEB RESOURCE

Several websites offer information and resources on stock options. Among them are:

- **www.mystockoptions.com.** This comprehensive site includes articles, a glossary, a tax guide, a tax calculator, and answers to frequently asked questions.
- **www.stockoptionscentral.com.** This site provides calculators that allow you to run "what if" scenarios to determine option value and taxability.
- **www.stockopter.com.** While primarily a site for financial professionals, it does include a glossary and explanatory articles.

Points to Include in the Options Agreement

While there are few firm rules about how to handle stock options in divorce cases, certain points ought to be covered in any agreement between you and your spouse. Here is a checklist of issues to include in your analysis:

- **Identify the grants.** Make sure you have all paperwork pertaining to the options as furnished by the employer, specifying the terms of the grant, the grant price, key dates, and all rules that apply.
- **Decide who will bear the exercise fees.** An employee typically pays a fee when an option is exercised. At divorce, the fee is usually paid by whoever exercises the option, whether that's the employee or the ex-spouse.
- **Agree to follow court orders and employer conditions.** Both spouses should agree that the options must be exercised in a way that adheres to both the judge's ruling and the employer's terms.
- **Clarify when options change or terminate.** The agreement should be clear about the duration of options periods and what happens to the options if the employee leaves the company. The employer should be responsible for providing information to the parties about any changes in conditions or timetables for exercising options.
- **Protect personal identification information.** Do not include in any court order personal information, such as address, date of birth, or Social Security number. Such information, should be provided to the employer and other parties by mail, certified with return receipt requested.

Nonqualified Deferred Compensation Plans

Nonqualified deferred compensation is one method by which companies have traditionally rewarded their highest-ranking executives. Far less common than stock options, these plans are generally found only in large corporations and are limited to a few executives, such as the chief executive officer, chief financial officer, president, and vice presidents.

Under such a plan, the employer defers part of an executive's income now—thereby reducing the executive's current tax burden—and provides the income later, when the executive will presumably be in a lower tax bracket. Like stock options, these plans are legally complex and difficult to value and divide at divorce, or even to evaluate to determine what is income available for support.

There are many kinds of nonqualified deferred compensation plans. Some look like 401(k)s, some resemble defined benefit plans (see Chapter 14), and others are merely promises by the employer to pay income to the employee in the future. A common type of plan is the SERP (Supplemental Executive Retirement Plan), or "top hat" plan, which provides additional retirement benefits to a select group of high-level managers through the use of various formulas.

If you are divorcing a spouse with a nonqualified deferred compensation plan, you or your lawyer will want to request documentation from the employer. This might include the plan summary, full plan documents, an employment contract, and an actuarial analysis.

Two factors reduce the value of these plans in divorce. Because such plans are often unfunded, there is a risk that the employee will never receive the benefits if the company goes bankrupt, reorganizes, or is taken over by another company. Also, the benefit cannot be split between divorcing spouses under a QDRO. (A QDRO assures direct payment of certain benefits to the nonemployee spouse. See Chapter 14.)

Questions to Ask an Attorney or a Tax Specialist

1. How do our state courts treat all our employment benefits? As property? As income?
2. What do our state courts have to say about how stock options should be valued and apportioned in divorce?
3. After my divorce, will I be entitled to any medical benefits I would have received under my spouse's retirement plan had we stayed married? (This is especially likely if your spouse is or was employed by a federal, state, or local government agency.) ●

How Will We Divide Debts?

General Rules on Who's Responsible for Debt 360

If You Live in a Community Property State...362

Listing Your Debts...363

Marital Debts and Bankruptcy 363

    Terms to Know...369

    Types of Bankruptcies 371

Dividing Debts at Divorce 375

Dividing Debts When There's Nothing to Fight Over 376

Questions to Ask a Divorce Attorney..377

Questions to Ask a Bankruptcy Attorney...378

When you divorce, you divide not only property but debts as well.
It's crucial that you and your spouse reach an agreement about how debts
will be handled.

f you are willing to assume a debt, you might ask for something in return. You may be able to make trade-offs—perhaps getting the car or the piano in exchange for paying off credit card debts.

Before negotiating your agreement, you'll want to know what your state laws have to say about who is responsible for what debts. You also need to be aware of the possibility that your spouse will declare bankruptcy, default, or fail to live up to your agreement in some other way. Use this chapter to investigate your legal and financial position on debts and divorce.

General Rules on Who's Responsible for Debt

Some general rules for separated, divorced, and married people having financial problems follow. Study these rules to understand which debts you are obligated to pay and which property you risk losing if you don't pay:

- In noncommunity-property states, a person is responsible for paying only the debts that person incurred during marriage—that is, you can't be forced to pay the bills your spouse runs up—except that both spouses are usually responsible for paying debts incurred:
  - with joint accounts
  - where the creditor was looking to both spouses for repayment
  - for the family's necessities, such as medical care, food, clothing, and shelter, and
  - for the children's education.
- While your spouse's creditor may not be able to come after you directly, the court will take the debt of both parties into account when dividing property, so you may end up paying indirectly.

Be Aware of Medical Expenses

State statutes often expressly state whether one spouse is liable for the necessities of the other. Are medical expenses always considered necessities? The issue of liability for a spouse's medical debts commonly arises in court these days. And there's no clear answer. About half the courts say neither spouse is responsible for the other; the other courts say that both are responsible to each other.

- Usually, debts incurred after the separation date but before the divorce is final are the responsibility of the spouse who incurred them and must be paid by that person. There is an exception—both spouses are responsible for paying debts for the family's necessities and the children's education, regardless of who incurred them.
- A spouse is generally not responsible for paying the debts a mate incurred before marriage or after the divorce became final.
- If either you or your ex-spouse agreed to pay certain joint debts as part of a divorce settlement—or if a divorce decree signed by a judge requires that one spouse pay the joint debts—the agreement or decree is binding only on you and your ex. Because you were married when the joint debts were incurred, the creditor has a right to collect from both of you, and no agreement between you and your spouse can change that. If you pay bills your ex-spouse was supposed to pay, whether voluntarily or involuntarily, your remedy is to try to get your ex to reimburse you. Complaints to the creditor will get you nowhere.
- The separately owned property of one spouse usually cannot be taken by a creditor to pay the separate debts of the other. Separate property of one spouse can be taken to pay debts you incur together, and separate property of one spouse is always available to pay that spouse's separate debts. (Separate property is property owned prior to the marriage, property accumulated after divorce, and property received during marriage by one spouse only, by

gift or inheritance. In a few equitable distribution states, separate property also includes wages earned by a spouse during marriage.)

If You Live in a Community Property State

The community property states are Arizona, California, Idaho, Louisiana, Nevada, New Mexico, Texas, Washington, and Wisconsin. In those states, property acquired and debts incurred during marriage—called community property and community debts—are joint. In most instances, both spouses are liable for all debts incurred by either, although there are exceptions. Here are the rules:

- Both spouses are generally liable for all debts incurred during marriage, unless the creditor was looking to only one spouse for repayment or the purchase in no way benefited the "community." For example, if on a credit application a spouse who bought a kayak claimed to be unmarried or stated that the other spouse's income would not be used to pay the debt, the spouse would not be liable to pay for the kayak if the borrowing spouse defaults. Similarly, if a spouse charged a trip to the Bahamas with a lover, the spouse who stayed at home would not be liable, as this debt does not benefit the community.

- Spouses are generally not liable for the separate debts their mates incurred before marriage or after permanent separation. But a court order dividing debt between the parties at divorce *cannot* affect the rights of the creditor without the creditor's consent. Where the divorcing parties have joint credit, a creditor may collect from either party.

- Community property includes all income earned during marriage, except for income earned on property owned before the marriage. This would include revenue from an income-generating piece of property or payments from a vested pension.

- Separate property is limited to property owned prior to the marriage, property accumulated after divorce (after separation, in a few states), and property received during marriage by one spouse only, by gift or inheritance.

Listing Your Debts

Below is a chart on which you can list all your debts—yours, your spouse's, and your joint (marital) debts. The liabilities worksheet, which you filled out as part of your "Net Worth Statement" in Chapter 12 should contain all this data.

Using the examples as a guide, list in Column 1 all current debts you and your spouse are responsible for paying. This information can be taken from the "Assets and Liabilities Worksheet" in Chapter 12. In Column 2, write the balance owed on each debt. In Column 3, note whether the debt is your separate debt, your spouse's separate debt, or a marital (joint) debt. In Column 4, write in the name of the person who will assume responsibility for paying the debt. Only in unusual circumstances will one of you be responsible for the other's separate debt. You will need to divide the marital debts, however.

You may want to make several copies of this chart in the event you negotiate back and forth about who will be responsible for paying which debts.

Marital Debts and Bankruptcy

For many couples, debts and divorce go together the way love and marriage once did. Unmanageable debts may serve as a catalyst for the ending of the marriage in some cases. Other couples feel the pinch as their expenses double when one spouse moves out—they maintain two households, yet their incomes stay the same.

For couples or recently divorced individuals overwhelmed by their debt burden, declaring bankruptcy may provide a way out. In some cases, bankruptcy lets you erase your debts in exchange for giving up certain property. Every state declares specific assets, such as clothing, certain equity in a home, certain equity in a car, and several other items to be exempt. You get to keep exempt property when you file for bankruptcy.

Spousal Debt Chart

Debts—Example	Balance	Husband, Wife, or Joint	Who Will Pay
Car loan	$ 8,000	Joint	Wife
Franklin Bank Visa	$ 1,800	Husband	Husband
Marshall Bank Visa	$ 2,300	Joint	Husband
Personal loan	$ 5,000	Joint	Husband
Student loan	$ 10,000	Wife	Wife
USA MasterCard	$ 4,000	Joint	Husband

Debts	Balance	Husband, Wife, or Joint	Who Will Pay

What to Do If You Are Facing Bankruptcy

Bankruptcy is a serious step that should never be considered until all other options are exhausted. For one thing, a bankruptcy may show up for ten years on your credit report, potentially affecting your ability to get a loan or a job, or even to rent an apartment. What's more, the recent tightening of the bankruptcy laws puts additional burdens on those who file, so it's an even less desirable option than it used to be. And if one spouse files for bankruptcy with the aim of destroying the other, the process can turn into a supreme mess. While your anger may be running as high as your debts, it's important not to act rashly or out of vindictiveness.

Instead, understand that, in some cases, a bankruptcy, if absolutely needed, can help out both spouses if they file jointly. But that will require cooperation and compromise. Though the new federal law makes bankruptcy more difficult (especially a Chapter 7 filing), it is sometimes the best of a bad range of alternatives.

If you're motivated to explore the bankruptcy option, consider these steps:

1. **Get a handle on what you owe.** Set aside time to gather all your bills and tally how much they're costing you each month and what interest rates you're paying. Include secured debts, such as your home and car, and mandatory incidentals like power, phone, groceries, and insurance. Then tally those you want but could live without, such as cable TV, gym memberships, dinners out, and other optional purchases. Always pay the mortgage and the car loan first and keep them current. But beyond that, be tough. If things are so tight that you're considering bankruptcy, it's time to wield the axe ruthlessly.

2. **Work with your spouse, if at all possible.** A joint filing can lessen the pain. If you can find the inner resources to work as a team, you'll save yourself money and grief.

3. **Put yourself on a strict allowance.** Take only a minimal amount from your paycheck in cash and, of course, keep the credit card under wraps.

4. **Sell assets.** Get rid of things that have cash value but not sentimental value, such as some antiques, old clothes, or collectibles. Selling them at garage sales, eBay, or consignment stores can free up cash. Put the money in exempt assets, like retirement plans or other assets that state or federal laws exclude from bankruptcy.

What to Do If You Are Facing Bankruptcy (cont'd)

5. **Seek a second or part-time job.** Even flipping burgers may get a little more money coming in and keep a bad financial situation from getting worse.

6. **Try credit counseling.** Set up an appointment with the local affiliate of the National Foundation for Credit Counseling (www.nfcc.org) and work with them to draft a budget and negotiate a payback plan with your creditors. Or you can try your hand at negotiating. A delayed repayment plan may negatively affect your credit report, but not as much as a bankruptcy would.

7. **Get help if your spouse files.** If your spouse files for bankruptcy, immediately contact a bankruptcy lawyer. Failing to act promptly may leave you without relief.

Married couples can file for bankruptcy jointly. Married persons can also file individually, and single people can file on their own. Before deciding on filing as an individual or a couple, you must understand what the different outcomes will be.

If a married couple files for bankruptcy together, they can eliminate all separate debts of the husband, all separate debts of the wife, and all jointly incurred marital debts.

Some debts cannot be erased (discharged) in bankruptcy. These include student loans, restitution for a crime, and in many cases, debts obtained by fraud.

Support-related debts (such as child support and alimony) owed under a marital settlement or a divorce decree are protected from bankruptcy discharge. A tough new law that went into effect in October 2005 tightened the rules on child support and alimony so that they cannot be erased. In fact, under that new bankruptcy law, any debts or support owed to a spouse as part of a divorce proceeding will be difficult, if not impossible, to discharge in bankruptcy. (Previously, many creditor-spouses were left holding the bag on debts the debtor-spouse failed to

pay.) Not only are dissolution debts not dischargeable but they actually go to the head of the line, taking priority over the need to repay almost every other kind of debt besides the bankruptcy trustee's fees. So if you are filing jointly with your spouse, any support-related debts that you or your spouse has from a previous marriage will remain even after your bankruptcy discharge.

And if you find yourself in the situation where your spouse is filing for bankruptcy alone, you will be protected under the new law—at least for the support-related debts. However, a successful bankruptcy can shift some other debts from the ex-spouse who assumed them under the marital settlement agreement and place them on the nondebtor ex-spouse—but not before the nondebtor ex-spouse gets a chance to object to the discharge of certain debts.

Here's some background on how this all works. Family law is a matter for the state courts, but bankruptcy is governed by federal laws. When those issues overlap—for example, when one spouse files for bankruptcy before, during, or after a divorce settlement—the issues can be complex and confusing. The bankruptcy court's job is to allow the debtor to have a fresh start after bankruptcy while at the same time protecting the rights of the nondebtor spouse and any minor children.

When one spouse files for bankruptcy, the family court can continue to hear and decide issues relating to establishing amounts for child or spousal support (alimony). But it cannot split up the family home, apportion investments, or distribute other property until it receives permission from the bankruptcy court. So if a debt is classified as a support obligation, it must be paid. But if it's judged to be a property settlement claim, it can remain in limbo until the bankruptcy case is settled. And the bankruptcy court may rule that the debt is owed by the nondebtor ex-spouse, regardless of what the marital settlement agreement states.

After a divorce, an ex-spouse who's assigned responsibility for paying joint debts may file for bankruptcy. If the petition is successful, the bankruptcy court may release that ex-spouse from having to pay the assigned debts. The creditor can no longer collect from that person—but can still seek payment from the other spouse even though the latter was supposed to be excused from responsibility for those debts. There's not much this innocent spouse can do to avoid having to pay, or to get reimbursement from the defaulting ex-spouse. If you're in this situation, you have three bad choices:

- You can file for bankruptcy yourself.
- You can refuse to pay the ex-spouse's debt and take the hit to your own credit record.
- You can pay the debt and chalk it up to experience.

Fortunately, you can reduce the likelihood of facing this unfortunate set of choices by structuring your separation agreement—and, ultimately, your divorce judgment—to take the possibility of bankruptcy into consideration. If you believe there's any chance that your spouse may file for bankruptcy protection, be sure to discuss it with your attorney. The lawyer can ask for a clause in your property settlement or divorce judgment that limits the impact of a bankruptcy—for example, by giving a nonbankrupt spouse the right to reopen the case if the other spouse files for bankruptcy.

The rules are more complicated for debts that aren't in the nature of support. Recent changes to the Bankruptcy Code made nonsupport obligations created in connection with a divorce nondischargeable if erasing the debt would harm the nondebtor spouse more than it would benefit the debtor. If nonsupport obligations are involved in the divorce, the nondebtor spouse must file a timely action to keep the debt from being discharged in bankruptcy.

Property settlements are kept on hold until the proposed bankruptcy is resolved. A potentially big issue that must be handled with great care is credit card debt. When spouses obtain a credit card, usually they sign a contract that says both spouses are "jointly and severally" liable. That means if one spouse dies or files for bankruptcy, the other is liable for the entire credit card debt.

For this reason among others, it's best to pay off the balance on all joint credit cards from marital assets before the divorce goes before a judge. If there's a sale of the marital home, credit cards can be paid off at the closing. In fact, in some cases, credit card companies will accept a reduced amount if the obligation is paid in one lump sum. If you do that, make sure you get a settlement letter from the credit card firm or collection agency, confirming the terms of the settlement.

If you don't retire your credit card debt when you sell your marital home or other assets, and one spouse files for bankruptcy, the credit card company will go after the other spouse for payment of jointly owned cards. The credit card companies don't care if this is fair; they care about getting what's owed them.

Do not assume that your soon-to-be ex-spouse will pay the credit card debts assigned by the divorce decree. That's doubly true if a bankruptcy is added to the mix. Any property settlement should contain provisions that address how credit card debt will be handled if one spouse files for bankruptcy.

Terms to Know

This brief glossary may aid your understanding of the ins and outs of bankruptcy law.

Automatic stay. An injunction (order) that automatically stops lawsuits, foreclosures, garnishments, and all collection activity against the debtor as soon as a bankruptcy petition is filed.

Bankruptcy estate. All property in which the debtor has an interest at the time of the bankruptcy, including property owned or held by another person and property the debtor acquires or becomes entitled to acquire within 180 days after filing as a result of a settlement agreement or divorce decree.

Chapter 7. This chapter of the Bankruptcy Code provides for prompt liquidation of assets and distribution of the proceeds to creditors. Generally, the new law limits Chapter 7 to individuals who earn the median income in their state and who take an approved course in financial management before the debts are discharged. The debtor must

pay in full all domestic support obligations that come due after the bankruptcy is filed.

Chapter 11. Most often used by businesses, this form of bankruptcy seeks to keep the business alive and pay creditors over time. Individuals can also seek a Chapter 11 bankruptcy, but it's an expensive and complex path.

Chapter 13. Under this chapter, the debtor retains property owned and must pay the debts over time, usually three to five years. The debtor must pay in full all domestic support obligations that come due after the bankruptcy is filed.

Discharge. The release of a debtor from personal liability for certain debts. The discharge also prevents the creditors to whom the debts are owed from taking any action against the debtor to collect.

Domestic support obligation. The Bankruptcy Code defines this term broadly. It includes child or spousal support, as well as obligations of one spouse or ex-spouse to the other in a property settlement or divorce decree.

A Carefully Worded Agreement May Protect You From an Ex-Spouse's Bankruptcy

To increase your chance that marital debts won't be discharged in the event your ex-spouse files for bankruptcy, your marital settlement agreement, order, or judgment should characterize the debts as "additional child support," "alimony," or "in the nature of support." This will provide guidance to a bankruptcy judge who is asked to determine whether or not the ex-spouse can avoid having to pay the debts. The language of the agreement won't be binding on the judge, who looks behind the agreement to the true nature of the debts, but it will be evidence to the judge that you and your ex-spouse intended the debts to be nondischargeable.

Exemptions/exempt property. Certain property that the law allows the debtor to keep from unsecured creditors. For example, in some states, the debtor may be able to exempt all or a portion of the equity in a home or some or all of the tools that contribute to the debtor's livelihood. State law governs the amount and availability of exemptions.

Means test. A test of a debtor's financial viability to determine eligibility to file for Chapter 7. If the debtor is determined to be not eligible, the petition for bankruptcy may be dismissed or converted into a Chapter 13 filing.

Nondischargeable debt. A debt that cannot be eliminated in bankruptcy. Examples include a home mortgage, alimony and child support, certain taxes, government-backed student loans, and criminal restitution owed.

Types of Bankruptcies

The two most common types of bankruptcy are called Chapter 7 and Chapter 13, after the sections of the law that are being followed. Chapter 7 bankruptcy allows you to discharge all your debts and be free of them forever. Chapter 13 is a "reorganization" bankruptcy where you pay some amount of each of your debts, in an order set by the court.

Being broke got tougher under the Bankruptcy Abuse Prevention and Consumer Protection Act of 2005. Previously, about two-thirds of those seeking bankruptcy protection opted for Chapter 7 instead of Chapter 13 because the former allows you to discharge all debts. But the new law makes it more difficult to get a Chapter 7 discharge.

How so? For one thing, a complicated means test, administered by your attorney, will determine whether you'll be allowed the more forgiving Chapter 7. If your income is above your state's median income, you must file under Chapter 13 unless the bankruptcy court rules your circumstances are extraordinary. Even those who earn less than the state's median income must file for Chapter 13 if they can pay more than $100 a month on their unsecured debt over the next five years.

The new law also requires debtors to take a credit counseling class before a bankruptcy can be filed, and a personal financial-management class before a bankruptcy is complete. Both involve a fee paid by the debtor.

In addition, a wide range of other provisions make filing for bankruptcy more difficult than it was in the past. For instance, miss one filing deadline and your bankruptcy case could be dismissed. What's more, your collateral, such as furniture and cars, will be assessed at a higher value—and if you're worth more, your creditors potentially can get more out of you. Further, it'll be harder to get out from under car loans, student loans, and credit card debt.

In a Chapter 7 bankruptcy, any debt arising from a separation agreement or divorce, or in connection with a marital settlement agreement, divorce decree, or other court order, can be considered dischargeable. If an ex-spouse wants to challenge the discharge of the debt, the bankruptcy court will hear the challenge. The court will allow the debt to be discharged unless either of the following applies:

- the ex-spouse who filed bankruptcy has the ability to pay the debt from income or property that is not reasonably necessary for support and not reasonably necessary to continue, preserve, and operate a business, or
- discharging the debt would harm the nonfiling ex-spouse or a child more than it would benefit the filing ex-spouse.

Most courts have said that an ex-spouse who files for bankruptcy and claims inability to pay certain debts or greater harm if they're not erased must prove the claim in court. (See, for example, *In re Hill*, 184 B.R. 750 (N.D. Ill. 1995).) At least one court, however, has held the opposite: that it's the nonfiling ex-spouse who must convince the court otherwise. (See *In re Butler*, 186 B.R. 371 (D. Vt. 1995).)

In a Chapter 13 bankruptcy, marital debts will be discharged after the debtor completes a repayment plan. This means that all, some, or none of the debts will be discharged, depending on the terms of the repayment plan approved by the bankruptcy court.

Sometimes in Chapter 7 bankruptcies, a court discharges certain debts that the debtor might have remained responsible for, if only an ex-spouse had challenged the discharge in court. On the other hand,

sometimes when an ex-spouse does challenge a discharge, the court finds in favor of the debtor. How can an ex-spouse get protection? One way is to make sure that when a spouse assumes a certain obligation upon divorce, the obligation is made subject to a security interest in that spouse's property.

There are several ways to protect yourself: 1) don't give up assets in exchange for your soon to be ex-spouse's assuming a debt, 2) consider the transfer of an exempt asset (for example, an IRA, certain retirement plans, governmental or qualified plans from an employer, or other statutorily exempt assets set forth by your state) in exchange for the debt assigned to the ex-spouse, and 3) don't assign assets in exchange for the promise to pay the debt.

> **EXAMPLE:** Jackie and Art divorce. Jackie receives a larger share of the property and gives Art a promissory note for $35,000 as an even-up payment. Jackie files for bankruptcy and asks the court to erase her obligation to make good on the note. Art fails to contest the discharge of the debt and therefore falls $35,000 short of getting his share of the marital property. Art could have protected himself by taking a promissory note secured by the house. This kind of note would allow him to force the sale of the house if Jackie doesn't pay. Also, this kind of note can't be erased in bankruptcy if the security interest (the note securing the house) is created at the same time the house transfers from the two spouses to the one keeping it after the marriage. (*Farrey v. Sanderfoot*, 500 U.S. 291 (1991).)

Bankruptcy isn't for everyone. A bankruptcy stays on your credit record for ten years, although you can take steps to rebuild your credit and probably obtain credit in two or three years after filing. Nonetheless, bankruptcy is always a possibility for you or your spouse, unless you've already filed within the last six years. If you're overwhelmed by your debt burden, don't immediately discount bankruptcy as an option. If your spouse assumes a large share of the marital debts or owes you a large sum

How to Choose a Credit Counselor

Whether you're facing bankruptcy or just need help with overwhelming debt, you can use these tips to help choose a reputable credit counseling agency.

Avoid promises of a quick fix. It probably took you many years to build a large debt. Significantly reducing it also can take time. Don't sign up for any plan until you have all the specifics in writing and understand what's being offered.

Do your homework. Shop around by talking to at least two or three agencies before deciding which one to use. Ask how much training their counselors receive. Avoid agencies that provide only a debt-management program, which typically involves the counselor's negotiating a payback and paying your bills for you in exchange for a fee. Select one that offers a variety of services that include things like face-to-face help with budgeting and savings.

Beware of high or up-front fees. Carefully read all documents and ask for a clear explanation about the cost for the services. You should not be asked for an up-front fee. Remember, just because an organization is nonprofit doesn't mean it's affordable or even legitimate. Check it out with the Better Business Bureau.

Ask about the impact on your credit. A bankruptcy will stay on your credit report for up to ten years, raising the interest rates you're charged for loans and possibly affecting your job prospects. Just undergoing counseling or coming up with a debt-management plan through a credit counselor should not. But be sure to ask.

Consider do-it-yourself negotiation. Remember, you have the ability to contact your creditors yourself to negotiate delays in payment or discounted rates if you're facing extreme hardship. While you may get a deeper discount if you work with a credit counselor, make sure you take into account that you'll also be paying a fee to the counselor.

Be wary of pitches about consolidating your debts. While consolidating bills into one large loan can lower monthly payments, don't do it if you may be tempted to use the money you save each month to go on another shopping spree. By stretching out the term of the loan, you'll pay far more total interest.

to even up the property division, be on the lookout for a bankruptcy filing, especially if your ex starts to complain loudly about being broke.

Make sure your divorce agreement doesn't leave you vulnerable. Promissory notes, especially unsecured notes (notes that don't give you a lien on property), are almost always wiped out in bankruptcy. You may have a little more protection with secured notes. You may be able to protect yourself by specifying in the settlement agreement that the promissory note or other payments are made in lieu of alimony. If you want to file for bankruptcy, be sure you know what's involved, and whether or not you can wipe out your debts. What's most important is that you get sound legal advice from a bankruptcy attorney or other bankruptcy resources available to you.

RESOURCE

See How to File for Chapter 7 Bankruptcy, by Stephen R. Elias, Albin Renauer, and Robin Leonard, Chapter 13 Bankruptcy: Keep Your Property & Repay Debts Over Time, by Kathleen Michon and Stephen R. Elias, or The New Bankruptcy: Will It Work for You? by Leon D. Bayer and Stephen R. Elias, all from Nolo. The first book gives all the information necessary to decide whether to file and the forms for filing a Chapter 7 bankruptcy case. A chapter on nondischargeable debts has information on when marital debts can and cannot be wiped out. The second book contains the same information for Chapter 13 bankruptcy cases. The third will explain the new bankruptcy rules and help you assess whether bankruptcy is appropriate for your situation.

Dividing Debts at Divorce

In Chapter 12, you had the chance to complete "Net Worth" and "Cash Flow Statements." If you skipped that material, you may want to go back and fill the charts out now. If you have already completed the "Net Worth" and "Cash Flow Statements" in Chapter 12, use this opportunity to update them. You might leave many items blank on these statements, but at least you'll get an idea of what you actually own and

owe, and how much it costs you to live. This information is crucial when you divide your debts at divorce.

You have four basic options in allocating your marital debts:

- You and your spouse can sell joint property to raise the cash to pay off your marital debts.
- You can agree to pay the bulk of the debts yourself; in exchange, you get a greater share of the marital property or a corresponding increase in alimony.
- Your spouse can agree to pay the bulk of the debts; in exchange, your spouse gets a greater share of the marital property or a corresponding increase in alimony.
- You and your spouse can divide joint property equally and divide the debts equally—that is, each of you gets half of the property and each of you agrees to pay half of the debts.

From your perspective, the first two options are much less risky than the second two possibilities. Remember the caution we have given throughout this book: *When you're connected to someone financially, you're at risk.* What happens if your spouse agrees to pay all or even half of the debts and then gets fired, refuses to pay, or files for bankruptcy? Your credit rating will be damaged, or you will be stuck with paying debts. The likelihood of getting reimbursed or getting a share of the marital property originally given to your spouse will be slim or nil. So if possible, sell your joint property and pay your joint debts, or agree to pay the debts yourself.

Dividing Debts When There's Nothing to Fight Over

As your financial picture gets clearer, you may come to the unhappy conclusion that your bills and liabilities far outweigh your assets and income. Don't feel alone. Not only do most Americans live above their means, but many people going through divorce find they simply do not have enough money to make ends meet. Unlike the bitter battles for property that have stereotyped divorce cases on television, you and your partner may admit in despair that there's "nothing to fight over."

If you're in such a position, it's critical for you to take an honest look at your needs and protect your own interests. Even if you have to borrow

money to get through your divorce, it could be well worth it to prevent problems later. Hiring a lawyer or financial professional now to review your situation could cost less than a huge tax bill or other debts in the future.

Extensive information on debts and creditors can be found in *Solve Your Money Troubles: Debt, Credit & Bankruptcy*, by Margaret Reiter and Robin Leonard (Nolo).

> **CAUTION**
>
> **The U.S. Supreme Court's June 2013 decision to strike down part of DOMA gives same-sex married couples living in states that legally recognize their marriages the full spectrum of federal protections, including the protections of the Bankruptcy Code.** That said, remember that some of these protections did not fully extend to same-sex couples living in the 37 states that have not legalized same-sex marriage, even if they married elsewhere or have registered as domestic partners or entered civil unions. As of 2013, unprotected domestic partners must stand in the same line as other creditors, even when their claims are for alimony and property division debts related to a marriage or marriage-like relationship.

Questions to Ask a Divorce Attorney

Debts and the fear of bankruptcy can cause you to lose sleep during divorce. Ask your attorney the following questions so that you know the best ways to protect yourself financially.

1. Who will be legally responsible to pay which debts?
2. What can be done legally to make sure my spouse pays debts according to our agreement?
3. Am I entitled to reimbursement or credit for our money spent on educational training and expenses, child and spousal support obligations from another relationship, payment of joint debts after filing for divorce, improvement or additions to my spouse's separate property made without my consent, or my spouse's gambling, drug, or alcohol expenses, including rehab?

4. If my spouse makes noise about filing for bankruptcy before our divorce becomes final, should I file, too?

5. How can I stop my spouse from spending marital assets until the divorce is final?

6. If only my spouse signed loan documents while we were married, is that still my debt?

7. If just my spouse signed the loan documents, does the creditor have a right to come after my assets once we're divorced?

8. If I believe my spouse is going to file for bankruptcy, what can I do to protect my child and spousal support?

Questions to Ask a Bankruptcy Attorney

If you do find yourself in a bankruptcy situation, whether it's you or your spouse who's filing, consider asking a bankruptcy attorney these questions:

1. What's the effect of filing for bankruptcy before, during, or after divorce?

2. Which assets are exempt from a bankruptcy estate? Which are included?

3. What debt is included in a bankruptcy estate? Which debts are not discharged?

4. How can I protect myself if I don't want to file for bankruptcy?

5. If my spouse files for bankruptcy after our divorce is final, would our property settlement protect me?

WEB RESOURCE

There are many good sites, government and private, for information about debt, credit, and other consumer issues. Typically, they include search mechanisms that allow you to home in on the issue most relevant for you. These sites include:

- **www.ftc.gov** (Federal Trade Commission)
- **www.publications.usa.gov** (Citizen Information Center of U.S. General Services Administration)
- **www.aba.com** (American Bankers Association)
- **www.bankrate.com**, and
- **www.marketwatch.com.**

Child Support and Alimony: What Might I Pay or Receive?

Child Support—Legal, Financial, and Emotional Realities.................................381

   Legal Realities..382

   Financial Realities...383

   Emotional Realities..383

Steps to a Settlement..384

   Who Will Have Custody of the Children?..385

   What Does It Cost to Rear Your Children?...386

   Health Insurance Coverage..392

   How Much Child Support Do You Need—Or Can You Afford to Pay?.......393

   What Are the Guidelines for Child Support in Your State? 396

   Methods of Computing Child Support 397

   Payment Practicalities...399

   What Is Your Bottom-Line Decision Regarding Custody and Child
      Support? 407

Alimony—Legal, Financial, and Emotional Realities 408

   Legal Realities..408

   Financial Realities...409

   Emotional Realities..410

Steps to a Settlement..410

   How Much Do You Need to Live On? 410

   What, If Any, Are the Financial Schedules for Alimony in Your Area? 411

   How Much Alimony Do You Need—Or Can You Afford to Pay?...............412

   Are There Alternatives to Paying or Receiving Monthly Alimony?.............424

   What Is Your Bottom-Line Decision on Paying or Receiving Alimony?.....428

Questions to Ask an Attorney...428

Child support and alimony payments can be the most hotly contested issues in a divorce. These payments are often necessary when the money that used to support a single household must stretch to support two.

Unemployment checks and insurance claims can help a couple survive the loss of a job or damage to a house. But who pays the bill when the financial disaster of divorce strikes?

Somehow, household expenses must be paid. Someone must come up with the "alimonia"—a Latin word for "food" or "support." Today, the word is alimony, although it is increasingly called spousal support, spousal maintenance, or rehabilitative support. The names differ depending on where you live, but the impact seems to be universal: The payer feels the amount is far too much, while the recipient knows it's not enough.

Both parties may be right.

Breaking one household into two costs more than anyone thinks it will. Besides doubling the basics—mortgage or rent, utilities, and phone bills—divorcing couples must admit that the choices they made as a couple may no longer be realistic. A wife who left her job to care for the children cannot retrace her steps and cover the ground she lost when she dropped out of the workforce. A husband who refused a transfer so his wife could pursue her entrepreneurial ambitions may be stuck in a dead-end job that he can't leave without jeopardizing his pension. When you are "fired" from the job of husband or wife, no one offers you an unemployment check. Is it any wonder, then, that alimony often becomes a major battleground in divorce?

If you have children, the struggle to make ends meet can be even more draining. Whether you will pay or receive child support, you'll need to know what it costs. But how can parents put a price tag on their children?

They can't and don't. It's therefore not surprising that the custodial parent (recipient) never feels the support payment is enough and the noncustodial parent (payer) feels hounded for increased support.

This chapter first takes you through the legal, financial, and emotional realities of alimony and child support, and then outlines your steps to a settlement.

SKIP AHEAD

If you have no children, skip ahead to the section on alimony.

Protect Your Parental Rights

Before making any legal moves that affect your children, make sure you are informed about your rights as a parent. Ask an attorney about what can happen when custody is disputed. Make sure you know when and where all hearings or mediation sessions will take place. Better to ask too many questions than to lose time with your children because you misunderstood the consequences of your actions. Further, because the court will examine your custody arrangements, be prepared for any stunt your spouse may try to pull in the courtroom. Don't be lulled into a false sense of security because of informal agreements you've made with your spouse. And never move out of the family residence without a written agreement about custody. For more about parenting and custody rights, see *Nolo's Essential Guide to Divorce*, by Emily Doskow.

Child Support—Legal, Financial, and Emotional Realities

Children, unfortunately, are frequently the emotional pawns of divorce. Parents express their resentments about their mates by degrading the other parent in front of the children and by fighting over custody and child support issues. We urge you not to do this. Fighting scares the children and makes them feel guilty. It also wastes time, money, and resources that could have been spent more positively on your children.

As difficult as it sounds, try to keep your emotions in check. This doesn't mean you should hide your feelings about your divorce from your children. It does mean that you should not take your anger out on your kids. Instead, try to include them in the process in constructive ways. By maintaining a healthy emotional attitude, you will be better equipped to deal with the legal and financial realities of child support.

Legal Realities

All parents have an obligation to support their children, no matter what the status of their adult relationships may be. Suppose one parent has primary custody, traditionally called "physical custody," and the other has secondary custody, or "visitation rights." (See "Legal Language Alert," below.) The parent with less custody time is usually ordered to pay some child support to the custodial parent. It is assumed that the custodial parent is supporting the children by providing their day-to-day care. For parents with joint physical custody, the support obligation of each is based on a number of factors, such as the ratio of each parent's income to their combined incomes, the percentage of time the child spends with each parent, and who pays necessary expenses for the child.

While divorcing spouses are permitted to decide virtually all the terms of their divorce without court intervention, the court will insist on examining the child support arrangements. If the judge approves of the arrangement, the court will include it in the judgment. Some of the factors evaluated by courts in setting child support include:

- the needs of the child
- each parent's ability to earn and pay support, and
- the amount of time the child spends with each parent.

Legal Language Alert: Custody by Any Other Name...

Traditionally, terms such as "custodial parent" or "noncustodial parent" have been used by courts and among people in general when talking about parent-child relationships after a divorce. In light of changing family patterns, however, courts are beginning to recognize shared parenting styles and other such arrangements. Legal language is therefore changing, and terms such as "primary custody" and "secondary custody" or even "residential" and "nonresidential" parents are beginning to be used.

For the time being, however, we will continue to refer to "custodial" or "noncustodial" parents until the new terms become more commonplace.

Society has become less tolerant of parents who fail to comply with child support obligations, and the federal government requires all state governments to enact laws to facilitate enforcement of child support orders.

Financial Realities

Any legal award of child support is probably going to be less than the actual amount necessary to meet the needs of growing children, meaning that you will have to devise strategies for finding the rest of the money you need to live on.

Even if you receive an adequate child support award, financial reality demands that you recognize the possibility that your spouse won't pay. If you are the recipient, what will you do when the check does not arrive or comes late? And if you are the payer, what happens when your paycheck is suddenly garnished—or worse, you are threatened with jail for noncompliance? Don't dwell on these negative questions, but at least confront them for your children's sake.

Also keep in mind two new, but sad, trends emerging from the financial realities of divorce. Some parents have reached the painful conclusion that they cannot retain custody of their children because they do not have the resources to support them. At the same time, other parents are fighting for joint custody so they can pay less child support. To avoid these situations, you must squarely face the true costs generated by children and prepare your strongest position for negotiating your settlement and protecting the interests of your children.

Emotional Realities

Your children's experience of your divorce can have a major effect on your emotional state, but the most important thing is that you make sure that your emotional state doesn't affect your children's experience. On all issues concerning your children—child support, custody, and visitation—you and your mate need to stay focused on what's best for your children.

Be aware of your own behavior, and don't get caught up in a power struggle or try to use the children to get back at the other parent. You will end up causing more damage to the children than to your spouse.

RESOURCE

There are lots of good books that can help you deal with the emotional realities of parenting during divorce. Check any of the websites listed in Chapter 1 for resource lists. Below are a few titles to get you started.

Building a Parenting Agreement That Works: Child Custody Agreements Step by Step, by Mimi L. Zemmelman (Nolo), helps you negotiate and prepare a parenting agreement with your spouse.

Being a Great Divorced Father: Real-Life Stories From a Dad Who's Been There, by Paul Mandelstein (Nolo), offers practical advice for dads.

Mom's House, Dad's House: Making Two Homes for Your Child, Revised Edition, by Isolina Ricci, Ph.D. (Simon & Schuster), is the classic text on dealing with shared custody.

Helping Kids Cope With Divorce the Sandcastles Way, by M. Gary Neuman, L.M.H.C., with Patricia Romanowski (Random House), helps parents understand what their kids—of any age—are experiencing during divorce and offers practical advice for communicating with children.

There are many books on the market written specifically for children of divorcing parents. Ask a local bookseller or visit online book forums for recommendations. You won't be lacking for choices, no matter what the age of your children. Take the time to find titles that speak best to your child's interests and reading abilities.

Steps to a Settlement

As you move through your divorce, you will find that your choices about child support are controlled by local financial guidelines. Answering the following questions may help you prepare to understand these guidelines and understand what's probable in terms of paying or receiving child support:

- Who will have custody of the children?
- What does it cost to rear your children?

- How much child support do you need—or can you afford to pay?
- What are the guidelines for child support in your state?
- What is your bottom-line decision regarding custody and child support?

Who Will Have Custody of the Children?

There are two types of custody: the legal authority to make decisions about the medical, educational, health, and welfare needs of a child (legal custody), and physical control over a child (physical custody). In most states, courts now tend to award joint legal custody, which lets divorced parents share in making decisions about their children's health, education, and welfare. The most common arrangement is to share legal custody, with one parent having primary physical custody and the other secondary physical custody or liberal visitation.

You can work out whatever arrangement you want regarding custody. If you can't agree, you may be able to use mediation to work things out. Mediation is a nonadversarial process in which a neutral person (a mediator) meets with parents to help them resolve problems. To reach a mediated solution, both parents must agree. The mediator cannot impose a solution on the parties. Many states require parents to participate in mediation before bringing a custody dispute to court. If you still can't agree and you leave the custody decision to the court, the legal standard the court will use is "the best interests of the child." The factors courts consider in determining the best interests usually include:

- the child's age, sex, maturity, temperament, and strength of attachment to each parent
- special needs of the child and parents
- the child's relationship with siblings and friends
- the distance between the parents' households
- the flexibility of both parents' work schedules and the child's schedule
- child care arrangements
- transportation needs
- the ability of the parents to communicate and cooperate

- the child's and parents' cultural and religious practices
- the willingness of a previously uninvolved parent to provide adequate supervision
- each parent's ability and willingness to learn basic caregiving skills such as feeding, changing, and bathing a young child, preparing a child for day care or school, or taking responsibility for helping a child with homework, and
- each parent's ability to care for the child's needs.

What Does It Cost to Rear Your Children?

Mentioned in this section are items you should consider in figuring the cost of rearing children. As you review these items, calculate the average amount you spend each month. You can make this calculation by totaling the three most recent months and then dividing by three. Check the monthly expenses column of your "Cash Flow Statement" (see Chapter 12), look back through your checkbook, or make an educated guess to find your monthly outlays.

After filling in each number, total the expenses at the end.

$ _____   *Child care*

If both parents work, child care costs may be significant. If you have relied in the past on informal babysitting arrangements with friends or relatives, you cannot and should not consider those arrangements permanent. If you don't know the going rate for child care in your area, investigate. You must factor the cost of competent care into your children's future. The court will order that child care expenses be shared by the parents to allow the custodial parent to work outside the home.

$ _____   *Education or college funds*

Even after divorce, it's possible for parents to work together to realize the joint goal of providing for a child's private schooling or college education. It takes planning and cooperation, but it can be done. Besides college, your children may want to receive special training for

The Dos and Don'ts of Parenting During Divorce

Do:

Initiate the child's contact with the other parent on a regular basis by phone, letter, audio and Skype, email, and other forms of communication.

Maintain predictable schedules.

Be prompt and have children ready at exchange time.

Avoid communication that may lead to conflict at exchange time.

Ensure smooth transitions by assuring the children that you support their relationship with the other parent and trust the other parent's parenting skills.

Allow the children to carry important items such as clothing, toys, and security blankets with them between the parents' homes.

Follow similar routines for mealtime, bedtime, and homework time in each home.

Handle rules and discipline in similar ways in each home.

Support contact with grandparents and other extended family so the children do not experience a sense of loss.

Be flexible so the children can take advantage of opportunities to participate in special family celebrations or events.

Give as much advance notice as possible to the other parent about special occasions.

Provide an itinerary of travel dates, destinations, and contact information when children are on vacation.

Establish a workable, businesslike method of communication.

Plan vacations around children's regularly scheduled activities.

The Dos and Don'ts of Parenting During Divorce (cont'd)

Don't:

Make your child choose between mom and dad.

Question your child about the other parent's activities or relationships.

Make promises you do not keep.

Argue with or criticize the other parent when your child is present or can hear you.

Discuss your personal problems with the child or where the child can hear you.

Use your child as a messenger, spy, or mediator.

Withhold visitation because of the other parent's failure to pay support.

Get professional help from an attorney or a mental health practitioner, court services mediator, domestic abuse agency, or social services agency if there are issues of:
- child abuse or neglect
- serious mental or emotional disorders
- drug or alcohol abuse or criminal activity
- domestic violence, or
- continuous levels of very intense conflict.

a profession or the arts. These costs can sometimes be staggering. It's necessary to anticipate them now, before your settlement is negotiated.

Agreements between divorcing spouses about post-high-school education should include:
- acceptable schools, location, and accreditation
- what GPA is expected of the child in high school and college
- what contributions each parent will make to the student's living expenses—such as allowances, car expenses, and traveling-home expenses—and whether money will be paid directly to the student, to a third party, or to the custodial parent
- student's application for financial aid

- academic load (full-time or part-time)
- whether the student will be required to work while in school, and
- methods to resolve disagreements about post-secondary education.

Rising College Costs

If you're already calculating how much it'll cost for your child to attend college, don't forget inflation. Tuitions have skyrocketed over the past few decades, often doubling the general inflation rate. But, even if they held steady for a decade, inflation means that the dollar you save today will be worth less by the time your child enrolls in college.

Tallying the impact of inflation can be a shock. Suppose tuition currently averages $10,000 annually (private institutions are much higher). If inflation is 4% a year and your seven-year-old enrolls as a freshman in 10 years, you will need $16,010 of today's dollars to pay that $10,000 tuition.

You must also consider the fact that college tuition itself has become much more expensive in recent decades. Suppose tuition increases at 7% per year: you will need $19,672 to cover freshman year a decade from now. There's much debate about how tuition might be reined in, but you shouldn't count on that happening. You will need to put your child's college funds in an investment vehicle that grows at a rate faster than inflation. Don't forget about grants, scholarships, public education and other alternatives that may help defray the cost of a college degree.

$ _____   *Emergency savings*

Financial planners make a standard recommendation that you have an emergency cash reserve equal to three to six months of fixed expenses. Be sure that reserve includes your children's normal expenses. You can also make your postdivorce life easier if you include a cash cushion for the emergencies that children can create.

$ _____   *Health and dental care*

Divorce affects your children's health care coverage. You and your spouse must reach an agreement as to who will carry the children on a health plan, who will buy coverage, and how you will pay for medical expenses not covered by the plan or co-payments and deductibles. If you cannot agree, a court will make the order for you. Some states mandate that if neither parent has insurance covering the children, one parent must pay for reasonable medical insurance. Also be aware of what benefits are covered and how long they last. For more details on health care coverage for youngsters, see "Are There Alternatives to Paying or Receiving Monthly Alimony?" below.

$ _____   *Hobbies*

Are your children involved in sports, music, or ballet? Do they act in community theater? Do they want to go to summer camp? Do they expect to as they grow older? Check how much you've paid for special lessons and hobbies, or talk to parents whose children are involved in similar pursuits to get an idea of future costs. You will want your agreement to indicate who will pay for which activities and hobbies.

$ _____   *Birthdays and holidays*

Do you splurge on special occasions and birthdays? Do your children have high expectations about gifts? These costs can add up. You may need to readjust your budget and let your children know about the change. Under the best circumstances, you and your soon-to-be ex-spouse might agree to a spending limit so that one partner does not use gifts to curry favor.

$ _____   *Total expenses for children*

Checklist of Issues to Consider When Negotiating Child Support

When you sit down to talk seriously about child support, you will need to take into account all of the considerations listed above, and you'll also need to look at factors that might cause a court to deviate from the standard support guidelines. Here's a checklist that will help you remember to cover all the bases when looking at child support issues. Consider things like:

☐ children's special needs, including physical, emotional, and health needs, educational needs, day care costs, or needs related to the child's age

☐ shared physical custody arrangements

☐ support obligations of either parent to a former household, including alimony

☐ an extraordinarily low- or high-income parent

☐ the intentional suppression, reduction, or hiding of income

☐ other support, such as lump-sum payments, use of a separate property residence, payment of a mortgage or medical expenses, or provision of health insurance coverage

☐ a parent's own extraordinary needs, such as unusual medical expenses

☐ deferred sale of family residence, in which the rental value exceeds the mortgage payments, homeowner's insurance, and property taxes

☐ travel expenses required for custodial time with one parent

☐ how the child support payment will be secured, including a requirement of life or disability insurance

☐ employee benefits, including medical savings plans, cost and quality of medical, dental, and life insurance, and

☐ other fairness factors.

Health Insurance Coverage

You may be worried about how your children's health care needs will be covered after your divorce. It may help you to know that a federal law (P.L. 103-66, Omnibus Reconciliation Act of 1993) allows custodial parents to obtain a court order for children's health insurance coverage through the noncustodial parent's employment insurance plan (as long as the plan isn't self-insured). Coverage cannot be denied by the noncustodial parent, that parent's employer, or the insurance company simply because the child does not live with that parent, is not claimed as a dependent, or lives outside of the plan's service area.

The court order for health care for your children is called a QMCSO, a qualified medical child support order. A QMCSO can require that:

- an employee's child is covered under an employer-provided health care plan
- the premium is deducted from the employee's paycheck
- reimbursements are made directly to the nonemployee parent if that parent pays the provider
- the nonemployee parent receives information regarding the health care plan and reimbursements, and
- other health coverage is provided as specifically directed.

You will want to use a QMCSO when:

- the employee parent is or may be a recalcitrant payer
- the employee parent's regular medical plan determines that the children are ineligible for dependent coverage if they are living with the other parent
- the custodial parent prefers to deal directly with the plan administrator for information, forms, and reimbursement
- the employee parent selects a medical plan that is not suitable for the child—for example, if the employed parent chooses a plan with a very high deductible (not appropriate when you have children who will inevitably visit the doctor) the QMCSO gives the other parent the option of selecting a plan with a very low or no deductible, or

- the employee parent wants to take advantage of favorable "family coverage" premiums.

To qualify as a QMCSO, the medical support order must:

- create the child's right to receive benefits under the group plan
- specify the employee parent's name and last known mailing address and also that of each child covered by the plan
- specify each plan and time frame to which the order applies, and
- not require any additional benefits not actually provided in the plan.

While the QMCSO can be valuable in making sure your children's health care is protected, it may include some of the following:

- limited coverage for a child residing outside the area served by the plan—although the plan cannot deny coverage for this reason, it can limit the coverage it offers, and
- limits on a child's ability to obtain medical coverage different from the coverage provided by the employee parent.

If you elect to use a QMCSO, the custodial parent should obtain copies of the following:

- a medical plan and summary plan description
- election forms highlighting the possible elections and open enrollment periods
- information on current beneficiary designations, and
- written material concerning plan requirements for terms of a QMCSO.

A sample QMSCO appears at the end of this chapter.

How Much Child Support Do You Need— Or Can You Afford to Pay?

In figuring out how much child support you need or can afford to pay, you will consider the same three items you considered with alimony— income needs, taxes, and payment practicalities.

Income Needs

The difference between legal reality and financial reality defines your income needs for child support. By legal reality, we mean the amount set by your state's child support formula. This is the guideline amount that you are likely to pay or receive. Your lawyer or local agencies can help you determine this figure (see "What Are the Guidelines for Child Support in Your State?" below).

The financial reality is the actual cost every month of rearing your children.

For recipients. If your actual child-raising costs are higher than the amount in the legal guideline and you have no way to make up the gap, you'll have to negotiate for a higher amount. If your spouse is aware that the guidelines are lower than the amount you are asking for, you may have little leverage. Nevertheless, you can still attempt to push up the amount of support.

For payers. Recognize that the gap between what your children cost and what you have to pay has to be closed somehow. If the custodial parent has no other source of revenue, keep in mind that you may need to consider paying more than guideline support. You and your spouse may be able to save yourselves time and trouble by negotiating child support from a realistic financial base right from the start. Often, divorcing couples find there is simply not enough money to meet these costs, and both spouses have to pare down their lifestyles.

Taxes

Child support payments have no tax consequences. They are neither deductible to the payer nor taxed as income to the recipient.

Children, however, have plenty of tax consequences. First and foremost, they are considered dependency exemptions. Parents must decide who will claim the children at tax time. Often, the noncustodial parent claims the children in exchange for paying higher support. The custodial parent must relinquish claim to the dependency exemption by signing IRS Form 8332, *Release of Claim to Exemption for Child of Divorced or Separated Parents*. When the parents have joint physical

custody, they may each claim a child (if they have more than one) or they may alternate the claim year to year. No matter who claims the children as dependents, both parents must list the children's Social Security numbers on their tax returns so the IRS knows that only one parent is claiming the children.

The following tax benefits are among those available to parents to offset the cost of raising children:

- earned income credit
- child care credit
- medical expense deductions
- head of household filing status
- dependency exemption
- child tax credit (for three or more children)
- deduction for interest on qualified education loans
- American Opportunity and Lifetime Learning credits for tuition costs, and
- education IRAs.

Not all of these benefits are available to all parents, but you need to be aware of these potential tax benefits as you make your custody and child support agreements.

Only a custodial parent is entitled to claim the child care tax credit. In general, an employed custodial parent of a dependent child under the age of 13 is eligible for the credit for child care expenses incurred so that the parent can earn an income. As the custodial parent's income increases, however, the credit phases out. Keep in mind the availability of the tax credit when negotiating your agreement about who will pay child care expenses.

Any taxpayer can claim a deduction for medical expenses actually paid, but only if those medical expenses exceed 10% of the taxpayer's adjusted gross income (there are a few exceptions to this that will apply through 2016). If your total medical expenses are high enough, you may want to allocate them to the lower wage earner so that parent can take the deduction.

Only a parent who has physical custody more than half the time can file as head of household. If the parents have joint legal and physical

custody (and physical custody is divided 50-50), neither can file as head of household, because the dependent child resides with neither parent for more than 50% of the year.

If you have more than one minor child and share physical custody, you can specify your arrangement as 51% for one child with one parent and 51% for the other child with the other parent. Because each parent has a dependent child in the home more than 50% of the year, each parent can file as head of household.

What Are the Guidelines for Child Support in Your State?

The Family Support Act of 1988 required all states to adopt a statewide uniform child support formula. Parents may be able to agree on a child support amount different from that required by the state formula, and as long as the parents can show that the children's needs will be met by the below-guideline payments, courts will generally approve the amount the parents have agreed on. However, neither the court nor the children are bound by the agreement if it is for less than the state minimum. Remember: *All* parents are obligated to support their children, and a custodial parent cannot give up permanently *the child's* right to receive support.

What If Your Spouse Takes a Lower-Paying Job to Reduce the Child Support Obligation?

Courts are displaying little tolerance for spouses who take lower-paying jobs than those for which they are qualified. A court may base child support on imputed income—a higher amount of income than the spouse is actually earning. Also, to determine support payments, a court may factor in benefits from deferred compensation, fringe benefits, and certain depreciation deductions for the purpose of determining business income. California and New York courts lead the way in making it difficult for individuals to play games with income in order to hold down alimony or child support payments.

Methods of Computing Child Support

There are two primary methods that courts use to compute child support. Which one applies to you depends on where you live. The two methods are called "income shares" and "percentage of income," and are discussed in more detail below. If you want to find out which formula is used where you live, check with a family law attorney or check one of the websites or support calculation programs described below.

The Income Shares Method

The income shares method calculates support by evaluating the proportional income of each parent and considering the cost of work-related day care and premiums for the children's medical insurance, among other factors.

The income shares model aims to provide the child with the same proportion of parental income—and thus the same lifestyle the family enjoyed—that the child would have received as a member of an intact household.

The Percentage of Income Model

The percentage of income model requires that either a flat percentage or varying percentage of the noncustodial parent's gross or net income be paid to the custodial parent.

Under the *flat percentage* model, the percentage of income devoted to child support remains constant at all income levels. Under the *varying percentage* model, however, the percentage of income devoted to child support varies according to the level of income, with the support percentage decreasing as income increases. Under either variation, the percentage to be applied is determined by the number of children and, in some states, by the ages of the children.

Calculating Support

Whichever system your state uses, the easiest way to get a general idea of what your support obligations will be is to ask a lawyer who is

familiar with your financial situation. But if you want to get a ballpark of what you might have to pay, or what you will be owed, or if you are representing yourself, in most states there are websites—either state sanctioned or commercial—that will help you calculate an estimate. See "How to Find and Use an Online Support Calculator," below.

How to Find and Use an Online Support Calculator

Many states have state-sponsored websites that use the state's guidelines to help you to make ballpark child support calculations for free. There are also numerous commercial websites for specific states that will help you make the calculations for a fee. Finally, www.alllaw.com offers support calculators for every state.

Often, there is a website specifically designed for your state that says it is the "official" one used by the courts in your state. Now that courts are required to have statewide guidelines for support, judges use computer programs to calculate support—and often you can find the same system, simplified for a layperson to use, online for a fee.

The fee that the websites charge will vary from about $3 to $10 per calculation. Some might offer a monthly fee, but most sites will charge you per calculation. You might think that you only need to do one calculation, but often you will want to do more—for example, if you aren't yet sure how much time your kids are going to spend with each parent, you'll need to do a calculation for each of the different possibilities for time sharing, which can affect support.

To use the support calculators, you'll need to know each parent's gross and net income, deductions, and exemptions—it helps to have a copy of a recent pay stub for each parent. Most calculators will ask for information that will allow for adjustments based on child care expenses, extraordinary medical or educational expenses, and other special circumstances.

For more on how to find state-sponsored websites and other resources relating to child support calculators, check out *Nolo's Essential Guide to Divorce*, by Emily Doskow.

The results you get from a support calculator will be estimates, but they will generally be in the ballpark, and you can use them to negotiate about support. If you have a dispute, though, the court will have the final say, using its official computer program.

Payment Practicalities

The practical matter of getting child support payments from one person to the other can create highly charged problems. Below are questions and suggestions to help prevent these problems. Whether you deal with these items during your negotiations or in the final settlement, use the spaces provided to jot down your ideas. You can refer to these points when needed. Put a check mark in front of the question once it is resolved.

How and When Child Support Will Be Paid

_____ *How will child support be paid?*

_____

_____

_____

_____

You and your spouse can agree that support will be paid in the form of a monthly check sent directly by the payer to the recipient. The court, however, may want to set up a different process. All child support orders automatically contain a wage withholding provision. An automatic wage withholding order works quite simply. When the court orders you to pay child support, the court—or your child's other parent—sends a copy of the court order to the payer's employer. The custodial parent's name and address is included with this order. At each pay period, the employer withholds a portion of the pay and sends it on to the custodial parent. Only if both parents agree—and the court allows it—can you avoid the wage withholding and make payments directly.

If you don't receive regular wages but do have a regular source of income, such as payments from a retirement fund, an annuity, or Social

Security, the court can order the child support withheld from those sources of income. Instead of forwarding a copy of the order and the custodial parent's name and address to an employer, the court sends the information to the retirement plan administrator or public agency from which you receive your benefits.

Mandatory wage withholding operates differently among the states. In Texas and Vermont, for example, all current orders include automatic wage withholding, regardless of the parent's payment history. The rationale is that by not distinguishing between parents with poor payment histories and parents who have paid regularly, no parent is stigmatized.

In California, wage withholding orders are automatic as well. But parents who show a reliable history of paying child support may be exempt. And in all states, employers must withhold wages if the payer is one month delinquent in support.

A few states have set up other mechanisms to collect child support as it becomes due. For example, a judge may order a noncustodial parent to pay child support to the state's child support enforcement agency, which in turn pays the custodial parent. This program is often used where automatic wage withholding is not in effect. It may also be used for noncustodial parents without regular income (the self-employed) or when parents agree to waive the automatic wage withholding.

Some other states let judges order noncustodial parents to make payments to court clerks or court trustees who in turn pay the custodial parents.

_____ *On what day of the month will payments be made?*

_____

_____

_____

_____

Most noncustodial parents find it convenient to pay child support shortly after being paid themselves. If you are a recipient, and payments

will not arrive until after your child's monthly child care bill or health insurance premium is due, you'll need to make different arrangements. Try saving money from the previous month or contacting the creditor and negotiating a new payment schedule. Try to get your payments in sync with your cash flow. You don't need to tell the creditor you're waiting for your child support check. Creditors may be surprisingly cooperative when you contact them in advance and honestly negotiate so they are assured of getting their payments.

How long will child support payments be made?

_____

_____

_____

_____

You must pay child support for as long as your child support court order says you must. If the order does not contain an ending date, in most states you must support your children at least until they reach age 18. A few states, however, extend the time and require you to support your children:

- until they finish high school
- until they reach 21 (for example, this is the requirement in New York)
- until they complete college, or
- if they are disabled, for as long as they are dependent.

Your child support obligation will end, however, if your child joins the military, gets married or moves out of the house to live independently, or if a court declares your child emancipated.

Child support obligations extend beyond the death of the paying parent—if that parent leaves an estate, the estate must continue paying support.

If you failed to make some of your child support payments and you owe unpaid support, it doesn't go away when your child turns 18 (or whatever age you are no longer liable for support). Many states give a custodial parent ten or 20 years to collect back child support—with interest.

_____ *How will you handle child support modifications?*

_____

_____

_____

_____

A court retains the power to make orders involving child support. If a noncustodial parent can't afford the payments, that parent must take the initiative to change the child support order. Or, if a custodial parent can't live on the current payment, it's that parent's responsibility to ask for more. As a first step, the parent wanting the change should call the other parent and try to work out an agreement. If you reach an agreement, be sure to get it in writing. You'll then need to get a judge's signature.

If you can't make satisfactory arrangements with your ex-spouse, the parent wanting the change will have to file papers with the court, schedule a hearing, and then show the judge that the support amount is too little to live on—or too much to pay. The court won't retroactively decrease child support, even if the noncustodial parent was too sick to get out of bed during the affected period. Once child support is owed and unpaid, it remains a debt until it is paid.

To get a judge to change a child support order, you must show a significant change of circumstance since the last order. What constitutes a significant change depends on your situation. Generally, the condition must not have been considered when the original order was made, and must affect your—or your child's or the other parent's—current standard of living.

Changes that qualify for a *noncustodial* parent to seek a reduction in support include:

- your income has substantially decreased
- you have increased expenses, such as a new child
- a raise you expected—which was the basis of the last order—didn't materialize
- the custodial parent received a large inheritance, an increase in compensation, or income-producing assets, or

- your child's needs have decreased—for example, she is no longer attending private school.

Changes that qualify for a *custodial* parent to obtain an increase in the support amount include:

- you have a substantial decrease in income
- you have increased expenses, such as a new child
- the noncustodial parent received a raise, large inheritance, or income-producing assets, or
- your child's needs have increased—for example, ongoing counseling.

Dealing With Unpaid Child Support

How will you handle nonpayment of child support?

_____

_____

_____

_____

Unpaid child support (sometimes referred to as "being in arrears") accumulates when the parent doesn't pay what is owed. A parent who is owed a great deal in arrears can ask a judge to issue a judgment for the amount due.

States' child support enforcement agencies, custodial parents, judges, and district attorneys use several different methods to collect child support.

Court hearings. Failing to follow a court order is called "contempt of court." A parent who is owed child support can schedule a "show cause" hearing before a judge. The other parent must be served with a document requiring attendance at the hearing, and then must explain why the support hasn't been paid. If the noncustodial parent is a no-show, the court can issue an arrest warrant, which could lead to a night or two at the county jail.

If the noncustodial parent attends the hearing, the judge can still require jail time for violations of the support order. To stay out of jail,

the parent must first show why the money wasn't paid and then explain what prevented the parent from requesting a modification hearing.

Federal law mandates that all states provide for wage withholding for child support orders.

Interception of tax refunds or lottery winnings. One of the most powerful collection methods available is an interception of the payer's federal income tax refund, state income tax refund, or sales tax refund. The custodial parent can ask the district attorney to call the Treasury Department for help. States also intercept tax refunds to satisfy child support debts. Also, many states with lotteries let custodial parents apply to the state for an interception of the other parent's winnings.

Property liens. In some states, a custodial parent can place a lien on the payer's real or personal property. The lien stays in effect until the payer pays up, or until the custodial parent agrees to remove the liens. The custodial parent can force the sale of the noncustodial parent's property, or can wait until the property is sold.

Posting bonds or assets to guarantee payment. Some states allow judges to require parents with child support arrears to post bonds or assets, such as stock certificates, to guarantee payment. In California, for example, if a parent misses a child support payment and the custodial parent requests a court hearing, the court may order the noncustodial parent to post assets (such as putting money into an escrow account) equal to one year's support or $6,000, whichever is less.

Reports to credit bureaus. If the custodial parent owes more than $1,000 in child support, that information may find its way into a credit file maintained by a credit bureau. Federal law requires child support enforcement agencies to report known child support arrears of $1,000 or more. Many child support enforcement agencies automatically send information about the child support order to credit bureaus, even though payments are current.

Reporting to "most wanted" lists. States are encouraged to come up with creative ways to embarrass parents into paying the child support they owe. Most states now publish "most wanted" lists of parents who owe child support.

Refusal to renew professional licenses. Some states will not renew contractor's or professional licenses of people behind on child support. These may include business, recreational, personal, or professional, such as attorney, contractor, cosmetologist, doctor, or hunting or fishing licenses. Without a license, the payer cannot legally practice the profession in the state.

Refusal to renew or suspension of driver's license. Parents in some states are not allowed to possess a driver's license if they have not met child support obligations.

Wage and property garnishments. Child support arrears can often be collected by a wage garnishment. A wage garnishment is similar to a wage withholding—a portion of the noncustodial parent's wages is removed from the paycheck and delivered to the custodial parent. They differ, however, in one way: the amount withheld. The amount of a wage withholding is *the amount of child support ordered each month*. The amount of a wage garnishment is *a percentage of the payer's paycheck*. The amount the payer was originally ordered to pay is irrelevant.

If the wage garnishment does not cover what's owed, the custodial parent may try to get other property to cover that debt. Common property targets include bank accounts, cars, motorcycles, boats and airplanes, houses and other real property, stock in corporations, and accounts receivable. It's also possible that unemployment compensation can be withheld in some states.

The federal government now requires states to set up agencies to oversee this collection. If the parties use such an agency, the employer of the paying parent makes the child support payment to this state agency, which then writes a check to the custodial spouse.

District attorney assistance. In most states, a local D.A. will help—and in some states is required to help—a custodial parent who hasn't received child support.

Passport denial. In cases where arrears exceed $2,500, the nonpaying parent may lose passport privileges.

Tapping Pension Plans to Pay Spousal and Child Support

A Qualified Domestic Relations Order (QDRO, pronounced "quadro") is a court order to the administrator of a retirement plan spelling out how the plan's benefits are to be assigned to each spouse in a divorce. (See Chapter 14 for details.) A QDRO can be used to collect child support or alimony payments from the funds in a retirement plan under certain conditions.

A QDRO is most useful when the payer is self-employed and has rights to benefits from a prior employer, or when the payer changes jobs a lot. If the payer is not yet eligible for benefits, you may be able to secure payments from a defined contribution plan, such as a profit-sharing or 401(k) plan. These plans are a particularly good source for support arrears because the plan administrators normally want to make lump sum payments, not series of monthly payments.

If your spouse's pension plan qualifies, you can use the QDRO to collect child support and alimony payments. The QDRO must clearly specify how much will be paid, to whom, and for how long, and the benefits must be currently available to an alternate payee (nonemployee spouse) under the terms of the plan.

If the retirement plan currently pays benefits, the QDRO operates like a garnishment of wages—that is, the money is removed from the benefits check and sent directly to the recipient. If retirement benefits are not yet being paid, the structure of the QDRO depends on the situation. If the payer is eligible to receive benefits under the plan (but has chosen to continue working), the QDRO could require that the plan make the monthly support payments. If the payer is still working, however, it is usually easier to use a wage garnishment.

Tax Problems With Using a QDRO

Child support, unlike alimony, is not taxable. Using a QDRO to collect child support from a pension plan, however, makes it taxable income to the recipient, because the tax code requires that any payments from a QDRO to a former spouse be taxed. To compensate, you can try to get the child support payments increased to allow for withholding of taxes. Otherwise, do not use this option unless no reasonable alternative exists to collect the payments.

Finding a Missing Parent

It's hard to collect back (or current) child support when you can't find the person who owes it to you. If your ex-spouse is missing, the best place to start searching is to contact your state child support enforcement agency (or the family support division of the district attorney's office). Be prepared to provide the following information about the missing parent:

- name, address, Social Security number, date of birth, physical description, father's name, and mother's maiden name
- names of friends and relatives and names of organizations to which the missing parent might belong
- name and address of current or recent employer, and
- information about the parent's income and assets, pay stubs, tax returns, bank accounts, investments, or property holdings.

You'll also need to give the enforcement agency the following information about yourself and the child(ren) who are owed support:

- children's birth certificates
- child support order
- divorce decree or separation agreement
- records of past child support received, and
- information about your income and assets.

For more information about collecting past-due child support or locating a parent, go to the website of the Office of Child Support Enforcement at www.acf.hhs.gov. Click on the link for "child support" and look up "Frequently Asked Questions About Child Support Enforcement."

What Is Your Bottom-Line Decision Regarding Custody and Child Support?

By this point, you've worked through questions on custody and child support. Use the space below to record your decision by writing out what

you want to accomplish regarding custody and support in the settlement negotiations.

_____

_____

_____

_____

_____

_____

_____

_____

Alimony—Legal, Financial, and Emotional Realities

It helps to think of alimony as a separate issue, distinct from your property settlement (except if your property earns income). Before reviewing the steps toward a settlement, take a few minutes to read about the legal, financial, and emotional realities surrounding alimony.

Legal Realities

Until the 1970s, alimony was a natural extension of the financial arrangement in traditional marriages, where the husband was the breadwinner and the wife stayed home, caring for the house and children. The amount of alimony was determined by a number of factors—the needs of the spouses, their status in life, their wealth, and their relative "fault" in ending the marriage. For example, if a husband was committing adultery or treating his wife cruelly, he would pay a relatively large amount. If the wife was having an affair or treating her husband cruelly, she would receive little or no alimony.

Today, the following factors are generally considered in determining alimony:

- the recipient's needs
- the payer's ability to pay

- the age, health, and standard of living of the spouses
- the length of the marriage
- each spouse's ability to earn
- the recipient's nonmonetary contributions to the marriage
- the recipient's ability to be self-supporting, and
- tax advantages and disadvantages.

Except in marriages of long duration (usually ten years or more) or in the case of an ill or ailing spouse, alimony today usually lasts for a set period of time, with the expectation that the recipient spouse will become self-supporting and eventually no longer need it.

In an effort to make alimony awards more uniform, some individual counties have adopted financial schedules to help judges determine the appropriate level of support. Of course, spouses can make arrangements outside of court that differ from the schedules. However, if you cannot agree on the amount, a judge will decide how much alimony will be awarded. Leaving such a decision to a judge may create a far different result from what you anticipated.

In some states, especially noncommunity-property states, the division of property can affect how much alimony is awarded. For example, if one spouse gets a house with no mortgage, plus rental properties that produce a good income, a court is not likely to award a high amount of alimony. You will have an opportunity to examine the interplay between alimony and your property division in Chapter 19.

Financial Realities

Generally, court-ordered alimony does not provide adequate money to live on. Often there's a big gap between the amount that's needed to live on and what the recipient actually gets.

To understand the financial reality of alimony, you must know what it really costs for you to live. Then you need to develop strategies for making up the difference between true living expenses and the amount you will receive in alimony. As you work through this section, you will look at the tax consequences of alimony as well as practical matters such as when, where, and how payments should be made. Remember, alimony

income is risky. It can be stopped, reduced, or increased by the court any time circumstances change.

Emotional Realities

Since the advent of no-fault divorce, fewer couples vent their anger in knock-down, drag-out fights in court. That doesn't mean the anger isn't there; it just manifests itself differently. That repressed anger may lead to fights over alimony, an item directly tied to a person's survival. You may be moving along nicely in the negotiations only to find that when you get to the alimony, your spouse suddenly makes unreasonable demands or starts playing tricks. Focusing on the legal and financial questions below will help you guard against emotional sabotage.

Steps to a Settlement

As you move through your divorce, you may find that your choices about alimony are guided by the local court's financial schedules. For instance, if a schedule shows that a person in your situation should get $500 a month in alimony, your spouse will probably use that number, at least as a starting point.

While you need to be aware of any schedule used in your area, don't rely on those figures in determining the amount you will pay or receive. By answering the following questions, you can analyze the financial consequences of paying or receiving alimony.

1. How much do you need to live on?
2. What, if any, are the financial schedules for alimony in your area?
3. How much alimony do you need—or can you afford to pay?
4. Are there alternatives to paying or receiving monthly alimony?
5. What is your bottom-line decision on paying or receiving alimony?

How Much Do You Need to Live On?

If you've completed the "Cash Flow Statement" in Chapter 12, you have already tracked your living expenses. If not, go back now and use the

statement to outline these costs. Whether you will pay or receive alimony, your goal is to get a realistic picture of your household expenses. If your spouse demands more than you feel you can afford to pay—or offers less than you feel you need to live on—you will have to document the cost of your lifestyle. Yes, that lifestyle will probably change after divorce, but for now you need to establish a baseline for expenses.

While you're documenting your own cost of living, estimate your spouse's expenses as well. You will need this information when you are negotiating. If you can't settle, the court will require that you each submit documentation of your income and expenses at trial.

If You Have Children

In calculating your expenses, don't overlook the special costs generated by children, such as allowances, piano lessons, child care, braces, Little League uniforms, and the like. You will need to make a list of ordinary and extra expenses. Some of those expenses may be listed on your "Cash Flow Statement" from Chapter 12. Other expenses are listed in "What Does It Cost to Rear Your Children?" above.

Check the monthly average figures from your "Cash Flow Statement" in Chapter 12 and other documentation containing your marital costs to estimate the following:

$ _____   *My personal monthly expenses*

$ _____   *My spouse's monthly expenses*

What, If Any, Are the Financial Schedules for Alimony in Your Area?

As mentioned previously, a number of counties have adopted informal financial schedules to help judges—and spouses—set alimony. Be sure

to ask your lawyer whether your county follows such a financial schedule or guideline.

More likely, alimony is determined based on the factors listed above. In addition, there's a trend for courts to follow principles such as these:

- If a marriage lasted less than five years and the couple has no children, many courts will deny alimony altogether, or award very short-term alimony.
- If there are children below school age, most courts will order some alimony.
- If a marriage lasted at least eight years, most courts make an alimony award to try to assure a continuity in the standard of living for both spouses.
- In marriages that lasted at least ten years, most courts will award alimony indefinitely, or at least until the recipient dies, remarries, cohabits, or no longer needs support.

Based on those factors, judges—and attorneys who regularly appear before those judges—estimate the amount of alimony. There's no universal formula or rule that applies. The amount of an alimony award, perhaps more than any other decision a judge can make, is seldom subject to challenge but often leaves ex-spouses very angry.

To get an estimate of the amount of alimony you may have to pay or are likely to receive if the issue is presented to a judge for decision, you will need to speak to an attorney or a mediator to determine local practice and custom. And remember, if you can't agree, you leave your fate to a judge whose decision may not meet your expectations.

How Much Alimony Do You Need— Or Can You Afford to Pay?

Figuring out how much alimony you need or can afford to pay centers on three concerns—income needs, taxes, and payment practicalities.

Income Needs

Only a hard look at the income-alimony gap can reveal potential problems. If you come to terms with this gap, you have a better chance of closing it. Here are a few suggestions if you, like many divorcing people, find that you run out of money before you run out of month:

- Cut expenses to reduce your debt load as much as possible.
- Do not use credit cards to finance your lifestyle.
- Take on a second job.
- Rent out part of your home to offset mortgage costs.
- Sell an asset and use the money to pay off debts.

Calculate how you would fare today if you had no option other than to pay or receive the alimony amount mandated by local custom or financial schedule. Then brainstorm about possible plans for resolving any budget problems you may face. Write your answers below.

Don't despair. In negotiating the overall settlement, you may get enough property or other assets to replace the loss of income caused by the divorce. You can explore this possibility in Chapter 19.

Alimony Recipients	
Monthly expenses:	$
Estimated alimony I will receive and my other net income (after taxes):	$
Difference:	$
How I will make up any shortfall:	

Alimony Payers	
Monthly net income:	$
Estimated alimony I will pay and my monthly expenses:	$
Difference:	$
How I will make up any shortfall:	

Factors to Consider in Making Alimony Nondeductible

A payer might not want to deduct alimony in any of the following situations:

- The payer can't use the alimony deduction—that is, if taxable income is low or income comes from nontaxable sources, such as tax-exempt bonds, Social Security benefits, nontaxable pension benefits, or nontaxable annuity payments.
- The payer expects other deductions (such as net operating loss carry-overs from a business) to equal or exceed gross income.
- The recipient is in a higher tax bracket than the payer.
- After the divorce, the recipient sold property to the payer and does not want the proceeds to be considered income.

Taxes

To understand taxes and alimony, remember this basic rule: Alimony is tax deductible to the payer and taxable as income to the recipient, but only if the alimony is based on an agreement in writing (such as your

divorce settlement papers) or is ordered by a court and very stringent rules are followed. If you do *not* want alimony to be taxed as income for the spouse receiving it, or to be a tax deduction for the spouse paying it, you can spell out that preference in your marital settlement or separation agreement, or make sure it's specified in a court order.

If your taxes are particularly complicated, you will need to consult with an accountant to fully understand alimony's tax consequences. Ideally, you and your spouse (or your respective attorneys or accountants) should figure the tax consequences for each of you and fashion an agreement in which each person can receive the greatest number of after-tax dollars. Many attorneys and financial planners have computer programs that can make these calculations. In addition, consider the following points:

Recipients. All too often, the person receiving alimony treats it as "free" money, as in "tax free." On April 15, however, you may find yourself scrambling to pay income taxes on it. Be sure you manage your money in such a way that you put aside enough to cover this bill by making extra estimated tax payments.

Payers. Keep careful records of the payments so you can deduct them properly. And obviously, you want to be sure the alimony award is in writing or is ordered by the court.

You can deduct alimony even if you do not itemize your deductions. You must provide your ex-spouse's Social Security number on your tax return (and you must use Form 1040—not Form 1040A or 1040EZ). If you fail to do so, you may lose your entire alimony deduction or be penalized for each failure to report per return.

To be deductible, alimony payments must meet the following conditions:

- There must be no obligation to pay after the death of the recipient.
- The payments must be made under a "divorce or separation instrument." These terms refer to a decree of divorce or separate maintenance, a written marital settlement or separation agreement, or a court order or temporary decree.
- The payer and recipient cannot file a joint income tax return.

- The payments must be in cash for a definite amount. Cash paid to cover the rent, mortgage, taxes, tuition, health insurance, or medical expenses of the recipient also qualify.
- Payments cannot be made to an ex-spouse living under the same roof after a final decree of divorce or legal separation.
- Payments cannot be disguised as child support.

The following payments qualify as alimony for tax purposes, assuming all of the preceding conditions are met:

- premium payments on term or whole life insurance if the recipient of alimony is also the owner of the policy
- payments of the recipient spouse's attorneys' fees
- payments of the recipient spouse's rent
- one-half of the payments for mortgage (principal and interest), taxes, utilities, and insurance on a residence co-owned by the payer and the recipient and held as tenants in common
- payments for mortgage (principal and interest), taxes, and insurance on a residence owned solely by the recipient
- payments made after a remarriage of either spouse as long as the divorce agreement specifies that payments will continue
- payments made under an annulment decree, provided your state treats alimony after an annulment the same as it treats alimony after divorce, and
- some allotments for those in the military. See a military adviser for more details on this possible deduction.

The following payments are not deductible as alimony for income tax purposes, even if all of the above conditions are met:

- the fair rental value of property owned by the payer and used by the recipient
- payments (including mortgage, taxes, and insurance premiums) to maintain property owned by the payer and used by the recipient. (These payments may qualify as a homeowner deduction for the payer, however.)

Alimony or Property Settlement Payments?

When your divorce settlement is being written, it is important to understand how alimony payments are structured. In some cases, the court may consider these payments as property being divided in the divorce. That matters to you because the payments may be subject to tax laws that differ from those involving a property transfer.

To qualify as alimony, the payments need not cease at remarriage, though they must terminate at death. In one case (*Prater v. Commissioner,* 95-1USTC § 50,271 (10th Cir. 1995)), payments made by one former spouse to the other out of proceeds from an oil and gas lease were considered alimony even though there was no mention of these payments ceasing upon remarriage. The court held that even though the payments did not cease upon remarriage, the provision that they cease upon the recipient's death meant that the payments should be characterized as alimony.

In another case (*Hoover v. Commissioner* (Tax Court Memo 1995 § 95,183)), the divorce decree awarded the wife the home, furnishings, and "alimony as division of equity." That was because the final divorce decree did not contain a provision that all payments would cease upon death or remarriage, although the husband was to provide medical insurance until his former wife died or remarried. Because the liability for these payments did not cease upon the death of the ex-wife, payments were deemed a property settlement and not alimony, in spite of what the divorce decree called them.

Payments are regarded as part of the property settlement when:

- the sum is fixed
- the payments are not related to the payer's income
- the payments are to continue regardless of the recipient's death or remarriage
- in exchange for the payments, the recipient gave up more property than the payer or gave up some other valuable right in the property settlement, and
- the payer has put up security to ensure payment.

The IRS Is Watching for "Frontloading"

Alimony is deductible by the payer in the year it's paid, as long as it hasn't been "frontloaded."

Frontloading does not describe a type of clothes dryer. It's a method some divorcing couples use to load up on tax-deductible alimony. Sometimes one spouse will pay a great deal of alimony the first year and then reduce it dramatically in later years. Or, a spouse will give the other some property, but call it alimony. In both cases, the IRS will want its share of money in taxes.

EXAMPLE 1: Biff will be paying Kit a total of $66,000 in alimony. He decides to pay her $44,000 the first year and $11,000 the second and third years. Biff is clearly frontloading his payments—paying most of the alimony in the first year to get a bigger tax break. The IRS will recompute Biff's taxes at the end of the third year and slap him with a bill to recapture the tax break he got the first year.

EXAMPLE 2: This time, Biff pays Kit $12,000 in cash and $30,000 for her new BMW in the first year. Even though only $12,000 was for the alimony itself, Biff deducts all $42,000 and calls all of it alimony. Biff plans to pay Kit $12,000 a year for the next two years to come up with the total $66,000. Again, Biff is clearly frontloading, and the IRS will hand him a bill at the end of the third year.

Biff could have avoided the frontloading charge by simply paying Kit $22,000 per year for each of the three years—without dramatically decreasing the amount or trying to pass off the transfer of the BMW as alimony. In general, the IRS rules against frontloading kick in if, during the first three years, the payments in one year exceed the payments in a succeeding year by more than $15,000.

Suppose, however, that Biff pays a large amount of alimony during the first year—more than $15,000—then loses his job and is able to have his payments greatly reduced. Even though he was not trying to trick the IRS, the frontload rules could nevertheless be triggered, and he could owe substantial taxes. Biff can protect himself with a tax indemnification agreement from Kit, which would make her responsible for such unintended tax consequences.

- payments made to a recipient as the beneficiary of the payer's estate, if the payer dies. The recipient, however, must report the payments as income.
- monthly allotments that represent an allowance for military quarters and are nontaxable. In situations in which a court orders alimony paid from these allotments, the military person cannot deduct the payments. But alimony deducted from regular pay is deductible and taxable.
- alimony paid under a court order retroactively reclassifying equalization payments (to even up the property division) as alimony, and
- payments made from a trust (alimony trust) set up to pay support, though the recipient spouse must include it as taxable income.

Payment Practicalities

If you will receive or pay alimony, you must be careful not to overlook the details involved in continuing a financial connection with your ex-spouse. Below are questions and suggestions to cover payment practicalities. Whether you deal with these items as your negotiations progress or wait until the final settlement, use the spaces provided to jot down your ideas, which you can refer to later when needed. Put a check mark in front of the question once it is resolved.

_____   *How will alimony payments be made?*

_____

_____

_____

_____

If you're the recipient, you basically have four choices:
- You can wait for the proverbial "check in the mail."
- You can set up a wage deduction through your ex's employer, then have that money automatically deposited (through direct deposit plans) into your bank account.

- You can set up a voluntary account deduction through your ex's bank, then have that money automatically deposited into your bank account.
- You can have your lawyer prepare a QDRO (qualified domestic relations order) specifying that payments come directly from your ex's retirement plan.

Most recipients wait for the check, but it is becoming more common for courts to order wage assignments or garnishments, especially when the payer is also paying child support. Whatever you arrange, be sure the details are clearly outlined in writing.

_____ *On what days of the month will alimony payments be made?*

_____

_____

_____

Payers usually find it most convenient to pay alimony shortly after receiving their salary checks or other income. If you are a recipient and payments won't arrive until after your mortgage, rent, or some other large monthly bill is due, you'll need to rearrange your budget. Try to manage your cash flow so there's some money left over from the previous month. If that won't work, contact the lender and ask that your payment date be pushed ahead to correspond with your cash flow—you don't need to tell the lender you're waiting for your alimony check unless the lender pushes the question. Most lenders will make arrangements to accommodate payments.

_____ *How long will alimony payments be made?*

_____

_____

_____

Records to Keep When You Pay or Receive Support

You must keep adequate records if you are paying or receiving alimony. This point cannot be overemphasized. Frequently, after a divorce, the spouses dispute, or the IRS challenges, the amounts that were actually paid or received. Without documentation, the payer may lose the alimony tax deduction and be ordered to pay back support if the other spouse makes a claim in court.

Payer

Here are suggestions of records to keep:

- a list showing each payment (date, check number, place where sent)
- original checks used for payments or bank copies of your cancelled checks showing both the front and the back (keep in a safe place, such as a safe-deposit box)—be sure to note on each check (before you send it) the month for which the support is being paid, and
- a receipt signed by the recipient, if you pay in cash.

Be sure to keep these records for at least three years from the date you file the tax return deducting the payments. You might want to keep them even longer, if, for example, your ex-spouse has more than three years to sue and claim that you owe back support.

Recipient

Make a list that shows each payment received. Include the following information:

- date payment was received
- amount received
- check number or other document identifying the number (for example, the number of the money order)
- account number on which any check was written
- name of bank on which check is drawn or money order issued
- a photocopy of the check or money order, and a copy of any signed receipt for cash payments.

It's becoming uncommon for alimony payments to be ordered indefinitely. And even in indefinite arrangements, alimony almost always ends if the recipient remarries or the payer retires. It usually ends if the payer dies, too. Sometimes, alimony lasts for a set number of years—often for the length of time it takes for the recipient to get training to reenter the workforce or, sometimes, to allow the recipient to stay home with young children.

Whatever your arrangement, pay close attention to the termination date. Suppose you are to receive alimony until September 30 of a certain year, based on the assumption that you will return to school, graduate in June of that year, and use the summer to find a job. What will you do if your education and career path don't work out as planned? You can usually go back to court and ask for an extension of the alimony, *but only if you make your request before the scheduled termination date.* If you wait until after the September 30 termination date, you're probably out of luck, because the court will have "reserved jurisdiction"—kept the power to make decisions—only until September 30 of that year.

When there is no termination date, alimony is generally modifiable upward or downward depending on a former spouse's continued ability to pay and the other spouse's need.

How will you handle future economic changes, such as inflation or salary increases?

_____

_____

_____

_____

If your alimony payment does not change, the amount it can actually buy year to year is sure to decrease. That's inflation. It's possible to take inflation into account in your alimony agreement with a COLA—cost of living adjustment—clause. Some couples tie COLA increases to the rise in the Consumer Price Index (CPI); others simply choose a specific dollar amount or percentage.

Similarly, couples can avoid extended bitterness over alimony arrangements if they face the question of salary increases in advance. Will the

recipient get a share of the payer's raise? Will the recipient's raise mean the alimony payment decreases? Be sure you know ahead of time how these questions will be handled.

_____ *What life or disability insurance policies will cover your ex-spouse?*

_____

_____

_____

If you are dependent on someone else for your income, what will you do if that person becomes disabled or dies? If you will be receiving alimony, you'll want to make sure your ex-spouse is covered by adequate insurance. You could purchase a life insurance policy on your ex. Or your ex could buy life and disability insurance—naming you as the beneficiary—to protect your income stream. To get a tax deduction on the insurance premiums, the payer can pay the premium money to the recipient by increasing the alimony and having the recipient make the insurance payments. The alimony recipient must be the absolute owner of the policy. (Remember, however, an increase in alimony may increase the recipient's income taxes.)

_____ *What will happen if the payer tries to modify the alimony payments or stops paying altogether?*

_____

_____

_____

If you are the payer and you cannot afford the amount of alimony you've been ordered to pay, you must file for modification of the order with the court. At your hearing, you must show a material change in circumstances since the last court order. Your ex-spouse's cohabitation with someone of the opposite sex may qualify.

Another common reason for changing alimony is your decreased income. Where you have voluntarily decreased your income, however—for example, you quit your job as a doctor to work in a pastry shop—the judge may consider your ability to earn, not your actual earnings.

If you are the recipient and your ex stops paying, you will need to go to court. Most likely, you will have to schedule a hearing in which your ex-spouse is ordered to come to court and explain why your payments aren't coming. You can also ask the court to grant you a judgment for the amount owed, or "in arrears." You can then use judgment collection methods, such as wage attachments and property liens, to try to collect the arrears.

Alimony and Bankruptcy

Alimony cannot be erased in bankruptcy. If your ex owes a substantial amount and tries to have it wiped out in bankruptcy, you will be protected so long as you follow bankruptcy laws and procedures. For more on bankruptcy and divorce, see Chapter 17.

Are There Alternatives to Paying or Receiving Monthly Alimony?

While financial schedules and legal guidelines help establish the amount of alimony, they don't require any particular payment arrangement. Although alimony is traditionally paid monthly, it is possible to pay it in a lump sum, periodically, or together with child support.

Lump Sum Alimony

A lump sum settlement of alimony upon divorce is sometimes called an alimony buyout. Before agreeing to a buyout, consider its pros and cons.

For the Recipient

There are four major advantages to accepting a lump sum settlement:

- You reduce the risk of depending on someone else—who could default, become disabled, or die—for income. Reducing the risk of

depending on someone who could default is important—alimony default rates exceed 50% in many places.

- You can use the lump sum for a down payment on a house or other purposes, such as paying bills.
- If you have no children, you can terminate any further contact with your ex-spouse.
- If the lump sum is characterized as nontaxable alimony, it is not taxable to you as the recipient.

There are also disadvantages to a buyout:

- You lose a steady flow of income.
- Once you accept a lump sum payment, you cannot go back to court to reinstate alimony.
- You may receive less than you would have in monthly payments. You and your soon-to-be ex-spouse may agree on a lower amount just to get rid of each other. Also, had you written a COLA (cost of living adjustment) clause into your settlement, your payments would have risen over time, a consideration usually not factored into a lump sum payment.
- You run the risk of losing some or all of the lump sum payment through bad investments or other unforeseen circumstances, such as a debilitating illness.

For the Payer

There are three major advantages to making a lump sum settlement:

- Because writing a monthly alimony check to a former spouse can be a grating experience that creates resentment and ill will after the divorce, a lump sum may be easier emotionally.
- If you have no children, you can end the marriage with a strong sense of finality.
- Because a buyout compresses all future payments into one present payment, the total is often discounted for taxes and is less than what would have been paid over a set number of years. For example, Mitch agrees to pay Sara $1,000 per month for the next five years, for a total of $60,000. Mitch figures inflation is around 4%, and in today's dollars, $60,000 is worth only $54,480. Mitch offers that amount less taxes in a present-day lump sum settlement.

There are also disadvantages to a buyout:

- You have to come up with a large sum of money.
- Under standard alimony plans, payments cease if the recipient dies, remarries, or, in some cases, cohabits with a new partner. You could pay a lump sum at divorce only to watch your ex remarry in a year.
- Because a buyout cannot be renegotiated, you could end up paying more in the lump sum than might have been required. For instance, if your ex obtains a high-paying job or your income drops a great deal, you might have been able to reduce the monthly alimony payments.
- Your payment might trigger the frontloading rules, or it could look like a property settlement payment if it's paid over less than three years and the annual amount exceeds $15,000. If you are not going to pay the entire buyout in one year, the IRS could interpret the arrangement as frontloading and assess taxes at the end of the third year.
- If the lump sum settlement is less than $15,000, it would be tax deductible. Otherwise, it may trigger the frontloading rules, and a portion of it would not be tax deductible.

Transferring Assets to the Recipient in Exchange for Alimony

Certain assets, such as the house, stocks, or a pension plan, could be given to your spouse in exchange for a release from the obligation to pay alimony. You may want to consider this option if you do not relish the idea of making monthly alimony payments or if you are strapped for cash and the payments would make your budget tighter.

You might also consider this option if you don't want to hold assets on which you may have to ultimately pay capital gains taxes, or if you want to sever all financial ties with your soon-to-be ex-spouse.

Exchanging property for alimony, however, means you would not get the tax deduction you get from making alimony payments. You still may want to consider this option when negotiating the final settlement.

Special Problems When Alimony and Child Support Are Combined Into One Payment of Family Support

If a family support payment is reduced, and the reduction is related to any of the events listed below, the amount by which the support is reduced will be treated like child support, not alimony. This can be a problem. The IRS may classify your former payments as child support, not alimony, and recompute your taxes to disallow the alimony deduction. The Internal Revenue Code does not recognize the concept of family support, classifying payments for tax purposes as either child support or alimony.

Events that can cause family support payments to be classified as child support include:

- a child reaching a specified age
- the death of a child
- a child leaving school
- a child leaving the household of the custodial parent, and
- a child becoming employed.

To avoid having the payments treated like child support (and losing the tax deductibility of the payments), don't base the reduction of family support on any contingency related to the child's age or needs. If you have more than one child, again, don't tie in the reduction to the date when each child reaches a certain age. Finally, don't call your support family support if your children are near the age of majority (usually 18 or 21).

Family Support

Sometimes, alimony and child support are combined into one monthly payment called family support. If the payments meet all the requirements of alimony, they may be fully taxable to the recipient and tax deductible to the payer.

Recipients of family support enjoy one advantage if a payer defaults. In that situation, the recipient has numerous child support collection techniques available even though part of the amount being collected is alimony. If the payer defaults on alimony payments, the only collection

method in most states involves taking the payer to court. Be warned, however, that a growing number of courts are disallowing family support, not wanting to burden the child support collection bureaus with collecting what is really alimony.

What Is Your Bottom-Line Decision on Paying or Receiving Alimony?

You may have little choice about the amount of alimony payments, but you must nevertheless make decisions about alimony. Will you negotiate for a lump sum buyout? Will you use alimony as leverage in negotiating for other assets? Use this space to describe your bottom-line decision on alimony payments so you can easily refer to this information in Chapter 19 when you bring all your decisions to the negotiating table.

_____

_____

_____

Questions to Ask an Attorney

1. If you live in a community property state: Will my payment of alimony before our divorce is final be deductible, or is it considered an allocation to my spouse of community income? If it isn't deductible, can a separation agreement or final divorce decree be written so as to make it deductible?
2. What conduct might jeopardize my right to custody of my children? Am I permitted to move away if I have custody? If I don't have custody and my ex-spouse wants to move with the children, can I fight for custody or block the move?
3. What formula is used to calculate child support in my state?
4. Does child support in my state cover all child-related expenses, including medical, clothing, allowances, and child care? If not, what expenses does child support cover, and who pays for the other costs?

5. What will I do if my spouse falls behind in child support or alimony?
6. How does the court enforce child support and alimony orders?
7. Is it wise to arrange for family support payments instead of child support?
8. How will alimony or child support be affected if a parent receives government welfare benefits?
9. If my ex-spouse remarries or takes a new partner, does that affect the obligation to pay or receive child support or alimony?

Sample Qualified Medical Child Support Court Order

A. _____[Name of party]_____ is ordered to maintain in full force and effect all health insurance coverage, including medical, dental, hospital, and health maintenance, available at reasonable cost through employment and/or other group health insurance plan, and to pay premiums on that coverage.

_____[Name of party]_____ is also ordered to maintain the minor child(ren) _____[specify]_____as alternate recipient(s), entitled to enroll in the group insurance plan, until such minor child(ren) attain majority, are emancipated, die, or marry, or until further order of this Court.

B. In the event that such health insurance coverage is not available at the entry of this judgment but subsequently becomes available at no cost or reasonable cost to _____[name of party]_____, ___[name of party]_____ is ordered to notify the Court and opposing party and to apply for that coverage.

C. The group health insurance plan that is the subject of this order is available through _____[employer or other group]_____. Coverage is provided by _____[name of insurance company]_____and coverage is for _____[description of the type of coverage provided by the plan]_____.

D. The employer of the plan participant (if the group health insurance is available through employment) is ordered to enroll the child(ren) as alternate recipients in the group health plan.

E. The insurance company is ordered to provide coverage to the child(ren) of the plan participant as alternate recipients even if the child(ren) were born out of wedlock, do not live with plan participant, are not claimed as dependents on the plan participant's federal income tax return, or live outside of the plan's service area.

F. The insurance company is ordered to reimburse the custodial parent directly for covered medical expenses paid by the custodial parent on behalf of the alternate recipients. The insurance company shall not pay such reimbursement to the plan participant.

G. The name and last known mailing address of the party ordered to maintain health insurance (the "plan participant") are:

_____

_____

H. The name and mailing address of the alternate recipient(s) are:

_____

_____

Negotiating and Finalizing the Best Possible Settlement

Have You Done Your Financial Homework? 433

Tallying Your Marital Balance Sheet 433

How Are the Offers and Counteroffers Made? 438

How Do You Finalize the Settlement? 441

Divorce Ceremonies 443

Until this stage of the divorce, you have focused on your personal, individual decisions—that is, the things you "want" in your divorce. But how will you get what you want? Through negotiation.

Negotiating a divorce settlement is a process of offers and counteroffers. You and your spouse may be able to compromise and reach a settlement easily, or your demands and those of your spouse may run headlong into each other. Some couples state their terms and reach a settlement in hours, while others go through years of bitter fighting.

Your settlement negotiations will probably bear an uncomfortable resemblance to the dynamics of your marriage. If your spouse has been rigid and demanding throughout your relationship, do not expect that you'll suddenly encounter flexibility and reasonableness in a settlement conference. Similarly, an uncommunicative mate will probably give you the silent treatment as you attempt to complete the settlement. Anticipating your spouse's most likely behavior can make it easier for you to accept and move through the negotiations.

Your decisions—and actions—become crucial in this phase, because you are playing for keeps. Once the settlement is final, it is costly, time-consuming, and almost impossible to modify it. Better to make changes to the settlement before you complete it, rather than try to alter it later.

In this chapter, you will address three basic issues regarding negotiating the settlement:

- Have you done your financial homework?
- How are the offers and counteroffers made?
- How do you finalize the settlement?

Remember that negotiation is the middle step in a longer process. You must communicate your needs and preferences clearly before negotiation begins, and you must be willing to compromise. Your negotiation should be guided by the acceptance that no settlement is perfect, for you (or for your spouse). Your goal is a settlement or judgment you can live with.

Have You Done Your Financial Homework?

Imagine walking into a room and sitting across a table from your spouse and your spouse's attorney. The attorney smiles while going through a long list of items that your spouse is prepared to "give" you if you will simply forfeit any claim to the items that are on your spouse's much shorter list.

The attorney points out that the dollar amount on the bottom line of both lists is equal. Surely you should agree to such a fair and equitable offer.

But should you?

You can't know whether or not the offer is fair unless you've done your homework. If you've worked through the previous chapters in this book, you should be able to accept or reject your spouse's offer relatively quickly. If you haven't read the prior chapters—analyzing your assets and debts, evaluating tax consequences, and understanding alimony and child support—you're apt to accept an offer that is a bad financial deal for you, regardless of how "equal" the dollar amounts on the bottom line appear to be.

You must know where you stand financially before you negotiate the settlement. We cannot stress this point enough. If you have not already done so, read the material relevant to your divorce from Chapters 12 (property and expenses), 13 (house), 14 (retirement plans), 15 (investments), 16 (employee benefit plans), 17 (debts), and 18 (alimony and child support). Even if your spouse has made an offer—or is preparing to counter an offer you've made—take no further steps until you have analyzed the financial consequences of any proposed settlement.

Tallying Your Marital Balance Sheet

In Chapter 12, you filled out the detailed "Assets and Liabilities Worksheet." Now, as you reach the crucial point of your negotiations, that attention to detail will pay off.

You can now devise your "Marital Balance Sheet," a key tool in arriving at a settlement. With the information you've assembled,

achieving a fair division of the assets and debts—the goal of every divorce proceeding—should be infinitely easier.

Refer back to the "Assets and Liabilities Worksheet" in Chapter 12. Make note of what property is jointly owned or owned by the spouses individually. Note also the equity or legal value in the last column. Then enter that information in the appropriate columns below. This will be a bit of a chore, but it's important. This is what you've been working toward.

Once you've entered that information, take a moment to study the overall picture. Think about which assets you want to keep and which ones you want to be rid of or are willing to give up. Bear in mind that what each person gets will also be an important element in deciding what, if any, amount of child support and alimony is ordered.

Begin by preliminarily dividing the property as you would ideally like to see it divided. First, consider which property each spouse would prefer to have and "deserves" to have. For example, if the equity or legal value of your house is $150,000, and you and your spouse each contributed equal amounts of separate property for the down payment, you would logically have a 50-50 share of the equity, or $75,000 apiece before taxes. But if you contributed unequal amounts for the down payment, that figure may need to be adjusted. Similarly, try to determine the stake each of you has in other assets based on who acquired them and with what funds.

The assets and liabilities listed in Chapter 12 are very detailed and specific. Here you're going to be grouping them—investments, collections, real estate, and so on—to get an overall sense of how they could be divided. Inequities will naturally develop. Allocate enough of the value of joint property to the spouse with the shortfall to roughly equalize the respective values. Your negotiations will determine who actually gets what. This exercise will help prepare you for those negotiations.

Look at specific items of jointly owned property to identify those that can be transferred from joint names into the sole name of one party to equalize the values. Consider trading property interests of similar value—say, the wife's interest in the husband's pension for the husband's interest in the family home.

When you get to the bottom of each portion of the "Marital Balance Sheet," check how close you've come to having equal allocations. Make note of any equalizing payment that would be required to bring the allocations into balance.

Dividing Miscellaneous Property

The bulk of your divorce may focus on splitting the miscellaneous items, such as furniture, books, electronic equipment, appliances, kitchenware, tools, and the like. The easiest way to divide these items is to make a list of *everything you jointly own*—this includes wedding gifts and joint inheritances—and write your name or your spouse's name next to each item. Give the list to your spouse. Ask your spouse to note those items on which he or she disagrees with your assignment of ownership. Then sit down together and review the items in dispute. If you can't resolve your differences, here are three suggestions:

- Flip a coin. The winner gets first choice of the disputed items. Alternate selecting until the list is exhausted.
- Put the disputed items on your settlement chart and divide them as you negotiate the settlement.
- Find a mediator to help you divide what's still on the list.

This property may not be worth the time you spend fighting over it. Do not let the issue of dividing the silverware and the CD collection interfere with your negotiations over more financially significant assets, such as the house or retirement plans.

Marital Balance Sheet

Division of Jointly Owned Property

Equity or Legal Value (from Chapter 12)	Allocation to Husband	Allocation to Wife
Assets:		
Marital home		
Rental house		
Real estate		
Investments		
Stock account		
Savings		
Checking		
Money market fund		
Mutual funds		
CDs		
Automobile 1		
Automobile 2		
Furnishings		
Art works		
Collections		
Debts:		
Credit cards		
Credit line		
Bank loan		
Family loan		
Other debt		
Jointly Owned Property Divided		
Equalizing payment, if any		
Husband's Property Divided **Equity or Legal Value (from Chapter 12)**	Allocation to Husband	Allocation to Wife
Husband's property		
Equalizing payment, if any		
Wife's Property Divided **Equity or Legal Value (from Chapter 12)**	Allocation to Husband	Allocation to Wife
Wife's property		
Equalizing payment, if any		

Summary of Property Division

Equity	To Husband	To Wife
Joint property		
Husband's property		
Wife's property		
Total division		
Overall equalizing payment, if any		

When Not to Keep the House

After Janet divorced, she found herself paralyzed, though not physically. Because of decisions she made during her divorce, she could not afford to move from her home.

Janet owed $60,000 to her uncle—he lent her the money so that she could buy out her husband. If Janet were to sell the house, she'd first have to pay her uncle. Then she'd have to pay taxes on any profit from the sale that exceeded $250,000. Janet wants a small condo worth less than her house. Between payments to her uncle, taxes she would owe on the profit, and the lower market price of her house due to slumping real estate values in her area, she would have barely enough left for a down payment on a new place.

Janet's goal is to keep her monthly house payment low. In her current home, she pays only $950 per month. After Janet repays her uncle and puts money aside for the taxes, she will have a little money left over from the sale of the house that she can use as a down payment on the condo. She'll still have to finance the deal. If she took a 30-year loan at 7.5% interest, her monthly payment would be $1,250—$300 more per month for housing than she pays now. And she simply cannot afford it on her current salary.

"It's ironic," Janet now muses. "I fought so hard to keep this house, and now I have to keep it whether I want to or not. I'm so busy at work that I hardly have time to mow the lawn, and I've had to rent out two rooms just to keep everything going. If I had it to do over again, I'd have sold the house during the divorce and started fresh."

How Are the Offers and Counteroffers Made?

Once you have a good grasp of your "Marital Balance Sheet," you can begin the negotiations. In many settlements, the spouses will agree on several issues. To resolve the remaining issues, the couple—perhaps with the help of a mediator or their attorneys—meet or correspond until all issues are settled. Once the details are settled, they are recorded in a settlement agreement. The settlement agreement is then taken to the court for a judge's approval. Unless an agreement contains a clause leaving the children unsupported, a judge will approve the agreement as a formality.

Don't Be Discouraged

Even after doing your financial homework and the hard work of preparing for settlement negotiations, you may still go through numerous settlement conferences in attempting to hammer out an agreement. Some cooling-off time may be needed between offers and counteroffers because of charged emotions. Take the time you need and don't give in just because you never want to see your ex again.

Your settlement may not follow the scenario described above exactly. Here are several other possibilities:

Scenario 1:

You and your spouse agree to terms ➤ *file your own papers with the court* ➤ *a final decree of divorce is issued* ➤ *the marriage ends.*

This kind of simple divorce usually takes place after a short marriage with no children or major assets to divide. If you fall into this situation, check to see whether there's a good do-it-yourself divorce book written for your state. (See the resources below.) You can also use the services of an independent paralegal to help you type up your papers. Check your phone book under "attorney support" or "typing services."

Scenario 2:

You and your spouse agree to terms on some issues ➤ *meet with a mediator to resolve remaining issues* ➤ *file papers with the court on your own or using an attorney, attorney-mediator, or typing service* ➤ *a final decree of divorce is issued* ➤ *the marriage ends.*

Mediators are discussed in Chapter 7. Remember that a mediator is not an advocate, but a person who helps you and your spouse arrive at a settlement you can live with. Once you settle the issues, a mediator will probably not help you file your court papers unless the mediator is also an attorney. You (or your spouse) can use an attorney for this, but make sure you find a lawyer who will respect the mediation process and your decisions. Otherwise, the attorney may want to get more involved—find out all the issues of your case, figure out why you settled it as you did, and possibly even counsel you to resolve it differently. This can mean you end up starting over.

Scenario 3:

You and your spouse—with your attorneys—agree on certain issues ➤ *through your attorneys' meeting or corresponding, you resolve remaining issues* ➤ *your attorneys file papers with the court* ➤ *a final decree of divorce is issued* ➤ *the marriage ends.*

If you and your spouse cannot settle your case, you'll probably turn to the services of attorneys. The cost will rise and the issues may seem more complicated, but your interests may be better protected. Nevertheless, it is imperative that you retain absolute control over your case. Having come this far in this book, you understand the financial realities of keeping certain assets, assuming (or having your spouse assume) debts, and paying or receiving alimony or child support. Be clear with your attorney as to what is an acceptable and unacceptable settlement. Give your lawyer a copy of the "Marital Balance Sheet" you completed in this chapter.

Understand What You're Getting

Judy and her husband went through an arduous negotiation process. When they couldn't agree on how to divide their property, they scheduled a trial. A few minutes before they were to go before the judge, they settled the issues in the courthouse cafeteria. Their attorneys initialed papers stipulating the terms of their agreement.

When Judy received her copy of the stipulations, she noticed that her husband's attorney had added language describing her husband's equalization payment—money owed for receiving a greater share of marital property—as a distribution of retirement benefits subject to a Qualified Domestic Relations Order (QDRO). Instead of receiving tax-free payments, Judy would be getting the payment in the form of a fully taxable distribution from the retirement plan. Judy could roll the money into an IRA to defer taxes, but she did not want a taxable asset when she and her husband had already agreed that she would receive the nontaxable equalization payment.

Fortunately, Judy reviewed the settlement agreement carefully and asked her attorney about it. Judy then demanded—and got—a correction. The equalization payment was separated from the retirement benefits, as was originally intended. Her alertness saved her almost $40,000 in taxes.

Scenario 4:

You and your spouse—with your attorneys—agree on certain issues ➤ through your attorneys' meeting or corresponding, you resolve some, but not all, remaining issues ➤ your attorneys prepare for trial ➤ a trial takes place, the judge makes rulings ➤ a final decree of divorce is issued ➤ the marriage ends.

Throughout this book, we have warned you against letting your fate be decided by a judge and letting your marital assets be eaten up in lawyers' fees. If, however, all else fails and you must go to trial, then be prepared. Don't be afraid to ask your lawyer what will happen—and how much the lawyer's bill will be. And don't forget that you'll probably have to pay experts—actuaries, appraisers, financial planners, and the like—if you're disputing asset values. If you can't agree on custody,

you'll probably need to bring counselors, psychologists, and others into court to testify. Be prepared for these bills.

If You Must Go to Trial

In some divorces, all attempts at negotiation break down and couples end up going to court. Before a trial, you will want to:

- Ask your attorney about the pros and cons of your position and what other courts have decided in cases like yours.
- Ask which judge you will most likely get and what biases that judge tends to hold.
- Reevaluate the financial and tax implications of your options.
- Reconsider the proposals made by your spouse. Think through their merits and drawbacks.
- Think about the benefits of settling without going to court, such as saving vast legal fees, court costs, and other expenses. What if you pay these expenses and lose anyway?

RESOURCE

If you live in California or Texas, check out *How to Do Your Own Divorce in California* **or** *How to Do Your Own Divorce in Texas*, **both by Ed Sherman (Nolo Occidental).** And wherever you live, you can probably find useful information in *Nolo's Essential Guide to Divorce*, by Emily Doskow (Nolo).

How Do You Finalize the Settlement?

As mentioned earlier, once the issues are resolved and reduced to a settlement agreement, a judge must approve the agreement. Unless it leaves the children unsupported, a judge will sign the agreement.

Usually, when the agreement is taken to court, so are the papers requesting an end of the marriage. Before that, only the complaint or petition requesting that the marriage end—and an answer or response from the other spouse—have been filed. Once the couple resolves the

property, debt, custody, alimony, and child support issues, they also usually end their marriage. In some cases, however, you may actually be divorced before you conclude your settlement. In these cases, the divorce process is "bifurcated"—divided in two parts—so that the part ending the marriage is resolved first, and then the issues of property division, debt allocation, custody, alimony, and child support are settled.

Tips for Negotiating Your Settlement

Here are several tips to bear in mind during your negotiations:

- You may not get what you deserve, but you can get what you negotiate.
- Know the least you are willing to get, the most you are willing to give, and the bottom line you are willing to agree on.
- Answer questions clearly and concisely—and without blame.
- Put yourself in your spouse's position and ask what you'd do if you were your spouse.
- Don't give in to pressure for the sake of an immediate response.
- Release tension during negotiation sessions; take deep breaths and let them out slowly.
- When difficulties or conflicts arise, use phrases that can help move things forward:

  "I appreciate what we've done so far..."

  "I'd like to settle this on the basis of principle, not power..."

  "Help me understand how you reached that conclusion..."

  "I'm trying to see the point you are making; would you mind clarifying it for me?"
- Deliver minor concessions early to set a positive tone. Focus on making immediate compromises that are easy for you and that your spouse might find valuable.
- Clearly label all major concessions, saying something such as "This is going to cost me dearly, but I'm giving this to you because...." This may encourage reciprocity.
- Prepare as if this negotiation is very important. Because it *is*.

Bifurcation is usually used when one party wants to remarry and doesn't want to wait the many months or years it may take to settle the unresolved issues in the divorce. Our experience is that couples who bifurcate their divorce have little incentive to untangle their finances. Financial discussions tend to drag on, and making a clean break and planning for your own financial future may become more difficult.

Do not expect a grand parade, or even a telegram announcing the end of your marriage. All you're apt to receive is a notice from the court. After all the drama of divorce, the end can be nothing more than an anticlimactic notice in the mail. You might even need to call your attorney's office or the courthouse to make sure the marriage is really over.

You can, of course, throw yourself a party to mark the occasion—or sleep in and stay under the covers, depending on how you feel about being divorced.

Divorce Ceremonies

Some couples craft beautiful and touching divorce ceremonies to end their marriages and begin their lives anew. An important part of any divorce ritual is to have the leader or clergyperson facilitate the ex-spouses in expressing gratitude for the best parts of their marriage. They recognize aloud what was good about the marriage and about each other. They may speak with pride and joy. Children in attendance can tell their parents how they feel about them and what their expectations are for the future. Friends can express their support and caring for both spouses.

Though it may not be for everyone, a divorce ceremony can be a powerful experience because it offers an opportunity to provide closure on the past and look forward to the future. ●

After the Divorce:
How Do I Get From "We" to "Me"?

How Do I Finish the Business of Divorce?...446

Can I—Or My Ex-Spouse—Change the Settlement?............................. 451

What Do I Want to Do With My Life? 453

    If You Have Money to Invest...453

    If Money Is Tight 453

    Postdivorce Financial Planning Tips..458

    Setting Goals After Divorce 461

If You Find a New Love, Protect Your Old Assets...and Your Alimony 461

    Cohabitation Can Be Costly..461

    Prenuptial Agreements...463

How Can I Move Beyond the Divorce? 465

Congratulations…you've made it. Your divorce is over, and it's probably a tremendous relief. As far as your financial life is concerned, however, the watchword after divorce is caution—not celebration.

Just as it was important to avoid spending sprees during the divorce, it's wise to keep your expenses down once the divorce is final. You need time to adjust to your new life. The last thing you want to do is run up big bills that you can't pay.

Similarly, don't let your guard down while you wrap up the details and get ready for life as a single person—or single parent. First, you must be sure to follow through—or see that your spouse follows through—on the settlement you've worked so hard to secure. Be on the lookout, too, for people who try to pressure you into investing your money one way or another. Sit tight. You'll have plenty of time to decide what to do once you've made sure there are no loose ends from your divorce that could trip you up in the future.

How Do I Finish the Business of Divorce?

Forgetting to follow through on details after divorce is one of the easiest things in the world to do. It's also one of the costliest.

During the divorce, you were under tremendous stress. Now, you may be more exhausted than you realize, and that can lead you to overlook important tasks. Even if you're feeling exhilarated instead of exhausted, that state of excitement can lead to overconfidence and a lack of focus on the business at hand.

You could literally lose the property you fought to keep because of simple errors in divorce paperwork. One man lost his vacation home to his ex-wife when she sold it before the deed transferring ownership to him was recorded. Another woman had to pay off debts her ex-husband incurred because she forgot to close a joint account.

Even after the judgment is entered, you must take steps to implement the provisions of the divorce decree and make sure your newly single status is reflected in your financial and legal dealings. Not all items on

the following checklist will apply to everyone, but read it all carefully and attend to all the ones that do apply to you.

Checklist: Required Postdivorce Actions

Name change

_____ See that your name is removed from any debts or loans that are no longer your responsibility.

_____ Contact the Social Security Administration by filing Form SS-5 at a local office or downloading the form from the agency's website, www.ssa.gov. It usually takes two weeks to verify the change. (A mismatch between a name on a tax return and a Social Security number could increase a tax bill or reduce the size of a refund.)

_____ Tell your employer, including the human resources department and retirement plan administrator or custodian.

_____ Notify all your creditors.

_____ Apply for credit in your new name.

_____ Change the name on your bank accounts and checks, or open new accounts in your new name.

_____ Apply for a name change on your passport; the application is available online at www.travel.state.gov/passport/correcting/changename/changename_851.html.

_____ Notify medical care providers, insurers, the registrar of voters, utilities, the post office, and the state department of motor vehicles.

A Rose by Any Other Name...

You may feel by now that you've gone through enough changes in your divorce. But there is one item which you may not mind changing: your name. Changing your name as part of the divorce process can be more efficient—and less expensive—than waiting to do it later, when you will probably have to pay court fees. You can use your birth name, use a name from a prior marriage, or, in some cases, simply make up a new name altogether. If you're in California, you can consult _How to Change Your Name in California_, by Lisa Sedano and Emily Doskow (Nolo).

Real estate and cars

Divide the property as set forth in the divorce decree. Specifically:

_____ Execute a quitclaim deed to transfer title to real estate, as required by the decree. New deeds must be recorded at the county recorder's office. Contact your tax assessor's office for forms to sign to keep property taxes from increasing because of the transfer.

_____ Fill out new registration forms changing title to cars, boats, motor homes, or other vehicles and file with the state department of motor vehicles. (Failure to do so could result in trouble or expense for you if the vehicle no longer belonging to you is later involved in a mishap or receives a citation.)

_____ Notify your auto insurer of any changes in automobile drivers, ownership, and addresses. (Also, if needed, apply for a new driver's license with your new name and address.)

_____ Remove your former spouse's name from your mortgage or lease.

Insurance

Review your health, dental, vision, life, and disability policies. Be certain you and your children have adequate coverage. In addition:

_____ Make sure the beneficiaries named in your policies are who you want them to be and are consistent with the divorce decree.

_____ Execute all necessary COBRA documents to ensure continued health insurance coverage, and make sure that your ex-spouse has done the same.

Estate planning

During your marriage you may have drafted a will, created a trust, or taken other steps to determine who will receive your property after you die. Review those and change them to reflect your new status. Also, create a new power of attorney for health care, naming the person you want to make medical decisions if you become incapacitated.

Debts

Verify that all joint accounts are closed and that you have paid or have a plan to pay all debts you agreed to cover. At the same time, contact the creditors your ex-spouse agreed to pay and make sure they're receiving

payments. (If they aren't, you may wish to minimize damage to your own credit by paying those bills and then seek reimbursement from your ex-spouse.)

Deposit accounts

Check your deposit accounts—checking, savings, money markets, certificates of deposit, and Treasury bills—to make sure the names on, and the amounts in, the accounts are consistent with the decisions reached in your settlement.

Changing Joint Deposit Accounts

To change joint deposit accounts, write to the financial institution where you have your account. Here's some sample language to use:

"Please be advised that as of _____[date]_____ , account number _____, in the names of _____, is to be closed. All assets from that account are to be transferred into a new account in the name of _____."

You and your ex-spouse must sign this form. Some banks and brokerage firms may ask for a copy of your divorce decree, but do not send it unless you're specifically requested to do so.

Investment accounts

Check investment accounts to see that ownership of stocks, bonds, and mutual funds is properly listed.

Children

If you share custody of children, you should have access to their school reports, medical records, and other information. Also, make sure you understand your custody and child support agreements and that you fulfill your obligations. Specifically:

_____ Pay all support when due. Keep detailed records of each payment made or received.

_____ Keep your scheduled visitation times with your children.

—— If you change jobs, notify your former spouse and your new employer. (Do what's right, even if a judgment or order doesn't require it.)

Tips for Completing the Business of Divorce

Here are a few tips for wrapping up the loose ends:

- **Review settlement details.** A few days after your divorce is final—after you've recovered from the euphoria or the exhaustion—meet with your lawyer or financial advisers to go over the details of your settlement. Make lists of the items that you, your attorney, your ex-spouse, or others are responsible for completing. Don't leave these items to memory. Make sure the Qualified Domestic Relations Order (QDRO) is entered and implemented, if required by the divorce decree.

- **Keep records of all payments you make or receive—for alimony, child support, or property exchange.** Make copies of all checks you send or receive. Also, use a calendar or log book to show payment amounts and the date payments are made or received. You need to record these details in case you ever go to court to raise or lower support payments or to collect unpaid support.

- **Make copies of your final judgment and settlement agreement and record appropriate documents.** Get at least one certified copy of your final judgment and settlement agreement. Make copies as needed. In addition, ask your attorney: "What documents do I need to record if my spouse owes me child or spousal support or money to equalize our property division?" Then record those documents with the county. Put the originals in a safe place, such as a safe-deposit box.

- **Start a postdivorce file for essential papers.** Keep your payment records and the duplicate of your final judgment and the settlement agreement in the file. If you experience problems, or need to refer to divorce papers for investment or tax purposes, everything will be located in an easy-to-find spot. On an emotional level, hunting for divorce documents can be frustrating. Avoid triggering old angers and reduce stress by keeping these documents within easy reach.

Can I—Or My Ex-Spouse—Change the Settlement?

Your settlement agreement contains several clauses. The main ones probably deal with property division, debt allocation, alimony, child support, and custody. You may be wondering whether you can change any of those provisions, or whether your spouse might try the same. The two of you can *voluntarily* change the terms of your agreement. Any change should be in writing, signed by both parties, dated, and preferably notarized. (Changes agreed to verbally may be unenforceable if you later have different memories about what you decided. In such cases, the courts generally rely on whatever is in writing.) But if your ex-spouse won't cooperate with a change you want, you will have to go to court. Ask your attorney for the specific rulings in your state and what's likely in your case, but below are some general guides.

Property division. It's highly unusual for a court to approve a request by one ex-spouse to change the terms of the property division after the divorce. Such a request might be approved if the ex-spouse who is asking to change the settlement agreement can show that the other ex-spouse hid assets—such as accounts receivable that should have been used to value a business. The ex-spouse will have to file a motion, schedule a court hearing, and then provide convincing evidence at the hearing of the other party's wrongdoing.

Debt allocation. If your ex-spouse doesn't pay debts according to your agreement, you will no doubt need the court's assistance. You can file a motion and ask the court to order your ex to pay—and to ensure that you're paid, ask the court to have your ex turn money over to the court. To avoid damaging your own credit, however, you should go ahead and pay the bills if you can, and ask the court for an order requiring reimbursement from your ex. The court can order the reimbursement by changing any equalization payment or property distribution that hasn't yet taken place. The court might also order a wage garnishment from your ex's paycheck, or require that other funds belonging to your ex be turned over to you.

Alimony. If you knowingly gave up alimony, you usually cannot go back to court and try to get some. Once it's waived, it's gone forever. If you are receiving or paying alimony, pay careful attention to the

termination date. If you are the recipient and need money beyond that date, you can usually go back to court and ask for an extension of the alimony, *but you usually must make your request before the scheduled termination date.*

If you are the payer and you can't afford to pay the amount you've been ordered to pay in the settlement agreement, you must file a motion for modification with the court. See Chapter 18.

Child support. Child support is always subject to modification by a court because the court retains power over child support until the children reach age 18 or 21 (in some states), or until whatever age your agreement specifies support must be paid. If you are the payer and can't afford the payments, or if you are the recipient and you can't live on the current payment, you must file papers with the court, schedule a hearing, and show the judge that you cannot afford to pay—or to live on—the current support. See Chapter 18.

Custody or visitation. Like child support, custody and visitation are always subject to the court's modification. The parent wanting to change custody or visitation must file papers with the court, schedule a hearing, and show the judge a significant "change of circumstance" since the original order. Changing custody is usually a difficult and emotionally draining process. The court uses the standard of the "best interests of the child" and does not like to change the status quo. Unless you can show that the current custody arrangement is detrimental to your children— for example, their grades are suffering or they are depressed—you are not likely to win your court hearing.

Get It in Writing

When you want to change an existing court order—whether it pertains to alimony, child support, custody, visitation, or any other provision—you can ask your ex-spouse to agree to the changes informally. If you do reach an agreement, it's important that you put the change in writing, have it signed by both parties, and have the signatures notarized. Failure to do so could result in a later finding that the modification wasn't valid.

What Do I Want to Do With My Life?

When your divorce is over, it's more crucial than ever to reassess where you're going financially. Rarely does anyone's postdivorce life resemble predivorce expectations. You need to adapt to changes you didn't expect.

If you have used goal-setting tools throughout your divorce, you may simply need to make minor adjustments to your basic plans. If, however, your financial position after divorce is drastically different from what you had anticipated, you may need to reevaluate your objectives, scale down your spending, or take a crash course on investing wisely. Whatever your situation, don't jump into risky ventures with your share of the settlement.

If You Have Money to Invest

Unless you're already a successful investor, it's best to put your money in a safe parking place until you get established—personally and professionally—in your new life. It's ideal to have three to six months' worth of cost-of-living expenses in a money market fund or a certificate of deposit (CD) as an emergency fund before you do any investing.

When you feel you have your postdivorce life under control and your future plans are relatively clear, you can move some of your assets into investments where your money can work harder for you—that is, earn more return than from a money market or CD.

You may start slowly with "dollar cost averaging." With this method, you put the same amount of money into an investment on a regular basis over a long period of time. For example, you might put $200 a month into a mutual fund. Your money will buy more shares when the market is down and fewer shares when the market is high. Over time, your average cost per share will be less than your average price per share, and you'll come out ahead.

If Money Is Tight

No matter how well you've planned, you may find yourself with limited resources following your divorce. Cutting costs, then, will be your major

priority. Some recently divorced people find it easier to make ends meet by pooling resources. You may be able to connect with others in similar circumstances by contacting singles clubs, Parents Without Partners, or groups sponsored by religious or community organizations. Setting up a baby-sitting exchange or bartering for other services with members of the group can help everyone lower expenses.

Also consider buying groceries in bulk or purchasing necessities from consumer cooperatives or warehouse distributors. To lower your housing costs, you might rent out a room or part of your home.

RESOURCE

To learn more about sharing resources of all kinds, see *The Sharing Solution: How to Save Money, Simplify Your Life, and Build Community,* **by Janelle Orsi and Emily Doskow (Nolo).**

Avoid taking cash advances or running up the balance on credit cards if you can help it. Most credit cards carry high interest charges—and that interest is not tax deductible.

You can also try to raise cash.

Borrow From Family or Friends

Do not feel discouraged if you have to borrow money from family or friends. Many people find they need short-term loans to get them through this transition period. In times of financial crises, some people are lucky enough to have friends or relatives who can and will help out. Be sure to give them a promissory note for the loan specifying the amount, interest rate, and payment schedule.

Sell a Major Asset or Many Minor Items

One of the best ways you can raise cash and keep associated costs to a minimum is to sell a major asset—or hold a garage or yard sale. Few people realize how much money may be lying around their house. Selling a hardly used car you were saving until your oldest child turns 16 or a computer system you no longer use could help you through a rough period.

RESOURCE

Find the forms you need. For promissory notes, bills of sale, and even child care agreements, look to *101 Law Forms for Personal Use*, by Ralph Warner and Robin Leonard (Nolo).

Cash in a Tax-Deferred Account

If you have an IRA or other tax-deferred account into which you've deposited money, consider cashing it in. You could have to pay the IRS a penalty—10% of the money you withdraw unless you qualify for the exceptions to this penalty. (See Chapter 14.) In addition, you'll owe income taxes on the money you take out. But paying these penalties to the IRS may be better than struggling to get by.

Get a Home Equity Loan

Many banks, savings and loans, credit unions, and other lenders offer home equity loans, also called second mortgages. Lenders usually lend between 50% and 80% of the market value of a house, less what is still owed on it. Home equity loans have advantages and disadvantages. Be sure you understand all the terms before you sign up for one.

Advantages

- You can obtain a closed-end loan—you borrow a fixed amount of money and repay it in equal monthly installments for a set period. Or you can obtain a line of credit—you borrow as you need the money, drawing against the amount granted when you opened the account, and repay according to your agreement.
- The interest you pay is fully deductible on your income tax return.
- Federal law requires that interest rates on adjustable rate home equity loans be capped—meaning that the rate can increase only a set amount each year as well as over the life of the loan.

Disadvantages

- You are obligating yourself to make another monthly or periodic payment. If you are unable to pay, you may have to sell your house or, even worse, face the possibility of the lender's foreclosing.

Before you take out a home equity loan, be sure you can make the monthly payment.

- While interest is deductible and capped for adjustable rate loans, it can still be high.
- You may have to pay an assortment of up-front fees for such costs as an appraisal, credit report, title insurance, and points, that can run to $1,000 or more. In addition, many lenders charge a yearly fee of $25 to $50.

Home Equity Loans in the Current Economy

Because of the recent economic downturn, it's not as easy as it used to be to obtain a home equity loan. For one thing, lenders have tightened standards. For another, falling home prices mean that many homeowners have lost equity in their homes. So what can you do? Try smaller banks or credit unions, many of which haven't been as quick to shut off the valves on home equity loans. They will, though, take a hard look at the appraisal and perhaps require more than one.

You can also take steps to boost your credit score. (See Chapter 5.) A better score can improve your chances of getting that loan. And you also may need to accept the reality that in burst-bubble states like California, Nevada, and Florida, home equity loans are now rare, regardless of the borrower's credit history.

Get Your Tax Refund Fast

Sometimes, getting a tax refund quickly will help you through a cash-flow crisis, especially if the IRS owes you a lot. Each IRS district office has a Taxpayer Advocate's Office (TAO); the offices can help callers get their refunds early. Local offices are listed in the government listing of the phone book. New electronic filing procedures can also help the refund process go much faster, so take advantage of that if you can.

Still Having Serious Trouble Paying Your Bills?

If you're getting dunning notices from creditors or are beginning to hear from debt collectors, you may be worried about losing whatever assets you managed to hold onto. If so, you need to take firm steps to deal with this potential crisis.

Develop a budget. Begin by listing all your expenses, both fixed (such as mortgage, car, and insurance payments) and those that vary, such as entertainment and clothing. (You may want to use updated figures from your portion of the "Cash Flow: Income and Expenses" statement in Chapter 12.) Writing down all your expenses will help you track your spending patterns, identify necessary expenses, and prioritize the rest. The aim is to make sure you can make ends meet on the basics, such as housing, food, and health care.

Contact your creditors. Tell them why you're having difficulties and try to work out a modified payment plan. Don't wait until your accounts have been turned over to a debt collector. If your creditors already have assigned your case to debt collectors, familiarize yourself with the Fair Debt Collection Practices Act, the federal law that details how and when debt collectors may contact you. Go to the Federal Trade Commission website at www.consumer.ftc.gov/articles/0149-debt-collection.

Consider credit counseling. If you're not disciplined enough to create a workable budget and stick with it and you aren't able to work out a repayment plan with your creditors, you may want to contact a credit counseling agency. Many are nonprofit and have certified counselors who will advise you on managing your money and debts. But beware: Scams abound in the "credit repair" industry. Ask for referrals from your bank, local consumer protection agency, or friends and family, or go to the National Foundation for Credit Counseling site at www.nfcc.org.

Study debt consolidation. You may be able to lower your cost of credit by consolidating your debt through a second mortgage or a home equity line of credit. But remember that these loans require you to put up your home as collateral. If you can't make the payments, or make them on time, you could lose your house. Also, check the cost—in addition to interest on a debt consolidation loan, you may be asked to pay "points"—each point being one percent of the amount you borrow.

Still Having Serious Trouble Paying Your Bills? (cont'd)

Bankruptcy. Personal bankruptcy should be considered the debt management option of last resort. It stays on your credit report for a decade and can make it difficult to obtain credit, buy a home, get insurance, or even get a job. Still, it's a legal procedure that can offer a fresh start for a debtor who receives a "discharge"—a court order that exempts him or her from repaying certain obligations. Recent changes to the law involve hurdles you must clear even before filing for bankruptcy. For instance, you must get credit counseling from a government-approved organization within six months before you file for bankruptcy relief. See Chapter 17 for more about bankruptcy.

For comprehensive information about credit problems, see *Solve Your Money Troubles: Debt, Credit & Bankruptcy*, by Margaret Reiter and Robin Leonard (Nolo).

Postdivorce Financial Planning Tips

Financial planning strategies are often based on the idyllic picture of a couple who moves from struggling to make ends meet in their 20s to a happy retirement in their 60s. This picture hardly resembles the norm for divorced people who must redesign their plans. Below are a few tips to help you do that.

Update Your Net Worth and Cash Flow Statements

Your net worth and cash flow have changed because of the property you received or exchanged during the divorce and payments you now make or receive that you didn't before. You can't realize your financial goals in the future unless you know where you're starting from. Update the "Net Worth Statement" and "Cash Flow Statement" you completed in Chapter 12 to reflect your new status.

Organize Your Bill-Paying Procedure

Pay off all debts you agreed to cover as part of your divorce settlement as quickly as possible. Create a payment schedule to match your pay periods. If you're paid on the first and 15th of the month, pay your bills then. Keep all bills in a folder. On the first of the month, take out only

those bills that you will pay; write your checks, mail the bills, and put the folder away until the 15th. Then do the same on the 15th. By using this system, you won't constantly think about your debts.

Make Sure You Have an Emergency Fund and an Emergency Plan

As we've recommended, you should have an "emergency fund" representing three to six months' worth of cost-of-living expenses. Make sure you can get to this money when you need it.

You may also want to make a contingency plan that takes into account your nonfinancial resources. Could you move in with friends or relatives—including your adult children—if you had to? Knowing you have some kind of cushion of support makes it easier to face life. Once you've been divorced, you may find yourself dwelling on the "worst case" you can imagine. Having a contingency plan can dispel fears and let you move on.

Reevaluate Your Insurance

Most people fall into one of two extremes—either they have too little insurance or they have too much of the wrong kind. Now that your divorce is over, reevaluate your insurance coverage. Read consumer publications and check with several agents to get a good overall update on your insurance position. Consider what your new needs are with respect to health, life, disability, auto, homeowner's or renter's, and personal liability coverage. If you are receiving alimony or child support, be sure to consider getting life insurance on your ex-spouse.

Keep Up With Changes in Your Tax Status

When your marital status changes, so does your tax status. Not only might you be in a different tax bracket, but tax laws themselves may change. If you're in a higher tax bracket, you need investments that provide tax-deferred or tax-free income. More likely, however, you are in a lower tax bracket and will want to take advantage of two of the best tax planning opportunities most people have: owning a house and contributing to an IRA or a 401(k) through your work.

Reevaluate Your Retirement Program

How much will it cost you to maintain your current lifestyle once you retire? If you and your ex-spouse divided retirement plans during divorce, be sure to do the follow-up work to secure any payments you are supposed to receive. If you have no retirement plan, go to a brokerage house and ask for a retirement projection. Do not let yourself be pushed into any investment you are not ready to make. Once you have an accurate calculation of postretirement living costs, investigate several types of plans before choosing one that is right for you.

Plan Your Estate

Many people mistakenly assume that they do not have enough property to plan their estates. If you do not plan, however, the state will do it for you after you die. And estate planning encompasses quite a few issues beyond just giving away your property at your death, including medical directives, guardianships, and beneficiary statements. It's important to deal with all of these after your divorce. You will want to:

- Prepare a new will. You may need only a simple will, or you may want to investigate a living trust. Not everyone writes the ex-spouse out after divorce, but it's important that you decide whether you want to or not, and make a new will either way. If you die without a will, the state government will provide a standard one according to its laws, and that may be quite different from what you intended. Make sure your new will doesn't include property that you no longer have after your property settlement.

- If you have minor children, name a guardian for them in your will or be sure the one you previously named is still appropriate. Be sure to talk to potential guardians before naming them; they may be reluctant to accept the responsibility. And designate an alternate in case the guardian you select can't serve.

- Change the beneficiary statements on your retirement plans, brokerage accounts, and insurance policies. These are independent of the wishes expressed in your will.

- Make a new durable power of attorney for financial matters—that is, naming someone to handle your money if you cannot. This

can avoid prolonged, costly court proceedings should you become incapacitated.

- Prepare a medical directive, also known as a living will even though it has nothing to do with giving away property at death. The medical directive declares your preference for medical treatment if you become seriously ill. It's often accompanied by a durable power of attorney for health care, which authorizes a trusted person to make the critical decisions if you're near death.

> **RESOURCE**
> **Nolo publishes a number of books and software programs that can help you take care of your estate.** They are described in Chapter 10.

Setting Goals After Divorce

Right now is the perfect time to look at your financial goals for the short and long term. Use the chart titled "Setting Goals," below, to open up ideas about where you'd like to go in the future.

If You Find a New Love, Protect Your Old Assets... and Your Alimony

For some people, going through a divorce leads them to swear off the institution of marriage altogether. For others, however, a divorce represents the freedom to be with the person they truly love. If you fall into that category, be sure that you don't let Cupid distract you from the financial realities that you will face. Read the next two sections if you are considering living with or marrying someone new.

Cohabitation Can Be Costly

Once the divorce is over, you may be thinking about moving in with your new partner. Cohabitation, or living together without legally marrying, has risks and it's important that you understand them.

Obviously, you will have to deal with the emotional side of your own and possibly your children's reactions to such a move. But you must recognize that you are taking legal and financial risks as well.

Here are a couple of cautionary tales that will illustrate why it is important to be financially aware when you live with someone.

We know of a couple who cohabited for four years in Minnesota. During that time, the man used his money to purchase the home in which they lived. The title to the home was taken in "joint tenancy," which means that if one owner died before the other, the survivor automatically inherited the deceased's share of the property, even if a will or trust said something else. (See Chapter 6.) When the couple split up, the woman sued for half the home. The man showed that he used his money to buy it. The woman claimed that he gave her a gift of half the house. The court ruled against her, stating that the house was his unless a written contract stated otherwise.

In a California case, a woman met and moved in with a new man while she was divorcing her first husband. She and her new companion bought a condominium together using joint funds. To avoid complications while negotiating her divorce property settlement, she and her companion put the title to the condominium in his name only. After her divorce was final, the woman married her companion and they moved into a larger home, renting out the condominium—for which title remained in her second husband's name alone. Twenty-seven years later, they were divorcing. The judge ruled that the condo was the property of the second husband, because title was in his name only and she could not prove any ownership interest.

The moral of these stories is, if you and your new love acquire any assets together, be sure the title document (deed or registration) properly reflects your intentions.

In addition to getting property titled the right way, remember that one companion could sue the other for a share of the assets or for alimony-like support. Given these legal and financial realities, it is in your interest to ask a lawyer's advice regarding cohabitation laws in your state before moving in with someone. We also suggest that you draft an agreement that details your understanding of your respective

ownership in the property. You can incorporate this understanding into a prenuptial agreement if you later marry.

CAUTION

In many states, you can lose your alimony if you move in with someone new. Several states presume that someone who cohabits with a person of the opposite sex has less need for alimony. (One or two courts have applied the "spirit" of the law to cut off alimony to a woman who entered a lesbian relationship, but others have refused to do that, and in virtually all other cases, alimony terminated after heterosexual cohabitation.)

RESOURCE

If you will move in with your new partner, we recommend that you get a copy of either *Living Together: A Legal Guide for Unmarried Couples*, **by Ralph Warner, Toni Ihara, and Frederick Hertz (Nolo), or** *A Legal Guide for Lesbian & Gay Couples*, **by Denis Clifford, Emily Doskow and Frederick Hertz (Nolo).** Both books cover the financial, legal, parenting, estate planning, and practical concerns of couples whose living arrangements are not covered by state marital laws.

Prenuptial Agreements

Many couples marrying for the second time (or more) consider a prenuptial agreement before taking the plunge. A prenuptial contract can be a valuable planning tool if one spouse has sizable wealth and wants to preserve it for children from a prior marriage. While such agreements have a reputation of being unromantic, they can actually prevent a great deal of heartache.

Engaged couples use prenuptial agreements to define the terms of their financial agreements, rather than letting state law control those terms. They might use a prenup to protect a family business, or to state the nature of certain assets brought into the marriage or to be acquired in the future. Couples can also use prenuptial agreements to ensure that

the laws of the state where they were married will govern in the event of a divorce, no matter where they live in the future.

If a couple gets a divorce but does not have a prenuptial agreement, assuming they can't otherwise agree, their assets will be divided in accordance with state laws. This means that in community property states (Arizona, California, Idaho, Louisiana, Nevada, New Mexico, Texas, Washington, and Wisconsin), marital property will be divided equally; in equitable distribution states (the rest), property will be split "fairly" according to the dictates of the court. If your new spouse contributed to the appreciation of assets that belonged to you alone prior to the marriage, those assets could ultimately be considered marital property subject to division at divorce.

In some cases, prenuptial agreements are challenged by unhappy spouses and struck down in court. This can happen if the terms are too favorable to one spouse or if the agreement was signed under what the court considers unfair circumstances. If you are considering a prenuptial agreement, it is more likely to stand up if it meets these conditions:

- The terms are fair.
- It is in writing (some states require acknowledgment and attorney certification as well).
- There is a clause stating that if any provision of the agreement is invalidated, the rest of the agreement still remains in effect.
- There is a listing attached showing each spouse's assets and liabilities.
- Each spouse has had the agreement reviewed by a separate lawyer.
- It includes a clause stating that all agreements between the prospective spouses are included in the prenuptial agreement.

In addition to a prenuptial agreement, you can protect your separate property through more sophisticated legal tools including an irrevocable trust, revocable living trust, or family limited partnership. For more information on these and other estate planning tools, see one of the many Nolo estate planning publications listed in Chapter 10.

> **RESOURCE**
>
> **If you are considering remarrying and want to prepare a prenuptial agreement, check out** *Prenuptial Agreements: How to Write a Fair & Lasting Contract,* **by Katherine F. Stoner and Shae Irving (Nolo).**

How Can I Move Beyond the Divorce?

By the time your divorce ends, you may not be sure who won or who lost. You need to give yourself time to reassess what has happened in your life. You may be so tired of the divorce process that you don't want to hear another word about money. Worse, you may feel bitter or gun-shy about moving forward. As we've said throughout this book, you must manage your money and your emotions well if you are to end the relationship with your spouse successfully.

Once you've followed up on the legal and financial details of your settlement, check your emotional state to determine what unresolved feelings you are carrying with you as you move into singlehood. If things are not what you anticipated, don't spend time and money in a continuing connection with your ex-spouse. Recognize your gains and cut your losses so you can get on with your new life.

These questions are designed to help you sort through the emotional and economic fallout of ending a marriage. Figuring out your answers can help you move on.

What changes in my financial life did I not anticipate before the final settlement?

_____

_____

_____

_____

What will be different in my children's lives now that the divorce is over?

_____

_____

_____

_____

What regrets do I still have about the divorce?

_____

_____

_____

_____

What financial steps do I feel good about, and what could I have done differently?

_____

_____

_____

_____

How can I continue to expand the positive steps I've taken?

_____

_____

_____

_____

What can I do to avoid repeating the financial mistakes I made?

_____

_____

_____

_____

We wish you the best of luck and a speedy recovery from the process of divorce.

Setting Goals

*Instructions: Use this matrix to write your short- and long-term goals. Begin now to focus on the future—**your** future—and the memories of the past will begin to fade.*

My professional goals are:

Next six months: Next one year: Next five years:

My educational goals are:

Next six months: Next one year: Next five years:

My recreation-travel-entertainment goals are:

Next six months: Next one year: Next five years:

Setting Goals (cont'd)

My plans for my children are:

Next six months:	Next one year:	Next five years:

My retirement goals are:

Next six months:	Next one year:	Next five years:

My other goals are:

Next six months:	Next one year:	Next five years:

Appendix

A

Law Libraries...470

   Finding a Law Library..470

   Finding a State Law That Affects Your Divorce.......................470

   Going Beyond State Laws..471

   Using Background Resources 471

Lawyers and Typing Services 472

   Finding a Lawyer...472

   What to Look for in a Lawyer..474

Online Legal Resources 475

Additional Resources 475

   Attorneys, Mediators, and Therapists 476

   Child Support 476

   Custody...477

   Domestic Violence..477

   Appraisers...477

   Tax Assistance 478

   Miscellaneous Financial Assistance...478

Present Value Factors 480

Future Value Factors 486

List of Professional Advisers..492

This book helps you through the divorce process by giving you strategies for evaluating your assets and debts, your likelihood of paying or receiving alimony or child support, and your children's custody. But you may need help beyond this book. The three best sources of legal help are a law library, lawyer, or typing service.

Before discussing additional resources in detail, here's a general piece of advice: Make all decisions yourself. By reading this book, you've taken on the responsibility for getting information necessary to make informed decisions about your legal and financial affairs. If you decide to get help from others, apply this same self-empowerment principle—shop around until you find an adviser who values your competence and intelligence and recognizes your right to make your own decisions.

Law Libraries

Often, you can handle a legal problem yourself if you're willing to do some research in a law library. Here, briefly, are the basic steps to researching a legal question. For more detailed, but user-friendly, instructions on legal research, see *Legal Research: How to Find & Understand the Law*, by Stephen R. Elias and Susan Levinkind (Nolo).

Finding a Law Library

To do legal research, you need to find a law library that's open to the public. Public law libraries are often housed in county courthouses, public law schools, and state capitals. If you can't find one, ask a public library reference librarian, court clerk, or lawyer.

Finding a State Law That Affects Your Divorce

Laws passed by state legislatures are occasionally referred to in this book. To find a state law, or statute, you need to look in a multivolume set of

books called the state code. State codes are divided into titles. States divide their titles by number or by subject, such as the civil code, family code, or finance code.

To read a law, find the state codes in your law library, locate the title you need, turn to the section number, and read. If you already have a proper reference to the law—called the citation—finding the law is straightforward. If you don't have a citation, you can find the law by referring to the index in the code you're using.

After you read the law in the hardcover book, turn to the back of the book. There should be an insert pamphlet (called a pocket part) for the current or previous year. Look for the statute in the pocket part to see if it has been amended since the hardcover volume was published.

Going Beyond State Laws

If you want to find the answer to a legal question, rather than simply looking up a law, you will need some guidance in basic legal research techniques. Resources that may be available in your law library are:

- *Legal Research: How to Find & Understand the Law*, by Stephen R. Elias and Susan Levinkind (Nolo)
- *Legal Research in a Nutshell*, by Kent C. Olsen and Morris Cohen (West), and
- *Finding the Law* (12th ed.), by Robert Berring and Elizabeth Edinger (West).

Using Background Resources

If you want to research a legal question related to family law but don't know where to begin, one of the best resources is the *Family Law Reporter*, published by the Bureau of National Affairs (BNA). This very thorough, four-volume publication covers all 50 states and the District of Columbia, and is updated weekly. It highlights and summarizes cases, new statutes, and family law news. It also includes a guide to tax laws affecting family law, a summary of each state's divorce laws, and a sample marital settlement agreement.

Most law libraries will have the *Family Law Reporter*. If you can't find it, however, you will need to look at materials written specifically for your state. Ask a law librarian for help.

Lawyers and Typing Services

A lawyer can provide you with information, guidance, or legal representation. A typing service can act as a "legal secretary" if you need to have documents prepared and filed in court. Typing services *cannot* give legal advice, but they charge far less than lawyers for their services. Typing services are covered in Chapter 7.

Finding a Lawyer

Before explaining how to find a lawyer, let's first eliminate the types of lawyers you are not looking for:

- the expensive, flamboyant lawyer who grandstands but fails to deliver promised services and passes your case on to a recent law school graduate who works in the office
- the associate or partner at a giant law firm that represents big businesses. These lawyers charge high fees, and few know much about divorce cases or keeping costs down.
- the lawyer who won't keep you informed about plans for your case and who wants to make all decisions without consulting you. These lawyers are annoyed—and intimidated—by clients who know anything about the law. What they want is a passive client who doesn't ask a lot of questions and pays the bill on time each month.

What you do want is a dedicated, smart, and skilled lawyer who regularly handles family law and divorce cases. The lawyer should understand that your input must be sought for every decision. This being said, here are several ways to find a lawyer:

Typing services. The best referrals will probably come from an independent paralegal listed in the yellow pages under "paralegals" or "typing services." Almost daily, independent paralegals refer their clients to lawyers and get feedback on the lawyers' work.

Personal referrals. This is the most common approach. If you know someone who was pleased with the services of a lawyer, call that lawyer first. If that lawyer doesn't handle divorces or can't take your case, ask for a recommendation to someone else. Be careful, however, when selecting a lawyer from a personal referral. A lawyer's satisfactory performance in one situation does not guarantee that the person will perform the same way in your case.

Group legal plans. Some unions, employers, and consumer action organizations offer group plans to their members or employees who can obtain comprehensive legal assistance free or for low rates.

Prepaid legal insurance. Prepaid legal insurance plans offer some services for a low monthly fee and charge more for additional work. Participating lawyers may use the plan as a way to get clients who are attracted by the low-cost basic services, and then sell them more expensive services. If the lawyer recommends an expensive course of action, get a second opinion before you agree.

But if a plan offers extensive free advice, your initial membership fee may be worth the consultation you receive. You can always join a plan for a specific service and then not renew.

There's no guarantee that the lawyers available through these plans are of the best caliber; sometimes they aren't. Check out the plan carefully before signing up. Ask about the plan's complaint system, whether you get to choose your lawyer, and whether or not the lawyer will represent you in court.

Lawyer referral panels. Most county bar associations will give out the names of attorneys who practice in your area. But bar associations often fail to provide meaningful screening for the attorneys listed, which means those who participate may not be the most experienced or competent.

Lawyer directories. There are lists of attorneys on most of the divorce websites. Nolo's lawyer directory at www.nolo.com has detailed profiles that let you see a lawyer's background, education, and fee structure.

What to Look for in a Lawyer

No matter what approach you take to finding a lawyer, here are three suggestions on how to make sure you have the best possible working relationship with your attorney.

First, fight the urge you may have to surrender your will and be intimidated. You should be the one who decides what you feel comfortable doing about your legal and financial affairs. You're hiring the lawyer to perform a service for you; shop around if the price or personality isn't right.

Second, you must be as comfortable as possible with any lawyer you hire. When making an appointment, ask to talk directly to the lawyer. If you can't, this may give you a hint as to how accessible the lawyer is.

If you do talk directly to the lawyer, ask some specific questions. Do you get clear, concise answers? If not, try someone else. If the lawyer says little except to suggest you turn over the problem—with a substantial fee—watch out. You're talking with someone who doesn't know the answer and won't admit it, or someone who pulls rank on the basis of professional standing. Don't be a passive client or hire a lawyer who wants you to be one. If the lawyer admits to not knowing an answer, that isn't necessarily bad. In most cases, the lawyer must do some research.

Also, pay attention to how the lawyer responds to the fact that you have considerable information. If you read this book, you know more about divorce and money than the average person. Does the lawyer seem comfortable with that? Does the lawyer give straightforward answers to your questions—or does the lawyer want to maintain an aura of mystery about the legal system? Pay attention to your own intuition. Many lawyers are threatened when the client knows too much—or, in some cases, anything.

Once you find a lawyer you like, make an hour-long appointment to discuss your situation fully. Your goal at the initial conference is to find out what the lawyer recommends and how much it will cost. Go home and think about the lawyer's suggestions. If they don't make complete sense or if you have other reservations, call someone else.

·Finally, keep in mind that the lawyer works for you. Once you hire a lawyer, you have the absolute right to switch to another—or to fire the lawyer and handle the matter yourself—at any time, for any reason.

Online Legal Resources

A growing number of basic legal resources are available online. Nolo's legal resources at www.nolo.com include a vast amount of free legal information for consumers on a wide variety of legal topics.

In addition, a wide variety of secondary sources intended for both lawyers and the general public has been posted by law schools and firms. A good way to find these sources is to visit any of the following websites, each of which provides links to legal information by specific subject:

- www.divorcenet.com. This site provides divorce information for all 50 states and a lawyer directory as well.
- www.findlaw.com. This legal search engine is an excellent resource for do-it-yourself legal research.
- www.law.cornell.edu. This site is maintained by Cornell Law School's Legal Information Institute. You can find the text of some state court decisions. You can also search for material by topic.
- www.law.indiana.edu. This site is maintained by Indiana University's School of Law at Bloomington. You can search by state governments and law journals, or by topic.
- www.divorce360.com. This site generates original content and tools along with links to technical material about divorce.

Additional Resources

Below are names, addresses, and phone numbers of organizations that may be able to offer additional assistance.

Attorneys, Mediators, and Therapists

American Academy of Matrimonial Lawyers
150 N. Michigan Avenue, Suite 1420
Chicago, IL 60601
312-263-6477 (phone)
312-263-7682 (fax)
www.aaml.org

American Arbitration Association
1633 Broadway, 10th Floor
New York, NY 10019
212-716-5800 (phone)
www.adr.org

American Association for Marriage and Family Therapy
112 S. Alfred Street
Alexandria, VA 22314-3061
703-838-9808 (phone)
703-838-9805 (fax)
www.aamft.org

Association for Conflict Resolution
12100 Sunset Hills Road, Suite 130
Reston, VA 20190
703-234-4141 (phone)
703-435-4390 (fax)
www.acrnet.org

Child Support

National Child Support Enforcement Association
1760 Old Meadow Road, Suite 500
McLean, VA 20447
703-506-2880 (phone)
www.acf.hhs.gov/programs/css

Administration for Children and Families
U.S. Department of Health and Human Services
Child Support Enforcement
370 L'Enfant Promenade, SW
Washington, DC 20201
202-401-9383 (phone)
202-401-5559 (fax)
www.acf_services.html#cse

Custody

Alliance for Noncustodial Parents Rights
www.ancpr.com

Domestic Violence

Futures Without Violence
100 Montgomery Street, The Presidio
San Francisco, CA 94129
415-678-5500 (phone)
415-529-2930 (fax)
www.futureswithoutviolence.org

Appraisers

American Society of Appraisers
(business appraisers)
11107 Sunset Hills Rd, Suite 310
Reston, VA 20190
703-478-2228 (phone)
703-742-8471 (fax)
www.appraisers.org

Appraisal Institute
(real estate appraisers)
200 W. Madison
Suite 1500
Chicago, IL 60606
888-756-4624 (phone)
312-335-4400 (fax)
www.appraisalinstitute.org

Tax Assistance

Internal Revenue Service (IRS)
(to obtain tax forms)
800-829-1040 (phone)
www.irs.gov

National Association of Enrolled Agents
(for a referral to an enrolled agent)
1120 Connecticut Avenue, NW, Suite 460
Washington, DC 20036
202-822-6232 (phone)
202-822-6270 (fax)
www.naea.org

Miscellaneous Financial Assistance

Institute for Divorce Financial Analysts
2224 Sedwick Dr, Suite 102
Durham, NC 27713
800-875-1760 (phone)
888-527-7657 (fax)
www.institutedfa.com

Financial Planning Association
(for a referral to a financial planner)
7535 E. Hampden Ave., Suite 600
Denver, CO 80231
800-322-4237 (phone)
303-759-0749 (fax)
www.fpanet.org

National Foundation for Credit Counseling
(to find a consumer credit counseling office)
2000 M Street, NW, Suite 505
Washington, DC 20036
202-677-4300
www.nfcc.org

Social Security Administration
(to confirm work and benefits history)
800-772-1213 (phone)
www.ssa.gov
(to get information pertaining to Social Security and women)
www.ssa.gov/women

Federal Trade Commission
202-326-2222 (phone)
www.ftc.gov

Present Value Factors

This chart shows the time value of money and can help you estimate the future value of benefits, such as retirement plans. The left-hand column represents years, while the percentages across the top are rates of inflation. For example, to find the value of a $10,000-a-year payment 20 years from now with 4% inflation, follow the 20-year line to 4%. The present value factor is 0.4564. Multiply that by $10,000, and you get $4,564, the purchasing power that $10,000 will have in two decades at that rate of inflation.

Present Value Factors

Year	1%	2%	3%	4%	5%	6%	7%
1	0.9901	0.9804	0.9709	0.9615	0.9524	0.9434	0.9346
2	0.9803	0.9612	0.9426	0.9246	0.9070	0.8900	0.8734
3	0.9706	0.9423	0.9151	0.8890	0.8638	0.8396	0.8163
4	0.9610	0.9238	0.8885	0.8548	0.8227	0.7921	0.7629
5	0.9515	0.9057	0.8626	0.8219	0.7835	0.7473	0.7130
6	0.9420	0.8880	0.8375	0.7903	0.7462	0.7050	0.6663
7	0.9327	0.8706	0.8131	0.7599	0.7107	0.6651	0.6227
8	0.9235	0.8535	0.7894	0.7307	0.6768	0.6274	0.5820
9	0.9143	0.8368	0.7664	0.7026	0.6446	0.5919	0.5439
10	0.9053	0.8203	0.7441	0.6756	0.6139	0.5584	0.5083
11	0.8963	0.8043	0.7224	0.6496	0.5847	0.5268	0.4751
12	0.8874	0.7885	0.7014	0.6246	0.5568	0.4970	0.4440
13	0.8787	0.7730	0.6810	0.6006	0.5303	0.4688	0.4150
14	0.8700	0.7579	0.6611	0.5775	0.5051	0.4423	0.3878
15	0.8613	0.7430	0.6419	0.5553	0.4810	0.4173	0.3624
16	0.8528	0.7284	0.6232	0.5339	0.4581	0.3936	0.3387
17	0.8444	0.7142	0.6050	0.5134	0.4363	0.3714	0.3166
18	0.8360	0.7002	0.5874	0.4936	0.4155	0.3503	0.2959
19	0.8277	0.6864	0.5703	0.4746	0.3957	0.3305	0.2765
20	0.8195	0.6730	0.5537	0.4564	0.3769	0.3118	0.2584

Present Value Factors

Year	1%	2%	3%	4%	5%	6%	7%
21	0.8114	0.6598	0.5375	0.4388	0.3589	0.2942	0.2415
22	0.8034	0.6468	0.5219	0.4220	0.3418	0.2775	0.2257
23	0.7954	0.6342	0.5067	0.4057	0.3256	0.2618	0.2109
24	0.7876	0.6217	0.4919	0.3901	0.3101	0.2470	0.1971
25	0.7798	0.6095	0.4776	0.3751	0.2953	0.2330	0.1842
26	0.7720	0.5976	0.4637	0.3607	0.2812	0.2198	0.1722
27	0.7644	0.5859	0.4502	0.3468	0.2678	0.2074	0.1609
28	0.7568	0.5744	0.4371	0.3335	0.2551	0.1956	0.1504
29	0.7493	0.5631	0.4243	0.3207	0.2429	0.1846	0.1406
30	0.7419	0.5521	0.4120	0.3083	0.2314	0.1741	0.1314
31	0.7346	0.5412	0.4000	0.2965	0.2204	0.1643	0.1228
32	0.7273	0.5306	0.3883	0.2851	0.2099	0.1550	0.1147
33	0.7201	0.5202	0.3770	0.2741	0.1999	0.1462	0.1072
34	0.7130	0.5100	0.3660	0.2636	0.1904	0.1379	0.1002
35	0.7059	0.5000	0.3554	0.2534	0.1813	0.1301	0.0937
36	0.6989	0.4902	0.3450	0.2437	0.1727	0.1227	0.0875
37	0.6920	0.4806	0.3350	0.2343	0.1644	0.1158	0.0818
38	0.6852	0.4712	0.3252	0.2253	0.1566	0.1092	0.0765
39	0.6784	0.4619	0.3158	0.2166	0.1491	0.1031	0.0715
40	0.6717	0.4529	0.3066	0.2083	0.1420	0.0972	0.0668
41	0.6650	0.4440	0.2976	0.2003	0.1353	0.0917	0.0624
42	0.6584	0.4353	0.2890	0.1926	0.1288	0.0865	0.0583
43	0.6519	0.4268	0.2805	0.1852	0.1227	0.0816	0.0545
44	0.6454	0.4184	0.2724	0.1780	0.1169	0.0770	0.0509
45	0.6391	0.4102	0.2644	0.1712	0.1113	0.0727	0.0476
46	0.6327	0.4022	0.2567	0.1646	0.1060	0.0685	0.0445
47	0.6265	0.3943	0.2493	0.1583	0.1009	0.0647	0.0416
48	0.6203	0.3865	0.2420	0.1522	0.0961	0.0610	0.0389
49	0.6141	0.3790	0.2350	0.1463	0.0916	0.0575	0.0363
50	0.6080	0.3715	0.2281	0.1407	0.0872	0.0543	0.0339

Present Value Factors

Year	8%	9%	10%	11%	12%	13%	14%
1	0.9259	0.9174	0.9091	0.9009	0.8929	0.8850	0.8772
2	0.8573	0.8417	0.8264	0.8116	0.7972	0.7831	0.7695
3	0.7938	0.7722	0.7513	0.7312	0.7118	0.6931	0.6750
4	0.7350	0.7084	0.6830	0.6587	0.6355	0.6133	0.5921
5	0.6806	0.6499	0.6209	0.5935	0.5674	0.5428	0.5194
6	0.6302	0.5963	0.5645	0.5346	0.5066	0.4803	0.4556
7	0.5835	0.5470	0.5132	0.4817	0.4523	0.4251	0.3996
8	0.5403	0.5019	0.4665	0.4339	0.4039	0.3762	0.3506
9	0.5002	0.4604	0.4241	0.3909	0.3606	0.3329	0.3075
10	0.4632	0.4224	0.3855	0.3522	0.3220	0.2946	0.2697
11	0.4289	0.3875	0.3505	0.3173	0.2875	0.2607	0.2366
12	0.3971	0.3555	0.3186	0.2858	0.2567	0.2307	0.2076
13	0.3677	0.3262	0.2897	0.2575	0.2292	0.2042	0.1821
14	0.3405	0.2992	0.2633	0.2320	0.2046	0.1807	0.1597
15	0.3152	0.2745	0.2394	0.2090	0.1827	0.1599	0.1401
16	0.2919	0.2519	0.2176	0.1883	0.1631	0.1415	0.1229
17	0.2703	0.2311	0.1978	0.1696	0.1456	0.1252	0.1078
18	0.2502	0.2120	0.1799	0.1528	0.1300	0.1108	0.0946
19	0.2317	0.1945	0.1635	0.1377	0.1161	0.0981	0.0829
20	0.2145	0.1784	0.1486	0.1240	0.1037	0.0868	0.0728
21	0.1987	0.1637	0.1351	0.1117	0.0926	0.0768	0.0638
22	0.1839	0.1502	0.1228	0.1007	0.0826	0.0680	0.0560
23	0.1703	0.1378	0.1117	0.0907	0.0738	0.0601	0.0491
24	0.1577	0.1264	0.1015	0.0817	0.0659	0.0532	0.0431
25	0.1460	0.1160	0.0923	0.0736	0.0588	0.0471	0.0378

Present Value Factors

Year	8%	9%	10%	11%	12%	13%	14%
26	0.1352	0.1064	0.0839	0.0663	0.0525	0.0417	0.0331
27	0.1252	0.0976	0.0763	0.0597	0.0469	0.0369	0.0291
28	0.1159	0.0895	0.0693	0.0538	0.0419	0.0326	0.0255
29	0.1073	0.0822	0.0630	0.0485	0.0374	0.0289	0.0224
30	0.0994	0.0754	0.0573	0.0437	0.0334	0.0256	0.0196
31	0.0920	0.0691	0.0521	0.0394	0.0298	0.0226	0.0172
32	0.0852	0.0634	0.0474	0.0355	0.0266	0.0200	0.0151
33	0.0789	0.0582	0.0431	0.0319	0.0238	0.0177	0.0132
34	0.0730	0.0534	0.0391	0.0288	0.0212	0.0157	0.0116
35	0.0676	0.0490	0.0356	0.0259	0.0189	0.0139	0.0102
36	0.0626	0.0449	0.0323	0.0234	0.0169	0.0123	0.0089
37	0.0580	0.0412	0.0294	0.0210	0.0151	0.0109	0.0078
38	0.0537	0.0378	0.0267	0.0190	0.0135	0.0096	0.0069
39	0.0497	0.0347	0.0243	0.0171	0.0120	0.0085	0.0060
40	0.0460	0.0318	0.0221	0.0154	0.0107	0.0075	0.0053
41	0.0426	0.0292	0.0201	0.0139	0.0096	0.0067	0.0046
42	0.0395	0.0268	0.0183	0.0125	0.0086	0.0059	0.0041
43	0.0365	0.0246	0.0166	0.0112	0.0076	0.0052	0.0036
44	0.0338	0.0226	0.0151	0.0101	0.0068	0.0046	0.0031
45	0.0313	0.0207	0.0137	0.0091	0.0061	0.0041	0.0027
46	0.0290	0.0190	0.0125	0.0082	0.0054	0.0036	0.0024
47	0.0269	0.0174	0.0113	0.0074	0.0049	0.0032	0.0021
48	0.0249	0.0160	0.0103	0.0067	0.0043	0.0028	0.0019
49	0.0230	0.0147	0.0094	0.0060	0.0039	0.0025	0.0016
50	0.0213	0.0134	0.0085	0.0054	0.0035	0.0022	0.0014

Present Value Factors

Year	15%	16%	17%	18%	19%	20%
1	0.8696	0.8621	0.8547	0.8475	0.8403	0.8333
2	0.7561	0.7432	0.7305	0.7182	0.7062	0.6944
3	0.6575	0.6407	0.6244	0.6086	0.5934	0.5787
4	0.5718	0.5523	0.5337	0.5158	0.4987	0.4823
5	0.4972	0.4761	0.4561	0.4371	0.4190	0.4019
6	0.4323	0.4104	0.3898	0.3704	0.3521	0.3349
7	0.3759	0.3538	0.3332	0.3139	0.2959	0.2791
8	0.3269	0.3050	0.2848	0.2660	0.2487	0.2326
9	0.2843	0.2630	0.2434	0.2255	0.2090	0.1938
10	0.2472	0.2267	0.2080	0.1911	0.1756	0.1615
11	0.2149	0.1954	0.1778	0.1619	0.1476	0.1346
12	0.1869	0.1685	0.1520	0.1372	0.1240	0.1122
13	0.1625	0.1452	0.1299	0.1163	0.1042	0.0935
14	0.1413	0.1252	0.1110	0.0985	0.0876	0.0779
15	0.1229	0.1079	0.0949	0.0835	0.0736	0.0649
16	0.1069	0.0930	0.0811	0.0708	0.0618	
17	0.0929	0.0802	0.0693	0.0600	0.0520	
18	0.0808	0.0691	0.0592	0.0508	0.0437	
19	0.0703	0.0596	0.0506	0.0431	0.0367	
20	0.0611	0.0514	0.0433	0.0365	0.0308	
21	0.0531	0.0443	0.0370	0.0309	0.0259	
22	0.0462	0.0382	0.0316	0.0262	0.0218	
23	0.0402	0.0329	0.0270	0.0222	0.0183	
24	0.0349	0.0284	0.0231	0.0188	0.0154	
25	0.0304	0.0245	0.0197	0.0160	0.0129	

Present Value Factors

Year	15%	16%	17%	18%	19%
26	0.0264	0.0211	0.0169	0.0135	0.0109
27	0.0230	0.0182	0.0144	0.0115	0.0091
28	0.0200	0.0157	0.0123	0.0097	0.0077
29	0.0174	0.0135	0.0105	0.0082	0.0064
30	0.0151	0.0116	0.0090	0.0070	0.0054
31	0.0131	0.0100	0.0077	0.0059	0.0046
32	0.0114	0.0087	0.0066	0.0050	0.0038
33	0.0099	0.0075	0.0056	0.0042	0.0032
34	0.0086	0.0064	0.0048	0.0036	0.0027
35	00075	0.0055	0.0041	0.0030	0.0023
36	0.0065	0.0048	0.0035	0.0026	0.0019
37	0.0057	0.0041	0.0030	0.0022	0.0016
38	0.0049	0.0036	0.0026	0.0019	0.0013
39	0.0043	0.0031	0.0022	0.0016	0.0011
40	0.0037	0.0026	0.0019	0.0013	0.0010
41	0.0032	0.0023	0.0016	0.0011	0.0008
42	0.0028	0.0020	0.0014	0.0010	0.0007
43	0.0025	0.0017	0.0012	0.0008	0.0006
44	0.0021	0.0015	0.0010	0.0007	0.0005
45	0.0019	0.0013	0.0009	0.0006	0.0004
46	0.0016	0.0011	0.0007	0.0005	0.0003
47	0.0014	0.0009	0.0006	0.0004	0.0003
48	0.0012	0.0008	0.0005	0.0004	0.0002
49	0.0011	0.0007	0.0005	0.0003	0.0002
50	0.0009	0.0006	0.0004	0.0003	0.0002

Future Value Factors

This chart lets you calculate an estimate of the future value of an asset, such as a house. The figures in the first column represent the holding period, or number of years you would expect to retain the asset. The percentages across the top indicate the inflation rate. When you follow the two lines until they intersect, you have the future value factor. For instance, if you have a $200,000 house that you expect to hold for another seven years and if you assume a 4% inflation rate, the FVF is 1.3159. Multiply that by $200,000, and you have an estimated future value of $263,180 for your house in seven years.

Future Value Factors							
Year	**1%**	**2%**	**3%**	**4%**	**5%**	**6%**	**7%**
1	1.0100	1.0200	1.0300	1.0400	1.0500	1.0600	1.0700
2	1.0201	1.0404	1.0609	1.0816	1.1025	1.1236	1.1449
3	1.0303	1.0612	1.0927	1.1249	1.1576	1.1910	1.2250
4	1.0406	1.0824	1.1255	1.1699	1.2155	1.2625	1.3108
5	1.0510	1.1041	1.1593	1.2167	1.2763	1.3382	1.4026
6	1.0615	1.1262	1.1941	1.2653	1.3401	1.4185	1.5007
7	1.0721	1.1487	1.2299	1.3159	1.4071	1.5036	1.6058
8	1.0829	1.1717	1.2668	1.3686	1.4775	1.5938	1.7182
9	1.0937	1.1951	1.3048	1.4233	1.5513	1.6895	1.8385
10	1.1046	1.2190	1.3439	1.4802	1.6289	1.7908	1.9672
11	1.1157	1.2434	1.3842	1.5395	1.7103	1.8983	2.1049
12	1.1268	1.2682	1.4258	1.6010	1.7959	2.0122	2.2522
13	1.1381	1.2936	1.4685	1.6651	1.8856	2.1329	2.4098
14	1.1495	1.3195	1.5126	1.7317	1.9799	2.2609	2.5785
15	1.1610	1.3459	1.5580	1.8009	2.0789	2.3966	2.7590
16	1.1726	1.3728	1.6047	1.8730	2.1829	2.5404	2.9522
17	1.1843	1.4002	1.6528	1.9479	2.2920	2.6928	3.1588
18	1.1961	1.4282	1.7024	2.0258	2.4066	2.8543	3.3799
19	1.2081	1.4568	1.7535	2.1068	2.5270	3.0256	3.6165
20	1.2202	1.4859	1.8061	2.1911	2.6533	3.2071	3.8697

Future Value Factors

Year	1%	2%	3%	4%	5%	6%	7%
21	1.2324	1.5157	1.8603	2.2788	2.7860	3.3996	4.1406
22	1.2447	1.5460	1.9161	2.3699	2.9253	3.6035	4.4304
23	1.2572	1.5769	1.9736	2.4647	3.0715	3.8197	4.7405
24	1.2697	1.6084	2.0328	2.5633	3.2251	4.0489	5.0724
25	1.2824	1.6406	2.0938	2.6658	3.3864	4.2919	5.4274
26	1.2953	1.6734	2.1566	2.7725	3.5557	4.5494	5.8074
27	1.3082	1.7069	2.2213	2.8834	3.7335	4.8223	6.2139
28	1.3213	1.7410	2.2879	2.9987	3.9201	5.1117	6.6488
29	1.3345	1.7758	2.3566	3.1187	4.1161	5.4184	7.1143
30	1.3478	1.8114	2.4273	3.2434	4.3219	5.7435	7.6123
31	1.3613	1.8476	2.5001	3.3731	4.5380	6.0881	8.1451
32	1.3749	1.8845	2.5751	3.5081	4.7649	6.4534	8.7153
33	1.3887	1.9222	2.6523	3.6484	5.0032	6.8406	9.3253
34	1.4026	1.9607	2.7319	3.7943	5.2533	7.2510	9.9781
35	1.4166	1.9999	2.8139	3.9461	5.5160	7.6861	10.6766
36	1.4308	2.0399	2.8983	4.1039	5.7918	8.1473	11.4239
37	1.4451	2.0807	2.9852	4.2681	6.0814	8.6361	12.2236
38	1.4595	2.1223	3.0748	4.4388	6.3855	9.1543	13.0793
39	1.4741	2.1647	3.1670	4.6164	6.7048	9.7035	13.9948
40	1.4889	2.2080	3.2620	4.8010	7.0400	10.2857	14.9745
41	1.5038	2.2522	3.3599	4.9931	7.3920	10.9029	16.0227
42	1.5188	2.2972	3.4607	5.1928	7.7616	11.5570	17.1443
43	1.5340	2.3432	3.5645	5.4005	8.1497	12.2505	18.3444
44	1.5493	2.3901	3.6715	5.6165	8.5572	12.9855	19.6285
45	1.5648	2.4379	3.7816	5.8412	8.9850	13.7646	21.0025
46	1.5805	2.4866	3.8950	6.0748	9.4343	14.5905	22.4726
47	1.5963	2.5363	4.0119	6.3178	9.9060	15.4659	24.0457
48	1.6122	2.5871	4.1323	6.5705	10.4013	16.3939	25.7289
49	1.6283	2.6388	4.2562	6.8333	10.9213	17.3775	27.5299
50	1.6446	2.6916	4.3839	7.1067	11.4674	18.4202	29.4570

Future Value Factors

Year	8%	9%	10%	11%	12%	13%	14%
1	1.0800	1.0900	1.1000	1.1100	1.1200	1.1300	1.1400
2	1.1664	1.1881	1.2100	1.2321	1.2544	1.2769	1.2996
3	1.2597	1.2950	1.3310	1.3676	1.4049	1.4429	1.4815
4	1.3605	1.4116	1.4641	1.5181	1.5735	1.6305	1.6890
5	1.4693	1.5386	1.6105	1.6851	1.7623	1.8424	1.9254
6	1.5869	1.6771	1.7716	1.8704	1.9738	2.0820	2.1950
7	1.7138	1.8280	1.9487	2.0762	2.2107	2.3526	2.5023
8	1.8509	1.9926	2.1436	2.3045	2.4760	2.6584	2.8526
9	1.9990	2.1719	2.3579	2.5580	2.7731	3.0040	3.2519
10	2.1589	2.3674	2.5937	2.8394	3.1058	3.3946	3.7072
11	2.3316	2.5804	2.8531	3.1518	3.4785	3.8359	4.2262
12	2.5182	2.8127	3.1384	3.4985	3.8960	4.3345	4.8179
13	2.7196	3.0658	3.4523	3.8833	4.3635	4.8980	5.4924
14	2.9372	3.3417	3.7975	4.3104	4.8871	5.5348	6.2613
15	3.1722	3.6425	4.1772	4.7846	5.4736	6.2543	7.1379
16	3.4259	3.9703	4.5950	5.3109	6.1304	7.0673	8.1372
17	3.7000	4.3276	5.0545	5.8951	6.8660	7.9861	9.2765
18	3.9960	4.7171	5.5599	6.5436	7.6900	9.0243	10.5752
19	4.3157	5.1417	6.1159	7.2633	8.6128	10.1974	12.0557
20	4.6610	5.6044	6.7275	8.0623	9.6463	11.5231	13.7435
21	5.0338	6.1088	7.4002	8.9492	10.8038	13.0211	15.6676
22	5.4365	6.6586	8.1403	9.9336	12.1003	14.7138	17.8610
23	5.8715	7.2579	8.9543	11.0263	13.5523	16.6266	20.3616
24	6.3412	7.9111	9.8497	12.2392	15.1786	18.7881	23.2122
25	6.8485	8.6231	10.8347	13.5855	17.0001	21.2305	26.4619

Future Value Factors

Year	8%	9%	10%	11%	12%	13%	14%
26	7.3964	9.3992	11.9182	15.0799	19.0401	23.9905	30.1666
27	7.9881	10.2451	13.1100	16.7386	21.3249	27.1093	34.3899
28	8.6271	11.1671	14.4210	18.5799	23.8839	30.6335	39.2045
29	9.3173	12.1722	15.8631	20.6237	26.7499	34.6158	44.6931
30	10.0627	13.2677	17.4494	22.8923	29.9599	39.1159	50.9502
31	10.8677	14.4618	19.1943	25.4104	33.5551	44.2010	58.0832
32	11.7371	15.7633	21.1138	28.2056	37.5817	49.9471	66.2148
33	12.6760	17.1820	23.2252	31.3082	42.0915	56.4402	75.4849
34	13.6901	18.7284	25.5477	34.7521	47.1425	63.7774	86.0528
35	14.7853	20.4140	28.1024	38.5749	52.7996	72.0685	98.1002
36	15.9682	22.2512	30.9127	42.8181	59.1356	81.4374	111.8342
37	17.2456	24.2538	34.0039	47.5281	66.2318	92.0243	127.4910
38	18.6253	26.4367	37.4043	52.7562	74.1797	103.9874	145.3397
39	20.1153	28.8160	41.1448	58.5593	83.0812	117.5058	165.6873
40	21.7245	31.4094	45.2593	65.0009	93.0510	132.7816	188.8835
41	23.4625	34.2363	49.7852	72.1510	104.2171	150.0432	215.3272
42	25.3395	37.3175	54.7637	80.0876	116.7231	169.5488	245.4730
43	27.3666	40.6761	60.2401	88.8972	130.7299	191.5901	279.8392
44	29.5560	44.3370	66.2641	98.6759	146.4175	216.4968	319.0167
45	31.9204	48.3273	72.8905	109.5302	163.9876	244.6414	363.6791
46	34.4741	52.6767	80.1795	121.5786	183.6661	276.4448	414.5941
47	37.2320	57.4176	88.1975	134.9522	205.7061	312.3826	472.6373
48	40.2106	62.5852	97.0172	149.7970	230.3908	352.9923	538.8065
49	43.4274	68.2179	106.7190	166.2746	258.0377	398.8813	614.2395
50	46.9016	74.3575	117.3909	184.5648	289.0022	450.7359	700.2330

Future Value Factors

Year	15%	16%	17%	18%	19%	20%
1	1.1500	1.1600	1.1700	1.1800	1.1900	1.2000
2	1.3225	1.3456	1.3689	1.3924	1.4161	1.4400
3	1.5209	1.5609	1.6016	1.6430	1.6852	1.7280
4	1.7490	1.8106	1.8739	1.9388	2.0053	2.0736
5	2.0114	2.1003	2.1924	2.2878	2.3864	2.4883
6	2.3131	2.4364	2.5652	2.6996	2.8398	2.9860
7	2.6600	2.8262	3.0012	3.1855	3.3793	3.5832
8	3.0590	3.2784	3.5115	3.7589	4.0214	4.2998
9	3.5179	3.8030	4.1084	4.4355	4.7854	5.1598
10	4.0456	4.4114	4.8068	5.2338	5.6947	6.1917
11	4.6524	5.1173	5.6240	6.1759	6.7767	7.4301
12	5.3503	5.9360	6.5801	7.2876	8.0642	8.9161
13	6.1528	6.8858	7.6987	8.5994	9.5964	10.6993
14	7.0757	7.9875	9.0075	10.1472	11.4198	12.8392
15	8.1371	9.2655	10.5387	11.9737	13.5895	15.4070
16	9.3576	10.7480	12.3303	14.1290	16.1715	
17	10.7613	12.4677	14.4265	16.6722	19.2441	
18	12.3755	14.4625	16.8790	19.6733	22.9005	
19	14.2318	16.7765	19.7484	23.2144	27.2516	
20	16.3665	19.4608	23.1056	27.3930	32.4294	
21	18.8215	22.5745	27.0336	32.3238	38.5910	
22	21.6447	26.1864	31.6293	38.1421	45.9233	
23	24.8915	30.3762	37.0062	45.0076	54.6487	
24	28.6252	35.2364	43.2973	53.1090	65.0320	
25	32.9190	40.8742	50.6578	62.6686	77.3881	

Future Value Factors

Year	15%	16%	17%	18%	19%
26	37.8568	47.4141	59.2697	73.9490	92.0918
27	43.5353	55.0004	69.3455	87.2598	109.5893
28	50.0656	63.8004	81.1342	102.9666	130.4112
29	57.5755	74.0085	94.9271	121.5005	155.1893
30	66.2118	85.8499	111.0647	143.3706	184.6753
31	76.1435	99.5859	129.9456	169.1774	219.7636
32	87.5651	115.5196	152.0364	199.6293	261.5187
33	100.6998	134.0027	177.8826	235.5625	311.2073
34	115.8048	155.4432	208.1226	277.9638	370.3366
35	133.1755	180.3141	243.5035	327.9973	440.7006
36	153.1519	209.1643	284.8991	387.0368	524.4337
37	176.1246	242.6306	333.3319	456.7034	624.0761
38	202.5433	281.4515	389.9983	538.9100	742.6506
39	232.9248	326.4838	456.2980	635.9139	883.7542
40	267.8635	378.7212	533.8687	750.3783	1051.6675
41	308.0431	439.3165	624.6264	885.4464	1251.4843
42	354.2495	509.6072	730.8129	1044.8268	1489.2664
43	407.3870	591.1443	855.0511	1232.8956	1772.2270
44	468.4950	685.7274	1000.4098	1454.8168	2108.9501
45	538.7693	795.4438	1170.4794	1716.6839	2509.6506
46	619.5847	922.7148	1369.4609	2025.6870	2986.4842
47	712.5224	1070.3492	1602.2693	2390.3106	3553.9162
48	819.4007	1241.6051	1874.6550	2820.5665	4229.1603
49	942.3108	1440.2619	2193.3464	3328.2685	5032.7008
50	1083.6574	1670.7038	2566.2153	3927.3569	5988.9139

List of Professional Advisers

Attorney:

Name: _____

Address: _____

_____

Phone: _____

Fax: _____

Email: _____

Financial Planner:

Name: _____

Address: _____

_____

Phone: _____

Fax: _____

Email: _____

Accountant or Other Tax Professional:

Name: _____

Address: _____

_____

Phone: _____

Fax: _____

Email: _____

Stockbroker/Money Manager:

Name: _____

Address: _____

_____

Phone: _____

Fax: _____

Email: _____

Banker:

Name: _____

Address: _____

_____

Phone: _____

Fax: _____

Email: _____

Mortgage Broker:

Name: _____

Address: _____

_____

Phone: _____

Fax: _____

Email: _____

Retirement Planning Adviser:

Name: _____

Address: _____

_____

Phone: _____

Fax: _____

Email: _____

Title Company Agent:

Name: _____

Address: _____

_____

Phone: _____

Fax: _____

Email: _____

Real Estate Broker/Home:

Name: _____

Address: _____

_____

Phone: _____

Fax: _____

Email: _____

Insurance Agent/Home:

Name: _____

Address: _____

_____

Phone: _____

Fax: _____

Email: _____

Insurance Agent/Auto:

Name: _____

Address: _____

_____

Phone: _____

Fax: _____

Email: _____

Insurance Agent/Life:

Name: _____

Address: _____

_____

Phone: _____

Fax: _____

Email: _____

Other:

Name: _____

Address: _____

_____

Phone: _____

Fax: _____

Email: _____

Other:

Name: _____

Address: _____

_____

Phone: _____

Fax: _____

Email: _____

Other:

Name: _____

Address: _____

_____

Phone: _____

Fax: _____

Email: _____

Other:

Name: _____

Address: _____

_____

Phone: _____

Fax: _____

Email: _____

Index

A

AAA. *See* American Arbitration Association (AAA)
ABA. *See* American Bar Association (ABA)
Accidental death or dismemberment insurance, 344
Accountants, 103, 104
  certified public accountants (CPAs), 102, 103
  enrolled agents (EAs), 103, 104
  forensic accountants, 103, 119
Account statements, 115
Acquisition costs
  collectibles, 334
  as tax basis for house, 229
ACR. *See* Association for Conflict Resolution (ACR)
Actuaries, 108
Aftermath of divorce, 445–468
  automobiles, 448
  change of settlement by self or ex-spouse, 451, 452
  children, 449, 450
  debts, 448, 449
  deposit accounts, 449
  estate planning, 448
  financial difficulties, 453–458
  financial planning tips, 458–461
  finishing business of divorce, 446–450
  goals, setting, 467, 468
  insurance, 448
  investment accounts, 449
  investments, 453
  joint accounts, 449
  moving beyond the divorce, 465, 466

  name change, 447
  new relationship, 461–465
  postdivorce actions, 447–450
  real estate, 448
  settlement detailed, reviewing, 450
  tips for completing business of divorce, 450
Agent's commission, real estate, 226, 227, 325, 326
Alimony, 408–429
  alternatives to monthly alimony, 424–428
  amount can afford to pay, 412–424
  amount needed, 410–424
  attorney, questions to ask, 428, 429
  bankruptcy and, 366, 367, 424
  Cash Flow Statement and, 410, 411
  change of after divorce, 451, 452
  children, considering needs of, 411
  cost of living adjustments (COLAs), 422
  day of month payments made, 419
  decision regarding paying receiving, 428
  disability insurance, 423
  duration of, 420, 422
  economic changes, future, 422, 423
  emotional realities, 410
  family support (one payment in combination with child support), 427, 428
  financial realities, 409, 410
  "frontloading," 418
  generally, 380
  house kept in exchange for release of, 246
  how payments will be made, 419
  income needs, 413, 414
  inflation and, 422, 423
  legal realities, 408, 409

life insurance, 423

lump sum alimony, 424–427

modification of payments by payer, 423, 424

moving out by one spouse and, 67

new relationship and, 463

nondeductibility, factors to consider, 414

nonpayment of, 423, 424

payers, chart, 414

payers, taxes and, 415, 419

payments, practical issues, 419, 420, 422–424

property settlement payments and, 417

recipients, chart, 413

recipients, taxes and, 415

recordkeeping, 421

schedules for in your area, 411, 412

settlement, steps to, 410–428

taxes and, 414–419

temporary, 142, 143

transferring of assets to recipient in exchange for, 426

when paid, 419

"All at once" syndrome, avoiding, 44, 45

Alternatives to traditional litigation

arbitration, 6

collaborative divorce, 7

limited-scope representation, 7

mediation, 6

private judging, 6

Ambivalence, 46

American Arbitration Association (AAA), 100

American Bar Association (ABA), 89, 90

American Taxpayer Relief Act, 165

Annual mortgage interest, 219

Annual property taxes, 219, 220

Annuities, 330

Annulments, joint tax returns and, 168

Appraisal Institute, 221

Appraiser Qualifications Board (AQB), 221

Appraisers

qualifications, 221

websites, 477, 478

Arbitration, described, 6

Arbitrators, 99, 100

Arizona

covenant marriages, 20

hidden assets, 118

Arkansas, covenant marriages, 20

Assets

bankruptcy, sale of assets, 365

cashing, holding of on, 7

classification as, 342

"easy-to-forget," 118–122

handling of, options, 295–297

hiding of by spouse, 116–118

house kept in exchange for, 246

income, differences, 191, 192

sale of, 142, 454

tax losses, 295–297

See also Assets and Liability Worksheet; Net Worth Statement Assessing Assets

Assets and Liability Worksheet, 179

debt, dividing, 363

negotiating settlement and, 433, 434

Association for Conflict Resolution (ACR), 95

Attitude

financial decisions and, 36, 37

mental attitude, financially-focused, 44

Attorneys

affording, 16

directories, 473

as employees, 17

finding, 472–475

free sessions with, 87

gender and, 92

group legal plans, 473

interest in case, 93

knowledge of, 93

personal referrals, 473

prepaid legal insurance, 473

rapport, 94

referral panels, 473

retainers, 89

selection of, 89, 93, 94

strategy of, 94

team, availability of, 94

websites, 476

what to look for, 93, 94, 474, 475

working with, 47, 91–94

See also Attorney's fees; Attorneys, questions to ask

Attorney's fees

house, sale of, 227

real estate, 326

Attorneys, questions to ask

alimony, 428, 429

bankruptcy attorney, 378

bankruptcy, divorce attorney, 377, 378

child support, 428, 429

employee benefits, 357

financial fact-finding, 128

house, decisions regarding, 247, 248

joint property, 83

moving out, 69, 70

property and expenses, 195

retirement plans, 288–290

risks, protection against, 155

stock options, 357

Automatic stay, bankruptcy, 369

Automobiles, postdivorce actions, 448

B

Background resources (law libraries), using, 471

Balance Sheet Summary, 190

Bankers, 104

Bankruptcy, 363, 365–375

after divorce, 458

alimony and, 366, 367, 424

assets, sale of, 365

automatic stay, 369

bankruptcy attorney, questions to ask, 378

bankruptcy estate, 369

Chapter 7, 369–373

Chapter 11, 370

Chapter 13, 370–372

child support and, 366, 367

credit card debt and, 369

credit counselor, choosing, 374

discharge, 370

divorce attorney, questions to ask, 378

domestic support obligation, 370

exemptions, 371

of ex-spouse, wording of agreement pertaining to, 370

means test, 371

nondischargeable debt, 371

one spouse filing for, 367, 368, 370

protecting self, 373

retirement plans and, 285

second job, *see*king, 366

terms to know, 369–371

types of, 371–375

websites, 378

what to do if facing, 365, 366

Bankruptcy Code, 370

changes to, 368

Defense of Marriage Act (DOMA) and, 377

Bankruptcy estate, 369

Bifurcation, 443

Bill-paying procedure, organizing post-divorce, 458, 459

Birthdays, child support and, 390

Bonds, settlement and, 311, 312
    corporate bonds, 315
    EE U.S. Savings Bonds, 315
    financial value (after tax/after sale), 314–316, 319–321
    HH bonds, 315
    municipal bonds, 315
    tax basis, 319–321
    zero coupon bonds, 315

Bonuses, collusion to delay, 119

Broker, sample letter to, 80

Budgets after divorce, 457

Business assets, moving out by one spouse and, 67

Business contacts, delay in signing, 119

Business continuation coverage, 152

Buyer (of house), thinking like, 224

Buy out of spouse's share of house, 210–212, 240

C

Cafeteria plans, 346

California
    child support, liens on property for, 404
    hidden assets, 118
    house as separate property debt for former spouse, 228
    retirement plans, separation date, 268

Capital gains and losses
    house, sale of, 211

    tax returns, 125, 126

Capital gains exclusion, sale of house, 212

Capital gains taxes, 60

Capital improvements, real estate, 324
    as tax basis for house, 229–231

Cash
    assets, holding off on cashing, 7
    financial value (after tax/after sale), 314
    preference for, 31, 32
    retirement plans and access to, 282, 283
    risk-free nature of, 32
    settlement, cash and cash equivalents, 311, 314
    See also Cash flow; Cash flow statements

Cash flow, 193–195
    income and expenses, 196–205

Cash flow statements
    adding up income, 194
    estimating expenses, 194
    updating post-divorce, 458

Cash value policies, insurance, 330
    life insurance, 338

Certified financial planners (CFPs), 102

Certified public accountants (CPAs), 102, 103

CFPs. *See* Certified financial planners (CFPs)

Chapter 7 bankruptcy, 369–373

Chapter 11 bankruptcy, 370

Chapter 13 bankruptcy, 370–372

Checking accounts, joint, 76, 77

Child care, child support and, 386

Child custody
    change of after divorce, 452
    decision regarding, 407, 408
    determination of custodial parent, 385, 386

moving out by one spouse, custody rights and, 57

"nonresidential" parents, 382

parental rights, protecting, 381–408

"primary custody," 382

"residential" parents, 382

"secondary custody," 382

websites, 477

Children

  cost of, 193

  discussions regarding money, 41

  estate protection, 154

  moving out of state, 60

  postdivorce actions, 449, 450

  tax deduction, claiming as, 8

  See also Child custody; Child support

Child support, 381–408

  amount can afford to pay, 393–396

  amount needed, 393–396

  attorney, questions to ask, 428, 429

  bankruptcy and, 366, 367

  birthdays, 390

  calculating support, 397–399

  change of after divorce, 452

  checklist of issues to consider, 391

  child care, 386

  college funds, 386, 388, 389

  computing methods, 397–399

  costs to rear children, 386–391

  court hearings regarding unpaid support, 403, 404

  day of month payments made, 400, 401

  decision regarding, 407, 408

  dental care, 389, 390

  do's and don'ts of parenting during divorce, 387, 388

  duration of, 401

  education, 386, 388, 389

emergency savings, 389

emotional realities, 381, 383, 384

failure to pay, 3

family support (one payment in combination with alimony), 427

financial realities, 381, 383

flat percentage model, 397

flexibility with, 7

generally, 380

health care, 389, 390

health insurance coverage, 392, 393

hobbies, 390

holidays, 390

how paid, 399, 400

income needs, 394

income shares method, 397

legal realities, 381–383

lower-paying job taken by spouse to reduce obligation, 396

medical expenses, deduction, 395

modifications to, 402, 403

"most wanted lists," regarding nonpayment of, 404

online calculator, finding and using, 398

payments, practical issues, 399–407

percentage of income model, 397

physical custody, 382

settlement, steps to, 384–408

state guidelines, 396

tax benefits, 395

taxes, filing and paying, 171, 172

varying percentage model, 397

visitation rights, 382

websites, 476, 477

when paid, 399–401

  See also Child support, nonpayment of

Child support, nonpayment of

  credit bureau reports, 404

dealing with, 403–406

district attorney assistance, 405

driver's licenses, refusal to renew, 405

garnishments, 405

liens on property, 404

lottery winnings, interceptions, 404

missing parent, finding, 407

passport denial, 405

pension plans, tapping of, 406

professional licenses, refusal to renew, 405

property garnishments, 405

tax refunds, interceptions of, 404

wage garnishments, 405

Closing costs

real estate, 326

sale of house, 227

COBRA (Consolidated Omnibus Budget Reform Act), 148

Cohabitation, 461–463

COLAs. See Cost of living adjustments (COLAs)

Collaborative divorce, 97–99

described, 7

pros and cons of, 98, 99

Collectibles

acquisition costs, 334

financial value (after tax/after sale), 333–337

improvements, 334

purchase price, 334

selling price, 334

settlement and, 313

tax brackets, 334–337

wholesale value, 334

College funds, child support and, 386, 388, 389

Colorado, hidden assets, 118

Commingled property, 176

Commissions, real estate, 226, 227, 325, 326

Community property

gains and losses, 297

Net Worth Statement Assessing Assets, 182

state laws and, 175

See also Community property states

Community property states

debt, dividing, 362

prenuptial agreements, 464

taxes, filing and paying, 162

Compensation

"Sources of Compensation," 191, 192

Consolidating debt, 374

Contributory retirement plans, income tax considerations, 270

Corporate bonds, 315

Cost approach, real estate, 324

Cost of living adjustments (COLAs)

alimony and, 422

retirement plans, 282

Costs

costly nature of divorce, 28, 29

unpredictability of, 26

Counseling, 52

credit counseling, 106, 374, 457

marriage counselors, 45, 46

Covenant marriages, 20

CPAs. See Certified public accountants (CPAs)

Credit

credit counselor and, 374

moving out by one spouse, 62

websites, 378

See also Credit cards; Credit cards, joint accounts; Credit reports

Credit cards

debt, bankruptcy and, 369

expenses, listing, 195
use of, 142
See also Credit cards, joint accounts
Credit cards, joint accounts, 72–75
   closing, sample letter, 73
   moving out by one spouse, 62, 64
   paying off balance, 74
Credit counseling, 457
   choosing counselors, 106, 374
Creditors, contacting, 457
Credit reports, 73
   building better report, 65–67
   child support, nonpayment of, 404
   credit file, checking, 63, 64
   credit score, defined, 65
   financial fact-finding, 115
   inaccurate, negative information, 66
Credit score, defined, 65
Current balance, Net Worth Statement
   Assessing Assets, 182, 183

D

Date of marriage, 61
Date of purchase/acquisition, Net Worth
   Statement Assessing Assets, 182
Date of separation, 61
Debt
   allocation, change of after divorce, 451
   concerns regarding payment of, 3
   consolidating debt, 374
   moving out by one spouse, 61, 62
   Net Worth Statement Assessing Assets,
   183
   postdivorce actions, 448, 449
   on property, 324
   responsibility for, 177
   websites, 378

See also Bankruptcy; Debt collection
   after divorce; Debt consolidation; Debt,
   dividing
Debt collection after divorce, 64
Debt consolidation, 457
Debt, dividing, 359–378
   Assets and Liability Worksheet, 363
   community property states, 362
   at divorce, 375, 376
   listing debts, 363, 364
   medical expenses, 361
   Net Worth Statement Assessing Assets,
   363
   nothing to fight over, 376, 377
   responsibility for debt, general rules,
   360–362
   separately owned property, 361, 362
   spousal debt chart, 364
Decision making, financial fact-finding, 113
Deductions, fees paid to professionals, 108
Deed, obtaining blank, 78
Default on mortgage as option, 214, 215
Defense of Marriage Act (DOMA), 9, 377
Deferred annuities
   described, 257
   tax basis of, 271
Defined benefit plans
   defined contribution plans, distinguished,
   253
   described, 256
   division of benefits, 287
   financial value of, 275, 276
   legal value of, 265, 266
Defined contribution plans
   defined benefit plans, distinguished, 253
   division of benefits, 286
   employer-sponsored plans, 254
   financial value of, 273, 274

legal value of, 265
Dental care, child support and, 389, 390
Dental expenses, tax returns, 124
Dependent care assistance programs, 345, 346
Deposit accounts
  joint accounts, 76, 77
  postdivorce actions, 449
Deposition, 178
Depreciation, 31, 32
  home office, 232
  real estate, 324
Depression, 38
Disability insurance, 152
  alimony and, 423
Disability plans, 343, 344
Discharge, bankruptcy, 370
Discovery
  deposition, 178
  documents, request for production of, 178
  hidden assets, 117, 118
  inspection, request for, 178
  interrogatories, 178
  subpoenas, 178
District attorney assistance, nonpayment of child support, 405
Dividends
  financial value (after tax/after sale), 321
  tax returns, 124
Dividing debt. See Debt, dividing
Division of benefits, retirement plans, 283–287
  bankruptcy filing and, 285
  consequences of, 283
  defined benefit plans, 287
  defined contribution plans, 286
  future (deferred) division, 284
  IRAs, 285, 286

present division, 284
  survivor benefits, 284
  TSAs, 286
Division of financial investments. See Financial investments, division of
Division of property
  after divorce, 451
  miscellaneous property, division of, 435
  preliminary division, 434
  summary of, 437
Divorce ceremonies, 443
Divorce financial analysts, 102, 103
Documents, request for production of, 178
Do-it-yourself divorce services, 109, 110
DOMA. See Defense of Marriage Act (DOMA)
Domestic support obligation, bankruptcy, 370
Domestic violence, websites, 477
Driver's licenses, refusal to renew for child support nonpayment, 405
Dues as assets, 122

E

Early withdrawal penalties, 314
  retirement plans, 272, 273
EAs. See Enrolled agents (EAs)
"Easy-to-forget" assets, 118–122
  dues, professional, 122
  frequent flyer points, 120, 121
  magazine subscriptions, 122
  prepaid insurance, 121
  property taxes, 121
  season tickets, 121
  stock options, 121
  tax refunds, 121
  time shares, 121

vacation pay, 121

Ebb and flow of emotions, management of, 45–47

EDGAR (Electronic Data Gathering Analysis and Retrieval), 350

Educational assistance programs, 347

Education, child support and, 386, 388, 389

EE U.S. Savings Bonds, 315

Email, reliance on, 40

Emergency funds
    child support and, 389
    postdivorce, 459

Emergency plans, postdivorce, 459

Emotional aspects of financial issues, 35–49
    "all at once" syndrome, avoiding, 44, 45
    ambivalence, 46
    attitude, 36, 37
    attorneys, working with, 47
    depression, 38
    ebb and flow of emotions, management of, 45–47
    "getting back together," hopes for, 37
    "getting it over with," 37, 46
    grief, stages of, 45, 46
    hardest time of divorce process, psychologists' view of, 46, 47
    marriage counselors, 45, 46
    mental attitude, financially-focused, 44
    overwhelmed, feelings of being, 47, 48
    "sore spots," 42
    spiritual level of divorce, 49
    stress reduction techniques, 39, 40
    threats from spouse, 43
    tracking feelings, 48
    well-being (emotional), safeguarding, 40, 41
    worst-case scenarios, preparation for, 42–44

Emotional realities
    alimony, 410
    child support, 381, 383, 384

Employee assistance programs, 347

Employee benefits, 342–347
    accidental death or dismemberment insurance, 344
    attorneys, questions to ask, 357
    cafeteria plans, 346
    dependent care assistance programs, 345, 346
    disability plans, 343, 344
    educational assistance programs, 347
    employee assistance programs, 347
    group legal services, 347, 473
    group term life insurance, 344
    health insurance, 343, 344
    health savings accounts (HSAs), 344, 345
    severance plans, 346
    tax specialists, questions to ask, 357
    top hat plans, 357

Employee stock ownership plans (ESOPs), 254

Employee stock purchase plans, 352

Enforcement of law, difficulty of, 19

Enrolled agents (EAs), 103, 104

Equitable distribution of property, 175, 176

Equity
    credit lines, joint accounts, 75, 76
    Net Worth Statement Assessing Assets, 183
    value of each investment, 310–313
    value of house, 224–226

ESOPs. See Employee stock ownership plans (ESOPs)

Estate planning
    continued home ownership and, 243
    financial fact-finding, 115

postdivorce, 448, 460, 461
tools, 156
See also Estate protection
Estate protection
children, planning for, 155
estate planning tools, 156
trusts, 154
wills, 154
Exempt property, bankruptcy, 371
Expenses, property and. *See* Property and
expenses

F

Fact-finding, financial. *See* Financial fact-
finding
Fair Debt Collection Practices Act, 457
Fair market value, real estate, 323, 324
Family home. *See* House, decisions regarding
Family necessities, 61
Family support (alimony and child support
combined), 427, 428
Fannie Mae, 223
Federal Trade Commission, 457
Fees
attorney's fees, 227, 328
credit counselor and, 374
professionals, deductibility of fees paid to,
108
redemption fees, 314
refinancing and selling proceeds to selling
spouse, 245
Finalizing settlement, 441–443
Financial advisers, questions to ask regarding
house, 248
Financial analysts, 102, 103
Financial assistance, websites, 478, 479
Financial commitments, anticipated, 137

Financial difficulties after divorce
assets, sale of, 454
bankruptcy, 458
budgets, 457
credit counseling, 457
creditors, contacting, 457
debt consolidation, 457
home equity loans, 455, 456
items, sale of, 454
loans from family or friends, 454
tax-deferred accounts, cashing in, 455
tax refunds, 456
Financial fact-finding, 111–133
account statements, 115
attorneys, questions to ask, 128
checklist, 128–133
credit reports, 115
decision making, 113
disorganized persons, advice for, 113–116
estate plans, 115
gifts given by spouse, 127
hiding of assets by spouse, 116–118
information analysis, 113
information gathering, 113
insurance policies, 115
loan applications, 115
professionals, finance, 115
retirement plans, 115
tax returns, 114, 122–127
title reports, 115
Financial investments, division of, 291–340
after-sale value of each investment,
313–337
after-tax value of each investment,
313–337
assets, options, 295–297
bonds, 311, 312, 314–316, 319–321
cash and cash equivalents, 311, 314

cash value life insurance, 338
collectibles, 313, 333–337
concepts to consider, 293–301
dividends, 321
equity value of each investment, 310–313
growth and, 297, 298
how investments will be divided, 338–340
income and, 297, 298
inflation, 301
insurance, 312, 330–332
investment chart, 303–309
investments held by spouse and self, determination of, 310
legal value of each investment, 310–313
limited partnerships, 312, 313, 332, 333
mutual funds, 312, 316–321
real estate, 322–330
reinvestments, 321
risk considerations, 293–295
security and, 297, 298
settlement, steps to, 302–338
single, difference of investments made while, 293
stocks, 311, 312, 316–321
tax basis, 298–300, 302
total return, looking to, 301, 302
trusted professionals, 295
Financially troubled times, managing divorce during, 5–8
alternatives to traditional litigation, 6, 7
family home, 5
recession of 2008, 5
Financial planners, 100–102
Financial Planning Association, 101
Financial planning tips, postdivorce, 458–461
bill-paying procedure, organizing, 458, 459
cash flow statements, updating, 458
emergency funds, 459
emergency plans, 459
estate planning, 460, 461
insurance, reevaluating, 459
net worth, updating, 458
retirement program, reevaluating, 460
tax status, keeping up with changes in, 459
Financial realities, 25–34
acceptance of costliness, 29
alimony, 409, 410
cash, preference for, 31, 32
child support, 382
costly nature of divorce, 28, 29
facing, 53
inflation, 32
math, mastering of, 34
problems, unexpected, 33
risks of financial connections to another person, 29, 30
settlements, financial angles of, 26, 27
tax laws, 27
time value of money, 32
unpredictability of costs, 26
See also Legal vs. financial realities of divorce
First-in-first out, stock, 320
Fix-up costs
real estate, 326–329
sale of house, 227
Flat percentage model of computing child support, 397
Flexibility, 7, 8
Foreclosure auctions, 214
Forensic accountants, 103, 119
401(k) plans, loans from, 273
Fraud, loans, 76

Freddie Mac, 223
Frequent flyer points, 120, 121
"Frontloading," alimony, 418
Future, planning for, 135–143
    financial commitments, anticipated, 137
    goals that will cost money, 138–140
    life events, major upcoming, 136, 137
    where money will come from, 141, 142
Future Value Factor table, 235–239

G

Gains and losses
    community property, 297
    stock options, 351
    See also Capital gains and losses; Tax
        losses
Garbage in/garbage out (GIGO), 34
Garnishments, nonpayment of child
    support, 405
Gender, working with attorneys and, 92
"Getting back together," hopes for, 37
"Getting even," 36
"Getting it over with," 37, 46
Gifts given by spouse, 127
GIGO (garbage in/garbage out), 34
Goals
    costliness of, 138–140
    setting after divorce, 467, 468
Government-sponsored retirement plans,
    256
Grief, stages of, 45, 46
Group legal plans and services, 347, 473
Group term life insurance, 344
Growth, balancing security and income,
    297, 298

H

Hardest time of divorce process,
    psychologists' view of, 46, 47
HARP. *See* Home Affordable Finance
    Program (HARP)
Health care, child support and, 389, 390
Health insurance, 343, 344
    child support and, 392, 393
    COBRA (Consolidated Omnibus Budget
        Reform Act), 148
    Qualified Medical Child Support Orders
        (QMCSOs), 147, 392, 393
Health savings accounts (HSAs), 344, 345
HELOCs. *See* Home equity lines of credit
    (HELOCs)
Help, getting, 85–110
    accountants, 103, 104
    actuaries, 108
    arbitrators, 99, 100
    attorneys, 89
    bankers, 104
    business appraisers, 108
    collaborative divorce, 97–99
    competence of person, considerations
        regarding, 86, 87
    credit counselors, 106
    divorce financial analysts, 102, 103
    financial planners, 100–102
    insurance agents, 104, 105
    limited-scope representation, 89–91
    mediators, 94–97
    money managers, 103
    objective evaluation of professional, 88
    online do-it-yourself divorce services, 109,
        110
    paralegal services, 107
    payment of professional, 87

private investigators, 109

private judges, 99, 100

questions to consider, 86–88

real estate agents, 105

selection of professionals, 88–110

services, considering best use of, 87, 88

stockbrokers, 103

suitability of person, considerations regarding, 86, 87

therapists, 105, 106

typing services, 107

"unbundling," 89–91

HH bonds, 315

Hidden assets, 116–119

bonuses, collusion to delay, 119

business contacts, delay in signing, 119

"discovery," 117, 118

lottery winnings, 117

searching for, 119

by spouse, 116–119

Hobbies, child support and, 390

Holidays, child support and, 390

Home Affordable Finance Program (HARP), 223

Home environment, controlling, 40

Home equity lines of credit (HELOCs), 64

Home equity loans, 455, 456

Home mortgage interest, tax returns, 124

Home office depreciation, 232

Hope for marriage, 51, 52

House, decisions regarding, 207–248

attorney, questions to ask, 247, 248

financial advisers, questions to ask, 248

financial vs. legal realities, 209

not keeping, when recommended, 437

questions to ask, 210

state laws and, 208, 209

values of, 5, 213, 221–224

See also House, steps toward selling; Living arrangements

House, steps toward selling, 210–247

acquisition costs as tax basis for house, 229

agent's commission, 226, 227

appraisers, qualifications, 221

attorney's fees, 227

best option, 240–243

buyer, thinking like, 224

buy out of spouse's share, 210–212, 240

capital gains, 211

capital gains exclusions, 212

capital improvements as tax basis for house, 229–231

closing costs, 227

costs involved in selling, 226, 227

defaulting on mortgage as option, 214, 215

equity value of house, 224–226

estate planning, continued home ownership and, 243

fair market value of house, 221–224

fix-up costs, 227

foreclosure auctions, 214

future sale of house, financial value, 234–239

Future Value Factor table, 235–239

joint ownership with spouse in order to sell in future, 212, 213, 242, 243

loan modification programs, 223

purchase price as tax basis for house, 229

purchasing house from spouse or former spouse, 243–247

real estate sites, 222

sale with spouse, financial value of house, 232–234

selling house with spouse and splitting proceeds, 212, 241

separate property debt for former spouse, 228

share of house, 239, 240

spouse, selling share of house to, 212

tax basis for house, 229–231

tax benefits of selling, 231

third person, sale to during year when divorce is final, 241

total housing cost per month, determining, 215–220

transferring share of house to spouse, 241

value of house, finding, 213

Housing cost per month, determining, 215–220

HSAs. *See* Health savings accounts (HSAs)

I

Improvements
    collectibles, 334
    real estate, 324

Incentive stock options (ISOs), 352

Income
    assets, differences, 191, 192
    cash flow statement adding up income, 194
    child support and income needs, 394
    classification as, 342
    income taxes and, 67

Income approach, real estate, 324

Income properties, real estate, 322, 323

Income shares method of computing child support, 397

Income taxes, 67
    retirement plans, 269–273

Indemnity clause in divorce agreement as protection from IRS, 168

Individual retirement accounts (IRAs)
    division of benefits, 285, 286
    early withdrawal penalties, 272, 273
    employer-sponsored, 255
    financial value of, 273, 274, 279
    legal value of, 265
    rollovers from ex-spouse's IRA, 279
    tax basis of, 271
    traditional IRAs, 257

Inflation, 32, 301
    alimony and, 422, 423

Information analysis, 113

Information gathering, 113

Innocent Spouse Relief, 163, 164

Inspection, request for, 178

Installment loans, sale of house, 245

Institute for Divorce Financial Analysts, 102, 103

Insurance, 146–152
    accidental death or dismemberment insurance, 344
    annuities, 330
    cash value policies, 330
    financial value (after tax/after sale), 330–332
    group term life insurance, 344
    health insurance, 146–148
    policies, 115
    postdivorce actions, 448
    prepaid insurance as assets, 121
    prepaid legal insurance, 473
    reevaluating, postdivorce, 459
    settlement and, 312
    See also Business continuation coverage; Disability insurance; Health insurance; Life insurance

Insurance agents, 104, 105

Interest, tax returns, 124

Internal Revenue Service (IRS), 30, 31, 108, 122

    alimony and, 418

    dishonesty of spouse, 163, 164

    tax basis and, 302

    tax returns, 114

    See also Taxes; Tax laws

Interrogatories, 178

Investment accounts, postdivorce actions, 449

Investment chart, 303–309

Investments

    after divorce, 453

    moving out by one spouse and, 67

    See also Financial investments, division of; Investments, joint property

Investments, joint property, 79, 80

    broker, sample letter to, 80

IRAs. *See* Individual retirement accounts (IRAs)

IRS. *See* Internal Revenue Service (IRS)

IRS Form 1099-MISC, 127

IRS Form 2119, 229

IRS Form 2848, 116

IRS Form 4506, 159

IRS Form 4506-T, 114, 116

IRS Form 4506-T-EZ, 114

IRS Form 8332, Release of Claim to Exemption for Child of Divorced or Separated Parents, 394

"Is marriage over" question, 51–53

ISOs. *See* Incentive stock options (ISOs)

J

Joint accounts, 72–83

    checking accounts, 76, 77

    credit card accounts, 62, 64, 72–75

    deposit accounts, 76, 77

    equity credit lines, 75, 76

    joint title to property, 77–79

    liquidation of by spouse, 77

    postdivorce actions, 449

    savings accounts, 76, 77

Joint property, 71–83

    attorneys, questions to ask, 83

    community property, 175

    equitable distribution of property, 175, 176

    investments, 79, 80

    marital balance sheet and, 435

    safe-deposit boxes, joint, 80, 81

    shared property, 81, 82

    special collections, 81, 82

    state laws and, 175, 176

    See also Joint accounts

Joint tax returns, 161, 165–168

    annulments, 168

    dividing liability, 168–171

    indemnity clause in divorce agreement as protection from IRS, 168

    liability, 166

    refunds, 168–171

    refusal of spouse to sign, 167

    responsibility when signing, 166, 167

    settlement of tax bill by spouse, 167, 168

Joint tenancy, 77, 78

    Net Worth Statement Assessing Assets, 181

    title documents, 153

Joint title to property, 77–79

    change in ownership, 79

    deed, obtaining blank, 78

    joint tenancy, 77, 78, 153

tenancy in common, 78, 79, 153
Judges, private, 99, 100

K

Key financial issues, flexibility on, 7, 8

L

Land, raw, 321
Law libraries, 470–472
Lawyers. *See* Attorneys
Legal discovery. *See* Discovery
Legal realities
    alimony, 408, 409
    child support, 381–383
    facing, 53
    See also Legal vs. financial realities of
      divorce
Legal services, employee benefits, 347
Legal value, Net Worth Statement Assessing
  Assets, 183
Legal vs. financial realities of divorce, 11–23
    attorneys, affording, 16
    comparison, 12, 13
    covenant marriages, 20
    enforcement of law, difficulty of, 19
    house, decisions regarding, 209
    legal reality, 13–20
    local nature of divorce law, 15, 16
    out of court, settlement of divorces, 14, 15
    professionals hired as employees, 17, 18
    questions, asking, 17
    settlement agreement, defined, 19
    stages of divorce, 21–23
    strategy of thinking financially, 20
    See also Financial realities

Liens
    child support, nonpayment of, 404
    on house, 225, 228
Life events, planning for major upcoming,
  136, 137
Life insurance, 149–151
    alimony and, 423
    cash value life insurance, 338
    group term life insurance, 344
    spouse, maintaining on, 150
Limited partnerships
    financial value (after tax/after sale), 332,
      333
    settlement and, 312, 313
    tax losses, 296
Limited-scope representation, 7
Living arrangements, rethinking, 7
Loan modification programs, 223
Loans, 142
    applications, 115
    from family or friends, 454
    401(k) plans, from, 273
    fraudulent, 76
    home equity loans, 455, 456
Local nature of divorce law, 15, 16
Lottery winnings, 117
    interceptions for nonpayment of child
      support, 404
Louisiana, covenant marriages, 20
Lump sum alimony, 424–427
    combination with child support, 427
    payer, 425, 426
    recipient, 424, 425

M

Magazine subscriptions as assets, 122
Maine, hidden assets, 118

Making Home Affordable Program (MHA), 223

Marital Balance Sheet, 179, 433–437
  Assets and Liability Worksheet and, 433, 434
  house, when not to keep, 437
  jointly owned property, specific items of, 435
  miscellaneous property, division of, 435
  preliminary division of property, 434
  sample, 436
  summary of property division, 437

Marital property, state laws and, 174–177
  jointly owned property, 175, 176

Market data approach, real estate, 324

Market value, Net Worth Statement
Assessing Assets, 182, 183

Marriage counselors, 45, 46

Math, mastering of, 34

Means test, bankruptcy, 371

Mediation
  described, 6
  when to use private mediation, 96, 97

Mediators
  selecting, 94–97
  websites, 476
  when to use private mediation, 96, 97

Medical expenses
  child support and, 395
  debt, dividing, 361
  tax returns, 124

Medicare taxes, 165
  returns, 123

Mental attitude, financially-focused, 44

Merrill Lynch, 102

MHA. See Making Home Affordable
Program (MHA)

Missing parent, finding for nonpayment of child support, 407

"Money crazies." See Emotional aspects of financial issues

Money managers, 103

Money purchase plans, 254

Monthly housing costs, determining
  gross monthly housing costs, 217
  listing costs, 216, 217
  net monthly housing costs, 217–220

Morgan Stanley, 102

Mortgage
  annual mortgage interest, 219
  defaulting on as option, 214, 215

Movers' costs, 60

Moving out by one spouse, 55–70
  alimony, 67
  attorneys, questions to ask, 69, 70
  copies of documents, 69
  credit, 62
  credit file, checking, 63, 64
  credit report, building better, 65–67
  custody rights, 57
  debt collection after divorce, 64
  debts, 61, 62
  family necessities and, 61
  income taxes and, 67
  investments, 67
  milestone days, 61
  movers' costs, 60
  before moving, 69
  out of state, moving of children, 60
  separation agreement, written, 59
  separation date, 57–64
  Social Security benefits after divorce, 68, 69

Municipal bonds, 315

Mutual funds

financial value (after tax/after sale), 316–321
net asset value (NAV), 312
settlement and, 312, 316–321
tax basis, 298–300, 319–321

N

Name change, 447
National Association of Enrolled Agents, 104
National Foundation for Credit Counseling, 106, 457
Negotiating settlement, 431–443
  Assets and Liability Worksheet, 433, 434
  bifurcation, 443
  finalizing settlement, 441–443
  financial home prior to, 433
  marital balance sheet, 433–437
  offers and counteroffers, 438–441
  tips for, 442
  trial, going to, 441
Net asset value (NAV), mutual funds, 312
Net monthly housing costs, 217–220
  annual mortgage interest, 219
  annual property taxes, 219, 220
  points, 217, 218
  tax bracket, 220
Net worth, 179–190
  Assets and Liability Worksheet, 179
  Balance Sheet Summary, 190
  forms, filling out, 180
  Marital Balance Sheet, 179
  Net Worth Statement Assessing Assets, 180–183
  Net Worth Statement Calculating Liabilities, 183–190

updating postdivorce, 458
Net Worth Statement Assessing Assets, 180–183
  community property, 182
  current balance, 182, 183
  date of purchase/acquisition, 182
  debt, 183
  debt, dividing, 363
  equity, 183
  joint tenancy, 181
  legal value, 183
  market value, 182, 183
  separate title, 182
  source of assets, 182
  tenancy in common, 181, 182
  tenancy in the entirety, 181
  title (or owner), 180, 182
Net Worth Statement Calculating Liabilities, 183–190
  worksheet, 184–189
New relationship, postdivorce, 461–465
  alimony and, 463
  cohabitation, 461
  prenuptial agreements, 463–465
Nondischargeable debt, bankruptcy, 371
Nonqualified deferred compensation plans, 356, 357
Nonqualified deferred compensation plans chart, 258
Nonqualified plans, tax returns, 123
"Nonresidential" parents, 382
Nonstatutory (or nonqualified) stock options (NQSOs), 353
NQSOs. *See* Nonstatutory (or nonqualified) stock options (NQSOs)

O

Offers and counteroffers in negotiating settlement, 438–441
 trial, going to, 441
 understanding what you are getting, 440
Omnibus Reconciliation Act of 1993, 392
Online do-it-yourself divorce services, 109, 110
Online legal resources, 475
Options. *See* Stock options
Options agreement, points to include, 356
Ordinary dividends, tax returns, 124
Out of court, settlement of divorces, 14, 15
Out of state, moving of children, 60
Overwhelmed, feelings of being, 4, 47, 48

P

Paralegal services, 107
Parental rights, protecting, 381–408
Partnership income, tax returns, 126, 127
Passport denial, nonpayment of child support, 405
Payment of professional, 87
Penalties. retirement plans, 272, 273
Pension benefits and plans, 66
 nonpayment of child support, tapping for payment of, 406
 See also Employee benefits
Percentage of income model of computing child support, 397
Personal annuities
 described, 257
 legal value of, 265
Personal retirement plans, 257
Phases of splitting up, 51

Points
 net monthly housing costs, 217, 218
 refinancing and selling proceeds to selling spouse, 245
Postdivorce actions
 automobiles, 448
 children, 449, 450
 debts, 448, 449
 deposit accounts, 449
 estate planning, 448
 insurance, 448
 investment accounts, 449
 joint accounts, 449
 real estate, 448
 settlement detailed, reviewing, 450
 tips for completing business of divorce, 450
Prenuptial agreements, 463–465
Prepaid insurance
 as assets, 121
 legal insurance, 473
Present Value Factors, retirement plans, 267, 480–485
"Primary custody," 382
Private investigators, 109
Private judges, 6, 99, 100
Problems, unexpected, 33
Professional advisers, list of, 492–495
Professional licenses, refusal to renew for nonpayment of child support, 405
Professionals, finance, 115
Professionals hired as employees, 17, 18
Profit or loss from business, tax returns, 125
Property alimony payments, 417
Property and expenses, 173–205
 assets and income, differences between, 191, 192

attorneys, questions to ask, 195

cash flow, 193–195

cash flow income and expenses, 196–205

cash flow statement adding up income, 194

cash flow statement estimating expenses, 194

children, cost of, 193

commingled property, 176

credit card expenses, listing, 195

debts, 177

discovery, 177, 178

equitable distribution of property, 175, 176

jointly owned property, 175, 176

marital property, state laws and, 174–177

net worth, 179–190

separately owned property, 176

"Sources of Compensation," 191, 192

Property division. *See* Division of property

Property garnishments, nonpayment of child support, 405

Property protection, 152, 153

title documents, 152, 153

Property, sale of, 142

Property taxes

annual property taxes, 219, 220

as assets, 121

Purchase price

collectibles, 334

as tax basis for house, 229

Purchasing house from spouse or former spouse

alimony, keeping house in exchange for release of, 246

assets, house kept in exchange for, 246

installment loans, 245

nonresident's interest, sale to third person, 245

refinancing and selling proceeds to selling spouse, 244, 245

retirement plan, borrowing from to finance buyout, 246, 247

Q

Qualified Domestic Relations Orders (QDROs), 261–263

cash, and access to, 282

defined, 261

described, 261, 262

division of benefits and, 283

generally, 253

interim QDROs, 262, 263

offers and counteroffers in negotiating settlement, 440

tax problems with using, 406

Qualified Medical Child Support Orders (QMCSOs)

generally, 147, 392, 393

sample, 430

Questions, asking, 17. *See also* Attorneys, questions to ask; Tax adviser, questions to ask; Tax specialists, questions to ask

R

Raising money, 77

Real estate

agent's commission, 325, 326

attorney's fees, 326

capital improvements, 324

closing costs, 326

cost approach, 324

debt on property, 324

depreciation, 324
fair market value, 323, 324
financial value (after tax/after sale), 322–330
fix-up costs, 326–329
improvements, 324
income approach, 324
income properties, 322, 323
land, raw, 321
market data approach, 324
postdivorce actions, 448
real estate investment trusts (REITS), 322
rental properties, 322, 323, 330, 339, 340
rental, tax losses, 296
sale, cost of, 325
tax basis, 298, 299
tax bracket, 325
title documents, 152, 153
values, 325
Real estate agents, 105
Real estate investment trusts (REITS), 322
Real estate taxes, returns, 124
Recordkeeping, alimony, 421
Redemption fees, 314
Refinancing and selling proceeds to selling spouse, 244, 245
fees, 245
points, 245
rates, 244
Reinvestments, financial value (after tax/after sale), 321
REITs. See Real estate investment trusts (REITS)
Rental properties, 322, 323, 330
problems with dividing, 339, 340
Rental revenue, tax returns, 126
"Residential" parents, 382

Restraining orders
joint property and, 79, 82
when inadvisable, 79
Restrictive stock, 353
Restrictive stock units (RSUs), 353
Retirement
benchmarks to, 264
planning for, 252
See also Retirement plans
Retirement plans, 115, 249–290
attorneys, questions to ask, 288–290
bankruptcy filing and, 285
buyout of house, borrowing from to finance, 246, 247
calculating financial value, 277, 278
cash, access to, 282, 283
cash, preference for, 31
charges, 272, 273
cost of living adjustments (COLAs), 282
division decisions, 261
division of benefits, 283–287
double payments, 282
early distributions, 273
financial value of, 268–279
future worth of, 267
income tax considerations, 269–273
legal value of, 265–28
letter to spouse's employer, 288–290
marital portion of, 66, 266
nonqualified plans, tax returns, 123
payout status, 281
penalties, 272, 273
Present Value Factors, 267, 480–485
reevaluating, postdivorce, 460
separation date, effect on value, 268
state income taxes, 269
survivor benefits, 284

tax basis of, 270, 271
tax brackets, 269
tax breaks, 259
terms used, 260
understanding plans, 253–261
value of, 58
variations of, 259
vesting, 280, 281
See also specific type of plan
Returns. *See* Tax returns
Risks
    investments, 293–295
    See also Risks, protection against
Risks, protection against, 145–156
    attorneys, questions to ask, 155
    estate protection, 154–156
    insurance, 146–152
    property protection, 152, 153
Rollovers from ex-spouse's IRA, 279
Roth IRAs
    described, 257
    tax basis of, 271
RSUs. *See* Restrictive stock units (RSUs)

S

Safe-deposit boxes, joint, 80, 81
Sale of real estate, 325
    See also House, steps toward selling
Same-sex marriage, states allowing, 9, 10
SAR-IRAs, 255
Sate of divorce, 61
Savings accounts, joint accounts, 76, 77
Schedule A, 124
Schedule B, 124
Schedule C, 125
Schedule D, 125–126

Schedule E, 126, 127
S corporation income, tax returns, 126, 127
Season tickets as assets, 121
"Secondary custody," 382
Secondary markets, limited partnerships, 312
Security, income and growth, balancing, 297, 298
Selling price, collectibles, 334
Separate Liability Election, 163, 164
Separately owned property, 176
    debt, dividing, 361, 362
Separate tax returns, 161–165
    dishonesty of spouse with IRS, 163, 164
Separate title, Net Worth Statement Assessing Assets, 182
Separation agreement, written, 59
Separation date
    defined, 58
    moving out by one spouse, 57–64
Separation, temporary, 53
SEP-IRAs, 255
SERPs. *See* Supplemental Executive Retirement Plans (SERPs)
Settlement
    alimony, 451, 452
    change by self or ex-spouse after divorce, 451, 452
    child custody, 452
    child support, 451
    debt allocation, 451
    financial angles of, 26, 27
    property division, 451
    visitation rights, 452
    in writing, 452
    See also Negotiating settlement; Settlement, steps to
Settlement agreement, defined, 19

Settlement, steps to, 302–338
  account charges, 314
  after-sale value of each investment, 313–337
  after-tax value of each investment, 313–337
  alimony, 410–428
  bonds, 311, 312, 314–316, 319–321
  cash and cash equivalents, 311, 314
  child support, 384–408
  collectibles, 313, 333–337
  date of divorce, 311
  dividends, 321
  equity value of each investment, 310–313
  insurance, 312, 330–332
  investment chart, 303–309
  investments held by spouse and self, determination of, 310
  legal value of each investment, 310–313
  limited partnerships, 312, 313, 332, 333
  mutual funds, 312, 316–321
  real estate, 322–330
  reinvestments, 321
  stocks, 311, 312, 316–321
Severance plans, 346
Sex offenders, checking proximity of, 242
Shared property, 81, 82
SIMPLE IRAs, 255
Social Security Administration (SSA), 68, 69
Social Security benefits, 258
  after divorce, 68, 69
"Sore spots" (financial issues), 42
Source of assets, Net Worth Statement Assessing Assets, 182
"Sources of Compensation," 191, 192
Special collections, joint, 81, 82
Spiritual level of divorce, 49
Spousal debt chart, 364

Spousal support, flexibility with, 7
SSA. See Social Security Administration (SSA)
Stages of divorce, legal vs. financial, 21–23
Start-up businesses, 342
State law affecting divorce, finding, 470, 471
Statutory stock options, 352
Stipulations, 143, 171
Stock
  average basis, 321
  financial value (after tax/after sale), 316–321
  first-in-first out, 320
  settlement and, 311, 312, 316–321
  specific identification, 320, 321
  tax basis, 298–300, 319–321
Stockbrokers, 103
Stock options, 348–356
  as assets, 121
  attorneys, questions to ask, 357
  cashless exercise, defined, 351
  documents needed, 350
  employee stock purchase plans, 352
  exercise date, defined, 351
  exercise, defined, 351
  expiration date, defined, 351
  gain, defined, 351
  grant, defined, 351
  holding period, defined, 351
  incentive stock options (ISOs), 352
  lingo, 351
  nonqualified deferred compensation plans, 356, 357
  nonstatutory (or nonqualified) stock options (NQSOs), 353
  options agreement, points to include, 356
  personal identification information, protection of, 356

restrictive stock, 353

restrictive stock units (RSUs), 353

similarities between different types, 350, 352

spread, defined, 351

statutory stock options, 352

Supplemental Executive Retirement Plans (SERPs), 357

tax specialists, questions to ask, 357

types of, 349–353

vesting, defined, 351

websites, 355

worth of at divorce, 354, 355

Stress reduction techniques, 39, 40

Subpoenas, 178

Supplemental Executive Retirement Plans (SERPs), 357

Survivor benefits, retirement plans, 284

T

Tax adviser, questions to ask, 172

Tax assistance, websites, 478

Tax basis

acquisition costs as tax basis for house, 229

bonds, 319–321

capital improvements as tax basis for house, 229–231

deferred annuities, 271

financial investments, division of, 298–300, 302

individual retirement accounts (IRAs), 271

mutual funds, 298–300, 319–321

real estate, 298, 299

retirement plans, 270, 271

stock, 298–300, 319–321

Tax bracket, 220

collectibles, 334–337

real estate, 325

retirement plans and, 269

Tax-deferred accounts, cashing in, 455

Taxes

alimony, 414–419

capital gains taxes, 60

child, claiming as deduction, 8

child support and, 394, 395

income taxes, 67

Medicare, 165

Qualified Domestic Relations Orders (QDROs) and, 406

"taxable event," divorce as, 30, 31

See also Income tax; Tax bracket; Taxes, filing and paying; Tax law; Tax refunds; Tax returns

Taxes, filing and paying, 157–172

child support, 171, 172

community property states, 162

dishonesty of spouse with IRS, 163, 164

estimate of tax bill, obtaining, 158, 159

Innocent Spouse Relief, 163, 164

joint filing, 161, 165–171

separate filing, 161–165

Separate Liability Election, 163, 164

status to use when filing, 160

tax adviser, questions to ask, 172

tax rates, 164

temporary alimony, 171, 172

written tax agreements, 172

Tax laws, 27

Tax losses, 295–297

Taxpayer Advocate's Office (TAO), 456

Taxpayer Relief Act of 1997, 323

Tax rates, 164

Tax refunds

as assets, 121

calculating share of, joint returns, 169–171

interceptions of for nonpayment of child support, 404

joint tax returns, 168–171

quick receipt of, 456

Tax returns, 114

capital gains and losses, 125, 126

dental expenses, 124

dividends, 124

home mortgage interest, 124

interest, 124

medical expenses, 124

Medicare, 123

nonqualified plans, 123

ordinary dividends, 124

partnership income, 126, 127

profit or loss from business, 125

real estate taxes, 124

rental revenue, 126

retention of, how long, 127

S corporation income, 126, 127

tips, 122, 123

wages, 122, 123

Tax-sheltered annuities (TSAs), 254

division of benefits, 286

financial value of, 273, 274

income tax considerations, 269

Tax specialists, questions to ask

employee benefits, 357

stock options, 357

Tax status, postdivorce changes, 459

Temporary alimony, 142, 143

taxes, filing and paying, 171, 172

Temporary separation, 53

Tenancy in common, 78, 79

Net Worth Statement Assessing Assets, 181, 182

title documents, 153

Tenancy in the entirety, Net Worth Statement Assessing Assets, 181

Therapists, 105, 106

websites, 476

Threats from spouse, 43

Time shares as assets, 121

Tips, tax returns, 122, 123

Title documents, 152, 153

Title reports, 115

Title to property, joint, 77–79

Top hat plans, 357

Tracking feelings, 48

Trading fees, 314

Treasury Department, 223

Trial, going to when negotiating settlement, 441

"Triple-squeeze" generation, 252

Trusts, 154

TSAs. *See* Tax-sheltered annuities (TSAs)

Typing services, 107

finding, 472

U

"Unbundling," 89–91

U.S. Department of Housing and Urban Development, 223

V

Vacation pay as asset, 121

Value

of house, 5, 213, 221–224

real estate, 325

Varying percentage model of computing child support, 397
Vesting
    retirement plans, 280, 281
    stock options, 351
Visitation rights, 382
    change of after divorce, 452
Voicemail, reliance on, 40

W

W-2 form, 122–127
Wages
    garnishments for nonpayment of child support, 405
    tax returns, 122, 123
Websites
    American Arbitration Association (AAA), 100
    Appraisal Institute, 221
    appraisers, 477, 478
    Association for Conflict Resolution (ACR), 95
    attorneys, 476
    attorneys specializing in working with particular gender, 92
    bankruptcy, 378
    child support, 476, 477
    child support, online calculators, 398
    credit, 378
    credit bureaus, 63
    custody, 477
    debt, 378
    domestic violence, 477
    Federal Trade Commission, 457
    financial assistance, 478, 479
    Financial Planning Association, 101
    Institute for Divorce Financial Analysts, 103
    Internal Revenue Service (IRS), 31, 122
    lawyer directories, 473
    legal resources, 475
    Making Home Affordable Program (MHA), 223
    mediators, 476
    missing parent, finding, 407
    National Association of Enrolled Agents, 104
    National Foundation for Credit Counseling, 106, 457
    Nolo, 21, 62, 110, 473, 475
    real estate, 222
    retirement plan information, 259
    stock options, 355
    tax assistance, 478
    therapists, 476
Well-being (emotional), safeguarding, 40, 41
Wholesale value, collectibles, 334
Wills, 154
Witnesses, threats from spouse and, 43
Worst-case scenarios, preparation for, 42–44
Written separation agreement, 59

Z

Zero coupon bonds, 315

NOLO® Online Legal Forms

Nolo offers a large library of legal solutions and forms, created by Nolo's in-house legal staff. These reliable documents can be prepared in minutes.

Create a Document

- **Incorporation.** Incorporate your business in any state.
- **LLC Formations.** Gain asset protection and pass-through tax status in any state.
- **Wills.** Nolo has helped people make over 2 million wills. Is it time to make or revise yours?
- **Living Trust (avoid probate).** Plan now to save your family the cost, delays, and hassle of probate.
- **Trademark.** Protect the name of your business or product.
- **Provisional Patent.** Preserve your rights under patent law and claim "patent pending" status.

Download a Legal Form

Nolo.com has hundreds of top quality legal forms available for download—bills of sale, promissory notes, nondisclosure agreements, LLC operating agreements, corporate minutes, commercial lease and sublease, motor vehicle bill of sale, consignment agreements and many more.

Review Your Documents

Many lawyers in Nolo's consumer-friendly lawyer directory will review Nolo documents for a very reasonable fee. Check their detailed profiles at **Nolo.com/lawyers**.

△△ NOL

...org

...ext order

~ Reg... ...n w... ...you a
coupon for 15% off your next Nolo.com order!

Nolo.com/customer-support/productregistration

On Nolo.com you'll also find:

Books & Software

Nolo publishes hundreds of great books and software programs for consumers and business owners. Order a copy, or download an ebook version instantly, at Nolo.com.

Online Legal Documents

You can quickly and easily make a will or living trust, form an LLC or corporation, apply for a trademark or provisional patent, or make hundreds of other forms—online.

Free Legal Information

Thousands of articles answer common questions about everyday legal issues including wills, bankruptcy, small business formation, divorce, patents, employment, and much more.

Plain-English Legal Dictionary

Stumped by jargon? Look it up in America's most up-to-date source for definitions of legal terms, free at nolo.com.

Lawyer Directory

Nolo's consumer-friendly lawyer directory provides in-depth profiles of lawyers all over America. You'll find all the information you need to choose the right lawyer.

DIMO11